Uses and Effects of Cultured Fishes in Aquatic Ecosystems

Support for the International Symposium and Workshop on the Uses and Effects of Cultured Fishes in Aquatic Ecosystems and publication of this proceedings was provided by

Sport Fish in Restoration Act Funds

administered by the

U.S. Fish and Wildlife Service, Division of Federal Aid

Conference Steering Committee
Delano R. Graff, *Chair*

Program Committee and Session Chairs

Gary J. Carmichael
Vincent A. Mudrak
Wayne J. Daley
Thomas W. Gengerke
Churchill B. Grimes

Harold L. Kincaid
Mark Konikoff
Donald D. MacKinlay
Franklin T. McBride
J. Holt Williamson

Arrangements Committee
Gary J. Carmichael (Albuquerque)
Donald L. Horak (Denver)

Uses and Effects of Cultured Fishes in Aquatic Ecosystems

Edited by

Harold L. Schramm, Jr.
Robert G. Piper

American Fisheries Society Symposium 15

Proceedings of the International Symposium and Workshop on the Uses and Effects of Cultured Fishes in Aquatic Ecosystems

Held in Albuquerque, New Mexico
12–17 March 1994

American Fisheries Society
Bethesda, Maryland
1995

The American Fisheries Society Symposium series is a registered serial.
Suggested citation formats follow.

Entire book

Schramm, H. L., Jr., and R. G. Piper, editors. 1995. Uses and effects of cultured fishes in aquatic ecosystems. American Fisheries Society Symposium 15.

Article within the book

Campton, D. E. 1995. Genetic effects of hatchery fish on wild populations of Pacific salmon and steelhead: what do we really know? American Fisheries Society Symposium 15:000–000.

© Copyright by the American Fisheries Society, 1995

All rights reserved. Photocopying for internal or personal use, or for the internal or personal use of specific clients, is permitted by AFS provided that the appropriate fee is paid directly to Copyright Clearance Center (CCC), 222 Rosewood Drive, Danvers, Massachusetts 01923, USA; phone 508-750-8400. Request authorization to make multiple photocopies for classroom use from CCC. These permissions do not extend to electronic distribution or long-term storage of articles or to copying for resale, promotion, advertising, general distribution, or creation of new collective works. For such uses, permission or license must be obtained from AFS.

Library of Congress Catalog Card Number: 95-060393

ISBN 0-913235-91-1   ISSN 0892-2284

Printed in the United States of America on recycled, acid-free paper.

American Fisheries Society
5410 Grosvenor Lane, Suite 110
Bethesda, Maryland 20814-2199, USA

Contents

SYMPOSIUM PROCEEDINGS

Preface xi

Acknowledgments xiii

Keynote Address: Fish Genetics, Fish Hatcheries, Wild Fish, and Other Fables
   G. C. Radonski and A. J. Loftus 1

PODIUM PRESENTATIONS

FISHERIES MANAGEMENT NEEDS: SPORT FISH RESTORATION AND ENHANCEMENT

Enhancement of Sportfishing in New York Waters of Lake Ontario with Hatchery-Reared Salmonines
   R. E. Lange, G. C. LeTendre, T. H. Eckert, and C. P. Schneider 7

Use of Cultured Salmonids in the Federal Aid in Sport Fish Restoration Program
   J. McGurrin, C. Ubert, and D. Duff 12

Contribution of Nonnative Fish to California's Inland Recreational Fishery
   D. P. Lee 16

A Review of Chinook Salmon Resources in Southeast Alaska and Development of an Enhancement Program Designed for Minimal Hatchery–Wild Stock Interaction
   W. Heard, R. Burkett, F. Thrower, and S. McGee 21

Changes in Stocking Strategies for Atlantic Salmon Restoration and Rehabilitation in Maine, 1871–1993
   J. R. Moring, J. Marancik, and F. Griffiths 38

Fish Stocking Programs in Wyoming: A Balanced Perspective
   M. D. Stone 47

An Evaluation of the Genetic Integrity of Ohio River Walleye and Sauger Stocks
   M. M. White and S. Schell 52

Native and Nonnative Fish Species Used in State Fisheries Management Programs in the United States
   D. Horak 61

FISHERIES MANAGEMENT NEEDS: THREATENED AND ENDANGERED SPECIES

Genetic Risks and Hazards in Hatchery Operations: Fundamental Concepts and Issues
   C. A. Busack and K. P. Currens 71

Captive Broodstocks for Recovery of Snake River Sockeye Salmon
   T. A. Flagg, C. V. W. Mahnken, and K. A. Johnson 81

Hybridization and Introgression between Introduced and Native Fish
   R. F. Leary, F. W. Allendorf, and G. K. Sage 91

An Augmentation Plan for Razorback Sucker in the Upper Colorado River Basin
T. Modde, A. T. Scholz, J. H. Williamson, G. B. Haines, B. D. Burdick, and F. K. Pfeifer 102

Apache Trout Management: Cultured Fish, Genetics, Habitat Improvements, and Regulations
G. J. Carmichael, J. N. Hanson, J. R. Novy, K. J. Meyer, and D. C. Morizot 112

A Restoration Plan for Pez Blanco in Lake Patzcuaro, Mexico
A. Chacon-Torres and C. Rosas-Monge 122

A Program for Maintaining the Razorback Sucker in Lake Mohave
G. Mueller 127

FISHERIES RESTORATION AND ENHANCEMENT: STOCKING CRITERIA AND GOALS

British Columbia's Trout Hatchery Program and the Stocking Policies that Guide It
B. Ludwig 139

Species-Specific Guidelines for Stocking Reservoirs in Oklahoma
E. Gilliland and J. Boxrucker 144

Salmon Stock Restoration and Enhancement: Strategies and Experiences in British Columbia
E. A. Perry 152

Beneficial Uses of Marine Fish Hatcheries: Enhancement of Red Drum in Texas Coastal Waters
L. W. McEachron, C. E. McCarty, and R. R. Vega 161

A Responsible Approach to Marine Stock Enhancement
H. L. Blankenship and K. M. Leber 167

An Ecological Framework for Evaluating the Success and Effects of Stocked Fishes
D. H. Wahl, R. A. Stein, and D. R. DeVries 176

FISHERIES RESTORATION AND ENHANCEMENT: GENETIC CRITERIA AND GOALS

An Evaluation of Inbreeding and Effective Population Size in Salmonid Broodstocks in Federal and State Hatcheries
H. L. Kincaid 193

Use of DNA Microsatellite Polymorphism to Analyze Genetic Correlations between Hatchery and Natural Fitness
R. W. Doyle, C. Herbinger, C. T. Taggart, and S. Lochmann 205

Genetic Monitoring of Life-History Characters in Salmon Supplementation: Problems and Opportunities
J. J. Hard 212

Distribution of Largemouth Bass Genotypes in South Carolina: Initial Implications
J. Bulak, J. Leitner, T. Hilbish, and R. A. Dunham 226

Fitness and Performance Differences between Two Stocks of Largemouth Bass from Different River Drainages within Illinois
D. P. Philipp and J. E. Claussen 236

Incorporating the Stock Concept and Conservation Genetics in an Illinois Stocking Program
K. D. Cottrell, S. Stuewe, and A. Brandenburg 244

Conservation of Genetic Diversity in a White Seabass Hatchery Enhancement Program in Southern California
D. M. Bartley, D. B. Kent, and M. A. Drawbridge 249

FISH PRODUCTION TO MEET NEEDS: CAPABILITIES AND LIMITATIONS

Adult Production of Fall Chinook Salmon Reared in Net-Pens in Backwaters of the Columbia River
J. W. Beeman and J. F. Novotny 261

Maintenance of Stock Integrity in Snake River Fall Chinook Salmon
R. M. Bugert, C. W. Hopley, C. A. Busack, and G. W. Mendel 267

Supplementation: Panacea or Curse for the Recovery of Declining Fish Stocks?
E. C. Bowles 277

Status of Supplementing Chinook Salmon Natural Production in the Imnaha River Basin
R. W. Carmichael and R. T. Messmer 284

Quality Assessment of Hatchery-Reared Spring Chinook Salmon Smolts in the Columbia River Basin
W. W. Dickhoff, B. R. Beckman, D. A. Larsen, C. V. W. Mahnken, C. B. Schreck, C. Sharpe, and W. S. Zaugg 292

Terminal Fisheries and Captive Broodstock Enhancement by Nonprofit Groups
J. A. Sayre 303

A Review of Seminatural Culture Strategies for Enhancing the Postrelease Survival of Anadromous Salmonids
D. J. Maynard, T. A. Flagg, and C. V. W. Mahnken 307

EVALUATION OF STOCKED FISH: IN HATCHERIES AND IN THE WILD

Use of a National Fish Hatchery to Complement Wild Salmon and Steelhead Production in an Oregon Stream
D. E. Olson, B. C. Cates, and D. H. Diggs 317

The Contribution of Hatchery Fish to the Restoration of American Shad in the Susquehanna River
M. L. Hendricks 329

Genetic Effects of Hatchery Fish on Wild Populations of Pacific Salmon and Steelhead: What Do We Really Know?
D. E. Campton 337

Assessment of Season of Release and Size at Release on Recapture Rates of Hatchery-Reared Red Drum
S. A. Willis, W. W. Falls, C. W. Dennis, D. E. Roberts, and P. G. Whitchurch 354

The Effect of Hatcheries on Native Coho Salmon Populations in the Lower Columbia River
T. A. Flagg, F. W. Waknitz, D. J. Maynard, G. B. Milner, and C. V. W. Mahnken 366

Marine Enhancement with Striped Mullet: Are Hatchery Releases Replenishing or Displacing Wild Stocks?
K. M. Leber, N. P. Brennan, and S. M. Arce 376

ROLE OF EXOTIC SPECIES: PAST AND FUTURE USES IN FISHERIES MANAGEMENT

Introduced Species as a Factor in Extinction and Endangerment of Native Fish Species
D. R. Lassuy 391

Coldwater Fish Stocking and Native Fishes in Arizona: Past, Present, and Future
J. N. Rinne and J. Janisch 397

Problems and Prospects for Grass Carp as a Management Tool
J. R. Cassani 407

The Case for Caution with Fish Introductions
W. R. Courtenay, Jr. 413

From Sportfishing Bust to Commercial Fishing Boon: A History of the Blue Tilapia in Florida
M. M. Hale, J. E. Crumpton, and R. J. Schuler, Jr. 425

Splake as a Control Agent for Brook Trout in Small Impoundments
J. R. Satterfield, Jr. and K. D. Koupal 431

An Evaluation of Aquatic Vegetation Control by Triploid Grass Carp in Virginia Ponds
R. Eades and E. Steinkoenig 437

Introduction and Establishment of a Successful Butterfly Peacock Fishery in Southeast Florida Canals
P. L. Shafland 443

Establishment and Expansion of Redbelly Tilapia and Blue Tilapia in a Power Plant Cooling Reservoir
J. U. Crutchfield, Jr. 452

HATCHERIES, HABITAT, AND REGULATIONS: PAST AND FUTURE USES IN FISHERIES MANAGEMENT

A Common Sense Protocol for the Use of Hatchery-Reared Trout
R. W. Wiley 465

Development of an Optimal Stocking Regime for Walleyes in East Okoboji Lake, Iowa
J. G. Larscheid 472

Evaluation of the Florida Largemouth Bass in Texas, 1972–1993
A. A. Forshage and L. T. Fries 484

Accomplishments and Roadblocks of a Marine Stock Enhancement Program for White Seabass in California
D. B. Kent, M. A. Drawbridge, and R. F. Ford 492

Using Genetics in the Design of Red Drum and Spotted Seatrout Stocking Programs in Texas: A Review
T. L. King, R. Ward, I. R. Blandon, R. L. Colura, and J. R. Gold 499

History, Genetic Variation, and Management Uses of 13 Salmonid Broodstocks Maintained by the Wyoming Game and Fish Department
C. B. Alexander and W. A. Hubert 503

The Roles of Hatcheries, Habitat, and Regulations in Wild Trout Management in Idaho
A. R. Van Vooren 512

Use of Cultured Fish for Put-Grow-and-Take Fisheries in Kentucky Impoundments
B. Kinman 518

Better Roles for Fish Stocking in Aquatic Resource Management
R. J. White, J. R. Karr, and W. Nehlsen 527

POSTER PRESENTATIONS

Use of Gene Marking to Assess Stocking Success of Red Drum in Texas Bays
R. Ward, T. L. King, I. R. Blandon, and L. W. McEachron 551

Multivariate Analysis of Red Drum Stocking in Texas Bays
R. Ward, L. W. McEachron, B. E. Fuls, and M. Muoenke 552

Predation and Cannibalism on Hatchery-Reared Striped Bass in the Patuxent River, Maryland
L. L. Andreasen 553

The Use of Hatchery-Produced Striped Bass for Stock Restoration and Validation of Stock Parameters
S. P. Minkkinen, C. P. Stence, and B. M. Richardson 555

American Shad Handling, Rearing, and Marking Trials
B. M. Richardson and S. P. Minkkinen 557

Long-Term Survival of Three Size-Classes of Striped Bass Stocked in the Savannah River, Georgia–South Carolina
J. E. Wallin and M. J. Van den Avyle 560

The Development of PCR-Based Genomic DNA Assays to Assess Genetic Variation in the Striped Bass Santee–Cooper River Population, South Carolina
M. Diaz, G. M. LeClerc, B. Ely, and J. S. Bulak 561

Development of a Hatchery Broodstock for Recovery of Green Lake Strain of Lake Trout
H. L. Kincaid 562

Hidden Falls Hatchery Chum Salmon Program
B. A. Bachen and T. Linley 564

Separation of Hatchery-Reared from Wild Red Drum Based on Discriminant Analysis of Daily Otolith Growth Increments
J. J. Isely, C. B. Grimes, and A. W. David 566

A Captive Broodstock Approach to Rebuilding a Depleted Chinook Salmon Stock
J. B. Shaklee, C. Smith, S. Young, C. Marlowe, C. Johnson, and B. B. Sele 567

The Assessment of Marine Stock Enhancement in Southern California: A Case Study Involving the White Seabass
M. A. Drawbridge, D. B. Kent, M. A. Shane, and R. F. Ford 568

Broodstock Development Plan for the Fluvial Arctic Grayling in Montana
W. P. Dwyer and R. F. Leary 570

Interactive Effects of Stocking-Site Salinity and Handling Stress on Short-Term Survival of Striped Bass Stocked in the Savannah River, Georgia–South Carolina
J. E. Wallin and M. J. Van den Avyle 571

Restoration of the Savannah River Striped Bass Population
M. J. Van den Avyle, J. E. Wallin, and C. Hall 572

Genetic Contribution of Hatchery Fish to Walleye Stocks in Saginaw Bay, Michigan
T. N. Todd and R. C. Haas 573

Homing Propensity in Transplanted and Native Chum Salmon
W. W. Smoker and F. P. Thrower 575

Advances to Increase Restocking of Native Fishes in Lake Patzcuaro, Mexico
C. Rosas-Monge and A. Chacon-Torres 577

Is Genetic Change from Hatchery Rearing of Anadromous Fish Really a Problem?
  R. R. Reisenbichler and G. Brown 578

Development of a Regional Policy for the Prevention and Control of Nonnative Aquatic
Species: The Chesapeake Basin Experience
  D. E. Terlizzi, R. J. Klauda, and F. P. Cresswell 580

AFS Section Position Statements

Fish Culture Section
  Presented by Kirby D. Cottrell, Section Delegate 585
Fisheries Management Section
  Presented by Harold L. Schramm, Jr., President 587
Genetics Section
  Presented by Fred M. Utter, President 588
Early Life History Section
  Presented by Fred P. Binkowski, Section Delegate 589
Fisheries Administrators Section
  Presented by Fred Harris, Section Delegate 590
Introduced Fish Section
  Presented by Dennis R. Lassuy, President 591
Bioengineering Section
  Presented by Susan Baker, President 592
Marine Fisheries Section
  Presented by Churchill B. Grimes, President 593

Symposium Summary: Melding Science and Values in Pursuit of Ethical Guidelines for the
  Uses of Cultured Fishes in Aquatic Ecosystems
  Paul Brouha 595

WORKSHOP RESULTS

Considerations For The Use of Cultured Fishes in Fisheries Resource management

  Introduction 601

  Considerations 603

  Appendix: Workshop Participants 607

Preface

As fisheries resource management has developed and expanded, the use of and need for cultured fishes has increased. Always a tool but never a panacea, stocking cultured fishes has been both supported and challenged. The use of cultured fishes in fisheries management was addressed at "The Role of Fish Culture in Fisheries Management" symposium in 1985. Increases in the scope of fisheries management responsibilities and declines in fisheries resources, coupled with advances in fisheries science and knowledge about fishes and fisheries systems, led to a clear need to examine again, scientifically, the uses and effects of cultured fishes.

The American Fisheries Society addressed this significant issue with a two-step process: (1) by making scientific information available to the diverse body of people that make fisheries resource management decisions, and (2) by asking representatives of resource management agencies to determine recommendations for the use of cultured fishes. To accomplish the first step, a symposium "Uses and Effects of Cultured Fishes in Aquatic Ecosystems" was convened in Albuquerque, New Mexico, 12–17 March 1994. The second step was accomplished by inviting all North American fisheries resource management agencies to send a representative to a facilitated workshop in Denver, Colorado, 29–30 July 1994, to develop comprehensive considerations for the use of cultured fishes.

This process could not have been accomplished without the dedicated support of many American Fisheries Society (AFS) subunits and individual members. The "Uses and Effects of Cultured Fishes in Aquatic Ecosystems" symposium was initiated by the Fish Culture Section and cosponsored by the Fisheries Management, Genetics, Fisheries Administrators, Introduced Fishes, Marine Fisheries, Bioengineering, Physiology, and Early Life History sections of AFS. The symposium Steering Committee generously provided their time, energy, and abilities to ensure a successful symposium and follow-up workshop.

This proceedings contains technical papers, abstracts of posters, and position statements of the sponsoring AFS Sections which were presented at the symposium, as well as results of the facilitated workshop. The poster and podium presentations are a blend of scientific research, case histories, and management programs. These presentations have all been peer reviewed and found to be acceptable for publication. The AFS Section position statements evolved during the symposium as new information was discussed. The "Considerations for the Use of Cultured Fishes" represents the collective thoughts of the participants at the July workshop and addresses biological, ecological, social, and economic issues associated with culturing and stocking fishes. The considerations have been reviewed by the symposium Steering Committee (who served as facilitators for the July workshop) to ensure that the final document accurately represents the message of the workshop participants.

The diligence of the Steering Committee and the workshop participants has resulted in a product that is technically informative and operationally useful. From the conception of this process, the Steering Committee's purpose has been to provide information that can be used to better manage fisheries resources. We believe this purpose has been accomplished; however, the utility of the material presented in this volume must be evaluated by researchers and managers who incorporate the information into fisheries management activities. We encourage all people responsible for fisheries resources to be responsive to future needs and opportunities to build on this information and continually improve our ability to wisely manage fisheries resources.

HAROLD L. SCHRAMM, JR.
ROBERT G. PIPER
Coeditors

Acknowledgments

The dedicated efforts of many people are required to plan and stage a major symposium, to conduct a facilitated workshop, and to prepare a proceedings of this size and scope. The symposium, "Uses and Effects of Cultured Fishes in Aquatic Ecosystems," sponsored by the American Fisheries Society, was a success due to the hard work and commitment of the Steering Committee and many individual volunteers. Gary J. Carmichael and Vincent A. Mudrak (U.S. Fish and Wildlife Service) compiled an excellent program that spanned 5 days, 68 podium papers, and 23 poster presentations. The eight technical sessions were moderated by Steering Committee members Vincent A. Mudrak, J. Holt Williamson (U.S. Fish and Wildlife Service), Wayne J. Daley (Kramer, Chin & Mayor, Inc.), Thomas W. Gengerke (Iowa Department of Natural Resources), Harold L. Kincaid (National Biological Service), Mark Konikoff (University of Southwest Louisiana), Donald D. MacKinlay (Canada Department of Fisheries and Oceans), and Franklin T. McBride (North Carolina Wildlife Resources Commission). Churchill B. Grimes (National Marine Fisheries Service) moderated a ninth program session during which AFS Section position statements were presented. He also organized a poster session and several lively point–counterpoint discussion sessions that followed each day of podium presentations. Fred P. Binkowski (University of Wisconsin) and Martin T. Marcinko (Pennsylvania Fish and Boat Commission) handled conference registration superbly. Wayne J. Daley was responsible for the overall budget and Dennis C. Ricker (Pennsylvania Fish and Boat Commission) provided scrupulous financial management. Several people at the Pennsylvania Fish and Boat Commission deserve recognition. Jule Weaver served as the on-site computer expert and assisted authors, attendees, and staff with various functions; Mary Ellen McMahon provided registration and clerical support for the symposium; Ted Walke designed and produced the printed symposium program on a very tight schedule; and the Commission Bureau of Education and Information Executive Director Peter Colangelo contributed the program book. The Steering Committee is grateful to Kim Marggraf of Marggraf Meetings who provided registration expertise. Gary J. Carmichael did an outstanding job coordinating the local arrangements for the symposium. Special thanks are also extended to the many regional office and field station personnel of the U. S. Fish and Wildlife Service Southwest Regional Office who worked hard to ensure that the symposium ran smoothly.

A follow-up workshop was convened four months after the symposium in Denver, Colorado. Local arrangements were skillfully negotiated and managed by Thomas G. Powell, Donald D. Horak (Colorado Division of Wildlife), and J. Michael Stempel (U.S. Fish and Wildlife Service). The three-day meeting was facilitated by Terry Radcliff who also provided training to the Steering Committee and who made it possible for a large group of people to work diligently to develop a consensus document that outlines considerations for the use of cultured fishes in aquatic ecosystems.

Harold L. Schramm, Jr. (Mississippi Cooperative Fish and Wildlife Research Unit) and Robert G. Piper (Piper Technologies) contributed enormous amounts of time, technical expertise, and editorial skill to evaluating and editing manuscripts from the symposium for this volume. Harold Schramm also prepared the workshop results for final publication. Many people provided peer reviews that strengthened the quality of this proceedings; their names are listed on the following page. Robert L. Kendall guided the editorial process for the American Fisheries Society and Beth D. Staehle coordinated the editing and production of this volume. Eva M. Silverfine, Amy E. Moore, and Janet E. Harry all made important contributions to the quality and production of this proceedings.

Financial support for the symposium and this publication was coordinated by AFS Executive Director Paul Brouha and was provided by the Federal Aid in Sport Fish Restoration Fund.

DELANO R. GRAFF, Chairman
Steering Committee

Reviewers

B. Barton
D. Barwick
J. Beeman
K. Bestgen
P. Bettoli
B. Bigler
N. Billington
H. Blankenship
J. Borawa
J. Boxrucker
R. Bugert
J. Bulak
C. Burger
C. Busack
D. Campton
G. Carmichael
L. Claggett
W. Clarke
D. Cloutman
L. Cofer
K. Cottrell
W. Courtenay, Jr.
T. Crawford
J. Crutchfield
W. Daley
L. Deaton
D. Degan
W. Devick
D. DeVries
W. Dickhoff
C. Dohner
M. Drawbridge
W. Dwyer
R. Eades
J. Epifanio
D. Fielder
S. Filipek
T. Flagg
R. Ford
A. Forshage
W. Fredenberg
J. Fries
G. Garrett
T. Gengerke

E. Gilliland
C. Goudreau
C. Grimes
M. Hansen
J. Hard
F. Harris
W. Heard
M. Hendricks
D. Hendrickson
W. Herschberger
J. Hightower
W. Hubert
C. Hunter
J. Isley
R. Jackson
D. Kent
J. Kerwin
H. Kincaid
T. King
P. Klerks
M. Konikoff
J. Koppelman
R. Lange
R. Langton
M. Larkin
J. Larscheid
D. Lassuy
R. Leary
D. Lee
S. Leider
B. Ludwig
M. Martin
E. Maughan
B. May
D. Maynard
M. McAfee
G. McMichael
T. McTigue
J. Meade
T. Modde
S. Mopper
J. Moring
V. Mudrak
G. Mueller

M. Mueller
B. Murphy
R. Muth
K. Nelson
J. O'Leary
E. Perry
R. Peterson
S. Phelps
D. Philipp
R. Pine
B. Potter
D. Propst
M. Ray
R. Reisenbichler
J. Rensel
J. Rice
B. Rosenlund
J. Satterfield, Jr.
R. Schultz
P. Shafland
R. Simons
M. Stempel
R. Stickney
M. Stone
D. Tave
T. Tiersch
J. Tomasso
F. Utter
R. Valdez
S. Van Horn
M. Van den Avyle
A. Van Vooren
G. Vaughn
L. Visscher
D. Wahl
R. Waples
M. Watson
M. White
R. White
R. W. Wiley
H. Williamson
D. Willis
S. Willis
G. Winans
P. Wingate

KEYNOTE ADDRESS
Fish Genetics, Fish Hatcheries, Wild Fish, and Other Fables

GILBERT C. RADONSKI

37 Pepper Tree Court, Warrenton, Virginia 22186, USA

ANDREW J. LOFTUS

American Sportfishing Association
1033 North Fairfax Street, Suite 200, Alexandria, Virginia 22314, USA

The central theme of this symposium is fish husbandry: the culture of fishes. Usually such a symposium would focus on husbandry techniques and the near-miraculous advances in technology would be chronicled. But instead we are gathered here to discuss the uses and effects of cultured fishes in aquatic ecosystems, that is, the ethical use of the products of fish husbandry. Fishery managers and biologists have debated this subject among themselves for decades. Now the discussion has moved to the broader societal forum, the environmental community.

As we are all well aware, the controversy over the role of cultured fishes in natural environments has increased in intensity during recent years. All aspects of fish husbandry, in the broader term, aquaculture, are in question, including facility siting, effluents, genetics, animal rights, and a host of social and economic impacts. It is a quantum leap from the early days of fish culture in the United States, described by Bowen (1970) as the product of fisheries destroyed by America's early eighteenth-century industries:

> While the decline of the inland fishery resources was not understood, an aroused public demanded corrective action. Therefore, the climate was created for the development of fish culture in America.

Fishery management in general is undergoing dramatic evolution as our knowledge of fish stocks, habitat relationships, and the effects of human activities are better understood. This dynamic is a normal process of most disciplines as technology and application evolve and the results of tested theories add to the store of knowledge.

Fishery managers generally have viewed fish culture as one of many management tools used to ameliorate the effects identified as limiting factors to reach a fishery management objective. Hence the hatchery product has been used to introduce species into new and existing waters, to supply year-classes when natural reproduction is absent or has failed, to create or maintain recreational and commercial fisheries, or to produce protein. Like the situation described by Bowen (1970), fishery management objectives are largely determined by public demand.

Cultured Fishes in Dynamic Systems

It is axiomatic that aquatic ecosystems are dynamic, in the classical sense, moving on a continuum from oligotrophy to eutrophy with species assemblages and quantity defined by location on the continuum. The practice of purposeful interruption of the dynamic by retarding succession in order to retain habitat that will support a particular species or species assemblage is antithetic to modern fish and wildlife management as expressed in the term "ecosystem management." At no time in history has the evolution and development of ecosystems and organisms been static nor shall it ever be; it is only the rate of change over time that varies. Even management efforts that interrupt the process are temporary in the relentless march of time.

Common definitions of the currently popular concept of biodiversity, as were reflected in several papers in a 1992 issue of the American Fisheries Society publication *Fisheries* (17[3]:6–38), are simply stated as the variety of life and its processes, with an inherent understanding that life refers to only native organisms. Indeed, Karr and Dudley (1981) described a comparable concept of biological integrity of communities as "species composition, diversity, and functional organization comparable to the natural habitat of the region." Thus, under these definitions, nonnative cultured fishes, and possibly even native cultured fishes that are genotypically different, would have no place in systems managed for biodiversity.

However, there are few, if any, contemporary areas that truly can be called natural in a static sense. This was eloquently characterized in 1993 during a presentation on ecological risk assessment at Tulane University by Dr. Robert Lackey of the Environmental Protection Agency. Dr. Lackey stated,

1

Many people, perhaps most I think, have a view of ecosystems characterized by Ansel Adams photographs–natural ecosystems are 'perfect'–an equilibrium condition in which all the pieces operate in a predictable, desirable way. Most views of ecosystems are of 'natural, unspoiled' panoramas. They are frozen in time and any deviation from this timeless condition is 'degradation.' This is not the way the natural world is.

Commenting on the nature and long-term dynamics of ecosystems in the context of human life spans and mortality he said,

> contrary to individual humans who die, and in most cases people think that is a bad situation, ecosystems change dramatically over time, have no optimal condition, and are only healthy when compared to some desired state specified by humans. Ecosystem 'health' is strictly an anthropocentric term.

Therefore, it is entirely unreasonable to wholeheartedly discount the value of using introduced species (usually cultured) based solely on the premise that people can hope to maintain systems in a purely static or historic state.

Appropriate Uses of Cultured Fishes

This is not to say that cultured species are universally appropriate. Partly as a result of public willingness to accept readily available cultured fishes as compensation for a variety of human-induced destruction of fisheries resources and habitats, much of the public, and possibly some managers, often view fish stocking as a panacea to the problem of declining populations. Again, that was clearly the case as described by Bowen (1970). Hatchery products should not be used as an expeditious surrogate that would justify the wanton destruction or taking of natural fish production systems. When the hatchery product should be used to mitigate purposeful habitat loss is a serious question and should be decided only through an open, public, deliberative process.

As Dr. Lackey's comments so appropriately outlined, possibly the one factor over which fishery managers often have the least control is habitat. Societal decisions that result in changing land use, water diversions, atmospheric deposition of nutrients, contaminants, and other materials, as well as increasing urbanization may force managers to compensate for changes in habitat through the use of cultured species. As habitats evolve, perhaps at an accelerated pace due to human influence, they may become unsuitable to maintain species and ecological processes as these species and processes occurred there in the past. Ironically, recognizing the reality that people have greatly altered most aquatic habitats, many arguments for preserving vestigial stocks of native species continue to be premised on the theory that humans cannot reengineer evolution and that native species are better adapted to the environment or to maintain historical biological diversity. Yet these very species evolved in an environment that is completely different from what presently exists. Present environmental conditions, had they developed over an epochal period of time rather than in the rapid pace so common under human-influenced conditions, may have led to a completely different evolutionary pathway and, ultimately, species composition than may be present.

Although there are numerous examples of management to achieve historic conditions, an appropriate case history lies in the attempted reestablishment of populations of Arctic grayling *Thymallus arcticus* in northern Michigan. Primarily due to irreversible habitat changes caused by logging, constructed barriers to movement, and other human-induced impacts, efforts to reestablish this species have met dismal failure. In cases such as this, managers must face a decision on whether to use cultured species for public benefit or to leave these streams with altered ecosystems to provide little or no public benefit.

In some instances, habitat reclamation may be technologically feasible, but the economic cost to society or public willingness to pay is prohibitive. In the case of some hydroelectric projects, it may be possible to remove dams and restore ecosystems to a state that somewhat resembles their original condition. However, society as a whole (the people component) has chosen not to undertake the cost or suffer the sacrifices to do this. The only suitable solution in these cases may be to use native or nonnative cultured species.

In other cases, the alteration of habitat has been accompanied by severe and irreversible changes in the entire aquatic community. There is no better example of a large-scale case such as this than the Great Lakes. Once relatively low in diversity, a tremendous variety of nonnative aquatic species has now been established in the Great Lakes either unintentionally or by managers trying to restore effective predator–prey assemblages while providing public benefits in a highly altered ecosystem. The entire fish assemblage and habitat have been changed during the past century so that it is not likely from a technological standpoint, and certainly not from an economic standpoint, that the lakes could be returned to the condition that existed hundreds of years ago. Even so, the aquatic community of the lakes is more diverse under current conditions and provides tremendously greater benefits to society than they had historically. In cases such as this, of which there are many throughout the United States, managers must be ready to utilize fish culture as a reclamation tool—both in the ecological and economic context.

Role of Cultured Fishes in Restoring Native Species

Using cultured fishes to restore native species that have been extirpated is an important part of several ongoing restoration efforts, including those for economically important species such as lake trout *Salvelinus namaycush*, Atlantic salmon *Salmo salar*, and Pacific salmon *Oncorhynchus* spp. The U.S. Fish and Wildlife Service, for example, maintains 28 hatcheries that are involved with the restoration of Threatened or Endangered fishes (including those on the proposed and candidate lists). In several cases, managers are faced with rebuilding entire stocks from remnant populations that represent a greatly reduced gene pool. In restoring specific runs of Pacific salmon in the northwestern United States, managers must consider the feasibility of maintaining the genetic integrity of hundreds of stocks of the same species. In the case of restoration of Atlantic salmon in the northeastern United States, most individual stock-specific gene pools have long since vanished, leaving managers with the task of restoring Atlantic salmon's presence in entire river systems from a very limited, nonnative gene pool. Cultured fishes will play an important role in restoring historic fish assemblages' phenotypes, although these fishes will not be genotypically native.

The Paradigm

In the future, fish culture will be affected by evolving environmental standards, and fishery managers and fish culturists will have to justify every aspect of the stocking and production process. It is said that those who do not learn from history are destined to relive it. There is a fishery with a significant aquaculture component that has raised the specter of what the future holds—the Atlantic salmon fishery.

Few fish have been as romanticized as the "silver swimmer." Its name, *Salmo salar*, means mighty leaper. It is legendary among anglers and is at the top of the list of gourmets. Its life history of traversing great oceanic distances to feed and then returning to its natal stream to spawn held naturalists in awe. The Atlantic salmon was extirpated from southern New England rivers by the industrial revolution of the late eighteenth and early nineteenth centuries. The industrial revolution was fueled by the construction of dams to harness water power. The mighty Atlantic salmon was deprived of its spawning grounds in many streams.

Interest in the culture of Atlantic salmon led to the first production hatchery built on the Rhine River in Germany in 1852. By the latter part of that same century every country with Atlantic salmon populations had a salmon hatchery and attempted to restore lost runs. To mitigate the loss of access to, or despoilment of, spawning habitat, Atlantic salmon were cultured and flushed down the rivers to accomplish little more than salve consciences. However, the New England Atlantic salmon hatcheries were so successful in producing fish that soon excess fish were being loaded into federal fish railroad cars and indiscriminately stocked from New England to Minnesota. Early in my (G. C. Radonski) career as a fishery biologist in northeastern Wisconsin, it was rare to review a lake or stream survey record without seeing an entry that Atlantic salmon had been stocked at the turn of the twentieth century. It is a wonder that we are not up to our ears in Atlantic salmon. Or, are we?

Due to diminishing quantity and quality of spawning habitat, and serious commercial overexploitation, wild Atlantic salmon in the market became limited and expensive in the best tradition of supply and demand. Natural stocks were producing about 10,000 tons annually. In the early 1970s the Norwegians began farming Atlantic salmon. In those early years the farmed production almost equaled the natural production. But cultural techniques and market demand for the high-quality cultured product produced a rapidly growing cultural infrastructure to the point where production in 1992 was in excess of 225,000 tons! That rapid growth unfolded the wide array of problems facing aquaculture including, but not limited to, genetic pollution, competition with wild stocks, aesthetics associated with facility siting, spread of parasites and diseases, water quality impairment, and interruption of the economic stability of other fisheries (such as Pacific salmon). There is a litany of benefits that offset the problems. However, benefits are enjoyed, problems must be dealt with.

The international body that coordinates the management of Atlantic salmon is the North Atlantic Salmon Conservation Organization, headquartered in Edinburgh, Scotland. The North Atlantic Salmon Conservation Organization has diligently followed the growth of Atlantic salmon culture, and we recommend their numerous studies and publications on the impacts of Atlantic salmon aquaculture for your review.

The Pendulum Swings

As with any issue, opinions regarding the use of cultured species span a wide spectrum. The arguments can often become very heated and polarized between the "greens"—those who would like to see absolutely no use of cultured fishes at all—and those who would prefer to return to the old days when cultured fish were freely dispensed with little or no regard for historic ranges, complex ecological effects, disease, or other factors. The current political climate among policy makers, particularly in Washington, D.C., is toward the greener side of the issue. Interpretations of responsibilities under federal mandates such as the Non Indigenous Aquatic Nuisance Prevention and Control Act (16 U.S.C.A. §§4701 to 4751), the Endangered Species Act (16 U.S.C.A. §§1531 to 1544), and others tend to be made through a romanticized vision of returning to a more natural state, which excludes or severely restricts the use of cultured species. President Carter's Executive Order (Number 11987, 24 May 1977), which was written to reduce introductions of species "not naturally occurring either presently or historically in any ecosystem of the United States," is currently being interpreted in some cases as prohibiting further stocking of species that have occurred for decades in some areas.

Although this is the mood of some of the current political factions, there is a question as to whether the American public as a whole is supportive. Recent articles in popular outdoor magazines are showing a backlash to such restrictions. In other arenas, the public is beginning to question the cost in terms of dollars and the sacrifice of alternative uses of the resources to maintain fish populations without the aid of artificial propagation where it is appropriate.

In Conclusion

Nine years ago, I (G. C. Radonski) keynoted the AFS-sponsored symposium, *Fish Culture in Fisheries Management*. My presentation was titled, "Fish Culture is a Tool, Not a Panacea" (Radonski and Martin 1986). We were tempted merely to recycle that presentation for this symposium. Upon review of that paper, we found it chronicled the past, and the past does not change. It described many fish culture success stories while noting that there were things done wrong in the developing fish culture science. As problems were identified, they were addressed and usually corrected. Most were corrected because of a strong and abiding ethic on the part of the fish culture practitioners toward the fishery resource and aquatic habitats. Other problems were dealt with in response to outside public pressure. In no case that we know of were these problems ignored. The process has not changed but the criteria have in response to evolving public fishery policy. We characterized this change in criteria as "modern fishery management." Modern fishery management would have natural systems with native biota and the presence of optimal biodiversity. We are not sure that we understand that in its fullest context, but we see the direction in which it is going. On the fishery management spectrum that goes from no management to intensive management, public fishery policy is moving toward no management. This symposium will play a role in the formation of that public policy. Neither end of the spectrum will prevail in an absolute sense. With hope, we will learn from each other in the process and the end result will be sound public policy that captures the social and economic benefits which can be obtained from the renewable common property fishery resource.

References

Bowen, J. T. 1970. A history of fish culture as related to the development of fishery programs. Pages 71–72 *in* N. G. Benson, editor. A century of fisheries in North America. American Fisheries Society Special Publication 7.

Karr, J. R., and D. R. Dudley. 1981. Ecological perspective on water quality goals. Environmental Management 5:55–68.

Radonski, G. C., and R. G. Martin. 1986. Fish culture is a tool not a panacea. Pages 7–13 *in* R. H. Stroud, editor. Fish culture in fisheries management. American Fisheries Society, Fish Culture Section and Fisheries Management Section, Bethesda, Maryland.

PODIUM PRESENTATIONS

Fisheries Management Needs:

Sport Fish Restoration

and Enhancement

Enhancement of Sportfishing in New York Waters of Lake Ontario with Hatchery-Reared Salmonines

ROBERT E. LANGE

New York State Department of Environmental Conservation
50 Wolf Road, Albany, New York 12233, USA

GERARD C. LETENDRE, THOMAS H. ECKERT, AND CLIFFORD P. SCHNEIDER

New York State Department of Environmental Conservation
Post Office Box 292, Cape Vincent, New York 13618, USA

Abstract.—Two centuries of habitat modification, overfishing, and the establishment of nonnative organisms has resulted in severely decreased abundance of native offshore fish species in Lake Ontario. The extirpation of the native open-water piscivores, Atlantic salmon *Salmo salar* and lake trout *Salvelinus namaycush*, resulted in a truncated trophic structure characterized by abundant alewife *Alosa pseudoharengus*, a nonnative planktivore. Sportfishing opportunities were limited to nearshore species such as smallmouth bass *Micropterus dolomieu* and yellow perch *Perca flavescens*. In 1968 New York began stocking nonnative salmonines to control alewife abundance and to provide sportfishing opportunities based on a put-grow-and-take management strategy. New York stocked 85 million hatchery-reared trout and salmon into Lake Ontario during 1968–1993. The cost of rearing and transporting these fish was US$788,000 in 1993. The fishery created by this stocking program has generated an estimated annual average of 885,000 angler trips over the past 9 years. Trout and salmon anglers spent an estimated $110 million traveling to and fishing in Lake Ontario in 1988. The presence of chemical contaminants in Lake Ontario fish, which has resulted in restrictive fish consumption advisories, remains a persistent problem. The Lake Ontario populations of alewife and rainbow smelt *Osmerus mordax*, the principal prey of trout and salmon, are in jeopardy of collapse due to predation and declining productivity related to decreased phosphorous loading. Concerned that the existing fishery may not be sustainable, the New York State Department of Environmental Conservation and the Ontario Ministry of Natural Resources has reduced stocking. The increasing emphasis on ecosystem structure and function highlights the need to broaden the scope of fishery management without losing sight of human demands for fishery resources.

The offshore fish community in Lake Ontario prior to European colonization of North America was typical for large oligotrophic lakes in north temperate zones (Christie 1973). The dominant fish species included blackfin cisco *Coregonus nigripinnis*, shortnose cisco *C. reighardi*, bloater *C. hoyi*, kiyi *C. kiyi*, lake herring *C. artedi*, lake whitefish *C. clupeaformis*, slimy sculpin *Cottus cognatus*, deepwater sculpin *Myoxocephalus thompsoni*, lake trout *Salvelinus namaycush*, Atlantic salmon *Salmo salar*, and burbot *Lota lota*.

By the mid-1960s, the offshore fish community was dominated by three species: alewife *Alosa pseudoharengus*, rainbow smelt *Osmerus mordax*, and slimy sculpin (O'Gorman et al. 1987). The alewife and rainbow smelt are nonnative species that successfully colonized Lake Ontario. Of the historic offshore fish community, only the slimy sculpin persisted in abundance. Populations of lake herring, lake whitefish, and burbot remained in Lake Ontario, albeit at diminished levels of abundance. The other species that formerly dominated the offshore fish community were either extirpated or nearly so by 1965.

The decline of the native offshore fish community in Lake Ontario resulted from an interaction of overexploitation, habitat degradation, and predation by the sea lamprey *Petromyzon marinus*. The new fish community offered little of value to the human population of the Lake Ontario region. The alewife, in particular, presented recurring problems from beach fouling following periodic mass mortalities suffered in the virtual absence of predatory control.

Small-scale commercial fisheries for lake herring and lake whitefish persisted in the Province of Ontario waters of Lake Ontario, but in New York State waters the commercial fishery that formerly targeted offshore species disappeared. Sportfishing was limited to nearshore species such as smallmouth bass *Micropterus dolomieu* and yellow perch *Perca flavescens*. Economic activity related to fishery resources was inconsequential (Eckert 1989).

An attempt was made in the mid-1950s to rehabilitate lake trout with limited stockings of hatchery-reared fish. However, survival to sexual maturity was poor, presumably due to sea lamprey predation and bycatch in gill nets fished for remaining coregonines. Little of the spawning habitat historically used by Atlantic salmon remained, and the limited abundance of this species worldwide made acquisition of eggs for hatching and rearing in hatcheries problematic. Thus, prospects for restoring native species were dim in the mid-1960s.

Similar problems arising from comparable fish community changes in Lake Michigan resulted in a unique fishery management initiative by the Michigan Department of Natural Resources in the mid-1960s. Their approach emphasized stocking hatchery-reared coho salmon *Oncorhynchus kisutch* and chinook salmon *O. tshawytscha* to consume overabundant alewives. This approach reduced beach fouling and converted alewife and rainbow smelt biomass into large piscivores attractive to anglers. This management strategy was soon emulated by other Great Lakes fishery management agencies, including the New York State Department of Environmental Conservation (NYSDEC).

Management of Lake Ontario 1968–1993

Coho salmon were introduced into New York waters of Lake Ontario in 1968, followed by chinook salmon in 1969. Hatchery-reared steelhead *O. mykiss* and brown trout *Salmo trutta* were added to the stocking program in 1973, followed by rainbow trout (nonanadromous *O. mykiss*) in 1974. Limited Atlantic salmon stocking began in 1983 to restore the presence of this native fish to Lake Ontario, encourage whatever natural reproduction could occur in the remaining limited habitat, and provide trophy fishing opportunities. The resulting sport fishery has become the most significant inland fishery in New York and an important economic asset.

Following the initiation of sea lamprey control in Lake Ontario by the Great Lakes Fishery Commission (GLFC) in 1972, lake trout stocking to restore a self-sustaining population was resumed in 1973. In 1974 the U.S. Fish and Wildlife Service began providing lake trout reared in federal hatcheries and, since 1977, has reared all of the lake trout stocked into U.S. waters of Lake Ontario.

The Province of Ontario conducts a similar and complementary program for its Lake Ontario waters. The activities of NYSDEC and the Ontario Ministry of Natural Resources (OMNR) pertaining to Lake Ontario fishery resources of common interest are closely coordinated through the Lake Ontario Committee, which operates under the auspices of the GLFC and provides a framework for cooperative management.

Fishing regulations for salmonine species other than lake trout and Atlantic salmon are intended to protect recently stocked fish and distribute the catch evenly among anglers, because natural reproduction of the nonnative species is not central to management goals. For these species there is a 12-in-minimum-size limit and, on the open lake, a daily limit of five trout and salmon in aggregate; in tributaries the daily limit is three in aggregate. There is no closed season. More stringent regulations apply for lake trout and Atlantic salmon to promote natural reproduction.

Since 1968, NYSDEC has stocked 85 million hatchery-reared trout and salmon into New York waters of Lake Ontario (Figure 1). The sharp decline in 1976 and 1977 resulted from a suspension of most stocking after the pesticide mirex and other chemical contaminants were found in the flesh of trout and salmon at levels sufficient for human consumption concerns. Kretser and Klatt (1981) found that most anglers preferred advisories as a basis for making decisions about fish consumption rather than a ban on fish possession and cessation of stocking. As a result, trout and salmon stocking was resumed in 1978. Restrictive fish consumption advisories are still in effect for all species of trout and salmon in Lake Ontario.

The New York annual stocking target of 5.2 million trout and salmon was first met in 1983 following the opening of the new Salmon River Hatchery in 1982. Stocking peaked at 6.6 million in 1984 and then declined toward the 5.2 million target as managers became more conscious of the need to maintain a predator–prey balance. Stocking was decreased to 3.45 million in 1993 to reduce predator demand significantly following indications of stress in the alewife and rainbow smelt populations (O'Gorman et al. 1993).

The estimated cost of rearing and transporting the trout and salmon stocked into New York waters of Lake Ontario during 1993, exclusive of lake trout, was US$788,000 (G. Seeley, NYSDEC, personal communication). Since the inception of the program the cost has increased at a rate of 4% annually. The stocking reduction effected in 1993 saved little money because the species involved were lake trout, which are reared by the U.S. Fish and Wildlife Service, and chinook salmon, which are stocked at a small size and cost relatively little to rear and stock.

The offshore boat fishery for trout and salmon averaged 435,000 angler trips per year during 1985–1993 (Eckert 1994). Based on the relationship between boat angler trips and shore and tributary

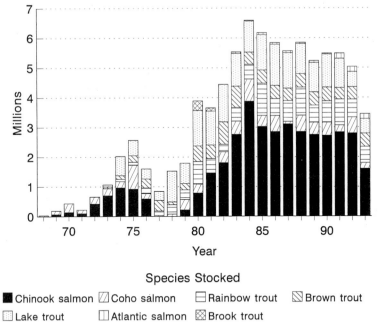

FIGURE 1.—Numbers of salmonines stocked into New York State waters of Lake Ontario, 1968–1993.

angler trips estimated in 1984 (NYSDEC, no date), we estimate that the Lake Ontario trout and salmon fishery supports an additional 450,000 angler trips at shore fishing sites and tributaries.

Connelly et al. (1990) developed estimates of economic indicators from a statewide angler survey in 1988. Trip-related expenditures for anglers fishing in Lake Ontario and its major tributaries were $110 million in 1988. Just over $1 billion in capital expenditures by anglers were attributable to freshwater fishing in New York State during 1988; based on 17% of the statewide fishing effort occurring in Lake Ontario, we conservatively estimate that Lake Ontario accounted for $170 million in capital expenditures. The estimated economic value of the Lake Ontario sport fishery, as determined by willingness to pay, was $69 million above expenditures.

Management Issues

By most accounts, the Lake Ontario trout and salmon fisheries created and maintained by a put-grow-and-take stocking program since 1968 have been successful. Gale (1987) described the development of Great Lakes trout and salmon fisheries as a "resource miracle." However, these fisheries have also resulted in significant issues for fishery managers.

The most longstanding vexation of Lake Ontario fisheries has been the persistent presence in the ecosystem of chemical contaminants such as mirex and polychlorinated biphenyls (PCB) that bioaccumulate in large piscivores. Although concentrations of these substances in fish have generally decreased over the last 20 years, restrictive consumption advisories are still required. This issue is a continuing concern for anglers and those with economic interests in Lake Ontario sportfishing. Stocking fish into a contaminated environment has been criticized (IJC 1990). However, the fish are not contaminated when they are stocked and reflect a broader ecosystem problem, as would self-sustaining native species had they persisted. The Lake Ontario ecosystem could not be considered healthy with the presence of excessive levels of chemical contaminants or the virtual absence of large piscivores.

The introduction of coho and chinook salmon into Lake Ontario resulted in the annual migration of many large fish into tributaries. Because habitat for natural reproduction is generally lacking or inaccessible, and because these species die naturally after spawning (or their spawning migration), snagging (foul-hooking) was permitted for anglers to take coho and chinook salmon in tributaries. Although seemingly pragmatic, this was a departure from acceptable angling methods. By the mid-1980s, some Lake Ontario tributaries were afflicted with an overcrowded, circus-like atmosphere during spawning runs that resulted in social problems among anglers and within riparian communities. Also, angling experience contributed to the growing knowledge that coho and chinook salmon could be caught by traditional angling techniques in tributaries even though they ceased feeding. Furthermore, snagging techniques spread to other fisheries in which snagging had never been permitted. In 1992 NYSDEC enacted regulations that would have phased out snagging after 1993. However, a successful legal challenge initiated by persons with an economic interest in the large number of people attracted by legal snagging resulted in a continuation of snagging in 1994. The court's decision was appealed by NYSDEC and remains the subject of ongoing litigation.

The status of alewife and rainbow smelt populations in Lake Ontario has received considerable attention from fishery managers because these species comprise most of the prey for stocked trout and salmon (Brandt 1986). Most of this attention has focused on the dynamics between salmonine predators and their prey. Jones et al. (1993) concluded that a tenuous balance existed between trout and salmon and their prey in the absence of a winter-induced mass mortality of alewives such as occurred in 1976–1977 (O'Gorman and Schneider 1986).

As trout and salmon stocking increased through the 1970s and early 1980s (Figure 1), nutrient abatement programs were reducing phosphorous levels in Lake Ontario. Open lake total phosphorous levels decreased from 20–25 µg/L during the mid-1970s to less than 10 µg/L by the early 1990s (Luckey 1994). Therefore, productivity in open-water food webs is declining.

Indications of stress in alewife and rainbow smelt populations, such as declining condition and population biomass (O'Gorman et al. 1993), prompted the Lake Ontario Committee to convene a scientific panel to evaluate the state of the Lake Ontario ecosystem as it pertained to the trout and salmon fishery. The panel concluded that prey fish populations were in a state of transition to lower production levels and biomass, and that predator demand at current stocking levels would likely lead to further reductions in prey fish abundance and decreased growth and survival of predators (Jones et al. 1992). The panel attributed these changes to declining productivity resulting from nutrient reduction, possibly exacerbated by the feeding activity of zebra mussels *Dreissena polymorpha*, a nonnative mollusk that has recently proliferated in the Great Lakes. Rand et al. (1994) concluded that the number of prey fish required in the daily ration of chinook salmon to maintain a constant growth rate more than doubled between 1978 and 1990 due to decreases in the mean weights of prey.

In response to the ecosystem changes occurring in Lake Ontario, NYSDEC and OMNR elected to reduce predator demand by 50% through stocking cuts. However, stocking reductions will not significantly reduce predator demand until 1995, and the future status of the alewife and rainbow smelt populations remains questionable. Should the alewife and rainbow smelt populations collapse, the potential response of the Lake Ontario fish community is uncertain, given the dearth of remaining native prey species.

Hartig et al. (1991) questioned the compatibility of water quality and fishery management programs in the Great Lakes and urged closer cooperation and coordination between program managers in the future. Recent ecosystem changes in Lake Ontario and fishery management responses represented a convergence of water quality and fishery management programs that was not widely anticipated and which amplified the admonitions of Hartig et al. (1991). A more ecosystem-oriented approach is indicated for both programs.

The difficulty in effecting timely adjustments to stocking rates in response to ecosystem changes is an example of the type of issue that has led to questions about the sustainability of fisheries supported with hatchery-reared fish. The GLFC (1992) acknowledged the necessity to maintain some Great Lakes fisheries with hatchery-reared fish, but only until self-sustaining populations could be restored. This view is based on the premise that wild fish will more readily persist in a changing environment and is related to emerging concepts of ecosystem integrity. The Revised Great Lakes Water Quality Agreement of 1978 (IJC 1988) established ecosystem integrity as a goal, and the concept is frequently encountered in ecosystem-level planning exercises. A benchmark of ecosystem integrity is the self-maintenance or self-sustainability of ecological systems (IJC 1991). Thus, the maintenance of fish populations with hatchery-reared fish implicitly indicates some lack of ecosystem integrity.

In the case of Lake Ontario, we think there is a strong likelihood that the fish community which existed in 1965, and the virtual absence of fishery benefits accruing from it, would have been little changed today had a put-grow-and-take stocking program not been implemented in 1968. Whatever the limitations and uncertainties associated with that stocking program, we believe the alternative was a continued scarcity of large piscivores and dominance of the offshore fish community by nonnative planktivores. To the extent that the productive potential of Lake Ontario can be fulfilled with self-sustaining, preferably native, predators, it should be. To the extent that it cannot, the use of hatchery-reared fish remains appropriate and beneficial.

References

Brandt, S. B. 1986. Food of trout and salmon in Lake Ontario. Journal of Great Lakes Research 12:200–205.

Christie, W. J. 1973. A review of the changes in the fish species composition of Lake Ontario. Great Lakes Fishery Commission Technical Report 23, Ann Arbor, Michigan.

Connelly, N. A., T. A. Brown, and B. A. Knuth. 1990. New York statewide angler survey. New York State Department of Environmental Conservation, Albany.

Eckert, T. H. 1989. Strategic plan for fisheries management in New York waters of Lake Ontario, 1989–2005. New York State Department of Environmental Conservation, Albany.

Eckert T. H. 1994. New York's 1993 Lake Ontario fishing boat census. Pages 59–111 in 1993 Bureau of Fisheries Lake Ontario Unit Annual Report to the Lake Ontario Committee and the Great Lakes Fishery Commission. New York State Department of Environmental Conservation, Albany.

Gale, R. P. 1987. Resource miracles and rising expectations: a challenge to fishery managers. Fisheries (Bethesda) 12(5):8–13.

GLFC (Great Lakes Fishery Commission). 1992. Strategic vision of the Great Lakes Fishery Commission for the decade of the 1990s. Great Lakes Fishery Commission, Ann Arbor, Michigan.

Hartig, J. H., J. F. Kitchell, D. Scavia, and S. B. Brandt. 1991. Rehabilitation of Lake Ontario: the role of nutrient reduction and food web dynamics. Canadian Journal of Fisheries and Aquatic Sciences 48:1574–1580.

IJC (International Joint Commission). 1988. Revised Great Lakes water quality agreement of 1978. IJC, Windsor, Ontario.

IJC (International Joint Commission). 1990. Fifth biennial report on Great Lakes water quality. IJC, Windsor, Ontario.

IJC (International Joint Commission). 1991. A proposed framework for developing indicators of ecosystem health for the Great Lakes region. IJC, Windsor, Ontario.

Jones, J. M., and six coauthors. 1992. Status of the Lake Ontario offshore pelagic fish community and related ecosystems in 1992. Report to the Lake Ontario Committee. New York State Department of Environmental Conservation, Albany.

Jones, M. J., J. F. Koonce, and R. O'Gorman. 1993. Sustainability of hatchery-dependent salmonine fisheries in Lake Ontario: the conflict between predator demand and prey supply. Transactions of the American Fisheries Society 122:1019–1030.

Kretser, W. A., and L. E. Klatt. 1981. 1976–77 New York angler survey Final Report. New York State Department of Environmental Conservation, Albany.

Luckey, F. 1994. Lakewide impacts of critical pollutants on United States boundary waters of Lake Ontario. New York State Department of Environmental Conservation, Albany.

NYSDEC (New York State Department of Environmental Conservation). No date. 1984 New York State Great Lakes angler survey, volume 1. New York State Department of Environmental Conservation, Albany.

O'Gorman, R., R. A. Bergstedt, and T. H. Eckert. 1987. Prey fish dynamics and salmonine predator growth in Lake Ontario, 1978–84. Canadian Journal of Fisheries and Aquatic Sciences 44:390–403.

O'Gorman, R., R. W. Owens, C. P. Schneider, and T. H. Eckert. 1993. Status of major forage fish stocks in U.S. waters of Lake Ontario, 1992. Report to the Lake Ontario Committee. National Biological Survey, Oswego, New York.

O'Gorman, R., and C. P. Schneider. 1986. Dynamics of alewives in Lake Ontario following a mass mortality. Transactions of the American Fisheries Society 115:1–14.

Rand, P. S., B. F. Lantry, R. O'Gorman, R. W. Owens, and D. J. Stewart. 1994. Energy density and size of pelagic prey fishes in Lake Ontario, 1978–1990: implications for salmonine energetics. Transactions of the American Fisheries Society 123:519–534.

Use of Cultured Salmonids in the Federal Aid in Sport Fish Restoration Program

JOSEPH MCGURRIN AND CHRISTOPHER UBERT

Trout Unlimited, 1500 Wilson Boulevard, Arlington, VA 22209, USA

DONALD DUFF

*USDA Forest Service, Wasatch–Cache National Forest
Room 8236-Federal Building, 125 South State Street
Salt Lake City, UT 84138, USA*

Abstract.—Declines in native trout and salmon throughout the United States have raised a number of issues about the impacts of cultured fish in coldwater ecosystems. The Federal Aid in Sport Fish Restoration Program (Program) has a significant effect on fisheries activities, particularly salmonid restoration. We examine the current use of cultured salmonids in the Program and make recommendations for improving Program accountability and, ultimately, the future use of all species of cultured fish. From 1985 to 1991, over 1.1 billion fish (adults and fingerlings) of all species were stocked under the auspices of the Program and accounted for 14.5% of total expenditures. For salmonid species, funding of fish culture activities accounted for an even greater percentage of Program expenditures; hatchery-related activities, including direct hatchery support, stocking activities, and stocked fish assessments and studies, accounted for 42% of the total funds spent for salmonid management. While some of the expenditures for hatcheries were dedicated to native trout and salmon restoration, this component of hatchery work was relatively small (less than 10% of the total hatchery projects and expenditures). The use of cultured fish under the Program, guided by responsiveness to angling constituencies, can benefit or harm native salmonids. To more fully evaluate this issue from a national perspective, the following is recommended. (1) The national database on Program activities should be expanded to include more detail on the scientific and technical aspects of stocking activities. (2) Surveys of future Program funding needs should include more specific questions about agency priorities for the use of cultured fish in fishery development and wild and native fish restoration. (3) Based on the analyses of present activities and future priorities, specific national guidelines for the use of cultured fish should be considered for adoption in the existing Program rules and policies.

The Federal Aid in Sport Fish Restoration Program (Program) is a cornerstone of the nation's fishing and boating community and will play a major role in the future of America's trout and salmon fisheries. In 1991 alone, states spent over US$35 million in Program funds for trout, salmon, char, and grayling projects. This money went toward salmonid fisheries that involve approximately 11.4 million anglers (age 16 and over) who spent nearly 100 million days and $3.5 billion on trout and salmon fishing (U.S. Fish and Wildlife Service 1993a). With the continuing crises involving salmon and native inland trout species, the potential role of the Program in restoring salmonids can be significant. One aspect of salmonid restoration that has been particularly controversial is the use of cultured salmonids in fishery conservation and management. This paper examines the current use of cultured salmonids in the Program and makes recommendations for improving Program accountability and, ultimately, the future use of all species of cultured fish.

Program Operation, Rules, and Stocking Regulations

The Program is a cooperative effort of federal and state government agencies, the sportfishing industry, and the angling public. It is designed to improve sportfishing and boating through the investment of anglers' and boaters' tax dollars in state sport-fishery conservation projects. After creating the Program in 1950 through the Federal Aid in Sport Fish Restoration Act (also known as the Dingell–Johnson Act; 16 U.S.C.A. §§ 777 to 777k), Congress expanded it in 1984 through the Wallop–Breaux Amendments to the Dingell–Johnson Act.

The U.S. Fish and Wildlife Service Division of Federal Aid (Federal Aid) serves as the central administrative office for the Program. One of the primary functions of the office is to formulate and establish program regulations and policies. The office translates the broad statements of the Federal Aid in Sport Fish Restoration Act into specific regulations that guide the Program. The standard criterion that Federal Aid uses to evaluate a proposed project is that it must be "substantial in character and design" (Federal Aid in Sport Fish Restoration Act). "Substantial in character" generally means that the state must demonstrate a substantial need to undertake a project. "Substantial in design" generally means that a project must be technically sound and competently designed.

In the first few years after Congress approved the Wallop–Breaux Amendments, there were few changes in the legislative or regulatory aspects of the Program. However, in 1991 a regulatory change allowed funds to be used for put-and-take stocking. The previous regulation only allowed for put-grow-and-take stocking and was intended to focus fish stocking on waters where the habitat could at least support fish survival and growth. The proponents of the change argued that the new regulation would foster the creation of more fishing opportunities. Opponents charged that such a regulation would only hurt the already weak Program emphasis on long-term restoration of fish stocks. Despite these objections, the Program change was adopted with little notoriety. Since that time, a variety of issues ranging from increasing native fish extinctions to overall Program accountability have combined to bring the use of cultured salmonids in the Program back into the forefront of policy debates.

Impacts of Stocking on Fish Biodiversity

Today, 10 years after the passage of the Wallop–Breaux Amendments, federal funding for state fisheries programs is approximately $200 million annually. While part of this money has been successfully used to create new fishing opportunities through fish stocking, questions have been raised about the impacts of stocking on fish biodiversity. Miller et al. (1989) reported on factors that led to the extinctions of three genera, 27 species, and 13 subspecies of North American fishes over the past 100 years. In most cases, multiple factors caused the declines. Habitat degradation (73% of extinctions) and introduced species (68% of extinctions) were the most frequently cited factors. Two-thirds of the extinctions due to introduced species were the result of intentional introductions primarily associated with sportfishing development. Two well-documented cases involve pure native salmonid species—Alvord cutthroat trout *Oncorhynchus clarki* ssp. and silver trout *Salvelinus agassizi*—that could have provided unique sportfishing opportunities today (Miller et al. 1989; Behnke 1980). In both of these cases, nonnative salmonids were introduced to diversify fishing opportunities. The stocking resulted in interspecific competition and hybridization and the elimination of the native fish.

A more recent study of Threatened and Endangered fish also highlights the potential for conflict between fish stocking and the protection of fish biodiversity. In a review of 69 Threatened and Endangered fish species by the Non-Indigenous Species Task Force, the effects of introduced species were cited as a cause of decline in 40 cases (58%) and a potential threat in eight others (12%) for a total of 70% of the cases reviewed (Office of Technology Assessment 1993). Most of these introductions were again associated with sportfishing development. The study noted that fish stocking activities under the Sport Fish Restoration Program could be a possible threat to the future maintenance of aquatic biodiversity. While the Program cannot be expected to solve all the problems of native fish extinctions, it should be able to foster activities that do not damage biodiversity and also play an active role in wild and native salmonid restoration.

National Perspectives

To examine the potential scope and impacts of fish stocking in the Program, it is helpful to look at the Federal Aid Information Reporting System (FAIRS), a national database on Program activities and expenditures. A draft supplemental environmental impact statement (EIS) incorporated FAIRS as part of an overall review of past Program activities (U.S. Fish and Wildlife Service 1993b). As part of the analysis, activities were grouped in general categories (e.g., hatcheries, habitat, research, education) and the expenditures for each category were compiled over the years from 1985 to 1991. During these years, over 1.1 billion sport fish (adults and fingerlings) were stocked for maintenance or restoration of fisheries and accounted for 14.5% of total Program expenditures.

The International Association of Fish and Wildlife Agencies (IAFWA), the organization that represents state fisheries agencies on national policy and program issues, also used a national perspective in a Survey of Future Sport Fish Restoration Fund Needs 1993–2003 (Hussey and Reeff 1994). The authors noted that the survey was not intended to be a comprehensive "accounting type" of Program review. Instead, state agency personnel were asked to estimate what level of funding in the year 2003 would be required to maintain current levels of different fisheries activities. Stocking needs were included, but were lumped in a broad category of "fisheries improvements" that also encompassed habitat management, fish attractors and reefs, and chemical reclamation of waters. While the survey indicated that the amount of funding needed for fisheries improvements would double by the year 2003, no specific analysis was provided for stocking expenditures.

Although the EIS and IAFWA efforts did not include specifics about the types and results of stocking activities, they provide the first national perspectives about the amount of current Program stocking and the importance of determining future needs. With a more detailed understanding of the current scope of stocking activities and needs, it will be possible to discuss the implications of the Program for both fishery development and native fish restoration. One approach to obtaining more specific information is to examine Program stocking activities in terms of individual species or species groups such as salmonids.

Use of Cultured Salmonids in the Program

We used FAIRS to examine the extent of activity and the amount of Federal Aid funding dedicated to the use of cultured salmonids. The assessment included all the salmonid project data found in FAIRS as of 1 January 1994 (Table 1). The analysis of salmonid restoration activities in 36 states between 1989 and 1993 revealed a substantial emphasis on hatchery- and stocking-related activities. Hatchery-related activities including direct hatchery support, stocking activities, and stocked fish assessments and studies accounted for 40% of all projects reported and received 42% of the total funds allocated to salmonid restoration. Although some of the expenditures for hatcheries were dedicated to native trout and salmon restoration, this component of hatchery work was relatively small (less than 10% of the total hatchery projects and expenditures). Habitat-related activities such as acquisition and improvement accounted for only 3% of salmonid restoration projects. Although lumped into one broad category for this study, all other individual salmonid program elements, including aquatic education, access projects, management planning and regulation, and research and surveys, individually accounted for far less Program resources than the cultured fish programs.

The percentage of total Program expenditures for salmonid fish culture (42%) was much higher than found in previous studies for all species (14.5%) of cultured fish (U.S. Fish and Wildlife Service 1993b). This is probably due to differences in methodology and differences in species resource problems. In terms of methodology, the EIS analysis included expenditures only related to facility development and fish production. The Trout Unlimited study included hatchery planning, surveys, and maintenance costs in addition to facility development and fish production. More importantly, part of the difference in expenditures also was probably due to differences in fish species and resource problems. Given that most of the historical salmonid habitat has been altered and degraded to the point that it can no longer support wild production, salmonid recreational fisheries are more dependent on cultured fish than are other species.

TABLE 1.—Average expenditures of Federal Aid in Sport Fish Restoration Program funds for salmonid fisheries management activities in 36 states, 1989–1993.

Activity	Projects		Funding[a]	
	No.	%	Program	%
Hatchery-related	264	40	$27.43	42
Habitat-related	16	3	$ 0.84	1
All other studies	372	57	$36.93	57
Totals	652	100	$65.20	100

[a] Dollars are in millions.

Recommendations

Program stocking activities, guided by responsiveness to angling constituencies, can benefit or harm wild and native salmonid fisheries. If the use of cultured salmonids and other cultured fish in the Program is to be effectively guided in the future, the current national assessments of the Program need to go beyond basic surveys of activities and expenditures and examine the results and impacts of stocking programs. Three recommendations will help attain this goal.

1. The data on stocking programs should be expanded to include more detail on the scientific and technical aspects of activities. The present FAIRS database is an appropriate system for compiling this information. The results of this work will give a better national picture of the past and present roles of cultured fish in the Federal Aid in Sport Fish Restoration Program.
2. Surveys of future Program funding needs should include more specific questions about agency priorities for the use of cultured fish in fishery development and wild and native fish restoration. The IAFWA Survey of Future Sport Fish Restoration Fund Needs is an established vehicle for compiling information about state agency activities and could be used for this purpose.

State agencies should remain the prime source of "on-the-ground" information about the problems and opportunities associated with the Program. This information is the key feedback mechanism for future planning for the use of cultured fish in sport-fish restoration.

3. Based on the analyses of present activities and future priorities, specific national guidelines for the use of cultured fish should be considered for adoption in the existing Program rules and policies. The development of guidelines should be a joint effort of the U.S. Fish and Wildlife Service and the states in consultation with fisheries constituencies and the public.

The above recommendations are made in the light of the ongoing battles over hatchery and wild fish management. The conflicts have already spilled over to arguments about the overall effectiveness of the Program. The final resolution of wild and hatchery fish issues will not be won by using fragmented and anecdotal information. Improvements in the use of cultured fish in the Program will only come about by collecting and analyzing evidence from a variety of sources and incorporating the information into effective Program policies.

References

Behnke, R. J. 1980. A systematic review of the genus *Salvelinus*. Pages 441–480 *in* E. K. Balon, editor. Charrs, salmonid fishes of the genus *Salvelinus*. Dr. W. Junk, Netherlands.

Hussey, S. L., and M. Reeff. 1994. Survey of future Sport Fish Restoration Fund needs: 1993–2003. International Association of Fish and Wildlife Agencies, Washington, DC.

Miller, R. R., J. D. Williams, and J. E. Williams. 1989. Extinctions of North American fishes during the past century. Fisheries (Bethesda) 14(6):22–38.

Office of Technology Assessment. 1993. Harmful nonindigenous species in the United States. U.S. Congress, Office of Technology Assessment, Washington, DC.

U.S. Fish and Wildlife Service. 1993a. National survey of fishing, hunting, and wildlife-associated recreation. U.S. Fish and Wildlife Service, Washington, DC.

U.S. Fish and Wildlife Service. 1993b. Supplemental environmental impact statement for the Federal Aid in Sport Fish and Wildlife Restoration Program. U.S. Fish and Wildlife Service, Washington, DC.

Contribution of Nonnative Fish to California's Inland Recreational Fishery

DENNIS P. LEE

California Department of Fish and Game
1701 Nimbus Road, Suite C, Rancho Cordova, California 95670, USA

Abstract.—California's native inland fish fauna includes only 10 species typically classified as game fish. During the past 125 years, 30 nonnative fish species were introduced into California to enhance recreational fishing opportunities. Many of these species subsequently developed naturalized populations. In addition, water development in California altered natural environments, impacted native fishes, and created new habitats. Nonnative fish species were introduced into many of these altered and new habitats to enhance angling opportunities. Fishing for nonnative species such as black basses *Micropterus* spp., catfishes *Ictalurus* spp., sunfishes *Lepomis* spp., and striped bass *Morone saxatilis* was 42–77% of the angling effort reported in periodic angler surveys conducted during 1936–1991. Whereas salmonids, principally trout, are the preferred species sought by the majority of California anglers, fishing for nonnative fishes is popular and has contributed to California's inland angling effort.

California native inland cool and coldwater fishes categorized as game fish include four anadromous species, white sturgeon *Acipenser transmontanus*, green sturgeon *A. medirostris*, chinook salmon *Oncorhynchus tshawytscha*, and coho salmon *O. kisutch*; two species with both anadromous and resident populations, rainbow trout *O. mykiss* and cutthroat trout *O. clarki*; and three resident species, golden trout *O. aguabonita*, bull trout *Salvelinus confluentus*, and mountain whitefish *Prosopium williamsoni*. California's only native warmwater game fish is the Sacramento perch *Archoplites interruptus*.

California's early fisheries management was limited to new species introductions (Shebley 1917; Evermann and Clark 1931), harvest regulations, and trout and salmon stocking. It was guided by the premise that prey were abundant and natural recruitment of game fish was low. To enhance recreational and commercial fishing opportunities, 30 different species were introduced by the California Fish Commission prior to 1900. Shebley (1917) reported that the success of many of these introductions was regarded as among the greatest achievements in fish culture and acclimatization. Game-fish species introduced included black basses *Micropterus* spp., sunfishes *Lepomis* spp., crappies *Pomoxis* spp., catfishes *Ictalurus* spp. and bullheads *Ameiurus* spp., trout and salmon *Salmo*, *Salvelinus*, and *Oncorhynchus* spp., American shad *Alosa sapidissima*, striped bass *Morone saxatilis*, white bass *Morone chrysops*, walleye *Stizostedion vitreum*, northern pike *Esox lucius*, and muskellunge *E. masquinongy*. Of these, only walleye, northern pike, and muskellunge did not establish self-sustaining (naturalized) populations.

Following dam construction on most of California's larger rivers, many native species were unable to maintain viable populations. Dams blocked access to spawning and nursery habitats, reservoir habitats were often unsuitable, and riverine habitats were affected by changes in flow patterns and physical characteristics. Hatchery propagation of trout and salmon was and still is used to mitigate and enhance affected populations. Nonnative species such as black basses, sunfishes, crappies, and catfishes were introduced into reservoirs and lower elevation streams to enhance fishing opportunities. Brown trout *Salmo trutta* and brook trout *Salvelinus fontinalis* were stocked throughout California into lakes and streams with suitable coldwater habitat. Rainbow trout, channel catfish *I. punctatus*, largemouth bass *Micropterus salmoides*, and striped bass have been and presently are reared in California as commercial food fishes and are stocked into private and public waters to enhance fishing.

Management of nonnative game fish remains a high priority for the California Department of Fish and Game (CDFG). Despite the apparent benefits from these species, little information has been published on their contribution to California's inland recreational fishery. In this report I review information on California's inland recreational catch and effort for native and nonnative species.

Methods

From 1936 to 1974, CDFG conducted periodic angling surveys to estimate California's inland recreational fish catch and effort (Curtis 1940; Calhoun 1950, 1951, 1953; Skinner 1955; Ryan 1959; Seeley et al. 1963; McKechnie 1966; Emig 1971; Pelzman

1973; Lal 1979). During 1936–1957, inland species were grouped as steelhead (anadromous rainbow trout), other trout, salmon (riverine), striped bass, black basses, crappies, sunfishes, catfishes, and no preference. After 1957, sunfishes and crappies were grouped together as panfishes, and kokanee (lacustrine sockeye salmon *O. nerka*) were included with trout (Ryan 1959). During 1936–1948, angling effort was based on fishing licenses sold (Calhoun 1950). Beginning in 1953, angling effort was based on anglers who fished (i.e., anglers who purchased a license but did not fish were excluded from estimated fishing effort; Lal 1979).

Results of the U.S. Fish and Wildlife Service (USFWS) national fishing and hunting surveys for California provided estimated angling effort for crappies, panfishes, white bass, black basses, striped bass, catfishes and bullheads, trout, salmon, steelhead, and other fishes in 1985 (USFWS 1989) and 1991 (USFWS 1994). A telephone survey conducted in 1988 identified California angler preferences (Fletcher and King 1988). In this survey the types of fish sought by inland anglers were grouped into species categories that included trout, black basses, striped bass, catfishes, steelhead, panfishes, sturgeons, corvinas *Cynoscion* spp., tilapias *Tilapia* spp., shads, croakers (Sciaenidae), other fishes, and no favorite fish.

Survey information did not distinguish between native and nonnative species. For my review, I grouped fish categories that included trout, salmon, steelhead, sturgeons, orangemouth corvina *Cynoscion xanthulus*, and bairdiella *Bairdiella icistia* as native species; this group included both native and nonnative trout and salmon. I grouped categories that included black basses, striped bass, white bass, catfishes, bullheads, crappies, panfishes, sunfishes, and American shad as nonnative species. Sacramento perch, a native species, is grouped with panfishes in the surveys.

Results

Fishing has been a popular recreational activity in California. The number of licensed California anglers steadily increased from 1935 to 1967 and reached a peak in 1981 (Figure 1). Since then sales of fishing licenses have dropped steadily until 1994 when sales showed a marked increase (CDFG 1994). Numbers of anglers fishing for nonnative species has varied annually and has accounted for approximately 42–77% of California inland anglers since 1936 (Figure 2).

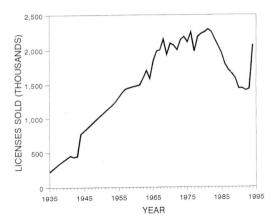

FIGURE 1.—California inland angling license sales, 1935–1994.

Discussion

Surveys indicate that anglers fishing for native species such as chinook and coho salmon, rainbow trout, and white and green sturgeons, continue to contribute to recreational fishing in California. A large part of this effort is in California's rivers and streams. Fletcher and King (1988) indicated that the majority of anglers preferred to fish for trout and salmon in 1988 and that 36% of urban and 43% of rural California inland anglers most often fished in rivers and streams.

Based on CDFG and USFWS surveys, at least 50% of inland fishing effort in California is for nonnative fishes. Most of these species have become naturalized in California waters. Attempts to enhance angling for species such as black basses, channel catfish, and striped bass through hatchery propagation have been discontinued by the CDFG. However, hatchery-reared, catchable-size catfish are stocked in a few lakes and reservoirs by private aquaculturists.

The percentage of anglers targeting specific species or groups varied in the surveys. Striped bass fishing has steadily declined, most likely as a result of declining Sacramento–San Joaquin Delta striped bass populations. Sunfishes and crappies were plentiful following construction of numerous multipurpose impoundments from 1950 to 1970. After inclusion of panfishes in the surveys in 1954, panfish anglers constituted up to 24% of inland anglers. However, in 1988 panfishes were reported to be one of the least preferred groups of fish by California anglers (Fletcher and King 1988).

The contribution of nonnative salmonids is un-

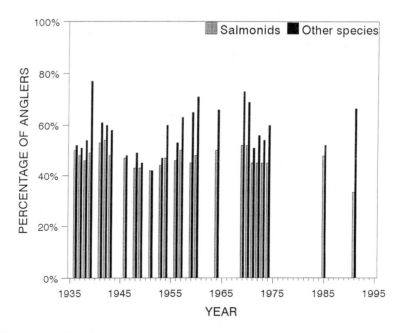

FIGURE 2.—Percentages of California inland anglers fishing for native (salmonids) and nonnative (other) species, 1936–1991.

known because they are not distinguished from native species in the surveys. However, wild and cultured rainbow trout are the most important fish in California's inland coldwater management program. Up to 12 million catchable-size rainbow trout are stocked annually in California waters.

California freshwater anglers expended an estimated US$1.2 million in 1985 (USFWS 1989) and $1.7 million in 1991 (USFWS 1994). The total economic value of recreational angling to California's economy most likely exceeds these estimates. At least half of these values could be attributed to fishing for nonnative fishes.

California's only warmwater game fish, the Sacramento perch, is considered rare in its native habitats (McGinnis 1984). Moyle (1976) reported the species eliminated in its original range due to habitat alteration and competition with nonnative fishes; however, the Sacramento perch's existence does not appear to be in jeopardy based on the success of populations established outside its native range. The CDFG has continued efforts to establish new populations where feasible. The contribution of Sacramento perch to California's recreational fishery is minor.

In California, development has had major effects on natural aquatic habitats through dam construction, water diversions, and flow alteration. California has 1,229 reservoirs that support recreational fisheries (Barrett and Cordone 1980). Fletcher and King (1988) reported that 57% of California's inland anglers fished in lakes and reservoirs. Most of California's higher-elevation reservoirs are generally infertile and provide coldwater habitat for trout and salmon. Lower-elevation multipurpose reservoirs provide habitat for both warmwater and coldwater species. Native fishes are often unable to maintain populations in these habitats (Murphy 1951; Courtois and Tippets 1979; Mills and Mamika 1980; Miller et al. 1989). Cordone and Nicola (1970) reported that a major cause of poor fishing for trout in reservoirs was inadequate natural reproduction. Recreational fisheries in these waters are maintained by stocking hatchery-reared fingerling and catchable-size trout, including rainbow, brown, and brook trout, lake trout *Salvelinus namaycush*, chinook and coho salmon, and kokanee. In lower-elevation reservoirs with suitable habitat, black basses, panfishes, catfishes, and striped bass have been introduced to enhance fishing opportunities. Nonnative forage species such as threadfin shad *Dorosoma petenense* and wakasagi *Hypomesus nipponensis* were also introduced to provide forage (Lee 1993). Today, nonnative fishes and hatchery-produced rainbow trout and chinook salmon support the majority of California's lake and reservoir fisheries.

Although data are not available, published information suggests most native populations have been reduced (Moyle et al. 1986; Miller et al. 1989; Moyle and Williams 1990). Declines or eliminations of native California fishes and problems associated with nonnative species have been reported (Mills and Mamika 1980; Courtois and Tippets 1979; Steinhart 1990). Nehlsen et al. (1991) reported that activities such as hydropower generation, fishing, logging, mining, agriculture, and urban growth have caused extensive losses to salmon and steelhead populations and their habitats. Some resident rainbow trout populations have been reduced by hydropower development and loss of habitat, whereas new populations have been established or are maintained by stocking programs. Golden trout, Lahonton cutthroat trout *O. clarki henshawi*, and mountain whitefish have been affected by introductions of nonnative fishes through hybridization and competition (A. Cordone, California Department of Fish and Game, personal communication).

In California, nonnative fish are popular with anglers and provide economic and recreational benefits. In altered habitats, nonnative fishes and hatchery-reared species have contributed to and enhanced recreational fisheries. Unfortunately, nonnative fishes have also exacerbated declines in native fishes resulting from habitat alteration. Nonetheless, it is doubtful that California's native fish fauna could provide similar recreational opportunities provided by nonnative fishes.

Acknowledgments

This work was funded by the Federal Aid in Sport Fish Restoration Act California Project F-51-R and the California Department of Fish and Game.

References

CDFG (California Department of Fish and Game). 1987. Final Environmental Impact Report—white bass management program. California Department of Fish and Game, Sacramento.

CDFG (California Department of Fish and Game). 1994. Combined Annual Report July 1, 1991–June 30, 1993. California Department of Fish and Game, Sacramento.

Barrett, J. G., and A. J. Cordone. 1980. The lakes of California. California Department of Fish and Game, Inland Fisheries Administrative Report 80-5, Sacramento.

Calhoun, A. J. 1950. California angling catch records from postal card surveys: 1936–1949; with an evaluation of postal card nonresponse. California Fish and Game 36:177–234.

Calhoun, A. J. 1951. California statewide angling catch estimates for 1949. California Fish and Game 37:69–75.

Calhoun, A. J. 1953. Statewide California angling estimates for 1951. California Fish and Game 39:103–113.

Cordone, A. J., and S. J. Nicola. 1970. Harvest of four strains of rainbow trout, *Salmo gairdneri*, from Beardsley Reservoir, California. California Fish and Game 56:271–287.

Courtois, L. A., and W. Tippets. 1979. Status of the Owens pupfish, *Cyprinodon radiosus* (Miller), in California. California Department of Fish and Game, Inland Fisheries Endangered Species Program Special Publication 79-3, Sacramento.

Curtis, B. 1940. Angler's catch records in California. Transactions of the American Fisheries Society 69:125–131.

Emig, J. W. 1971 California inland angling survey for 1969, with corrections for the 1964 survey. California Fish and Game 57:99–106.

Evermann, B. W., and H. W. Clark. 1931. A distributional list of species of freshwater fishes known to occur in California. California Fish and Game Fish Bulletin 35, Sacramento, California.

Fletcher, J. E., and M. King. 1988. Attitudes and preferences of inland anglers in the state of California. Final Report to California Department of Fish and Game, Sacramento.

Lal, K. 1979. California inland angling survey for 1971, through 1974. California Fish and Game 65:4–22.

Lee, D. P. 1993. Recreational fishery management strategies for California reservoirs. Proceedings of the American Fisheries Society, Western Division, 1993 Annual Meeting, Sacramento, California.

McGinnis, S. M. 1984. Freshwater fishes of California. University of California Press, Berkely.

McKechnie, R. J. 1966. California inland angling survey for 1964. California Fish and Game 52:293–299.

Mills, T. J., and K. A. Mamika. 1980. The thicktail chub, *Gila crassicauda*, an extinct California fish. California Department of Fish and Game, Inland Fisheries Endangered Species Program Special Publication 80-2, Sacramento.

Miller, R. R., J. D. Williams, and J. E. Williams. 1989. Extinctions of North American fishes during the past century. Fisheries (Bethesda) 14(6):22–38.

Moyle, P. B., H. W. Li, and B. A. Barton. 1986. The Frankenstein effect: impact of introduced fishes in North America. Pages 415–426 *in* R. H. Stroud, editor. Fish culture in fisheries management. American Fisheries Society, Fish Culture Section and Fisheries Management Section, Bethesda, Maryland.

Moyle, P. B. 1976. Inland fishes of California. University of California Press, Berkeley.

Moyle, P. B., and J. E. Williams. 1990. Biodiversity loss in the temperate zone: decline of the native fish fauna of California. Conservation Biology 4:275–284.

Murphy, G. I. 1951. The fishery of Clear Lake, Lake County, California. California Fish and Game 37:439–484.

Nehlsen, W., J. E. Williams, and J. A. Lichatowich. 1991. Pacific Salmon at the crossroads: stocks at risk from California, Oregon, Idaho, and Washington. Fisheries (Bethesda) 16(2):4–21.

Pelzman, R. J. 1973. California inland angling survey for 1971. California Fish and Game 59:100–106.

Ryan, J. H. 1959. California inland angling estimates for 1954, 1956, and 1957. California Fish and Game 45: 93–109.

Seeley, C. M., R. C. Tharratt, and R. L. Johnson. 1963. California inland angling surveys for 1959 and 1960. California Fish and Game 49:183–190.

Shebley, W. H. 1917. History of the introduction of food and game fishes into the waters of California. California Fish and Game 3:3–12.

Skinner, J. E. 1955. California statewide angling catch estimates for 1953. California Fish and Game 41:19–32.

Steinhart, P. 1990. California's wild heritage—threatened and endangered animals in the golden state. California Department of Fish and Game, Sacramento.

USFWS (U.S. Fish and Wildlife Service). 1989. 1985 National survey of fishing, hunting and wildlife-associated recreation—California. USFWS, Washington, DC.

USFWS (U.S. Fish and Wildlife Service). 1994. 1994 National survey of fishing, hunting and wildlife-associated recreation—California. USFWS, Washington, DC.

A Review of Chinook Salmon Resources in Southeast Alaska and Development of an Enhancement Program Designed for Minimal Hatchery–Wild Stock Interaction

WILLIAM HEARD

National Marine Fisheries Service
Auke Bay Laboratory, Alaska Fisheries Science Center
11305 Glacier Highway, Juneau, Alaska 99801, USA

ROBERT BURKETT

Alaska Department of Fish and Game
Division of Commercial Fisheries and Fisheries Management and Development
Post Office Box 3-2000, Juneau, Alaska 99802, USA

FRANK THROWER

Auke Bay Laboratory

STEVE MCGEE

Alaska Department of Fish and Game

Abstract.—Chinook salmon *Oncorhynchus tshawytscha* in southeast Alaska are characterized by diverse origins and life histories. Compared with other salmon, endemic stocks spawn in a limited number of streams. Complex mixtures of stocks of variable ages, originating from southeast Alaskan streams and more southerly non-Alaskan streams, occupy marine waters of the area. Chinook salmon hatcheries in southeast Alaska increased from 2 in 1976 to 15 in 1992. Adult production from hatcheries in 1991, 1992, and 1993 was 112,000, 79,000, and 73,000, fish respectively. Objectives for enhancing this species were to provide more salmon for Alaskan fisheries and help reduce fishing pressure on depressed wild stocks. Enhancement was developed to minimize adverse effects on wild populations. Specific approaches regarding siting of hatcheries, broodstock selection, disease control, and genetic protocols were taken. No hatcheries were built on streams with spawning populations of chinook salmon and most were 50 to 240 km from wild stocks. Five hatchery stocks, from five discrete wild stocks in different geographic parts of the region, were developed for use at specific hatcheries. Evaluation of contributions to fisheries and interactions with wild stocks was based on tagged hatchery smolts. Examination of 40,762 adults in 8 wild chinook salmon systems during 1979–1993 revealed 123 hatchery-origin fish (0.3%). However, in the Farragut River, hatchery strays exceeded 8% of adults examined, and in 1 year represented 26% of adults sampled. Straying to the Farragut River was decidedly nonrandom and raises questions about location and culture practices of certain hatcheries and streams that attract hatchery fish. Wild chinook salmon spawners in regional streams have increased or remained stable since the enhancement program began in the 1970s. Hatcheries provided an additional 42,000–70,000 chinook salmon annually to recreational and commercial fisheries during the years 1990–1992. Other than possibly in the Farragut River, deleterious wild stock-hatchery stock interactions of chinook salmon in southeast Alaska appear minimal.

Chinook salmon *Oncorhynchus tshawytscha*, the least abundant salmon in southeast Alaska, is a highly prized species by both commercial and sport fisheries. Coastwide declines in wild chinook salmon stocks and mixed-stock interception fisheries for this species in coastal waters of Canada and the United States were among the major factors leading to signing of the Pacific Salmon Treaty (Treaty) in 1985.

Mixed-stock fisheries for chinook salmon along the Pacific coast are the result of commingled oceanic migration patterns. Smolts originating from stocks in Oregon, Washington, British Columbia, and Alaska migrate north along the coastline (Mason 1965; Major et al. 1978; Healey 1983, 1991), and adults are caught in fisheries far from natal waters. Various fisheries probably have contributed to declining runs of chinook salmon; however, the loss and deterioration of natural habitats due to industrialization, urbanization, other land-use practices, and, especially, the damming of rivers in southern parts of the range are thought to be the main factors in the coastwide decline of many stocks (Netboy 1980; CRMJS 1992).

Beginning in 1978, commercial fisheries in southeast Alaska were managed under a fixed harvest quota. The quota, initially a harvest range established by the Alaska Department of Fish and Game (ADF&G) and the North Pacific Fishery Management Council (NPFMC), was set well below historical harvests (NPFMC 1978). These quotas became part of a 15-year stock-rebuilding program, begun in 1981, for natural stocks that spawn in southeast Alaska and transboundary rivers that originate in Canada. In addition to management restrictions on commercial fisheries, enhancement programs were started that included operation of hatcheries and rehabilitation efforts on natural runs. Except for harvest management of wild stocks, the term enhancement, as used in this paper, includes any action taken to increase the numbers of chinook salmon of local origin available to common-property fisheries in southeast Alaska; it does not include allocated harvest of wild stocks. Rehabilitation, a form of enhancement, relates to efforts to increase the production of chinook salmon in streams with natal runs. Common-property fisheries are general fisheries open to commercial fishers or anglers; common-property fisheries contrast with cost-recovery fisheries that allow a restricted proprietary harvest of hatchery fish in terminal migration areas near a hatchery.

Catch limits for chinook salmon in Alaska were adopted as quotas by the Treaty in 1985 and also were implemented in parts of British Columbia. Under Treaty provisions, stock-rebuilding programs began on a coastwide basis. In southeast Alaska, quotas for chinook salmon were established for commercial and sport fisheries.

Objectives for enhancing this species in the region were focused on providing more Alaska-origin fish for local fisheries and reducing fishing mortality on depressed wild stocks. Rehabilitation of natural runs included two approaches. First, natural escapements were increased by restricting or eliminating fisheries known to target specific stocks in marine waters near the mouths of spawning rivers. Second, a few depressed runs were stocked with indigenous-origin hatchery-produced fry. In addition, policies were implemented prohibiting wide-scale releases of hatchery fry or smolts into wild stocks within the region. The principal enhancement effort involved building a number of hatcheries to help minimize social and economic impacts of quota-restricted fisheries by providing supplemental catches during the stock-rebuilding process. In this paper we examine the chinook salmon resources in southeast Alaska and describe development of an enhancement program to produce more chinook salmon for the region while attempting to minimize hatchery stock-wild stock interactions.

Chinook Salmon in Southeast Alaska

Part of the Alexander Archipelago, southeast Alaska is composed of large and small islands along a narrow mainland of coastal fjords and mountains. Over 2,000 primary watersheds (watersheds flowing into salt water) on islands and the mainland support populations of pink salmon *O. gorbuscha*, chum salmon *O. keta*, and coho salmon *O. kisutch*. Many lake systems in the region also support runs of sockeye salmon *O. nerka*.

Chinook Salmon Resources

Less than 40 watersheds support spawning populations of chinook salmon. The largest rivers with chinook salmon in this region, which also support the largest spawning populations of chinook salmon in the region, are the Taku, Stikine, and Alsek rivers (Figure 1). These rivers originate in drier, more subarctic environments of northern British Columbia and Yukon Territory; most spawning in these rivers occurs in Canada. The other streams supporting spawning populations are shorter (generally less than 50 km) streams along the mainland coast, and many produce only a few hundred spawners annually (Kissner 1986; Pahlke 1993). A small extant stock in the King Salmon River at the northern end of Seymour Canal on Admiralty Island may be the only viable chinook salmon run on islands in the region.

Populations in the region have been grouped into categories based on the approximate numbers of spawners. Naturally spawning chinook salmon in the region are found in 22 to 24 minor stream systems (<1,500 spawners), 8 medium systems (1,500 to 10,000 spawners), and 3 major systems (>10,000 spawners) (Holland et al. 1983; Kissner and Hubartt 1987; Pahlke 1993). Many stream systems are remote and have sections of glacial meltwater that are difficult to survey. Some may not have chinook salmon spawning in them every year. Estimates of spawner escapements for some stocks in the region show an overall increase in numbers of spawners in recent years (Table 1). Escapement is defined as the number of maturing salmon in a particular year that arrive on spawning grounds or return to a hatchery.

Freshwater Biology

Adult chinook salmon in southeast Alaska migrate into spawning streams from May through July,

TABLE 1.—Average annual spawning escapements of chinook salmon in southeast Alaska river systems between 1975 and 1993 (adapted from Table 1 in McGee et al. 1993). Escapement goals were established by the Alaska Department of Fish and Game.

River system	Goal	1975–1980	1981–1985	1986–1990	1991–1993
Major systems					
Alsek	7,300	4,028	2,632	3,315	3,322
Taku	36,515	12,440	14,923	23,440	31,223
Stikine	21,200	7,887	15,374	19,258	30,177
Medium systems					
Andrew	750	379	495	1,012	1,194
Blossom	1,280	163	739	1,156	369
Chickamin	1,440	338	1,149	1,579	652
Keta	800	407	1,004	1,214	454
Situk	600	1,299	995	1,238	1,022
Unuk	2,880	1,469	1,993	2,427	1,386
Other medium systems	3,875	2,028	3,187	4,313	2,538
Minor systems					
King Salmon	250	92	213	210	177
Other minor systems	5,250	1,925	4,473	4,410	3,717
Total	82,140	32,455	47,177	63,572	76,231

and peak spawning usually occurs from late July to mid-August (Kissner 1978). Most are stream-type chinook salmon (juveniles rear in freshwater for one or more years before migrating to the ocean) (Meehan and Siniff 1962; Healey 1983; Kissner and Hubartt 1987; Van Alen et al. 1987). Healey (1991) noted ocean-type chinook salmon (juveniles migrate to sea during their first year of life) are predominant south of the Alaska–British Columbia border. Although stream-type chinook salmon are predominant in southeast Alaska, ocean-type salmon have been documented in the Situk River (Kissner 1986; Johnson et al. 1992a). Kissner (1988) thought the proportion of age-0 and age-1 smolts in the Situk River might vary from year to year depending on annual temperature regimes. Based primarily on early emergence timing of fry and an extended growing season, Johnson et al. (1992a) suggested that Situk River chinook salmon have life history characteristics intermediate between typical stream- and ocean-type populations. Ocean-type fish probably constitute less than 10% of total chinook salmon production from southeast Alaskan streams.

The timing of downstream migration for age-1 chinook salmon smolts in southeast Alaska is from late April to mid-June with a peak in mid-May (Meehan and Siniff 1962). Johnson et al. (1992a) found age-0 smolts migrating to the lower Situk River rearing areas in late June and July, entering the main estuary by early August, and migrating to sea by September.

Marine Biology

Biology of chinook salmon in marine waters of southeast Alaska involves both rearing and migration patterns of stocks that originate within the region and a mixture of stocks that originate from areas to the south. Each of these groups has different behavior patterns as to when, where, and at what life history stage it occupies specific marine waters of the region (Orsi et al. 1987; Van Alen et al. 1987; Orsi 1988).

Healey (1983) suggests that ocean-type chinook salmon remain relatively close to the coastline, often in sheltered marine waters around island complexes, whereas stream-type fish move offshore. Orsi et al. (1987) and Orsi (1988) found complex temporal and spatial behavior patterns for stream- and ocean-type fish during their first years of marine life; specific ages of these fish were found in outside coastal and inside waters of southeast Alaska. Outside coastal waters are those contiguous with the Gulf of Alaska along the coastline of outer islands, whereas inside waters are fjords, inlets, and straits among the islands and along parts of the mainland.

Ocean-type chinook salmon of marine ages-.1 and -.2 were 54% of the fish collected in different

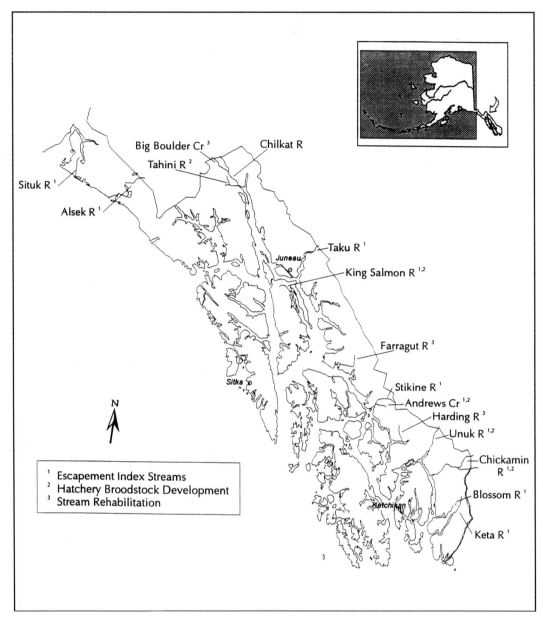

FIGURE 1.—Location of several southeast Alaska chinook salmon stocks used for escapement level indices, hatchery broodstock development, and stream rehabilitation projects.

parts of southeast Alaska by experimental trolling with small lures (Orsi 1987). These fish, of non-Alaskan origin, were found primarily in outside coastal waters (Orsi et al. 1987). No ocean-type marine age-.0 fish were collected, indicating these chinook salmon do not migrate into southeast Alaska waters during their first summer (Orsi 1988). In contrast, marine age-.0 stream-type chinook salmon were present primarily in inside waters during late summer and early fall. Whereas some of these stream-type juveniles may have been migrating toward offshore oceanic waters, as the life history pattern reported by Healey (1983) would suggest, data from Kissner and Hubartt (1987) indicate that at least some southeast Alaska stream-type stocks remain in inshore waters during much of their marine life. The presence of immature chinook salmon of different life history types in inside and outside marine waters of southeast Alaska illustrates the importance this region has as a primary nursery area for this species (Orsi et al. 1987).

Regional Chinook Salmon Fisheries

Fisheries for chinook salmon in the region are managed by gear types, seasons, and location. Two fisheries, commercial trolling and marine sportfishing, target this species. Purse-seine and drift gill-net fisheries directed at other species of salmon are permitted to retain a limited number of incidentally caught chinook salmon.

Chinook salmon harvest in the region is limited by quotas established by the Pacific Salmon Commission under auspices of the Treaty. The present Treaty quota for southeast Alaska is 263,000 fish, and the Alaska Board of Fisheries allocates this quota among fisheries. Chinook salmon produced by southeast Alaskan hatcheries in excess of pre-Treaty levels do not count toward Treaty quotas. Fish from non-Alaskan hatcheries, however, generally are included in Treaty quotas because most hatchery production of this species from British Columbia and Pacific Northwest states was in place before the Treaty was signed.

Most chinook salmon harvested in the region are caught in the commercial troll fishery (Parker and Kirkness 1956; Gunstrom 1980). Traditionally this fishery is concentrated along outside coastal waters where the majority of chinook salmon caught are from non-Alaskan origins (NPFMC 1978). Prior to 1978 the commercial troll fishery was essentially open all year. More recently this fishery has been managed to allow fishing during a 6-month winter period, mid-October to mid-April, and intermittent shorter periods in June, July, and August (McGee et al. 1993). Intermittent fisheries in June are directed at mature fish from southeast Alaskan hatcheries, whereas those during July and August are directed at keeping overall catches within treaty limits and Alaska Board of Fisheries allocations. Since catch quotas were implemented, the summer commercial troll fishery for chinook salmon has declined from 169 d in 1978 to 4.5 d in 1992.

Although chinook salmon are available for sportfishing year-round in marine waters of southeast Alaska, most of the sport-fish catch occurs during a 5-month period from mid-April to mid-September. Average annual sport harvest in the region increased from 21,000 fish during 1977–1987 to 40,000 fish during 1988–1992. The largest sport harvest of chinook salmon occurred in 1991 when over 60,000 fish were caught (Suchanek 1994). The recent rapid growth in sport harvest of chinook salmon has caused conflicts over allocations within the Treaty quota and over catch limits imposed on different user groups of fishers. Increased sport-fish harvests of chinook salmon in recent years have coincided with increased production from enhancement efforts.

Sport catches of chinook salmon vary considerably in different parts of the region depending on seasonal abundance of fish. In spring and early summer, anglers target maturing fish returning to mainland rivers or regional hatcheries. At other times, marine sport angling is focused on immature fish. A minimum legal size limit of 71 cm total length is enforced on both commercial troll and marine sport fisheries. Freshwater fishing for chinook salmon in the region is restricted to a few streams.

History, Development of Guidelines, and Current Policy

Enhancement of chinook salmon in southeast Alaska was based on a different set of circumstances than those associated with similar programs in other states. First, spawning populations in wild chinook salmon stocks throughout the region were still relatively abundant. Information, however, on many of the smaller stocks was and remains minimal. Second, spawning and rearing habitats for chinook salmon in this region were essentially unaltered; there was no need to build large hatcheries to compensate for irrevocable habitat losses. Third, growing concerns about adverse interactions between hatchery stocks and wild stocks in other areas provided a framework for more conservative approaches to enhancement. There were clear examples of past mistakes in other programs that could be avoided in southeast Alaska. Paramount among guidelines for chinook salmon enhancement was a focus on preserving wild stocks and preventing adverse hatchery–wild stock interaction.

Chinook Salmon Plan

In 1983 an interagency planning group, under authority of the Commissioner of ADF&G, codified policy guidelines for the management and enhancement of chinook salmon that developed within southeast Alaska during the previous decade. This plan (Holland et al. 1983) considered many related issues including management of wild and cultured fish, stock transport, sensitive and nonsensitive zones (relative to locations of wild chinook salmon stock), genetic diversity among hatchery stocks, tagging and evaluation of both enhanced and wild stocks, hatchery siting, and alternative enhancement strategies. An important feature of developing this plan was active participation by a team of scientists to update the original plan and document progress and change. The most recent update (McGee et al. 1993) includes a comprehensive overview of the current status of chinook salmon enhancement and management in the region.

Genetics

Fish geneticists and other scientists established a formal ADF&G genetics policy to protect wild stocks while developing enhancement programs with Alaskan salmonids (Davis et al. 1985). The policy prohibits both interstate transport of live salmonids, including gametes, into Alaska and interregional transport of salmonids within the state (Davis and Burkett 1989). Furthermore, intraregional transport is permitted only after careful review of associated issues. The policy also includes guidelines for maintaining genetic diversity in both wild and hatchery stocks, preventing inbreeding in hatchery stocks, minimizing introgression of hatchery genes into wild stocks, establishing wild stock sanctuaries, and rehabilitating depressed endemic stocks. Rehabilitation involving fry stocking into natural chinook salmon streams can be done only with fry from the indigenous wild stock, either parental or first filial progeny (Davis et al. 1985).

Disease and Fish Health Issues

Fish disease and fish health management in Alaska operate under detailed policies set forth in comprehensive regulations and formal state statutes (SPRC 1988). Statutes and regulations include guidelines for wild fish transplants, broodstock screening, transfers of fish or eggs between hatcheries, diagnostic procedures, disease history of juvenile fish prior to release, and the general use of chemical disinfectants and therapeutic drugs (SPRC 1988). All chinook salmon hatcheries in southeast Alaska are inspected routinely by certified pathologists for compliance with fish health policies.

Extensive propagation of sockeye salmon in Alaska has led to a detailed protocol for infectious hematopoietic necrosis (IHN) virus (Meyers et al. 1990), a rhabdovirus that occurs widely in Alaskan sockeye salmon and can infect other salmonids including chinook salmon. State fish health policy prohibits, except under special conditions, raising sockeye and chinook salmon in the same hatchery. Although chinook salmon are susceptible to IHN, some Alaskan stocks may have immunity to some isolates of the virus (Wertheimer and Winton 1982).

Siting of Hatcheries

No hatcheries in southeast Alaska were built on streams with natural runs of chinook salmon. All chinook salmon hatcheries in the region (Figure 2) are located at or near tidewater. Most are 50 to 240 km from any endemic stock and are located on islands in designated nonsensitive zones for the species. Many hatcheries in the region are situated on water sources isolated by barrier waterfalls and without runs of any anadromous salmonids. Locations of hatcheries in this region may influence marine migration patterns of the chinook salmon smolts they release (Holland et al. 1983). Many hatchery sites were also selected to have areas for terminal fishing for hatchery fish where wild stocks would not be harvested.

Hatchery Broodstock Diversity

A genetics policy established by different agency scientists encourages genetic diversity among hatchery stocks and recommends that a single donor stock not be used to establish or contribute to more than three hatchery broodstocks (Davis et al. 1985). In addition, this policy and other similar recommendations advise maximizing diversity within hatchery broodfish by using 400 or more founder parents as an effective population size from donor stocks (Hynes et al. 1981; Allendorf and Ryman 1987). Because it was necessary to select broodstocks from native populations that have relatively few spawners each year, some elements of these guidelines were not met at all hatcheries.

Broodstock development generally involves collecting gametes from a donor stock for several years to secure several hundred parent fish during that period. Numbers of eggs removed from wild stocks in a given year depends on escapement. Removing eggs from wild populations is initiated only after a predetermined number of spawners enters the stream and thereafter some portion of all segments of the run is allowed to spawn naturally. In most cases gametes from different run segments, including different age components of the run, are included in broodstock founder parents. No spawners are removed from wild stocks when escapements are below threshold levels established by ADF&G.

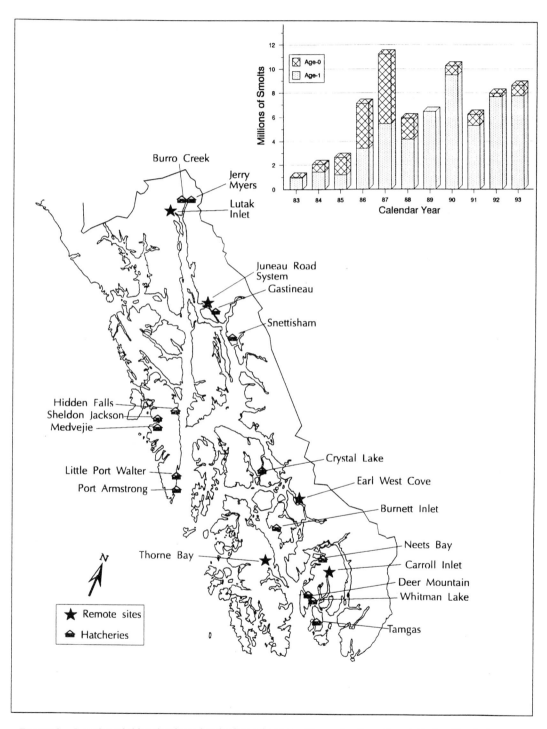

FIGURE 2.—Location of chinook salmon hatcheries and remote release sites in southeast Alaska. (Smolts from some hatcheries are held in marine net-pens at remote sites prior to release.) Insert shows number of smolts released by age and year, 1983–1993. (Smolt data were modified from McGee et al. 1993.)

TABLE 2.—Chinook salmon hatcheries in southeast Alaska—the operating agency, initial year of chinook salmon culture, principal broodstock, and origin of principal broodstocks.

Operating hatchery (code)	Principal agency[a]	Initial year	Stock	Origin
Burro Creek (BC)	PNP	1990	Tahini River	Wild donor and HR[b] from HF
Burnett Inlet (BI)	PNP	1987	Andrew Creek	HR from CL
Crystal Lake (CL)	ADF&G	1971	Andrew Creek	Wild donor
Deer Mountain (DM)	ADF&G	1977	Unuk River	Wild donor
Gastineau (G)	PNP	1988	Andrew Creek[c]	HR from CL
Hidden Falls (HF)	NSRAA	1981	Andrew Creek	HR from CL
Little Port Walter[d] (LPW)	NMFS	1976	Chickamin River	Wild donor
			Unuk River	Wild donor
			King Salmon River	Wild donor and HR from S
Medvejie (M)	NSRAA	1982	Andrew Creek	HR from CL and HF
Neets Bay (NB)	SSRAA	1981	Unuk River	Wild donor and HR from LPW
Port Armstrong (PA)	PNP	1985	Unuk River[c]	HR from LPW
Jerry Meyers (JM)	PNP	1985	Tahini River	HR from HF
Sheldon Jackson (SJ)	PNP	1984	Andrew Creek[c]	HR from CL
Snettisham (S)	ADF&G	1977	Andrew Creek[c]	HR from CL
			King Salmon River	Wild donor
Tamgas Creek (TC)	MIC	1982	Unuk River	HR from DM and LPW
Whitman Lake (WL)	SSRAA	1980	Chickamin River	Wild donor and HR from LPW

[a] Private nonprofit, PNP; Alaska Department of Fish and Game, ADF&G; Northern Southeast Regional Aquaculture Association, NSRAA; National Marine Fisheries Service, NMFS; Southern Southeast Regional Aquaculture Association, SSRAA; Metlakatla Indian Community, MIC.
[b] Hatchery return (HR) from indicated facility.
[c] Temporary surrogate stock; developments are underway for a change in the principal stock used at this hatchery.
[d] Primarily a research hatchery operated cooperatively with ADF&G. The three stocks are maintained separately, all smolts are tagged with coded wire tags before release, and all adults are decoded before spawning.

Five hatchery broodstocks derived from five discrete wild runs of chinook salmon in different geographic parts of southeast Alaska are now used in regional programs (Figure 1). Sensitive and nonsensitive zones are established to define where each stock can be used in hatcheries. In hatcheries on or close to the mainland, where most wild runs are located, three zones—northern, central, and southern—are designated as areas for using brood derived from wild stocks within the corresponding zone. Hatcheries located along the outside coastline of the outer islands are designated as nonsensitive zones where other stocks from within the region can be used.

Three stocks—Andrew Creek, Unuk River, and Chickamin River (Figure 1)—are now used at several hatcheries (Table 2). A King Salmon River broodstock is under development for eventual use at two or three hatcheries presently using another stock from surplus returns at other facilities. Also, a stock from the Tahini River now has limited use in two small hatcheries in Lynn Canal in the northern part of the region (McGee et al. 1993). Boulder Creek, Farragut River, and Harding River broodstocks presently are used for rehabilitation programs on those streams.

Enhancement Programs

Hatchery Development

In 1971, the Alaska state legislature created the Fisheries Rehabilitation Enhancement and Development Division (FRED) within ADF&G. The FRED goal was to ensure continued availability of fisheries resources to the state. Because salmon runs then were at all time lows, hatchery development and other enhancement activities were viewed as a way to help increase abundance to allow continued harvest. Options for enhancement expanded in 1974 when legislation was enacted permitting the operation of private nonprofit (PNP) hatcheries and the formation of regional aquaculture associations under ADF&G oversight. The primary goal for PNP hatcheries was to make contributions to common-property fisheries.

The number of hatcheries raising chinook salmon in southeast Alaska increased from one in 1971 to 15 in 1992 (Table 2). Presently these 15 hatcheries are operated by three governmental entities, two regional aquaculture associations, and six other PNP corporations (Table 2).

In addition to incubating and rearing chinook salmon, some facilities release smolts at marine net-pen sites located away from the hatchery (Figure 2). The main purpose of these remote releases is to create adult returns in areas, away from the hatchery, for specific commercial or recreational fisheries. Remote releases are approved only after careful evaluation of stocks involved and release locations relative to wild stocks. To maximize successful imprinting and final prerelease growth, smolts usually are released at remote sites after a 4- to 8-week culture period in marine net-pens (Josephson and Kelley 1993).

Smolt Production Levels

Combined production capacity for all hatcheries in the region in 1992 totaled about 12 million age-1 and 3 million age-0 smolts (McGee et al. 1993). Releases of smolts for the 11-year period, 1983–1993, have steadily increased from 1 million in 1983 to over 11 million in 1987 and have exceeded 6 million per year since 1986 (Figure 2). By the late 1980s Crystal Lake, Hidden Falls, Medvejie, Snettisham, Tamgas Creek, and Whitman Lake hatcheries were each routinely releasing 1 million or more smolts annually.

Broodstocks used in regional hatcheries were derived from wild stream-type stocks, and most smolt releases in recent years were age-1 smolts. Releases of age-0 smolts were made, partly for economic reasons, experimentally at several hatcheries (Heard 1985; Denton 1988a; Amend 1988). The cost of raising age-1 chinook salmon smolts in southeast Alaskan hatcheries can be four times the cost of producing age-0 smolts (Amend 1988). However, age-0 smolt releases have not been cost effective due to much lower marine survivals than for age-1 smolts. Marine survivals of age-0 smolts generally have been less than 1% (Denton 1988a; ADF&G, National Marine Fisheries Service, and Southern Southeast Regional Aquaculture Association, unpublished data).

Adults from Hatcheries

Annual chinook salmon production from southeast Alaskan hatcheries was 25,000–112,000 adults during 1985–1992 (Figure 3). These estimates included the number of fish caught in various common-property fisheries, the number of adults returning to hatcheries, and the number of adults recovered in proprietary cost-recovery fisheries conducted at some PNP hatcheries. Cost-recovery harvest and sale of a portion of the adult salmon returning to PNP hatcheries is allowed under Alaskan statutes to help cover operational costs and repay facility indebtedness. Such fisheries are only

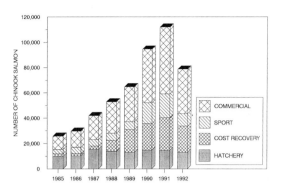

FIGURE 3.—Total southeast Alaskan hatchery production of chinook salmon based on catches in the commercial, sport, and proprietary (cost-recovery) fisheries and returns to the hatchery, 1985–1992.

allowed in terminal areas where returning hatchery adults are separate from other stocks. Cost-recovery harvest of chinook salmon now approximates one fourth of total hatchery production in the region (Figure 3).

In 1993, nine facilities in southeast Alaska recorded 1,000 or more adult chinook salmon either harvested in fisheries or returning to the hatchery, and three hatcheries each accounted for over 10,000 fish (Table 3). Crystal Lake Hatchery has been the consistent top producer of hatchery chinook salmon caught in common property fisheries in the region. The percentage contribution of southeast Alaskan hatcheries to total chinook salmon harvest in southeast Alaska since 1984 has ranged from less than 2% to almost 20% (Figure 4).

Annual harvest rate of chinook salmon produced by regional hatcheries in commercial and sport fisheries varies widely among hatcheries and between years. This rate is the total number of hatchery fish caught in fisheries divided by the sum of the number caught plus the number returning to the hatchery in a given year. With reduced fishing days in the commercial troll fishery, along with other management restrictions, harvest rates of Alaskan hatchery chinook salmon have declined in recent years. In 1992 the harvest rate from individual hatcheries ranged from 20 to 85% and averaged 50% across all hatcheries (McGee et al. 1993).

Marine survival of smolts released from southeast Alaskan hatcheries have varied widely between hatcheries and among years (Table 4). Several studies have been done in the region to measure effects of different freshwater culture methods and hatchery procedures on marine survival of chinook

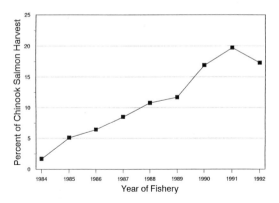

FIGURE 4.—Percent contribution of southeast Alaskan hatcheries to total common-property commercial and sport-fishery harvest of chinook salmon in southeast Alaska, 1984–1992 (adapted from Table 3 *in* McGee et al. 1993).

salmon smolts. These studies include the effects of different stocks on survival and maturation schedules (Hard et al. 1985), size at release on survival (Denton 1987), freshwater culture density on survival (Denton 1988b; Martin and Wertheimer 1989), and different diet formulations on growth and survival (Denton 1986; Martin and Leon, in press). Measured marine survivals for some hatcheries and broods have ranged from less than 0.1 to 14.9% (Table 4).

TABLE 3.—Estimated number of chinook salmon produced by southeast Alaska hatcheries, 1988–1993[a].

Hatchery	1988	1989	1990	1991	1992	1993
Burnett Inlet	0	0	0	93	1,109	763
Crystal Lake[b]	15,550	13,793	25,013	29,903	18,129	16,327
Deer Mountain	539	1,515	1,219	1,324	1,002	938
Gastineau	0	0	138	762	175	858
Hidden Falls	566	515	899	2,819	3,298	2,583
Little Port Walter	7,592	5,782	6,507	7,587	3,026	2,905
Medvejie	409	476	2,522	6,427	16,272	17,975
Neets Bay	17,320	25,669	15,657	9,470	8,908	11,058
Port Armstrong	286	277	1,074	804	1,651	2,385
Jerry Myers	0	34	56	91	32	57
Sheldon Jackson	0	174	352	490	467	880
Snettisham	2,100	1,284	3,736	2,704	3,644	7,056
Tamgas Creek	1,821	2,470	4,275	8,617	7,233	2,848
Whitman Lake[c]	4,873	8,889	29,150	32,610	10,582	3,786
Unallocated[d]	1,761	1,774	6,927	8,220	3,545	2,230
Totals	52,817	62,652	97,525	111,921	79,073	72,649

[a]Modified from data in Alaska Department of Fish and Game annual chinook salmon plan annex reports.
[b]Includes returns to Earl West Cove.
[c]Includes returns to Carroll Inlet.
[d]Primarily sport-caught chinook salmon unallocated by hatchery.

TABLE 4.—Marine survival of southeast Alaska hatchery chinook salmon from yearling smolt release to their recovery at the hatchery or in a fishery.

Hatchery and inclusive brood years	N^a	Percent survival		Source[b]
		Mean	Range	
Crystal Lake				
1979	3	9.3		1
1980	1	5.6		1
1981	1	4.4		1
1982	3	4.9	3.9–6.3	1
1983	3	2.7		1
1984	4	3.6		1
1985	3	1.4	0.8–2.8	1
1986	1	4.4		1
Deer Mountain				
1977	2	1.3		1
1978	4	3.9		1
1979	5	4.3	3.5–6.3	2
1980	2	2.2		2
1981	3	3.5	0.9–6.3	3
1984	6	4.2		1
1985	1	2.2		1
1986	1	1.7		1
Little Port Walter				
1976	8	8.5	2.2–14.9	4
1977	6	0.6		5
1978	8	1.6	0.7–3.0	6
1979	2	8.8	7.1–10.6	5
1980	2	3.4	2.9–3.8	5
1981	7	5.5	3.3–6.3	5
1982–1983	47	5.9	0.9–10.2	5
1984–1985	77	1.6		5
1986–1987	54	3.4	1.0–8.4	5
Medvejie				
1982–1985	9	1.1	0.1–2.8	7
1986–1987	4	2.9	2.5–3.4	7
Neets Bay				
1981	2	6.1		8
1982	3	8.8		8
1983	1	4.6		8
1984	5	4.4		8
1985	4	0.9		8
1986	2	1.0		8
1987	2	1.6		8
Whitman Lake				
1980	1	8.0		9

[a]Individual tag groups of smolts for which marine survival was determined.
[b]Sources are (1) ADF&G, unpublished; (2) Denton 1987; (3) Denton 1988b; (4) Hard et al. 1985; (5) National Marine Fisheries Service, unpublished; (6) Martin and Wertheimer 1989; (7) Northern Southeast Regional Aquaculture Association, unpublished; (8) Southern Southeast Regional Aquaculture Association, unpublished; and (9) Freitag 1988.

Other Enhancement Methods

In addition to raising and releasing smolts from hatcheries, other enhancement efforts for southeast Alaska chinook salmon include stocking fry for rehabilitating or enhancing wild runs, and stocking

TABLE 5.—Incidence of hatchery-produced chinook salmon in wild chinook salmon spawning rivers in southeast Alaska, 1979–1993.

River system	Number of chinook salmon examined	Survey years	Number of hatchery tags recovered[a]	Expanded number to account for unmarked hatchery fish	Percent hatchery origin
Chickamin	1,209	1985–1990	3	27	2.23
Chilkat	3,759	1983–1987, 1989–1993	7	7	0.19
Farragut	617	1983–1985, 1989, 1991–1993	34	51	8.27
Harding	363	1986, 1989–1993	2	4	1.10
King Salmon	459	1979, 1981–1992	0	0	
Stikine	10,260	1979–1992	4	8	0.08
Taku	18,383	1979–1990	0	0	
Unuk	5,791	1985–1993	6	26	0.45
Totals	40,841		56	123	0.30

[a] These data do not include eight hatchery tags (expanded to 10 fish) from Taku River and three hatchery tags (equal 3 fish) from Stikine River because the total number of fish examined was not documented.

fry into streams or lakes with no chinook salmon and with underutilized salmonid rearing environments (Coombs 1982; Hard 1986; Killinger et al. 1988).

Rehabilitation projects are currently underway on Harding River, Farragut River, and Boulder Creek in the Chilkat River system (Figure 1). These projects temporarily increase fry recruitment into natal streams with the ultimate goal of greater smolt and adult production levels. This method of enhancement consists of collecting eggs from a limited number of adults, incubating the eggs in a controlled environment, briefly rearing early-stage fry (in some cases), then returning the fry to the natal stream. The intent is to maximize numbers of fry in the natal rearing environment with minimal effects from the short-term hatchery experience.

Hatchery–Wild Stock Interactions

Between 1979 and 1993 over 40,000 adult chinook salmon were examined for hatchery returns (hatchery releases were marked with coded wire tags) in rivers and on spawning grounds of most of the principal wild chinook salmon systems within the region (Table 5). Many of these surveys were made to collect biological data or to find adults from earlier juvenile tagging programs in the same streams. These surveys also provided a measure of straying by tagged hatchery adults into wild chinook salmon systems or of straying of tagged wild stocks between systems. Fifty-six tagged chinook salmon from regional hatcheries were recovered in wild systems representing, when expanded for tagging ratios, 123 fish (0.3% of adults examined).

The greatest percentage of hatchery strays into a wild system was in the Farragut River where 8.3% of the 617 adult chinook salmon examined over a 10-year period were from regional hatcheries (Table 5). For 2 years, no strays were found in this river, but in 1991, 26% of the adults examined were from hatcheries. Of 51 strays in the Farragut River, all but one were from Hidden Falls, Little Port Walter, and Port Armstrong hatcheries. These hatcheries are located along the east side of Baranof Island between 120 and 200 km from the stream.

Straying from these three hatcheries into Farragut River appears decidedly nonrandom, because overall straying rates from these and other hatcheries into other wild chinook salmon systems were low. These findings raise questions about how the location and culture practices at specific hatcheries affect straying and about features of streams that attract strays. A similar phenomenon in the lower Columbia River was found when both spring chinook salmon (stream-type) and fall chinook salmon (ocean-type) strays were attracted disproportionately to Lewis and Kalama rivers (Quinn and Fresh 1984; Quinn et al. 1991).

Discussion

Chinook salmon enhancement efforts in southeast Alaska during the past 15 years followed a planned format to minimize adverse effects on wild stocks. No hatcheries were built on wild chinook salmon systems. Conservative genetic, fish transport, and fish disease policies were put in place. Genetic diversity among hatcheries was achieved by developing different broodstocks from separate wild stocks in different parts of the region. Also, several smaller hatcheries were used rather than a few large hatcheries. An appropriate question would be, "Has the program provided more chinook salmon for regional fisheries while minimizing hatchery impacts on wild populations?" To attempt to answer this question, it is instructive to examine several issues.

Based on harvests of chinook salmon produced in regional hatcheries in recent years, the program has been successful. Since 1984, regional hatcheries have made steadily increasing contributions to the total commercial and recreational harvest of chinook salmon, reaching almost 20% in 1991. Adult chinook salmon produced from southeast Alaskan hatcheries, in recent years, have equaled or exceeded numbers of wild adults spawning in regional streams. Recent estimates of the contribution from wild stocks in medium stream systems in southeast Alaska (exclusive of transboundary rivers) to chinook salmon fisheries in the region have been less than 1% of total harvest (CTC 1993). Similar estimates for minor and major stream systems are not available. Several coastwide issues, including continued depressed stock-rebuilding programs, future Treaty negotiations, and actions required by the U.S. Endangered Species Act (ESA; 16 U.S.C.A. §§ 1531 to 1544) may necessitate future reductions in chinook salmon harvest levels. With further reductions in Treaty-managed harvest of chinook salmon, production from southeast Alaskan hatcheries will increase in importance for fisheries in this region.

There is some concern about the relation between appropriate escapement goals and current escapement levels (CTC 1993) in a few wild index stocks (stocks monitored to assess the effectiveness of stock-rebuilding programs) in southeast Alaska. Research is underway to define habitat requirements of juveniles better, determine the amounts of available habitat in streams, and refine spawner-recruitment relations. Stocks of particular concern involve four in the Behm Canal area (Blossom, Chickamin, Keta, and Unuk rivers) where escapements have fluctuated in recent years. Precise causes for fluctuations in escapements are unknown but may relate to harvest or to variations in natural mortality in freshwater and marine environments. Another index stock below the current escapement goal is the transboundary Alsek River. Here rapid geological changes in the watershed, especially postglacial uplift and siltation of the estuary (Celewycz 1988), very likely have reduced rearing habitats and may have rendered the present escapement goals, based on historic levels, unrealistically high.

Because many wild stocks in southeast Alaska spawn and rear in remote, glacial streams where visibility is poor, there often is a paucity of accurate information on spawner-escapement levels and available habitat for spawning and rearing. Gaining accurate information for such stocks and systems, where foot or aerial survey methods are ineffective, is difficult and expensive. For example, in one glacially occluded system, Chilkat River, a radio-tagging study indicated escapements were an order of magnitude greater than initially thought (Johnson et al. 1992b; CTC 1993). Similar studies are needed on other wild chinook salmon streams in the region; however, because radio-tagging studies are expensive they can not be routinely conducted on most streams during a given spawning season.

Contributions to Fisheries

Although southeast Alaskan hatcheries contribute to common-property chinook salmon harvest in the region, the present structure of the fisheries limits the level of hatchery contribution. Fisheries in the region are dictated by harvest quotas on wild and non-Alaskan hatchery fish, gear-group allocations, complex seasonal time and area fishing periods, and behavioral features of different stocks. Exploitation rates on Alaskan hatchery fish currently average only 50%, thus surplus numbers of adults return to many hatcheries. Although experimental commercial troll fisheries targeting hatchery chinook salmon in terminal areas are conducted to catch more hatchery fish, they have only slightly raised harvest rates due, in part, to reduced feeding behavior as the adult salmon approach sexual maturity.

Large returns of chinook salmon to hatcheries, in turn, have increased levels of cost-recovery harvest at PNP facilities. Whereas cost-recovery harvest may help to partly defray hatchery cost, most hatcheries would prefer that more chinook salmon were caught in common-property fisheries where per-fish value is greatest.

Harvest of southeast Alaskan hatchery chinook salmon in commercial and sport fisheries depends on many factors, including marine nursery areas and migration timing of hatchery stocks. For example, up to 24% of chinook salmon caught during the 6-month winter commercial troll fishery have been southeast Alaskan hatchery fish (K. Crandall, ADF&G, personal communication). This fishery is restricted primarily to stocks rearing in inside waters within the surf line of the outer coast. By contrast, the shorter summer troll fishery (4.5 d in 1992) is concentrated on the outer coast beyond the surf line and has much higher catch rates of non-Alaskan stocks. Alaskan hatchery stocks compose less than 5% of the summer troll fishery. Because of Treaty quotas, specific gear allocations for harvesting chinook salmon, and migration behavior of southeast Alaska stocks, management options for increasing common-property harvest of hatchery fish are limited.

The contribution of southeast Alaskan hatchery fish in the Sitka, Juneau, and Ketchikan urban areas has averaged 12, 20, and 45%, respectively, of chinook salmon caught by recreational anglers in those areas in recent years (Suchanek 1994). Up to 60% of the chinook salmon caught in peak sportfishing periods in the Ketchikan area have come from local hatcheries. Hatcheries and off-station release sites located near urban centers provide better access to hatchery fish for sport fisheries than do facilities in remote areas.

Influence of Life History

The stream-type life history, predominant in southeast Alaskan stocks and the principal culture method used in regional hatcheries (i.e., production of age-1 smolts), is important relative to marine migration patterns and the ability of hatchery fish to contribute to local fisheries. Some southeast Alaskan stocks have ocean migration patterns quite different from those described for stream-type fish by Healey (1983). Based on recovery patterns of juvenile chinook salmon tagged in short coastal systems (Chickamin and Unuk rivers) and in upriver transboundary systems (Taku and Stikine rivers), Kissner and Hubartt (1987) concluded that only the more distant upriver stocks exhibit the classic offshore oceanic migratory behavior of stream-type fish. This conclusion was based on the recovery of upriver tagged juveniles only as mature fish either on or migrating near spawning grounds. Many juveniles tagged in the shorter coastal systems, in contrast, were recovered in inshore waters in southeast Alaskan fisheries as subadults throughout their marine-life history, a behavior typical of ocean-type stocks. Because life history behavior in the marine environment, especially ocean migration pattern, directly affects the ability of southeast Alaska hatchery and wild chinook salmon stocks to contribute to regional fisheries, an important question is why do some stream-type stocks in this region have ocean-type migration behavior?

Gharrett et al. (1987), examining genetic relationships among Alaska chinook salmon stocks, provide a possible explanation for this paradox. They proposed two colonizing movements of chinook salmon into the region following Wisconsin glaciation: (1) from the Bering Refuge via Yukon River and headwater-stream capture into the Taku and Stikine rivers; and (2) from the Pacific Refuge via coastwide movement from the south into shorter coastal streams inside the barrier mountains. Given these possible differences in origins, it seems reasonable to equate the more restrictive coastal migration behavior of short coastal-stream stocks in southeast Alaska with the likely ocean-type invaders that colonized these streams. The shift to a stream-type life history in fresh water could have been dictated by climate and environment, whereas marine migration patterns remained true to initial genetic origins.

Variation in Marine Survival

Causes of variations in marine survival of chinook salmon smolts released from regional hatcheries both between hatcheries and across years are poorly understood. Differences in stocks, hatchery location, and details of fish culture, including size and time of smolt releases, all are likely to play key roles in interannual differences in marine survival. This variation determines, to a great extent, year-class strength and the ultimate ability of hatchery fish to enhance fisheries. Another cause of variation in survival in both hatchery and wild smolts from this region is thought to occur within the marine environment.

Stocks residing in outside coastal waters, for example, may encounter significant differences in marine conditions from stocks residing in inside waters. This inference is based, in part, on recovery patterns of Medvejie Hatchery chinook salmon in regional fisheries. Although this hatchery is releasing Andrew Creek stock (stream-type) smolts, the physical location in Sitka Sound on the outside coast of Baranof Island presumably is the reason most Medvejie Hatchery fish are caught in outside waters, in contrast to inside hatcheries releasing smolts from the same stock. Few Medvejie Hatchery fish are recovered, either as immature subadults or maturing fish, in inside waters. Recent improved survival of smolts released from this hatchery support the notion that variable marine conditions differentially affect survival of inside- and outside-rearing stocks. During the same period, survival of smolts released from other hatcheries, where significant rearing occurs in inside waters, has been low. This relationship also probably applies to wild stocks, as suggested by recently improved escapements of upriver Taku and Stikine river spawners rearing in outside marine waters and poor escapements of some coastal stocks rearing mostly in inside waters. Better understanding of ocean migration patterns and the effect of different marine conditions on chinook salmon survival is crucial to the effective management of fisheries with intermingled wild and hatchery stocks in this region.

Are Hatchery Fish Adversely Affecting Wild Stocks?

The high incidence of hatchery fish recovered in the Farragut River may be a cause for concern about hatchery–wild stock interaction. It is puzzling that all but one of the tags recovered were from three east Baranof Island hatcheries and none were recovered from Crystal Lake Hatchery, which is located much closer to Farragut River. Furthermore, it is not known if any hatchery chinook salmon in Farragut River are spawning either with siblings or with wild cohorts. All recovered strays were removed from the river before spawning.

Even if spawning between hatchery strays and wild Farragut River chinook salmon is taking place, the level of introgression may be small or undetectable. Recent studies on coho salmon by Fleming and Gross (1993) document the competitive inferiority of hatchery strays when attempting to spawn with wild fish. Hatchery males were less aggressive, more submissive, and were denied access to spawning females; hatchery females spawned smaller portions of their eggs than did wild females and lost more eggs to nest destruction by other females.

Farragut River chinook salmon had lower levels of genetic variability than did other Alaskan stocks examined by Gharrett et al. (1987). Introgression from hatchery strays into the wild Farragut River genome may be detectable through analyses of baseline genetic material from hatchery stocks and their founder wild stock populations. Given the suggestion by Gharrett et al. (1987) that short coastal runs of chinook salmon in southeast Alaska are, geologically, from a common source and that hatchery broodstocks from nearby streams are similar to Farragut River, it is likely that effects from introgression between hatchery and wild fish in this stream, if it is occurring, will be difficult to measure.

In addition to nonrandom hatchery strays into Farragut River, another hatchery stock–wild stock concern involves fishing pressure and fishery patterns in the southern part of the region near Ketchikan. Whereas local area hatcheries provide up to 60% of the chinook salmon catch in some fisheries in this vicinity (Suchanek 1994), high harvest levels of hatchery fish have a potential to also overharvest adjacent or commingled small or weak wild runs. Conversely, with similar levels of fishing effort, catches of hatchery stocks can provide a protective buffer against overharvest of wild fish.

Except for straying into the Farragut River and possibly the development of fisheries with high hatchery harvest rates near wild stocks, interactions between hatchery fish and wild stocks of chinook salmon in southeast Alaska appear minimal. This conclusion is based on the low overall recovery rate (0.3%) of hatchery fish from 40,841 adults examined in most of the major wild systems in the region plus the steadily increasing numbers of spawners occurring in most wild stocks. However, many important and difficult questions on possible hatchery–wild stock interactions in southeast Alaska chinook salmon remain unanswered and need attention.

Recommendations

We believe the primary goals of chinook salmon enhancement and the hatchery program for this species in southeast Alaska, with qualified exceptions, are being successfully met. In spite of complex fishery constraints, enhancement in the region is now producing significant numbers of chinook salmon that increase the harvest of this species in both commercial and recreational fisheries. Many, although not all, wild stocks concurrently have shown increases in spawning escapements as a result of the stock-rebuilding program.

Because quotas limit harvest levels, except for fish produced from regional hatcheries, and because of growing demands for increased catches in both commercial and recreational fisheries, the long-range need for some level of hatchery production of chinook salmon in southeast Alaska will probably continue. The ability to increase contributions of hatchery fish to common-property fisheries significantly, however, may be limited as evidenced by the growing number of adults harvested in cost-recovery fisheries by hatchery operators.

Research should be directed at filling gaps in basic knowledge on wild chinook salmon stocks in the region and on possible avenues of interactions with hatchery stocks. Better ways to manage fisheries to increase common property harvest of hatchery fish are needed. More knowledge on exploitation rates of wild stocks and optimal escapement levels is essential to the success of stock-rebuilding programs. This, in turn, requires comprehensive data on available spawning and rearing habitat in wild systems. Factors influencing straying of east Baranof Island hatchery chinook salmon into Farragut River and reasons for this stream having a special attraction for certain hatchery fish should be examined in detail. Better knowledge on the marine ecology of wild and hatchery chinook salmon and the complex oceanography of this region is needed. A program for genetic monitoring of wild and hatchery stocks should be coupled with research to examine effects of any introgression.

In conclusion we note that strict harvest management coupled with protection of habitat in southeast Alaska is providing robust stocks of wild chinook salmon, and an effective and responsible enhancement program is providing additional fish to local fisheries. Coastwide, chinook salmon management is growing more difficult due to habitat losses and population declines in many stocks, complex Treaty issues, and ESA concerns, especially in the Columbia River drainage. The future of chinook salmon hatcheries and enhancement of this species will continue to evolve. Plans for any expansion of hatchery production in southeast Alaska are uncertain, and some reduction in current levels of production may occur. Regardless of what future hatchery-production levels of chinook salmon in this region may be, conservative policies favoring minimal hatchery–wild stock interaction as detailed in this paper and consistent with new research findings should be continued.

Acknowledgments

We acknowledge contributions made by Marianne McNair, Ron Josephson, Karen Crandell, Bob Zorich, Carrita Morris, Jerry Taylor, and Adrian Celewycz in assisting with graphics or data compilation. Bruce Bachen, James Borawa, Alex Wertheimer, Doug Mecum, and four anonymous reviewers provided many useful comments to earlier drafts of the manuscript.

References

Allendorf, F. W., and N. Ryman. 1987. Genetic management of hatchery stocks. Pages 141–159 in N. Ryman and F. Utter, editors. Population genetics and fishery management. University of Washington Press, Seattle.

Amend, D. F. 1988. Enhancement of chinook salmon. Pages 137–151 in W. R. Heard, rapporteur. Report of the 1987 Alaska chinook salmon workshop. Northwest and Alaska Fisheries Center Processed Report 88-06, Juneau, Alaska.

Celewycz, A. 1988. Alsek River juvenile chinook salmon studies. Pages 113–123 in W. R. Heard, rapporteur. Report of the 1987 Alaska chinook salmon workshop. Northwest and Alaska Fisheries Center Processed Report 88-06, Juneau, Alaska.

Coombs, C. I. 1982. Preliminary results of transplanting chinook salmon fry to three locations of Carroll River in southern southeast Alaska. Southern Southeast Regional Aquaculture Association, Inc., Ketchikan, Alaska.

CTC (Chinook Technical Committee). 1993. Pacific Salmon Commission Report TCCHINOOK (93)-2, Vancouver, British Columbia.

CRMJS (Columbia River Management Joint Staff). 1992. Columbia River fish runs and fisheries 1938–92. Oregon Department of Fish and Wildlife and Washington Department of Fisheries, Status Report, Olympia, Washington.

Davis, B., and R. Burkett. 1989. Background of the genetic policy of Alaska Department of Fish and Game. Alaska Department of Fish and Game FRED (Fisheries Rehabilitation, Enhancement and Development) Report 95, Juneau.

Davis, B., and seven coauthors. 1985. Alaska Department of Fish and Game genetic policy. Genetic Alaska Department of Fish and Game, Juneau.

Denton, C. 1986. Effects of supplementary dietary salt on chinook salmon at Deer Mountain Hatchery. Alaska Department of Fish and Game FRED (Fisheries Rehabilitation, Enhancement and Development) Report 63, Juneau.

Denton, C. 1987. Deer Mountain Hatchery chinook salmon 1979 brood year completion report: a size at release study. Alaska Department of Fish and Game FRED (Fisheries Rehabilitation, Enhancement and Development) Report 76, Juneau.

Denton, C. 1988a. Overview of age-0 smolt program at Deer Mountain Hatchery. Pages 185–190 in W. R. Heard, rapporteur. Report of the 1987 Alaska chinook salmon workshop. Northwest and Alaska Fisheries Center Processed Report 88-06, Juneau, Alaska.

Denton, C. 1988b. Marine survival of chinook salmon, Oncorhynchus tshawytscha, reared at three densities. Alaska Department of Fish and Game FRED (Fisheries Rehabilitation, Enhancement and Development) Report 76, Juneau.

Fleming, I. A., and M. R. Gross. 1993. Breeding success of hatchery and wild coho salmon (Oncorhynchus kisutch) in competition. Ecological Applications 3(2): 230–245.

Freitag, G. 1988. Contribution patterns and returns of Unuk stock chinook salmon released from Whitman Lake Hatchery. Pages 198–206 in W. R. Heard, rapporteur. Report of the 1987 Alaska chinook salmon workshop. Northwest and Alaska Fisheries Center Processed Report 88-06, Juneau, Alaska.

Gharrett, A. J., S. M. Shirley, and G. R. Tromble. 1987. Genetic relationships among populations of Alaskan chinook salmon (Oncorhynchus tshawytscha). Canadian Journal of Fisheries and Aquatic Sciences 44: 765–774.

Gunstrom, G. 1980. The troll fishery in Southeast Alaska. Alaska Department of Fish and Game Fishery Information Pamphlet 1, Juneau.

Hard, J. J. 1986. Production and yield of juvenile chinook salmon in two Alaskan lakes. Transactions of the American Fisheries Society 115:305–313.

Hard, J. J., A. C. Wertheimer, W. R. Heard, and R. M. Martin. 1985. Early male maturity in two stocks of chinook salmon (*Oncorhynchus tshawytscha*) transplanted to an experimental hatchery in southeastern Alaska. Aquaculture 48:351–359.

Healey, M. C. 1983. Coastwide distribution and ocean migration patterns of stream- and ocean-type chinook salmon (*Oncorhynchus tshawytscha*). Canadian Field-Naturalist 97:427–433.

Healey, M. C. 1991. Life history of chinook salmon (*Oncorhynchus tshawytscha*). Pages 311–393 in C. Groot and L. Margolis, editors. Pacific salmon life histories. University of British Columbia Press, Vancouver.

Heard, W. R. 1985. Chinook salmon fisheries and enhancement in Alaska: a 1982 overview. NOAA (National Oceanic and Atmospheric Administration) Technical Report NMFS (National Marine Fisheries Service) F/NWC 27.

Holland, J., B. Bachen, G. Freitag, P. Kissner, and A. Wertheimer. 1983. Chinook salmon plan for southeast Alaska. Alaska Department of Fish and Game, Juneau.

Hynes, J. D., E. H. Brown, Jr., J. H. Helle, N. Ryman, and D. A. Webster. 1981. Guidelines for the culture of fish stocks for resource management. Canadian Journal of Fisheries and Aquatic Sciences 38:1867–1876.

Johnson, R. J., R. P. Marshall, and S. E. Elliott. 1992b. Chilkat River chinook salmon studies, 1991. Alaska Department of Fish and Game Fish Data Series 92-49, Juneau.

Johnson, S. W., J. F. Thedinga, and K. V. Koski. 1992a. Life history of juvenile ocean-type chinook salmon (*Oncorhynchus tshawytscha*) in the Situk River, Alaska. Canadian Journal of Fisheries and Aquatic Sciences 49:2621–2629.

Josephson, R., and S. Kelley. 1993. Sport fisheries rehabilitation, enhancement and development. Alaska Department of Fish and Game, Federal Aid in Sport Fish Restoration, Project F-32-2, Juneau.

Killinger, G., D. Logan, and J. McNair. 1988. Habitat utilization, movement and survival of chinook salmon fry stocked in a barriered stream. U.S. Forest Service, Habitat Hotline 88-2, Juneau, Alaska.

Kissner, P. D. 1978. A study of chinook salmon in Southeast Alaska. Alaska Department of Fish Game Studies AFS-41-5, volume 18, Juneau.

Kissner, P. D. 1986. Status of important native chinook salmon stocks in southeast Alaska. Alaska Department of Fish Game Studies AFS-41-12, volume 26:1–57, Juneau.

Kissner, P. D. 1988. Situk River age-0 chinook smolts. Pages 19–25 in W. R. Heard, rapporteur. Report of the 1987 Alaska chinook salmon workshop. Northwest and Alaska Fisheries Center Processed Report 88-06, Juneau, Alaska.

Kissner, P. D., and D. J. Hubartt. 1987. A study of chinook salmon in Southeast Alaska. Alaska Department of Fish Game Studies AFS-41-13, volume 27:26–124, Juneau.

Major, R. L., J. Ito, S. Ito, and H. Godfrey. 1978. Distribution and abundance of chinook salmon (*Oncorhynchus tshawytscha*) in offshore waters of the North Pacific Ocean. International North Pacific Fisheries Commission Bulletin 48:1–54.

Martin, R. M., and K. A. Leon. In press. Freshwater performance of Alaskan chinook salmon fed diets with 10, 20, and 30% moisture. Progressive Fish-Culturist.

Martin, R. M., and A. Wertheimer. 1989. Adult production of chinook salmon reared at different densities and released at two smolt sizes. Progressive Fish-Culturist 51:194–200.

Mason, J. E. 1965. Chinook salmon in offshore waters. International North Pacific Fisheries Commission Bulletin 16:41–73.

McGee, S., and six coauthors. 1993. 1993 Annex: chinook salmon plan for southeast Alaska. Alaska Department of Fish and Game, Juneau.

Meehan, W. R., and D. B. Siniff. 1962. A study of the downstream migrations of anadromous fishes in the Taku River, Alaska. Transactions of the American Fisheries Society 91:399–407.

Meyers, T. R., J. B. Thomas, J. E. Follett, and R. R. Saft. 1990. Infectious hematopoietic necrosis virus: trends in prevalence and the risk management approach in Alaskan sockeye salmon culture. Journal of Aquatic Animal Health 2:85–98.

NPFMC (North Pacific Fishery Management Council). 1978. Fishery management plan and environmental impact statement for the high seas salmon fishery off the coast of Alaska east of 175 degrees east longitude. NPFMC, Anchorage, Alaska.

Netboy, A. 1980. The Columbia River salmon and steelhead trout: their fight for survival. University of Washington Press, Seattle.

Orsi, J. A. 1987. Small versus large trolling lures for sampling juvenile chinook salmon and coho salmon. Transactions of the American Fisheries Society 116:50–53.

Orsi, J. A. 1988. Size, age, origin, and distribution of juvenile chinook salmon in the marine waters of southeastern Alaska. Pages 50–57 in W. R. Heard, rapporteur. Report of the 1987 Alaska chinook salmon workshop. Northwest and Alaska Fisheries Center Processed Report 88-06, Juneau, Alaska.

Orsi, J. A., A. C. Celewycz, D. G. Mortensen, and K. A. Herndon. 1987. Sampling juvenile chinook salmon (*Oncorhynchus tshawytscha*) and coho salmon (*O. kisutch*) by small trolling gear in the northern and central regions of Southeastern Alaska, 1985. NOAA (National Oceanic and Atmospheric Administration) Technical Report NMFS (National Marine Fisheries Service) F/NWC-115.

Pahlke, K. A. 1993. Escapements of chinook salmon in southeast Alaska and transboundary rivers in 1992. Alaska Department of Fish and Game Fishery Data Series 93-46, Juneau.

Parker, R. R., and W. Kirkness. 1956. King salmon and the ocean troll fisheries of southeastern Alaska. Alaska Department of Fisheries Research Report 1, Juneau.

Quinn, T. P., and K. Fresh. 1984. Homing and straying in chinook salmon (*Oncorhynchus tshawytscha*) from Cowlitz River Hatchery, Washington. Canadian Journal of Fisheries and Aquatic Sciences 41:1078–1082.

Quinn, T. P., R. S. Nemeth, and D. O. McIsaac. 1991. Homing and straying patterns of fall chinook salmon in the lower Columbia River. Transactions of the American Fisheries Society 120:150–156.

SPRC (State Pathology Review Committee). 1988. Regulation changes, policies and guidelines for Alaska fish and shellfish health and disease control. Alaska Department of Fish and Game FRED (Fisheries Rehabilitation, Enhancement and Development) Special Report, Juneau.

Suchanek, P. 1994. Overview of the sport fishery for chinook salmon in Southeast Alaska. Report to the Alaska Board of Fisheries Alaska Department of Fish and Game, Douglas.

Van Alen, B. W., K. A. Pahlke, and M. A. Olsen. 1987. Abundance, age, sex and size of chinook salmon (*Oncorhynchus tshawytscha* Walbaum) catches and escapements in southeastern Alaska in 1985. Alaska Department of Fish Game Technical Data Report 215, Juneau.

Wertheimer, A. C., and J. R. Winton. 1982. Differences in susceptibility among three stocks of chinook salmon, *Oncorhynchus tshawytscha*, to two isolates of infectious hematopoietic necrosis virus. NOAA (National Oceanic and Atmospheric Administration) Technical Report NMFS (National Marine Fisheries Service) F/NWC22.

Changes in Stocking Strategies for Atlantic Salmon Restoration and Rehabilitation in Maine, 1871–1993

JOHN R. MORING

*National Biological Service, Maine Cooperative Fish and Wildlife Research Unit
University of Maine, 5751 Murray Hall, Orono, Maine 04469, USA*

JERRY MARANCIK AND FREDERICK GRIFFITHS

*U.S. Fish and Wildlife Service, Maine Fish Program Office
Craig Brook National Fish Hatchery, East Orland, Maine 04431, USA*

Abstract.—The culture of Atlantic salmon *Salmo salar* in Maine began in 1871, when 72,300 eggs were obtained from captured Penobscot River fish and fertilized and cultured in the basement of an old mill at Craig's Brook. Subsequently, a hatchery was built (Craig Brook National Fish Hatchery) that continues in operation today. Culture and stocking strategies and philosophy have changed during this 120-year period. Initially, most Atlantic salmon were reared to the eyed egg or fry stage, then stocked in southern New England rivers where runs were severely depleted. In 1890, the stocking strategy changed to the release of fingerlings (parr). Most runs of Atlantic salmon in Maine were extirpated in the twentieth century. After the establishment of the Maine Atlantic Sea Run Salmon Commission, the passage of several water quality laws, and the construction of fish passage facilities, restoration efforts began with a hybrid strain of Atlantic salmon derived from Canadian and native Maine river stocks. In the 1960s and 1970s, 2-year-old smolts were stocked. Later, with the use of heated water during culture, 1-year-old smolts dominated the program. In the 1990s, more emphasis is being placed on fry stocking and the use of river-specific stocks to maintain the remaining genetic integrity of the river populations in eastern Maine, where Atlantic salmon have never been extirpated. Other techniques used, with varying degrees of success, have included upstream trucking, reducing sportfishing harvest, and stocking in tributary systems rather than main-stem river locations. In future years, restoration success will be influenced by recent restrictions of commercial fisheries on the high seas, improved fish passage, genetic considerations, and refinements in habitat protection. The restoration and rehabilitation efforts in Maine waters would not have been possible without the stocking of hatchery-reared fish. However, high numbers of fish stocked, particularly at the smolt stage, have not resulted in large runs or higher angler catches. Other factors related to habitat, dams, and ocean mortality may be critical. In all likelihood, stocking fish will be necessary into the twenty-first century. Thorough evaluations of management strategies are critical to future management decisions.

Most estimates taken from fragmented reports by the first colonists concluded that indigenous runs of Atlantic salmon *Salmo salar* were large. Captain George Weymouth, aboard the ship *Archangel*, was looking for a northwest passage to India when he first viewed the Kennebec River. When he entered the mouth of the river in 1605, he reported seeing "Plenty of salmon and other fishes of great bigness." There is some question as to the magnitude of the runs before 1600 (Carlson 1988), but Stolte (1986) estimated the size of the New England spawning stocks at the time of colonization by Europeans as 300,000, based on available habitat.

Most runs of Atlantic salmon in New England began to decline in the 1850s. By 1866, many runs of Atlantic salmon in southern New England had been extirpated, and eggs were obtained from the Penobscot River, Maine, and elsewhere to stock in southern New England waters (Moring 1986). At the turn of the century, the Penobscot River was still considered the premier Atlantic salmon river of the country, despite the increased pollution from some 250 sawmills as early as 1837 (Netboy 1968). Commercial catches in the river and bay exceeded 15,000 fish in 1872; but by the end of the nineteenth century Atlantic salmon were declining rapidly in many rivers of Maine, primarily due to dams, pollution, and commercial fishing (Baum 1983; Moring 1987). Most runs of Atlantic salmon were extirpated in Maine by the mid-twentieth century. By 1947, only 40 salmon were commercially harvested in the Penobscot River (Netboy 1968). Salmon runs were eliminated in the state, with the exception of those on several smaller rivers of Downeast Maine, the easternmost portion of the state, where rivers were largely free of dams (Figure 1).

A separate state fisheries agency, the Maine Atlantic Sea Run Salmon Commission (Salmon Commission), was established in 1947 to restore Atlantic salmon to their former range in state waters. But it

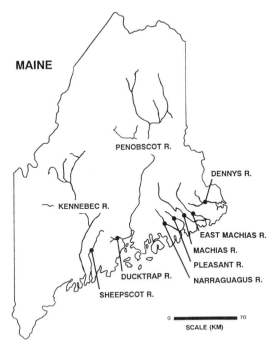

FIGURE 1.—Maine rivers with Atlantic salmon runs.

federal government is primarily concerned with restoration of extirpated runs and not introductions to new waters (U.S. Fish and Wildlife Service 1989). The state of Maine's highest priority is to maintain and enhance the Atlantic salmon populations and sport fisheries in rivers in eastern Maine that have never lost their populations of Atlantic salmon; their second priority is to restore populations and sport fisheries on rivers with extirpated runs (Beland 1984). In this review, we document the changes in management strategies of the cooperating agencies, discuss their successes and failures, and predict which actions may be necessary for successful restoration and rehabilitation in the future.

Historical Perspective

Due to the decline in Atlantic salmon and other anadromous fish species in New England, a hatchery was constructed in a converted mill at Craig's Brook (now the Craig Brook National Fish Hatchery) in 1871 to receive and incubate Atlantic salmon eggs for stocking. Later, the facility was remodeled and expanded; it continues operations today, along with Green Lake National Fish Hatchery, as the source of Atlantic salmon used in restoration and rehabilitation programs in Maine. During the last decades of the 1800s, hatchery superintendent Charles Atkins developed techniques for stripping eggs from adult salmon, and by the mid-1870s the hatchery was incubating about 3 million eggs annually.

The hatchery was closed from 1875 to 1878, but reports of adult Atlantic salmon returning to rivers as far south as the Delaware and Susquehanna prompted the U.S. Fish Commission to re-open the hatchery in 1879. Until 1890, the hatchery produced fry for stocking. After 1890, the stocking of fingerlings (parr) was emphasized. When the Salmon Commission was established in 1947 and the Penobscot River was later designated as a Model Restoration River by the federal government, restoration efforts concentrated on the culture of 2-year-old smolts in order to mimic the age of migrating smolts produced in the wild. By 1984, by use of accelerated growth techniques such as heated water and improved diets, 1-year-old smolts became the norm. Currently, stocking policy has returned to the emphasis of Atlantic salmon culturists of 100 years ago—stocking fry. Managers believe that stocking fry will allow fish to become more acclimated to conditions in the wild, thus increasing the ultimate survival and homing of returning adults.

Runs of Atlantic salmon in the Penobscot River

was not until the passage of the Maine Water Pollution Bill of 1965 and the Federal Water Pollution Control Act of 1972 (33 U.S.C.A. §§ 1251 to 1387) and the installation of more effective fish passage structures that the restoration began to show results. During the past 120 plus years of fish culture and stocking, management strategies for restoration and rehabilitation have shifted. However, changes were not always in response to strictly biological concerns; some had political or economic origins. Restoration is defined as the return of self-sustaining populations of fish to waters where they were extirpated, whereas rehabilitation involves the recovery of depressed populations.

Federal and state priorities differ somewhat in the management of Atlantic salmon. In the 1990s, restoration and rehabilitation of salmon is a cooperative function of the state of Maine, the federal government, and the Penobscot Indian Nation. Fish culture is the responsibility of the U.S. Fish and Wildlife Service, whereas management of Atlantic salmon is the responsibility of the Salmon Commission; these two agencies and the Penobscot Indian Nation work together on technical issues. In recent years, because of severe budgetary restrictions at the state level, federal agencies are temporarily playing a larger role in funding and monitoring. The

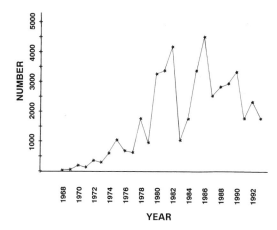

FIGURE 2.—Numbers of adult Atlantic salmon returning to the Penobscot River, Maine, 1968 to 1993. Data are from records of the Maine Atlantic Sea Run Salmon Commission.

steadily increased to a peak of 4,529 fish in 1986, but numbers of returning adults have declined since (Figure 2). Only 1,774 Atlantic salmon returned to the Penobscot River in 1993. This decline in numbers of returning fish led to a recent, temporary closure of the West Greenland commercial fishery and adoption of new management strategies, such as improved fry stocking, to increase the number of spawners.

Restoration and Rehabilitation Strategies

Stocking Eggs, Fry, and Fingerlings

Between 1866 and 1870, eggs from Atlantic salmon were brought from the Miramichi River, New Brunswick, and stocked into waters of several states, including rivers in Maine where salmon runs had declined (C. E. Atkinson, American Embassy-Tokyo, unpublished data). After Craig Brook National Fish Hatchery was constructed in 1871, eggs were obtained from Maine rivers. Eggs and fry were stocked until 1890, but the effectiveness of this program was never fully evaluated.

In 1890, there was a shift in focus to longer hatchery-rearing regimes and the stocking of parr (termed fingerlings at the time). The objective was to increase survival by eliminating the high mortality known to be associated with the vulnerable fry stage but, again, there was virtually no evaluation of stocking this life stage by federal or state agencies.

Development of the Penobscot Strain

Because the strains of Atlantic salmon native to the Penobscot and other major rivers in Maine were extirpated, restoration had to be accomplished using gametes acquired from elsewhere. The Penobscot River was the first system to be stocked using fish of nonnative origin (Moring 1993). The exact genetic composition of the Penobscot strain is unknown, but fragmented records indicate that it likely is a combination of genetic material from strains of fish taken from Canadian and Downeast Maine rivers (Ritter 1975; A. Meister, Salmon Commission, personal communication). As adults returned to the Penobscot River, their gametes provided the fish for subsequent stocking in the system, as well as in the Downeast rivers where runs were declining. Subsequently, eggs of the Penobscot strain were transported to the Merrimack, Connecticut, Pawcatuck, and other rivers in southern New England. Today, a portion of the returning adults are captured in a fish trap at the Veazie Dam on the Penobscot River (and previously at a trap on the Union River) and transported to Craig Brook National Fish Hatchery for broodstock. In 1993, 534 adults, including 283 females, produced 2.1 million eggs. This hybrid strain has now been stocked in the Penobscot drainage for more than six generations.

The slow pace of restoration may be attributable, in part, to the reliance on a nonnative, hybridized strain for restoring runs on the southern boundary of the species' historical range. As Reisenbichler and McIntyre (1986) have shown for *Oncorhynchus* spp. and Griffith et al. (1989) have shown for terrestrial animals, the farther the distance from the original source, the greater the amount of straying and lower the rates of survival of translocated animals.

2-Year-Old Versus 1-Year-Old Smolts

In the 1960s, emphasis shifted from parr stocking to the use of 2-year-old smolts. Most of Maine's remaining wild Atlantic salmon spend 2 years in freshwater before migrating to sea as smolts. Therefore, hatchery rearing was primarily directed at producing fish that were of sufficient size to undergo smoltification in the second year (some faster growing fish were often graded and released as 1-year-old smolts). By 1984, however, culturists used improved diets and heated water to accelerate the growth of juveniles so that the Atlantic salmon could attain near-optimal size for stocking as functional, 1-year-old smolts. In 1991, median size of 1-year-old smolts at Craig Brook National Fish Hatchery was 17.5 cm, and most were between 16.5 and 18.5 cm. Most 2-year-old smolts released in that year ranged from 19.5 to 25.5 cm (F. Griffith, U.S.

Fish and Wildlife Service, unpublished data). Fish that did not reach optimal size (approximately 17 cm in total length) in their first year were held over to release as 2-year-old smolts. In 1987, 718,200 smolts were stocked into Maine rivers, only 12% of them as 2-year-old fish.

Various questions remain unanswered concerning possible differences between the behavior of these two types of smolts. Downstream migration rates of 2-year-old smolts released in the lower river (Fried et al. 1978) were more rapid and continuous than those of 1-year-old smolts released 70 km upstream (Vanderpool 1992). The causes of the declines in numbers of adults returning each year since 1986 are complex and may include commercial harvest on the high seas, inadequate downstream fish-passage facilities, and riverine predation (Hosmer et al. 1979; U.S. Atlantic Salmon Assessment Committee 1991). However, declines in adult numbers may also be related to the age of smolts released. Although some of these factors have been evaluated, the effects of each independent variable have not been thoroughly addressed. Further, possible synergistic effects of different levels of several variables have not been evaluated.

Fry Stocking Versus Smolt Stocking

Smolt stocking has been a major component of Atlantic salmon restoration efforts in Maine since 1947. Culturing fish in hatcheries until the smolt stage has long been known to reduce mortality of juvenile salmonids compared with fish living in the wild (e.g., Cultus Lake studies, Foerster 1968). Stocking smolts, rather than a younger life stage, has been presumed to increase the returns of adults, which are a source of gametes as well as the source of political and financial support for the program.

Fry stocking has been the management practice in southern New England because of the poor survival of smolts used in those stocking programs and the low returns (averaging only 0.01 to 0.02% from stocked smolt to adult return). Survival of hatchery-reared smolts to returning adult stage in Maine rivers has averaged 0.3% since 1989, but models developed by the U.S. Fish and Wildlife Service predict that stocking larger numbers of fry, raised at less cost than smolts, will result in comparable or higher numbers of returns. Management theory today is that stocking fish at the fry stage allows fish to adapt to natural conditions and to home more adequately to suitable spawning habitat. As a consequence, management priorities in Maine since 1987 have resulted in increased emphasis on fry stocking.

TABLE 1.—Numbers (in thousands) of Atlantic salmon fry, parr, and smolts stocked into Maine rivers, 1970–1993. Data are from the Maine Atlantic Sea Run Salmon Commission.

Year	Fry	Parr	Smolts	Total
1970	25.0	0	51.0	76.0
1971	0	15.8	73.1	88.9
1972	129.0	0	110.1	239.1
1973	0	0	142.5	142.5
1974	0	44.2	137.0	181.2
1975	0	15.3	169.1	184.4
1976	0	83.8	310.0	393.8
1977	0	0	377.1	377.1
1978	0	126.8	295.3	422.1
1979	95.1	5.2	374.8	475.1
1980	0	0	682.2	682.2
1981	202.0	121.3	257.1	580.4
1982	349.0	375.3	408.7	1,133.0
1983	20.0	77.7	529.0	626.7
1984	134.0	56.9	838.0	1,028.9
1985	420.0	167.0	721.6	1,308.6
1986	125.0	53.6	778.5	957.1
1987	746.0	233.2	718.2	1,697.4
1988	376.0	39.9	935.1	1,351.0
1989	582.0	430.0	612.8	1,624.8
1990	963.0	538.5	677.3	2,178.8
1991	968.0	561.7	840.8	2,370.5
1992	1,178.0	523.2	895.2	2,596.4
1993	1,940.3	527.3	640.0	3,107.6

In 1993, 62% of the 3.1 million fish released into Maine rivers were fry. Large numbers of smolts are still being released, but there has been a substantial increase in the number of fry stocked into Maine waters (Table 1). If adult returns increase as the result of these additional fry stocked, we expect that the number of fish stocked as fry to continue to increase.

Miscellaneous Techniques

During the initial 15 years of restoration, many adult Atlantic salmon captured in the trap at the Veazie Dam were transported upstream and released above several dams on the river that had inadequate fish passage. Because of improved fish passage and concerns of the Penobscot Indian Nation that such procedures prevented returning Atlantic salmon from passing by tribal lands, trucking was eliminated and adult Atlantic salmon are now able to migrate upstream. Except for mortality and delay associated with upstream passage of adult fish around and over dams, this policy adheres to the long-term objectives of true restoration—the absence of human assistance.

Until the 1990s, most smolts were stocked directly into the main stem of the Penobscot and other Maine rivers, partly to increase numbers of returning adults. Survival is likely to be lower for fish stocked into upstream tributaries, where there is a longer downstream journey and more dams and predators. However, concentrated stocking has contributed to higher levels of predation by freshwater fish and birds (van den Ende 1993; Krohn et al., in press) and is not conducive to return migration to suitable spawning areas, which are primarily located in upriver tributaries (Baum 1983).

In the 1970s, the emphasis was on upper river stocking. That changed in the 1980s to a policy of more stocking in the lower river to maximize adult returns. In 1992, there was another change in management policy that eliminated main-stem stocking entirely in favor of upriver tributary stocking in an attempt to improve natal river imprinting and spawning success.

River-Specific Stocks

For rivers where populations have never been extirpated, it is appropriate to maintain whatever genetic uniqueness still exists, rather than dilute a natural run with fish from elsewhere. In 1992, the rehabilitation program began to shift from a policy of stocking smolts and fry from the Penobscot strain to one of river-specific stocking (i.e., only progeny from adults captured in a particular river will be stocked in that river).

To help clarify the potential genetic distinctions among river stocks, DNA samples were obtained from fish in the Downeast rivers in 1990, 1992, and 1993, and compared with samples taken from fish in the Gander River, Newfoundland; Craig Brook National Fish Hatchery and commercial aquaculture stocks; and from adults returning to the Penobscot River. Definitive results are not yet available; however, the river-specific broodstock program is progressing with the possibility that the nonextirpated runs in rivers of eastern Maine may still have unique genetic characteristics. Thus, separate stocking programs are appropriate.

This change in policy necessitated changes in facilities. Craig Brook National Fish Hatchery was converted during 1992–1993 from a smolt production facility with a single broodstock to a fry production facility with multiple broodstocks. The current program involves collecting wild parr from the different Downeast rivers and raising them to adult size in captivity. These broodfish will serve as the source for Atlantic salmon stocked in these rivers in the future.

Smolt Self-Release Ponds

Three circular, concrete-lined ponds (9.1 m diameter, 1.2 m deep) constructed by Bangor Hydro-Electric Company near the West Enfield Dam on the Penobscot River have been used for a portion of the smolt releases since 1989. Rather than stocking smolts directly into a river at a time deemed appropriate by humans, self-release ponds allow smolts to exit the ponds and enter the river at their own volition (Isaksson et al. 1978; Rottiers and Redell 1993).

A videotape system (J. Pippy, Department of Fisheries and Oceans, personal communication) monitored emigration to the river. An overnight rise in water temperature from 9 to 11°C on 7 May 1990 was followed by increased migration from the ponds. In 1991, a similar overnight rise in water temperature from 11.5 to 13°C on 12 May led to a similar peak in emigration on 13 May. Analysis of 1991 data showed positive correlations for the number of fish emigrating and water temperature (Spearman rank correlation coefficient, $r_s = 0.4547$; $P = 0.0047$) and hours of daylight ($r_s = 0.4204$; $P = 0.0096$), similar to the results found by Rottiers and Redell (1993). There was a marked 24-h periodicity to smolt emigration, with most movement between 2000 hours and midnight (Vanderpool 1992).

Despite these peaks in emigration from self-release ponds, a significant number of fish failed to leave by the time the water temperature reached 16°C in mid- to late-May, a temperature at which most riverine smolt migration had ceased. Fish only departed when the ponds were drained, leading to speculation that fish either were not true smolts or were unable to locate an exit. Out-migration delays also were experienced by Atlantic salmon held by Rottiers and Redell (1993). Smolts that did leave the ponds were often eaten by smallmouth bass *Micropterus dolomieu* at the pond exit tube (C. Fay, Penobscot Indian Nation, unpublished data). The ponds continue to be used to hold smolts that are stocked in mid-April as a logistical convenience during the period of peak stocking. All fish that have not departed by mid-May are released into the river when the ponds are drained.

Stocking Density Studies

From the earliest days of stocking fry, appropriate stocking densities have been questioned. Although some studies have shown higher survival of salmonids stocked at low densities (Hume and Parkinson 1984), stocking too few fry will underutilize habitat. Stocking too many fish will cause territorial interactions and emigration to increase (Gustafson-Greenwood and Moring 1990). Using a predictive model, Gibson (1992) found that stocking 40 fry per unit (1 unit = 100 m^2) of nursery habitat, as measured during detailed summer habitat surveys, was generally insufficient to saturate available habitat, and a density of 120 fry per unit was excessive. Based on a habitat classification system, stocking densities of 70 to 111 fry have been recommended for the Merrimack River (U.S. Atlantic Salmon Assessment Committee 1993). These densities are in accordance with studies in Vermont, where McMenemy (1989) found that low stocking densities of 20–40 fry per unit did not saturate the habitat, and in Maine, where 60 fry per unit is considered low (Fay 1990).

In Maine, fry are stocked by the Salmon Commission, which uses canoes to distribute the fish throughout an appropriate stretch of habitat. Currently, fry are stocked at a density of about 60 per unit, whereas age-0 parr are stocked at 17 to 24 per unit. Gibson (1994) recommended that optimal-quality habitat in a reach should be stocked with fry to the acceptable density before stocking fry in the next-best grade of habitat.

Angling Restrictions

Maine continues to be the only state in the country that has always had a legal sport fishery for sea-run Atlantic salmon. However, in the past decade, angling restrictions have been imposed to reduce angler harvest, allow more adult fish to reach spawning grounds, and hasten eventual restoration. Anglers removed 25% of the returning adults on the Penobscot River in the early 1980s, when bag limits were relatively liberal (Table 2). Season possession limits imposed in the late 1980s reduced angler harvest to about 11% of the run. When a limit of one adult Atlantic salmon per season was imposed for the 1992 season, sport catch on the Penobscot River was 497 fish (153 kept, 344 released), and angler harvest was reduced to 6.4% (U.S. Atlantic Salmon Assessment Committee 1993). In 1993, the angler catch on the river was 574 (124 kept, 450 released), reflecting a harvest rate of 7.0% (Salmon Commission, unpublished trap and rod records).

In 1994, anglers in Maine were allowed to keep only one grilse; all adult Atlantic salmon had to be released. This marks the first time that anglers have not been able to keep any large (two sea-winter) Atlantic salmon, a reflection of the growing acceptance of catch-and-release fishing and support of the restoration efforts, particularly by salmon angling clubs.

TABLE 2.—Angler catches of Atlantic salmon from rivers in Maine, 1985–1986 and 1992–1993. The former period was during the peak runs and the latter period represents the most recent angler records.

River	1985	1986	1992	1993
Androscoggin	0	0	3	0
Aroostook	0	0	9	0
Dennys	20	15	12	4
Ducktrap	15	5	0	0
East Machias	31	13	9	3
Machias	30	43	10	12
Narraguagus	61	46	62	27
Penobscot	625	778	497	574
Saco	85	2	0	12
Saint Croix	20	55	2	1
Sheepscot	5	11	7	14
Union	1	5	0	0
Others[a]	0	0	1	12
Totals	893	973	612	659

[a] Includes Kennebec, Pleasant, and upper St. John rivers.

Discussion

Maine is fortunate to have the largest number of Atlantic salmon runs, the only nonextirpated populations of Atlantic salmon, and the most successful Model Restoration River—the Penobscot—in the United States. The ultimate objective of state and federal management programs is to maintain (and enhance) the runs of Atlantic salmon in the Downeast rivers and to restore runs in rivers, such as the Penobscot, that Atlantic salmon once inhabited. Yet, since the peak run of 1986, numbers of returning adults to Maine rivers have steadily declined, as have angler catches. The stocking program has been the key element in restoration efforts. Without the use of hatchery-reared fish, there would be no hope for restoration, and little optimism for the rehabilitation of depressed populations in rivers of eastern Maine. It appears that stocking will continue to be essential for years to come.

A Monte Carlo simulation model, using the best available data on survival rates at various life stages, river section, and spawning and age determinations, has predicted that a run of 5,500 is necessary for achieving a self-sustained population of Atlantic salmon in the Penobscot River (Marancik 1988). In recent years, less than a third of the needed fish have returned to the Penobscot River.

Stocking smolts, long the technique of choice in Maine, has not resulted in increased returns to the Penobscot River (Figure 3). Rather, the increased

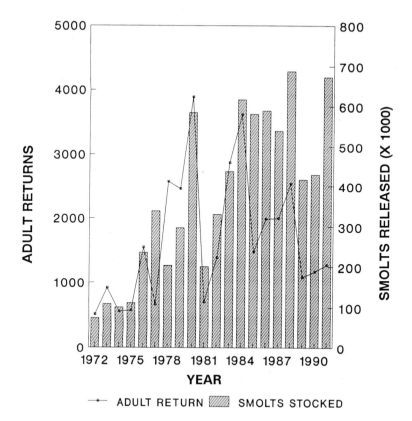

FIGURE 3.—Numbers of stocked Atlantic salmon smolts and the number of adults returning 2 years later, 1972–1991. Data are from records of the Maine Atlantic Sea Run Salmon Commission.

numbers of smolts stocked in the 1980s have been associated with declining numbers of adult fish returning to the river. In the wild rivers of eastern Maine, angler catches are inversely related to numbers of stocked smolts (Figure 4).

It would be easy to conclude that management practices have simply returned to the procedures used in the 1870s—stocking fry. However, the difference in approach between these two time periods is not related to the actual culture as much as it is to the disposition of the stocked fry. In the late 1980s and now in the 1990s, fry have been stocked with a more scientific approach toward density and habitat. In all likelihood, fry were stocked in the past at densities too low to be successful and in habitats that were inappropriate for instream residence. Stocking evaluation has just begun. Despite the years of marginal success from fry stocking in waters of southern New England, fisheries managers in Maine now are placing more emphasis on fry stocking. Reasoning that the steadily increasing incidence of fish spawning in the wild in Maine waters may be an indication of better habitat than that in southern New England, Maine managers are expecting better survival of stocked fry.

Stocking smolts has not resulted in continued improvement of runs in any of the Maine rivers, and producing smolts is considerably more expensive than producing fry. Whether stocking fry at appropriate densities will result in an improvement in adult returns awaits final evaluation.

Restoration and rehabilitation of Atlantic salmon runs in Maine has not been affected as much by cultural and stocking practices as by external factors. Dams, a culprit in the original demise of Atlantic salmon populations of New England, still have a major influence on fish passage and the susceptibility of Atlantic salmon to predators. More than 30% of the Penobscot River drainage that was originally accessible to Atlantic salmon is still unavailable to returning adults (Salmon Commission, unpublished data).

Atlantic salmon restoration in 1994 has reached an important milestone. Several management strategies have recently been implemented and show promise

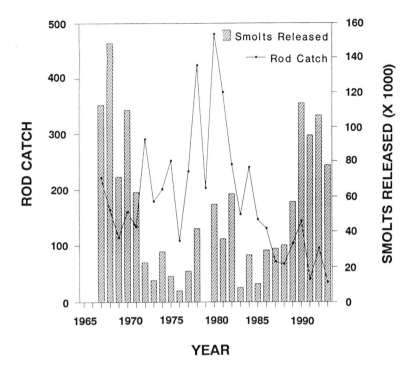

FIGURE 4.—Numbers of Atlantic salmon smolts stocked into the Dennys, East Machias, Machias, Narraguagus, Pleasant, and Sheepscot rivers and angler catches 2 years later, 1967–1993. Data are from records of the Maine Atlantic Sea Run Salmon Commission.

for increasing the returns of salmon to Maine rivers. Rather than a careful, stepwise series of scientific evaluations, the consequences of many actions will manifest themselves simultaneously. It will be difficult to assess which strategies may be effective and which may not. Short-term and long-term success will depend on several actions in the remaining years of the twentieth century: (1) the temporary closure of the West Greenland commercial fishery and negotiated agreements concerning Canadian fisheries must result in significantly higher oceanic survival and higher returns to the river; (2) the shift to a large fry-stocking program in tributaries must show positive results in terms of survival in the wild, higher numbers of resulting smolts, and a better reinforcement of homing back to natal rivers to increase natural reproduction; (3) angling restrictions will allow continued recreational fishing but will minimize sportfishing mortality of adult fish (*viz* catch-and-release only); and (4) the river-specific stocking program must secure sufficient broodfish each year without reducing the number of wild fish returning to spawn in Downeast rivers. Whatever the results in the next several years, future management strategies will continue to emphasize increasing the number of adult spawners returning to rivers, the numbers spawning in the wild, and the numbers of juveniles rearing in appropriate habitat.

Acknowledgments

We appreciate the contributions of biologists with the Maine Atlantic Sea Run Salmon Commission for providing unpublished records from the Atlantic salmon program, and Clem Fay, Penobscot Indian Nation, for providing additional records. William Krohn, National Biological Service, kindly reviewed the manuscript.

Cooperators of the Maine Cooperative Fish and Wildlife Research Unit are the University of Maine, Maine Department of Inland Fisheries and Wildlife, National Biological Service, and the Wildlife Management Institute.

References

Baum, E. T. 1983. The Penobscot River. Maine Atlantic Sea-Run Salmon Commission, River Management Report, Bangor.

Beland, K. F. 1984. Strategic plan for management of

Atlantic salmon in the state of Maine. Maine Atlantic Sea-Run Salmon Commission, Bangor.

Carlson, C. C. 1988. Where's the salmon? A reevaluation of the role of anadromous fisheries in aboriginal New England. Pages 47–80 *in* B. P. Nicholas, editor. Holocene human ecology in northeastern North America. Plenum Press, New York.

Fay, C. W. 1990. Evaluation of survival and growth of juvenile Atlantic salmon in selected sections of the Mattamiscontis Stream drainage. Penobscot Indian Nation, Final Report, Old Town, Maine.

Fried, S., J. D. McCleave, and G. LaBar. 1978. Seaward migration of hatchery-reared Atlantic salmon, *Salmo salar*, smolts in the Penobscot River estuary, Maine: riverine movements. Journal of the Fisheries Research Board of Canada 35:76–87.

Foerster, R. E. 1968. The sockeye salmon, *Oncorhynchus nerka*. Fisheries Research Board of Canada Bulletin 162.

Gibson, M. R. 1992. Analysis of fry stocking results in New England salmon restoration: effect of stocking density and environment on survival. U.S. Atlantic Salmon Assessment Committee, Working Paper 92/1, West Kingston, Rhode Island.

Gibson, M. R. 1994. Overwinter mortality of Atlantic salmon parr in Rhode Island streams in relation to density and age. U.S. Atlantic Salmon Assessment Committee, Working Paper 94/1, West Kingston, Rhode Island.

Griffith, B., J. M. Scott, J. W. Carpenter, and C. Reed. 1989. Translocation as a species conservation tool: status and strategy. Science 245:477–480.

Gustafson-Greenwood, K. I., and J. R. Moring. 1990. Territory size and distribution of newly-emerged Atlantic salmon (*Salmo salar*). Hydrobiologia 206:125–131.

Hosmer, M. J., J. G. Stanley, and R. W. Hatch. 1979. Effects of hatchery procedures on later return of Atlantic salmon to rivers in Maine. Progressive Fish-Culturist 41:115–119.

Hume, J. M. B., and E. A. Parkinson. 1984. The effects of various stocking strategies on the survival and growth of headwater stocked steelhead fry. Proceedings of the Western Association of Fish and Wildlife Agencies 64:274–287.

Isaksson, A., T. J. Rasch, and P. H. Poe. 1978. An evaluation of smolt releases into a salmon- and a non-salmon-producing stream using two release methods. Journal of Agricultural Research of Iceland 10:100–113.

Krohn, W. B., R. B. Allen, J. R. Moring, and A. E. Hutchinson. In Press. Double-crested cormorants in New England: population and management histories. Colonial Waterbirds 17.

Marancik, J. 1988. Projected impacts of the proposed Basin Mills Dam on the Penobscot River Atlantic salmon restoration program. U.S. Fish and Wildlife Service and Atlantic Salmon Working Group, Final Report, East Orland, Maine.

McMenemy, J. R. 1989. West River and tributaries: Atlantic salmon fry stocking evaluation. Final Report, Project F-12-R-22. Vermont Department of Fish and Wildlife, Springfield.

Moring, J. R. 1986. Stocking anadromous species to restore or enhance fisheries. Pages 59–74 *in* R. H. Stroud, editor. Fish culture in fisheries management. American Fisheries Society, Fish Culture Section and Fisheries Management Section, Bethesda, Maryland.

Moring, J. R. 1987. Restoration of Atlantic salmon in Maine: overcoming physical and biological problems in the estuary. Pages 3129–3140 *in* O. T. Magoon, H. Converse, D. Miner, L. T. Tobin, D. Clark, and G. Domurat, editors. Proceedings of Ocean Zone '87, the fifth symposium on coastal and ocean management. American Society of Civil Engineers, New York.

Moring, J. R. 1993. Anadromous stocks. Pages 553–580 *in* C. C. Kohler and W. A. Hubert, editors. Inland fisheries management in North America. American Fisheries Society, Bethesda, Maryland.

Netboy, A. 1968. The Atlantic salmon: a vanishing species? Houghton Mifflin Company, Boston.

Reisenbichler, R. R., and J. D. McIntyre. 1986. Requirements for integrating natural and artificial production of anadromous salmonids in the Pacific Northwest. Pages 365–374 *in* R. H. Stroud, editor. Fish culture in fisheries management. American Fisheries Society, Fish Culture Section and Fisheries Management Section, Bethesda, Maryland.

Ritter, J. A. 1975. The importance of stock selection to the restoration of Atlantic salmon in New England. Pages 129–131 *in* J. R. Bohne and L. Sochasky, editors. Proceedings of the New England Atlantic salmon restoration conference. International Atlantic Salmon Foundation, Special Publication Series 6.

Rottiers, D. V., and L. A. Redell. 1993. Volitional migration of Atlantic salmon from seasonal holding ponds. North American Journal of Fisheries Management 23:238–252.

Stolte, L. W. 1986. Atlantic salmon. Pages 696–713 *in* R. L. DiSilvestro, editor. Audubon wildlife report 1986. National Audubon Society, New York.

U.S. Atlantic Salmon Assessment Committee. 1991. Report No. 3—1990 activities. Report to the U.S. Section of the North Atlantic Salmon Conservation Organization, Woods Hole, Massachusetts.

U.S. Atlantic Salmon Assessment Committee. 1993. Report No. 5—1992 activities. Report to the U.S. Section of the North Atlantic Salmon Conservation Organization, Turners Falls, Massachusetts.

U.S. Fish and Wildlife Service. 1989. Restoration of Atlantic salmon in New England. Final environmental impact statement, 1989–2001. U.S. Fish and Wildlife Service, Newton Corner, Massachusetts.

van den Ende, O. 1993. Predation on Atlantic salmon smolts (*Salmo salar*) by smallmouth bass (*Micropterus dolomieu*) and chain pickerel (*Esox niger*) in the Penobscot River, Maine. Master's thesis. University of Maine, Orono.

Vanderpool, A. M. 1992. Migratory patterns and behavior of Atlantic salmon smolts in the Penobscot River, Maine. Master's thesis. University of Maine, Orono.

Fish Stocking Programs in Wyoming: A Balanced Perspective

MICHAEL D. STONE

Wyoming Game and Fish Department, 5400 Bishop Boulevard
Cheyenne, Wyoming 82006, USA

Abstract.—Introduction of nonnative salmonids into Wyoming began with stocking brook trout *Salvelinus fontinalis* and rainbow trout *Oncorhynchus mykiss* in 1880. Both good and bad experiences from the use of cultured fishes to manage sport fisheries have been documented. Fifty-three fish species are native to Wyoming; four of these are now considered extirpated from the state. None were extirpated by fish stocking. The Wyoming Game and Fish Department (WGFD) has developed a balanced fisheries management program that addresses both the ecological integrity of waters capable of sustaining native or wild fisheries and public demands for sport fisheries. Practices appropriate for one setting may be inappropriate for another. Maintenance of native subspecies of cutthroat trout *O. clarki* has been a management priority for over 40 years. Early stocking of brook trout and rainbow trout was detrimental to certain native trout stocks. Cultured trout have since been instrumental in cutthroat trout restoration. Most trout streams in Wyoming are managed as wild fisheries in accordance with WGFD guidelines. Low returns of stocked trout in streams, and the demonstrated ability of streams to support acceptable wild trout fisheries, lead to these guidelines. Reservoirs constitute 68.5% (159,800 acres) of the standing water acreage in Wyoming. Most reservoirs are managed with cultured fish due to habitat conditions that limit natural trout reproduction. Altered habitats present the greatest challenges in terms of overall fisheries management and use of cultured fish.

Cultured salmonids have been stocked in Wyoming since 1880 when brook trout *Salvelinus fontinalis* and rainbow trout *Oncorhynchus mykiss* were introduced. In 1884, production of hatchery trout began at the Red Buttes Hatchery near Laramie. In the following 110 years, fish hatcheries and rearing facilities have continued to operate for the purpose of improving sportfishing. The Wyoming Game and Fish Department (WGFD) currently operates eleven fish culture facilities. Rainbow trout and cutthroat trout *O. clarki* most commonly are reared.

Use of cultured salmonids to provide sportfishing opportunities has been both praised and maligned over the years. A balanced fisheries management program has been developed by the WGFD, and use of cultured fish represents one tool of this management program. The use of cultured fish continues to improve through stocking evaluations and evolution of management philosophy. This paper reviews fish stocking in Wyoming. Examples were selected to demonstrate both good and poor uses of cultured trout. Current management philosophy regarding the use of cultured trout also is reviewed.

Species Occurrence

Fifty-three species of fish are native to Wyoming (Table 1). Native salmonids were limited to cutthroat trout, mountain whitefish, and Arctic grayling. Four native fish species are now considered extinct in the state: the Colorado squawfish, razorback sucker, bonytail, and Snake River sucker. The Colorado squawfish, razorback sucker, and bonytail are federally listed Endangered species native to the Colorado River system. Construction of Fontenelle and Flaming Gorge reservoirs on the main-stem Green River effectively eliminated habitat for these fishes in Wyoming. Only one Snake River sucker was ever collected from Wyoming. This specimen was collected from Jackson Lake in 1927 (Baxter and Simon 1970). Reasons for the disappearance of this fish are unknown. None of the species extinctions in Wyoming can be attributed to stocking of cultured or nonnative species.

Twenty-nine introduced fish species have become established in Wyoming (Table 1). These include rainbow trout, brook trout, brown trout, lake trout, golden trout, kokanee, walleye, yellow perch, largemouth bass, smallmouth bass, bluegill, green sunfish, black crappie, and northern pike. Along with the native cutthroat trout and channel catfish, these introduced species are the most popular game fish in the state.

Numbers and Types of Waters Stocked

A database inventory maintained by the WGFD contains individual records for 4,694 streams and 5,360 standing waters (Table 2). Reservoirs constitute 68.5% of the 233,310 surface acres of standing waters in the state. Over half of the fishing effort occurs on these waters. Due to habitat conditions, reservoirs seldom are able to support naturally reproducing trout populations. Reservoirs receive the

TABLE 1.—Native and nonnative fish species established in Wyoming.

Common Name	Scientific Name
Native	
Shovelnose sturgeon	Scaphirhynchus platorynchus
Goldeye	Hiodon alosoides
Central stoneroller	Campostoma anomalum
Lake chub	Couesius plumbeus
Red shiner	Cyprinella lutrensis
Utah chub	Gila atraria
Leatherside chub	Gila copei
Bonytail[a]	Gila elegans
Roundtail chub	Gila robusta
Western silvery minnow	Hybognathus argyritis
Brassy minnow	Hybognathus hankinsoni
Plains minnow	Hybognathus placitus
Common shiner	Luxilus cornutus
Sturgeon chub	Macrhybopsis gelida
Pearl dace	Margariscus margarita
Hornyhead chub	Nocomis biguttatus
Bigmouth shiner	Notropis dorsalis
Sand shiner	Notropis stramineus
Suckermouth minnow	Phenacobius mirabilis
Finescale dace	Phoxinus neogaeus
Fathead minnow	Pimephales promelas
Flathead chub	Platygobio gracilis
Colorado squawfish[a]	Ptychocheilus lucius
Longnose dace	Rhinichthys cataractae
Speckled dace	Rhinichthys osculus
Redside shiner	Richardsonius balteatus
Creek chub	Semotilus atromaculatus
River carpsucker	Carpiodes carpio
Quillback	Carpiodes cyprinus
Utah sucker	Catostomus ardens
Longnose sucker	Catostomus catostomus
White sucker	Catostomus commersoni
Bluehead sucker	Catostomus discobolus
Flannelmouth sucker	Catostomus latipinnis
Mountain sucker	Catostomus platyrhynchus
Snake River sucker[a]	Chasmistes muriei
Shorthead redhorse	Moxostoma macrolepidotum
Razorback sucker[a]	Xyrauchen texanus
Black bullhead	Ameiurus melas
Channel catfish	Ictalurus punctatus
Stonecat	Noturus flavus
Cutthroat trout	Oncorhynchus clarki
Mountain whitefish	Prosopium williamsoni
Arctic grayling	Thymallus arcticus
Burbot	Lota lota
Plains topminnow	Fundulus sciadicus
Plains killifish	Fundulus zebrinus
Mottled sculpin	Cottus bairdi
Paiute sculpin	Cottus beldingi
Iowa darter	Etheostoma exile
Johnny darter	Etheostoma nigrum
Orangethroat darter	Etheostoma spectabile
Sauger	Stizostedion canadense
Nonnative	
Gizzard shad	Dorosoma cepedianum
Goldfish	Carassius auratus
Grass carp	Ctenopharyngodon idella
Common carp	Cyprinus carpio
Golden shiner	Notemigonus crysoleucas
Emerald shiner	Notropis atherinoides
Spottail shiner	Notropis hudsonius
Northern pike	Esox lucius
Golden trout	Oncorhynchus aguabonita
Rainbow trout	Oncorhynchus mykiss

TABLE 1.—Continued.

Common Name	Scientific Name
Nonnative	
Kokanee (lacustrine sockeye salmon)	Oncorhynchus nerka
Ohrid trout	Salmo letnica
Brown trout	Salmo trutta
Brook trout	Salvelinus fontinalis
Lake trout	Salvelinus namaycush
Western mosquitofish	Gambusia affinis
Guppy	Poecilia reticulata
Green swordtail	Xiphophorus helleri
Rock bass	Ambloplites rupestris
Green sunfish	Lepomis cyanellus
Pumpkinseed	Lepomis gibbosus
Bluegill	Lepomis macrochirus
Smallmouth bass	Micropterus dolomieu
Largemouth bass	Micropterus salmoides
White crappie	Pomoxis annularis
Black crappie	Pomoxis nigromaculatus
Yellow perch	Perca flavescens
Walleye	Stizostedion vitreum
Freshwater drum	Aplodinotus grunniens

[a] Believed to be extinct in Wyoming.

greatest emphasis for stocking. Over one-fourth (26.7%) of the reservoirs are stocked each year; 7.0% of the lakes and only 3.0% of the streams are stocked annually (Table 3). Less than 6% of the individual waters listed in the inventory are stocked with cultured fish. A total of 527 waters (5.2%) were scheduled for stocking in 1993 and 480 waters (4.8%) in 1994 (Table 3).

Native Trout Restoration

Maintenance of native cutthroat trout subspecies has been a management priority of the WGFD for more than 40 years. One subspecies, the Colorado River cutthroat trout *O. clarki pleuriticus*, is native to the Colorado River basin. In Wyoming it historically was restricted to the higher, cooler waters in the Green and Little Snake river basins. Habitat

TABLE 2.—Waters (with and without fisheries) contained in the Wyoming Game and Fish Department stream and lake inventory. Sizes of streams are given in miles; all other water types are given in acres.

Water type	Number	Total size
Streams	3,668	16,260
Nonfishery streams	1,026	3,410
Ponds	1,674	900
Lakes	1,894	51,210
Reservoirs	487	159,800
Nonfishery standing waters	1,305	21,400
Total	10,054	

TABLE 3.—Number of waters (percentage of total waters in parentheses) scheduled for stocking in Wyoming.

Water type	1993	1994
Streams	174 (3.7)	140 (3.0)
Ponds	90 (5.3)	77 (4.6)
Lakes	136 (7.2)	133 (7.0)
Reservoirs	127 (26.1)	130 (26.7)
Total	527 (5.2)	480 (4.8)

degradation and effects of introduced trout species have adversely affected the Colorado River cutthroat trout (Binns 1977). Brook trout displaced cutthroat trout at lower elevations and rainbow trout hybridized with cutthroat trout.

Colorado River cutthroat trout now occupy less than 1% of their historic range. Status reviews have been conducted under the Endangered Species Act (16 U.S.C.A. §§ 1531 to 1544), but this subspecies has not been listed. The WGFD is aggressively attempting to restore this sensitive subspecies and avoid listing. Formal cutthroat trout management plans were completed in 1987 and revised in 1994. Restoration efforts include habitat protection, habitat improvement, rehabilitation of streams containing nonnative salmonids, and stocking to expand the range and upgrade genetic purity of existing populations.

Sunrise Lake is a 28-acre alpine lake (10,460 ft elevation) located approximately 24 mi east of Pinedale on the Bridger–Teton National Forest in northwestern Wyoming. Sunrise Lake contained no fish until 1979, when Colorado River cutthroat trout from the WGFD Daniel Fish Hatchery were stocked to establish an additional Colorado River cutthroat trout population. Stocking consisted of 510 fish at 68/lb in 1979 and 1,250 fish at 48/lb in 1987. Gill-net sampling in 1992 collected fish ranging in size from 4.6 to 15.5 in and ages 1 to 4. In addition, more than 1,000 young-of-the-year Colorado River cutthroat trout were observed near the lake outlet, confirming natural reproduction.

The Sunrise Lake population of Colorado River cutthroat trout contributes to restoration efforts within the native range of this subspecies. Sunrise Lake trout stocking represents a successful and beneficial use of cultured fish.

Catchable-Size Trout Stocking

Catchable-size trout traditionally have been stocked in Wyoming to provide put-and-take fishing opportunities in waters that sustain heavy recreational use but lack the habitat features necessary to support sustained trout populations (Stone 1978). Put-and-take fisheries supported by stocking catchable-size trout are not considered natural resources by the WGFD but rather a means of providing recreational opportunity.

Catchable-size trout stocking has been evaluated statewide to refine the use of cultured trout. Wiley et al. (1993) reported return rates to anglers ranging from 4.4 to 89.6% for Wyoming lakes stocked with catchable-size trout from 1965 to 1988. Beginning in the 1980s, evaluation standards for successful catchable-size trout stocking required that at least 50% of the catchable-size trout stocked must be creeled and stocked fish must provide at least 25% of all game fish harvested. In addition, the receiving water must be incapable of providing a satisfactory fishery under another management strategy.

Tensleep Creek, located on the west slope of the Big Horn Mountains in north-central Wyoming, is an example of a stream where put-and-take stocking did not work successfully. Approximately half of the creek's 18-mile length is located on the Big Horn National Forest. Public access is also provided on the portion of Tensleep Creek associated with the WGFD Wigwam Rearing Station. Catchable-size rainbow trout have been stocked into Tensleep Creek from the Wigwam Rearing Station on a regular basis since 1957 to augment fisheries for wild rainbow and brown trout. A creel survey conducted in 1987 reported that only 26.7% of the stocked rainbow trout were harvested; wild trout were 55.8% of the trout harvested (Yekel 1990). The harvested stocked trout averaged 2 in longer than the wild trout. Population estimates obtained by electrofishing were 102 lb/acre in the stocked sections and 135 lb/acre in the section that was not stocked (Annear 1990).

Tensleep Creek failed to meet the criteria for continued catchable-size trout stocking. Instead, it met the criteria for management as a wild trout fishery. Stocking was reduced from 6,000 trout in 1987 to 4,000 in 1988, 3,000 in 1989, and 1,000 in 1990–1991. Stocking ceased in 1992. Monitoring will determine if trout stocking was detrimental to the wild trout population. Stocking in Tensleep Creek is an example of poor use of cultured fish.

Guidelines for Balanced Stocking Programs

Use of cultured fish represents only one tool in fisheries management. Although stocking has been both praised and maligned, such judgements are inappropriate. Stocking itself is neither good nor bad. As a tool, stocking can be used either properly or improperly. Managers must recognize that stocking can only address limitations in natural recruitment. Stocking is no substitute for correcting habitat limitations or preventing excessive harvest.

Evaluations of catchable-size trout stocking in Wyoming streams have provided variable results (Wiley et al. 1993). Return rates were generally unsatisfactory (<50%). Consequently, WGFD has eliminated catchable-size trout stocking in almost all streams. Social tradition and public expectations created by years of stocking must be considered when catchable-size trout stocking in streams is eliminated. Anglers are becoming highly educated with respect to fisheries management and are becoming more demanding for certain types of fishing experiences and resource agency accountability. These factors make possible changes that previously may not have been socially acceptable. Catchable-size trout stocking programs now emphasize smaller standing waters near population centers, which typically experience heavy use and lack habitat capability to provide fisheries under another management concept.

Maintenance of wild fisheries has been emphasized and stream stocking has been evaluated for years (Stone 1978; Wiley et al. 1993). Recent analyses of the trout standing stocks in streams provide additional evaluation tools. Salmonid standing stocks were 60 lb/acre or less in 55.5 to 96.2% of observations for western streams (Platts and McHenry 1988). Wiley (1992) analyzed 1,037 data entries for Wyoming streams. He found that 55% of the reported trout stocks were 60 lb/acre or less, 80% were 100 lb/acre or less, and 90% were 200 lb/acre or less. Streams over 50 ft wide usually had trout stocks 60 lb/acre or less. Habitat evaluation techniques (Binns and Eiserman 1979) and standing stock data permit comparison of observed stock size with expected stock potential and formulation of management recommendations. Trout stocking should not be expected to increase trout populations beyond natural production limits (Wiley 1992).

Prior to 1992, WGFD stocking guidelines stated that those streams managed as wild trout fisheries would not be stocked, except when deemed necessary to reestablish fish populations following disasters. Guidelines for stream stocking were revised in 1992; now all Wyoming streams with the potential to support wild trout populations will be managed as wild fisheries. To further this guideline, written rationale, management plans, or approved study plans will be developed for continued stocking of those waters where wild fisheries management is impossible.

Cultured fish play an important role in WGFD restoration programs for native species. Management strategies for the restoration of Colorado River cutthroat trout and Bonneville cutthroat trout *O. clarki utah* include a combination of habitat and fish population management measures. Watershed and riparian area management and habitat enhancement are used to address habitat concerns. Population management includes removal of competing trout and transplanting trout from other waters or stocking with genetically pure strains of cultured fish. Regulations are established as needed to limit harvest by anglers.

Cultured trout are especially important where depletion of native populations precludes sufficient transplants to establish new populations. If sufficient numbers of fish are available to establish a broodstock, the pace of introductions can be accelerated using cultured trout. Caution is needed, however, to establish broodstocks that reflect the genetic characteristics of the native stocks.

Game and forage fish reproduction in Wyoming reservoirs is typically limited by reservoir drawdown and lack of suitable habitat features. Over one-fourth of the reservoirs are stocked annually to provide recreational fisheries. Evaluations are important to determine the effectiveness of reservoir stocking practices. Comprehensive evaluations are expensive. Routine stocking evaluations are often in the form of angler catch rates, density indices, and satisfied anglers. Programmed creel surveys are conducted as budgets permit. Criteria for evaluating stocking success for individual waters are developed prior to conducting evaluations. Factors considered in development of evaluation criteria include the management goal and fisheries potential for each water.

Altered river habitats present great challenges in terms of overall management and use of cultured fish. Tailwater fisheries can be extremely productive, but many suffer from deficiencies ranging from flow depletion to channel down cutting or insufficient flushing flows. Past management activities focused on stocking to compensate for these habitat deficiencies. Too often the results have been unsatisfactory. Present WGFD management includes efforts to correct habitat deficiencies, stock select strains of cultured trout, and establish regulations when needed to limit harvest. Efforts to correct habitat deficiencies require extensive personnel commitments and patience. Such a balanced approach is warranted to manage altered river habitats effectively.

References

Annear, T. C. 1990. Tensleep Creek fisheries monitoring study. Wyoming Game and Fish Department, Fish Division Administrative Report, Cheyenne.

Baxter, G. T., and J. R. Simon. 1970. Wyoming Fishes. Wyoming Game and Fish Department, Bulletin 4, Cheyenne.

Binns, N. A. 1977. Present status of indigenous populations of cutthroat trout, *Salmo clarki*, in southwest Wyoming. Wyoming Game and Fish Department, Fisheries Technical Bulletin 2, Cheyenne.

Binns, N. A., and F. M. Eiserman. 1979. Quantification of fluvial trout habitat in Wyoming. Transactions of the American Fisheries Society 108:215–228.

Platts, W. B., and M. L. McHenry. 1988. Density and biomass of trout and char in western streams. U.S. Forest Service General Technical Report INT-241.

Stone, M. D. 1978. Wyoming's catchable trout program as defined by fisherman attitudes. Pages 13–17 *in* J. R. Moring, editor. Proceedings of the wild trout-catchable trout symposium. Oregon Department of Fish and Wildlife, Corvallis.

Wiley, R. W. 1992. Consideration of trout standing stocks in Wyoming waters. Wyoming Game and Fish Department, Fish Division Administrative Report, Cheyenne.

Wiley, R. W., R. A. Whaley, J. B. Satake, and M. Fowden. 1993. Assessment of stocking hatchery trout: a Wyoming perspective. North American Journal of Fisheries Management 13:160–170.

Yekel, S.A. 1990. Programmed creel census on Tensleep Creek, Washakie County, 1987. Wyoming Game and Fish Department, Fish Division Administrative Report, Cheyenne.

An Evaluation of the Genetic Integrity of Ohio River Walleye and Sauger Stocks

MATTHEW M. WHITE

Ohio University, Department of Biological Sciences
Athens, Ohio 45701, USA

SCOTT SCHELL

Ohio Department of Natural Resources, Division of Wildlife
360 East State Street
Athens, Ohio 45701, USA

Abstract.—An electrophoretic survey of populations of walleye *Stizostedion vitreum* and sauger *S. canadense* from the Ohio River was conducted to determine the patterns of genetic variation, population structuring, and the degree of hybridization between these two species and their stocked F_1 hybrid, the saugeye (female walleye × male sauger). Thirty-six presumptive structural loci were surveyed from the eye, liver, and muscle tissue of 500 sauger from nine locations and 222 walleyes from seven locations. Levels of variation in sauger were low and suggested limited population differentiation along the river. Levels of variation among walleye populations suggested a significant degree of population differentiation; however, no clear pattern of differentiation was observed. Two polymorphisms, not previously observed in walleye populations, are shared with sauger, suggesting past hybridization events or geographically unique alleles. Recombinant genotypes were detected in samples from three Ohio River pools, confirming that hybrid reproduction has occurred. These three pools are consecutively affected by one major river and four smaller watersheds that have received large numbers of stocked saugeyes. If maintaining the genetic integrity of the parental species is a concern, our data strongly suggest that saugeye should not be stocked where self-sustaining parental populations occur.

Walleye *Stizostedion vitreum* and sauger *S. canadense* are widespread and important components of the fish faunas of the Mississippi River drainage, the Great Lakes, and Canada, and are sympatric throughout much of their ranges (Pflieger 1975; Trautman 1981). Walleye displayed significant population differentiation and stock structuring based on allozyme data in several studies from the Great Lakes and Canada (Clayton et al. 1974; Ward et al. 1989; Todd 1990), and mitochondrial DNA (mtDNA) data (Billington and Hebert 1988; Ward et al. 1989; Billington et al. 1992). Despite the broad geographic area covered by these studies, there are no data documenting genetic variation within and among populations from the Ohio River. In contrast to the walleye, limited studies addressing geographic variation among sauger populations have observed relatively low levels of allozyme variation (Uthe et al. 1966; Uthe and Ryder 1970; Todd 1990).

Natural hybridization between walleye and sauger is known but appears to be rare (Trautman 1981; but see Ward 1992). The F_1 hybrid is readily identifiable, both phenotypically and electrophoretically (Uthe et al. 1966; Clayton et al. 1973; Billington et al. 1988). Billington et al. (1988) have also identified introgression between walleye and sauger based on mtDNA. The extent of introgression due to natural hybridization between sauger and walleye has not been evaluated.

The F_1 hybrid of female walleye × male sauger (saugeye) has become an important component of the fisheries management programs of several states. The saugeye is a unique hybrid in that it is very adaptable, exhibits remarkable heterosis, and is capable of reproduction. Saugeyes have been shown to survive in ponds on a *Lepomis* spp. forage base (Lynch et al. 1982). Evaluations of twelve Ohio reservoirs demonstrated poststocking, oversummer survival rates (35mm fingerlings) ranging from 0.7 to 9.2%, with a mean survival rate of 3.3% (Austin 1993). Siegwarth and Summerfelt (1990) found in a controlled growth comparison that juvenile saugeyes performed better than walleyes for every variable at two temperature regimes. Recent culture experiments revealed that introgressive hybridization had no influence on embryonic survival and hatching rate of eggs from either walleye or sauger females (Malison et al. 1990).

Widespread stocking of saugeye in Ohio River drainages has raised the issue of hybrid reproduction and the consequences of increased hybrid abundance in the presence of both parental species. Such reproduction and introgression could have

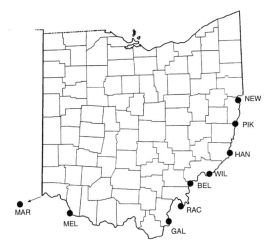

FIGURE 1.—Ohio River lockchamber sampling locations for species of *Stizostedion*, 1991–1993. (NEW, New Cumberland; PIK, Pike Island; HAN, Hannibal; WIL, Willow Island; BEL, Belleville; RAC, Racine; GAL, Galipolis; MEL, Meldahl; and MAR, Markland.)

consequences for stock integrity of the parental species within the Ohio River or anywhere that the saugeye coexists with either parental species.

This study was designed to evaluate the pattern of genetic variation and population structuring of walleye and sauger in the Ohio River. We also evaluated the extent of introgression with the parental species by detecting the presence of recombinant individuals based on electrophoretic pattern.

Methods

Walleyes and saugers were collected during 1991 and 1993 Ohio River lockchamber surveys (Figure 1). Additional walleye samples were collected in 1992 and 1993 by Ohio River creel clerks and from electrofishing surveys. Whole fish were placed on wet ice in the field and then stored at −20°C until they were processed for electrophoresis. Walleye fin tissue samples (single pelvic fin) from the creel survey were stored on ice and maintained at −20°C.

Homogenates of muscle, liver, and eye were prepared for each fish. Tissues (including fin clips) were homogenized in an equal volume of an aqueous buffer (0.1M Tris, 0.001M EDTA, 0.0001M NAD, 0.0001M NADP, pH 7.0). Homogenates were centrifuged at 12,000 rotations per minute at 4°C. The products of 36 presumptive structural gene loci were resolved by standard horizontal starch gel electrophoresis (Morizot and Schmidt 1990). Locus designations and electrophoretic conditions are shown in Table 1. Locus designations follow Shaklee et al. (1989). Stain recipes were adapted from Turner (1983), Billington et al. (1990), and Todd (1990).

Data were analyzed with the BIOSYS-1 computer package (Swofford and Selander 1981). Specific analyses included (1) comparisons with random mating expectations by a chi-square goodness-of-fit test, (2) extent of interpopulation genic heterogeneity and differentiation, and (3) magnitude of gene flow. Genic heterogeneity was evaluated using a chi-square test (Workman and Niswander 1970). The extent of genic differentiation in subdivided populations is measured by F-statistics (Wright 1978). The reduction in heterozygosity in the subpopulation and the total population due to inbreeding is measured by F_{IS} and F_{IT}. The reduction in heterozygosity due to drift among subpopulations is estimated by F_{ST} and is a measure of population differentiation. The significance of the genic heterogeneity chi-square determines the significance of the value of F_{ST} computed for that locus (Workman and Niswander 1970). Gene flow estimates were computed from

$$F_{ST} = 1/(4Nm + 1),$$

where the product Nm is the effective number of migrants per population per generation (Slatkin 1981). Substitution of F_{ST} into this equation yields an estimate of Nm. Values greater than 1 suggest gene flow is sufficient to overcome the effects of stochastic processes on allele frequencies (Slatkin 1981).

Hybrid reproduction was detected by the presence of recombinant genotypes. Because walleye and sauger exhibit fixed allele differences at two loci (*MDH-1**, *PGM-1**; Clayton et al. 1973; Todd 1990; this study), their hybrid would display heterozygous genotypes at both loci. Recombinants would be detected by observing individuals that were hybrid at one locus and parental at another (Avise and Van Den Avyle 1984; Campton 1987).

Results

Polymorphisms were detected at two loci for 500 saugers from nine populations (Table 2). The *EST** polymorphism was widespread in all sauger populations from the Ohio River. The *PGM** polymorphism was also widespread and newly identified in sauger. An *LDH** polymorphism was detected, but found in only 3 of 500 Ohio River saugers sampled. Of 18 comparisons of observed genotype frequencies with random mating expectations, 3 exhibited

TABLE 1.—Enzymes (designations follow Shaklee et al. [1989]) and electrophoretic conditions used in the analysis of *Stizostedion* species from the Ohio River. Tissue types analyzed were eye (E), pelvic fin (F), liver (L), and muscle (M). Buffers were (A) Tris-citrate pH 6.9 (Todd 1990); (B) Tris-borate-EDTA pH 8.6 (Turner 1983); and (C) Lithium hydroxide (Selander et al. 1971).

Enzyme	Enzyme number	Locus abbreviation	Tissue	Buffer
Acid phosphatase	3.1.3.2	ACP*	L	A
Aconitate hydratase	4.2.1.3	mAH*	L	A
	4.2.1.3	sAH*	L	A
Adenosine deaminase	3.5.4.4	ADA-A*	M	B
Alcohol dehydrogenase	1.1.1.1	ADH*	L	A
Aspartate aminotransferase	2.6.1.1	m-AAT*	M	B
	2.6.1.1	s-AAT*	M	B
Creatine kinase	2.7.3.2	CK-A*	M	B
	2.7.3.2	CK-C*	E	B
Dihydrolipoamide dehydrogenase	1.8.1.4	DLDH*	L	B
Dipeptidase	3.4.13.11	PEPA*	L	C
Esterase (Nathyl Acetate)		EST*	E, F	B
Fumarate hydratase	4.2.1.2	FH*	M	A
Glucose-6-phosphate isomerase	5.3.1.9	GPI-A*	M	A
	5.3.1.9	GPI-B*	M	A
Glucose-6-phosphate dehydrogenase	1.1.1.49	G6PDH*	L	B
Glycerol-3-phosphate dehydrogenase	1.1.1.8	G3PDH*	M	A
L-Iditol dehydrogenase	1.1.1.14	IDDH*	L	A
Isocitrate dehydrogenase	1.1.1.42	IDH-1*	M	A
	1.1.1.42	IDH-2*	M, F	A
L-Lactate dehydrogenase	1.1.1.27	LDH-A*	M	B
	1.1.1.27	LDH-B*	M	B
	1.1.1.27	LDH-C*	E	B
Malate dehydrogenase	1.1.1.37	MDH-1*	M, F	A
	1.1.1.37	MDH-2*	M	A
	1.1.1.37	MDH-3*	M	A
Malic enzyme	1.1.1.40	MEP*	M	A
Mannose-6-phosphate isomerase	5.3.1.8	MPI-A*	L	B
	5.3.1.8	MPI-B*	M	B
Muscle protein		MP*	M	B
Phosphoglucomutase	5.4.2.2	PGM-1*	M	A
	5.4.2.2	PGM-2*	M	A
Phosphogluconate dehydrogenase	1.1.1.44	PGDH*	L	A
Prolyl dipeptidase	3.4.13.8	PEPS*	L	C
Superoxide dismutase	1.15.1.1	SOD*	L	B
Triose-phosphate isomerase	5.3.1.1	TPI-A*	L	A
	5.3.1.1	TPI-B*	L	A
Tripeptide aminopeptidase	3.4.11.4	PEPB*	L	C

significant departures, but no pattern was observed and we found both heterozygote deficiencies and excesses.

Genic heterogeneity among sauger populations was significant at the *PGM** locus but not at the *EST** locus (Table 3). Overall heterogeneity was highly significant. Gene flow estimates for these two loci (12.2 for *PGM** and 3.48 for *EST**) suggest that gene flow was sufficient to overcome the effects of stochastic pressures on allele frequencies (Slatkin 1987). A graphical analysis of allele frequency and river position (Figure 2) suggests relative homoge-

TABLE 2.—Allele frequencies for nine samples of sauger from the Ohio River, 1991–1993. Population abbreviations (river position) are defined in Figure 1.

Locus, allele, and sample size	Population								
	NEW	PIK	HAN	WIL	BEL	RAC	GAL	MEL	MAR
*EST**									
*100	0.551	0.591	0.456	.0487	0.412	0.418	0.290	0.430	0.367
*105	0.449	0.409	0.544	0.513	0.588	0.582	0.710	0.570	0.633
N	78	11	57	115	57	49	23	50	60
*PGM-1**									
*100	0.603	0.818	0.967	0.851	0.699	0.805	0.900	0.838	0.853
*92	0.397	0.182	0.033	0.149	0.301	0.195	0.100	0.162	0.147
N	34	11	46	47	68	41	10	34	34

TABLE 3.—Genic heterogeneity (F_{IS}, F_{IT}, and F_{ST}) and estimates of gene flow (Nm) for sauger from the Ohio River, 1991–1993. The significance of F_{ST} is given by P.

Locus	F_{IS}	F_{IT}	F_{ST}	P	Nm
EST*	0.140	0.199	0.020	0.056	12.2
PGM-1*	0.414	0.454	0.067	<0.001	3.48
Mean	0.245	0.296	0.038	<0.001	

neity among populations and the absence of a cline or other obvious pattern to allele frequency variation.

Polymorphisms were observed at six loci for 222 walleyes from seven populations (Table 4). The EST* and IDDH* polymorphisms are newly observed in walleye. Three of 33 comparisons with random mating expectations were significant and indicated a deficit of heterozygotes. Five of the six loci exhibited significant interpopulation genic heterogeneity (Table 5), and the overall heterogeneity was highly significant. Gene flow estimates from these data ranged from 3.4 and 3.0 at MDH-3* and MP*, respectively, to 0.6 and 0.9 for IDDH* and ADH*, respectively. Although the Nm values are greater than 1, the data, and the interpopulation genic heterogeneity, suggest low levels of gene flow insufficient to overcome the differentiating effects of drift (Slatkin 1987). A plot of allele frequencies with river position for the walleye populations (Figure 3) suggests a much greater level of stock structuring, relative to sauger, along the river; however, the pattern of allele frequency variation appears to be random.

Electrophoretic data indicated the presence of recombinant genotypes at three Ohio River tailwaters. Willow Island (sample frequency = 30% in 1992, 4% in 1993), Belleville (10% in 1993), and Racine (25% in 1993). Recombinant genotypes were not documented from other locations. These recombinant individuals exhibited genotypes confirming hybrid reproduction, although locus variability is not sufficient to determine the exact nature or direction of reproduction. The recombinant individuals phenotypically appeared to be walleyes, suggesting that hybridization between walleye and saugeye is occurring.

Discussion

Levels of allozyme variation in sauger observed in this study were similar to those measured by Billington et al. (1990) but quite low relative to walleye and other percid fishes (Billington et al. 1989; Todd 1990). Although the analysis of genic heterogeneity among populations indicate significant differences among populations along the Ohio River, gene flow estimates suggest there is considerable exchange among populations. Spatial comparisons of allele frequencies also suggest that these populations are not very different. There are two conclusions that can be drawn from this: (1) sauger from the Ohio River (in the vicinity of Ohio) can be considered a single panmictic population and no stock structure is evident; and (2) the presence of locks and dams on the Ohio River do not constitute a significant barrier to gene flow among populations. The highly migratory behavior of sauger (Collette et al. 1977) likely contributes to the lack of population structure.

The allozyme variation exhibited by walleyes from the Ohio River is consistent with the patterns of variation observed among other walleye populations (Clayton et al. 1971; Ward et al. 1989; Todd 1990). Highly significant genic heterogeneity and low estimates of gene flow suggest that some degree of stock structuring occurs in the Ohio River; however, a pattern to the structuring was not evident. The localized nature of spawning habitat may contribute to this structuring. Extensive stocking of walleye (primarily of Lake Erie origin) in many lakes and rivers draining into the Ohio River only

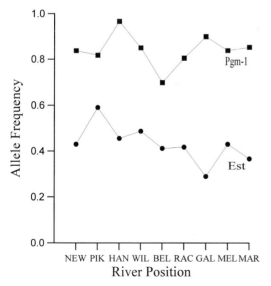

FIGURE 2.—Allele frequencies for polymorphic loci of saugers collected from Ohio River locks, 1991–1993. River locations are defined in Figure 1. Loci are defined in Table 1.

TABLE 4.—Allele frequencies for walleye populations from the Ohio River. All samples were collected during 1991 except for RAC, which was collected during 1993. Population abbreviations (river position) are defined in Figure 1.

Locus, allele, and sample size	Population							
	NEW	PIK	HAN	WIL	BEL	RAC	GAL	MEL
*MDH-3**								
*100	0.462	0.425	0.333	0.235	0.333	0.240	0.458	0.200
*120	0.538	0.575	0.667	0.765	0.667	0.760	0.542	0.800
N	53	53	9	17	39	25	12	5
*IDH-1**								
*120	0.355	0.375	0.750	0.450	0.404	0.532	0.708	0.500
*100	0.645	0.625	0.250	0.550	0.596	0.468	0.298	0.500
N	55	40	—	4	20	26	31	12	5
*ADH**								
*−100	1.00	0.912	0.750	1.00	0.950	0.896	0.929	0.600
*−60	0.0	0.088	0.250	0.0	0.050	0.104	0.071	0.400
N	15	17	4	20	10	24	7	5
*IDDH**								
*100	0.900	1.00	1.00	1.00	0.722	0.839	1.00	1.00
*10	0.100	0.0	0.0	0.0	0.278	0.161	0.0	0.0
N	20	17	4	20	10	31	7	5
*EST**								
*105	0.663	0.909	1.00	1.00	0.903	0.841	0.800	0.875
*100	0.337	0.091	0.0	0.0	0.097	0.159	0.200	0.125
N	46	22	7	20	31	22	5	4
*MP**								
*70	0.469	0.500	0.125	0.500	0.333	0.300	0.214	0.400
*100	0.531	0.500	0.875	0.500	0.667	0.700	0.786	0.600
N	16	16	4	11	9	30	7	5

complicates the interpretation of the observed genetic structure.

This study has also documented *EST** and *IDDH** polymorphisms not previously observed in walleye. The electrophoretic mobilities of the two alleles at both loci in walleye were identical to those observed in sauger. Todd (1990) has suggested that a rare polymorphism in Lake St. Clair walleye was due to a previous hybridization event with sauger. It is possible that the *EST** and *IDDH** polymorphisms in Ohio River walleye are due to such a hybridization event. It also could be that these electrophoretic variants are specific to Ohio River walleye and not due to a hybrid origin.

Recombinant genotypes confirming hybrid reproduction were observed at three localities. The observed recombinant frequencies must be viewed as conservative estimates in light of the recombinant ratios of Mendelian genetics and the limited number of polymorphic loci. The fish assayed in this study were phenotypically walleye, suggesting that walleye was one of the parents. However, electrophoresis does not permit us to determine the nature or direction of the hybrid reproduction in *Stizostedion*.

The four samples in which recombinants were observed originated from Belleville, Racine, and Gallipolis pools, which are downriver of the confluence of the Muskingum and Ohio rivers. In addition, several smaller drainages that affect these Ohio River pools have been stocked annually with saugeye since 1990. Two impoundments on the Hocking River and one impoundment on the Little Hocking River, which impact Belleville pool, have received saugeye stockings. Forked Run lake, which affects Racine pool, is less than one mile from the Ohio River. One impoundment on the Raccoon

TABLE 5.—Genic heterogeneity (F_{IS}, F_{IT}, and F_{ST}) and gene flow estimates (Nm) for walleye from the Ohio River, 1991–1993. The significance of F_{ST} is given by P.

Locus	F_{IS}	F_{IT}	F_{ST}	P	Nm
*MDH-3**	0.134	0.184	0.057	0.002	3.4
*IDH-1**	−0.091	−0.004	0.080	0.014	2.1
*ADH**	−0.323	−0.150	0.131	0.014	0.9
*IDDH**	−0.190	−0.008	0.153	0.001	0.6
*EST**	0.039	0.131	0.096	<0.001	1.6
*MP**	0.014	0.076	0.063	0.261	2.9
Mean	−0.031	0.055	0.083	<0.001	2.0

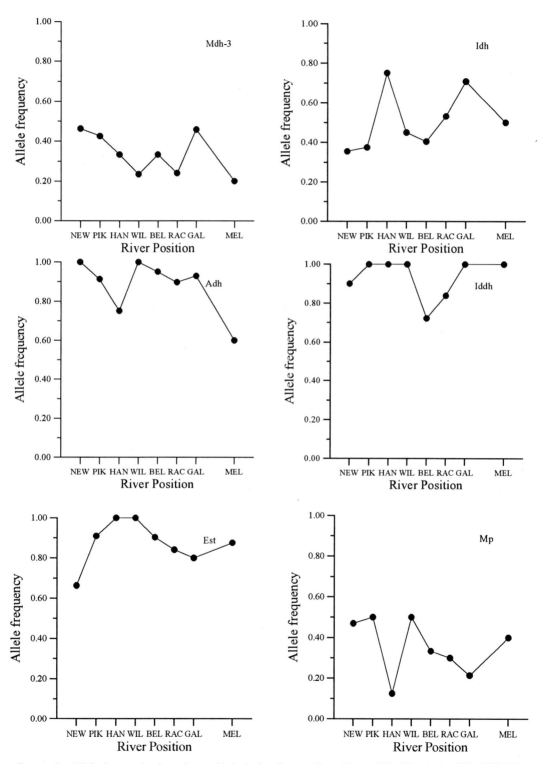

FIGURE 3.—Allele frequencies for polymorphic loci of walleyes collected from Ohio River locks, 1991–1993. River locations are defined in Figure 1. Loci are defined in Table 1.

Creek watershed, which drains into the Gallipolis pool, has also received annual saugeye stockings since 1990. Other pools farther upstream in which recombinants were not observed are not impacted by drainages receiving saugeye. The very remote possibility that the observed recombinant individuals were inadvertently propagated and stocked must be considered, since Belleville, Racine, and Gallipolis pools are traditional sources of sauger broodstock for saugeye propagation.

Management Implications

Recent mandates affirming part ownership of the Ohio River and a multistate management agreement have permitted the Ohio Department of Natural Resources Division of Wildlife to assess this fishery for the first time. Self-sustaining walleye and sauger populations have been documented in every Ohio River pool bordering Ohio. Existing strategic plans to manage saugeye fisheries did not include strategies for Ohio River percids.

Hybridization between walleye and sauger is rare in natural populations (Clayton et al. 1973; Trautman 1981; Billington et al. 1988, but see Ward 1992). Low levels of natural introgression are likely normal where walleye and sauger coexist, because both species are broadcast spawners. The lack of greater inbreeding between sympatric walleye and sauger populations is likely due to isolating mechanisms (Billington et al. 1988). Biological and ethological differences between walleye and sauger have been documented, but much needs to be learned about spawning isolating mechanisms. Walleye and sauger used different prespawn habitats in the upper Mississippi River (Siegwarth et al. 1993). Though walleye and sauger are reported to use similar rocky spawning habitat in large rivers (Hickman et al. 1989, Siegwarth et al. 1993) and reported spawning temperatures overlap (Scott and Crossman 1973), walleye begin spawning first. Whatever the isolating mechanisms are, walleye and sauger have coexisted for many years and remain distinct species.

Even less is known about the reproductive behavior of the hybrid. A recent impoundment saugeye telemetry study (Leeds 1991) revealed that 75% of the transmitted individuals moved upstream in spring (7.1–16.0°C water temperatures). Clearly, reproductive isolation among walleye, sauger, and their hybrids is not absolute.

Public demand for quality recreational fisheries combined with improved hatchery capabilities have driven Ohio's saugeye stocking program. During 1993, 10.6 million saugeye fingerlings and 1.9 million saugeye fry were stocked into 45 reservoirs and two river systems. These highly successful reservoir fisheries have led to river stockings in an effort to expand saugeye fishing opportunities. The Muskingum River has received annually increasing saugeye stockings of 126,000 fingerlings in 1990 to 750,000 fingerlings in 1993. The full effect of these introductions will not be realized until these fish mature.

The results of this genetic evaluation demonstrate that Ohio River walleye are experiencing introgression, though the evaluation did not confirm if the introgression is natural or a consequence of saugeye stocking. Introgressive hybridization can cause the genetic loss of an entire species, subspecies, or unique population (Campton 1987). The fact that the natural hybrids are rare raises the question of the possible long-term impacts on endemic Ohio River percids if hybrid abundance exceeds natural background levels. Increased saugeye abundance likely will increase existing levels of introgression in Ohio River walleye. Additional samples of walleye from Belleville, Racine, and Gallipolis pools and from Greenup, Meldahl, and Markland pools (downstream from the confluences of the Scioto and Miami rivers, where additional impoundment and river stockings are occurring) would help further clarify the frequency of these reproductive events. With continued saugeye stocking expected, stewardship of Ohio River native percid resources is a valid concern. It remains to be seen if sauger eventually will be affected as well or if natural isolating mechanisms prevail.

Habitat loss from damming, pollution, dredging, and corridor development have contributed to the past decline of Ohio River aquatic fauna. Introductions of nonnative walleye stocks have also occurred in the Ohio River drainage. Despite these combined stresses, naturally reproducing walleye populations persist in the Ohio River and are responding favorably to recent improvements in water quality (Sanders 1993). A unique mitochondrial marker has also been confirmed in some Ohio River walleye (N. Billington, Southern Illinois University, personal communication). Conservation of this resource must be considered in future management activities.

The results of this study suggest that in lieu of prestocking impact studies a more conservative approach to saugeye stockings is in order until more is known about their potential effects on endemic percid populations and these endemic populations can be identified. Recent water quality improvements may permit endemic walleye populations to reestablish in their former range. Propagation and stocking of endemic walleye strains may prove to be a cost-effective option, especially if natural reproduction results. Propagation of sterile saugeyes should also be seriously considered, especially for waters that directly impact the Ohio River. Development of a policy regarding the stocking of percid hybrids in Ohio must be considered.

Acknowledgments

We wish to thank Michelle Wright, Brenda Miller, Katie Richter, and Sandy Weaver for laboratory assistance. Gratitude is extended to Jerry Schulte of the Ohio River Valley Water Sanitation Commission, Bernie Dowler of the West Virginia Division of Wildlife Resources, Larry Lehman and Tom Flatt of the Indiana Division of Fish and Wildlife, numerous individuals from the Ohio Division of Wildlife, and Roy Stein from the Aquatic Ecology Laboratory at Ohio State University. This research was supported by the Ohio Division of Wildlife through Federal Aid in Sport Fish Restoration Project F-57-R-15.

References

Austin, M. R. 1993. Evaluation of percid success in Ohio. Ohio Department of Natural Resources, Federal Aid in Sport Fish Restoration, Project F-69-P, Annual Performance Report, Columbus.

Avise, J. C., and M. J. Van Den Avyle. 1984. Genetic analysis of hybrid white bass × striped bass in the Savannah River. Transactions of the American Fisheries Society 113:563–570.

Billington, N., and P. D. N. Hebert. 1988. Mitochondrial DNA variation in Great Lakes walleye (*Stizostedion vitreum*) populations. Canadian Journal of Fisheries and Aquatic Sciences 45:643–645.

Billington, N., P. D. N. Hebert, and R. D. Ward. 1988. Evidence of introgressive hybridization in the genus *Stizostedion*: interspecific transfer of mitochondrial DNA between sauger and walleye. Canadian Journal of Fisheries and Aquatic Sciences 45:2035–2041.

Billington, N., P. D. N. Hebert, and R. D. Ward. 1990. Allozyme and mitochondrial DNA variation among three species of *Stizostedion* (Percidae): phylogenetic and zoogeographical implications. Canadian Journal of Fisheries and Aquatic Sciences 47:1093–1102.

Billington, N., R. J. Barrette, and P. D. N. Hebert. 1992. Management implications of mitochondrial DNA variation in walleye stocks. North American Journal of Fisheries Management 12:276–284.

Campton, D. E. 1987. Natural hybridization and introgression in fishes: methods of detection and genetic interpretations. Pages 161–192 *in* N. Ryman and F. Utter, editors. Population genetics and fishery management. University of Washington Press, Seattle.

Clayton, J. W., R. E. K. Harris, and D. N. Tretiak. 1973. Identification of supernatant and mitochondrial isozymes of malate dehydrogenase on electropherograms applied to the taxonomic discrimination of walleye (*Stizostedion vitreum vitreum*), sauger (*S. canadense*), and suspected hybrids. Journal of the Fisheries Research Board of Canada 30:927–938.

Clayton, J. W., R. E. K. Harris, and D. N. Tretiak. 1974. Geographic distribution of alleles for supernatant malate dehydrogenase in walleye (*Stizostedion vitreum vitreum*) populations in western Canada. Journal of the Fisheries Research Board of Canada 31:342–345.

Collette, B. B., and seven coauthors. 1977. Biology of the percids. Journal of the Fisheries Research Board of Canada 34:1891–1897.

Hearn, M. C. 1986. Reproductive viability of sauger–walleye hybrids. Progressive Fish-Culturist 48:149–150.

Hickman, G. D., K. W. Hevel, and E. M. Scott. 1989. Density, movement patterns, and spawning characteristics of sauger (*Stizostedion canadense*) in Chickamauga Reservoir, Tennessee—1988. Tennessee Valley Authority, River Basin Operations, Chattanooga.

Larmann, P. W. 1978. Case histories of stocking walleyes in inland lakes, impoundments, and the Great Lakes—100 years with walleyes. American Fisheries Society Special Publication 11:254–260.

Leeds, L. G. 1991. Distribution, movement, and habitat preference of saugeye in Thunderbird Reservoir, Oklahoma. Federal Aid in Sport Fish Restoration, Project F-37-R, Study 17, Final Report, Norman.

Lynch, W. E., D. L. Johnson, and S. A. Schell. 1982. Survival, growth, and food habits of walleye × sauger hybrids (saugeye) in ponds. North American Journal of Fisheries Management 4:381–387.

Malison, J. A., T. B. Kayes, J. A. Held, and C. L. Amundson. 1990. Comparative survival, growth, and reproductive development of juvenile walleye, sauger, and their hybrids reared under intensive culture conditions. Progressive Fish-Culturist 52:73–82.

Morizot, D. C., and M. E. Schmidt. 1990. Starch gel electrophoresis and histochemical visualization of proteins. Pages 23–80 *in* D. H. Whitmore, editor. Electrophoretic and isoelectric focusing techniques in fisheries management. CRC Press, Boca Raton, Florida.

Pflieger, W. F. 1975. The fishes of Missouri. Missouri Department of Conservation, Jefferson City.

Sanders, R. E. 1993. Ohio's near shore fishes of the Ohio River (Year 2: 1992 results). Ohio Environmental Protection Agency for Ohio Department of Natural Resources, Division of Wildlife, Nongame and Endangered Wildlife Program, Columbus.

Scott, W. B., and E. J. Crossman. 1973. Freshwater fishes of Canada. Fisheries Research Board of Canada Bulletin 184.

Seigwarth, G. L., and R. C. Summerfelt. 1990. Growth comparisons between fingerling walleyes and walleye × sauger hybrids reared in intensive culture. Progressive Fish-Culturist 52:100–104.

Siegwarth, G. J., J. Pitlo, and D. Willis. 1994. Walleye and sauger spawning habitat survey in pool 16 of the upper Mississippi River. 56th Midwest Fish and Wildlife Conference (Abstracts), St. Louis, Missouri.

Selander, R. K., M. H. Smith, S. Y. Yang, W. E. Johnson, and J. B. Gentry. 1971. Biochemical polymorphism and systematics in the genus *Peromyscus*. I. Variation in the old-field mouse, *Peromyscus polionotus*. Studies in Genetics VI:49–90.

Shaklee, J. B., F. W. Allendorf, D. C. Morizot, and G. S. Whitt. 1990. Genetic nomenclature for protein coding loci in fish. Transactions of the American Fisheries Society 119:2–15.

Slatkin, M. 1981. Estimating levels of gene flow in natural populations. Genetics 99:323–335.

Slatkin, M. 1987. Gene flow and the geographic structure of natural populations. Science 236:787–792.

Swofford, D. L., and R. B. Selander. 1981. BIOSYS-1: a FORTRAN program for the comprehensive analysis of electrophoretic data in population genetics and systematics. Journal of Heredity 72:281–283.

Todd, T. N. 1990. Genetic differentiation of walleye stocks in Lake St. Clair and Western Lake Erie. U.S. Fish and Wildlife Service Fish and Wildlife Technical Report 28.

Trautman, M. B. 1981. The Fishes of Ohio. Ohio State University Press, Columbus.

Turner, B. J. 1983. Genetic variation and differentiation of remnant natural populations of the desert pupfish, *Cyprinodon macularius*. Evolution 37:690–700.

Uthe, J. F., E. Roberts, L. W. Clarke, and H. Tsuyuki. 1966. Comparative electropherograms of representatives of the families of Pteromyzontidae, Esocidae, Centrarchidae, and Percidae. Journal of the Fisheries Research Board of Canada 23:1663–1671.

Uthe, J. F., and R. A. Ryder. 1970. Regional variation in muscle myogen polymorphism in walleye (*Stizostedion vitreum vitreum*) as related to morphology. Journal of the Fisheries Research Board of Canada 27:923–927.

Ward, N. 1992. Electrophoretic and morphological evaluation of *Stizostedion* species collected from Lake Sakakawea, North Dakota. Master's thesis. South Dakota State University, Brookings.

Ward, R., B. Billington, and P. D. N. Hebert. 1989. Comparison of allozyme and mitochondrial DNA variation in populations of walleye, *Stizostedion vitreum*. Canadian Journal of Fisheries and Aquatic Sciences 46:2074–2084.

Workman, P., and J. Niswander. 1970. Population studies of southwest indian tribes. II. Local genetic differentiation in the Papago. American Journal of Human Genetics 1970:24–49.

Wright, S. 1969. Evolution and the genetics of populations, volume 2. University of Chicago Press, Chicago.

Native and Nonnative Fish Species Used In State Fisheries Management Programs in the United States

DON HORAK

Colorado Division of Wildlife, 317 West Prospect
Fort Collins, Colorado 80526, USA

Abstract.—A one-page survey was sent to state freshwater fishery chiefs to obtain current data on the use of native and nonnative fishes in their fisheries programs. Generally, Hawaii, western states, and northeastern states had the lowest number of native fish species and the fewest native sport fishes. Twelve states reported having 10 or fewer native sport-fish species. State recreational fishery programs used an average of 31 sport-fish species per state, 19 being native and 12 nonnative. The lowest number of sport fishes used in a state program was 16. Thirty-six percent of the states had fewer native than nonnative sport-fish species. On average, nonnative sport fishes provided 38% of recreational fishery use in state programs; two western states reported 99% use. Seventy-five percent of the nonnative sport-fish species have become naturalized (i.e., would continue to exist in an ecosystem without stocking). The extent of aquatic habitat alteration among the states averaged 13% for severe alteration, 42% for moderate, 29% for slight, and 16% for no alteration. The five states that cited the most severely altered habitat also had the lowest number of native sport-fish species and used double the average number of nonnative sport-fish species. The number of federally listed Threatened and Endangered fish species averaged four per state; 6% of the listings were attributed primarily to nonnative sport-fish introductions and 88% were attributed to habitat alteration. Nonnative sport fishes are a significant component of most states' recreational fishing programs; however, management agencies must assess their use of native and nonnative species and evaluate their recreational program accordingly.

The appropriate use of hatchery fish for recreational fisheries management has been debated for several decades. In the past, that debate generally focused on the appropriate uses of cultured fishes and the management of wild, self-reproducing populations. With passage of the Endangered Species Act (ESA) in 1973 (16 U.S.C.A. §§ 1531 to 1544) and the Nonindigenous Aquatic Nuisance Prevention and Control Act of 1990 (16 U.S.C.A. §§ 4701 to 4751), managing recreational fisheries with native and nonnative species (or lower taxa) came into a different focus. It is appropriate, therefore, to identify current fishery management practices, especially those of state fish and wildlife agencies.

Methods

A one-page survey (Table 1) was sent to state freshwater fishery chiefs in September 1993 to obtain data on the use of native and nonnative fish species in freshwater fisheries management programs. Definitions used in the survey were (1) native species—any species within its historic range (the area occupied at the time of European colonization of North America), (2) nonnative species—any species that occupies an ecosystem beyond its historic range, and (3) naturalized species—any nonnative species that has been caused to adapt and grow or multiply as if native (*Webster's New International Dictionary*, 3rd edition, s.v. "naturalized"). State fishery chiefs were asked to identify numbers of native and nonnative fish species, including those they consider sport fishes; numbers of naturalized species; and the percentage of their fishery program, estimated from recreation days, that was produced by native and by nonnative sport fishes. Questions about habitat degradation and federally listed Threatened and Endangered (i.e., listed under the ESA) fish species were also asked. Even though authority for listing species as Threatened and Endangered is a federal responsibility, views of state agencies provided pertinent information. All 50 states responded to the survey, although only 47 states answered all the questions on Endangered species and 45 states answered the questions on habitat alteration. The data contained numbers of freshwater and anadromous species; subspecies and lower taxa were treated as separate species if they were federally listed Threatened or Endangered species.

Results

The number of native fishes ranged from a low of 5 species in Hawaii to a high of 330 species in Georgia, with an average of 120 per state (Table 2). Generally, the lowest number of native fish species was found in western and northeastern states (Fig-

TABLE 1.—Survey sent to state freshwater fishery chiefs (September 1993) to obtain data about the use of native and nonnative (nonindigenous, exotic) species.

Survey of Native and Nonindigenous Fish Species by State
(Freshwater only)

State:

Definitions to be used:
1. Native species—any species within its historic range (the area occupied at the time of European colonization of North America)
2. Nonindigenous species—any species that occupies an ecosystem beyond its historic range
3. Naturalized species—any nonindigenous species in your state that has been caused to adapt and grow or multiply as if native

Number of native fish species:
* Number of native fish species that are considered sport fish:
Number of nonindigenous fish species:
* Number of nonindigenous fish species that are considered sport fish:
* Number of these nonindigenous fish species that are considered sport fish which you consider naturalized:

The percent of your recreational fisheries program that is supported by
1. Native sport fish:
2. Nonindigenous sport fish:

Estimation of your aquatic habitat base that is
1. Severely altered physically and/or chemically (would not provide a recreational fishery without yearly stocking):
2. Moderately altered physically and/or chemically (would provide some recreational fishery from natural recruitment but may require special regulations and/or periodic stocking to satisfy recreational pressures):
3. Slight alteration physically and/or chemically (same criteria as in #2 above):
4. No alteration or affect from human activities physically and/or chemically (same criteria as in #2 above):
5. The four percentages add up to: 100%

Number of fish federally listed as Threatened or Endangered:
Did introduction of "exotic" sport fishes play a major role in listing any of these fish?
If the answer is yes, how many?
For how many did habitat alteration play a major role in listing?
* Names of these fish species would be appreciated if they are readily available to you.

---

ure 1). Numbers of native species generally were high in southeastern states and those adjacent to the Mississippi River. The number of species can be misleading because lower taxa were not counted unless they were a federally listed species. For example, Nevada reported that they have 37 native species, but this number would increase to 94 if subspecies were included.

Numbers of native sport fishes ranged from 3 in Nevada to 42 in Illinois (Table 2). The mean number of native sport fishes per state was 19; 27 states had 19 or fewer native sport fishes and 23 states had 20 or more. Generally, Hawaii, western states, and New England states had few native sport-fish species (Figure 1). Twelve states reported having 10 or fewer native sport-fish species.

The average number of nonnative fishes for all states was 24, of which 12 were sport fishes (Table 2). All states except Louisiana reported at least one nonnative species of sport fish; Colorado had the largest number (30). States with the higher numbers of native sport fishes generally had fewer than 12 nonnative sport-fish species. The exceptions to this were Hawaii, Rhode Island, and Delaware, which reported low numbers of both native and nonnative sport fishes.

Another way of indexing the prevalence of nonnative sport fishes was by calculating the ratio of native sport fishes to nonnative sport fishes. Eighteen states (38%) had a ratio of less than 1.0, which demonstrates they have fewer native sport fishes than nonnative sport fishes (Table 2). Half of the states (25) had a ratio of less than 2.0. Mississippi had the highest ratio. A ratio could not be calculated for Louisiana, which reported no nonnative sport-fish species.

Total number of sport fishes per state was calculated by adding the cited numbers of native and nonnative sport-fish species. Alaska, Hawaii, and Louisiana each had the lowest number (16). Seven states had 20 or fewer total sport fishes, 14 states had less than 26, and 27 states had less than 31. The average for all states was 31, with a high of 53 in Illinois and Washington.

Ninety to 99% of recreational fishing days were provided by nonnative sport fishes in eight states (Table 2). Arizona and Colorado each reported that 99% of recreational fishing was for nonnative sport fishes. Eighteen states reported nonnative sport fish use of 50% or greater. Massachusetts and Montana reported 80% and 85%, respectively, nonnative sport fish use. Averaged for all states, nonnative sport fish provided 38% of recreational fishery use.

The average number of naturalized sport-fish species was nine per state (Table 2). When compared with the average of 12 nonnative sport fishes reported per state, 75% of the species had become naturalized. All nonnative species were naturalized in 11 states. More than half the states (32) reported all but three or fewer sport fishes had become naturalized.

Numbers of fish species reported as federally listed as Threatened or Endangered ranged from 0 in nine states to 25 in Nevada and averaged 4 per state (Table 3). Nevada, Tennessee, Arizona, Michigan, California, Alabama, New Mexico, and Oregon had more than seven listed species. Six states reported that nonnative ("exotic" used in the survey

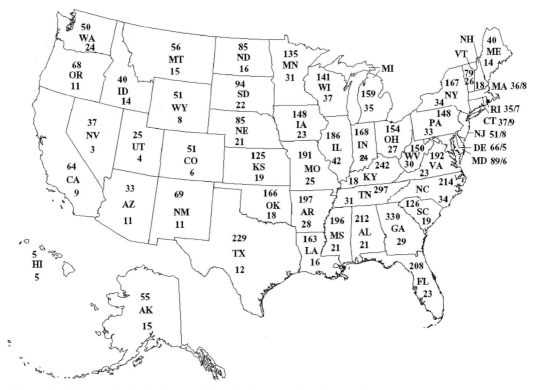

FIGURE 1.—Number of native freshwater fish species in each of the 50 states (upper number) and number of native freshwater sport-fish species (lower number).

form) sport fishes played a major role in the listing of Threatened or Endangered species. Three states did not report the number of species listings due to nonnative sport fishes. Averaged among the 47 responding states, 6% of federal listings were reported as due to nonnative sport fishes and 88% were attributed to habitat alteration; the remaining causes of listings were not reported.

Estimation of habitat alteration in the survey produced subjective information, as noted from comments on the survey forms, but it is still useful. For the 45 respondents, on average 84% of the habitat had been altered (Table 3). Some states had much lower values, especially South Carolina and Alaska, where only 3% and 6% had been altered. The extent of aquatic habitat alteration averaged 13% for severe alteration, 42% for moderate, 29% for slight, and 16% for no alteration.

Discussion

Numbers of native fish species cited in the survey were similar but not identical to numbers presented by Warren and Burr (1994). These authors presented fewer species in 28 states, the same number in 1 state, and slightly more species in the remaining states. Anadromous species were counted in the current survey, which may account for some differences.

It is difficult to determine a minimum number of sport fishes required for a recreational program because each state is managed differently. Theoretically, one could be the minimum number. However, the lowest number used in any state was 16, and the average number was 31. These numbers could represent minimum and optimum levels.

More than one-third of the states reported having fewer native than nonnative sport-fish species. Arizona and Colorado were essentially totally dependent on nonnative sport fishes. It is interesting that Massachusetts and Montana use nonnative sport fishes for 80% and 85%, respectively, of their recreational fisheries programs. Montana has 15 native sport fishes whereas Massachusetts has only 8. Congressional representatives from both states have

TABLE 2.—Native and nonnative freshwater fish species as reported by state freshwater fishery chiefs in 1993 survey.

State	Number of sport-fish species				Ratio of native to nonnative fish[a]	Number of nonnative sport-fish species		Total sport-fish species[a]	Percent recreational fishery program	
	Native		Nonnative							
	Fish	Sport fish	Fish	Sport fish		Naturalized	Not naturalized[a]		Native sport fish	Nonnative sport fish
Alabama	212	21	14	2	10.5	0	2	23	99	1
Alaska	55	15	2	1	15.0	1	0	16	99	1
Arizona	33	11	43	29	0.4	23	6	40	1	99
Arkansas	197	28	23	17	1.6	7	10	45	94	6
California	64	9	48	28	0.3	28	0	37	49	51
Colorado	51	6	62	30	0.2	18	12	36	1	99
Connecticut	37	9	35	20	0.5	17	3	29	10	90
Delaware	66	5	15	12	0.4	5	7	17	30	70
Florida	208	23	23	4	5.8	1	3	27	96	4
Georgia	330	29	9	2	14.5	2	0	31	96	4
Hawaii	5	5	50	11	0.5	11	0	16	5	95
Idaho	40	14	45	28	0.5	17	11	42	50	50
Illinois	186	42	13	11	3.8	10	1	53	90	10
Indiana	168	24	18	5	4.8	0	5	29	70	30
Iowa	148	23	8	6	3.8	4	2	29	90	10
Kansas	125	19	38	23	0.8	13	10	42	98	2
Kentucky	242	18	16	6	3.0	2	4	24	93	7
Louisiana	163	16	10	0	0.0	0	0	16	100	0
Maine	40	14	12	7	2.0	7	0	21	76	24
Maryland	89	6	21	15	0.4	10	5	21	30	70
Massachusetts	36	8	38	18	0.4	17	1	26	20	80
Michigan	159	35	14	7	5.0	7	0	42	65	35
Minnesota	135	31	14	8	3.9	4	4	39	95	5
Mississippi	196	21	7	1	21.0	1	0	22	99	1
Missouri	191	25	11	4	6.3	1	3	29	88	12
Montana	56	15	31	12	1.3	9	3	27	15	85
Nebraska	85	21	31	14	1.5	7	7	25	90	10
Nevada	37	3	27	24	0.1	23	1	27	5	95
New Hampshire	48	18	17	15	1.2	13	2	33	63	37
New Jersey	51	8	17	17	0.5	13	4	25	10	90
New Mexico	69	11	55	28	0.4	25	3	39	5	95
New York	167	34	13	11	3.1	8	3	45	74	26
North Carolina	214	34	8	4	8.5	4	0	38	85	15
North Dakota	85	16	10	4	4.0	1	3	20	80	20
Ohio	154	27	16	8	3.4	4	4	35	99	1
Oklahoma	166	18	18	8	2.3	2	6	26	69	31
Oregon	68	11	36	26	0.4	24	2	37	87	13
Pennsylvania	148	33	10	6	5.5	3	3	39	93	7
Rhode Island	35	7	12	10	0.7	7	3	17	25	75
South Carolina	126	19	24	11	1.7	11	0	30	75	25
South Dakota	94	22	18	13	1.7	2	11	35	80	20
Tennessee	297	31	18	4	7.8	4	0	35	85	15
Texas	229	12	18	6	2.0	2	4	18	96	4
Utah	25	4	82	26	0.2	22	4	30	4	96
Vermont	79	26	10	7	3.7	6	1	33	80	20
Virginia	192	23	13	6	3.8	6	0	29	70	30
Washington	50	24	34	29	0.8	24	5	53	50	50
West Virginia	150	30	25	6	5.0	3	3	36	70	30
Wisconsin	141	37	16	6	6.2	3	3	43	40	60
Wyoming	51	8	29	19	0.4	18	1	27	15	85
Mean	120	19	24	12	1.6	9	3	31	62	38

[a] Calculated from data reported in the survey.

been very influential in stressing the disadvantages of nonnative species.

Seventy-five percent of the nonnative sport-fish species in all states are reported to have become established and are considered naturalized. These figures suggest that a majority of the nonnative sport fishes would continue to exist in a state without additional stocking. The number of species that would not exist without continued stocking is low (three or less) in 32 states (64%). Therefore, should removal of nonnative sport fishes ever be judged necessary or required by law, the simple act of

TABLE 3.—The roles of habitat alteration and introduction of nonnative sport fishes in the federal listing of Threatened or Endangered species. Data collected in 1993 survey of freshwater fishery chiefs.

State	Alteration of aquatic habitat				Number of listed species[a]	Number of listed species due to		Percent listing from	
	Severe	Moderate	Slight	None		Nonnative sport fish	Habitat alteration	Nonnative sport fish[b]	Habitat alteration[b]
Alabama	0	90	10	0	11 (12)	0	11	0	100
Alaska	1	1	4	94	0				
Arizona					19 (17)				
Arkansas	6	11	3	80	3	0	3	0	100
California	12	33	40	15	17 (16)	5	14	29[c]	82[c]
Colorado	55	27	10	8	5	1	5	20[c]	100[c]
Connecticut	15	30	55	0	1	0		0	
Delaware	20	30	50	0	1	0	1	0	100
Florida	5	45	50	0	1 (3)	0	1	0	100
Georgia	1	40	58	1	6 (7)	0	6	0	100
Hawaii	10	50	20	20	0				
Idaho	10	45	25	20	3	0	3	0	100
Illinois	0	60	40	0	1	0		0	
Indiana	10	85	5	0	0				
Iowa	10	55	35	0	1	0	1	0	100
Kansas	5	92	3	0	2	0	2	0	100
Kentucky	1	10	10	79	2 (4)	0	0	0	0
Louisiana	1	2	72	25	2	0	0	0	0
Maine	10	10	30	50	0 (1)				
Maryland	10	45	40	5	1 (2)	0	1	0	100
Massachusetts	5	60	30	5	1	0	1	0	100
Michigan	0	30	65	5	19 (0)	0	12	0	63
Minnesota	20	40	30	10	0				
Mississippi	3	7	40	50	3	0	3	0	100
Missouri	1	69	25	5	4	0	4	0	100
Montana	5	60	30	5	1	0	1	0	100
Nebraska	3	10	40	47	2 (1)	0	2	0	100
Nevada	40	45	10	5	25 (22)	0	25	0	100
New Hampshire	5	5	90	0	0				
New Jersey	60	10	10	20	1	0	1	0	100
New Mexico	30	40	20	10	10	10	7	100[c]	70[c]
New York	10	50	30	10	1	0	1	0	100
North Carolina	5	78	15	2	4	0	4	0	100
North Dakota	20	70	8	2	1	0	1	0	100
Ohio					1	0		0	
Oklahoma	10	70	20	0	3	0	3	0	100
Oregon					10 (11)	3	10	30[c]	100[c]
Pennsylvania	15	75	10	0	1 (0)	0	1	0	100
Rhode Island					(1)				
South Carolina	1	1	1	97	1	0	1	0	100
South Dakota	15	75	9	1	1	0	1	0	100
Tennessee	10	40	45	5	23 (14)	0	23	0	100
Texas	0	10	90	0	7	0	7	0	100
Utah	70	23	6	1	7 (8)	3	4	43	57
Vermont	0	90	10	0	0				
Virginia	20	30	40	10	5 (6)	0	5	0	100
Washington					(3)				
West Virginia	20	50	30	0	0				
Wisconsin	10	40	40	10	0				
Wyoming	20	50	20	10	1 (3)	0	0	0	0
Mean	13	42	29	16	4			6	88

[a] Values in parentheses are numbers currently listed in the Endangered Species Act federal rules and are reported if different than numbers given in the 1993 survey.
[b] Some rows total more than 100% because both causes were listed for some species.
[c] Calculated from data presented in the survey.

curtailing stocking would eliminate only a few species. Population levels would probably be lower, but most species would still survive; some may even be impossible to eradicate.

Sterile hybrids, which were reported as nonnative sport fishes in the survey, would eventually be eliminated if stocking was curtailed. Even though nonnative, hybrids can provide many desirable attributes, especially sterility for population control. Regular stocking of sterile hybrids increases management costs, but it is a cheap option when managing a recreational fishery in high risk situations, such as habitats near Endangered species. Hybrids also express specific traits that are very popular with anglers (Bayless 1967; Tuncer et al. 1990).

The states of Utah, New Jersey, Colorado, Nevada, and New Mexico reported the highest percentages of severely altered habits. These states averaged 25 nonnative sport-fish species, much higher than the average of 12 for all states; further they averaged only 6 native sport-fish species compared with the average of 19 for all states. Although three states (Arkansas, Mississippi, and South Carolina) indicated high use of nonnative sport-fish species with little habitat alteration, these data suggest that the use of nonnative sport fish is often chosen when habitat alteration is severe and there are relatively few native species. Many nonnative sport-fish species, however, are also stocked simply to satisfy angler desire. The number of anglers, in turn, likely influences the amount of habitat alteration.

State fishery chiefs reported that they believed introduction of nonnative sport fishes was a major factor contributing to population declines for only 6% of federally listed Threatened and Endangered fishes. Lassuy (1993) reported 73% of listings were related to introductions for sport fishing. Lassuy reviewed ESA final rules and used those files that he believed contained adequate information (he did not analyze all listed species) to make a determination about the reason for a listing. From this partial data set on listed species, he concluded that species introductions related to sport fishing were the most common reason for a listing. The large difference in percentages may reflect that Lassuy's (1993) analysis showed that introductions only contributed to the listing rather than being the major cause. The 88% of listings fishery chiefs attributed to habitat alteration is very similar to the 91% reported in ESA listings. Last, it should be noted that most listed species are small fishes such as chubs, dace, darters, gambusia, madtoms, pupfish, shiners, and suckers—fish that are eaten by both native and nonnative predators.

Conclusion

Forty-nine of 50 state recreational fishery programs use nonnative sport-fish species, and some states are almost totally reliant on them to provide recreational fishing. Even where few species are used, they are important to the program. Fifty million Americans go fishing each year, and they generate over US$69 billion in economic output (Anonymous 1994). Significant social and economic losses would occur if managers stopped using nonnative sport fishes. However, nonnative fishes have negatively affected some native fish populations, causing nonnative organisms to come under increased scrutiny; they may soon be considered politically incorrect as evidenced by the recently used terms "alien" and "alien invasive" (Miller 1994). Therefore, it is important for each management agency to assess its use of native and nonnative species and evaluate its recreational program accordingly.

Extreme positions, such as encouraging the use of only native or only nonnative species are inappropriate. Many states lack an adequate number of native sport-fish species to support current recreational fishing expectations. In addition, the integrity of our aquatic ecosystems has been and continues to be severely, and in many cases irreversibly, altered by human activities. To compensate for these altered ecosystems and satisfy the desires of anglers, nonnative species are usually introduced. The influence of humans cannot be erased, and fisheries managers must do the best they can with the resources available to them. Consideration of recreational fishery programs should be included in reauthorization of the ESA. In addition, fishery managers must work hard to ensure that recreational fishing programs, even those using nonnative species such as the brown trout *Salmo trutta*, do not jeopardize ecosystem integrity and biodiversity.

References

Anonymous. 1994. SFI releases support on economics of sportfishing. SFI Bulletin 452(March–April):6. American Sportfishing Association, Alexandria, Virginia.

Bayless, J. D. 1968. Striped bass hatching and hybridization experiments. Proceedings of the Annual Conference Southeastern Association of Game and Fish Commissioners 21(1967):233–244.

Lassuy, D. R. 1993. Introduced species as a factor in endangered species listing. 123rd Annual Meeting of the American Fisheries Society (Abstract), Portland, Oregon.

Miller, K. 1994. National policy needed to fight alien species, experts say. The Coloradoan (March 12): A7.

Tuncer, H., R. M. Harrell, and E. D. Houde. 1990. Comparative energetics of striped bass and hybrid juveniles. Aquaculture 86:387–400.

Warren, M. L., Jr., and B. M. Burr. 1994. Status of freshwater fishes of the United States. Fisheries (Bethesda) 19(1):6–18.

Wiltzius, W. J. 1985. Fish culture and stocking in Colorado, 1972–1978. Colorado Division of Wildlife Division Report 12.

Fisheries Management Needs:

Threatened and

Endangered Species

Genetic Risks and Hazards in Hatchery Operations: Fundamental Concepts and Issues

CRAIG A. BUSACK

Washington Department of Fish and Wildlife
600 Capitol Way North, Olympia, Washington 98501, USA

KENNETH P. CURRENS

Oregon State University, Department of Fisheries and Wildlife
104 Nash Hall, Corvallis, Oregon 97331, USA

Abstract.—As concern over erosion of genetic diversity in fish stocks has increased over the years, so has concern about the role of hatcheries in influencing genetic change. Whereas past genetic concerns regarding hatchery operations have tended to emphasize effective population size of hatchery broodstocks, now hatchery managers need to consider a more comprehensive view of genetic risk. In this paper we present some basic concepts and associated issues in such a broad view. We recognize four fundamentally different types of genetic hazard: (1) extinction, (2) loss of within-population variability, (3) loss of among-population variability, and (4) domestication. The importance of type-2 hazards in hatchery operations has long been realized, but types 3 and 4 are controversial because of a scarcity of empirical data and because consideration of them has great ramifications for hatchery operations. Precise quantification of genetic impacts in terms of fitness depression is likely to remain a difficult if not impossible task. Ultimately, incorporation of genetic concerns into hatchery operations and other aspects of fisheries management will require managers to shift their perspective from one of managing fitness to one of managing genetic diversity.

During the last century, fishery managers increasingly have asked hatcheries to meet the demands of growing human populations for fish. This trend shows no signs of abating. In the Columbia River basin alone, more than 90 state and federal hatcheries raise and release 190 million juvenile Pacific salmon *Oncorhynchus* spp. annually, and even more hatcheries are planned (Anonymous 1990a). In the past, hatcheries produced fish primarily to augment fisheries. Now hatcheries in the Columbia River basin and elsewhere also are asked to help conserve and restore depleted natural populations. Implicit in this shift—from producing an exploitable commodity to conserving populations—is the need to protect the capacity of populations to persist and be productive.

The productivity of populations and their resilience to environmental change is a result of the genetic diversity they contain. In the last 20 years, fish geneticists increasingly have become aware of how hatchery operations can alter genetic diversity, and managing these changes is now a great concern (Bakke 1989; Hindar et al. 1991; Hilborn 1992; Meffe 1992). Unlike disease or nutritional problems that can be corrected in the next cycle of hatchery production, unwanted changes in appropriate forms or combinations of genes in populations can depress productivity for many years. Genes are transmitted over generations, and productive combinations of genes evolve in populations over many hundreds or thousands of generations (Dobzhansky et al. 1977).

Despite recent efforts to acquaint fishery managers and aquaculturists with genetic concepts (Kapuscinski and Jacobson 1987; Ryman and Utter 1987; Tave 1993), awareness of genetic concerns in hatchery operations varies widely among fishery professionals. Not surprisingly, many do not recognize genetic threats to the success of hatchery programs intended to conserve or restore natural populations. Our goal in this paper is to acquaint fishery professionals with both the basic concepts of genetic risk and the issues of risk management in the culture and uses of hatchery fish. We review the genetic vocabulary necessary to describe genetic risks and hazards and their relationship to genetic diversity. We classify the basic types of genetic hazards, describe their potential sources in hatchery programs, and discuss issues related to them. We close by recommending a shift from fitness-based to diversity-based management of genetic risk.

Basic Terms

Risk often is used ambiguously in describing a threat. Here, we distinguish between hazard and risk. A hazard is a potentially adverse consequence of an event or activity, whereas risk is the probability of the hazard occurring (Smith 1992). Thus, production of fewer offspring is a hazard of interbreeding with a foreign population; risk is the probability of producing few offspring.

The most general definition of genetic hazard is loss of genetic diversity. Genetic diversity is all the genetic differences contained within a population or group of populations. A population here means a group of interbreeding individuals. A gene is a hereditary unit of genetic information, which is contained at a site on a chromosome called a locus. A trait controlled by a single gene is called a single-locus trait. Each fish usually has two copies of each gene—one from each parent. Copies may be biochemically different forms of the gene, or alleles. The fish is homozygous if the copies are the same allele and heterozygous if they are different. Likewise, different fish may have different numbers of chromosomes. Genetic diversity then primarily consists of the quantity and variety of alleles, chromosomes, and arrangements of genes on the chromosomes that are present in the population(s).

Fundamental to how we manage genetic diversity is how we detect and measure it. Here the distinction between genes and their effects is very important. The genetic composition of a fish at one or more loci is its genotype. The biochemical, physiological, morphological, or behavioral expression of the genotype is the phenotype. Phenotypic differences may be the expression of genes at one to hundreds of loci. Expression of biochemical markers such as allozymes is typically controlled by one or two gene loci. Expression of traits related to fish performance is typically controlled by numerous loci. Such traits are called quantitative traits.

We easily can detect and quantify many single-locus genotypic differences through DNA and protein electrophoretic analyses. Well-established statistical methods exist for evaluating single-locus diversity within and among populations and higher taxa (e.g., Weir 1990). However, it is difficult to predict the effect of these differences on the performance of individuals.

In contrast, we often do understand the effects of phenotypic variation in quantitative traits, such as physiological, morphological, or behavioral differences. However, we usually can measure genotypic differences in quantitative traits only indirectly by examining differences in phenotype, and the phenotype is also influenced by the environment (Falconer 1981; Tave 1993). Consequently, to manage genetic differences within or among populations at quantitative traits, we need to understand potential variation caused by different genes in different environments, the cumulative effects of different alleles over all loci, the interaction of different alleles at each locus, and the interaction of alleles at different loci. This requires elaborate breeding experiments in controlled environments. Thus, an interesting dichotomy exists: measurement of genotypic diversity is straightforward at many traits that do not clearly relate to phenotype, but measurement of genetic diversity is extremely difficult at traits that do relate directly to phenotype. This dichotomy is at the root of many of the genetic risk issues described later in this paper.

Genetic Hazards

Any condition that has the potential to decrease either within- or among-population genetic diversity is a source of genetic hazard. Consideration of four major types of genetic hazards have been useful in planning multispecies hatchery programs for Pacific salmon (Clune and Dauble 1991): (1) extinction; (2) loss of within-population genetic variability; (3) loss of among-population genetic variability; and (4) domestication. When natural populations are used as sources of broodfish or artificially propagated fish are released into the wild, hazards occur at multiple geographical and genetic scales. For example, loss of a unique population means extinction at the population level but loss of among- and within-group diversity at the species level. Likewise, a threat to a hatchery population may also threaten wild populations but with a different kind or magnitude of hazard. Consequently, hazards need to be always defined relative to a reference population or scale.

Extinction

Definition.—Extinction is the complete loss of all genetic information. It is the most serious hazard, because once a population is gone, all the unique aspects of the diversity it contained also are lost. Because different populations have different gene pools, extinction of any population also reduces overall genetic diversity of the species.

Mechanisms.—Extinction significantly differs from the other hazards in hatcheries because it is primarily caused by nongenetic mechanisms. In most cases, the main causes of extinction have been grouped into three nongenetic sources of fluctuations in population size (Shaffer 1981): (1) demographic variation or random differences in reproductive success, (2) environmental variation, and (3) catastrophes. Genetic mechanisms that potentially reduce reproductive success, such as inbreeding in very small populations, in theory can also contribute to an "extinction vortex" (Gilpin 1987).

Sources.—Extinction is the primary focus of most risk assessment in conservation biology (e.g., Burgman et al. 1993), but it has been overlooked in hatchery programs. One of the attractions of artificial propagation is that it can reduce environmental variation, thereby reducing the risk of extinction. However, hatchery programs still can be abundant sources of uncontrolled demographic, environmental, and catastrophic changes. Broodfish may be taken from small, wild populations without replacement. Disease, power failures, and dewatering can be catastrophic to even the best hatchery programs. Ecological interactions between released hatchery fish and wild fish that may depress populations (e.g., Sholes and Hallock 1979; Nickelson et al. 1986; Hindar et al. 1991) are another uncontrolled source of demographic variation.

Unresolved issues.—Consideration of extinction as a hazard of hatchery operations is new. It is important because, more and more, hatcheries are being identified as a means of recovering populations in danger of extinction. If we can identify and remove sources of this hazard in hatcheries, and genetic function of the population has not been impaired, the risk of extinction may be reduced, and the population may be able to grow (but see Peterman 1987).

Methods for assessing risk of extinction are rapidly developing. However, it remains very much a theoretical and modeling process (Gilpin and Soule 1986; Goodman 1987; Shaffer 1987; Burgman et al. 1993), and is likely to remain so. Validation of theory requires extinctions to be carefully observed, which is not likely to occur when we are intervening to prevent extinction. Viability analyses of fish populations (e.g., Rieman and McIntyre 1993) are rare but will undoubtedly become more common.

Loss of Within-Population Variability

Definition.—Loss of within-population variability (diversity) is the reduction in quantity, variety, and combinations of alleles in a population. Quantity is the proportion of an allele in the population. Variety is the number of different kinds of alleles.

Mechanisms.—Two mechanisms of genetic change influence within-population diversity: random genetic drift and inbreeding. Genetic variability is lost in all populations through random genetic drift. It occurs because during spawning, many more gametes are produced by the parents than actually unite to become new zygotes. Each new generation, then, is a sample of the quantity and variety of alleles present in the gametes of the previous generation. Most of the time, the quantity and variety of alleles present in the progeny will not be an exact copy of the parents; the rarer the allele and the smaller the number of gametes that start the next generation, the more likely that the allele will not be represented exactly. Consequently, over time, variation will be lost, especially in small populations.

In the real world, sampling of gametes is not random. Gametes of some fish in the population will be better represented because there were an unequal number of males and females, some individuals reproduced more than once or at older ages, or more of their offspring survived to reproduce. All these variables make the genetically effective population size smaller than the census size. To compare rates that heterozygotes or alleles are lost in different populations, therefore, geneticists correct for several deviations from the ideal state: unequal sex ratio, age structure, differences in family size, and temporal fluctuations in population size (Falconer 1981). For example, a broodstock of 4 males and 100 females will lose as much variability due to drift as a population of 8 males and 8 females, everything else being equal.

Inbreeding is the breeding of related individuals. By itself, inbreeding does not lead to changes in frequency or variety of alleles in a population (Falconer 1981). Rather, inbreeding increases individual and population homozygosity, because more closely related individuals are more likely to have the same alleles than are less related individuals. This leads to changes in the frequency of phenotypes in the population. If selection then acts on these phenotypes, allele frequencies can also change.

Many studies have documented poor phenotypic performance associated with inbreeding (inbreeding depression) in captive fish populations (see Tave 1993; Waldman and McKinnon 1993). Inbreeding depression has two different genetic sources. First, it may result from increases in phenotypic expression of homozygous genotypes for rare, harmful alleles that are normally hidden in the population in heterozygotes. Second, if heterozygotes normally perform better than do homozygotes, then the decrease in heterozygosity will lead to a decrease in performance (Waldman and McKinnon 1993).

Sources.—The concept of effective population size can be used to identify potential sources of random genetic drift in hatchery programs and to recommend guidelines to minimize it (Kapuscinski and Jacobson 1987; Simon 1991; Tave 1993). Potential sources of smaller effective population size in artificial propagation systems are easily identified and many have been documented. These sources include using small numbers of broodfish, using more females than males (or the alternative) and pooling gametes, changing age structure, and allowing progeny of some matings to have greater survival than allowed others (Gharrett and Shirley 1985; Simon et al. 1986; Withler 1988).

The most important source of small effective population size is the variance of family size, or variation among families in the number of offspring that survive to reproduce (Falconer 1981). Simon et al. (1986) documented sources of this hazard in common hatchery procedures and noted that making simplified assumptions about family size variance could lead to large overestimates of effective population size. Likewise, simplifying assumptions about family size variance may also lead to serious overestimates of effective population size for natural populations in which hatchery fish mingle with wild fish when there is an overall survival difference between hatchery and wild fish (Ryman and Laikre 1991).

Reduced genetic diversity in hatchery stocks compared with their wild counterparts (Allendorf and Phelps 1980; Ryman and Stahl 1980; Waples et al. 1990) indicates the potential for random genetic drift in hatcheries. Leberg (1992) experimentally verified the loss of genetic diversity due to genetic bottlenecks—temporary reductions to very small effective population sizes (Nei et al. 1975)—predicted by genetic theory. Bottlenecking can have significant effects on the success of hatchery strains. For example, populations of eastern mosquitofish *Gambusia holbrooki* established from a small number of related founders grew at much lower rates than did those established from unrelated founders (Leberg 1990).

Unresolved issues.—Most questions pertain to two major issues. First, what are the critical thresholds for loss of within-population variability? Second, can this hazard be controlled?

When hatchery programs are judged based on the performance of the fish they produce, it is critical to know the biological threshholds that lead to reduced genetic diversity. However, threshholds, such as the degree of relatedness beyond which inbreeding depression becomes significant or the minimum effective population size that can be maintained over time before population growth suffers, are unknown and may be different for each population (Shields 1993). How large an effective population size should a population have? Lande and Barrowclough (1987) suggest that 500 individuals may be sufficient for conservation of genetic diversity underlying quantitative traits. What level of heterozygosity is desirable? This depends on whether fish performance declines because of an increase in the number of harmful alleles at homozygous loci or because of the loss of superior heterozygous loci (Lande 1988; Mitton 1993). How do we judge impairment? At what point do we view a population as genetically damaged? And what course of action should we take? Answers to these questions require detailed genetic knowledge of quantitative traits, which is nearly impossible for hatchery managers to monitor.

Total control over random loss of within-population genetic diversity is very difficult. Managing loss by controlling broodfish number, sex ratios, and age structure is possible, though logistically difficult. Because variance in family size is measured on adult progeny, it cannot be estimated directly without a pedigree, which is usually unavailable.

Loss of Among-Population Variability

Definition.—Loss of among-population variability is the reduction in differences in quantity, variety, and combinations of alleles among populations. As with the loss of within-population diversity, consequences of this hazard can be viewed from two different perspectives. At the multipopulation level, the potential evolutionary consequence is reduced ability of the species or group of populations to respond differently to environmental change. At the individual population level, at which most hatchery programs operate, this hazard is the loss of genetic uniqueness with a concurrent reduction in performance of the fish.

Mechanism.—The genetic mechanism for loss of among-population genetic diversity is gene flow at excessive levels or from nonnatural sources. For management purposes, we often consider populations as reproductively isolated units. In many fish species, however, naturally occurring gene flow is an important factor in maintaining genetic diversity. Consequently, the standard for judging gene flow is natural levels and from natural sources.

Excessive gene flow may reduce performance of individual populations (outbreeding depression) by disrupting their genetic organization (Shields 1982). Outbreeding depression has two possible genetic sources (Templeton 1986). The first is loss of adaptation. A population is adapted to a local environment if its gene pool contains high frequencies of alleles that help it do well there. Introduced alleles from other populations that have evolved in different environments may be less beneficial than are the native ones. Their presence automatically reduces the frequency of favorable alleles. The net result is that the population becomes less well adapted.

The second cause of outbreeding depression is the breaking up of favorable combinations of alleles called coadapted complexes. Recall the complex relationships between alleles and loci that underlie the expression of quantitative traits. As immigrating alleles replace existing alleles in the population, new less favorable allelic combinations may be formed, reducing performance. Whereas outbreeding depression caused by loss of adaptation can be expected to become evident the first generation after the gene flow occurs, outbreeding depression caused by breakdown of coadapted complexes may not be apparent until the second generation (Gharrett and Smoker 1991; Lynch 1991).

Sources.—Conditions that increase gene flow are common to past and present hatchery practices in this country (Philipp et al. 1993). In the Columbia River, for example, hatcheries routinely transfer eggs and fish from different populations between hatcheries to meet production needs (Howell et al. 1985). Likewise, stocking programs commonly release fish into streams outside the original distribution of the introduced fish, resulting in gene flow if stocked fish survive to reproduce with native fish. Many management programs have combined both practices by using hatchery stocks of mixed ancestry over wide geographical areas (Howell et al. 1985).

Evidence of loss of among-population genetic diversity as a result of hatchery programs is extensive, especially for North American salmonids. Numerous distinctive populations of western trout *Oncorhynchus* spp. have been lost by hybridization with introduced rainbow trout *O. mykiss* (Behnke 1992; Busack and Gall 1981; Campton and Johnston 1985). Other studies have noted that in some environments native genotypes may persist in spite of large levels of stocking (Wishard et al. 1984; Currens et al. 1990), presumably because introduced fish were poorly adapted to these environments.

Unresolved issues.—As with loss of within-population diversity, the major unanswered questions for managers revolve around two issues: identifying threshholds for managing hatchery operations based on fish performance and determining how loss of genetic diversity can be realistically controlled.

Although evidence exists for local adaptation, especially in salmonids (reviewed by Taylor 1993), we know of no empirical data on outbreeding depression in fish that involves anything but extremely distantly related populations (e.g., Gharrett and Smoker 1991). Thus questions about how much outbreeding depression can be expected under different circumstances remain unanswered. For management of hatcheries based on performance measures, there are few standards for monitoring the risks of this hazard and consequently few incentives. Most evidence of local adaptation, for example, is circumstantial (Taylor 1993). Many studies suggest adaptation, but definitive proof is difficult because natural selection is difficult to study. Theoretical models of outbreeding depression (Emlen 1991; Lynch 1991) may be helpful but will be difficult to verify. Other thresholds for managing based on performance include determining whether there is a maximum acceptable level of genetic or ecological distinctness between populations beyond which performance suffers with gene flow. If gene flow has already occurred, how fast can natural selection overcome outbreeding depression? In what cases will benefits of gene flow be expected to outweigh the temporary cost of outbreeding depression (Templeton 1994)?

Unlike measures of performance, genetic diversity among populations can be measured and monitored. Sources of the hazard can be eliminated, although it might be expensive and logistically difficult. Consequently, loss of genetic diversity among populations can be potentially managed. However, the critical question is how to measure natural levels and sources of gene flow that lead to patterns of among-population differences. This is especially important if gene flow is to be used as a means of rejoining fragmented populations.

Domestication

Definition.—Domestication is the changes in quantity, variety, or combination of alleles within a captive population or between a captive population and its source population in the wild as a result of selection in an artificial environment. This hazard is similar to loss of within-population diversity with two important differences. First, changes in genetic diversity by genetic drift are random in character, whereas diversity lost due to domestication is directly related to specific traits. Second, diversity is lost through random genetic drift at a rate inversely proportional to effective population size, whereas through domestication it is lost at a rate dependent on the genetic nature of the traits and selection intensity imposed. Domestication makes fish culture more effective, but it may also decrease the performance of hatchery fish and their descendants in the wild.

Mechanism.—Taking fish into an artificial environment for all or part of their lives imposes different selection pressures on them than does the natural environment. Decreased reproductive success of some genotypes in the hatchery environment leads to genetic changes in the population.

Sources.—Domestication can occur at single-locus traits, but in general it is expressed as changes in quantitative traits. We recognize three types of domestication selection: (1) intentional or artificial selection, (2) biased sampling during some stage of culture, and (3) unintentional selection. In practice, it may be nearly impossible to distinguish and control these separately.

Artificial selection is the deliberate effort to alter a population to suit management needs, such as development of rainbow trout stocks with specific spawning timing (e.g., Busack and Gall 1980). Artificial selection becomes a hazard when fish to be released into the wild perform well in the hatchery but poorly in the wild because of divergence from their source population at the intended traits or because of correlated changes in other traits. Additionally, if hatchery fish survive to reproduce in the wild, performance of the wild population may be reduced by outbreeding depression.

Biased sampling originates more from error than intent. It can happen during any stage of hatchery operation where genetic variability might be excluded. For example, a common source of biased sampling is broodstock collection. Ideally fish are chosen randomly. More often, however, fish are chosen to represent the distribution of spawning timing, size, age, or some other trait of the source population. If sampling errors are random or involve traits that do not respond strongly to selection, little or no genetic change results. But if sampling errors are systematic and involve traits that respond easily to selection, variability is lost. The potential for genetic change in hatchery operations because of sampling error has been demonstrated by Leary et al. (1986), who found that electrophoretically detectable allele frequencies in a rainbow trout hatchery stock varied over the course of a spawning season.

Unintentional selection is genetic change that results from uncontrolled differences in reproductive success imposed by the hatchery environment and rearing regimes. The fundamental reason for operating hatcheries is to achieve a survival advantage by altering the environment. Fish in hatchery environments may be exposed to higher densities or different food, drift, flow, substrate, protective structure, photoperiod, and so on. These changes in environment allow more fish to survive in the hatchery than survive in the wild, but they also produce the opportunity for genetic change.

The biggest obstacle to serious consideration of the hazard of domestication is a tenacious belief that hatcheries can not impose selection simply because they allow so many fish to survive. Rather than impose selection, the reasoning goes, they release fish from it. This is genetically naive for three reasons. First, what is important is survival to reproductive age, not juvenile survival. Mortality rates of stocked and wild fish are high. If the offspring of hatchery fish that survive to reproduce differ genetically from those in the wild, selection has occurred. Second, death of less fit individuals is not a prerequisite for selection. All that is required is for some genotypes to leave more adult offspring than others. If reproductive potentials among genotypes differ because of the hatchery environment, domestication selection has occurred. This fact points to the third flaw in the "benign environment" notion. Hatcheries may release fish from many of the selection pressures they would have encountered in nature, but this shift in the relative fitness of genotypes will also cause selection to occur. For example, release from competition for mates may be changing expression of secondary sexual characteristics in coho salmon *O. kisutch* (Fleming and Gross 1989). In summary, a natural population continually receiving hatchery introductions is essentially being simultaneously selected for performance in two different environments, with the possible outcome being reduced fitness in the wild. Consequently, potential exists for hatchery-dependent populations in which the spawner–recruit relationship is reduced by domestication to the point where populations are no longer self-sustaining in the wild.

As with the other types of genetic hazard, theoretical argument exceeds empirical evidence. This is not surprising. Domestication selection is measurable in quantitative rather than qualitative traits, and it is difficult to separate genetic and environmental effects on the phenotype (Hard 1995, this volume). Many arguments for domestication are based on evidence of selection regimes. Captive propagation of any organism poses very different selection regimes than does the wild (Frankham et al. 1986). Consequently, selective changes are expected to be fairly strong (Kohane and Parsons

1988). Doyle (1983), for example, showed that high selection differentials can easily exist in hatchery environments.

Many studies have demonstrated phenotypic differences between hatchery and wild fish, but in relatively few are the effects clearly genetic. The best study to date is that of Reisenbichler and McIntyre (1977), who compared early survival of a two-generation-old hatchery stock of steelhead (anadromous rainbow trout) with the wild stock from the same stream. Hatchery fish exhibited a statistically significant survival advantage over wild fish in hatchery environments; the situation was reversed in natural environments. Swain and Riddell (1990) noted that hatchery juvenile coho salmon exhibited more agonistic behavior than did wild juveniles. Also, differences in foraging behavior have been noted between wild and hatchery × wild steelhead juveniles (Johnsson and Abrahams 1991).

Unresolved issues.—Of all the issues surrounding genetic hazards and risk, probably the most controversial is domestication selection, because it strikes at the heart of hatchery technology. Hatchery situations can be envisioned in which the other types of hazards are controlled. Complete control of domestication, however, would require perfect random sampling of broodfish and eliminating differences between hatchery and natural environments. This is unimaginable. Hatcheries exist because they offer very different environments from nature, which allow higher juvenile survival. Like other hazards, the main issues for managing domestication are whether we know enough about biological thresholds to manage based on performance of the fish, and whether we can control for possible sources of loss of genetic diversity without such information.

Lack of empirical data on domestication is a major problem. If hazards in hatcheries are to be managed based on performance measures (e.g., survival rate to harvest and fecundity), there are no standards by which to monitor the risks and, consequently, little incentive to consider it. We believe domestication should be considered a ubiquitous phenomenon in hatchery operations until it is shown otherwise. It is one of the costs of using hatcheries. The challenge is to learn enough about the types and magnitude of the changes in hatcheries so that both short-term and long-term costs can be understood.

There are ways to reduce domestication. The way to reduce the intentional selection component is obvious: stop artificial selection or stop using the selected stock. The problem is that hatchery managers can only stop artificial selection of which they are aware. For example, using only the early spawners is obviously artificial selection and can be eliminated; but how much artificial selection results from routine culling that occurs during hatchery rearing?

Control of domestication due to biased sampling depends on the ability to incorporate random sampling into hatchery procedures and the kind of traits that are important. True random sampling is virtually impossible, however. Rigorous sampling methods can be developed for easily observed and readily measured traits, but random sampling for many traits of a population is impossible. Thus, some loss of diversity due to sampling seems inevitable. Intentional gene flow from wild populations might reduce this loss.

There are two obvious ways to reduce unintentional selection in hatcheries. First, selection potentials can be decreased by minimizing the time fish are exposed to the hatchery environment. For example, only wild fish can be used as broodstock so that hatchery fish are regularly cycled through the natural environment and the proportion of hatchery fish on the spawning grounds can be limited (Anonymous 1990b). Second, hatchery environments can be made more similar to the wild without loss of efficiency (Maynard et al. 1995, this volume). Recently Allendorf (1993) has suggested a third method that is applicable only in pedigreed populations: reducing selection potentials by equalizing family size.

The Future of Genetic Risk Management

Many fisheries scientists have concluded that the empirical data do not support current concerns about genetic risks and hazards. We disagree with such a conclusion. We are unaware of rigorous research designed to detect genetic impacts that has failed to find them. Such data would be very important. The data that do exist support current concerns.

Clearly, we need more research into genetic risk (see also Campton 1995, this volume). The single greatest need is a rigorous treatment of outbreeding depression, but work on domestication selection runs a close second. We also need studies of the effect of reducing effective population size from optimal to various lower, but not pathologically low, levels. For all three areas, we need to understand not only immediate, short-term consequences, but also recovery time from genetic impacts. Ideally these effects would be studied in the species of greatest management interest. However, constraints of money and time will make this very difficult in some species, such as Pacific salmon. In addition, legal protection may preclude research on certain species. Thus, we encourage research on small "laboratory" species (e.g., Leberg 1990, 1992). Although data collected from carefully controlled experiments is most desirable, other avenues of obtaining information should not be overlooked. Large-scale management research, perhaps via adaptive management (Walters and Hilborn 1978), is another alternative. Genetic monitoring of hatchery programs will also provide valuable information. Theoretical work, including modeling, is vital, both to provide new information and to provide management with some guidance in the absence of empirical data.

Although we enthusiastically support all these approaches to reducing the uncertainties surrounding genetic risk, we believe it is unrealistic to expect too much from these efforts. Certainly important illustrative examples will appear, and mechanisms will become better understood. But uncertainty will continue to be a fact of life in managing genetic risks simply because genetic impacts are often a function of chance. One of the biggest unknowns in predicting the magnitude of a genetic impact is the genetic composition of the population(s) involved. The genetic composition of a population at any given time is a product of its entire history of selection, mutation, gene flow, and drift. Stated more simply, any two fish populations subjected to the same genetic effect can be expected to respond differently. A recent sobering illustration of the dependence of genetic effect on genetic composition is provided by a study of inbreeding depression in mice *Peromyscus* sp. (Brewer et al. 1990). Theory predicts that chronically small populations should be less susceptible to inbreeding depression than are large or recently large ones. When mice from several populations of differing current and recent abundance levels were inbred, the relative levels of inbreeding depression were quite different from those expected.

In summary, we see the current situation in genetic risk management as follows. There are sound theoretical reasons for expecting genetic impacts from many common types of hatchery practices and operations, and the empirical record supports these. More research will make management of genetic risk easier, but there may be real limits to our ability to predict genetic effects. We can expect the relationship between genetic structure and function to become clearer, but certainly not as clear as we would like it to be.

Faced with present uncertainties and the possibility that additional research will not provide all the answers, we believe the only realistic approach to genetic risk management is to manage based on maintaining diversity rather than performance. We do not want to imply that diversity-based management is a second choice approach, however. It is a fundamentally sounder approach because it addresses the source of fitness.

Diversity-based management is not a new idea among conservation biologists (e.g., Meffe 1987; Ryman 1991), but the idea is fairly new to many managers. The basic precept is the same as that of ecosystem conservation—conserve function by conserving diversity (Meffe et al. 1994). Rather than managing to keep performance depressions within acceptable limits, we should manage instead to maintain diversity. Essential elements of such an approach are inventories of genetic diversity and programs to monitor it. These programs may be based on allozyme data, DNA analyses, and studies of quantitative genetic variation. Diversity-based programs would stress keeping effective population sizes high, allowing gene flow among closely related populations only, and minimizing domestication selection. They should also include creation of genetic refuge areas where no hatchery activities take place. These kinds of programs may be expensive or logistically difficult in the short term, but we see them as the only way to protect the productivity and resilience of populations for the future.

Acknowledgments

We thank the symposium organizers for giving us the opportunity to participate. Jim Shaklee, Fred Utter, Claribel Coronado, John Winton, and three anonymous reviewers provided helpful comments during manuscript preparation.

References

Allendorf, F. W. 1993. Delay of adaptation to captive breeding by equalizing family size. Conservation Biology 7:416–419.

Allendorf, F. W., and S. R. Phelps. 1980. Loss of genetic variation in a hatchery stock of cutthroat trout. Transactions of the American Fisheries Society 109:537–543.

Anonymous. 1990a. Review of the history, development, and management of anadromous fish production facilities in the Columbia River basin. Columbia Basin Fish and Wildlife Authority and U.S. Fish and Wildlife Service, Portland, Oregon.

Anonymous. 1990b. Yakima/Klickitat production project preliminary design report (with appendices). Bonne-

ville Power Administration, BP-00245, Portland, Oregon.
Bakke, B. 1989. Misincentives spur fish agencies to degrade wild fisheries. Forest Watch 10(3):15–20.
Behnke, R. J. 1992. Native trout of western North America. American Fisheries Society Monograph 6.
Brewer, B. A., R. C. Lacy, M. L. Foster, and G. Alaks. 1990. Inbreeding depression in insular and central populations of *Peromyscus* mice. Journal of Heredity 81:257–266.
Burgman, M. A., S. Ferson, and H. R. Akcakaya. 1993. Risk assessment in conservation biology. Chapman and Hall, New York.
Busack, C. A., and G. A. E. Gall. 1980. Ancestry of artificially propagated California rainbow trout strains. California Fish and Game 66:17–24.
Busack, C. A., and G. A. E. Gall. 1981. Introgressive hybridization in a population of Paiute cutthroat trout (*Salmo clarki seleniris*). Canadian Journal of Fisheries and Aquatic Sciences 38:939–951.
Campton, D. E., and J. M. Johnston. 1985. Electrophoretic evidence for a genetic admixture of native and nonnative rainbow trout in the Yakima River, Washington. Transactions of the American Fisheries Society 114:782–793.
Campton, D. E. 1995. Genetic effects of hatchery fish on wild populations of Pacific salmon and steelhead: what do we really know? American Fisheries Society Symposium 15:337–353.
Clune, T., and D. Dauble. 1991. The Yakima/Klickitat fisheries project: a strategy for supplementation of anadromous salmonids. Fisheries 16(5):28–34.
Currens, K. P., C. B. Schreck, and H. W. Li. 1990. Allozyme and morphological divergence of rainbow trout (*Oncorhynchus mykiss*) above and below waterfalls in the Deschutes River, Oregon. Copeia 1990: 730–746.
Dobzhanksy, T., F. J. Ayala, G. L. Stebbins, and J. W. Valentine. 1977. Evolution. Freeman, San Francisco.
Doyle, R. W. 1983. An approach to the quantitative analysis of domestication in aquaculture. Aquaculture 33: 167–185.
Emlen, J. M. 1991. Heterosis and outbreeding depression: a multilocus model and an application to salmon production. Fisheries Research 12:187–212.
Falconer, D. S. 1981. Introduction to quantitative genetics. Longman Group, New York.
Fleming, I. A., and M. R. Gross. 1989. Evolution of female life history and morphology in a Pacific salmon (coho: *Oncorhynchus kisutch*). Evolution 43:141–157.
Frankham, R., and six co-authors. 1986. Selection in captive populations. Zoo Biology 5:127–138.
Gharrett, A. J., and S. M. Shirley. 1985. A genetic examination of spawning methodology in a salmon hatchery. Aquaculture 47:245–256.
Gharrett, A. J., and W. W. Smoker. 1991. Two generations of hybrids between even- and odd-year pink salmon (*Oncorhynchus gorbuscha*): a test for outbreeding depression? Canadian Journal of Fisheries and Aquatic Sciences 48:1744–1749.
Gilpin, M. E. 1987. Spatial structure and population vulnerability. Pages 87–124 *in* M. E. Soule, editor. Viable populations for conservation. Cambridge University Press, New York.
Gilpin, M. E., and M. E. Soule. 1986. Minimum viable populations: processes of species extinctions. Pages 19–34 *in* M. E. Soule, editor. Conservation biology: the science of scarcity and diversity. Sinauer Associates, Sunderland, Massachusetts.
Goodman, D. 1987. The demography of chance extinction. Pages 11–34 *in* M. E. Soule, editor. Viable populations for conservation. Cambridge University Press, New York.
Hard, J. J. 1995. Genetic monitoring of life-history characters in salmon supplementation: problems and opportunities. American Fisheries Society Symposium 15:212–225.
Hilborn, R. 1992. Hatcheries and the future of salmon in the Northwest. Fisheries (Bethesda) 17(1):5–8.
Hindar, K., N. Ryman, and F. Utter. 1991. Genetic effects of cultured fish on natural fish populations. Canadian Journal of Fisheries and Aquatic Sciences 48:945–957.
Howell, P., K. Jones, D. Scarnecchia, L. Lavoy, W. Kendra, and D. Ortmann. 1985. Stock assessment of Columbia River salmonids, Volumes 1 and 2. Bonneville Power Administration Publication DOE/BP-12737, Portland, Oregon.
Johnsson, J. L., and M. V. Abrahams. 1991. Interbreeding with domestic strain increases foraging under threat of predation in juvenile steelhead trout (*Oncorhynchus mykiss*): an experimental study. Canadian Journal of Fisheries and Aquatic Sciences 48: 243–247.
Kapuscinski, A. R., and L. D. Jacobson. 1987. Genetic guidelines for fisheries management. University of Minnesota, Minnesota Sea Grant College Program, Sea Grant Research Report 17, Duluth.
Kohane, M. J., and P. A. Parsons. 1988. Domestication: evolutionary change under stress. Evolutionary Biology 23:31–48.
Lande, R. 1988. Genetics and demography in biological conservation. Science 241:1455–1460.
Lande, R., and G. F. Barrowclough. 1987. Effective population size, genetic variation, and their use in population management. Pages 87–124 *in* M. E. Soule, editor. Viable populations for conservation. Cambridge University Press, New York.
Leary, R. F., F. W. Allendorf, and K. L. Knudsen. 1986. Genetic differences among rainbow trout spawned on different days within a single season. Progressive Fish-Culturist 51:10–19.
Leberg, P. L. 1990. Influence of genetic variability on population growth: implications for conservation. Journal of Fish Biology 37 (Supplement A):193–195.
Leberg, P. L. 1992. Effects of population bottlenecks on genetic diversity as measured by allozyme electrophoresis. Evolution 46:477–494.
Lynch, M. 1991. The genetic interpretation of inbreeding depression and outbreeding depression. Evolution 45:622–629.
Maynard, D. J., T. A. Flagg, and C. V. W. Mahnken. 1995. A review of seminatural culture strategies for

enhancing the postrelease survival of anadromous salmonids. American Fisheries Society Symposium 15:307–314.

Meffe, G. K. 1987. Conserving fish genomes: philosophies and practices. Environmental Biology of Fishes 18:3–9.

Meffe, G. K. 1992. Techno-arrogance and halfway technologies: salmon hatcheries on the Pacific coast of North America. Conservation Biology 6:350–354.

Meffe, G. K., and C. R. Carroll, principals. 1995. Principles of conservation biology. Sinauer Associates, Sunderland, Massachusetts.

Mitton, J. B. 1993. Theory and data pertinent to the relationship between heterozygosity and fitness. Pages 17–41 in N. W. Thornhill, editor. The natural history of inbreeding and outbreeding. University of Chicago Press, Chicago.

Nei, M., T. Maruyama, and R. Chakraborty. 1975. The bottleneck effect and genetic variability in populations. Evolution 29:1–10.

Nickelson, T. E., M. F. Solazzi, and S. L. Johnson. 1986. Use of hatchery coho salmon (*Oncorhynchus kisutch*) presmolts to rebuild wild populations in Oregon coastal streams. Canadian Journal of Fisheries and Aquatic Sciences 43:2443–2449.

Peterman, R. M. 1987. Review of the components of recruitment of Pacific salmon. American Fisheries Society Symposium 1:417–429.

Philipp, D. P., J. M. Epifanio, and M. J. Jennings. 1993. Point/counterpoint: conservation genetics and current stocking practices—are they compatible? Fisheries (Bethesda) 18(12):14–16.

Reisenbichler, R. R., and J. D. McIntyre. 1977. Genetic differences in growth and survival of hatchery and wild steelhead trout (*Salmo gairdneri*). Journal of the Fisheries Research Board of Canada 34:123–128.

Rieman, B. E., and J. D. McIntyre. 1993. Demographic and habitat requirements for conservation of bull trout. U.S. Forest Service General Technical Report INT-302.

Ryman, N. 1991. Conservation genetics considerations in fishery management. Journal of Fish Biology 39(Supplement A):211–224.

Ryman, N., and G. Stahl. 1980. Genetic changes in hatchery stocks of brown trout (*Salmo trutta*). Canadian Journal of Fisheries and Aquatic Sciences 37:82–87.

Ryman, N., and F. M. Utter, editors. 1987. Population genetics and fishery management. Washington Sea Grant College Program, Seattle.

Ryman, N., and L. Laikre. 1991. Effects of supportive breeding on the genetically effective population size. Conservation Biology 5:325–329.

Shaffer, M. L. 1981. Minimum viable population sizes for species conservation. Bioscience 31:131–134.

Shaffer, M. L. 1987. Minimum viable populations: coping with uncertainty. Pages 69–86 in M. E. Soule, editor. Viable populations for conservation, Cambridge University Press, New York.

Shields, W. M. 1982. Philopatry, inbreeding, and the evolution of sex. State University of New York Press, Albany.

Shields, W. M. 1993. The natural and unnatural history of inbreeding and outbreeding. Pages 143–169 in N. W. Thornhill, editor. The natural history of inbreeding and outbreeding. University of Chicago Press, Chicago.

Sholes, W. H., and R. J. Hallock. 1979. An evaluation of rearing fall-run chinook salmon, *Oncorhynchus tshawytscha*, to yearlings at Feather River Hatchery, with a comparison of returns from hatchery and downstream releases. California Fish and Game 65:239–255.

Simon, R. C. 1991. Management techniques to minimize the loss of genetic variability in hatchery fish populations. American Fisheries Society Symposium 10:487–494.

Simon, R. C., J. D. McIntyre, and A. R. Hemmingsen. 1986. Family size and effective population size in a hatchery stock of coho salmon (*Oncorhynchus kisutch*). Canadian Journal of Fisheries and Aquatic Sciences 43:2434–2442.

Smith, K. 1992. Environmental hazards. Chapman and Hall, London, UK.

Swain, D. P., and B. E. Riddell. 1990. Variation in agonistic behavior between newly emerged juveniles from hatchery and wild populations of coho salmon, *Oncorhynchus kisutch*. Canadian Journal of Fisheries and Aquatic Sciences 47:566–571.

Tave, D. 1993. Genetics for fish hatchery managers, 2nd edition. AVI, Westport, Connecticut.

Taylor, E. B. 1993. A review of local adaptation in Salmonidae, with particular reference to Pacific and Atlantic salmon. Aquaculture 98:185–207.

Templeton, A. R. 1986. Coadaptation and outbreeding depression. Pages 105–116 in M. E. Soule, editor. Conservation biology: the science of scarcity and diversity. Sinauer Associates, Sunderland, Massachusetts.

Templeton, A. R. 1994. Coadaptation, local adaptation, and outbreeding depression. Pages 152–153 in G. K. Meffe, and contributors. Principles of conservation biology. Sinauer Associates, Sunderland, Massachusetts.

Waples, R. S., G. A. Winans, F. M. Utter, and C. Mahnken. 1990. Genetic monitoring of Pacific salmon hatcheries. NOAA (National Oceanic and Atmospheric Administration) Technical Report NMFS (National Marine Fisheries Service) 92.

Waldman, B., and J. S. McKinnon. 1993. Inbreeding and outbreeding in fishes, amphibians, and reptiles. Pages 250–282 in N. W. Thornhill, editor. The natural history of inbreeding and outbreeding. University of Chicago Press, Chicago.

Walters, C. J., and R. Hilborn. 1978. Ecological optimization and adaptive management. Annual Review of Ecology and Systematics 9:157–178.

Weir, B. S. 1990. Genetic data analysis. Sinauer Associates, Sunderland, Massachusetts.

Wishard, L., J. Seeb, F. M. Utter, and D. Stefan. 1984. A genetic investigation of suspected redband trout populations. Copeia (1984):120–132.

Withler, R. E. 1988. Genetic consequences of fertilizing chinook salmon (*Oncorhynchus tshawytscha*) eggs with pooled milt. Aquaculture 68:15–25.

Captive Broodstocks for Recovery of Snake River Sockeye Salmon

THOMAS A. FLAGG AND CONRAD V. W. MAHNKEN

National Marine Fisheries Service, Northwest Fisheries Science Center
2725 Montlake Boulevard East, Seattle, Washington 98112, USA

KEITH A. JOHNSON

Idaho Department of Fish and Game, Eagle Fish Health Laboratory
1800 Trout Road, Eagle, Idaho 83616, USA

Abstract.—The National Marine Fisheries Service and the Idaho Department of Fish and Game have established captive broodstocks to aid recovery of Snake River sockeye salmon *Oncorhynchus nerka* which is listed as Endangered under the U.S. Endangered Species Act. These efforts focus on protecting the last known remnants of this stock: sockeye salmon that return to Redfish Lake, Idaho at the headwaters of the Salmon River. Only one to eight sockeye salmon adults have returned to Redfish Lake in each of the last 6 years. Since 1991, all returning adults and a few residual sockeye salmon maturing in the lake have been captured and spawned, and their eggs have been retained for captive broodstocks. In addition, almost 900 outmigrating smolts have been captured and retained for rearing; some of these fish matured in 1993. Over 10,000 juvenile sockeye salmon from the fall 1993 captive spawnings were released into Redfish Lake in 1994. Within the next few years, hundreds of thousands of juveniles should be available from the captive broods for stocking in Redfish Lake and other lakes in the Sawtooth Basin. However, for these sockeye salmon to attain self-sustaining populations, conditions through the migratory corridor must be improved so enough adults can return from the ocean to allow natural recolonization of the habitat.

In December 1991, the National Marine Fisheries Service (NMFS) listed Snake River sockeye salmon *Oncorhynchus nerka* as Endangered under the U.S. Endangered Species Act (ESA; 16 U.S.C.A. §§1531 to 1544). Snake River sockeye salmon is a prime example of a species on the threshold of extinction (Waples et al. 1991a). The use of the term species in the context of the ESA can refer to true taxonomic species, subspecies, and distinct population segments. The definition of what constitutes a species under the ESA is addressed by Waples (1991a). The reader should note that the use of the term species in the context of ESA should be interpreted in the legal sense.

The last known remnants of this Snake River sockeye salmon stock return to Redfish Lake in the Sawtooth Basin of Idaho at the headwaters of the Salmon River (Figure 1). The NMFS is developing a recovery plan for Snake River sockeye salmon (Snake River Salmon Recovery Team [SRSRT], unpublished[1]). The goal of this plan will be to rebuild Snake River sockeye salmon within its historic range in order to delist the species.

Restoration mandates by the ESA are focused on natural populations and the ecosystems upon which they depend. Nevertheless, the ESA recognizes that conservation of listed species may be facilitated by artificial means while factors impeding population recovery are rectified (Hard et al. 1992). Frequently, restoration of severely depleted populations will be hindered by lack of suitable numbers of juveniles for effective supplementation (i.e., release of hatchery-propagated fish to increase natural production), even if factors impeding recovery could be corrected. For restoration of these populations to occur in a timely fashion, the full reproductive potential of such populations must be harnessed in the short-term to produce large numbers of juveniles. Often the only reasonable avenue to build populations quickly enough to avoid extinction will be through captive broodstock technology.

Only one to eight sockeye salmon adults per year have returned to Redfish Lake in each of the last 6 years. On the basis of these critically low population numbers, NMFS and the Idaho Department of Fish and Game (IDFG) recently implemented captive broodstocks as an emergency measure for Redfish Lake sockeye salmon (Flagg 1993; Johnson 1993). The Redfish Lake program is intended to be a temporary measure until migration habitat improvements can be implemented to increase survival.

In this paper we (1) identify general concepts and concerns regarding the use of captive broodstocks, (2) discuss the status of the NMFS–IDFG captive broodstock program for Endangered Redfish Lake

---

[1]Draft Snake River Salmon Recovery Plan (SRSRP). Available from National Marine Fisheries Service, 911 N.E. 11th Avenue, Room 620, Portland, Oregon 97232.

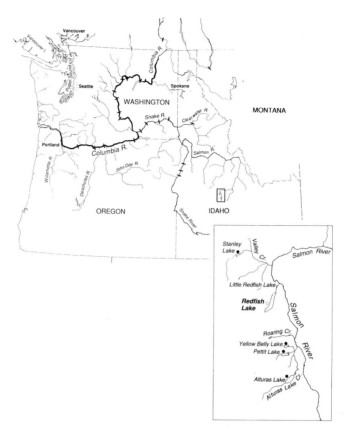

FIGURE 1.—Map showing location of Redfish Lake in the Columbia River basin. Lines across rivers indicate hydroelectric dams.

sockeye salmon through December 1993, and (3) discuss enhancement strategies for fish from our captive broodstocks. Our comments are directed toward Redfish Lake sockeye salmon as a case example. However, many other populations of Pacific salmon *Oncorhynchus* spp. are also in decline (Nehlson et al. 1991; Matthews and Waples 1991; Waples et al. 1991b). We believe information presented in this paper can be applied to all Pacific salmon.

Captive Broodstock Concepts and Concerns

Captive propagation of animals to maximize their survival and reproductive potential has won acceptance in endangered species restoration (Gipps 1991; Johnson and Jensen 1991; DeBlieu 1993; Olney et al. 1994). Currently, over 105 species of mammals, 40 species of birds, 12 species of reptiles, 29 species of fish, and 14 species of invertebrates are maintained or enhanced through forms of captive breeding (CBSG 1991). These efforts range from establishment of free-roaming breeding colonies on localized preserves to full-term captive rearing (Gipps 1991; Johnson and Jensen 1991; DeBlieu 1993; Olney et al. 1994). Each approach has merit. Full-term rearing of captive broodstocks maximizes production of juveniles for enhancement, whereas preserves provide natural influences that may aid in reducing effects of domestication. The highly migratory nature of some animals (e.g., Pacific salmon) effectively eliminates the breeding-colony-on-a-preserve approach favored by many conservation biologists and limits captive broodstock approaches for these species to full life-cycle culture.

Sockeye salmon normally live 4 to 6 years (Foerster 1968; Groot and Margolis 1991) and can have fecundities of about 2,500 eggs per female (Mullan 1986; Flagg et al. 1991). The relatively short generation time of these and other Pacific salmon and their potential to produce large numbers of offspring make them suitable for captive broodstock rearing. Survival advantages offered through protective culture can be significant. The potential advantages of captive culture over natural production can best be viewed in terms of two near-independent stages of anadromous Pacific salmon life history: egg to smolt and smolt to adult.

Pacific salmon generally have high natural mortality through their early life history stages. For instance, for Redfish Lake sockeye salmon, egg-to-smolt survival is generally less than 6% (Bjornn et al. 1968). In contrast, the protective environment of hatcheries will generally result in many more juveniles than are produced in the wild: egg-to-smolt survival for hatchery-reared sockeye salmon is generally 75% or greater (Mullan 1986; Flagg et al. 1991). Thus, successful hatchery rearing through juvenile stages alone can provide a 12-fold survival advantage compared with natural production.

Nevertheless, the potential survival advantages of protective culture offer the greatest benefits during the smolt-to-adult phase. Under current migration conditions, wild smolt-to-adult survival of anadromous sockeye salmon from Redfish Lake has been estimated at about 0.2% (IDFG, unpublished data). However, potential smolt-to-adult survival of Pacific salmon in protective captive culture may exceed 40% (T. Flagg and C. Mahnken, NMFS, unpublished data)—a 200-fold survival advantage over natural production during these life stages.

Theoretically, captive culture of Redfish Lake sockeye salmon through both egg-to-smolt and smolt-to-adult life stages could provide more than a 2,400-fold survival advantage over natural production. At present, expected natural adult returns from a spawning pair of Redfish Lake sockeye salmon is less than 0.3 fish (IDFG, unpublished data). In comparison, full-term captive culture of an equivalent number of eggs could result in over 750 adults. The substantial survival advantage for captive-reared fish provides potential to produce large numbers of juveniles to amplify the natural population during the second generation.

Even though captive broodstock theory is promising, artificial propagation and captive broodstock technologies are not without risks. Many past attempts to use artificial propagation to supplement natural populations of Pacific salmon have yielded, at best, mixed results (Miller 1990; Cuenco et al. 1993). Hatchery rearing programs have also been criticized as halfway technologies that do not address the underlying causes of endangerment (Frazer 1992; Meffe 1992). These philosophical concerns are similar to concerns about potential domestication and ecological interactions (Reisenbichler and McIntyre 1977; Nickelson et al. 1986; Hillman and Mullan 1989; Waples 1991b) that have been used by some authors to argue against continued use of fish hatcheries for enhancement (Goodman 1990; Hilborn 1992). Some authors argue that the primary course of recovery for depleted populations should be through habitat improvements, after which populations should be left to rebound naturally. Others point out the hazard of catastrophic loss of the portion of the gene pool in captivity through failure of the culture facility or disease outbreaks.

Captive breeding is also widely regarded as less cost-effective in the long-term than is *in situ* preservation (Magin et al. 1994). However, as Johnson and Jensen (1991) point out in presenting arguments for using captive broodstocks.

> If the gene pool is lost, no amount of habitat protection will help the species. [If] The species is almost extinct, only hands-on research will help us learn why the species is continuing to decline and how to counteract the problems it faces.

Kincaid (1993) furthers this argument by stating,

> The potential hazards of using captive culture (inbreeding, genetic drift, domestication, selection, behavioral conditioning, and exposure to disease) and the negative interactions of hatchery and wild fishes that affect the hatchery generation have been well documented (Hynes et al. 1981; Krueger et al. 1981; Kincaid 1983; Allendorf and Ryman 1987; Kapunscinski and Jacobson 1987; Waples 1991b). However, waiting for restoration of natural production is a more dangerous risk because the entire population is threatened. The continued decline in population size risks additional loss of genetic variability and possible extinction of the population.

It is apparent that, in many cases, the risk of waiting for natural recovery via habitat improvements is greater than the risk of husbandry intervention.

Pragmatically, captive broodstocks may offer the best chance for continued existence of endangered populations such as Snake River sockeye salmon. However, relatively few past attempts have been made to grow anadromous Pacific salmon to maturity in captivity, and little is known regarding techniques to maximize survival and reproduction. Expected levels of success of captive broodstocks for Pacific salmon are, as yet, undefined. A concern is that the species of Pacific salmon that are currently listed under ESA (i.e., chinook salmon *O. tshawytscha* and sockeye salmon) are the most difficult to rear to adulthood in captivity due to high mortality from disease (Harrell et al. 1984a, 1984b, 1985, 1987; Hopley 1986; T. Flagg and C. Mahnken, unpublished data; C. Wood, Canada Department of Fish and Oceans Pacific Biological Station, personal communication). In addition, gamete quality from captive-reared fish may be poor (McAuley 1983; Peterschmidt 1991; IDFG, unpublished data; T. Flagg and C. Mahnken, unpublished data).

We support the concept that captive broodstocks cannot be a substitute for restoring fish in the habitat. However, we think that it is unrealistic to rely solely on habitat improvements and harvest restrictions, especially when populations have reached critically low numbers. For severely depleted populations, captive broodstocks will often be the only method to rebuild numbers quickly enough to reduce inbreeding or avoid extinction. However, we recommend each year-class of captive broodstock should be maintained for only a single generation or a limited number of generations to help assure that genetic diversity and adaptability to native habitats are preserved.

We know of only five cases in which captive broodstocks are being implemented to aid recovery of Pacific salmon stocks: (1) California Sacramento River winter chinook salmon listed as Threatened under ESA (B. Wyatt, University of California, personal communication), (2) Washington White River spring chinook salmon (A. Appleby, Washington State Department of Fish and Wildlife, personal communication), (3) Washington Dungeness River chinook salmon (Shaklee et al. 1995, this volume), (4) Washington Hood Canal coho salmon *O. kisutch* (Sayre 1995, this volume), and (5) our program for Redfish Lake sockeye salmon. However, no Pacific salmon captive broodstock program has yet reached the point of fully evaluating success of juveniles released to the stock's historic habitat. At present, successful application of captive broodstock technology to Pacific salmon is uncertain and, as such, should be applied cautiously.

Redfish Lake Sockeye Salmon Captive Broodstock Program

The exact status of the Snake River sockeye salmon population was unknown at the time of ESA listing. Construction of impassable hydroelectric and irrigation dams on the Snake River system in the 1950s and 1960s had markedly reduced the geographic distribution of Snake River sockeye salmon to a single watershed in the Sawtooth Basin at the headwaters of the Salmon River in Idaho (Figure 1). In addition, barriers to upstream migration were installed in the 1950s at three of the four remaining salmon-producing lakes in the Sawtooth Basin and the lakes were poisoned. These alterations were made to promote fisheries for trout *Oncorhynchus* spp., but they further limited the range of Snake River sockeye salmon to a single lake—Redfish Lake (Figure 1). Eight major hydroelectric dams on the Columbia River system currently interfere with the almost 1,450-km migration to and from the ocean for Redfish Lake sockeye salmon.

No sockeye salmon returned to the Sawtooth Basin during 1990, the year of the ESA-mandated biological review. However, because redds (nests) were observed in Redfish Lake in 1988 and 1989, and assuming that juveniles could still be in the lake or at sea, the NMFS Biological Review Team decided that the Snake River sockeye salmon population could still exist (Waples et al. 1991a). Subsequent collections of outmigrating juveniles and returns of anadromous adult sockeye salmon to Redfish Lake in 1991, 1992, and 1993 confirmed the persistence of the species.

All three known forms of *O. nerka* occur in Redfish Lake. The anadromous form usually spends 1 to 2 years in its nursery lake before migrating to sea as a smolt during the spring of the year and remains at sea for an additional 2 to 4 years before returning to the natal area to spawn (Bjornn et al. 1968; Foerster 1968; Groot and Margolis 1991). Residual sockeye salmon are progeny of anadromous fish and remain in freshwater to mature and reproduce; they produce mostly anadromous offspring (Ricker 1938; Foerster 1968; Groot and Margolis 1991). It was theorized that residual sockeye salmon helped maintain the Redfish Lake sockeye salmon population during historic population lows (Waples et al. 1991a). The more distinct kokanee form appears to have diverged from anadromous stock in recent geological time and is fully adapted to freshwater (Foerster 1968; Groot and Margolis 1991). Anadromous and residual sockeye salmon in Redfish Lake were included together in the anadromous gene pool for ESA protection, whereas kokanee were excluded.

Because both anadromous and residual forms of sockeye salmon inhabit Redfish Lake along with kokanee, mechanisms were needed to differentiate them from kokanee when developing broodstocks. Sockeye salmon (anadromous and residual forms) and kokanee occupy overlapping habitats in lake environments (Foerster 1968; Groot and Margolis 1991); however, spatial and temporal spawning separation occur between sockeye salmon forms and kokanee in Redfish Lake. The sockeye salmon forms are shoal spawners that reproduce in the lake in late October, whereas kokanee spawn in a tributary to the lake in late August and early September (Spaulding 1993; Teuscher et al. 1994). Also, skin and flesh may be more red at spawning in kokanee than in the residual form, because kokanee, which have adapted to a carotenoid-poor forage environment, appear to be more efficient than are residual sockeye salmon at storing carotenoid (C. Wood, personal communication). In addition, recent investigations have indicated that anadromous and residual sockeye salmon can be genetically differentiated from kokanee by protein electrophoresis (R. Waples, NMFS, personal communication) and DNA analysis (Brannon et al. 1992). Recent information also suggests that because anadromous fish spend time in seawater, an environment rich in strontium, it is possible to distinguish the progeny of anadromous and nonanadromous parents based on the elevated strontium:calcium ratio in the primordial core of the progeny's otoliths (Kalish 1990; IDFG, unpublished data). All of the criteria described above were employed to differentiate kokanee from the anadromous and residual sockeye salmon gene pool when developing broodstocks.

Between 1991 and 1993, captive broodstocks were initiated from the following sources: (1) wild juveniles captured at a weir during their outmigration from Redfish Lake, (2) eggs taken from wild returning adults captured at a weir just below Redfish Lake, (3) eggs from wild adult residuals captured in nets in the lake, and (4) eggs from captive broodstocks reared and spawned in captivity (Table 1). Because Redfish Lake sockeye salmon are listed as Endangered under ESA, husbandry research has been deemed infeasible, and the fish are not routinely handled during rearing. This precludes documentation of variables such as growth except as an endpoint measurement. Therefore, survival (Table 1), primary causes of death, and percent maturation are the only quantities reported in this paper.

Mating strategies for Redfish Lake sockeye salmon broodstock were structured to maintain genetic diversity. These strategies included random pairing, pairing in as many different combinations as possible, avoidance of pairing between siblings, fertilization between different year-classes, and fertilization with cryo-preserved sperm from other generations as suggested by Hard et al. (1992). Genetic consequences of captive broodstock programs are beyond the scope of this paper. However, we think it is important to point out that Hard et al. (1992) cautioned that artificially amplifying only a portion of a population through propagation may

TABLE 1.—Status of Redfish Lake sockeye salmon captive broodstocks through December 1993.

Broodstock source	Agency[a]	Initial stage	Number	Months in culture	Average survival[b] (%)
Wild juvenile outmigrants					
Spring 1991	IDFG	Juveniles	759	32	39
Spring 1992	IDFG	Juveniles	79	20	88
Spring 1993	IDFG	Juveniles	48	8	48
Captive-reared adults					
Fall 1993	NMFS	Eggs	750	2	100
Fall 1993	IDFG	Eggs	450	2	100
Wild adult residuals					
Fall 1992	IDFG	Eggs	35	12	100
Fall 1993	IDFG	Eggs	120	1	100
Fall 1993	NMFS	Eggs	120	1	100
Wild adult returns[c]					
Fall 1991	NMFS	Eggs	991	26	59
Fall 1991	IDFG	Eggs	998	26	93
Fall 1993	NMFS	Eggs	1,200	3	100
Fall 1993	IDFG	Eggs	1,000	3	100

[a] Agency responsible for captive broodstock: Idaho Department of Fish and Game (IDFG) and National Marine Fisheries Service (NMFS).
[b] Captive broodstocks are being held as multiple discrete lots in multiple rearing containers. Survival percentage is approximate overall average for total number of months of culture.
[c] In fall 1991, one female and three male adult sockeye salmon returned to Redfish Lake and were captured and spawned; in fall 1992, one male returned, was captured, and its milt cryo-preserved; in fall 1993, two females and six males returned and were captured and spawned.

reduce effective population size (N_e) by dramatically increasing only a fraction of the available genotypes in the parent population. For our Redfish Lake sockeye salmon program, many of the potential adverse consequences of broodstock selection were avoided by capturing all returning adults and a large fraction (up to 25%) of the migrating juveniles. Nevertheless, it should be recognized that such heavy mining of a native population can only be justified in the face of otherwise certain extinction.

One of the primary obligations when maintaining an endangered species in protective culture is ensuring the highest possible survival. Full-term culture in pathogen-free freshwater has generally resulted in higher survival to spawning and higher percentages of viable gametes than has culture in seawater for Pacific salmon (McAuley 1983; Harrell et al. 1984a, 1984b, 1985, 1987; Peterschmidt 1991; C. Mahnken and T. Flagg, unpublished data; C. Wood, personal communication). Therefore, full-term freshwater rearing in pathogen-free water was chosen for these Endangered captive broodstocks.

Two separate captive populations of Redfish Lake sockeye salmon have been established to reduce the risk of catastrophic loss of these valuable gene pools. Most broodstocks obtained as eggs have been divided between IDFG hatcheries and NMFS facilities (Table 1). Because of health risks and regulations associated with interstate transfer of live fish, IDFG is maintaining all broodstocks obtained as juveniles. Captive broodstocks are cultured at the IDFG Eagle Hatchery near Boise, Idaho, in 13°C well water. Fish are also being reared in 10°C well water at a NMFS facility at the University of Washington's Big Beef Creek Research Station near Seabeck, Washington.

Fish are being reared using standard fish culture practices (Leitritz and Lewis 1976; Piper et al. 1982; Rinne et al. 1986). Each group is reared at low density in multiple containers at each facility. All fish are tagged with passive integrated transponders (PIT tags) to allow pedigree identification. Fish density in rearing tanks starts at less than 2 kg/m^3 and is maintained at less than 8 kg/m^3 during most of the culture period; however, fish density has risen to 25 kg/m^3 at maturity. Fish are fed a commercial diet (Biodiet) daily.

Most mortality for the IDFG groups captured as juvenile outmigrants from Redfish Lake in spring 1991 (Table 1) was attributed to standpipe failures, bacterial kidney disease (BKD) caused by *Renibacterium salmoninarum*, and an infection of *Aeromonas* sp. Mortality during rearing of fish from the NMFS portion of eggs obtained from wild returning adults in 1991 (Table 1) was primarily attributed to BKD. Most mortality during IDFG rearing of groups captured as juvenile outmigrants from Redfish Lake in spring 1993 (Table 1) was caused by standpipe failure.

Fish captured as juvenile outmigrants from Redfish Lake in spring 1991 were mostly yearlings (1989 brood), and a few fish (about 15%) matured in fall 1993 as age 4 fish. These were the first fish to mature from our captive broodstocks. Twenty-four maturing adults (12 males and 12 females) from this group were released into Redfish Lake in fall 1993 to spawn naturally. The remaining maturing fish were spawned at the Eagle Hatchery and produced about 10,000 viable eggs. Fish from these eggs hatched in early 1994. Most of these fish will be reared by IDFG and released to Redfish Lake in fall 1994. However, about 1,200 eggs were retained as safety-net captive broodstock to protect against loss of this gene component if initial fish stockings to Redfish Lake fail (Table 1).

Over the next few years, hundreds of thousands of progeny from the various age-classes of Redfish Lake sockeye salmon in our captive broodstocks should be available for use in recovery efforts. In addition, each year a few maturing adults from the captive broodstock will be available for release into Redfish Lake to spawn naturally. Release groups will be marked (with coded wire tags, otolith marks, and PIT tags) so performance can be evaluated.

Enhancement Strategies for Fish from Redfish Lake Captive Broodstocks

The goal of recovery programs for Snake River sockeye salmon listed as Endangered under ESA is the restoration of the species to viable self-sustaining numbers that will remove them from serious threat of extinction (SRSRT, unpublished). Captive broodstocks can be an integral part of these efforts. The production of juveniles from the captive broodstock programs described above should ensure against immediate extinction for Redfish Lake sockeye salmon. However, methods for the return of fish from captive broodstocks to the natural environment require careful consideration. Target areas and appropriate stocking levels must be properly identified if enhancement is expected to be successful.

All Sawtooth Basin lakes are being considered for use in recovery of Snake River sockeye salmon. However, initial efforts will be focused on enhancement at Redfish Lake. Redfish Lake has about one-half of the total estimated carrying capacity of the Sawtooth Basin lakes that historically supported anadromous sockeye salmon. Redfish Lake and other Sawtooth Basin lakes are situated at over 2,000 m in elevation in the Sawtooth Mountains of Idaho. These lakes are ultraoligotrophic and have limited carrying capacity. The historic production of sockeye salmon smolts from Redfish Lake probably never exceeded 100,000 fish annually (200,000 total from the Sawtooth Basin) (Bjornn et al. 1968; SRSRT, unpublished).

Spawning and rearing habitat for sockeye salmon in Redfish Lake is still in relatively good condition (Spaulding 1993; IDFG, unpublished data). However, overall nutrient loading in the lake has been reduced by the decline of returning adult salmon. Based on recent productivity estimates (Spaulding 1993), the current maximum annual carrying capacity for Redfish Lake is probably less than 50,000 yearling sockeye salmon (W. Wurtzbaugh, Utah State University, personal communication). Research is under way to determine the feasibility of lake fertilization as an enhancement strategy to boost production potential of Sawtooth Basin lakes (Teuscher at al. 1994). However, we anticipate that lake fertilization may only be able to increase natural sockeye salmon carrying capacity by a maximum of 50–150% per lake.

A few maturing adults from Idaho captive broodstocks will be released to spawn naturally in Redfish Lake each year. However, most captive broodstock will be spawned and eggs retained for culture. We are emphasizing strategies for release of juveniles from captive broodstock that will provide a natural rearing phase in the lake prior to outmigration. Some fish may be released into the lake as marked fry. Although fry releases would provide the longest rearing period in the natural habitat, fry-to-smolt survivals are often only 5–15% for sockeye salmon (Foerster 1968). Because of anticipated low survival, we plan to release only limited numbers of captive broodstock offspring as fry for experimental evaluations.

The majority of progeny from the captive broodstocks will be reared in net-pens in Redfish Lake for about 5 months (from June to October) and released. These juveniles should overwinter in the lake and outmigrate naturally as smolts the next spring. This strategy will use the positive attributes of culture while minimizing potential effects of domestication. In addition, because these fish should not be actively feeding over winter, presmolt releases in the fall could allow stocking of larger numbers of captive broodstock progeny into the lake.

There is a conflict between the estimated carrying capacity of Sawtooth Basin lakes and the numbers of smolts required to produce substantial adult returns under current conditions. It has been suggested that for cases such as Redfish Lake sockeye salmon a return of 1,000 or more adults per year may be necessary to ensure long-term persistence of the population (Thomas 1990; Thompson 1991; SRSRT, unpublished). However, under the currently estimated 0.2% smolt-to-adult survival for wild Redfish Lake sockeye salmon (IDFG, unpublished data) this would require that about 500,000 smolts outmigrate from Redfish Lake each year—almost 3.5 times the optimistic smolt production capacity of the lake and 10 times current estimated carrying capacity. These estimates should not be confused with population requirements for long-term recovery for Redfish Lake sockeye salmon: recovery has tentatively been suggested to require a 2:1 adult recruit:spawner ratio, with at least 500 adult spawning pairs per generation returning over at least two generations (SRSRT, unpublished).

Any differential survival between wild and captive-reared juveniles will increase the required stocking density to achieve adult returns. For instance, smolt-to-adult survival of hatchery chinook salmon smolts released in the Sawtooth Basin is typically one-half that of wild smolts (SRSRT, unpublished; IDFG, unpublished data). If progeny from Redfish Lake sockeye salmon captive brood survive at the same hatchery:wild fish ratio, a worst-case scenario might require stocking of a million fish (20 times estimated current carrying capacity) to achieve a desired 1,000 adult returns under current smolt-to-adult survival conditions.

To further complicate matters, as we described earlier an indigenous kokanee stock inhabits Redfish Lake along with anadromous sockeye salmon. This kokanee population appears healthy, and about 60,000 juveniles annually from 10,000 spawners apparently occupy most of the currently available sockeye salmon habitat in Redfish Lake (Spaulding 1993; Teuscher et al. 1994; IDFG, unpublished data). These fish will compete for food and territory with offspring from captive broodstocks (Teuscher et al. 1994). Therefore, carrying capacity estimates for sockeye salmon smolt in Redfish Lake need to be adjusted downward to accommodate for expected kokanee:sockeye ratios in the lake. This will further widen the gap between potential sockeye salmon smolt production from the lake and numbers required for substantial adult returns under current conditions. Similar situations undoubtedly exist for the other Sawtooth Basin lakes capable of producing sockeye salmon.

It is clear that an artificial propagation fix alone will not recover Redfish Lake sockeye salmon. Any scenario we can develop requires ecologically unsound overstocking of the lake to produce significant adult returns under current conditions. In any event, natural long-term increases in population size are required for effective recovery (SRSRT, unpublished). Even if artificial propagation could initially boost adult returns, if barriers to survival remain at the current level and returning adults are allowed to spawn naturally, the population would quickly decline again in numbers leading toward extinction.

Most of the severe barriers to survival for Snake River sockeye salmon are downstream of the spawning and rearing habitat. Columbia River system dams (Figure 1) are considered by many to be a major source of smolt-to-adult mortality for migrating fish (SRSRT, unpublished). However, other natural and artificial habitat alterations, harvest, and changes in ocean productivity may also have contributed to reduction in abundance of Snake River sockeye salmon (SRSRT, unpublished). Regardless, the cumulative effects of decreased smolt-to-adult survival have reduced the adult recruit:spawner ratio for Redfish Lake sockeye salmon to a current 0.15:1 (IDFG, unpublished data). Recovery to population equilibrium of 1:1 replacement would require over a 6-fold increase in survival from current conditions. Recovery to a 2:1 recruit:spawner ratio as suggested by the draft recovery plan (SRSRT, unpublished) would require an overall increase in natural survival of more than 13 fold.

Most management groups are focusing hopes for increased survival for Redfish Lake sockeye salmon on improvements in conditions through the migratory corridor to the ocean. Unfortunately, it is unlikely that smolt-to-adult survival increases of the magnitude which we suggest are necessary can be accomplished in the near future. It is probable that even aggressive habitat improvements will take several fish generations to complete. Therefore, captive broodstocks and artificial propagation will probably be key components in maintaining Redfish Lake sockeye salmon for years to come.

Conclusions

We conclude that captive broodstock technology for salmonids, although in its initial development stages, is sufficiently advanced to allow careful propagation with captive broodstock to proceed. However, the husbandry methodology for full-term culture of Pacific salmon in captivity is poorly developed and has not yet been thoroughly evaluated. Success of supplementation using offspring from captive broodstock is uncertain. Because the benefits and risks have not been determined through appropriate monitoring and evaluation, captive broodstock development should be considered an experimental approach and used with caution. Releases from captive broodstocks must be scaled to a level appropriate to reduce ecological risks to the population as a whole.

Captive broodstocks can be only one leg of a recovery program; they must be used hand in hand with habitat improvements to fully aid in recovery. As we have pointed out, primary consideration should be given to restoring fish in the habitat. Nonetheless, in some cases captive broodstocks may provide the only mechanism to prevent extinction of a stock and must be undertaken regardless of prospects for immediate habitat improvement. This is the case for Redfish Lake sockeye salmon. Because of the current low population replacement rate and the critically low population size, it is virtually certain that without the boost provided by our captive broodstock projects, Redfish Lake sockeye salmon would soon be extinct.

Acknowledgments

Support for this research came from the Bonneville Power Administration. Reference to trade names does not imply endorsement by the National Marine Fisheries Service.

References

Allendorf, F. W., and N. Ryman. 1987. Genetic management of hatchery stock: past, present, and future. Pages 141–159 in N. Ryman and F. Utter, editors. Population genetics and fisheries management. University of Washington Press, Seattle.

Bjornn, T. C., D. R. Craddock, and D. R. Corley. 1968. Migration and survival of Redfish Lake, Idaho, sockeye salmon, *Oncorhynchus nerka*. Transactions of the American Fisheries Society 97:360–375.

Brannon, E. L., A. L. Setter, T. L. Welsh, S. J. Rocklage, G. H. Thorgaard, and S. J. Cummings. 1992. Genetic analysis of *Oncorhynchus nerka*. Report (Contract DE-BI79-90BP12885) to Bonneville Power Administration, Portland, Oregon.

CBSG (Captive Breeding Specialist Group). 1991. Regional captive propagation programs worldwide. Captive Breeding Specialist Group News 2, World Conservation Union, Apple Valley, Minnesota.

Cuenco, M. L., T. W. H. Backman, and P. R. Mundy. 1993. The use of supplementation to aid in natural stock restoration. Pages 269–293 in J. G. Cloud and G. H. Thorgaard, editors. Genetic conservation of salmonid fishes. Plenum Press, New York.

DeBlieu, J. 1993. Meant to be wild: the struggle to save endangered species through captive breeding. Fulcrum Publishing, Golden, Colorado.

Flagg, T. A. 1993. Redfish Lake sockeye salmon captive broodstock rearing and research, 1991–1992. Report (Contract DE-AI79-92BP41841) to Bonneville Power Administration, Portland, Oregon.

Flagg, T. A., J. L. Mighell, T. E. Ruehle, L. W. Harrell, and Conrad V. W. Mahnken. 1991. Cle Elum Lake restoration feasibility study: fish husbandry research, 1989–1991. Report (Contract DE-AI79-86BP64840) to Bonneville Power Administration, Portland, Oregon.

Foerster, R. E. 1968. The sockeye salmon. Fisheries Research Board of Canada Bulletin 162.

Frazer, N. B. 1992. Sea turtle conservation and halfway technology. Conservation Biology 6(2):179–184.

Gipps, J. H. W., editor. 1991. Beyond captive breeding: reintroducing endangered species through captive breeding. Zoological Society of London Symposium 62. Clarendon Press, Oxford, UK.

Goodman, M. L. 1990. Preserving the genetic diversity of salmonid stocks: a call for federal regulation of hatchery programs. Environmental Law 20:111–166.

Groot, C., and L. Margolis, editors. 1991. Pacific salmon life histories. University of British Columbia Press, Vancouver.

Hard, J. J., R. P. Jones, Jr., M. R. Delarm, and R. S. Waples. 1992. Pacific salmon and artificial propagation under the Endangered Species Act. NOAA (National Oceanic and Atmospheric Administration) Technical Memorandum NMFS (National Marine Fisheries Service) NWFSC-2.

Harrell, L. W., T. A. Flagg, and A. J. Novotny. 1984a. Broodstock restoration programs at Manchester Marine Experimental Laboratory, Puget Sound, Washington. Proceedings of the 34th annual fish culture conference, Moscow, Idaho. (Available from Northwest Fisheries Science Center, 2725 Montlake Boulevard East, Seattle, Washington.)

Harrell, L. W., T. A. Flagg, T. M. Scott, and F. W. Waknitz. 1985. Snake River fall chinook salmon broodstock program, 1984. Report (Contract DE-AI79-83BP39642) to Bonneville Power Administration, Portland, Oregon.

Harrell, L. W., T. A. Flagg, and F. W. Waknitz. 1987. Snake River fall chinook salmon broodstock program (1981–1986). Report (Contract DE-AI79-83BP39642) to Bonneville Power Administration, Portland, Oregon.

Harrell, L. W., and six coauthors. 1984b. Status of the National Marine Fisheries Service/U.S. Fish and Wildlife Service Atlantic salmon broodstock program. Report to National Marine Fisheries Service, Northeast Region, Gloucester, Massachusetts.

Hilborn, R. 1992. Hatcheries and the future of salmon in the Northwest. Fisheries (Bethesda) 17(1):5–8.

Hillman, T., and J. Mullan. 1989. Effect of hatchery releases on the abundance and behavior of wild juvenile salmonids. Pages 265–284 in D. Chapman, editor. Summer and winter ecology of juvenile chinook salmon and steelhead trout in the Wenatchee River, Washington. Report to Chelan County Public Utility District, Wenatchee, Washington.

Hopley, W. 1986. Artificial production of south Puget Sound spring chinook salmon. Pages 41–59 in South Puget Sound spring chinook salmon technical committee report. Washington State Department of Fisheries, Olympia.

Hynes, J. D., E. H. Brown, J. H. Helle, N. Ryman, and D. A. Webster. 1981. Guidelines for the culture of fish stocks for resource management. Canadian Journal of Fisheries and Aquatic Sciences 38:1867–1876.

Johnson, J. E., and B. L. Jensen. 1991. Hatcheries for endangered freshwater fish. Pages 199–217 in W. L. Minckley and J. E. Deacon, editors. Battle against extinction. University of Arizona Press, Tucson.

Johnson, K. A. 1993. Research and recovery of Snake River sockeye salmon, 1991–1992. Report (Contract DE-BI79-91BP21065) to Bonneville Power Administration, Portland, Oregon.

Kalish, J. M. 1990. Use of otolith microchemistry to distinguish the progeny of sympatric anadromous and non-anadromous salmonids. U. S. National Marine Fisheries Service Fishery Bulletin, 88:657–666.

Kapuscinski, A., and L. Jacobson. 1987. Genetic guidelines for fisheries management. University of Minnesota Sea Grant Report 17, Duluth.

Kincaid, H. L. 1983. Inbreeding in fish populations used in aquaculture. Aquaculture 3:215–227.

Kincaid, H. L. 1993. Breeding plan to preserve the genetic variability of the Kootenai River white sturgeon. Report (Contract DE-AI79-93BP02886) to Bonneville Power Administration, Portland, Oregon.

Krueger, C. A., A. Garret, T. Dehring, and F. Allendorf. 1981. Genetic aspects of fisheries rehabilitation programs. Canadian Journal of Fisheries and Aquatic Sciences 38:1877–1881.

Leitritz, E., and R. C. Lewis. 1976. Trout and salmon culture (hatchery methods). California Department of Fish and Game Fish Bulletin 164.

Magin, C. D., T. H. Johnson, B. Groombridge, M. Jenkins, and H. Smith. 1994. Species extinctions, endangerment, and captive breeding. Pages 1–31 in P. J. S. Olney, G. M. Mace, and A. T. C. Feistner, editors. Creative conservation: interactive management of wild and captive animals. Chapman and Hall, London.

Matthews, G. M., and R. S. Waples. 1991. Status review for Snake River spring and summer chinook salmon. NOAA (National Oceanic and Atmospheric Administration) Technical Memorandum NMFS (National Marine Fisheries Service) F/NWC-200.

McAuley, W. C. 1983. DOMSEA coho broodstock program. Pages 23–24 in T. Nosho, editor. Salmonid broodstock maturation. University of Washington Sea Grant Publication WSG-WO 80-1, Seattle.

Meffe, G. K. 1992. Techno-arrogance and halfway technologies: salmon hatcheries on the Pacific coast of North America. Conservation Biology 6(3):350–354.

Miller, W. H., editor. 1990. Analysis of salmon and steelhead supplementation. Report (Contract DE-AI79-88BP92663) to Bonneville Power Administration, Portland, Oregon.

Mullan, J. W. 1986. Determinants of sockeye salmon abundance in the Columbia River, 1880s–1982: a review and synthesis. U.S. Fish and Wildlife Service Biological Report 86(12).

Nehlson, W., J. E. Williams, and J. A. Lichatowich. 1991. Pacific salmon at the crossroads: stocks at risk from California, Oregon, Idaho, and Washington. Fisheries (Bethesda) 16(2):4–21.

Nickelson, T. E., M. F. Solazzi, and S. L. Johnson. 1986. Use of hatchery coho salmon (*Oncorhynchus kisutch*) presmolts to rebuild wild populations in Oregon coastal streams. Canadian Journal of Fisheries and Aquatic Sciences 43:2443–2449.

Olney, P. J. S., G. M. Mace, and A. T. C. Feistner. 1994. Creative conservation: interactive management of wild and captive animals. Chapman and Hall, London.

Peterschmidt, C. J. 1991. Broodstock rearing and reproductive success of coho salmon (*Oncorhynchus kisutch*). Master's thesis. University of Washington, Seattle.

Piper, G. R., I. B. McElwain, L. E. Orme, J. P. McCraren, L. G. Fowler, and J. R. Leonard. 1982. Fish hatchery management. U. S. Fish and Wildlife Service, Washington, DC.

Reisenbichler, R. R., and J. D. McIntyre. 1977. Genetic differences in growth and survival of juvenile hatchery and wild steelhead trout, *Salmo gairdneri*. Journal of the Fisheries Research Board of Canada 34:123–128.

Ricker, W. E. 1938. Residual and kokanee salmon in Cultus Lake. Journal of the Fisheries Research Board of Canada 4:192–217.

Rinne, J. N., J. E. Johnson, B. L. Jensen, A. W. Ruger, and R. Sorenson. 1986. The role of hatcheries in the management and recovery of threatened and endangered fishes. Pages 271–285 *in* R. H. Stroud, editor. Fish culture in fisheries management. American Fisheries Society, Fish Culture Section and Fisheries Management Section, Bethesda, Maryland.

Sayre, J. 1995. Terminal fisheries and captive broodstock enhancement by nonprofit groups. American Fisheries Society Symposium 15:303–306.

Shaklee, J. B., C. Smith, S. Young, C. Marlowe, C. Johnson, and B. B. Sele. 1995. A captive broodstock approach to rebuilding a depleted chinook salmon stock. American Fisheries Society Symposium 15:567.

Spaulding, S. 1993. Snake River sockeye salmon (*Oncorhynchus nerka*) habitat/limnological research. Report (Contract DE-BI79-91BP22548) to Bonneville Power Administration, Portland, Oregon.

Teuscher, D., and six coauthors. 1994. Snake River sockeye salmon habitat and limnological research. Report (Contract DE-BI79-91BP22548) to Bonneville Power Administration, Portland, Oregon.

Thomas, C. D. 1990. What do real population dynamics tell us about minimum viable population sizes? Conservation Biology 4(3):324–327.

Thompson, G. G. 1991. Determining minimum viable populations under the Endangered Species Act. NOAA (National Oceanic and Atmospheric Administration) Technical Memorandum NMFS (National Marine Fisheries Service) F/NWC-198.

Waples, R. S. 1991a. Definition of species under the Endangered Species Act: application to Pacific salmon. NOAA (National Oceanic and Atmospheric Administration) Technical Memorandum NMFS (National Marine Fisheries Service) F/NWC-194.

Waples, R. S. 1991b. Genetic interactions between wild and hatchery salmonids: lessons from the Pacific Northwest. Canadian Journal of Fisheries and Aquatic Sciences 48(Supplement 1):124–133.

Waples, R. S., O. W. Johnson, and R. P. Jones, Jr. 1991a. Status review for Snake River sockeye salmon. NOAA (National Oceanic and Atmospheric Administration) Technical Memorandum NMFS (National Marine Fisheries Service) F/NWC-195.

Waples, R. S., R. P. Jones, Jr., B. R. Beckman, and G. A. Swan. 1991b. Status review for Snake River fall chinook salmon. NOAA (National Oceanic and Atmospheric Administration) Technical Memorandum NMFS (National Marine Fisheries Service) F/NWC-201.

Hybridization and Introgression between Introduced and Native Fish

ROBB F. LEARY, FRED W. ALLENDORF, AND GEORGE K. SAGE

University of Montana, Department of Biological Sciences, Missoula, Montana 59812, USA

Abstract.—The ability of stocked fish to mitigate loss of natural reproduction effectively is again being questioned. A major concern is that the introduced fish may hybridize with native populations and disrupt local adaptations. Numerous examples of interspecific hybridization and introgression exist. Hybridization between introduced fish and conspecific native populations is not as well documented, but likely reflects difficulty of detection rather than limited occurrence. Hybridization often adversely affects a number of attributes important to fitness (e.g., survival, growth, reproductive success, and development). These potential adverse effects of hybridization between introduced and native fish can best be minimized by reducing the role of introductions in mitigation programs. Initial emphasis should be placed on treating causes, such as degraded habitat, not on the symptom, reduced reproduction. If introductions are necessary, then a knowledge of the population genetic structure of the taxon, at least in the supplemented area, is essential to minimize potential genetic changes in native populations. Once a hybrid swarm has been formed, it is unlikely that natural processes will cause the hybrid swarm to revert back to a genetically pure population of the native taxon. This reversion can be accomplished by eradication of the hybrid swarm and reintroduction of the native taxon. This process is likely to be effective in only a small number of circumstances. When eradication is not feasible, genetic restoration can be attempted by introducing native fish into the hybrid swarm. If the native fish successfully reproduce, they will reduce the proportion of foreign genes in the population.

The ability of stocked fish to mitigate a loss of natural reproduction successfully was questioned 50 years ago (Schuck 1943). Despite this call for an increased emphasis on regulations and habitat management as a means of restoring natural reproduction, managers have continued to stock fish as the primary means of mitigation although little evidence supports its effectiveness.

The ability of stocking to replace lost natural reproduction effectively has again been questioned (e.g., Ferguson 1990; Krueger and May 1991; Waples 1991). A major concern is that introduced fish may hybridize with native populations. When fertile, these hybrids can backcross with the native fish and infuse, or introgress, genes from the introduced fish into the native population. This introgressive hybridization (hereafter, introgression) can result in potentially deleterious genetic changes. Introgression between different taxa results in extinction, from a genetics perspective, of the native population. Genes of the native fish still exist in the population but in novel combinations with genes of the nonnative taxon. The native genotypes no longer exist. Creation of these novel genotypic combinations also has the potential of disrupting local adaptations, thereby reducing the survival and reproductive capabilities of individuals and thus the productivity of a fishery. The net result of introgression, therefore, is reduced biological diversity and productivity.

In this paper, we review the evidence indicating that hybridization and introgression between introduced and native fish have been common among salmonids in western North America. Introduced fish in this paper are those individuals transported from their original place of existence to another, those released from hatcheries, and future descendants of such fish produced in the wild. Native populations are those that sustain themselves solely by natural reproduction and have attained access to the waters they inhabit without direct or indirect human intervention. We then consider the evidence indicating that introgression potentially can have deleterious effects on fitness. Finally, we recognize introductions will play an integral role in ameliorating past mistakes and consider steps that can be taken to minimize the genetic effects of these introductions on native populations.

Hybridization and Introgression

Hybridization between native bull trout *Salvelinus confluentus* and introduced brook trout *S. fontinalis* is common and widespread (Paetz and Nelson 1970; Cavender 1978; Leary et al. 1983a; Howell and Buchanan 1992). Electrophoretic analysis of proteins has revealed that the majority of hybrids are first generation hybrids, indicating that most are effectively sterile (e.g., Leary et al. 1983a). For example, we have electrophoretically identified 53 individual fish of hybrid origin from nine tributary streams to the Bitterroot River, Montana. Most of these fish were included in the sample because they were suspected to be hybrids based on field identification. Thus, the samples cannot be used to estimate the extent of hybridization in these streams, but they provide insight into the proportion of hybrids that successfully reproduce. Of the 53 hybrids, 50 were first generation hybrids, 2 appeared to be backcrosses to bull trout, and 1 a backcross to brook trout. Each backcross individual came from a different stream, indicating that hybrid reproduction is a rare and sporadic event.

Because brook and bull trout hybridization mainly produces sterile progeny and no individuals beyond the backcrosses have been detected, hybridization does not constitute a threat of introgression of brook trout genes into bull trout populations. This does not mean, however, that hybridization is benign with respect to the continued existence of bull trout populations. From a demographic perspective, the production of sterile progeny is equivalent to not reproducing; it is simply wasted reproductive potential.

We suspect this wasted reproductive potential can aid displacement of bull trout populations by brook trout when hybridization is frequent. Brook trout tend to mature earlier and have higher annual and lifetime fecundity than do bull trout (Scott and Crossman 1973). Thus, each hybrid produced constitutes relatively more reproductive loss to bull trout than to brook trout and could facilitate displacement of the former.

The South Fork of Lolo Creek, Montana, represents an example of rapid and almost complete displacement of bull trout by brook trout in which the initial phases were characterized by frequent hybridization (Leary et al. 1993). Brook trout first invaded the creek in the late 1970s. In 1982, bull trout (43.6%) and hybrids (35.9%) were the predominant fish in the sample. Brook trout predominated by 1987 (46.7%) and by 1990 were much more frequent (64.7%) than were bull trout (23.5%) or hybrids (11.8%). Frequent hybridization between brook and bull trout is not unique to the South Fork of Lolo Creek (Table 1) and is considered to be one of the factors threatening the continued existence of many bull trout populations (Howell and Buchanan 1992; Thomas 1992).

Hybridization of introduced rainbow trout *Oncorhynchus mykiss* and native subspecies of cutthroat trout *O. clarki* represents a dramatic example of widespread extinction of native populations by introgression. There appears to be little impediment to introgression among these fish and this

TABLE 1.—Numbers (percentages) of bull trout, brook trout, and their hybrids in samples from four streams in western North America. Each stream was sampled only once, so no temporal data on the relative abundance of the three fish are available.

Sample site	Trout sampled			
	Bull	Hybrid	Brook	Total
Dewey Creek, Idaho	16 (40.0)	9 (22.5)	15 (37.5)	40
Little Weiser River, Idaho	10 (43.5)	12 (52.2)	1 (4.3)	23
Slate Creek, Montana	12 (57.1)	5 (23.8)	4 (19.0)	21
Sun Creek, Oregon	10 (31.3)	11 (34.4)	11 (34.4)	32

often results in the formation of hybrid swarms (e.g., Leary et al. 1984; Gyllensten et al. 1985; Forbes and Allendorf 1991). Hybrid swarms are populations in which the genes from the parental taxa are randomly, or nearly so, distributed among fish in the population so that no fish is likely to be derived from the parental taxa unless it is a recent immigrant. From an evolutionary genetics perspective, hybrid swarms represent extinction. Genes of the native fish still exist in the population but in combination with the genes of the introduced taxon. The native genotypes no longer exist, and, in this case, the fish represent neither a rainbow nor a cutthroat trout.

Hybrid swarms between native westslope cutthroat trout *O. c. lewisi* and introduced rainbow trout or Yellowstone cutthroat trout *O. c. bouvieri* are common throughout the range of westslope cutthroat trout (Leary et al. 1984; Allendorf and Leary 1988). Westslope cutthroat trout inhabits only about 2.5% of its historic distribution in Montana (Liknes and Graham 1988). Introgression is considered to be the major factor responsible for this dramatic range reduction (Allendorf and Leary 1988; Liknes and Graham 1988). For example, the South Fork of the Flathead River is considered to be the stronghold of westslope cutthroat trout in Montana (Liknes and Graham 1988). The status of 32 lake and 17 stream populations in the drainage was determined using electrophoretic analysis of proteins at the loci known to distinguish westslope cutthroat, Yellowstone cutthroat, and rainbow trout (Leary et al. 1987). The results indicated the drainage is not a westslope cutthroat trout stronghold. Only 35% of the streams and 22% of the lakes sampled contained pure westslope cutthroat trout populations (Table 2). The remaining samples came from hybrid swarms or populations of introduced rainbow or Yellowstone cutthroat trout (Table 2). These populations threaten the continued existence of many of the remaining westslope cutthroat trout because of upstream or downstream migration.

We have obtained similar results for Yellowstone cutthroat trout. Electrophoretic analysis of 83 samples from 72 streams in the Yellowstone River drainage, Montana, indicated that 71% of the samples came from native Yellowstone cutthroat trout populations and the remainder from hybrid swarms with introduced rainbow or westslope cutthroat trout (R. Leary, F. Allendorf, and G. Sage, unpublished data). No evidence of introgression was detected in the 22 samples from 21 streams in the Shields River drainage, a tributary to the Yellowstone River, indicating this to be a true Yellowstone cutthroat trout stronghold. Outside the Shields River drainage only 61% of the samples came from native populations, and, because of the potential for migration, many of these are threatened with future introgression. Other studies demonstrating introgression between native and introduced rainbow or Yellowstone cutthroat trout in western North America are summarized in Table 3.

Introgression between some nonsalmonid taxa is also common. Guadalupe bass *Micropterus treculi* within its native range is considered seriously threatened by introgression with introduced smallmouth bass *M. dolomieu* (Whitmore 1983; Morizot et al. 1991). The range of the Pecos pupfish *Cyprinodon pecosensis* recently has been reduced about 60% by introgression with the introduced sheepshead minnow *C. variegatus* (Wilde and Echelle 1992).

Conspecific introgression is not nearly as well documented as interspecific introgression. Given that barriers to gene exchange are expected to be stronger between taxa than between conspecific populations, we do not consider sparse documentation to reflect sporadic occurrence: it more likely reflects difficulty of detection. Genetic differences between conspecific populations are usually much smaller than those commonly observed between taxa. Instead of differences being due mainly to complete or nearly complete divergence at certain loci, conspecific populations usually differ in frequencies of shared alleles at a number of loci. In these instances, conclusive demonstration of introgression requires a knowledge of the genetic characteristics of the native and introduced fish prior to the introductions; these data are almost invariably lacking.

Despite the lack of preintroduction data, there is some evidence of conspecific introgression in salmonid fishes. Substantially less genetic divergence exists among steelhead (anadromous rainbow trout) populations along the coast of California, Oregon, and Washington than among those in British Columbia (Reisenbichler and Phelps 1989; Reisenbichler et al. 1992). The introduction of hatchery fish, mainly from a single source, has been common in the former areas but rare in British Columbia. Thus, one explanation for the smaller amount of genetic divergence among the U.S. populations is that straying of hatchery fish has increased gene flow among populations and this has acted to homogenize the genetic characteristics of the steelhead populations. The same explanation has been offered for the smaller amount of genetic divergence observed among the heavily stocked Sacramento and San Joaquin rivers drainage populations of chinook salmon *O. tshawytscha* than among populations in other less heavily stocked rivers of California (Bartley and Gall 1990). These explanations, although reasonable, are speculative. The available data do not allow rejecting the possibility that in the areas of lesser genetic divergence historically there was more gene flow among the native populations than there was in the other areas, and that the hatchery operations have had little effect on the genetic characteristics of the native populations.

More conclusive evidence for conspecific introgression has come from situations in which a relatively large genetic difference between the native and introduced fish reasonably can be inferred. Substantial genetic differences at two enzyme loci exist between rainbow trout that spawn in waters west of the Cascade Mountains (coastal) and those that spawn east of the mountains (interior) (Allendorf and Utter 1979; Parkinson 1984). Until the last 10 years almost all rainbow trout broodstocks were of the coastal type. Noting that interior rainbow trout in the heavily stocked upper Yakima River drainage, Washington, had allele frequencies at these two loci shifted towards coastal rainbow trout, Campton and Johnston (1985) suggested introgression with the introduced fish had occurred. Large genetic differences between northern and southern brook trout populations (Stoneking et al. 1981; McCracken et al. 1993) and accurate stocking records enabled McCracken et al. (1993) to obtain evidence for extensive introgression between the native southern and introduced northern fish in Great Smoky Mountains National Park. Substantial biochemical genetic differences exist between Atlantic and Mediterranean populations of brown trout *Salmo trutta* (Krieg and Guyomard 1985). Analysis of Mediterranean populations supplemented with

TABLE 2.—Electrophoretically determined status of 49 trout populations in the South Fork Flathead River drainage, Montana.

Population	Location			Sample size
	Township	Range	Section	
Pure westslope cutthroat trout				
Big Salmon Lake (2 collections)	21N	14W	3	45
Cliff Lake	28N	18W	28	25
Doctor Lake	19N	15W	14	25
Jenny Lake	29N	19W	18	26
Lower Marshall Lake	18N	14W	19	27
Upper Marshall Lake	18N	14W	19	7
Squaw Lake	27N	18W	5	26
Danaher Creek	19N	12W	4, 5, 9	26
Lower Gordon Creek	19N	13W	5	26
Marshall Creek	18N	14W	13	25
Mid Creek	23N	14W	4	26
South Fork Flathead River	23N	14W	4, 9, 16	35
Middle Wheeler Creek	27N	17W	32	7
Westslope cutthroat trout × rainbow trout hybrids				
Bigelow Lake	26N	18W	1	25
Blackfoot Lake (4 collections)	28N	18W	19	100
Crater Lake	27N	18W	8	26
Doris Lake #2	29N	19W	6	32
Doris Lake #3	29N	19W	6	26
Fawn Lake	30N	19W	31	28
Lower Pilgrim Lake	26N	18W	1	22
Lower Seven Acres Lake	28N	18W	27	26
Aeneas Creek (2 collections)	27N	18W	3	55
Westslope cutthroat trout × Yellowstone cutthroat trout hybrids				
Bighawk Lake (2 collections)	28N	18W	14	97
Birch Lake	28N	18W	32	28
Clayton Lake (3 collections)	28N	18W	16	75
George Lake (2 collections)	19N	15W	26	55
Koessler Lake	19N	15W	15	26
Margaret Lake	27N	17W	19	26
Sunburst Lake	23N	16W	23	14
Upper Three Eagles Lake	27N	18W	10	26
Wildcat Lake (2 collections)	28N	19W	12	59
Clayton Creek (2 collections)	28N	18W	3, 8	52
Deep Creek	29N	17W	28	25
Upper Gordon Creek	19N	14W	7	10
Gorge Creek	24N	15W	35	25
Jones Creek (2 collections)	27N	18W	12	51
Upper Wheeler Creek	27N	18W	25, 27	30
Lower Wheeler Creek (3 collections)	27N	17W	22	69
Wildcat Creek (2 collections)	28N	18W	6	53
Westslope cutthroat trout × Yellowstone cutthroat trout × rainbow trout hybrids				
Black Lake (4 collections)	28N	18W	30	49
Handkerchief Lake (2 collections)	28N	18W	36	31
Graves Creek (2 collections)	28N	18W	35	53
Pure rainbow trout				
Jewel Lakes	28N	18W	19	10
Lena Lake	20N	15W	25	27
Lower Necklace Lake	20N	15W	17	8
Big Salmon Creek	20N	15W	10	2
Pure Yellowstone cutthroat trout				
Lick Lake	19N	15W	9	35
Tom Tom Lake (2 collections)	27N	18W	27	25
Rainbow trout × Yellowstone cutthroat trout hybrids				
Woodward Lake	20N	15W	18	2

TABLE 3.—Summary of protein electrophoretic studies providing evidence of introgression between introduced rainbow trout (RT) or Yellowstone cutthroat trout (YCT) and other trout taxa native to western North America. The extent of introgression is given as the number of introgressed populations/total populations sampled.

Native taxon	Extent of introgression	Reference
Apache trout *O. gilae apache*	19(RT)/31	Carmichael et al. 1993
	1(YCT)/31	
Alvord cutthroat trout *O. clarki* subsp.	extinct (RT)	Bartley and Gall 1991
Bonneville cutthroat trout *O. c. utah*	6(RT)/22	Martin et al. 1985
Coastal cutthroat trout *O. c. clarki*	2(RT)/23	Campton and Utter 1985
Colorado River cutthroat trout *O. c. pleuriticus*	1(RT)/9	Martin et al. 1985
	3(YCT)/9	
Gila trout *O. g. gilae*	2(RT)/6	Loudenslager et al. 1986
Paiute cutthroat trout *O. c. seleniris*	extinct in native range (RT)	Busack and Gall 1981

fish of Atlantic origin revealed introgression between the two fish was common (Barbat-Leterrier et al. 1989).

Not all studies exploiting known intraspecific genetic differences have provided results as conclusive as the above. Noting that there was a strong geographic pattern in the amount of genetic divergence among inland rainbow trout populations in southern Idaho, Wishard et al. (1984) suggested introgression with introduced coastal rainbow trout had not occurred. They argued that extensive introgression would lead to genetic homogenization and a destruction of historic patterns of differentiation. The geographic pattern of divergence suggested that the historic pattern still existed. Leary et al. (1983b) observed that some of the rainbow trout populations in this area contained rare electrophoretic variants present in a hatchery population used to stock the area. The populations in this area also tended to have unusually high amounts of genetic variation for inland rainbow trout. Thus, Leary et al. (1983b) suggested that introgression had occurred but cautioned this interpretation was not conclusive. The possibility that these populations naturally possessed rare variants and high amounts of genetic variation could not be rejected.

Phenotypic Effects of Hybridization and Introgression

Heterosis, or hybrid vigor, refers to the often observed situation in which crosses (hybrids) between highly inbred lines have higher performance ratings than do the lines themselves (e.g., Schuler 1954). Crossing the lines breaks down inbreeding depression and results in increased fitness or heterosis. Increased fitness, however, is not a general expectation when genetically divergent genomes from randomly mating populations are brought together by hybridization and introgression. The general expectation is intermediate or reduced fitness. Reduced fitness due to hybridization or introgression is generally called outbreeding depression (Templeton 1986).

There are two different, but not mutually exclusive, causes of outbreeding depression: breakdown of local adaptation and breakdown of coadapted gene complexes. Genetic changes caused by hybridization or introgression could disrupt local adaptations, such as thermal tolerance or homing behavior, in the native population. Genetic diversity results when genes exist in two or more allelic states or forms. When the form of a gene evolves in response to the form of other genes in the genome, selection will create groups of genes that act among themselves to create well-regulated physiological and developmental processes. These groups of genes are usually called coadapted gene complexes (Dobzhansky 1948). If populations are reproductively isolated for a sufficient amount of time, selection is expected to create different coadapted gene complexes among populations. Because these different gene complexes have not been selected to act together, bringing them together by hybridization and introgression can result in a disruption of physiological and developmental processes and lead to a reduction in fitness parameters such as fertility or survival. This process is generally referred to as a breakdown of coadapted gene complexes (Templeton 1986).

Although the phenotypic effects of hybridization and introgression have not been extensively investigated, the available data indicate that hybridization and introgression seldom results in an increase in fitness attributes. Northern largemouth bass *M. salmoides salmoides* and Florida largemouth bass *M. s. floridanus* represent a good example of a fitness reduction due to a breakdown of local adaptation. Hybrids had intermediate survival and growth compared with pure northern and Florida largemouth bass when raised in central Illinois ponds (Philipp and Whitt 1991). The extent to which the hybrids and Florida largemouth bass had lower growth and survival than did the northern largemouth bass was related to the severity of the winter; the more severe the winter, the poorer the performance of the hybrid and Florida largemouth bass. Likewise, populations with introgression between northern and Florida largemouth bass had lower growth rates compared with pure northern largemouth bass (Philipp 1991). These results are most easily explained by reduced tolerance to cold water in the Florida largemouth bass and hybridized fish.

There is compelling evidence that hybridization in pink salmon *O. gorbuscha* can disrupt local adaptations. Reduced returns, possibly because of reduced homing ability, were observed in first generation hybrids between native fish and fish from different river drainages compared with the pure native fish (Bams 1976; Withler and Morely 1982). Gharett and Smoker (1991) observed reduced returns of second generation hybrids between even- and odd-year pink salmon from the same river, but the possibility that this resulted from reduced survival and not from reduced homing ability could not be excluded.

We used the Arlee strain of westslope cutthroat trout and the Eagle Lake strain of rainbow trout in a hybridization experiment. The eggs from each of 24 female westslope cutthroat trout were divided into two groups of approximately equal number. One lot of eggs from each female was fertilized with sperm from a male westslope cutthroat trout, the other with sperm from a rainbow trout. The eggs from each cross were placed in a section of a partitioned Heath incubator tray. The percentage of eyed eggs tended to be higher in the hybrid crosses than in the westslope crosses (Table 4). Eleven of the 24 comparisons between half-sibling families were significantly different, and in 9 of these cases the hybrid cross had a higher percentage of eyed eggs. Similar results were obtained for hatching success. Nine of the comparisons were significantly different, and in seven of these cases the hybrid cross had a higher percentage of hatched eggs. The overall percentage of hatched eggs was also significantly higher in the hybrid crosses. After hatching, however, the hybrids grew more slowly and had poorer survival than did the westslope cutthroat trout. Thus, although hybrids had higher hatching percentage, overall they had reduced fitness because of poorer growth and posthatch survival.

There is other evidence that hybridization can disrupt developmental processes of salmonid fishes.

TABLE 4.—Developmental success, growth, and survival of westslope cutthroat trout (westslope) and westslope cutthroat × rainbow trout hybrids from 24 half-sibling families.

	Westslope	Hybrid
Developmental success		
Percent eyed eggs	76.4	79.8[a]
Percent hatched eggs	75.0	78.8[a]
Growth (mean total length, mm)		
89 d postfertilization	29.0	27.2[b]
112 d postfertilization	40.1	33.6[b]
Survival (percent population)		
Fertilization	52.6	47.4
Eyed egg	51.5	48.5
Hatch	51.4	48.6
89 d	61.1	38.9
112 d	94.0	6.0

[a] Values in row significantly different ($P < 0.001$) by chi-square analysis.
[b] Values in row significantly different ($P < 0.05$) by Student's t-test.

Hybrids between bull trout and brook trout and between rainbow trout and coastal, westslope, or Yellowstone cutthroat trout had a higher number of asymmetric meristic characters per individual than was observed in the parental species (Leary et al. 1985). The hybrids also tended to have higher instead of intermediate average values for the meristic traits than did the parental taxa (Leary et al. 1983a, 1985). These differences between hybrid offspring and parents were interpreted to be a consequence of disrupted development.

Comparison of interstrain rainbow trout hybrids with the parentals revealed that the hybrids generally had reduced developmental rate (Ferguson et al. 1985a) and higher meristic counts (Ferguson and Danzmann 1987) than did the parentals. Again, these deviations from intermediacy were considered to result most likely from a perturbation of development in the hybrids.

Ferguson et al. (1985b) did not find evidence of superior performance of hybrids between rainbow and cutthroat trout. Westslope and Yellowstone cutthroat trout had faster developmental rate, based on time of hatching, than did rainbow trout. Hybrids between rainbow trout and both cutthroat trouts had intermediate developmental rate. The cutthroat trout subspecies had a higher rate of yolk sac absorption after hatching than did the rainbow trout, and hybrids were similar to the cutthroat trout.

In contrast to the above results, Ferguson et al. (1988) reported some evidence for superiority of hybrids between non-inbred westslope and Yellowstone cutthroat trout. At 56 d after fertilization, the hybrids were larger and tended to have faster development, as measured by degree of yolk sac absorption, than did the parental taxa. The hybrids also had a lower mean number of asymmetric meristic traits than did the Yellowstone cutthroat trout but were similar to the westslope cutthroat trout.

Overall the above data indicate that superior performance associated with hybridization and introgression is the exception rather than the rule. Thus, interspecific introgression not only results in extinction of native genomes and a loss of genetic diversity due to the homogenization of once genetically distinct evolutionary lineages, but also tends to replace native populations with less adapted and productive populations. Conspecific introgression also results in homogenization and can disrupt development and local adaptation, potentially reducing productivity. This reduced productivity may lead to a call for increased stocking, which in turn may serve to exacerbate the damage already done. Even if introgression does not result in an immediate and detectable reduction in fitness, it still has deleterious evolutionary consequences. Homogenization results in evolutionary inertia: the diversity within and among many separate lineages produced by millions of years of evolution is exchanged for a new and widespread mongrel species. We conclude that the indiscriminate introduction of fish has resulted in widespread hybridization and introgression. By far, this has been a harmful and not beneficial management practice in terms of maintaining the viability and productivity of native populations.

Eradication and Genetic Restoration

Introduced and introgressed populations now threaten many of the few remaining native trout populations because of the possibility of gene flow. In many of these instances, we think the genetically prudent action would be eradication. Eradication by the removal of fish or the application of poisons is certainly the quickest means of eliminating introduced or introgressed populations. Poisons, however, can present problems. They may conflict with other conservation goals because of adverse affects on other aquatic life forms, including downstream native populations, if the chemicals are applied carelessly. Introduced and introgressed populations may exist in waters too remote or large for chemical treatment to be effective or practical.

When eradication is not considered feasible, an alternative is genetic restoration. By this, we mean the conversion of an introduced or introgressed population to one that is nearly a genetically pure population of the native taxon by stocking genetically pure conspecifics. If the stocked fish survive and reproduce, this will reduce the proportion of foreign genes in the target population. With repeated stockings, the frequency of foreign genes may be reduced to near zero.

Neither eradication nor genetic restoration can convert an introgressed population back to the native population unless genetically pure members of the latter were used to establish a hatchery broodstock or were stocked into a once fishless body of water—a very unlikely scenario. The purpose of eradication and genetic restoration is to admit past mistakes and to try to rectify these as best as circumstances permit. Instead of having waters within a taxon's range replete with introduced and introgressed fish, eradication and genetic restoration efforts will convert the range of the native taxon to a situation more similar to that prior to widespread introgression.

The efficacy of genetic restoration depends on how well the stocked fish survive and reproduce. Even with high survival and reproductive capabilities, however, the time and expense of genetic restoration may be substantial. In a closed system with discrete generations (parents reproduce once and at the same age), random mating, and equal parental contributions of the resident and stocked fish, the proportion of foreign genes in the next generation will be reduced by one-half. If the fish do not mature until age 3, then four successive annual stockings will be required to reduce the proportion of foreign genes in the next generation effectively by one-half because initially there are three immature age-classes and a group of reproductive adults in the population. Thus, if a population initially has 50% foreign genes under this scenario, 24 stockings will be required to reduce the percentage to less than one.

A restoration program in progress by the state of Montana in once fishless Tom Tom Lake in the South Fork Flathead River drainage indicates that other issues need to be considered before implementing a restoration program. In 1984, the lake contained an introduced population of Yellowstone cutthroat trout, and its outlet stream, Wheeler Creek, contained an introgressed population of westslope and Yellowstone cutthroat trout. Between 1985 and 1990, 6,000 westslope cutthroat trout fingerlings were stocked into the lake. A sample from the lake in 1990 indicated the introduced westslope cutthroat trout were producing pure westslope cutthroat trout and first generation hybrids. Samples from Wheeler Creek provided interesting and potentially disturbing results. There was a low percentage (1.3) of introgressed Yellowstone cutthroat trout genes in the population when it was sampled in 1983 and 1984. In 1991, the percentage of Yellowstone cutthroat trout genes (5.9) in the population had significantly ($P < 0.05$) increased. This increase was due to the presence of a single, apparently pure Yellowstone cutthroat trout in the sample. This fish was most likely a migrant from the lake population, which is by far the nearest source of pure nonnative Yellowstone cutthroat trout.

The above data pose the question of whether stocking of westslope cutthroat trout stimulated migration of Yellowstone cutthroat trout out of the lake. If so, these data suggest caution in implementing a genetic restoration program when a native population exists downstream from the restoration target. Eradication may be more prudent, or stocking levels must be low enough so as not to overpopulate the system and stimulate migration. Otherwise, migration may result in the opposite outcome desired—introgression and loss, not protection, of the native population.

Because of the time and expense required, we think restoration will be practical only for introgressed populations that contain a low proportion (10% or less) of foreign genes or those introgressed populations that do not pose an immediate threat to native populations. In waters with higher percentages of foreign genes or in which there is a high threat of introgression to an upstream or downstream native population, eradication should be considered the primary management option.

There likely will be no general consensus on the point at which a population has been effectively genetically restored. We suggest a level of 1% or less foreign genes. Levels of introgression this low are difficult to detect conclusively unless the population was previously known to be introgressed. Furthermore, we suspect this amount of introgression is unlikely to alter the biological characteristics of the fish population from those of the native taxon.

Whether eradication or genetic restoration are adopted as management strategies, genetic concerns must be addressed when considering the source of the fish to be stocked. If native populations exist in the drainage, then the stocked fish or their descendants have the potential of causing undesirable genetic changes in the remaining native populations through gene flow. Undesirable genetic changes are especially possible if the stocked fish are substantially different from the natives, thereby increasing the chances of a breakdown of local adaptation or outbreeding depression.

Obviously, the most effective means of guarding against deleterious changes would be to use native fish or their descendants in a stocking program where there is a potential for gene flow between stocked and native fish. This, however, may not always be economically or logistically practical. When a large number of waters are to be stocked, it will be impossible to maintain a separate broodstock for every perceived introduction. In such a situation, a compromise must be made. In order for the compromise to be genetically rational, knowledge of the population genetic structure of the taxon, at least in the region where introductions are to be made, is essential.

Consider Colorado River cutthroat trout in the North Fork Little Snake River drainage, Wyoming. Evidence of genetic variation was electrophoretically detected at 6 of 46 protein-coding loci in samples from four populations (Table 5). These data indicated a substantial amount of genetic divergence among populations: 26.4% of the total variation detected was due to differences among populations. Most of this divergence, however, was due to the presence of $GPI\text{-}B1*O$ in only the Roaring Fork population and at a high frequency. Differences between the Roaring Fork and the other populations accounted for 25.8% of the total genetic variation. A trivial amount of genetic divergence (0.6% of total) existed among the North Fork Little Snake River, Solomon Creek, and Ted Creek populations.

Because of the small amount of genetic divergence among the North Fork Little Snake River, Solomon Creek, and Ted Creek populations, they could be treated as a single management unit from a genetics perspective. That is, no population could be considered a genetically safer source of fish for introductions in the drainage than could the others. Furthermore, it would not be inappropriate to establish a captive, generalist broodstock from two or all three of these populations.

Conversely, the large amount of genetic divergence between the Roaring Fork and the other populations indicates that it would be wise to treat the Roaring Fork fish as a separate unit. The Roaring Fork fish inhabit an unusual habitat, in terms of high elevation and low summer flows, and apparently have been reproductively isolated from other populations in the drainage for a long time. These fish, therefore, could negatively effect other populations if introduced, and we do not recommend they be used in a hatchery program.

TABLE 5.—Allele frequencies at the polymorphic loci in four samples of Colorado River cutthroat trout from the North Fork Little Snake River drainage, Wyoming. A contingency table chi-square test for homogeneity of allele frequencies among samples was performed for each locus with all comparisons having three degrees of freedom. Only sMEP-1* allele frequencies were not significantly different ($P \geq 0.05$).

Locus[a] and alleles	Allele frequencies				Chi-square value
	North Fork Little Snake	Roaring Fork	Solomon Creek	Ted Creek	
CK-C1*					
*38	0.926	1.000	1.000	0.948	8.490*
*20	0.047			0.052	
G3PDH-1*					
*100	1.000	0.952	1.000	1.000	9.110*
*140		0.048			
GPI-B1*					
*100	1.000	0.323	1.000	1.000	151.689***
*0		0.677			
IDDH-1*					
*−63	1.000	0.903	1.000	1.000	18.446**
*−100		0.097			
sMDH-B1,2*					
*100	0.875	0.960	0.836	0.952	15.838**
*83	0.125	0.040	0.164	0.048	
sMEP-1*					
*100	1.000	0.919	0.938	0.887	6.813
*90		0.081	0.062	0.113	

[a] The loci code for creatine kinase, enzyme number 2.7.3.2 (IUBNC 1984) (CK-C1*); glycerol-3-phosphate dehydrogenase, 1.1.1.8 (G3PDH-1*); glucose-6-phosphate isomerase, 5.3.1.9 (GPI-B1*); L-iditol dehydrogenase, 1.1.1.14 (IDDH-1*); malate dehydrogenase, 1.1.1.37 (sMDH-B1,2*); and malic enzyme, 1.1.1.40 (sMEP-1*).

Yellowstone cutthroat trout in the Yellowstone River drainage, Montana, have a population genetic structure that is also easily incorporated into a genetically rational introduction program. We detected genetic variation at 14 of the 46 protein-coding loci analyzed in samples from 59 populations. At 13 of these loci the variant allele was found in only a few populations and at a frequency of less than 0.05. The only highly variable locus was sAAT-3,4* (aspartate aminotransferase, 2.6.1.1). This locus has three alleles, and all of them exist at frequencies greater than 0.10 in almost every population. Overall, only 5.5% of the total genetic variation detected was due to differences among populations. From a genetics perspective, therefore, all populations in the drainage could be considered a single unit.

A substantial proportion of the total genetic variation of westslope cutthroat trout throughout the taxon's range was due to genetic differences among populations from different river drainages (16.7%) and among populations within drainages (15.7%; Allendorf and Leary 1988). Much of this genetic divergence was due to alleles that were detected in only one or two populations but at high frequency. These data indicate that large genetic differences exist among populations of westslope cutthroat trout even over very short geographic distances and that the chances of local adaptations are high. In this situation, we recommend each population should be considered a separate management unit, and therefore a separate broodstock would have to be maintained for every introduction need. Given that this is likely impossible, establishing a generalist broodstock from a number of populations is probably a reasonable compromise. This homogenized broodstock probably will have, on the average, a smaller amount of genetic divergence from the native populations than the latter will have among themselves.

Conclusion

We do not advocate the use of introductions to supplement natural reproduction in waters inhabited by native populations. The reason for reduced reproduction often is degraded habitat. Thus, initial emphasis should be directed towards habitat restoration because this is the only way to cure the problem. If supplementation is considered necessary, a knowledge of the population genetic structure of the taxon is essential to minimize potential genetic changes in native populations. Furthermore, care must be taken when establishing and maintaining broodstocks to ensure that the genetic characteristics of the source population are adequately incorporated and perpetuated in the broodstock (e.g., Allendorf and Ryman 1987; Allendorf 1993). Otherwise, the broodstock may eventually serve as a means of causing large and rapid genetic changes in native populations.

Acknowledgments

Funding for our work presented was supplied by the Montana Department of Fish, Wildlife, and Parks, the U.S. Fish and Wildlife Service, the U.S. Forest Service, the U.S. Park Service, the Wyoming Department of Game and Fish, and the National Science Foundation. John Epifanio, Gary Garrett, and an anonymous reviewer made helpful comments for revision.

References

Allendorf, F. W. 1993. Delay of adaptation to captive breeding by equalizing family size. Conservation Biology 7:416–419.

Allendorf, F. W., and R. F. Leary. 1988. Conservation and distribution of genetic variation in a polytypic species, the cutthroat trout. Conservation Biology 2:170–184.

Allendorf, F. W., and N. Ryman. 1987. Genetic management of hatchery stocks. Pages 141–159 in N. Ryman and F. M. Utter, editors. Population genetics and fishery management. University of Washington Press, Seattle.

Allendorf, F. W., and F. M. Utter. 1979. Population genetics. Pages 407–454 in W. S. Hoar, D. J. Randall, and J. R. Brett, editors. Fish physiology, volume 8. Academic Press, New York.

Bams, R. A. 1976. Survival and propensity for homing as affected by presence or absence of locally adapted paternal genes in two transplanted populations of pink salmon (Oncorhynchus gorbuscha). Journal of the Fisheries Research Board of Canada 33:2716–2725.

Barbat-Leterrier, A., R. Guyomard, and F. Krieg. 1989. Introgression between introduced domesticated strains and Mediterranean native populations of brown trout (Salmo trutta L.). Aquatic Living Resources 2:215–223.

Bartley, D. M., and G. A. E. Gall. 1990. Genetic structure and gene flow in chinook salmon populations of California. Transactions of the American Fisheries Society 119:55–71.

Bartley, D. M., and G. A. E. Gall. 1991. Genetic identification of native cutthroat trout (Oncorhynchus clarki) and introgressive hybridization with introduced rainbow trout (O. mykiss) in streams associated with the Alvord Basin, Oregon and Nevada. Copeia (1991):854–859.

Busack, C. A., and G. A. E. Gall. 1981. Introgressive hybridization in populations of Paiute cutthroat trout (Salmo clarki seleniris). Canadian Journal of Fisheries and Aquatic Sciences 38:939–951.

Campton, D. E., and J. M. Johnston. 1985. Electrophoretic evidence for a genetic admixture of native and nonnative rainbow trout in the Yakima River, Washington. Transactions of the American Fisheries Society 114:782–793.

Campton, D. E., and F. M. Utter. 1985. Natural hybridization between steelhead trout (Salmo gairdneri) and coastal cutthroat trout (Salmo clarki clarki) in two Puget Sound streams. Canadian Journal of Fisheries and Aquatic Sciences 42:110–119

Carmichael, G. J., J. N. Hanson, M. E. Schmidt, and D. C. Morizot. 1993. Introgression among Apache, cutthroat, and rainbow trout in Arizona. Transactions of the American Fisheries Society 122:121–130.

Cavender, T. M. 1978. Taxonomy and distribution of the bull trout, Salvelinus confluentus (Suckley), from the American northwest. California Fish and Game 64:139–174.

Dobzhansky, T. 1948. Genetics of natural populations. XVIII. Experiments on chromosomes of Drosophila pseudoobscura from different geographical regions. Genetics 33:588–602.

Ferguson, M. M. 1990. The genetic impact of introduced fishes on native species. Canadian Journal of Zoology 68:1053–1057.

Ferguson, M. M., and R. G. Danzmann. 1987. Deviation from morphological intermediacy in interstrain hybrids of rainbow trout, Salmo gairdneri. Environmental Biology of Fishes 18:249–256.

Ferguson, M. M., R. G. Danzmann, and F. W. Allendorf. 1985a. Developmental divergence among hatchery strains of rainbow trout (Salmo gairdneri), part 2. Hybrids. Canadian Journal of Genetics and Cytology 27:298–307.

Ferguson, M. M., R. G. Danzmann, and F. W. Allendorf. 1985b. Absence of developmental incompatibility in hybrids between rainbow trout and two subspecies of cutthroat trout. Biochemical Genetics 23:557–570.

Ferguson, M. M., R. G. Danzmann, and F. W. Allendorf. 1988. Developmental success of hybrids between two taxa of salmonid fishes with moderate structural gene divergence. Canadian Journal of Zoology 66:1389–1395.

Forbes, S. H., and F. W. Allendorf. 1991. Associations between mitochondrial and nuclear genotypes in cutthroat trout hybrid swarms. Evolution 45:1332–1349.

Gharett, A. J., and W. W. Smoker. 1991. Two generations of hybrids between even- and odd-year pink salmon (Oncorhynchus gorbuscha): a test for outbreeding depression? Canadian Journal of Fisheries and Aquatic Sciences 48:1744–1749.

Gyllensten, U., R. F. Leary, F. W. Allendorf, and A. C. Wilson. 1985. Introgression between two cutthroat trout subspecies with substantial karyotypic, nuclear, and mitochondrial genomic divergence. Genetics 111:905–915.

Howell, P. J., and D. V. Buchanan, editors. 1992. Proceedings of the Gearhart Mountain bull trout workshop. American Fisheries Society, Oregon Chapter, Corvallis.

IUBNC (International Union of Biochemistry, Nomenclature Committee). 1984. Enzyme nomenclature. Academic Press, San Diego, California.

Krieg, F., and R. Guyomard. 1985. Population genetics of French brown trout (*Salmo trutta* L.): large geographical differentiation of wild populations and high similarity of domesticated stocks. Genetics, Selection, and Evolution 17:369–376.

Krueger, C. C., and B. May. 1991. Ecological and genetic effects of salmonid introductions in North America. Canadian Journal of Fisheries and Aquatic Sciences 48(Supplement 1):66–77.

Leary, R. F., F. W. Allendorf, and S. H. Forbes. 1993. Conservation genetics of bull trout in the Columbia and Klamath River drainages. Conservation Biology 7:856–865.

Leary, R. F., F. W. Allendorf, and K. L. Knudsen. 1983a. Consistently high meristic counts in natural hybrids between brook trout and bull trout. Systematic Zoology 32:369–376.

Leary, R. F., F. W. Allendorf, and K. L. Knudsen. 1983b. Genetic analysis of four rainbow trout populations from Owyhee County, Idaho. University of Montana Population Genetics Laboratory Report 83/6, Missoula.

Leary, R. F., F. W. Allendorf, and K. L. Knudsen. 1985. Developmental instability and high meristic counts in interspecific hybrids of salmonid fishes. Evolution 39: 1318–1326.

Leary, R. F., F. W. Allendorf, S. R. Phelps, and K. L. Knudsen. 1984. Introgression between westslope cutthroat and rainbow trout in the Clark Fork River drainage, Montana. Proceedings of the Montana Academy of Sciences 43:1–18.

Leary, R. F., F. W. Allendorf, S. R. Phelps, and K. L. Knudsen. 1987. Genetic divergence and identification of seven cutthroat trout subspecies and rainbow trout. Transactions of the American Fisheries Society 116:580–587.

Liknes, G. A., and P. J. Graham. 1988. Westslope cutthroat trout in Montana: life history, status, and management. American Fisheries Society Symposium 4:53–60.

Loudenslager, E. J., J. N. Rinne, G. A. E. Gall, and R. E. David. 1986. Biochemical genetic studies of native Arizona and New Mexico trout. The Southwestern Naturalist 31:221–234.

Martin, M. A., D. K. Shiozawa, E. J. Loudenslager, and J. N. Jensen. 1985. Electrophoretic study of cutthroat trout populations in Utah. Great Basin Naturalist 45:677–687.

McCracken, G. F., C. R. Parker, and S. T. Guffey. 1993. Genetic differentiation and hybridization between stocked hatchery and native brook trout in Great Smoky Mountains National Park. Transactions of the American Fisheries Society 122:533–542.

Morizot, D. C., S. W. Calhoun, L. L. Clepper, M. E. Schmidt, J. H. Williamson, and G. J. Carmichael. 1991. Multispecies hybridization among native and introduced centrarchid basses in central Texas. Transactions of the American Fisheries Society 120: 283–289.

Paetz, M. J., and J. S. Nelson. 1970. The fishes of Alberta. The Queen's Printer, Edmonton, Alberta.

Parkinson, E. A. 1984. Genetic variation in populations of steelhead (*Salmo gairdneri*) in British Columbia. Canadian Journal of Fisheries and Aquatic Sciences 41:1412–1420.

Philipp, D. P. 1991. Genetic implications of introducing Florida largemouth bass, *Micropterus salmoides floridanus*. Canadian Journal of Fisheries and Aquatic Sciences 48(Supplement 1):58–65.

Philipp, D. P., and G. S. Whitt. 1991. Survival and growth of northern, Florida, and reciprocal F_1 hybrid largemouth bass in central Illinois. Transactions of the American Fisheries Society 120:56–64.

Reisenbichler, R. R., J. D. McIntyre, M. F. Solazzi, and S. W. Landino. 1992. Genetic variation in steelhead of Oregon and Northern California. Transactions of the American Fisheries Society 121:158–169.

Reisenbichler, R. R., and S. R. Phelps. 1989. Genetic variation in steelhead (*Salmo gairdneri*) from the north coast of Washington. Canadian Journal of Fisheries and Aquatic Sciences 46:66–73.

Schuck, H. A. 1943. Survival, population density, growth, and movement of the wild brown trout in Crystal Creek. Transactions of the American Fisheries Society 73:209–230.

Schuler, J. F. 1954. Natural mutations in inbred lines of maize and their heterotic effect. I. Comparison of parent, mutant and their F_1 hybrid in a highly inbred background. Genetics 39:908–922.

Scott, W. B., and E. J. Crossman. 1973. Freshwater fishes of Canada. Fisheries Research Board of Canada, Bulletin 184, Ottawa, Ontario.

Stoneking, M., D. J. Wagner, and A. C. Hildebrand. 1981. Genetic evidence suggesting subspecific differentiation between northern and southern populations of brook trout (*Salvelinus fontinalis*). Copeia (1981):810–819.

Templeton, A. R. 1986. Coadaptation and outbreeding depression. Pages 105–116 *in* M. E. Soule, editor. Conservation biology, the science of scarcity and diversity. Sinauer Associates, Sunderland, Massachusetts.

Thomas, G. 1992. Status report: bull trout in Montana. Montana Department of Fish, Wildlife and Parks, Helena.

Waples, R. S. 1991. Genetic interactions between hatchery and wild salmonids: lessons from the Pacific northwest. Canadian Journal of Fisheries and Aquatic Sciences 48(Supplement 1):124–133.

Whitmore, D. H. 1983. Introgressive hybridization of smallmouth bass (*Micropterus dolomieu*) and Guadalupe bass (*M. treculi*). Copeia 1983:672–679.

Wilde, G. R., and A. A. Echelle. 1992. Genetic status of Pecos pupfish populations after establishment of a hybrid swarm involving an introduced congener. Transactions of the American Fisheries Society 121:277–286.

Wishard, L. N., J. E. Seeb, F. M. Utter, and D. Stefan. 1984. A genetic investigation of suspected redband trout populations. Copeia (1984):120–132.

Withler, F. C., and R. B. Morely. 1982. Use of milt from on-year males in transplants to establish off-year pink salmon (*Oncorhynchus gorbuscha*) runs. Canadian Technical Report of Fisheries and Aquatic Sciences 1139, Ottawa, Ontario.

An Augmentation Plan for Razorback Sucker in the Upper Colorado River Basin

TIMOTHY MODDE

U.S. Fish and Wildlife Service, Colorado River Fish Project
266 West 100 North, Suite 2, Vernal, Utah 84078, USA

ALLAN T. SCHOLZ

Eastern Washington University, Department of Biology
Cheney, Washington 99004, USA

J. HOLT WILLIAMSON

U.S. Fish and Wildlife Service, Dexter National Fish Hatchery and Technology Center
Post Office Box 219, Dexter, New Mexico 88230, USA

G. BRUCE HAINES

U.S. Fish and Wildlife Service, Colorado River Fish Project
266 West 100 North, Suite 2, Vernal, Utah 84078, USA

BOB D. BURDICK AND FRANK K. PFEIFER

U.S. Fish and Wildlife Service, Colorado River Fish Project
764 Horizon Drive, South Annex A, Grand Junction, Colorado 81506, USA

Abstract.—The Endangered razorback sucker *Xyrauchen texanus* historically occupied the major tributaries of the Colorado River system. Habitat loss due to water development, predation and competition from nonnative fish, and contaminants have resulted in decreases in population size such that razorback suckers in the upper Colorado River basin are consistently collected in only the middle Green River. An augmentation plan is proposed that integrates habitat improvement (*viz*, flooded bottomland enhancement) with reintroduction. The goal of an augmentation program for razorback sucker is the establishment of viable (i.e., persistent), locally adapted populations. The basic genetic conservation unit was defined as the local spawning population and was delineated by river tributaries. A decision-making process, based on the threats of inbreeding depression and genetic drift, to determine whether a rare population should be augmented is presented. Populations that have sufficient size to prevent the probability of inbreeding ($N > 250$) and to reproduce successfully should not be augmented. Rather, resources should be directed toward habitat restoration. Genetic management incorporates the establishment of a refugium stock represented by the offspring of a spawning matrix of a minimum of five wild males and five wild females. In addition to providing refugia to the genetic diversity in wild populations, these fish may serve as broodstock for future reintroduction efforts. If reintroduction is necessary, synthetic imprinting will be incorporated to influence behavior and distribution of stocked fish following stocking.

The razorback sucker *Xyrauchen texanus* is a relatively large, monotypic catostomid endemic to the larger tributaries of the Colorado River drainage. Once abundant, this fish has been extirpated from most of its historical range and is now found consistently in only the middle Green River and Lake Mohave (Minckley et al. 1991). The widespread decline of this fish and the continuation of threats to its viability have resulted in listing it as an Endangered Species (USFWS 1991). The major factor associated with the decline of the razorback sucker is the interruption of natural water flow patterns via water diversions and upstream impoundments that alter both the timing and quantity of historical flows (Tyus and Karp 1989, 1990; Minckley et al. 1991).

These alterations have decreased historic habitat and created main-channel barriers that block migration to historical spawning sites (Minckley 1983). Additionally, predation and competition from nonnative fishes (Behnke and Benson 1980; Marsh and Brooks 1989) and toxic heavy metal contamination (Stevens et al. 1992; Ono et al. 1983) have contributed to the decline of razorback sucker.

The factors described above have impaired the ability of the razorback sucker to recruit throughout its range (McAda and Wydoski 1980; Tyus 1992). The remaining populations are composed of older nonrecruiting adults (Lanigan and Tyus 1989; Marsh and Minckley 1989), with most individuals 30 to 40 years old (McCarthy and Minckley 1987). In response to the need to recover the razorback sucker, as well as the remaining three Endangered fishes of the main-stem upper Colorado River basin (Colorado squawfish *Ptychocheilus lucius*, humpback chub *Gila cypha*, and bonytail *G. elegans*), a multi-agency recovery program was established among state and federal agencies and private interest groups in the upper Colorado River basin to accommodate water development and fish recovery simultaneously (Wydoski and Hamill 1991). The goal of this recovery program is to establish and protect self-sustaining populations and natural habitat of the razorback sucker (USFWS 1987). Among several components of the program are habitat rehabilitation, including minimum-flow requirements, and artificial propagation. This paper outlines a strategy to recover razorback sucker in the upper Colorado River basin by enhancing habitat necessary for recruitment and developing a population augmentation program that includes a genetic management plan, a decision-making process to determine when stocking is necessary, and the use of chemical imprinting to increase management options following stocking.

The initial step in the development of a recovery plan is the identification of population units of concern. A baseline genetic survey that included all known wild and captive populations of razorback sucker was conducted throughout its range in both upper and lower Colorado River basins. Two methods of genetic analysis were chosen—mitochondrial DNA, by means of an analysis of restriction fragment length polymorphisms, and allozyme variation, by means of starch gel protein electrophoresis. Preliminary results support provisional division of the upper Colorado River basin razorback sucker populations into Green River, Colorado River, and San Juan River stocks. In addition to genetic findings, behavioral and distributional data support this hypothesis.

The razorback sucker occupying the Green River represents the largest population of this species occupying riverine habitat (Minckley et al. 1991). Lanigan and Tyus (1989), assuming no recruitment or tag loss over a nine-year period, estimated the middle Green River razorback sucker population to be 948 fish (95% confidence interval, 758–1,138). Using the same data, plus 4 years of additional data, Modde, Burnham, and Wick (unpublished data) estimated the population to be 524 fish (95% confidence interval 351–696). The latter study reported limited recruitment, an adult survival rate of 71%, and an unspecified rate of tag loss.

The razorback sucker in the middle Green River has been observed at two spawning sites (Figure 1), one approximately 16 km below Split Mountain Canyon (defined as the Jensen site by Tyus and Karp 1990) and the other at the mouth of the Yampa River just before it enters the Green River. During the spawning season, razorback suckers can be collected regularly at both spawning sites. Some of these fish migrate 30–100 km and show fidelity to selected spawning sites in the Green and Yampa rivers in Colorado and Utah (Holden and Stalnaker 1975; Tyus 1987; Tyus and Karp 1990). Previous investigators have speculated that homing of razorback suckers to these sites may be related to natal imprinting (Wick et al. 1982; Tyus and Karp 1990).

Suspected larval razorback suckers were collected below the two spawning sites by Tyus (1987) and again in 1992, 1993, and 1994 by the U.S. Fish and Wildlife Service (D. Snyder and R. Muth, Larval Fish Laboratory, Colorado State University, personal communication). In 1993, 700 suspected larval razorback suckers were collected between the known spawning sites and the Colorado River arm of Lake Powell (E. Wick, National Park Service, personal communication). The dates when these fish were collected and their sizes suggested they could have originated from the middle Green River, as well as other unknown spawning sites in the lower Green River. Despite the relative abundance of the razorback sucker larvae in the Green and Colorado rivers, less than 10 juvenile razorback suckers have been collected in the upper Colorado River basin during the last 10 years (near the "Mineral Bottoms" by the Utah Division of Wildlife Resources in 1991, and on the Ouray National Wildlife Refuge in 1993).

In the upper Colorado River, razorback sucker adults were commonly collected during the 1970s (Minckley et al. 1991). Despite extensive sampling, very few adult razorback suckers have been collected in the upper Colorado River main stem since 1986, and only seven razorback sucker adults have been collected from riverine habitats since 1988 (McAda et al. 1994). Recently a gravel pit near Grand Junction, Colorado, was found to support several hundred razorback suckers (F. Pfeifer, U.S. Fish and Wildlife Service, personal communication). However, genetic analysis (A. Echelle, D. Philipp, and F. Allendorf, Recovery Program Genetics Advisory Panel, personal communication) personal communication) of these fish indicated that most were either the offspring of a limited number of parents that spawned in the pond during the high-flow years of 1983 and 1984 or the offspring of hybridization between razorback sucker

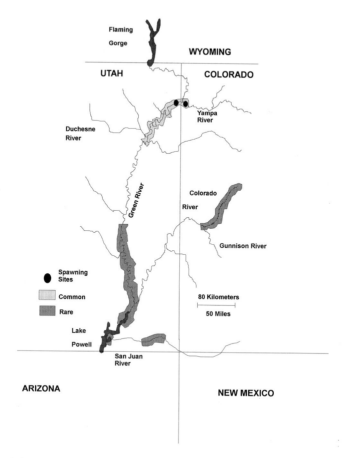

FIGURE 1.—Map of the upper Colorado River basin showing the abundance and the known spawning locations of razorback sucker.

and flannelmouth sucker *Catostomus latipinnis*. No riverine razorback sucker spawning sites in the upper Colorado River have been identified recently above Grand Junction.

Information on the historic distribution of the razorback sucker in the San Juan River is sparse. Anecdotal reports from as late as 1960 indicated razorback sucker as far upstream as Farmington, New Mexico. Earlier reports (1891) of "humpies" in the Animas River near Durango, Colorado, suggested that the razorback sucker may have occupied this tributary of the San Juan River as well (Platania 1990). In 1961 preimpoundment rotenone applications in the Navajo Dam area killed fish 64 km downriver to Farmington. However, no razorback suckers were documented among those fish killed (Olson 1962). The first verified collection of a razorback sucker in the San Juan River basin was in 1976 when two adult fish were seined from a pond near Bluff, Utah. According to local residents, 100–250 juvenile razorback suckers were stranded and died when a nearby pond was drained in 1976. These ponds were connected to the river via a canal (Minckley et al. 1991). Between 1987 and 1989, 16 adult razorback suckers were collected from the San Juan River arm of Lake Powell. In 1988, one adult razorback sucker was captured in the San Juan River near Bluff, Utah, close to the 1976 pond capture site (Platania 1990). No larval or juvenile razorback suckers have ever been documented in the San Juan River and no potential spawning sites have been identified. Intensive riverwide sampling of the river fish community for all life stages from 1991 to 1993 has failed to collect razorback sucker (D. Ryden and F. Pfeifer, U.S. Fish and Wildlife Service, unpublished data).

Razorback sucker adults are flexible in their habitat use. Although these fish apparently evolved in large river systems, adults have survived well in both lacustrine (Wallis 1951; Marsh and Langhorst 1988) and riverine (Tyus 1987) habitats. However, because few immature fish have been captured, little is known of their habitat needs. Most suspected larval razorback suckers have been collected in slack-water habitats in both the Green and Colorado rivers (R. Muth and D. Snyder, personal communication). Tyus and Karp (1990) observed that razorback sucker spawns during increasing and peak spring flows suggested that flooded bottomlands are necessary for recruitment. Modde, Burnham, and Wick (unpublished data) reported limited recruitment of the razorback sucker in the middle Green River and found a significant positive correlation between the number of small (<475 mm total length) razorback suckers collected during 1980 and 1992 and spring flow magnitude 5 years prior to their collection. Discovery in 1991 of several hundred razorback suckers of the same apparent age in Etter Pond (F. Pfeifer, personal communication), which is adjacent to the Colorado River and had been isolated since the 1984 flood, suggests these flooded bottomlands adjacent to the main channel are important to recruitment. Larval razorback suckers stocked into isolated coves in Lake Mohave (Mueller et al. 1993) and a gravel pit adjacent to the Colorado River (Osmundson and Kaeding 1989) demonstrated rapid growth and relatively high survival. Rearing of razorback suckers in hatchery ponds at the Dexter National Fish Hatchery and Technology Center support this hypothesis. Given the limited information available, it appears recruitment of razorback sucker is likely associated with high-flow events, most notably with the availability of flooded bottomlands.

Recovery Plan

Habitat Enhancement

Floodplain enhancement in the upper Colorado River basin is a necessary component to any augmentation plan if self-sustaining populations of razorback sucker are desired. Wetlands needed for razorback sucker recruitment can be provided by removing barriers to historic bottomlands and by providing sufficient flow to inundate bottomlands in a manner that approximates the natural hydrograph. Because the peaks and duration of high flows have decreased following construction of main-stem impoundments, habitat and water flow manipulations will be necessary to regain connectivity of wetlands at the lower flows. Management (e.g., draining and water elevation control) of flooded bottomlands will be needed during the early phases of the recovery process to reduce the abundance of nonnative fish predators and to capture and mark razorback suckers produced in these nursery sites. Marking of fish produced in rehabilitated bottomlands is imperative to monitor the importance of these sites to recovery. Specific bottomland management strategies that will maximize recruitment have yet to be developed. In isolated coves in Lake Mohave, Mueller et al. (1993) have successfully reared razorback suckers to nearly adult size (>400 mm total length) in a single growing season. In the upper Colorado River basin, two growing seasons may be required before vulnerability of razorback suckers to predators is reduced substantially. Flooded bottomland management strategies that will be effective in the upper Colorado River are in the process of being developed and will be determined by available floodplain resources. Experimental evaluation of flooded bottomland management is currently underway at Ouray National Wildlife Refuge, Ouray, Utah.

Captive Propagation Strategy

The goal of the captive propagation program for razorback sucker in the upper Colorado River basin is to avoid imminent extinction and preserve genetic diversity of the species, particularly the heritable component associated with fitness, adaptation, and long-term survival. The intent is to rely on inherent genetic characteristics that will increase the probability that fish reintroduced into suitable habitat will survive, reproduce, and recruit over successive generations, ultimately resulting in a naturally sustaining population.

Genetic conservation units (GCU) are local riverine stocks defined by spawning location, movement and distribution, and genetic information, in that order (Simpson 1961; Mayr 1977; Utter 1981; Dizon et al. 1992; Ruggiero et al. 1994). Dowling and Minckley (1993) suggested, based on haplotype similarity, the razorback sucker in the entire Colorado River drainage may be panmictic. In view of repeated homing and fidelity of razorback sucker to spawning sites previously determined from a tagging and telemetry study (Tyus 1987), we have taken a conservative approach of identifying each major tributary as a unique GCU until conclusively proven otherwise.

The first captive propagation objective is development of artificial genetic refuges (AGRs), production broodstocks, and progeny of presumed stocks. The second captive propagation objective is

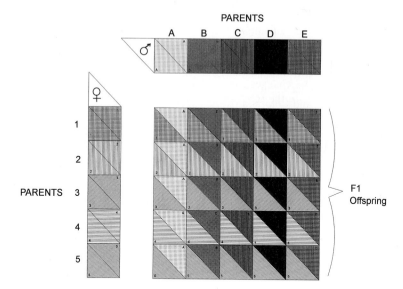

FIGURE 2.—Diagram of the di-allele crossbreeding strategy for maintaining genetic diversity and developing razorback sucker broodstock in the upper Colorado River basin.

production and reintroduction of progeny into suitable habitat to augment declining or restore self-sustaining razorback sucker populations. Managing captive populations to maximize genetic diversity preserves future options for effective management and recovery of the taxon (Ralls and Ballou 1992). Razorback sucker populations most threatened by extinction provide a management opportunity and priority for establishing AGRs or gene banks. AGRs provide a refuge of genetic material in the event a natural catastrophe destroys a majority of the wild stock.

Except for Lake Mohave in the lower basin and the middle Green River in the upper basin, accurate population estimates do not exist for razorback sucker populations. Nevertheless, monitoring data suggest that the razorback sucker population in the upper Colorado River basin is small. Therefore, a di-allele crossbreeding system (Williamson and Wydoski 1994) is being used to develop AGR populations for the upper Colorado River basin (Figure 2). This crossbreeding system is particularly valuable when the total number of adults available is small and the contribution of each individual is critical to maintaining population characteristics. A di-allele mating strategy ensures production of every genotypic combination possible among the few fish. Each fish is genetically represented in the next generation in five unique lots, thus greatly increasing the likelihood of preserving allelic diversity in the offspring of the captive stock (Williamson and Wydoski 1994). The matrix produces 25 half-sibling family lots; the five lots along the diagonal are unique (i.e., unrelated). They provide the matrix core. The remaining family lots provide security of parental contribution through time in the event of a loss of a unique lot. The number of adults used in the matrix is arbitrary; however, any number much beyond five males and five females becomes unwieldy and the utility of the di-allele matrix questionable. After the F_1 broodstock has been developed from the five by five cross, additional wild fish will be used to supplement the genetic component of the original founders when available.

Stocking

The goal of razorback sucker stocking is to assist the establishment of naturally sustainable populations while posing a minimum risk to the endemic fish population. The widespread stocking of millions of razorback sucker in the lower Colorado River basin during 1981–1987 has produced disappointing results (Minckley et al. 1991), and the stocking of 6,400 advanced fingerling razorback suckers in the middle Green River during 1987–1989 has failed to produce a single recapture (U.S. Fish and Wildlife Service, Colorado River Fish Project, unpublished data). However, these stocking efforts were initiated without habitat modifications and a stocking evaluation plan. The failure of these reintroduction efforts indicated that the continued decline of the razorback sucker is not solely a problem of stocking, but probably a wider problem associated with environmental changes. Thus, any stocking program to reintroduce the razorback sucker must be complimented with some form of habitat enhancement.

After the habitat has been enhanced to allow razorback sucker recruitment, the decision of whether or not to stock should be made relative to genetic and demographic criteria. The question of when to initiate stocking is linked with how many individuals are necessary to persist, i.e., population viability (Salwasser et al. 1984). The often-used numbers of individuals necessary to maintain sufficient genotypic variation in order to avoid inbreeding depression and genetic loss due to drift have been 50 and 500, respectively (Franklin 1980). Although this recommendation for short- and long-term variation has been widely cited (Simberloff 1988), these criteria were intended only as guidelines. The lower number was based on agricultural breeding guidelines that allow 1% loss of heterozygosity per generation and the higher number on loss of alleles within a single-locus trait in the absence of selection in a single species (*Drosophila*). Lande (1988) suggested that more complicated genetic variation may require larger populations to prevent losses due to drift. Boyce (1993) stated that the 50/500 rule was arbitrary and capricious and indicated that very few empirical studies define the size of a minimum viable population. Several bird populations of approximately 200 individuals have persisted through time (Thomas 1990). Despite the variation in the numbers suggested for minimum viability, there appears some uniformity among recommendations in the order of magnitude between 100 and 1,000 individuals (Thomas 1990). Within this range, variation in species responses and the inability of current science to predict persistence leaves managers with no useful management rules (Simberloff 1988).

Despite concern for genetic variation and the ability of a population to adapt to environmental changes, Lande (1988) suggested that most populations become extinct due to demographic rather than genetic factors. He stated that demographic and environmental stochasticity are highly integrated. Similarly, Boyce (1992) concluded that demographic modeling is more likely to be of practical significance in determining viability. Thus, sufficient numbers of individuals need to exist to ensure growth and survival of a population. Environmental factors are important in determining reproductive success, growth rates, and thereby, survival. It is demographic collapse that results in population declines that cause lack of genetic variation (Rabb and Lacey 1990). Because of the high longevity of razorback sucker (McCarthy and Minckley 1987), demographic stochasticity is tempered against rapid fluctuations in generation size. Therefore, given representation of most age-classes, a smaller population of razorback sucker would be needed to provide demographic stability than would be needed by a shorter lived species.

In an effort to integrate genetic and demographic criteria, a process will be used to assist in deciding whether or not to stock razorback sucker (Figure 3). The logic presented assumes that habitat restoration or enhancement has been accomplished to support the numerically depressed population. A population size of 250 is arbitrary, but serves as a guideline for the lowest number that represents genetic viability. If it is assumed that the effective population size, N_e (Wright 1931; Kimura and Crow 1963), is approximately 0.2 of a natural stock (Mace and Lande 1991), a total population of 250 should approximate an effective population size of 50 (Mace and Lande 1991). If a natural population has greater than 250 individuals and is successfully spawning and recruiting, then sufficient genetic variation should exist to prevent inbreeding depression and stocking should not be initiated. Rather, management resources should be directed toward habitat rehabilitation or enhancement to remove existing environmental constraints and improve the existing dynamics. If a population exists that has greater than 250 individuals and is not recruiting but maintains a locally adapted genetic stock, stocking should be avoided until genetic risk to that population can be assessed and an appropriate course of action defined. The accumulation of native broodstock will be necessary to provide a broodstock for reintroduction if natural recruitment cannot be attained. The introduction of fish from outside the local stock could potentially establish a population that replaces the local genome. If a population is functionally extirpated (i.e., no opportunity for recovery), and the habitat provides the carrying capacity and necessary conditions for survival, then stocking should be initiated. Determination of the stock used to support reintroduction should represent those genotypes that evolved under the most similar environment, i.e., nearest neighbor population. Stocking should be terminated once natural recruitment has been renewed, but monitoring should be continued.

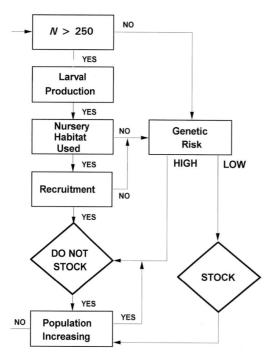

FIGURE 3.—Decision logic for determining whether or not to stock razorback sucker in the upper Colorado River basin. The population size (N) of 250 has been selected arbitrarily to achieve an effective population size of 50. Genetic risk is that to the wild population.

Given the rational described above, we suggest that razorback sucker should not be stocked into the middle Green River, but rather habitat should be modified to improve the existing dynamics of the population. The demographic status of the Green River population most closely fits the criteria of the high or moderate risk categories suggested by Mace and Lande (1991). If the population continues to decline, spawning adults are rarely collected during the spawning season, and larval production and recruitment ceases, then the decision on whether to stock needs to be revisited. The maintenance of a refugium population would allow a genetic source of the original population that would support augmentation efforts.

In contrast to the Green River, fish in the Colorado and San Juan rivers have populations smaller than 250 breeding individuals and no spawning or recruitment is known. These populations fit the very high risk category defined by Mace and Lande (1991), and stocking of razorback sucker, in accordance with habitat development, should proceed with care taken to match genetic and life history characteristics of local populations as closely as possible. If the populations are functionally extinct, AGRs and broodstocks from the nearest neighbor population should be considered. In addition, sites historically occupied by razorback sucker in the upper Colorado River, such as the Gunnison and Uncompangre rivers or the Colorado River above Moab, also should be considered for reintroduction using the same criteria (Williamson and Wydoski 1994).

Imprinting

The use of artificial imprinting to attract razorback suckers to target spawning sites will improve the probability of successful reintroduction programs. The use of imprinting will expand management options following stocking and can be used to test the hypotheses that nonnative fishes and loss of wetland habitats are primary factors in causing the decline of this species. The hypotheses that nonnative fish and loss of backwater nursery habitat are primary factors causing decline could be tested by introducing fish into a relatively predator free system at a site upstream from a good nursery habitat. The successful application of imprinting could also keep stocked fish separate from wild spawners, thereby reducing genetic risks to endemic populations, and attract adult hatchery fish to spawning locations that would enable resulting larvae to colonize new nursery sites.

Based on evidence of homing to selected spawning sites by native fishes in the Green and Yampa rivers (Holden and Stalnaker 1975; Tyus 1987; Tyus and Karp 1990), and the presence of water seeps at known spawning sites that might provide distinctive odors required for olfactory recognition (Tyus 1985), it is thought that natal imprinting may be involved in selection of spawning sites by razorback sucker (Wick et al. 1982; Tyus 1985, 1990; Tyus and Karp 1990). Scholz et al. (1992b) and Tilson et al. (1994) found elevated levels of thyroxine during two periods of imprinting (alevin–swim-up and smolt stages) for kokanee *Oncorhynchus nerka* (lacastrine sockeye salmon). Similar physiological conditions have been found for razorback sucker; peaks in whole-body thyroxine content of eggs and larvae from Green River adults and hatchery broodstock occurred at hatching and at swim-up (Scholz et al. 1992a, 1993). After swim-up thyroxine content declined to very low levels (Scholz et al. 1992a, 1993). If razorback suckers experience chemical imprinting, as these results indicate, the critical period is at hatch or swim-up stages. It is noteworthy that the thyroxine peaks occurred prior to the time larvae would normally drift downstream away from the sites where they were spawned.

A principle advantage of employing synthetic chemical imprinting would be the potential to reduce risk to wild fish by decoying hatchery fish to spawning areas that are remote from wild spawners. Results of the imprinting experiments described above can be used to determine the percentage of fish (p) that can be expected to be attracted to a synthetic chemical. Thus, 1 minus p is the stray rate or the percentage of hatchery stocked fish that could potentially be expected to spawn with wild fish. These results could then be used to evaluate genetic risk to wild stocks from stocking hatchery fish. The genetic assessment model would first calculate Hardy–Weinberg equilibrium for selected razorback sucker genotype frequencies observed in the population. A simulation run could then be made assuming that a certain number of hatchery fish breed with the wild population. The number of stocked fish that could potentially breed with the wild population can be determined by knowing the number of fish stocked, their allele frequencies, and multiplying this value by the above calculated stray rate (assuming that fish with different alleles stray in proportion to their frequencies). Using simulation modeling techniques, the number of hatchery fish could be adjusted up or down so as not to effect gene flow by more than a predetermined amount. In this manner the maximum number of hatchery fish to be stocked could be determined.

Conclusion

The goal of the augmentation plan presented is persistence of the razorback sucker in the upper Colorado River basin. The razorback suckers in each major tributary in the basin are assumed to be offspring of isolated spawning activity; this plan addresses the viability of populations in each major tributary. Because loss of habitat is considered a primary element in the decline of this species, augmentation efforts should be coincident with environmental enhancement. A captive breeding program will be necessary to preserve genetic variability of wild populations, especially those populations for which reproductive success is undocumented. If populations are not recruiting, augmentation will be necessary to recover these populations. When necessary, augmentation should proceed according to a well-established plan using the offspring of pedigree broodstock to maximize the effective population size of the reintroduction effort. All reintroductions should be carefully monitored to determine the causes of success and failure. Synthetic imprinting of stocked fishes will increase management options following stocking and provide insight into the factors affecting the decline of the species.

References

Behnke, R. J., and D. E. Benson. 1980. Endangered and threatened fishes of the upper Colorado River basin. Extension Service Bulletin 503A. Colorado State University, Fort Collins, Colorado.

Boyce, M. S. 1992. Population viability analysis. Annual Review of Ecology and Systematics 23:481–506.

Boyce, M. S. 1993. Population viability analysis: adaptive management for threatened and endangered species. Transactions of the North American Wildlife and Natural Resources Conferences 58:520–527.

Dizon, A. E., C. Lockyer, W. F. Perrin, D. P. DeMaster, and J. Sisson. 1992. Rethinking the stock concept: a phylogenetic approach. Conservation Biology 6:24–36.

Dowling, T. E., and W. L. Minckley. 1993. Genetic diversity of razorback sucker as determined by restriction endonuclease analysis of mitochondria DNA. Final Report (Contract 0-FC-40-0950-004) to the U.S. Bureau of Reclamation. Arizona State University, Tempe.

Franklin, I. R. 1980. Evolutionary change in small populations. Pages 135–149 in E. Soule and B. A. Wilcox, editors. Conservation biology. Sinauer, Sunderland, Massachusetts.

Holden, P. B., and C. B. Stalnaker. 1975. Distribution and abundance of mainstream fishes of the middle and upper Colorado River basins, 1967–1973. Transactions of the American Fisheries Society 104:217–231.

Kimura, M., and J. F. Crow. 1963. The measurement of effective population number. Evolution 17:279–288.

Lande, R. 1988. Genetics and demography in biological conservation. Science 241:1455–1460.

Lanigan, S. H., and H. M. Tyus. 1989. Population size and status of the razorback sucker in the Green River basin, Utah and Colorado. North American Journal of Fisheries Management 9:68–73.

Mace, G. M., and R. Lande. 1991. Assessing extinction threats: toward a reevaluation of IUCN threatened species categories. Conservation Biology 5:148–157.

Marsh, P. C., and J. E. Brooks. 1989. Predation by ictalurid catfishes as a deterrent to re-establishment of hatchery reared razorback sucker. Southwestern Naturalist 34:188–195.

Marsh, P. C. and D. R. Langhorst. 1988. Feeding and fate of wild larval razorback suckers. Environmental Biology of Fishes 21:59–67.

Marsh, P. C., and W. L. Minckley. 1989. Observations on recruitment and ecology of razorback sucker: lower

Colorado River, Arizona-California-Nevada. Great Basin Naturalist 49:71–78.

Mayr, E. 1977. Population, species, and evolution. Belknap Press, Harvard University, Cambridge, Massachusetts.

McAda, C. W., B. Bates, S. Cranney, T. Chart, B. Elmblad, and T. Nesler. 1994. Interagency standardized monitoring program: summary of results, 1986 though 1992. Recovery Implementation Program, upper Colorado River basin. U.S. Fish and Wildlife Service, Denver, Colorado.

McAda, C. W., and R. S. Wydoski. 1980. The razorback sucker, Xyrauchen texanus, in the upper Colorado River. U.S. Fish and Wildlife Service Technical Papers 99:1–15.

McCarthy, M. S., and W. L. Minckley. 1987. Age estimation for razorback sucker from Lake Mohave, Arizona and Nevada. Journal of the Arizona-Nevada Academy of Science 21:87–97.

Minckley, W. L. 1983. Status of the razorback sucker, Xyrauchen texanus (Abbott), in the lower Colorado River Basin. Southwestern Naturalist 28:165–187.

Minckley, W. L., P. C. Marsh, J. E. Brooks, J. E. Johnson, and B. L. Jensen. 1991. Management toward recovery of the razorback sucker. Pages 303–358 in W. L. Minckley and J. E. Deacon, editors. Battle against extinction. University of Arizona Press, Tucson.

Mueller, G., T. Burke, and M. Horn. 1993. A program to maintain razorback sucker in a highly modified riverine habitat. Proceedings of the U.S. Committee on Irrigation and Drainage.

Olson, H. F. 1962. State-wide rough fish control: rehabilitation of the San Juan River. New Mexico Department of Game and Fish, Federal Aid in Sport Fish Restoration, Project F-19-D-4, Job C-16-4, Job Completion Report, Santa Fe.

Ono, R. D., J. D. Williams, and A. Wagner. 1983. Vanishing fishes of North America. Stonewall Press, Inc., Washington DC.

Osmundson, D. B., and L. R. Kaeding. 1989. Colorado squawfish and razorback sucker grow-out pond studies as part of conservation measures for the Green Mountain and Ruedi Reservoir water sales. U.S. Fish and Wildlife Service, Colorado River Fish Project, Final Report. Agreement 6-AA-60-00150, Grand Junction, Colorado.

Platania, S. P. 1990. Biological summary of the 1987 to 1989 New Mexico–Utah ichthyofaunal study of the San Juan River. U.S. Bureau of Reclamation, Salt Lake City, Utah.

Rabb, G. B., and R. Lacey. 1990. Endangered species biology. Science 249:612.

Ralls, K., and J. D. Ballou. 1992. Managing genetic diversity in captive breeding and reintroduction programs. Transactions of the North American Wildlife and Natural Resources Conference 57:263–282.

Ruggiero, L. F., G. D. Hayward, and J. R. Squires. 1994. Viability analysis in biological evaluations: concepts of population viability analysis, biological population and ecological scale. Conservation Biology 8:364–372.

Salwasser, H., S. Mealey, and K. Johnson. 1984. Wildlife population viability—a question of risk. Transactions of the North American National Resources Conferences 49:421–439.

Scholz, A. T., R. J. White, S. A. Horton, and V. A. Koehler. 1992a. Measurement of egg and larval thyroxine concentration as an indicator of the critical period for imprinting in razorback suckers [Xyrauchen texanus (Abbott)]: implications for endangered stocks in the Colorado River Basin. Colorado River Fisheries Project, Technical Report 1 (Cooperative Agreement 2-FC-40-11830) to U.S. Department of Interior, Bureau of Reclamation, Salt Lake City, Utah.

Scholz, A. T., R. J. White, S. A. Horton, and V. A. Koehler. 1992b. Measurement of thyroxine concentrations as an indicator of the critical period for imprinting in kokanee salmon (Oncorhynchus nerka): implications for operating Lake Roosevelt kokanee hatcheries. U.S. Department of Energy, Bonneville Power Administration, Supplemental Report for Project 88-63 (Contract DE-B179-88B91819), Portland, Oregon.

Scholz, A. T., R. J. White, S. A. Horton, M. B. Tilson, C. I. Williams, and B. Haines. 1993. Thyroxine concentrations and chemical imprinting of razorback sucker (Xyrauchen texanus) and Colorado squawfish (Ptychocheilus lucius) eggs and larvae reared at Dexter National Fish Hatchery, NM. Colorado River Fisheries Chemoreception Project, Technical Report 3 (Cooperative Agreement 2 FC-40-11830) to U.S. Department of Interior, Bureau of Reclamation, Salt Lake City, Utah.

Simberloff, D. 1988. The contribution of population and community biology to conservation science. Annual Review of Systematics and Ecology 19:473–511.

Simpson, G. G. 1961. Principles of animal taxonomy. Columbia University Press, New York.

Stevens, D. W., B. Waddell, L. A. Peltz, and J. B. Miller. 1992. Detailed study of selenium and selected elements in water, bottom sediment, and biota associated with irrigation drainage in the Middle Green River Basin, Utah, 1988–90. U.S. Geological Survey, Water-Resources Investigations Report 92-4084. Salt Lake City, Utah.

Thomas, C. D. 1990. What do real population dynamics tell us about minimum viable population sizes? Conservation Biology 4:324–327.

Tilson, M. B., A. T. Scholz, R. J. White, and H. Galloway. 1994. Assessment of smolt transformation tendency and critical period for olfactory imprinting in kokanee salmon: 1993 Annual Report. U.S. Department of Energy, Bonneville Power Administration, Project 88-63 (Contract DE-8179-88BP-91819), Portland, Oregon.

Tyus, H. M. 1985. Homing noted for Colorado squawfish. Copeia 1985:213–215.

Tyus, H. M. 1987. Distribution, reproduction, and habitat use of the razorback sucker in the Green River, Utah, 1979–1986. Transactions of the American Fisheries Society 116:111–116.

Tyus, H. M. 1990. Potamodromy and reproduction of Colorado squawfish in the Green River basin, Colorado and Utah, 1979–1986. Transactions of the American Fisheries Society 119:1035–1047.

Tyus, H. M. 1992. Razorback sucker listed as endangered: a further decline in the Colorado River system. Endangered Species Update 9(5&6):1–3.

Tyus, H. M., and C. A. Karp. 1989. Habitat use and streamflow needs of rare and endangered fishes, Yampa River, Colorado and Utah. U.S. Fish and Wildlife Service Biological Report 89(14):1–27.

Tyus, H. M., and C. A. Karp. 1990. Spawning and movements of razorback sucker, *Xyrauchen texanus*, in the Green River basin of Colorado and Utah. The Southwestern Naturalist 35:427–433.

USFWS (U.S. Fish and Wildlife Service). 1987. Recovery implementation program for endangered fish species in the upper Colorado River basin. USFWS, Region 6, Denver, Colorado.

USFWS (U.S. Fish and Wildlife Service). 1991. Endangered and threatened wildlife and plants: the razorback sucker (*Xyrauchen texanus*). Determined to be an endangered species. Federal Register 56 (205):54957–54967.

USFWS (U.S. Fish and Wildlife Service). 1993. Draft, Section 7 consultation, sufficient progress, and historic projects agreement and recovery action plan. Recovery implementation program for endangered fish species in the upper Colorado River basin. USFWS, Region 6, Denver, Colorado.

Utter, F. M. 1981. Biological criteria for definition of species and distinct intraspecific populations of anadromous salmonids under U.S. Endangered Species Act of 1973. Canadian Journal of Fisheries and Aquatic Sciences 38:1626–1635.

Wallis, O. L. 1951. The status of the fish fauna of the Lake Mead National Recreational Area, Arizona-Nevada. Transactions of the American Fisheries Society 80:84–92.

Wick, E. J., C. W. McAda, and R. V. Bulkley. 1982. Life history and prospects for recovery of the razorback sucker. Pages 120–126 in W. H. Miller, H. M. Tyus, and C. A. Carlson, editors. Fishes of the upper Colorado River system: present and future. American Fisheries Society, Western Division, Bethesda, Maryland.

Williamson, J. H., and R. S. Wydoski. 1994. Genetics management guidelines. Recovery implementation program for endangered fish species in the upper Colorado River basin. U.S. Fish and Wildlife Service, Denver, Colorado.

Wright, S. 1931. Evolution in Mendelian populations. Genetics 16:97–159.

Wydoski, R. S., and J. Hamill. 1991. Evolution of a cooperative recovery program for endangered fishes in the upper Colorado River basin. Pages 123–135 *in* W. L. Minckley and J. E. Deacon, editors. Battle against extinction. University of Arizona Press, Tucson.

Apache Trout Management: Cultured Fish, Genetics, Habitat Improvements, and Regulations

GARY J. CARMICHAEL

U.S. Fish and Wildlife Service, Southwest Region, Division of Fisheries
PO Box 1306, Albuquerque, New Mexico 87103, USA

JAMES N. HANSON

U.S. Fish and Wildlife Service, Arizona Fisheries Resources Office
PO Box 39, Pinetop, Arizona 85935, USA

JAMES R. NOVY

Arizona Game and Fish Department, Regional Office
HC 66, Box 57201, Pinetop, Arizona 85935, USA

KELLY J. MEYER

White Mountain Apache Tribe, Game and Fish Department
PO Box 220, Whiteriver, Arizona 85941, USA

DONALD C. MORIZOT

The University of Texas, M. D. Anderson Cancer Center, Research Division
PO Box 389, Smithville, Texas 78957, USA

Abstract.—The Apache trout *Oncorhynchus apache* historically was an important food and sport fish in Arizona. Through past management practices that included stocking cultured fishes, and through habitat loss and alteration, the Apache trout became imperiled, was recognized as such in the 1950s by the White Mountain Apache Tribe, and was listed as Endangered under the Endangered Species Act of 1973. Hybridization of Apache trout with rainbow trout *O. mykiss* and cutthroat trout *O. clarki* has been shown using isozyme locus polymorphisms in aggregate as diagnostics for taxon discrimination. Predation by and competition with brown trout *Salmo trutta* and brook trout *Salvelinus fontinalis* have been identified as range-limiting factors for Apache trout populations. Land-use practices have reduced the amount of suitable habitat for Apache trout in some watersheds. A combination of modern fish culture practices, genetic analyses and management, habitat improvements, and management regulations are needed to sustain Apache trout in the southwestern United States. Recent genetic analyses have confirmed the identity of an additional population of Apache trout previously considered extirpated. Ecological surveys of the watersheds in the White Mountains will be required to determine habitat-specific management efforts such as fish stocking, barrier installation, and stream renovation to ensure survival of Apache trout populations. Based on genetic analyses and watershed surveys of trout populations and habitat, no single management option is adequate to address the recovery of Apache trout populations. We recommend a management strategy that combines habitat-specific objectives, fish stocking, and regulations and that ranges from no intentional human use or impact to stocking of nonnative fishes and Apache trout into altered habitats.

Many White Mountain area settlers reported the presence of native trout which they referred to as yellow-bellied, speckled trout. Although known since 1873, this fish was not formally described as Apache trout *Oncorhynchus apache* until a century later (Miller 1972). Apache trout inhabited most of the streams in east-central Arizona's White Mountains, and historically, the Apache trout was the only salmonid resident in the Black, White, Blue, and Little Colorado river drainages.

Loss of habitats, hybridization with other trouts, and predation and competition from other trouts have greatly reduced the distribution and abundance of Apache trout (USFWS 1979, 1983). Stocking cultured fish in Apache trout ecosystems has been used, has affected, and will play a role in Apache trout conservation. Predation by introduced brown trout *Salmo trutta* is the major threat to Apache trout today. Fish stocking practices have been changed, and hybridization is now less of a threat. Habitat alterations increased siltation and reduced the range of Apache trout, but current habitat management strategies mitigate habitat loss. Habitat improvements, management regulations, and genetic analyses are all part of Apache trout ecosystem management. Several biological and genetic surveys have been conducted, which have led to development of regulations that limit fishing and further habitat alterations.

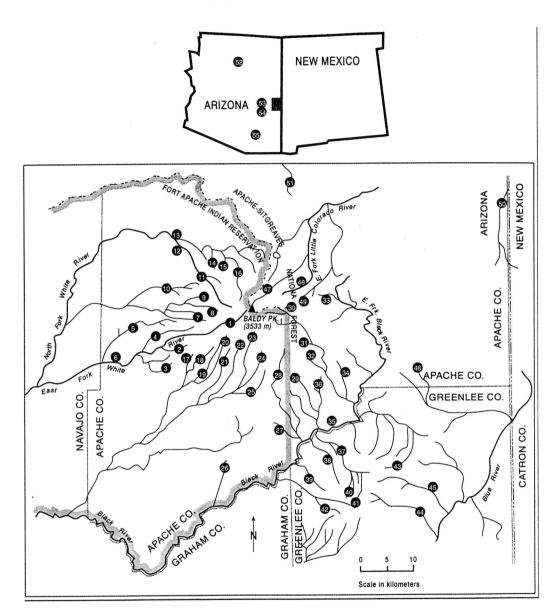

FIGURE 1.—Collection and stocking sites of Apache, brown, brook, and hybrid trout in the Fort Apache Indian Reservation, Apache Sitgreaves National Forest, and nonhistorical sites. Site numbers correspond to map locations. The map is expanded from Carmichael et al. (1993). (1) East Fork, White River; (2) Elk Canyon Creek; (3) Rock Creek; (4) Deep Creek; (5) Firebox Creek; (6) Lower East Fork, White River; (7) Sun and Moon creeks; (8) South Fork, Diamond Creek; (9) Diamond Creek; (10) Little Diamond Creek; (11) Paradise Creek; (12) Lower Paradise Creek; (13) North Fork, White River; (14) Bear Cienega Creek; (15) Smith Creek; (16) Ord Creek; (17) Little Bonito Creek; (18) Crooked Creek; (19) Boggy Creek; (20) Lofer Cienega Creek; (21) Flash Creek; (22) Squaw Creek; (23) Big Bonito Creek; (24) Hughey Creek; (25) Hurricane Creek; (26) Paddy Creek; (27) Soldier Creek; (28) Reservation Creek; (29) Boggy Creek; (30) West Fork, Black River and Thompson and Burro creeks; (31) Stinky Creek; (32) Hay Creek; (33) East Fork, Black River; (34) Home Creek; (35) Black River; (36) Centerfire Creek; (37) Fish Creek; (38) Conklin Creek; (39) Snake Creek; (40) Double Cienega Creek; (41) Corduroy Creek; (42) Bear Wallow Creek; (43) Hannagan Creek; (44) KP Creek; (45) Grant Creek; (46) Coleman Creek; (47) West Fork, Little Colorado River; (48) East Fork, Little Colorado River; (49) Lee Valley Creek; (50) Coyote and Mamie creeks; (51) Mineral Creek; (52) North Canyon Creek; (53) Cibeque Creek; (54) Canyon Creek; (55) Grant Creek.

Informed management actions need to set the direction for Apache trout conservation as well as for delisting under the Endangered Species Act (16 U.S.C.A. §§ 1531 to 1544). These actions include (1) identification and genetic characterization of Apache trout populations; (2) renovation[1] of candidate Apache trout streams (Rinne et al. 1981); (3) survey, restoration,[2] and improvement of previously altered Apache trout habitats; (4) stocking of hatchery-reared Apache trout for fisheries recovery and enhancement[3]; and (5) establishment of fisheries management policies for critical habitats and other streams to ensure Apache trout conservation.

The purpose of this paper is to provide the conservation community with a synthesis of scientific information that will lead to conservation, recovery, and possibly delisting of the Apache trout. In this paper we present (1) Apache trout distribution data; (2) conservation and enhancement efforts, including those enhancement actions that promote Apache trout as a game fish by use of stocked fish in streams and impoundments not described as critical[4]; and (3) delisting recommendations. We report progress and recommend actions that should be initiated or continued to reduce habitat loss, and we discuss cessation of stocking of other fishes in streams that contain Apache trout.

Apache Trout Distribution

The Apache trout was originally found in the upper Salt River drainage, including the Black and White rivers, the San Francisco River drainage, including the Blue River, and the headwaters of the Little Colorado River (Miller 1972) (Figure 1). The range of Apache trout was thought to have dropped to a low of about 48 stream kilometers or less in the mid-1900s, reduced from an estimated original range of about 965 km (Harper 1978). The present distribution is a minimum 141 km. According to our surveys, the headwaters of the White and Black rivers on the Fort Apache Indian Reservation (FAIR) now contain the largest populations of Apache trout. Larger stream systems such as Bonito Creek and the East Fork of the White River could contain several thousand Apache trout. Postwinter populations in small tributary streams may be less than 100 individuals, and fall populations are usually four to five times greater (Harper 1978).

Many of the watersheds formerly inhabited by Apache trout have been stocked with rainbow trout *O. mykiss*, cutthroat trout *O. clarki*, brook trout *Salvelinus fontinalis*, or brown trout for more than 50 years (USFWS 1983). Cutthroat trout are believed to have been first stocked into Apache trout streams in the late 1800s (USFWS 1983). Cutthroat and rainbow trout were spread by stocking extensively over the entire range of Apache trout, although natural barriers to upstream migration prevented species mixture in some watersheds (USFWS 1983). Hatchery and management records from Williams Creek National Fish Hatchery, the U.S. Fish and Wildlife Service (USFWS) Arizona Fisheries Resources Office, and the Arizona Game and Fish Department (AGFD) indicate that more than 1.2 million cutthroat trout were stocked into at least 53 White Mountain streams from 1920 to 1942. Similar records indicate that rainbow trout were widely stocked between 1934 and 1954. Some rainbow, brown, brook, and cutthroat trout stocking still occurs today, although into streams and lakes that are marginal Apache trout habitat. These stockings are conducted under auspices of Section 7 of the Endangered Species Act, which regulates impacts on Endangered species. Streams critical to recovery of Apache trout are not stocked with other trout and are or will be protected from the influence of introduced fish by either natural or artificial stream barriers that prevent immigration. Some nonnative salmonids outcompete Apache trout for food, displace them, or may prey upon them. Competition with brown and brook trout has been identified as a cause of the decline of Apache trout (Rinne and Minckley 1985). Rainbow and cutthroat trout have hybridized with the Apache trout, and biochemical genetic surveys have identified populations with high proportions of hybrids (Loudenslager et al. 1986; Dowling and Childs 1992; Carmichael et al. 1993).

Habitat alterations due to logging, sand and gravel mining, grazing, water impoundment, and road construction have reduced the quality and amount of Apache trout habitat. These alterations are more prominent at lower elevations in the White Mountains (USFWS 1983). In the southwestern United States, livestock grazing is practiced on public land, and water is a key to livestock distribution. As a result, many streams, springs, and wetlands have been altered and riparian vegetation has been reduced or eliminated. Because grazing on public lands (including wilderness areas) is provided by law, it is an activity that must be coordinated or regulated in order to maintain or improve stream, riparian, and watershed conditions in Apache trout habitat.

Survey records from the 1980s (Rinne 1985; Rinne and Minckley 1985; Loudenslager et al. 1986; Dowling and Childs 1992; Carmichael et al. 1993) indicate that populations of Apache trout still remain in a few streams on the FAIR and Apache Sitgreaves National Forest (ASNF). Meristic and morphometric variation and biochemical genetic variations were characterized in these surveys, and generally, data were in agreement. Isozyme locus polymorphisms, in aggregate diagnostic for discrimination of Apache, rainbow, and cutthroat trout, indicated only 11 potentially unhybridized populations of Apache trout (Carmichael et al. 1993). Rainbow trout introgression was documented in 19 of 31 populations, including at least 2 in which all individuals sampled were hybrids. Hybridization between cutthroat and Apache trout was detected in four populations; two of these contained individuals with alleles from all three species (Carmichael et al. 1993). In 19 of 20 hybridized populations sampled, a trend of backcrossing toward Apache trout was demonstrated. No pure rainbow or cutthroat trout were found in the populations sampled (Carmichael et al. 1993).

In 1988, we conducted additional surveys in 35 White Mountain streams on the FAIR and ASNF to identify streams containing Apache and other trouts (Table 1). At least 2 of the 11 unhybridized populations are threatened with brown trout competition (Boggy Creek on FAIR[5] and Flash Creek) (Table 1). Additional streams containing Apache trout populations that are in danger due to brown trout predation include Little Diamond, Little Bonito, Hughey, Sun, and Moon creeks. Brown trout, but few Apache trout, were found in Big Bonito and Squaw creeks, two streams that historically contained Apache trout populations. Ord Creek contained Apache trout (Carmichael et al. 1993), but brook trout were more abundant (Table 1).

Other headwaters and remote streams throughout the White Mountain area may contain additional Apache trout populations—many streams support populations that display both Apache and rainbow trout characteristics (Carmichael et al. 1993). Presence of these hybrids suggests that additional isolated headwater areas may contain as yet unknown Apache trout populations.

Conservation Efforts and Fish Culture

Conservation of Apache trout was first attempted by the White Mountain Apache Tribe (WMAT) in the late 1940s and 1950s. By 1950, the only known Apache trout populations occurred on the FAIR. In 1955, the WMAT closed sportfishing on all Mount Baldy streams (Figure 1) that contained suspected populations of Apache trout and later closed all streams on the FAIR to fishing for Apache trout.

TABLE 1.—Number of fish collected by electrofishing in 35 creeks on the Fort Apache Indian Reservation (FAIR) and the Apache Sitgreaves National Forest (ASNF), Arizona, 1988. No rainbow or cutthroat trout were located in any of the streams.

Creek	Number of fish collected		
	Apache trout	Brown trout	Brook trout
Bear Cienega	11	38	7
Big Bonito	1	37	0
Boggy (ASNF)	35	10	0
Boggy (FAIR)	32	25	0
Centerfire	50	13	0
Conklin	30	0	0
Corduroy	90	0	0
Crooked	22	0	0
Deep	79	0	0
Double Cienega	110	0	0
Elk Canyon	33	0	0
Firebox	55	0	0
Fish	55	0	0
Flash	2	94	0
Hannagan	79	0	0
Hughey	43	17	0
Hurricane	15	7	0
Little Bonito	2	11	0
Little Diamond	20	142	0
Lofer Cienega	2	0	0
Moon	4	31	0
North Fork Diamond	26	0	0
Ord	58	0	125
Paddy	43	0	0
Paradise	24	37	0
Reservation	220	1	0
Rock	2	13	0
Smith	15	24	0
Snake	50	7	0
Soldier	15	0	0
South Fork Diamond	24	0	0
Squaw	1	74	0
Sun	33	55	0
Thompson	0	42	2
West Fork Black	0	33	1

In the 1960s Apache trout status was determined by the USFWS, WMAT, and the AGFD using fishery surveys. Federal, state, and tribal representatives all agreed that management actions must be implemented to secure the survival of Apache trout (USFWS 1983). The AGFD, in cooperation with the WMAT and USFWS, entered into a controlled propagation (hatchery) program. Apache trout were collected from Ord Creek in 1962 and were successfully propagated at the AGFD Sterling Springs Hatchery. Separate hatchery broodfish populations were created from Apache trout taken from Ord Creek and Crooked Creek in 1963. Resulting progeny were introduced into Christmas Tree, Bear Canyon, and Becker lakes, Lee Valley Reservoir, and streams on the Apache Sitgreaves, Kaibab, Tonto, and Coronado National Forests. The stocking continued from 1965 through 1974.

Some stocking sites were outside the historic range and were chosen because they lacked salmonids. All sites were intended to provide the general public a chance to fish for this imperiled trout.

In 1964, the Apache trout was recommended for inclusion in the Secretary of the Interior's list of Rare and Endangered species and was listed as Endangered in 1969 (Miller 1969). In a 1964 WMAT resolution, the Tribe adopted a management plan proposed by the USFWS that called for the construction of fish barriers and the reclamation of streams and lakes for the reintroduction of Apache trout. As part of this plan, the WMAT renovated Sun and Moon creeks and constructed an impoundment (Christmas Tree Lake) at their confluence. In 1965 the WMAT closed Ord Creek, the upper reaches of East Fork of the White River, Paradise Creek and its tributaries to fishing. Christmas Tree Lake filled in the early spring of 1967, and Apache trout were stocked from Ord Creek, Firebox Creek, and Deep Creek. In addition, fry from Ord Creek broodfish held at the AGFD's Sterling Springs Hatchery were introduced. Ord Creek fish were moved to North Canyon, Coyote, Mamie, and Mineral creeks on the ASNF, and Grant, Horton, Sun, and Moon creeks. Ord Creek was renovated in 1977 and restocked with Apache trout from Paradise Creek in 1980 (Rinne et al. 1981).

The Endangered Species Act was passed in 1973, and the Apache trout was brought under its protection. FAIR streams and others in Arizona were closed to taking of Apache trout in 1974. A recovery team was formed, and in 1975 the Apache trout was one of the first species to be downlisted from Endangered to Threatened status. Public waters in Arizona were reopened to fishing, but the FAIR remained closed. Fishing is still prohibited for wild Apache trout on greater than 80% of FAIR streams, although there is a substantial fishery for Apache trout introduced into noncritical habitats. The WMAT also implemented regulations that created a 100-m buffer zone on the banks of Apache trout streams, which prevents heavy equipment from damaging aquatic habitats.

Population surveys completed in the 1980s identified streams that could be renovated to achieve recovery goals without losing genetic diversity in Apache trout. Fish barriers have been constructed on 14 streams to protect restored populations following stream renovation. Apache trout populations have been established in Coleman, Hayground, Home, Lee Valley, Wildcat, and Hurricane creeks. More populations will be established in the near future.

Since 1987, an interagency (ASNF and AGFD) team has been conducting extensive surveys of ASNF streams, where degraded stream and riparian habitat has been recognized as a major problem. All Apache trout streams on the ASNF, with the exception of two (KP Creek and Grant Creek), have been inventoried. Aquatic habitat surveys identified degraded streams and reaches that were in need of habitat restoration (Table 2). Streams on the ASNF containing Apache trout were given priority in an implementation schedule for habitat restoration in a habitat recovery plan. Management actions affecting six ASNF Apache trout streams were identified in 1993 (USFS 1993a; 1993b). The remaining Apache trout streams on the ASNF in need of habitat management will be addressed in 1994 or 1995.

Apache trout are presently being reared by the USFWS at the Williams Creek National Fish Hatchery. Several hundred Apache trout were collected by electrofishing from the East Fork of the White River on the FAIR (type locality for the species) and were spawned on site. Embryos were transferred to the Williams Creek National Fish Hatchery and 944 fish reared to maturity. First generation hatchery broodfish were produced in 1983 and 1984. Fish culture technology and facilities have been established for the successful production of hatchery-reared Apache trout (David 1990; R. E. David, G. J. Carmichael, and D. C. Morizot, USFWS, unpublished data). As of 1988, fish used for controlled reproduction at the Williams Creek National Fish Hatchery were still genetically representative of Apache trout located in the East Fork of the White River (Carmichael et al. 1993).

Apache trout, which rarely exceed 300 mm in streams, grew fast at the hatchery, and females typically achieved a length exceeding 600 mm and a weight of 2,200 g within 4 years. Egg production by captive Apache trout represented a 2–10-fold increase in fecundity over that observed in the wild and greatly assists propagation efforts at the hatchery. Production of Apache trout was successful and increased from 1,200 fish in 1986 to 550,000 in 1990 (David et al. 1993).

Many sites have been stocked with Apache trout from Williams Creek National Fish Hatchery, creating a very popular fishery. In 1991, a rainbow trout stocking program on the FAIR and ASNF was replaced with over 500,000 catchable-size, subcatchable-size, and fingerling Apache trout. Apache trout have been stocked into 1,200 ha of lakes and up to 100 km of streams on the FAIR.

TABLE 2.—Streams involved in Apache trout management and delisting of Apache trout as Endangered. Population numbers 22–36 are replicate populations; source populations are in parentheses. Management actions needed include stream renovation (renovate), construction of fish passage barriers (barrier), genetic sampling (genetics), collection of larger genetic sample sizes (sample), habitat restoration[a] (habitat), and collection of fish with alleles to be incorporated into hatchery broodfish (*PGK*a*, *CK-B*c*, or *PGM*d*; defined in text).

Population number	Stream or stream complex	Confirmed hybridization with rainbow trout	Managed for reproducing Apache trout population	Management action needed
1	Big Bonito	No	Yes	Renovate, genetics
2	Boggy on FAIR	No	Yes	Sample, *PGK*d*
3	Boggy and Centerfire	Yes	No	Habitat, genetics
4	Coyote and Mamie[b]	No	Yes	Sample, *CK-B*c*
5	Conklin	Yes	No	Habitat, genetics
6	Corduroy, Double Cienega, and Fish	Yes	No	Habitat, genetics
7	Crooked	No	Yes	Sample, barrier
8	Deep	No	Yes	Sample, *PGM*a*
9	East Fork White River	No	Yes	Sample
10	Elk Canyon	No	Yes	Sample
11	Firebox	No	Yes	Sample
12	Flash	No	Yes	Sample, barrier, renovate
13	Hannagan	Yes	No	Barrier, habitat, genetics
14	Hughey	Yes	No	Genetics
15	Little Bonito	No	Yes	Renovate
16	Little Diamond	Yes	Yes	Genetics
17	Lofer Cienega[c]	No	No	Sample, habitat
18	Mineral[b]	No	Yes	Genetics, habitat
19	North Canyon[b]	No	Yes	Sample
20	Reservation	No	No	Genetics
21	Soldier	No	Yes	Sample, habitat
22	Bear Wallow (Crooked)	No	Yes	Renovate, habitat
23	Coleman (Soldier)	No	Yes	Genetics
24	Grant[b]	No	No	Genetics
25	Grant (Elk Canyon)	Yes	Yes	Renovate
26	Hayground (White)	No	Yes	Genetics, habitat
27	Home (White)	No	Yes	Genetics, habitat
28	K.P. (Deep)	No	Yes	Renovate
29	Lee Valley (White)	No	Yes	Genetics, habitat
30	Ord (North Canyon)	No	Yes	Renovate
31	Paddy (Soldier)	No	Yes	Renovate
32	Paradise (Coyote)	No	Yes	Renovate
33	Snake (Flash)	No	Yes	Renovate
34	Squaw (Boggy on FAIR)	No	Yes	Renovate
35	Stinky (Firebox)	No	Yes	Renovate, habitat
36	Wildcat (White)	No	Yes	Genetics, habitat

[a] Habitat restoration measures to be implemented on Boggy Creek on ASNF and Centerfire, Hayground, Home, Stinky, and Wildcat creeks have been issued by the U.S. Forest Service (1993). Remaining streams are currently under analysis by the ASNF, and recommendations for habitat restoration measures are expected in 1994 or 1995.
[b] Stocked with fish from Ord Creek during previous conservation efforts. Population 24, Grant Creek, is located in Coronado National Forest.
[c] A replicate population will be created if enough fish can be located.

North Canyon and Coyote Creeks

In order to determine the success of a previous conservation effort that transferred fish from Ord Creek to others (USFWS 1983), we collected 30 Apache trout from both North Canyon and Coyote creeks by means of electrofishing and angling. All fish collected in Coyote Creek were Apache trout (Carmichael et al. 1993). Fish from North Canyon Creek were handled and analyzed using methods of Morizot and Schmidt (1990) and Carmichael et al. (1993). The electrophoretic analyses of the sample indicated that North Canyon Creek fish are Apache trout. No other trouts or hybrids were present. The North Canyon Creek and Coyote Creek populations were originally created using fish from Ord Creek. As the Ord Creek population has been extirpated, North Canyon Creek and Coyote Creek fish are more representative of Ord Creek than any other extant Apache trout population and can be used to repopulate Ord Creek and other streams.

Management and Delisting Recommendations

The high number of populations that contain hybrids, brown trout, and brook trout (Table 1) hinder recovery efforts for Apache trout. Delisting of Apache trout will be possible only by means of various fisheries management techniques or actions; specific management actions are listed in Table 2. Impediments to recovery and enhancement include manageable problems of hybridization, competition, predation, and habitat improvement. We recommend that to delist the Apache trout under the Endangered Species Act, the following management procedures should be exercised.

Establish Self-Sustaining Apache Trout Populations

To delist Apache trout it will be necessary to establish or maintain 18–21 self-sustaining discrete populations of Apache trout (Table 2). Also, 28 creeks throughout its historic range should be managed to contain reproducing populations of Apache trout without competition or hybridization. Unless population sizes become too small to be self-sustaining, no hatchery or other Apache trout should be introduced into these streams. Priority should be given to protecting the populations for which no hybridization with rainbow trout has been detected (Carmichael et al. 1993). In these populations, sample sizes for genetic analyses should be increased to at least 50 individuals each, if possible, to verify virtual absence of rainbow trout introgression. These populations are the most critical ones sampled for recovery. Every effort should be made to maintain whatever barriers appear to prevent introduction of other trouts. The streams should be monitored to determine self-sustaining status of Apache trout populations and lack of predators and competitors.

Replicate Existing Apache Trout Populations

Apache trout populations that are self-sustaining and show no hybridization should be replicated in other streams after the elimination of other species by stream renovation and construction and maintenance of barriers to prevent recolonization of other trouts (Table 2). Hatchery-reared fish can be used if others of close ancestry are not available.

Conduct Genetic Surveys

Biochemical genetic analyses should be conducted on all populations suspected to contain Apache trout. Recently established populations should be sampled and analyzed within the next 5 years. The eight Apache trout-predominant streams that are considered partially hybridized but still important to Apache trout recovery should be monitored and evaluated. Hybrid-predominant populations should be used as experimental systems for testing recovery techniques as recommended in Carmichael et al. (1993). Given the very high proportions of hybrids in these populations, we consider that renovation may be the most effective action for establishment of Apache trout populations. Five populations are considered to be Apache trout predominant (>50% apparent Apache trout): Little Diamond, Sun, Moon, Hughey, and Paddy creeks. Barriers to rainbow trout immigration should be maintained and sample sizes for genetic study increased. If brown trout competition or predation becomes limiting to Apache trout, these populations should be renovated and restocked. Resampling Paddy Creek (which contains approximately 50% hybrids at present) in 5 years would yield an estimate of the efficacy of a do-nothing approach to Apache trout recovery; the other four populations, in which more than 86% of the fish sampled were Apache trout, could be considered for other management practices, including reassessment after a period of nonintervention, stocking with hatchery-produced Apache trout, or stocking with Apache trout from nonhybridized populations. One population may be reserved for estimation of the rate of loss of rainbow trout alleles over time. All populations should be sampled and fish genetically analyzed prior to movement of fishes to other streams.

Use of Hatchery-Reared Fish

Hatchery-reared fish should be used as a management strategy to assist delisting and to enhance fisheries by establishing sportfishing populations that also serve as Apache trout gene pool reserves. Hatchery-derived sport fisheries for Apache trout will be useful to develop public support for management in critical streams. Criteria for selection, maintenance, and acquisition of broodfish should be developed. These should address minimized domestication, inbreeding, disease susceptibility, hybridization, and maintenance of genetically distinct populations. Guidelines for the disposal of excess fish should be developed. Only Crooked, Deep,

Boggy on FAIR, and Firebox creeks, and the East Fork of the White River should be considered sources of Apache trout for introduction or for development of broodfish (Carmichael et al. 1993). Coyote Creek and North Canyon Creek fish, originally derived from Ord Creek, are also acceptable as Apache trout sources for introductions. Fish from populations containing three Apache trout alleles detected in surveyed populations should be brought into the hatchery (creatine kinase [enzyme number 2.7.3.2: IUBNC 1984], *CK-B*a*, from Coyote Creek; phosphoglucomutase [5.4.2.2: IUBNC 1984], *PGM*d* from Deep Creek, and phosphoglycerate kinase [2.7.2.3: IUBNC 1984], *PGK*a* from Boggy Creek on FAIR) (Carmichael et al. 1993). Crosses among these populations can rapidly yield populations carrying the Apache trout polymorphic alleles. Hatchery-reared fish to be stocked for enhancement, therefore, will be representative of the Apache trout genome and will serve as gene pools in addition to those populations now occurring in headwater streams.

Sport fisheries should be created using hatchery-reared Apache trout and other salmonids. Several streams should be considered as Apache trout enhancement streams to be used for sportfishing and as habitat used to hold populations of Apache trout that need not be self-sustaining (Hurricane, Sun, Moon, Canyon, Cibeque, Diamond, East Fork Diamond, Thompson, Burro, and lower Paradise creeks, and lower East Fork White, North Fork White, East Fork Black, West Fork Black, Black, East Fork Little Colorado, and West Fork Little Colorado rivers). Future studies should address and assess reproductive performance of stocked Apache trout, and consideration should be given to the efficacy of swamping hybrids by repeated Apache trout introductions or selective removal of probable hybrids. Such swamping on a sustained basis could develop a competitive advantage for Apache trout due to its greater abundance than other trouts. These studies can be carried out in enhancement streams.

The Williams Creek National Fish Hatchery will produce Apache trout required to replace rainbow trout stocked in streams on the FAIR. As Apache trout numbers increase, a corresponding number of other trouts, especially rainbow trout, could be deleted from the fisheries management commitment. Apache trout should be substituted for rainbow trout whenever Apache trout provide an acceptable sportfishing opportunity.

Monitor and Develop Habitat

Habitat conditions should be monitored in all Apache trout streams at least once every 5 years to determine that Apache trout have not declined to dangerously low levels because of alterations of ecosystem variables. Surveys should include detailed fish population estimates, water quality characterization, benthic macroinvertebrate analysis, and riparian condition analysis. Physical barriers should be assessed and maintained by conducting annual inspections.

Conservation actions should include the collection and evaluation of scientific data, improvement and maintenance of habitats, development and application of grazing strategies, and the planting of native vegetation within riparian zones.

Develop, Implement, and Enforce Regulations

Policies and regulations in place should continue to be implemented, such as those addressing riparian buffer zones and those preventing further stocking of rainbow trout and other nonnative species in streams where Apache trout predominate. Headwater streams on the FAIR that contain populations of Apache trout should be either not opened to fishing or open only to closely regulated fishing. Conservation efforts should include development and application of logging and silviculture regulations, recreation standards, and road construction standards, and development and continuation of efforts to evaluate buffer zones for undergrowth vegetation. Public support of the Apache trout program should be further developed. An explanatory pamphlet covering the laws and regulations concerning Apache trout should be distributed to the fishing and conservation public.

Streams and impoundments capable of supporting Apache trout should be divided or categorized into those that (1) can support self-sustaining populations; (2) are suitable for put-grow-and-take fisheries; or (3) are suited for only put-and-take fisheries. Once categorized, management techniques will be applied accordingly. Streams and impoundments that offer high quality sportfishing experiences should be available to the fishing public. Fishing should continue to be regulated in a manner that provides a variety of fishing experiences for anglers. Management techniques may include a standard creel limit (e.g., 6–10 fish/d), trophy fishing, catch and release, and a combination of the three. Enforcement of all federal, state and tribal laws and regulations should be ensured to protect the Apache trout.

Discussion

When the above management goals have been achieved and viable fisheries created, the Apache trout should be delisted under the Endangered Species Act. Public demand for Apache trout as a game and food fish has already been strongly established.

Future research on Apache trout should investigate (1) habitat requirements of Apache and other trouts; (2) competition effects from other trouts on Apache trout; (3) genetics of unsampled populations and the efficacy of swamping; (4) reproductive performance of Apache trout; (5) growth dynamics of Apache trout in various habitats; (6) behavioral characteristics of Apache trout; (7) physiological needs and characteristics of Apache trout; and (8) chronic temperature and pH tolerances of Apache trout in various habitats.

The fisheries in Apache trout streams can serve as examples of the beneficial as well as the harmful uses and effects of cultured fishes on aquatic ecosystems. Past stocking of brown, rainbow, cutthroat, and brook trout created viable fisheries at the expense of Apache trout. The Apache trout can be delisted and viable fisheries can be managed by means of hatchery-reared Apache trout and nonnative salmonids. As knowledge and awareness of Apache trout genetics increases, resource managers can continue to use cultured fish to improve the prospects of Apache trout and provide quality fisheries. Resource managers cannot, however, rely solely on cultured fish. We must manage various altered habitats, and we must improve habitats for Apache trout and nonnative fisheries.

Acknowledgments

D. Parker, C. Williams, S. Leon, L. Wirtenen, R. David, J. Caid, and many game rangers from the White Mountain Apache Tribe provided invaluable support and assistance.

Endnotes

[1] Renovation consists of building physical barriers in streams to prevent upstream movement of other species, removing all fishes upstream of the barrier, usually with piscicides, and restocking desired fishes.

[2] Restoration is defined as management actions taken to improve habitat that has been negatively altered due to past actions, including renovations.

[3] Enhancement populations are created when Apache trout are stocked outside the critical streams. They are created to be used as recreational fisheries and to serve as gene pools in addition to populations in critical streams.

[4] Critical streams are defined as streams or habitat that contain nonhybridized, reproducing Apache trout populations, which are necessary for delisting under the Endangered Species Act.

[5] There are two Boggy Creeks in Arizona, one on FAIR and another on ASNF.

References

Carmichael, G. J., J. N. Hanson, M. E. Schmidt, and D. C. Morizot. 1993. Introgression among Apache, cutthroat, and rainbow trout in Arizona. Transactions of the American Fisheries Society 122:121–130.

David, R. E. 1990. Apache trout culture: an aid to restoration. Endangered Species Update 8. U.S. Fish and Wildlife Service, Washington, DC.

Dowling, T. E., and M. R. Childs. 1992. Impact of hybridization on a threatened trout of the southwestern United States. Conservation Biology 6:355–364.

Harper, K. C. 1978. Biology of a southwestern salmonid, *Salmo apache* (Miller 1972). Pages 99–111 in J. R. Moring, editor. Proceedings of the wild trout-catchable trout symposium. Oregon Department of Fish and Wildlife, Eugene.

IUBNC (International Union of Biochemistry, Nomenclature Committee). 1984. Enzyme Nomenclature. Academic Press, San Diego, California.

Loudenslager, E. J., J. N. Rinne, G. A. E. Gall, and R. E. David. 1986. Biochemical genetic studies of native Arizona and New Mexico trout. Southwestern Naturalist 31:221–234.

Miller, R. R., editor. 1969. Red data book IV—Pisces. International Union for Conservation of Nature and Natural Resources, World Conservation Union, Morges, Switzerland.

Miller, R. R. 1972. Classification of the native trouts of Arizona with the description of a new species, *Salmo apache*. Copeia 1972:401–422.

Morizot, D. C., and M. E. Schmidt. 1990. Starch gel electrophoresis and histochemical visualization of proteins. Pages 23–80 in D. H. Whitmore, editor. Electrophoretic and isoelectric focusing techniques in fisheries management. CRC Press, Boca Raton, Florida.

Rinne, J. N. 1985. Variation in Apache trout populations in the White Mountains Arizona. North American Journal of Fisheries Management 5:1465–158.

Rinne, J. N., and W. L. Minckley. 1985. Patterns of variation and distribution in Apache trout (*Salmo apache*) relative to co-occurrence with introduced salmonids. Copeia 1985:285–292.

Rinne J. N., W. L. Minckley and J. N. Hanson. 1981. Chemical treatment of Ord Creek, Apache County, Arizona, to re-establish Arizona trout. Journal of the Arizona–Nevada Academy of Science 16:74–78.

USFS (U.S. Forest Service). 1993a. Decision notice and finding of no significant impact, revised allotment management plan (West Fork Allotment) Apache County Arizona. USFS, Alpine Ranger District, Alpine, Arizona.

USFS (U.S. Forest Service). 1993b. Decision notice on a finding of no significant impact, West Fork of the Black River watershed and fisheries restoration projects and Burro Creek/Hayground/Reservation allotments, allotment management plans, Apache County, Arizona. USFS, Springerville Ranger District, Springerville, Arizona.

USFWS (U.S. Fish and Wildlife Service). 1979. Recovery plan for Arizona trout, *Salmo apache*, Miller, 1972. USFWS, Albuquerque, New Mexico.

USFWS (U.S. Fish and Wildlife Service). 1983. Recovery plan for Arizona trout, *Salmo apache*, Miller, 1972. USFWS, Albuquerque, New Mexico.

A Restoration Plan for Pez Blanco in Lake Patzcuaro, Mexico

ARTURO CHACON-TORRES

Escuela de Biología
Universidad Michoacána de San Nicolás de Hidalgo
Michoacán, Mexico

CATALINA ROSAS-MONGE

Dirección General de Pesca
Gobierno del Estado de Michoacán, Michoacán, Mexico

Abstract.—Lake Patzcuaro, Mexico, supports a typical artisanal fishery that involves numerous small-scale operations, a large need of human labor, low capital costs, and rudimentary technology. The fish fauna in Lake Patzcuaro consists of 10 native and 4 nonnative species. The native fauna includes species of Atherinidae, Cyprinidae, and Goodeidae. Atherinids are the most commercially important fish group of Central Mexico. Management of the fishery for pez blanco (a Mexican silverside) *Chirostoma estor estor*, the most valuable fish product in the country (US$25/kg), must consider ecological, socioeconomic, and technological issues. Ecological factors include the piscivorous role of pez blanco in the aquatic ecosystem. Environmental deterioration and the introduction of nonnative species, such as largemouth bass *Micropterus salmoides*, common carp *Cyprinus carpio*, grass carp *Ctenopharyngodon idella*, and tilapias *Oreochromis* spp., have modified habitat quality and trophic relationships. Socioeconomic factors include the role of the fishery in the local economy, market relationships, and the social organization of P'urhépecha Indian communities. Overfishing, including the use of small-mesh nets that capture smaller silverside species and juveniles of pez blanco, omission of closed seasons, and resource mismanagement have contributed to the collapse of the pez blanco fishery. It is necessary to develop a fishery management program that includes habitat restoration, closed fishing seasons, and elimination of nonselective gear. Simultaneously, a program to increase regional aquaculture and restocking of pez blanco is needed. Aquaculture activities should include artificial fertilization, incubation of eggs in floating cages, and grow-out in artificial channels confluent with the lake. These activities can allow continued harvest and prevent resource extinction.

Current knowledge of Mexican lakes and reservoirs and their biodiversity is limited and fragmentary. The area of Mexican freshwater lakes and reservoirs is approximately 1,200,000 ha (INEGI 1991); most of these bodies are sources for drinking water, irrigation, hydroelectric power, and fish production. Many of these lakes and reservoirs have been severely affected by humans. Disturbance and degradation of waters and their watersheds and overexploitation of fishery resources have negatively affected rural economies and food production. Processes responsible for environmental degradation and resource deterioration in these inland waters are poorly understood, and little information is available to provide a basis for strategic management plans for these threatened environments, most of which contain unique faunas of ecological and commercial importance (Chacon-Torres 1993).

Lake Patzcuaro is one of the most important lakes in Mexico. Its native fish fauna and the P'urhépecha Indian settlements around its shores make the lake an important and valuable resource. However, the lake has become severely degraded. This lake represents the habitat of pez blanco (a Mexican silverside) *Chirostoma estor estor*, an endemic species with high commercial value. Its natural populations are surviving under conditions of intensive overfishing, ecological competition, and habitat deterioration. Andrade (1990), Ledesma (1990), and Oseguera (1990) found hybridization among four species of *Chirostoma* inhabiting Lake Patzcuaro and suggested that habitat deterioration could possibly encourage the hybridization of wild populations.

This paper presents analyses of factors affecting the fishery of pez blanco and discusses strategies for species restoration and fisheries management.

Study Area

Lake Patzcuaro is a tropical, high-altitude (2,035 m), freshwater lake (130 km^2 surface area) located on the Mexican plateau in west-central Mexico. The lake has no outlet; inflow is from temporary streams during the rainy season. Maximum depth is 12 m in the northern part of the lake; extensive shallow areas are developing in the southern part (Chacon-Torres 1993).

The annual water balance is controlled by differences among rainfall, seepage, watershed runoff, and evaporation. A mass-balance analysis indicates that hydrological inputs from the watershed are being reduced as a result of watershed deterioration. High turbidities result from volcanic silt, raw sewage, and increasing erosion loads to the lake. Lake Patzcuaro is light limited due to scattering and attenuation from suspended inorganic materials. The lake has high concentrations of magnesium, sodium, carbonates, total phosphorus, chlorophyll a, and suspended solids (Chacon-Torres 1993).

Lake Patzcuaro is eutrophic with some less productive areas. Proposed lake management strategies focus on the watershed and include erosion control and reforestation, basinwide sewer systems and wastewater treatment, and preservation of the native fauna (Chacon-Torres 1993).

Pez Blanco Biology

Pez blanco is adapted to limnetic waters rich in oxygen (5.0–7.0 mg/L). This species has a relatively low temperature tolerance (15–23°C). Pez blanco occupies neutral or slightly alkaline waters (pH 7.0–8.5) with low ammonia concentration (<0.125 mg/L), low suspended solids (<120 mg/L), and a low concentration of organic matter (5 d biological oxygen demand <5.0 mg/L) (A. Chacon-Torres, unpublished data).

Small individuals (<150 mm standard length) primarily feed on littoral zone invertebrates, including cladocerans, ostracods, insects, and crustaceans such as *Hyallela azteca* and *Cambarellus montezumae*; small fish are 25% of their diet (Garcia 1984). Larger individuals (>150 mm standard length) primarily are piscivorous. Other food items in the diet include chironomids, crustaceans, and other invertebrates. Pez blanco feeds in limnetic waters, particularly those close to submerged vegetation (Garcia 1984).

Although Garcia (1984) indicated that the pez blanco spawns throughout the year, we found the highest spawning activity during the end of winter and beginning of spring (January–June), with a peak between February and April. The species is a multiple spawner and becomes sexually mature at age 2–3; the minimum standard length at sexual maturity is 102 mm for males and 140 mm for females. Fecundity increases with fish size (Figure 1).

Sexually mature fish move inshore and spawn in littoral areas. Eggs are deposited on filamentous green algae along the shoreline that is exposed to wave action and oxygenation. The upper tempera-

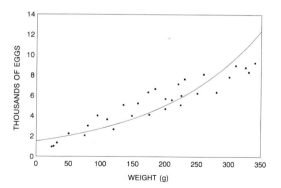

FIGURE 1.—Relationship between fecundity and total fish weight for pez blanco.

ture limit for normal embryonic development is near 24°C. Incubation is about 210 h at 20°C, and larvae are about 1.5 mm long at hatching. The yolk sac is absorbed when the larvae are 2.5 mm (4 d after hatching at 20–22°C) (Morelos et al. 1994).

Lake Patzcuaro Fishery

For centuries fishing activity has been the basis of the subsistence economy of the P'urhépecha Indian villages surrounding Lake Patzcuaro. Traditionally, the fishery has been based on the capture of endemic species, including pez blanco, "charales" (a suite of smaller silversides *Chirostoma grandocule*, *C. attenuatum*, and *C. patzcuaro*), acumara *Algansea lacustris*, and the neotenic amphibian achoque *Ambystoma dumerilii*. However, the yields of these native species have greatly decreased in comparison with capture of introduced species. Thus, native species, which are more favored by regional consumers, are declining and becoming endangered.

The average annual catch between 1980 and 1993 was 1,239 metric tons (95.3 kg/ha). This fishery was composed of the exotic common carp *Cyprinus carpio* and grass carp *Ctenopharyngodon idella* (21.4% of weight harvested), acumara (19.7%), charales (19.5%), largemouth bass *Micropterus salmoides* (16.6%), tilapias *Tilapia* spp. (9.3%), "chehuas" (a suite of native goodeids *Goodea atripinnis*, *Allophorus robustus*, *Neoophorus diazi*, *Allotoca vivipara*, and *Skiffia lermae*; 7.9%); and pez blanco (5.6%). Pez blanco has the highest market value (up to US$25/kg). Since 1992, the largemouth bass fishery has collapsed with a maximum annual catch of less than 1 metric ton. Reasons for this decline are not totally understood; however, habitat destruction by dredging, competition with other exotic species, and deterioration of lake water quality are potential causes for the collapse of the largemouth bass fishery.

The commercial fishing operations on Lake Patzcuaro are sustained by 24 communities located around its shores and islands. In 1981, an estimated 1,100 people fished on Lake Patzcuaro; by 1991 that estimate had increased to 1,500. Some people combine their fishing activities with agriculture, handcrafting, and trade. Since fishing is not a full-time activity in some villages, there are uncertainties about the total number of people who use the lake, which makes evaluation of the total fishing effort difficult.

Fishing methods include bag seines, gill nets, and dip nets. Seines, known as "chinchorros," typically are 200–300 m long and 5 m deep and have 1–2-cm-square mesh. The middle section of chinchorros is made of smaller mesh in the form of a pocket, or "bolsa." Gill nets, or "cheremekuas," are made of two mesh sizes—a small (0.5–0.8-cm-square) mesh for the charales and a larger (up to 8 cm) mesh for pez blanco and largemouth bass. Butterfly-shaped dip nets are known as "mariposa" or "guaromutacua." Although fishing with the guaromutacua was a common practice in the past, this net has almost been abandoned for more efficient fishing gear. Other fishing methods in the lake include rod-and-reel and spearing with the "atlatl" or "fisga."

We estimated that there were 17,250 gill nets and 130 seines being used in Lake Patzcuaro in 1993. If the average gill net is 50 m long and the seine is 300 m, the lake is potentially fished with 876 km of gill nets and 26 km of seines.

There are two different types of fishing boats—artisanal canoes ("cayucos") made of wood, and fiberglass boats with outboard motors. In 1993 there were about 1,890 cayucos and 4 fiberglass boats used on Lake Patzcuaro.

Increased fishing effort also has resulted from the change from cotton to nylon nets. The nylon nets can be used more frequently and for longer time periods. Consequently, gill nets can remain in the water for days and even for weeks.

A large increase in fish catch has been observed during the last decade (Figure 2). This may be the result of the establishment of exotic species of carp and tilapia, as well as increased fishing efficiency made possible by nylon nets with small mesh size. However, since 1990 there has also been a decline in fish yields, a reduction in fish size, and a lower frequency of native fish species in the catch. This suggests that local fish populations are currently

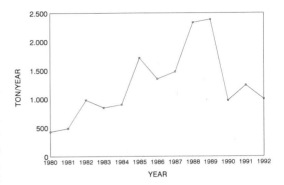

FIGURE 2.—Total fish catch in Lake Patzcuaro, Mexico, 1980–1992.

overexploited and that fishing effort is beyond the sustainable maximum.

The trend in the fishery of the pez blanco indicates a progressive decline over time. The maximum recorded annual harvest was 113 metric tons in 1982. In 1993 the harvest was less than 15 metric tons, a reduction of over 95 metric tons. Pez blanco was 11.5% of the total harvest in 1982 but less than 1.5% in 1993.

A regression analysis using 14 years of fish catch data indicates a possible collapse of the fishery by the end of the twentieth century (Figure 3). Economic value of the pez blanco fishery in 1982 was approximately $2,260,000; in 1993 it had declined to only $300,000. This represents an economic loss of about $2 million.

Some of the reasons for the decline of pez blanco in Lake Patzcuaro are habitat deterioration, competition with exotic species, overfishing, omission of

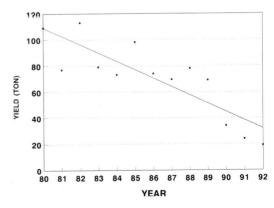

FIGURE 3.—Decline of the pez blanco fishery in Lake Patzcuaro, Mexico, 1980–1992. Yield = $12,845 - 6.43$ (year); $r^2 = 0.70$; $F = 28.82$; $P < 0.0001$.

closed seasons, and harvest of juvenile pez blanco using small (<2.0-cm-square) mesh nets.

Given the high commercial value and the progressive decline of captures of pez blanco, attempts to reproduce and culture this species have been made but only on an experimental scale (Armijo and Sasso 1979; Rosas 1970; Lara 1974). There have been several attempts to elucidate the basic biology of some of the native species in Lake Patzcuaro, but most of these studies have been insufficient and isolated. Hence, the biology of the native fish fauna of the lake area remains poorly understood, and the potential for aquaculture of these native species is restricted. The slow growth rate, nutritional problems, high mortality rate during larval stages, and pelagic behavior of pez blanco have been identified as the major reasons for the lack of progress.

Mexican freshwater aquaculture, which remains a small-scale enterprise, has been oriented mainly to the culture of exotic species such as rainbow trout *Oncorhynchus mykiss*, channel catfish *Ictalurus punctatus*, carps, and tilapias. However, the culture of these species does not satisfy the demand of the majority of Mexicans for fish protein. Rainbow trout and channel catfish are cash-crop species that are either consumed by a minority of upper social classes or exported to U.S. markets, and carp and tilapia are rarely included in the diet of the Mexican people as a function of tradition. Hence there is a need for development of an appropriate fish farming scheme to meet the national needs. This fish farming program must be designed in accordance with cultural preferences of the people and the potential size of the local market. An example of the culture of Mexican native fish species has been provided recently by the Centro de Investigaciones y de Estudios Avanzados del Instituto Politecnico Nacional in southeast Mexico, where the Mexican "mojarra" (Mayan cichlid *Cichlasoma urophthalmus*) has been integrated successfully into regional aquaculture (Martinez 1987; Ross 1988; Ross and Martinez 1990).

A Plan for Restoration

Because the Lake Patzcuaro regional economy depends greatly on fish productivity in the lake, development of reliable sources of fry of the native fish species is important. This would be of value not only for the restoration of an endangered natural resource, but also for the use of these valuable native fish species in a regional aquaculture program. At present, given the increasing eutrophication of the lake and the effects of introduced species, aquaculture programs for restocking purposes are recommended under an appropriate habitat restoration and fishery management strategy.

Because the pez blanco fishery has high social and economic values, a program oriented towards the restoration and experimental aquaculture of this species was established. This program includes obtaining necessary biological information about pez blanco. We evaluated hatching and survival rates at different incubation temperatures in three different culture systems—conventional aquaria, water recirculation systems, and floating net cages. An incubation temperature of 20°C in the water recirculation system resulted in up to 95% survival of pez blanco fry. A complete description of embryonic development obtained during these trials (Morelos et al. 1994) will be useful for future identification of embryonic stages of wild and cultured pez blanco.

We are now studying larval feeding and growth in floating net cages. Groups of pez blanco fingerlings are fed five different diets: wild zooplankton from Lake Patzcuaro, high-protein commercial salmon starter, an experimental artificial diet, a local commercial artificial diet for rainbow trout, and nauplii and adults of *Artemia* spp. All diets are particle size graded to be compatible with the size of the pez blanco mouth.

The Michoacan Directory of Fisheries has established a portion of Lake Patzcuaro as an experimental fish reserve to recover the pez blanco and other valuable native species. Pez blanco research activities at this reserve include artificial fertilization using wild broodstock; incubation in floating cages using water hyacinth *Eichornia* spp. as a substrate for fertilized eggs; feeding of larvae and fry with natural and artificial food until they reach 5–7 cm in length; and release of juveniles to the environment to increase recruitment. To date, we have attained an egg to fry survival rate of 75%; causes of mortality are fungi *Saprolegnia* spp. infestations, copepod and insect predation, and occasional poor water quality.

Several habitat restoration actions were initiated in the area of Lake Patzcuaro that was to become the fish reserve. Submerged aquatic vegetation was removed to increase water circulation and oxygenation; the area was dredged to increase water depth and remove excess organic matter; and channels confluent with the lake were constructed and closed with nets at each end. These channels will be stocked with pez blanco fingerlings for grow-out and domestic broodstock production as well as for reintroduction.

It is necessary to protect breeding pez blanco from harvest to increase recruitment. We will do this by restricting fishing at spawning sites. These sites will be located and mapped by use of water quality data and remote sensing, specifically satellite multispectral imagery. Fishing permits will be issued to people involved with recovery actions such as artificial fertilization, hatchery activities, and juvenile grow-out programs.

Regional aquaculture can be established by using floating cages and earthen ponds close to the shoreline. To prevent hardship and discouragement among P'urhépecha Indian communities, the controlled rearing of easily cultured exotic species like carp and tilapias could be combined with that of native species to ensure, simultaneously, fish production during closed seasons and native fish species restoration.

References

Andrade, T. E. 1990. Desarrollo embrionario y larval de *Chirostoma patzcuaro* Meek 1902 y de los hibridos obtenidos por fecundacion artificial con *Chirostoma grandocule* Steindachner 1894 (Pisces: Atherinidae) del lago de Patzcuaro, Michoacan, Mexico. Bachelor dissertation, Escuela de Biologia, Universidad Michoacana de San Nicolas de Hidalgo, Michoacan, Mexico.

Armijo, O. A., and L. Sasso. 1979. Observaciones preliminares en acuarios sobre incubación y alevinaje de aterínidos (*Chirostoma* spp.) del lago de Patzcuaro. Pages 149–153 *in* R. S. W. Pullin and W. A. Dill, editors. Advances in Aquaculture. Fishing News Books, Farnham, Surrey, UK.

Chacon-Torres, A. 1993. Lake Patzcuaro, Mexico: watershed and water quality deterioration in a tropical high-altitude Latin American lake. Lake and Reservoir Management 8(1):37–47.

Garcia de, L. F. 1984. Ecologia pesquera, alimentacion y ciclo gonadico de *Chirostoma estor* Jordan y *Microp-*

terus salmoides Lacepede en el lago de Patzcuaro, Michoacan, Mexico. Bachelor's dissertation. Universidad Nacional Autonoma de Nuevo Leon, Nuevo Leon, Mexico.

INEGI. 1991. Datos basicos de la geografia de Mexico. Instituto Nacional de Estadistica, Geografia e Informatica, Mexico.

Lara, V. A. 1974. Aspectos de cultivo extensivo e intensivo del pescado blanco de Patzcuaro, *Chirostoma estor* Jordan 1879. Actas del Simposio sobre Acuicultura en America Latina. Documentos de Investigacion. Montevideo, Uruguay. FAO Informes de Pesca (159)1:113–116.

Ledesma, A. P. 1990. Analisis de las fases ontogenicas primarias y reconocimiento del hibrido obtenido por fecundacion artificial entre *Chirostoma attenuatum* y *Chirostoma patzcuaro* (Pisces: Atherinidae) del lago de Patzcuaro, Mexico. Bachelor's dissertation. Universidad Michoacana de San Nicolás de Hidalgo, Michoacán, Mexico.

Martinez, P. C. A. 1987. Aspects of the biology of *Cichlasoma urophtalmus* (Gunther) with particular reference to its culture. Doctoral dissertation. University of Stirling, Stirling, UK.

Morelos, L. M. G., V. Segura, and A. Chacon. 1994. Desarrollo embrionario del pez blanco de Patzcuaro *Chirostoma estor estor* 1879 (Pisces: Atherinidae). Rev. Zoología Informa, Instituto Politécnico Nacional, Mexico 8(27):22–46.

Oseguera, F. L. 1990. Caracterizacion morfologica de estadios embrionarios y juveniles de *Chirostoma grandocule* Steindachner (1896) y *Chirostoma attenuatum* Meek (1902) del lago de Patzcuaro, Michoacan, Mexico. Bachelor's dissertation. Universidad Michoacana de San Nicolás de Hidalgo, Michoacán, Mexico.

Rosas, M. M. 1970. Pescado blanco (*Chirostoma estor*) su fomento y cultivo en Mexico. Series de Divulgacion del Instituto Nacional de Pesca, Mexico.

Ross, L. G. 1988. *Cichlasoma urophtalmus* aquaculture. Cichlidae (Nottingham) 9(3):45–49.

Ross L. G., and C. A. Martinez. 1990. The biology and culture of *Cichlasoma urophtalmus*. A technical manual. Overseas Development Administration and Centro de Investigaciones y de Estudios Avanzados del Instituto Politecnico Nacional, UK.

A Program for Maintaining the Razorback Sucker in Lake Mohave

GORDON MUELLER

National Biological Service, Post Office Box 25007, Denver, Colorado 80225, USA

Abstract.—Lake Mohave, Arizona–Nevada, supports the last large population of the Endangered razorback sucker *Xyrauchen texanus*. Razorback suckers successfully spawn in Lake Mohave; however, predation by nonnative fish appears to restrict recruitment. Most razorback suckers are believed to be more than 40 years old and nearing the end of their life span. The population is expected to perish within the next few years unless steps are taken to ensure survival and recruitment of young.

Concerned biologists from seven state and federal agencies formed the Native Fish Work Group (NFWG) to maintain the razorback sucker in Lake Mohave. The NFWG has developed a program to replace the aging population with young adults that reflect the genetic divergence of the reservoir population. Larval fish spawned naturally in the reservoir are collected and stocked by themselves into small, isolated nursery areas. Fish are grown in these predator-free environments to 25 cm, a length believed large enough to evade most predators. The NFWG plans to release a minimum of 10,000 young adult razorback suckers back into Lake Mohave and monitor their survival. This represents the first step in a long-term management commitment.

Populations of razorback suckers *Xyrauchen texanus* are now restricted to less than 25% of their former range (Minckley et al. 1991). Nearly all razorback suckers collected during the past 2 decades have been adults. These fish are nearing extinction in the upper Colorado River basin (Tyus 1987; Osmundson and Kaeding 1991; USFWS 1991). The largest remaining population of razorback suckers is found in Lake Mohave, Arizona–Nevada, a regulatory reservoir formed by Davis Dam. These fish are old, and, as in the more pristine upper Colorado River basin, reproduction occurs but few young survive to reach adulthood. The Lake Mohave population has declined 60% (59,500 to 23,300) in the past 5 years (Marsh 1994). Populations elsewhere are also declining and extinction will occur this decade unless steps are taken immediately to augment populations (Minckley et al. 1991; Burdick 1992).

The species decline is attributed to habitat degradation and competition and predation by nonnative species. Razorback suckers do successfully spawn at several locations throughout the basin. Eggs incubate and hatch and larval razorback suckers are produced, but young longer than 25 mm total length (TL) rarely are found. Today young razorback suckers are vulnerable to a large host of predators not present a century ago. Over 40 nonnative fish species have been successfully introduced into the Colorado River basin. Implications of these introductions are difficult to quantify; however, Marsh and Brooks (1989) showed that small razorback suckers stocked in Arizona streams were eaten by channel catfish *Ictalurus punctatus* and other nonnative species. Predation on larval razorback suckers, although not quantified, is considered the single most important factor limiting recruitment in Lake Mohave (Minckley 1983; Marsh and Langhorst 1988; Minckley et al. 1991).

Efforts to recover the razorback sucker and other native fishes began in 1976 with the creation of the Colorado River Fishes Recovery Team (K. D. Miller 1982). The Colorado River Fishery Project was initiated by the U.S. Fish and Wildlife Service (USFWS) in 1979. Its purpose, along with a substantially larger recovery program for the upper Colorado River basin that followed in 1987, was to recover the razorback sucker, bonytail *Gila elegans*, humpback chub *G. cypha*, and Colorado squawfish *Ptychocheilus lucius* in a manner that allows further water development. The cost for the remaining 10 years of the 15-year recovery program is estimated to be US$84–134 million (USFWS 1993). A cooperative agreement between USFWS and Arizona Game and Fish Department delayed listing of the razorback sucker in order to reintroduce nearly 12 million small razorback suckers into Arizona streams between 1980 and 1989 (Johnson 1985; Minckley et al. 1991). This massive stocking attempt was unsuccessful in reestablishing self-sustaining populations, and the razorback sucker was listed as Endangered in 1990.

Unfortunately, 18 years of recovery effort have failed to slow, let alone reverse, the decline of this native fish (Tyus 1987; Marsh and Brooks 1989; Langhorst 1989; Osmundson and Kaeding 1991;

Minckley et al. 1991). Much effort has gone into research, habitat restoration, and legislative protection. Furthermore, recovery efforts have generally focused on the upper Colorado River basin where the razorback sucker is rare; little has been done to actively manage surviving populations or repress nonnative fish communities.

The continual decline of the razorback sucker prompted Lake Mohave field biologists to adopt an active management effort to maintain razorback suckers in Lake Mohave. The group adopted a quick response management approach, rather than developing a more conventional recovery program. The goal was simple: to replace Lake Mohave's old razorback sucker population before the relict population was lost. The purpose of this paper is to describe the development, operation, and results of a cooperative program to maintain the razorback sucker in Lake Mohave.

Approach

Cooperative Partnership

The Native Fish Work Group (NFWG) is a cooperative effort of the Arizona Game and Fish Department, Arizona State University, Bureau of Reclamation, USFWS, National Biological Service, National Park Service, and Nevada Division of Wildlife. The NFWG and the program to save the Lake Mohave razorback sucker were conceived, created, and implemented at the field level. Participants are local biologists, many directly responsible for managing the Lake Mohave resource. The NFWG has no specific budget; work is accomplished through the collective resources of participating agencies. The NFWG meets every 4 months, or as needed, to review past work and to plan and assign future activities. Decisions are made at the lowest possible management level. Tasks are assigned according to available expertise, resources, and function. For instance, environmental compliance and permits are usually handled by the National Park Service, Endangered Species Act (16 U.S.C.A. §§1531 to 1544) compliance is completed by the USFWS, construction activities are managed by the Bureau of Reclamation, and Arizona State University and the National Biological Service assist with research needs. Monitoring activities, fish collection, and site maintenance are generally accomplished by personnel from all the agencies.

The NFWG first drafted a research and management plan in 1990 that outlined the goals and potential methods for sustaining the Lake Mohave razorback sucker population. The plan is a working document that identifies and prioritizes goals while allowing for implementation flexibility. The plan is periodically updated as needed. Annual and specific task reports provide information regarding activities. The program has focused resources on actual implementation rather than administrative processes.

Factors influencing the program are limited resources, the rapid decline in razorback sucker numbers, and time. The existing population is dying of old age and could perish by the end of this decade. The program is being expedited by using the reproductive potential of thousands of reservoir spawners rather than the conventional culturing practice of mass-producing fish from a small captive broodstock.

Management Concept

Habitat degradation in the lower Colorado River basin has been extensive, but research suggests predation is the single most important factor for recruitment failure in Lake Mohave (Minckley 1983; Marsh and Langhorst 1988; Minckley et al. 1991). Razorback suckers do successfully spawn in reservoirs and other lentic bodies of water. Eggs incubate and hatch, but young razorback suckers apparently survive only in environments where nonnative predators are absent or rare (Minckley et al. 1991). Razorback suckers once flourished in several newly impounded reservoirs in the lower Colorado River basin (Minckley et al. 1991). Apparently, populations of razorback suckers were able to expand while reservoirs filled and before nonnative predators became established.

The ecology of these fish is not clearly understood. However, razorback suckers do spawn with the rising water from spring runoff. Prior to the channelization of the Colorado River and the creation of large reservoirs, larval razorback suckers were dispersed into large, newly flooded, and highly productive nursery areas (Minckley et al. 1991). Survival may have depended on a combination of high spawner fecundity, rapid larval growth, and the dispersal of a naturally small predator community contained by seasonal low river conditions. Today, conditions are much different. Reservoirs have inundated seasonally flooded nursery areas and have modified, but also expanded and stabilized, aquatic habitats. The early colonization of many of the lower Colorado River reservoirs by razorback suckers suggests the fish could tolerate some physical habitat changes. However, it appears the introduction and management of sport fishes may have been the decisive factor in the razorback sucker's decline.

Options available to the NFWG were few. Meaningful habitat restoration or the removal of nonnative fish from Lake Mohave and other upstream portions of the river basin are unlikely. We thought the only feasible approach was to stock fish large enough to ensure some survival. Razorback suckers are reared successfully and grow rapidly. Unfortunately, the costs of raising and transporting thousands of large razorback suckers from Dexter National Fish Hatchery (1,100 km) or other existing fish culture facilities would be high. Experiments conducted in the mid-1980s showed that razorback suckers spawned and produced young if provided habitats free of nonnative fishes (Minckley et al. 1991). We believed a razorback sucker of 25–30 cm TL was large enough to escape most predators, and we chose that size as the target for stocking. Fish would be grown to this length, implanted with passive integrated transponder (PIT) tags, and released into the reservoir. A minimum of 10,000 young razorback suckers would be released over 5 years. Arizona State University has conducted an ongoing reservoir monitoring program for nearly 20 years, and this continuing monitoring program will provide the information necessary to determine stocked fish survival. Stocking will be refined as data become available.

Development of Rearing Areas

Aerial photographs and surveys identified several potential rearing sites along Lake Mohave's shoreline. The reservoir has several naturally occurring backwaters that seasonally become isolated from the main reservoir. These ephemeral backwaters and coves that easily could be closed were visited, prioritized, and surveyed. Few coves were suitable for rearing because of their size or the expense to make them suitable for rearing razorback suckers. Alternative rearing sites away from the reservoir were also examined. Five specific types of potential rearing areas have been identified or are being used. They include ephemeral backwaters, backwaters closed by net, permanent backwaters, hatchery facilities, and outlying ponds.

Since 1990, nine small ephemeral backwaters (0.1–0.8 ha) have been developed as seasonal rearing areas. These shallow backwaters are seasonally closed to the main reservoir by gravel berms formed by wave-induced beach erosion and deposition. Berms prevent fish passage, but backwaters remain hydraulically connected to the reservoir. Backwaters are seasonally flooded and drained as the reservoir is operated within its 5-m-vertical-fluctuation zone. Backwaters are normally flooded from spring through late summer and drained in autumn as the reservoir is lowered to develop storage for spring runoff. Drainage prevents the establishment of resident fish populations and assists in the recovery of reared fish.

Growth-rate information provided by Dexter National Fish Hatchery suggested 25-cm razorback suckers could be raised in 18 to 24 months; therefore, permanent (year-round) rearing areas were desirable. Ephemeral backwaters would be used for rapid seasonal growth, but fish less than 25 cm would have to be moved to a deeper, permanent-water cove in autumn when the reservoir elevation dropped. Davis Cove provided a potentially excellent rearing site because of its size, shape, and depth. The cove has a maximum depth of 4 m at minimum reservoir pool, a narrow entrance, and a surface area of 1.2 ha. The cove's entrance was closed in 1991 using a 10-mm-bar-mesh net and a flotation boom. The net conformed to the entrance and accommodated the reservoir's 5-m fluctuation zone. This temporary barrier was used for 3 years and was replaced with an earthen berm in fall 1994.

Willow Beach National Fish Hatchery is a cold-water trout hatchery located on Lake Mohave, 16 km downstream from Hoover Dam. The facility is currently adding water heaters that will allow the hatchery to hatch and rear razorback sucker. Raceway operations are being modified to rear 5–15-cm (TL) razorback suckers. We plan to test the concept of stocking backwaters with larger juveniles in order to reduce the number of fish smaller than 25 cm at the end of the growing season. This approach could help optimize survival, production, and fish growth while reducing the need to transfer, hold, and handle fish smaller than 25 cm.

Razorback suckers have been reared successfully elsewhere (Minckley et al. 1991). The option of rearing fish away from Lake Mohave in city park ponds, golf course ponds, and housing development lakes is being explored. A Memorandum of Agreement was recently signed between the Bureau of Reclamation and City Manager of Boulder City, Nevada. Under the agreement, razorback suckers have been stocked in a municipal golf course pond.

Propagation and Rearing

The Lake Mohave project is unique among endangered fish stocking programs because of the range of methods available to produce young fish.

Some methods optimize production, whereas other techniques produce fewer fish but focus on genetic diversity. The following propagation methods have been or currently are being tested.

(1) Backwaters were stocked with mature adults from the reservoir prior to the spawning season. Adults were stocked at a ratio of one female to two males. Males were stocked in larger numbers because females normally spawn with multiple males at one time (Minckley 1973). Adults were recaptured and returned to Lake Mohave following spawning. A different spawning group was used each year.

(2) Another approach was examined to reduce spawner handling and stress associated with allowing fish to spawn in the backwaters and eventually recapturing and returning them to the reservoir. Spawning fish were collected from Lake Mohave in 1993, gametes were stripped, eggs were fertilized, and fish were returned to the reservoir. The egg contribution from each female was fertilized by at least two males. About 250,000 eggs from 24 females and about 60 males were collected; 200,000 were dispersed in Yuma Cove, and the remainder were hatched in the laboratory for larval research. Unfortunately, the reservoir dropped about 1 m shortly after the eggs were distributed in Yuma Cove.

(3) Larvae are phototactic and can be captured easily at night by using lights. Naturally spawned reservoir larvae have been collected for a number of years using handheld spotlights and small dip nets or light traps (Minckley et al. 1991; Mueller et al. 1993). These techniques are being refined to collect large numbers of larval razorback suckers for stocking into the backwaters.

(4) Dexter National Fish Hatchery is the USFWS warmwater, endangered-fish culturing facility. The hatchery maintains a broodstock of Lake Mohave razorback sucker and produces fish for scientific research and stocking. The hatchery supplied 10,000, 6-cm (TL) fish for Davis Cove on 28 June 1992.

All backwaters except Davis Cove were poisoned to remove nonnative fish. Davis Cove was not poisoned due to its close proximity to a large public marina and in order to test if razorback suckers could survive in a cove where nonnative fish were reduced but not eliminated. Here the nonnative fish community was reduced using gill and trammel nets and by electrofishing.

Stocking Criteria

Lake Mohave contains several piscivores, including striped bass *Morone saxatilis*, largemouth bass *Micropterus salmoides*, bluegill *Lepomis macrochirus*, and channel catfish. The stocking-size criterion was based on the assumption that most piscivores would have difficulty consuming a 25-cm razorback sucker. This criterion will be refined as survival rates of recaptured fish are developed. Currently, razorback suckers longer than 25 cm that are taken from the rearing areas are tagged with PIT tags and stocked directly into the reservoir. Smaller fish are moved to a permanent backwater and allowed to continue their growth.

Monitoring

Larval and juvenile fish are collected using lights and small dip nets, light traps, small-mesh seines, fish traps, small-mesh trammel nets, and electrofishing. Ephemeral backwaters and Davis Cove are sampled bimonthly to monitor fish survival, growth, and condition. Piscivorous fish collected in Davis Cove during sampling events are removed.

Arizona State University has been monitoring the reservoir's razorback sucker population for nearly 20 years. The reservoir is sampled specifically for razorback suckers during spring and Thanksgiving class breaks. This effort has been intensified during the past 8 years to estimate the population size. Sampling is conducted in March when spawners are concentrated in shallow water and are vulnerable to trammel nets, large beach seines, and electrofishing. Usually 1,000 to 2,000 adults are collected and tagged with PIT tags annually. The spring and fall monitoring efforts are providing the information required to determine the survival of backwater-reared fish.

Results

Propagation and Rearing

In January 1991, 100 (33 females, 67 males) adult razorback suckers were stocked into Yuma Cove just prior to spawning season. Larvae were collected during the spawn; however, for reasons unknown, no young fish survived. A similar effort was attempted in January 1992 with 88 adults (28 females and 60 males), and 296 young fish were recovered (Table 1). Juveniles collected in November 1992 averaged 35.4 cm TL (maximum 39.1 cm). One hundred fifty-three fish larger than 25 cm were tagged with PIT tags and released in Lake Mohave; the remainder (<25 cm) were placed in Davis Cove to continue their growth. Fifteen juveniles were sacrificed for genetic (mitochondrial DNA and allozyme) analyses to determine the number of females that actually contributed to the 1992 year-class. Results showed that although 28 females were introduced, the 15 juveniles were produced by 5 females, and 8 were produced by a single female (T. Dowling, Arizona State University, personal communication).

The 1993 fertilized egg experiment produced 17

TABLE 1.—Razorback sucker production and survival in ephemeral backwaters in Lake Mohave, Arizona–Nevada, during 1991–1994. The number stocked represents larvae unless otherwise noted.

Year and location	Surface area[a] (ha)	Number stocked (size, cm)	Number juveniles harvested (% survival)	Yield (number/ha)	Average size (cm) at harvest[b]
1991					
Yuma Cove	0.82	100 spawners	0	0	
1992					
Yuma Cove	0.82	88 spawners	296	361	35.4
1993					
Yuma Cove	0.82	200,000 eggs	17 (<0)	21	32
		420 (2.6)	386 (92)	470	23.2
Willow Cove	0.17	500 (1.5)	26 (5)	153	21.9
Nevada Larvae	0.10	2,000 (1.1)	250 (13)	2,500	13.8
Arizona Juvenile	0.17	2,010 (1.5)	198 (10)	1,165	17.4
		69 (13.5)	55 (80)	324	30.7
1994					
Yuma Cove	0.82	3,000 (2)	358 (12)	407	32.0
Willow Cove	0.17	1,000 (2)	160 (16)	941	20.8
Dandy Cove		1,000 (2)	562 (56)		19.8
Nevada Larvae	0.10	500 (2)	217 (44)	2,190	22.4
North Chemhuevie	0.16	1,000 (2)	812 (81)	4,476	19.4
South Sidewinder	0.05	500 (2)	201 (40)	3,960	20.1

[a] Surface area at reservoir elevation 195 m.
[b] Harvest dates ranged from August to November.

juvenile razorback suckers. After the water level fell, 420, 2.6-cm-laboratory-reared fry were stocked into Yuma Cove. This stocking produced 386 juveniles; survival was 91.5% (Table 1). Three other coves were stocked with laboratory-reared fish of various size groups. Fish stocked at lengths of 1.1–1.5 cm showed a 5.2–12.5% survival rate; the survival of stocked fish larger than 2.5 cm was 80% (Table 1). Four hundred eighty-seven juvenile fish were tagged with PIT tags and released into the reservoir; the remaining smaller fish were stocked into Davis Cove.

Intensive larval collections from the reservoir began in January 1994. Larval razorback suckers were collected each week at multiple locations until the last week in March. The 3-month effort yielded over 11,000 larvae. Larvae were held in laboratory tanks and fed until they reached 2 cm TL and then distributed to all nine ephemeral backwaters. Stocking rates ranged from 4,000 to 10,000 larvae/ha. The USFWS received 2,000 larvae to rear for future broodstock at Dexter National Fish Hatchery.

Six of the nine backwaters produced young razorback suckers. The absence of razorback suckers in three backwaters was attributed to berm failure (reconnection to the reservoir) or poor water quality. Approximately 2,200 young suckers were harvested, tagged with PIT tags, and released into Lake Mohave in the fall of 1994. Survival rates for fish in backwaters ranged from 12 to 76% (Table 1).

Davis Cove was intensively sampled prior to stocking to remove resident fish. Electrofishing and gill netting harvested over 600 kg of nonnative fish (largemouth bass, striped bass, green sunfish *Lepomis cyanellus*, bluegill, channel catfish, yellow bullhead *Ameiurus natalis*, and common carp *Cyprinus carpio*) from the 1.2-ha cove. Scuba divers removed additional fish with spear guns and concluded that few large (>30-cm-long) predators remained. The cove was stocked on 24 June 1992, with 10,000 juvenile razorback suckers from Dexter National Fish Hatchery and 143 fish reared in an ephemeral backwater. Although the hatchery- and cove-reared fish were young-of-the-year, hatchery fish were about half the size (average 6.8 cm) of cove-reared fish (>12 cm). Predation of smaller hatchery fish by *Lepomis* sp. and juvenile largemouth bass was witnessed by divers. Subsequent sampling during the following 18 months yielded only 2% of the hatchery fish and over 50% of the larger cove-reared fish.

Recovery of Cove-Reared Fish

The spring monitoring effort recovered 5 of the 153 cove-reared fish released into Lake Mohave during autumn 1992 (Table 2). These fish were collected as far as 34 km from their release points. All five fish were males. Release of 487 fish in 1993 brought the total number of fish stocked into Lake Mohave to 640. Ten 28–52.9 cm (TL) cove-reared

TABLE 2.—Growth of and distance traveled by razorback suckers released into and recaptured from Lake Mohave during 1992 through 1994.

Year class and sex	Release		Recapture		Distance traveled (km)
	Date	Length (cm)	Date	Length (cm)	
1992					
Male	Jan 15, 1993	28.3	Mar 17, 1993	28.4	3
Male	Oct 15, 1992	35	Mar 17, 1994	52.9	8
Male	Oct 23, 1992	36	Mar 14, 1993	36.3	8
Male	Nov 23, 1992	36.5	Mar 15, 1993	38	24
Male	Nov 23, 1992	34.2	Mar 15, 1993	36.1	29
Male	Nov 23, 1992	37.5	Mar 19, 1993	38	34
Male	Nov 23, 1992	34.5	Mar 15, 1994	46.9	2
Male	Nov 23, 1992	35.2	Mar 15, 1994	46.9	2
Male	Nov 23, 1992	35.2	Mar 14, 1994	43.7	13
Male	Nov 24, 1992	34.6	Mar 15, 1994	48	2
Male	Nov 24, 1992	35.3	Mar 15, 1994	44.5	2
Male	Nov 25, 1992	37.1	Mar 14, 1994	49	13
1993					
Immature	Jul 8, 1993	28.5	Mar 14, 1994	36	4
Immature	Aug 18, 1993	36	Mar 17, 1994	36.8	4
Immature	Sep 21, 1993	24.3	Mar 15, 1994	28	2

fish were recaptured during the 1994 spawning period. The majority of fish collected in both 1992 and 1993 were spawning males (Gustafson 1975). One male taken in 1993 had reached adult size (50–55 cm) in just 2 years.

Discussion

The NFWG recognized, as others before us, that any attempt to manage an Endangered species should also adhere to social, political, and economic constraints (Johnson 1977; Wydoski 1977, 1982). Four basic issues and their relationship to the Lake Mohave razorback sucker population provided guidance for our activities: recovery, management considerations, genetics, and economics.

One of the first questions asked was, "Is recovery of the razorback sucker (i.e., attainment of a self-sustaining population) in Lake Mohave a realistic goal?" The team agreed recovery would be nice but not realistic. Recovery of the razorback sucker would require habitat restoration or at least changes in resource management. The water resource of the lower Colorado River and its associated politics would prevent any rapid modifications to existing river operations. Reservoirs would not be drained nor would dams be removed before the existing population disappeared. Evidence showed that dam removal would not be necessary if nonnative fish could be eliminated (Minckley 1983; Minckley et al. 1991). The current fishery of Lake Mohave, Lake Mead, other upstream reservoirs and the Colorado River is dominated by nonnative fishes that constitute a valuable recreational fishery. The removal of this biological component would be undesirable to the general public and logistically impossible to accomplish.

State and federal agencies have legislative commitments under the Endangered Species Act; however, often political and environmental issues concerning recovery are complex and controversial. Administrative processes, inaction justified by uncertainty, inadequate resources, and conflicting management goals often lead to slow or ineffective recovery programs (Rohlf 1991; Tear et al. 1993). The rapid decline of the Lake Mohave razorback sucker population was warning us that we no longer had the luxury of time for further debate or research. Our choices were simple: either continue the debate and monitor their decline, or, for the present, accept something less than total recovery by actively managing the species.

Minckley and Deacon (1991) recently pointed out that technology and resources are normally available to sustain or replace endangered populations. The critical question is whether the affected agencies have the political conviction to do so. We found the agencies would support an active Endangered Species management program as long as the effort did not unreasonably conflict with other resource management objectives. Any major conflict would demand formal coordination, consultation, and, most importantly, time necessary for resolution. Recognizing that time was our greatest enemy, we concluded our best chance to implement a maintenance effort was through an active stocking and management program. We were confident stocking 25-cm razorback suckers was feasible, but we had to identify where, by whom, and how these fish were going to be raised.

No one agency volunteered the funds or facilities to accomplish this task; it became a cooperative effort. We were unable to identify any existing culturing facilities that would raise tens of thousands of large razorback suckers, nor did we have the resources to build new facilities. As an alternative to conventional culturing, we proposed to develop a low-cost, on-site rearing program as a method of producing 10,000, 25-cm razorback suckers. We thought if this concept worked, it could be used to reintroduce and maintain other Threatened or Endangered, long-lived fish species in different reservoir and riverine habitats.

The debate over the method of producing genetically acceptable razorback suckers has been an evolutionary process. Mitochondrial DNA diversity in the Lake Mohave population is high compared with other relict populations located farther upstream (Dowling and Minckley 1993). The reservoir population is composed of direct descendants of a very large, diverse population that inhabited the river prior to impoundment. Methods being used to produce 25-cm razorback suckers were reviewed in autumn 1993 (Dowling and Minckley 1993). Natural spawning (stocking reservoir adults in backwaters) was successful only 1 of 2 years. The progeny produced by these fish exhibited the greatest growth but resulted in only 296 juveniles for the 2-year effort. Genetic analysis suggested the majority of these fish came from very few females. We also suspected the use of early spawners may not have adequately represented the total spawning effort. Fertilized egg experiments were unsuccessful, partially because of unpredictability of reservoir operations. Light-trapping experiments showed that large numbers of razorback sucker larvae could be harvested (Mueller et al. 1993). Dowling and Minckley (1993) recommended, in order of priority, the following methods for producing razorback suckers for stocking into Lake Mohave: (1) collect naturally produced larvae, (2) artificially collect gametes (protocol would be developed), (3) stock backwaters with spawning adults, and (4) use hatchery-produced fish. The method of choice, collecting larvae naturally spawned in the reservoir, allows us to produce young adult razorback suckers that represent greater genetic diversity than those produced using other recommended matrix spawning techniques or hatchery facilities (Williamson and Wydoski 1994; Dowling and Minckley, in press). Rather than manipulate spawning, we are now taking advantage of the product of natural spawning as a means of conserving the population's genetic diversity.

Fish have survived nearly 18 months since their release into Lake Mohave. Return of 15 of 640 stocked fish from an 11,400-ha reservoir was higher than expected and represents the largest number of subadults collected from Lake Mohave, and possibly the entire Colorado River, in the last 20 years. We anticipate returns will increase in 1995; a total of 2,880 fish have been released and earlier-stocked females should become sexually active and more susceptible to capture.

Active management of relict populations is a critical component of recovery that many feel is being neglected or overlooked. The razorback sucker is following the same path toward extinction as other fish species. For instance, the bonytail, which co-inhabits the Colorado River, was federally listed as Endangered in 1980. A recovery plan was formalized in 1984 (revised in 1990) calling for the augmentation of wild populations through stocking (Colorado River Fishes Recovery Team 1984). Even though culturing facilities, broodstock, and a recovery plan have been in place for over a decade, managers remain reluctant to stock fish. Bonytail are now considered extirpated from the upper Colorado River basin, where less than 5 have been captured during the last 10 years (USFWS 1990; R. S. Wydowski, personal communication). Concerns have now shifted from recovery to preventing extinction.

The "hands off" recovery philosophy for the bonytail and the razorback sucker is failing while unique, irreplaceable biological components are being lost. The necessity for an active and long-term management commitment to maintain these populations was recognized 14 years ago (W. H. Miller 1982), but researchers have not identified any solutions. Management should be considered a practical safeguard to conserve remaining populations while recovery programs are further developed, implemented, and tested. Reversing environmental degradation will take a concerted and long-term commitment not obtainable in a 10- or 15-year recovery program. The Lake Mohave program falls short of recovery; however, it does represent a modest step toward conserving an existing population. Similar, proactive management approaches are needed to prevent further population declines and potential extinctions.

Acknowledgments

We gratefully acknowledge the assistance of the following individuals: Francisco Abarca, Frank Baucom, Brent Bristow, Bill Burke, Mike Burrell, John Davison, Tom Dowling, Greg Finnegan, Leslie Fitzpatrick, Victor Gamboa, Ross Haley, Jim Heinrich, Gene Hertzog, Mike Horn, John Hutchings, Joe Kahl, Jr., Tom Liles, Chuck Minckley, Wendell Minckley, Alan O'Neil, William Rinne, Tom Shrader, Jon Sjoberg, Kent Turner, Curt Young, Mary Webb, and many other biologists and volunteers. Special thanks go to Thomas Burke, George Devine, and Paul Marsh, key participants who also provided helpful comments and suggestions on this manuscript, and to Larry Visscher, Gary Carmichael, Holt Williamson, and Harold Schramm for their critical review.

References

Burdick, B. D. 1992. A plan to evaluate stocking to augment or restore razorback sucker in the upper Colorado River. U.S. Fish and Wildlife Service, Colorado River Fishery Project, Grand Junction, Colorado.

Colorado River Fishes Recovery Team. 1984. Bonytail chub recovery plan. Report to U.S. Fish and Wildlife Service, Denver, Colorado.

Dowling, T. E., and W. L. Minckley. 1993. Genetic diversity of razorback sucker as determined by restriction endonuclease analysis of mitochondrial DNA. Final Report (Contract 0-FC-40-09530-004) to the Bureau of Reclamation. Arizona State University, Tempe, Arizona.

Dowling, T. E., and W. L. Minckley. In press. Mitochondrial DNA variability in the endangered razorback sucker (*Xyrauchen texanus*): analysis of hatchery stocks and implications for captive propagation. Biological Conservation.

Gustafson, E. S. 1975. Capture, disposition, and status of adult razorback sucker from Lake Mohave, Arizona. Report (Contract No. 14-16-0002-3585) to U.S. Fish and Wildlife Service, Albuquerque, New Mexico.

Johnson, J. E. 1977. Realistic management of endangered species: progress to date. Proceedings of the Annual Conference Western Association of Game and Fish Commissioners 57:298–301.

Johnson, J. E. 1985. Reintroducing the natives: razorback sucker. Proceedings of the Desert Fishes Council 13:73–79.

Langhorst, D. R. 1989. A monitoring study of razorback sucker (*Xyrauchen texanus*) reintroduced into the Lower Colorado River in 1988. Final Report (Contract FG-7494) to California Department of Fish and Game, Blythe, California.

Marsh, P. C. 1994. Abundance, movements, and status of adult razorback sucker, *Xyrauchen texanus* in Lake Mohave, Arizona and Nevada. Proceedings of the Desert Fishes Council 25:35.

Marsh, P. C., and D. R. Langhorst. 1988. Feeding and fate of wild larval razorback suckers. Environmental Biology of Fishes 21:59–67.

Marsh, P. C., and J. L. Brooks. 1989. Predation by ictalurid catfishes as a deterrent to re-establishment of introduced razorback suckers. The Southwestern Naturalist 34:188–195.

Miller, K. D. 1982. Colorado River fishes recovery team. Pages 95–97 *in* W. H. Miller, H. M. Tyus, and C. A. Carlson, editors. Fishes of the upper Colorado River system: present and future. American Fisheries Society, Western Division, Bethesda, Maryland.

Miller, K. H. 1982. Closing remarks. Pages 130–131 *in* W. H. Miller, H. M. Tyus, and C. A. Carlson, editors. Fishes of the upper Colorado River system: present and future. American Fisheries Society, Western Division, Bethesda, Maryland.

Minckley, W. L. 1983. Status of the razorback sucker, *Xyrauchen texanus* (Abbott), in the lower Colorado River basin. The Southwestern Naturalist 28:165–187.

Minckley, W. L. 1973. Fishes of Arizona. Arizona Game and Fish Department, Phoenix.

Minckley, W. L., and J. E. Deacon, editors. 1991. Battle against extinction: native fish management in the American West. The University of Arizona Press, Tucson.

Minckley, W. L., P. C. Marsh, J. E. Brooks, J. E. Johnson, and B. L. Jensen. 1991. Management toward recovery of the razorback sucker. Pages 303–357 *in* W. L. Minckley and J. E. Deacon, editors. Battle against extinction: native fish management in the American West. The University of Arizona Press, Tucson.

Mueller, G., M. Horn, J. Kahl, Jr., T. Burke, and P. Marsh. 1993. Use of larval light traps to capture razorback sucker (*Xyrauchen texanus*) in Lake Mohave, Arizona—Nevada. The Southwestern Naturalist 38:399–402.

Osmundson, D. B., and L. R. Kaeding. 1991. Recommendations for flows in the 15-mile reach during October—June for maintenance and enhancement of endangered fish populations in upper Colorado River. U.S. Fish and Wildlife Service, Colorado River Fishery Project, Grand Junction, Colorado.

Rohlf, D. J. 1991. Six biological reasons why the Endangered Species Act doesn't work—and what to do about it. Conservation Biology 5:273–282.

Tear, T. H., J. M. Scott, P. H. Hayward, and B. Griffith. 1993. Status and prospects for success of the Endangered Species Act: a look at recovery plans. Science 262:976–977.

Tyus, H. M. 1987. Distribution, reproduction, and habitat use of razorback sucker in the Green River, Utah, 1979—1986. Transactions of the American Fisheries Society 116:111–116.

USFWS (U.S. Fish and Wildlife Service). 1990. Bonytail chub recovery plan. USFWS, Denver, Colorado.

USFWS (U.S. Fish and Wildlife Service). 1991. Newsletter—Spring 1991. Recovery implementation program for endangered fish species in the upper Colorado. USFWS, Denver, Colorado.

USFWS (U.S. Fish and Wildlife Service). 1993. Section 7

consultation, sufficient progress, and historic projects agreement and recovery action plan. Recovery implementation program for endangered fish species in the upper Colorado River basin. USFWS, Denver, Colorado.

Williamson, J. H. 1992. Development of refugia populations and captive broodstocks for threatened and endangered fishes in the upper Colorado River basin. March 1, 1990. Fisheries and Federal Aid, U.S. Fish and Wildlife Service, Denver, Colorado.

Williamson, J. H., and R. S. Wydoski. 1994. Genetics management guidelines: recovery implementation program for endangered fish species in the upper Colorado River basin. U.S. Fish and Wildlife Service, Denver, Colorado.

Wydoski, R. S. 1977. Realistic management of endangered species—an overview. Proceedings of the Annual Conference Western Association of Game and Fish Commissioners 57:273–286.

Wydoski, R. S. 1982. Concerns about the status and protection of endemic fishes, especially threatened and endangered species: synthesis of a panel discussion. Pages 127–129 *in* W. Miller and H. M. Tyus, editors. Fishes of the upper Colorado River system: present and future. American Fisheries Society, Western Division, Bethesda, Maryland.

Fisheries Restoration

and Enhancement:

Stocking Criteria and Goals

British Columbia's Trout Hatchery Program and the Stocking Policies that Guide It

BRYAN LUDWIG

Ministry of Environment, Lands and Parks
Victoria, British Columbia V8V 1X5, Canada

Abstract.—In British Columbia, management of anadromous and inland trout is the responsibility of the provincial government. The Fisheries Program Strategic Plan specifies key objectives for the program, including conserving wild stocks and serving the public interest. The latter involves providing a diversity of angling opportunities including harvest. The provincial government operates five hatcheries. In 1993, 10.1 million fish were released into over 1,000 lakes and streams. This release included 5.4 million rainbow trout *Oncorhynchus mykiss* and 1.1 million steelhead (anadromous rainbow trout). Over one-half of the rainbow trout eggs collected to supply the hatchery program come from wild parents. Most of the stocked lakes are high-productivity, high-use lakes in the southern portion of the province, and these lakes generally do not support wild populations. A number of safeguards have been built into the steelhead program to protect wild stocks including (1) marking all hatchery smolts, thus allowing a directed harvest; (2) collecting eggs from wild broodstock; (3) avoiding transplants of steelhead to nonnative systems; and (4) releasing hatchery smolts to the lower portion of the river. Formal policies have been described to guide the inland lakes program, including policies on fish stocking, conservation, wild indigenous fish, and waters designation.

British Columbia has an area of 366,000 square miles. The population, however, is made up of only 3.5 million people, most of whom live in the southern portion of the province. Angling is an extremely popular recreational activity. Almost 400,000 angling licenses are sold each year, and freshwater fishing annually generates Can$450,000,000 to the economy. The actual total number of lakes and streams in the province is unknown, but estimates of 25,000 lakes and 4,600 streams have been cited in various ministry documents (D. Narver, MELP, unpublished data).

In British Columbia, management of anadromous and inland trout is the responsibility of the provincial government. The Provincial Fisheries Program Strategic Plan for 1990–1995 (Anonymous 1990) specifies three key objectives: (1) conserve wild fish stocks, (2) protect and manage fish habitat, and (3) serve the public interest. The goal of the third objective is to provide a diversity of angling opportunities, which includes harvesting. Given the population growth and interest in angling, conserving wild fish stocks and allowing for the harvest of fish is one of the critical challenges for fisheries managers. The policies and procedures used to guide the operation of the hatchery program in British Columbia have also undergone an evolution towards protection of wild stocks.

This report describes: (1) the hatchery program, (2) stocking policies, (3) program success measurement, (4) problems, and (5) future direction of the Fisheries Program Strategic Plan.

Hatchery Program

Stocking records date back to the early 1900s, and it is clear that for many areas of the province there is a long legacy of indiscriminate stocking (Houston 1975). The program has evolved, however, from one that used primarily domesticated strains of fish to one that relies largely on wild stocks.

There are five hatcheries operated by the provincial government. All of these facilities are located in the southern portion of the province. In 1993, 10.1 million fish were released into approximately 1,000 lakes and streams (Table 1). These releases included 1.1 million steelhead and 0.2 million anadromous cutthroat trout *Oncorhynchus clarki* released into approximately 40 streams. Rainbow trout *O. mykiss* accounted for over one-half the production and are released annually to between 600 and 800 lakes and streams (Figure 1).

The eggs used for the rainbow trout program come from one of four broodstock sources (Figures 2 and 3). Wild stocks supply 28% of the eggs for the program; semiwild, 27%; mixed-strain, 35%; and domestic, 10%. Eggs from wild stocks are incubated and reared from 3 to 12 months prior to release and are not restocked in the lake of origin. Semiwild broodstock are derived from brood lakes that were stocked with wild juvenile fish that spent between 3 and 12 months in the hatchery; mixed-strain stocks are from hatchery-origin adults recycled in brood lakes (Figure 2); and domestic stocks (captive native, domestic) are held at the hatchery (Figure 3).

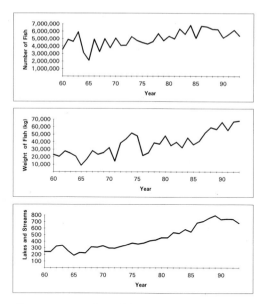

FIGURE 1.—Number of lakes and streams and number and weight of rainbow trout stocked in British Columbia from 1960 through 1993.

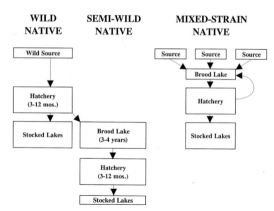

FIGURE 2.—Broodstock-source classification system, identifying types of wild stocks of rainbow trout used in the hatchery program in British Columbia.

Thus, over one-half of the rainbow trout eggs come from wild parents and of these, one-half come from lakes that are never restocked.

Stocked lakes vary from landlocked potholes to large bodies of water with inlet and outlet streams and some natural recruitment. Most of the lakes stocked are high-use, high-productivity lakes in the south-central portions of the province. These lakes generally do not support significant numbers of wild stocks. On Vancouver Island, and in the central and northern portions of the province, however, wild stocks are more common and must be protected from the potential effects of stocking hatchery fish.

The main reason for stocking fish is to provide a harvestable product; a high proportion of the stocked lakes have limited natural recruitment. Demand for nonconsumptive angling opportunities is increasing, and a number of lakes are stocked to provide this opportunity. A small number of lakes are stocked as broodstock lakes.

The steelhead portion of the hatchery program is characterized by the release of small numbers of fish into a large number of rivers. A typical smolt release into a stream on Vancouver Island is 25,000 fish. There are only two programs in the province where smolt releases exceed 100,000 fish. Most programs use fish stocks that are native to the release area. Only 15% of releases are transplanted, and these are to water systems where only remnants of the wild population remain.

Stocking Policies

There are a number of policies, procedures, and guidelines that steer our fish stocking program. Not

TABLE 1.—Salmonid species stocked in British Columbia in 1993.

Species	Number of fish	Number of lakes
Anadromous cutthroat trout *Oncorhynchus clarki*	236,000	45
Coastal cutthroat trout *O. clarki*	281,000	100
Rainbow trout *O. mykiss*	5,411,000	745
Steelhead (anadromous rainbow trout)	1,150,000	40
Kokanee (lacustrine sockeye salmon) *O. nerka*	2,102,000	13
Brown trout *Salmo trutta*	7,000	2
Brook trout *Salvelinus fontinalis*	817,000	118
Bull trout *Salvelinus confluentus*	110,000	12
Total	10,114,000	1,075

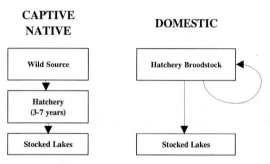

FIGURE 3.—Broodstock-source classification system, identifying types of domestic stocks of rainbow trout used in the hatchery program in British Columbia.

all of these are necessarily entrenched in formal policy documents.

Anadromous Trout Program

The principle factors that steer the anadromous trout program are for the most part guidelines. The objective of the anadromous trout program is to provide a harvestable product for anglers. In order to protect wild stocks a number of safeguards have been incorporated into the program: (1) all hatchery-produced smolts are marked with an adipose clip, which allows for a selective harvest of returning hatchery fish, and catch and release of wild stocks; (2) only wild adult fish are used as broodstock and are collected throughout the spawning season to preserve natural timing; (3) streams are classified as wild, augmented, or hatchery, and these classifications define the fish culture activities that may occur in that system (this is crucial for protection of wild stocks); (4) steelhead cannot be transplanted between water systems without the prior approval of the Federal-Provincial Transplant Committee (this committee reviews transplant requests on the basis of potential fish disease transmission and ecological and genetic impacts on wild stocks); (5) historic spawning-run sizes, impact on the wild population by anglers, accessibility to waters, and proximity to population centers are considered in determining smolt production goals for stocking (the production goal is adjusted if the hatchery fish to wild fish harvest ratio exceeds 1:1); and (6) transport and release of smolts is restricted to the lower portion of rivers to reduce the effect of residual smolts on wild parr. In some cases, rather than risk a problem with residual smolts, very small smolts are released to landlocked lakes.

Inland Lakes Program

In contrast with the anadromous trout program, the inland lakes program has been guided by more formal policies. There has been a progression toward increased protection of wild stocks in recent years.

Sources of eggs for fish culture (1983).—This policy states that the first priority of the ministry is to use only wild endemic stocks of salmonids for broodstock or first generation hatchery fish planted in lakes where natural selection may occur. The policy is somewhat open in that it allows for the use of hatchery fish for broodstock, if wild fish returns are too low.

Fish stocking (1984).—This policy describes the conditions under which stocking of public waters with hatchery fish may be carried out. These conditions include lakes, rivers, and streams in which a self-sustaining fish population does not exist or is at less than the biological carrying capacity. There is no specific protection for barren lakes or restrictions on the stocks that can be used.

Fish and aquatic invertebrate transplant and introduction (1984).—In British Columbia, approval of the Federal-Provincial Transplant Committee is required before fish can be transplanted. Transplant requests are reviewed on the basis of fish disease transmission and ecological and genetic risks to wild stocks. This committee provides an independent risk assessment for fish transplants.

Conservation (1993).—The Conservation Policy states that the long-term conservation of indigenous populations and their habitats is a primary objective of the ministry. The policy is not limited to fish, and the concept is to be recognized in all planning documents. The policy ensures that indigenous stocks are protected from the effects of exotic species. Indigenous stocks refer to those fish that occur in British Columbia, whereas exotic stocks are those that do not occur naturally in the watershed.

Wild indigenous fish (1993).—The Wild Indigenous Fish Policy gives first priority to the protection, maintenance, and enhancement of wild indigenous fish stocks. Stocking is permitted only if there has been a previous history of stocking, or if it can be carried out without compromising wild stocks. Introduction of exotic species or races of fish is not considered compatible.

Lake stocking of cultured fish (1993).—This policy establishes procedures for determining if a previously unstocked lake may be stocked. Unstocked lakes must be inventoried before fish stocking will be considered. If fish are present, a conservation data center is to be consulted to determine if rare or endangered species are present. In addition, a management plan is required to determine possible effects on wild stocks.

Waters designation (draft).—The policy states that all lakes and streams will be classified as wild indigenous, wild naturalized, augmented, or hatchery. Stocking is not permitted in waters with a wild designation. Waters that are not inventoried are assigned a wild indigenous classification by default. Barren lakes are to be considered wild naturalized. In order to change the classification to permit stocking, a fish inventory must be conducted and a management plan prepared that assesses the possible effects of stocking on existing fish stocks.

The effectiveness of these policies in protecting wild fish stocks can be determined by considering their policy application in a hypothetical situation.

Consider a lake in the northern portion of the province that has never been stocked but for which stocking is being considered. The Lake Stocking Policy will require that an inventory of the lake be conducted to determine if fish are present. If they are, the Conservation Data Center will be consulted to determine if any of the species are rare or endangered. A management plan will be prepared to determine the effect of the stocking program. The Waters Designation Policy requires that the management plan be reviewed to ensure compliance with existing policies. If the lake contains fish, the Conservation Policy provides that the genetic characteristics of the population must be preserved, and the Wild Indigenous Fish Policy requires that the stocking program not compromise wild stocks. Is it possible to carry out a stocking program in a lake without compromising wild stocks? This is not addressed in the policies.

Program Success Measurement

Given the large number of systems in which steelhead are released in the anadromous trout program, and the fact that very few of these systems have counting facilities, we rely on a mail-out harvest questionnaire to provide a measure of performance of the various hatchery programs and an index of how well the wild populations are doing. The questionnaire is mailed to one-half the anglers purchasing a license.

In 1993, the estimated total steelhead catch was 118,000, of which 38,000 were hatchery steelhead (MELP, unpublished data; Figure 4). This is a slight increase from previous years. The average catch was 0.59 fish/d. The steelhead catch was reported for over 200 streams.

It is more difficult to assess the inland lakes program. A complete survey of recreational freshwater fishing in Canada is conducted every 5 years. The last complete survey was conducted in 1990, and at that time there were 392,000 anglers who generated 4.6 million angler-days in British Columbia (B. Williams, MELP, unpublished data; Table 2). The total catch was 9.6 million fish, of which 46% were harvested. Rainbow trout made up 53% of the total catch.

Problems

While allowing fish stocking, the existing policies clearly state that wild native stocks are a priority. It has not been determined if it is possible to stock native fish without compromising wild stocks. Criteria have not been established to determine if a wild population is unique or important and thus should be protected from stocking. Should all wild

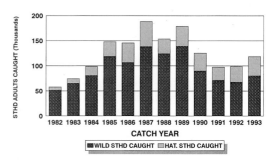

FIGURE 4.—Number of wild and hatchery (HAT) steelhead (STHD) adults caught in British Columbia during the years 1982–1993.

stocks be considered important no matter the size of the population? We have recognized the unique nature of some steelhead stocks and some lake fish stocks (i.e., Gerrard rainbow trout), but we have not clearly recognized less obvious characteristics that may be unique, such as behavior, temperature tolerance, or disease resistance (J. Ptolemy, MELP, unpublished data).

There is no specific protection for barren lakes in existing policies. These lakes may contain unique nonfish species that should be protected.

For many lakes, the inventory information is not adequate to make informed decisions. We know there are lakes being stocked that have wild populations. Although we are reasonably comfortable with the policies, we have yet to review the existing stocking programs intensively in terms of their effects on wild populations.

In some steelhead systems the wild stock is declining, and unless hatchery production is decreased, the hatchery harvest may exceed the wild harvest. For example, several years ago there were serious concerns about the declining wild catch in the Somass River (Figure 5). Although there has been an increase in the wild fish catch for 1993, there were record hatchery fish catches as well. The 1:1 ratio of hatchery to wild fish in the catch has been exceeded, and as a result, the stocking program needs further review.

TABLE 2.—Freshwater angling statistics in British Columbia for 1990.

Licenced anglers (N)	392,000
Juvenile anglers (N)	151,000
Angler-days (N)	4,600,000
Total fish caught (N)	9,600,000
Total fish released	54%
Rainbow trout catch	53%
Angler expenditures (Canadian $)	356,000,000

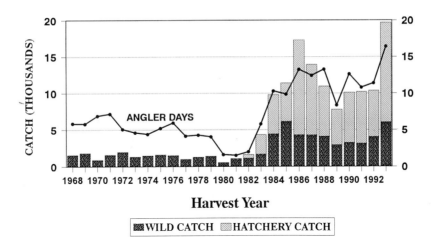

FIGURE 5.—Number of wild and hatchery-reared steelhead caught in the Somass River, British Columbia, and angler-days of effort, in thousands, during 1968–1993.

Future Direction

It was not until the early 1980s that we realized there were significant differences in the performance of various rainbow trout stocks. Concurrent with this, it became apparent that the Fish Culture Section was relying to a great extent on stocks that performed well in the hatchery but had some undesirable characteristics, such as early maturation, that influenced their performance after release.

Our research section, under the guidance of K. Tsumura, has been evaluating the use of (1) all-female populations of fish; (2) rainbow trout stocks from coarse-fish lakes, because rainbow trout from monoculture lakes do not perform as well in coarse-fish lakes as do rainbow trout stocks from lakes containing coarse fish (Blann et al. 1990); (3) rainbow trout stocks from alkaline lakes, because some lakes in the interior of the province have extremely high pH waters (on the order of 9.2; Toth and Tsumura [1993] have found that broodstocks from lakes with high pH have higher rates of survival in alkaline lakes than broodstocks from lakes with a more neutral pH); and (4) triploid stocks of rainbow trout and brook trout in situations in which early maturation is a problem and in which wild stocks must be protected. We anticipate being able to reduce the stocking rate significantly because fish will stay in the water system longer. The use of these special fish stocks could improve the performance of hatchery fish, but the rearing and release costs for the hatchery program may increase.

We now have the opportunity to make use of geographic information systems (GIS) to evaluate fish culture programs. For example, in an area such as Vancouver Island, the total number of lakes, the number of lakes with wild populations, and the number of lakes stocked can easily be cross-referenced using GIS. Many of the lakes in British Columbia have not been inventoried, but this technology ensures that maximum use can be made of available information.

The experience in British Columbia indicates that the key to protecting wild stocks is a waters classification system that includes a no-stocking designation and a system that requires all fish transplants to be reviewed by an independent committee. In British Columbia, although we have these systems in place and are comfortable with the direction we are taking in terms of new hatchery programs, we have yet to reconsider the effect of existing stocking programs on wild populations of fish.

Acknowledgments

D. Peterson, K. Tsumura, J. Ptolemy, B. Williams, and S. Billings contributed unpublished data. D. Peterson and M. Labelle commented on the manuscript. Thanks to S. Billings and D. Stanton for preparing the figures.

References

Anonymous. 1990. Conserving our resource. Fisheries Program Strategic Plan, 1991–1995. Ministry of Environment Lands and Parks, Victoria, British Columbia.

Blann, V. E., T. H. Godin, and K. Tsumura. 1990. Rainbow trout broodstocks for coarse fish lakes. Ministry of Environment Lands and Parks, Fisheries Project Report RD24, Vancouver.

Houston, C. 1975. Historical stocking record for British Columbia. Master's thesis. University of British Columbia, Vancouver.

Toth, B. M., and K. Tsumura. 1993. Alkaline lakes enhancement. Ministry of Environment Lands and Parks, Fisheries Project Report RD33, Vancouver.

Species-Specific Guidelines for Stocking Reservoirs in Oklahoma

EUGENE GILLILAND AND JEFF BOXRUCKER

Oklahoma Fishery Research Laboratory
500 East Constellation, Norman, Oklahoma 73072, USA

Abstract.—The Oklahoma Department of Wildlife Conservation established stocking guidelines in 1990 for largemouth bass *Micropterus salmoides,* smallmouth bass *M. dolomieui,* hybrid striped bass female *Morone saxatilis* × male *M. chrysops,* walleye *Stizostedion vitreum,* saugeye female *Stizostedion vitreum* × male *S. canadense,* and channel catfish *Ictalurus punctatus.* Committees composed of management, research, and hatchery biologists established systems by species that assigned points for various physical and biological criteria and fisheries characteristics of candidate reservoirs. Reservoirs were then ranked according to point totals and hatchery production was assigned in order of priority until fully allocated. Guidelines also included flowcharts to assist managers with stocking decisions, provided target values for the required evaluation sampling, and offered guidance on when to consider discontinuing stocking. Reservoirs that were stocked for three consecutive years and failed to meet evaluation criteria for the stocked species were removed from consideration, thus allowing new reservoirs to be considered based on their point rankings. The mean number of lakes for which stocking was requested (all species) decreased from 243/year for the 5 years before implementation of the guidelines to 186/year for the next 5 years. The mean number of reservoirs stocked declined by 30%, except those receiving smallmouth bass and saugeye, which increased by 85% and 150%, respectively. Mean numbers of fish stocked per year increased from 15,317,000 in 1985–1989 to 19,766,000 in 1990–1993. This increase was attributed to a large increase in walleye fry stockings that offset decreased numbers of stocked fingerlings of other species. The reservoir selection criteria were modified annually as the effectiveness of the system was tested. These guidelines proved useful to fishery managers by providing objective criteria to establish priorities for fish stocking. They also allow hatcheries to set realistic production goals by stocking only those reservoirs most likely to produce successful fisheries. The size and quality of stocked fish were also improved by shifting production demands.

Stocking hatchery-produced fish by the Oklahoma Department of Wildlife Conservation (ODWC) has produced many popular fisheries. Successful introductions of Florida largemouth bass *Micropterus salmoides floridanus* have increased trophy bass fishing opportunities (Horton and Gilliland, in press). Stockings of hybrid striped bass female *Morone saxatilis* × male *M. chrysops* produced quality fisheries in reservoirs where largemouth bass habitat was limited (Glass and Maughan 1985). Stocking walleye *Stizostedion vitreum* over a 30-year span produced several good fisheries, but stockings are now limited to those lakes where natural recruitment is insufficient to maintain fisheries (D. Driscoll, ODWC, personal communication). Saugeye female *Stizostedion vitreum* × male *S. canadense* was first introduced as a sport fish that might be better suited to Oklahoma reservoirs than walleye, but its utility as a biological control for crappie *Pomoxis* spp. (Horton and Gilliland 1990; Boxrucker 1994) has expanded the demand for these predators. Smallmouth bass *Micropterus dolomieu* produced from Tennessee broodstock exhibited excellent survival, growth, and reproduction in an Oklahoma reservoir (Gilliland et al. 1991). Catchable-sized channel catfish *Ictalurus punctatus* were stocked into numerous smaller bodies of water to increase angling opportunities in urban areas.

Before 1990, management biologists requested fish for reservoirs in their regions based on standardized sampling surveys and public demand. Requests were sent to administrators who subjectively ranked fish requests from across the state. Limited hatchery production of a species was often divided among all reservoirs for which requests were received. This approach often resulted in inadequate stocking rates with no guarantees that the reservoirs receiving the most fish were necessarily the best suited for a species. The wide variation in stocking rates and schedules made evaluation of stocking success difficult and precluded the establishment of optimum stocking protocols (Welcomme et al. 1983).

Faced with an increasing demand for cultured fish from management biologists, and wanting to better use existing hatchery production, ODWC Fisheries Division personnel set out to develop a method of establishing priorities for lakes to be stocked. The new system needed to offer objective

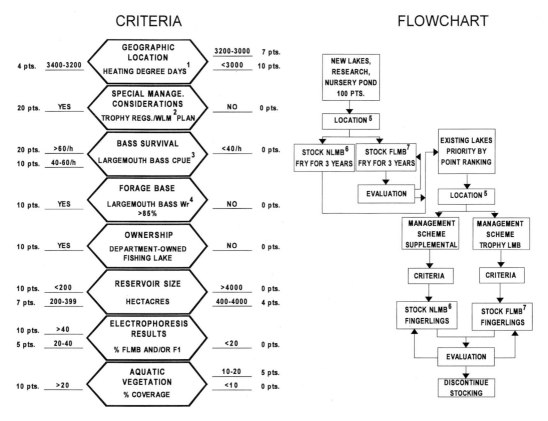

FIGURE 1.—Criteria for ranking Oklahoma reservoirs to be stocked with Florida largemouth bass and the flowchart of stocking protocol. [1] Heating degree days (HDD) is the sum over all days, fall to spring, of the difference between 65°F and the average daily temperature; [2] WLM is water level management of the reservoirs; [3] largemouth bass CPUE (electrofishing) is fish caught per hour; [4] W_r is the mean relative weight; [5] location—waters north of a diagonal boundary from the northeast corner of the state to the southwest corner roughly following the 3,800 HDD cline are stocked only with northern largemouth bass, waters south of the 3,800 HDD cline can be stocked with northern or Florida subspecies; [6] NLMB is the northern largemouth bass; and [7] FLMB is the Florida largemouth bass.

methods of selecting reservoirs for stocking, include a guide to lead biologists through stocking strategies (Turner 1989), and provide evaluation protocols to determine success of the stockings and make decisions regarding continuation of the program (Murphy and Kelso 1986).

Methods

In 1989, management and research biologists and hatchery managers were assigned to committees that were responsible for developing stocking guidelines for each major species cultured at ODWC hatcheries. The four committees included (1) Florida largemouth bass, northern largemouth bass *Micropterus salmoides salmoides,* and smallmouth bass; (2) hybrid striped bass; (3) walleye and saugeye; and (4) channel catfish. Each committee was responsible for conducting a literature review and compiling a summary of survey data used to define stocking guidelines and evaluation criteria. Specific criteria for assigning point values to candidate reservoirs were established so that fish requests could be ranked objectively. A flowchart to guide managers through stocking decisions was also developed for each species. Biologists and managers met periodically to critique the flowcharts and reservoir selection criteria from each committee. Consensus opinions were then incorporated into each successive draft. Personnel at the ODWC's Oklahoma Fishery Research Laboratory compiled the output from each committee and prepared working documents that followed standardized formats.

We compared the mean annual number of re-

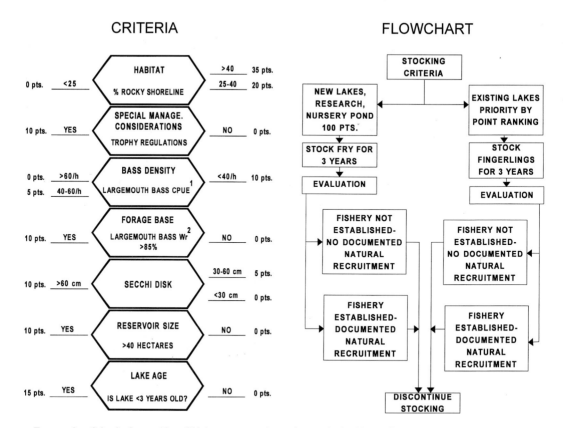

FIGURE 2.—Criteria for ranking Oklahoma reservoirs to be stocked with smallmouth bass and the flowchart of stocking protocol. [1] Largemouth bass CPUE (electrofishing) is fish caught per hour; [2] W_r is the mean relative weight.

quests for fish stocking by species, the mean number of reservoirs actually stocked with that species, and the mean number and weight of fish stocked for the 5 years prior to the implementation of the system (1985 through 1989, hereafter designated as BEFORE) with the 5 years following implementation (1990 through 1994, hereafter designated as AFTER) using Student's t-test with a P of 0.10. A liberal significance level was used because even relatively small differences in de-

TABLE 1.—Mean annual number of reservoirs for which stocking was requested, the number of reservoirs actually stocked (in parentheses), and the percent change; the mean number and kilograms of fish stocked per year in Oklahoma for the years 1985 through 1989 (BEFORE stocking guidelines were developed) and for 1990 through 1994 (AFTER stocking guidelines were developed) for Florida largemouth bass (FLMB), northern largemouth bass (NLMB), smallmouth bass (SMB), walleye (WALL), saugeye (SAUG), hybrid striped bass (HYB), and channel catfish (CCAT).

| | Number of reservoirs requesting and receiving stockings | | | Fish stocked | | | |
| | | | | Number (×1000) | | Kilograms (×1000) | |
Species	BEFORE	AFTER[a]	Percent change	BEFORE	AFTER	BEFORE	AFTER
FLMB	35 (41)[b]	32 (23)	−44	950	773	281	216
NLMB	18[b] (18)[b]	4 (10)	−44	586[b]	180	130	83
SMB	7 (7)	18[b] (13)[b]	+85	180	374[b]	11	49[b]
WALL	27[b] (30)[b]	12 (11)	−63	8,927	13,800[b]	100	94
SAUG	11 (8)	25[b] (20)	+150	469	1,212[b]	29	84[b]
HYB	20 (22)	18 (20)	−9	2,743	2,326	147	90
CCAT	125[b] (128)[b]	77 (82)	−36	1,462[b]	1,101	5,133	6,686
Total	243 (254)	186 (179)	−30	15,317	19,766	5,851	7,302

[a] Reservoirs stocked include only years 1990–1993.
[b] Mean significantly greater at $P = 0.10$.

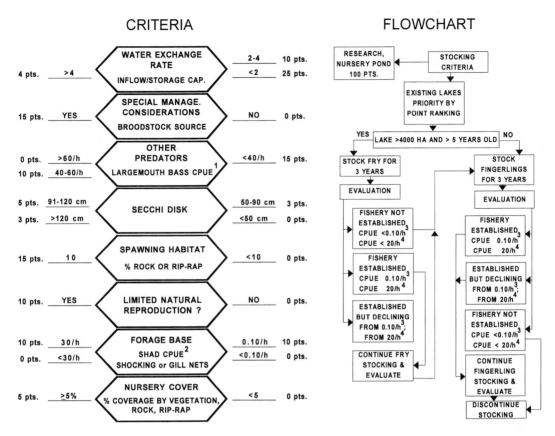

FIGURE 3.—Criteria for ranking Oklahoma reservoirs to be stocked with walleye and the flowchart of stocking protocol. [1] Largemouth bass CPUE (electrofishing) is fish caught per hour; [2] gizzard shad CPUE (electrofishing, ≤30/h, and gill netting, ≤0.10/h) is fish caught per hour; [3] CPUE ≥ 0.10/h is the minimum catch of walleye by gill netting; and [4] CPUE ≥ 20/h is the minimum catch of walleye by electrofishing.

mands on the hatchery system can result in significant operational changes.

Results

Criteria for Ranking Lakes

Criteria were prepared for each species (left side of Figures 1 through 6). Categories for each species' worksheet totaled 100 points and included physical, biological, and fishery characteristics of that reservoir. Several criteria were common to each species, including estimators of habitat quality and predator and forage abundance. Water quality and habitat parameters considered were mean secchi disc transparency (Figures 2 through 6) and abundance of protective cover or spawning substrate such as aquatic vegetation or rocky shoreline (Figures 1, 2, 3, 4, and 6). Willis and Stephen (1987) found an inverse relationship between storage ratio and stocking success of walleye (water exchange rate; Figures 3, 4, and 5). Forage abundance was estimated by catch of gizzard shad *Dorosoma cepedianum* in standardized electrofishing and gill netting (Figures 3, 4, and 5). Because of the inherent biases associated with gizzard shad density estimates using these gear types (Michaletz 1990), mean relative weights (W_r) of the target species or competing predators were used as an indirect measure of forage availability (Figures 1, 2, 5, and 6). Predator density has been shown to affect survival of stocked fish (Carline et al. 1984; Mosher 1987). Lower points were awarded for high largemouth bass electrofishing catch per unit effort (CPUE as fish/h) when stocking smallmouth bass, walleye, or saugeye (Figures 2, 3, and 4). More points were awarded to reservoirs for which largemouth bass were requested if they had largemouth bass catch rates indicating a more suitable environment for survival

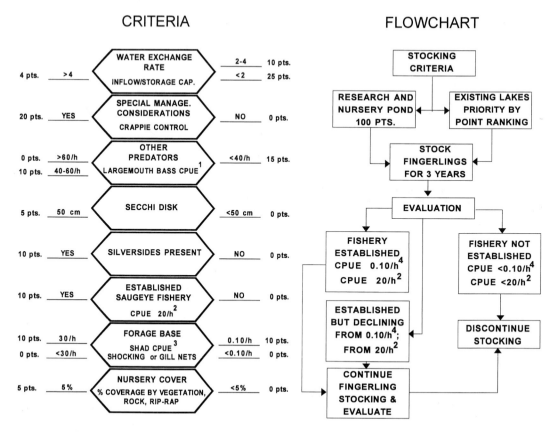

FIGURE 4.—Criteria for ranking Oklahoma reservoirs to be stocked with saugeye and the flowchart of stocking protocol. [1] Largemouth bass CPUE (electrofishing) is fish caught per hour; [2] saugeye CPUE (electrofishing) is fish caught per hour; [3] gizzard shad CPUE (electrofishing, $\leq 30/h$, and gill netting, ≤ 0.10) is fish caught per hour; and [4] CPUE $\geq 0.10/h$ is the minimum catch of saugeye by gill netting; the minimum catch by electrofishing is $\geq 20/h$.

(Figure 1). Pelagic predator density was used in channel catfish criteria (Figure 6). Special management considerations such as reservoirs with water level management plans or restrictive creel and trophy length limits received extra points (Figure 1), as did walleye, saugeye, and hybrid striped bass reservoirs with established tailwater fisheries (Figures 3, 4, and 5). Other special management considerations receiving higher points included introductions of saugeye for crappie control (Figure 4) and proximity to urban areas (Figures 5 and 6). New reservoirs, those with federal aid research projects dealing with the species in question, or nursery ponds automatically received 100 points and were given top priority on stocking lists for each species.

Flowcharts that led managers through stocking decisions and provided target values for the required evaluation sampling (right side of Figures 1 through 6) were created for each species. These flowcharts offered guidance for whether to continue or discontinue stockings because of failure to meet the established sampling criteria. Sampling to evaluate the success of each stocking was required for each species (Noble 1986). Gears were assigned for evaluation of each species after a consensus was reached among biologists as to the most effective means of collecting the fish at an appropriate age. Target values for evaluation sampling were from electrofishing or gill netting for smallmouth bass, walleye, saugeye, hybrid striped bass, and channel catfish. Percentage of Florida largemouth bass subspecific alleles in samples of age-1 largemouth bass as determined by starch gel electrophoresis were used to measure the success of Florida largemouth bass introductions. Details on sampling methods and target values were provided in a text accompanying each flowchart.

In practice, managers made stocking requests by

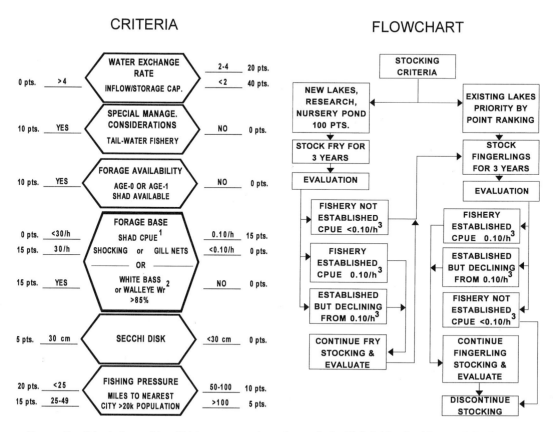

FIGURE 5.—Criteria for ranking Oklahoma reservoirs to be stocked with hybrid striped bass and the flowchart of stocking protocol. [1] Gizzard shad CPUE (electrofishing, ≤30/h, and gillnetting, ≤0.10) is the fish caught per hour; [2] W_r is the mean relative weight of walleye or white bass *Morone americana;* and [3] CPUE ≥0.10/h is the minimum catch of saugeye by gill netting.

completing the appropriate worksheet and submitting them to ODWC administrators. All worksheets were compiled by species, and reservoirs were ranked by point totals. Hatchery production was assigned to each reservoir in order until fully allocated. Once stocked, reservoirs were automatically included for 3 years to allow the required evaluation. After three consecutive years of stocking, 3-year mean catch rates were compared with target catch values. If a population failed to meet the target value, that reservoir was dropped from stocking and others further down the list were moved up in priority the following year.

Effects of the Guidelines on Stocking Requests

Total kilograms of fish produced by ODWC hatcheries has increased slightly in the past 10 years, as have numbers of fish produced (Table 1). The total number of requests for stockings decreased from 243/year BEFORE to 186/year AFTER, a decline of 23%; numbers of fish stocked increased from 15,317,000 BEFORE to 19,766,000 AFTER (Table 1).

Numbers of Florida largemouth bass requested for stocking declined from 950,000/year BEFORE to 773,000/year AFTER, in part because of a change dictated in the stocking guidelines to use larger fingerlings at a lower rate (Table 1). However, the mean number of reservoirs stocked with Florida largemouth bass dropped significantly from 41/year BEFORE to 23/year AFTER ($P = 0.09$; Table 1). Numbers of northern largemouth bass requests and numbers of reservoirs stocked declined (18/year requested and stocked BEFORE versus 4 requested and 10 stocked/year AFTER, $P = 0.01$ and 0.05, respectively). Numbers of northern largemouth bass stocked also decreased (586,000/year BEFORE versus 180,000/year AF-

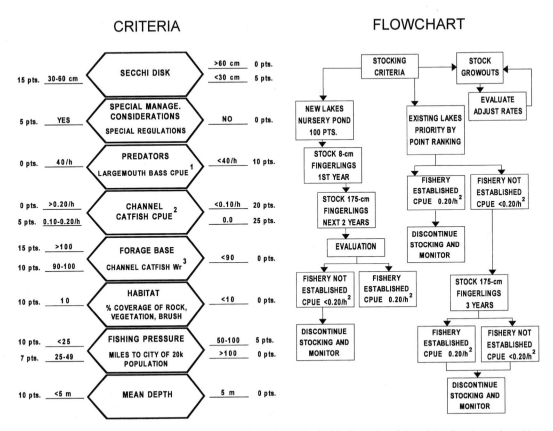

FIGURE 6.—Criteria for ranking Oklahoma reservoirs to be stocked with channel catfish and the flowchart of stocking protocol. [1] Largemouth bass CPUE (electrofishing) is the fish caught per hour; [2] channel catfish CPUE (gill netting) is the fish caught per hour; and [3] W_r is the mean relative weight.

TER, $P = 0.08$; Table 1). Smallmouth bass became more popular among anglers and managers when Tennessee-strain broodstock were brought into Oklahoma in 1985. Requests, reservoirs stocked, numbers, and weight of smallmouth bass stocked increased ($P = 0.01$, $P = 0.07$, $P = 0.03$, and $P = 0.03$, respectively; Table 1). Walleye requests and reservoirs stocked declined significantly ($P < 0.01$ and $P = 0.01$, respectively). Numbers of walleye stocked increased (8,927,000 fish/year BEFORE versus 13,800,000 fish/year AFTER, $P = 0.07$; Table 1), but biomass remained relatively constant because more requests were for fry than for fingerlings after 1990. The mean number of reservoirs for which saugeye was requested rose significantly from 11 BEFORE to 25 AFTER ($P = 0.02$), and numbers and weight stocked increased from 469,000 fish and 29 kg/year BEFORE to 1,212,000 fish and 84 kg/year AFTER ($P = 0.10$ and $P < 0.01$, respectively; Table 1). Although the number of reservoirs stocked with saugeye did not show a statistically significant increase, in practical terms the increase from 8 BEFORE to 20 AFTER was an important change. Channel catfish showed significant reductions in requests ($P = 0.01$), number of reservoirs stocked ($P = 0.01$), and numbers of fish stocked each year ($P = 0.09$; Table 1). Hybrid striped bass was the only fish that did not show a significant change in either number of requests, reservoirs stocked, or fish released.

Discussion

Hatchery priorities have changed because of the new system. Before 1990, pond space at the four ODWC hatcheries was allocated to species that would be stocked close to the hatchery; that is, driving distances for hatchery personnel were minimized. Production has since shifted to those facilities that obtained the best fingerling return for a species regardless of distances to receiving reservoirs. Stocking fish is now a cooperative effort among the entire Fisheries Division rather than relying primarily on hatchery personnel for delivery. Reduced numbers of requests for certain species has allowed hatchery managers to shift toward production of fewer but larger fingerling largemouth bass, smallmouth bass, and channel catfish as requested by managers.

The stocking guidelines and lake selection criteria have several strengths. They provide an objective method of ranking reservoirs that are to receive hatchery-produced fish, a standardized worksheet and text format for all species, and a written path for all managers to follow when stocking so that procedures are standardized.

Several weaknesses were identified, and ODWC personnel are researching methods to overcome these shortfalls. Effort is being made to find more reliable and accurate methods of assessing forage abundance, measure habitat variables that affect stocking success (survival and recruitment) more accurately, determine the best methods for breaking ties in point rankings among reservoirs, and come to a consensus on how much influence public opinion should have on setting stocking priorities.

Acknowledgments

We would like to thank the ODWC management, research, and hatchery personnel and administrators involved in creating, testing, and implementing this system. This effort was funded in part through Oklahoma Federal Aid in Sport Fish Restoration Project F-43-D and F-44-D.

References

Boxrucker, J. 1994. Results of concomitant predator and prey stockings as a management strategy in combatting stunting in an Oklahoma crappie population. Proceedings of the Annual Conference Southeastern Association of Fish and Wildlife Agencies 46(1992): 327–336.

Carline, R. F., R. A. Stein, and L. M. Riley. 1984. Effects of size at stocking, season, largemouth bass predation, and forage abundance on survival of tiger muskellunge. Pages 151–167 in G. E. Hall, editor. Managing muskies. American Fisheries Society Special Publication 15.

Gilliland, E., R. Horton, B. Hysmith, and J. Moczygemba. 1991. Smallmouth bass in Lake Texoma, a case history. Pages 136–142 in D. Jackson, editor. The first international smallmouth bass symposium. Mississippi Agricultural and Forestry Experiment Station, Mississippi State University, Mississippi State.

Glass, R. D., and O. E. Maughan. 1985. Concentrated harvest of striped bass × white bass hybrids near a heated water outlet. North American Journal of Fisheries Management 5:105–107.

Horton, R. A., and E. R. Gilliland. 1991. Diet overlap between saugeye and largemouth bass in Thunderbird Reservoir, Oklahoma. Proceedings of the Annual Conference Southeastern Association of Fish and Game Agencies 44(1990):98–104.

Horton, R. A., and E. R. Gilliland. In press. Monitoring trophy largemouth bass in Oklahoma using a taxidermist network. Proceedings of the Annual Conference Southeastern Association of Fish and Wildlife Agencies 47(1993).

Michaletz, P. 1990. Comparison of electrofishing and gill netting for sampling shad in two Ozark impoundments. Missouri Department of Conservation, Federal Aid in Sport Fish Restoration, Project F-1-R-039, Job 2, Final Report, Jefferson City.

Mosher, T. 1987. An assessment of walleye populations in small Kansas lakes with recommendations for further stocking. Kansas Fish and Game Commission, Federal Aid in Sport Fish Restoration, Project FW-9-P-5, Final Report, Pratt.

Murphy, B. R., and W. E. Kelso. 1986. Strategies for evaluating freshwater stocking programs: past practices and future needs. Pages 303–313 in R. H. Stroud, editor. Fish culture in fisheries management. American Fisheries Society, Fish Culture Section and Fisheries Management Section, Bethesda, Maryland.

Noble, R. L. 1986. Stocking criteria and goals for restoration and enhancement of warm-water and cool-water fisheries. Pages 139–146 in R. H. Stroud, editor. Fish culture in fisheries management. American Fisheries Society, Fish Culture Section and Fisheries Management Section, Bethesda, Maryland.

Turner, G. E. 1989. Codes of practice and manual of procedures for consideration of introductions and transfer of marine and freshwater organisms. EIFAC (European Inland Fisheries Advisory Commission) Occasional Paper 23.

Welcomme, R. L., C. C. Kohler, and W. R. Courtenay, Jr. 1983. Stock enhancement in the management of freshwater fisheries: a European perspective. North American Journal of Fisheries Management 3:265–275.

Willis, D. W., and J. L Stephen. 1987. Relationship between storage ratio and population density, natural recruitment, and stocking success of walleye in Kansas reservoirs. North American Journal of Fisheries Management 7:279–282.

Salmon Stock Restoration and Enhancement: Strategies and Experiences in British Columbia

E. A. PERRY

Department of Fisheries and Oceans, 555 West Hastings Street
Vancouver, British Columbia V6B 5G3, CANADA

Abstract.—The use of hatcheries as a resource management tool for chinook salmon *Oncorhynchus tshawytscha* and coho salmon *O. kisutch* in western Canada and United States is being questioned. In British Columbia other salmon species catches have increased substantially since the 1970s, but chinook salmon catch has decreased and many wild coho salmon stocks are declining in abundance. The future role of hatcheries in areas where there remains potential natural salmon production, will likely depend on whether they benefit wild stocks. Such benefits will depend on careful planning and implementation of hatchery projects and operations, avoidance of interactions between hatchery and wild stocks, acquisition of and response to new knowledge, and conservation-driven harvest management. This paper presents three examples of strategies being used and considered in British Columbia to meet these challenges.

Wild coho salmon stocks in the Strait of Georgia are in serious decline. The plan developed to rebuild these stocks is unique in our experience. It integrates harvest at rates sustainable by the wild stocks, freshwater habitat management, and specific stated criteria for hatchery expansion and operations. Stocks will be monitored for two cycles after harvest is reduced. Failure of a large number of stocks to respond will result in further harvest reduction or other measures, not expanded hatchery production. If only a few stocks do not rebuild after two cycles, hatchery augmentation may be considered. Hatchery releases will be limited to levels that will produce an escapement equal to 50% of the natural spawning target. Operational procedures will be in accordance with guidelines intended to minimize ecological and genetic interactions between wild and hatchery coho salmon. Finally, hatchery production will be reduced if there is evidence of negative interactions between wild and hatchery coho salmon in the marine environment.

Coho salmon fry augmentation programs on the Eagle and Coldwater rivers in interior British Columbia are reviewed to examine interactions between wild and hatchery coho salmon during freshwater rearing. The return per spawner for wild spawners and the survival of hatchery fry to catch and escapement decreased at higher spring fry abundance in the Eagle River but not in the Coldwater River. The fry augmentation program has been discontinued in the Eagle River. These results demonstrate that coho salmon fry augmentation may be a beneficial strategy for some stocks but that negative interactions between wild and hatchery coho salmon may occur. The results illustrate the need to monitor and adjust stocking programs. The success of augmentation work designed to rebuild wild stocks ultimately means the hatchery will no longer be required; this suggests portable or temporary hatchery facilities should be considered.

Adult returns from chinook and coho salmon smolts released from Fraser River and coastal hatcheries feeding into the Strait of Georgia are compared with release numbers. There is a relatively strong suggestion of density-dependent mortality for hatchery chinook salmon. Interpretation of the hatchery coho salmon data is more debatable. Progress and problems encountered in designing a large-scale experiment involving manipulation of hatchery-release numbers to test the density dependence hypothesis are described.

Hatchery and spawning channel production of salmon *Oncorhynchus* spp. has been an increasingly important component of fisheries resource management in British Columbia during the past 20 years. In the 1960s and 1970s salmon catches were low compared with the previous 4 decades. This decline provided impetus for the Canadian Salmonid Enhancement Program (SEP), started in 1977. In the 1990s Canadian salmon catch has risen to record levels, partly due to enhancement but primarily due to improved escapement management and favorable ocean conditions (Canadian Department of Fisheries and Oceans [DFO], unpublished data). Most of this recovery has been in sockeye salmon *O. nerka* and pink salmon *O. gorbuscha*, and to less extent chum salmon *O. keta*. The catch of coho salmon *O. kisutch* has been relatively stable during the past 30 years, but many wild stocks are at low levels. Catch of chinook salmon *O. tshawytscha* has declined dramatically.

Perhaps because hatcheries in western Canada and United States are typically associated with chinook and coho salmon, and abundance of many stocks of these species has recently declined, the use of hatcheries is being questioned. The technology is blamed per se, despite the success of hatchery-produced pink and chum salmon in British Columbia, Japan, and Alaska. In the extreme, and more often in the popular media than in scientific reports, hatcheries are blamed for many of the problems facing wild chinook and coho salmon, especially overfishing and failure to protect fish habitat.

It is clear that debate will not resolve this issue. There will undoubtedly continue to be a role for hatcheries as a final resort for conserving threatened stocks and for enhancing discrete fisheries in areas barren of wild stocks. The hatchery's role in areas where there remains significant potential for natural salmon production, however, will likely be dictated by whether or not hatcheries are shown to provide a net benefit for wild stocks. Success will depend on extreme vigilance in the way that hatchery programs are planned and implemented, with emphasis on avoidance of genetic and ecological interactions with wild stocks, acquisition of and response to new knowledge, and conservation-driven harvest management policies. Widespread recognition that hatcheries are a beneficial management tool will be the measure of success.

This report examines some of these requirements for planning and implementation from the perspective of the salmon hatchery program in British Columbia, with emphasis on chinook and coho salmon hatcheries. I describe the most recent example in the evolution of enhancement planning, the goal of which is to assist in the rebuilding of wild stocks. Data that indicate possible negative interactions between wild and hatchery stocks in fresh water and possible intraspecific interactions in the marine environment are presented, and our response to these data is described.

Enhancement Planning

The approach to enhancement planning is very different today than it was 20 years ago, reflecting changing goals. Although much improved, it is not flawless. There remains substantial progress to be made in integration of our enhancement, habitat, and harvest activities. We do not have stock production plans closely linked to harvest plans for most stocks, nor do we have perfect understanding of what is required to maintain genetic diversity when using a hatchery to rebuild wild stocks perfect. Regardless, planning and action are necessary, and we must use the best available information for these purposes.

Wild coho salmon stocks in the Strait of Georgia (Figure 1) were the subject of a recent large-scale planning exercise in British Columbia. Coastal and Fraser River wild stocks that contribute to the Strait of Georgia are declining in abundance. Overfishing and habitat loss were identified as the major causes for the decline. Recommendations from the planning exercise integrate harvest and habitat management, enhancement, and evaluation (DFO 1992). The key action points are to increase spawning escapement through harvest reduction and to increase habitat awareness, protection, and improvement. The SEP's responsibilities include habitat improvement and educational activities, so SEP is an integral part of the key habitat recommendations. What is new in this plan is the emphasis on wild stock harvest management, the highlighting of habitat issues, and stated criteria for hatchery expansion and operations.

New hatchery production is not viewed as part of the initial phase of the plan. Rebuilding will first be attempted using harvest reduction. If, after two cycles, most stocks have responded, hatchery augmentation may be considered for the remainder. Failure of a large number of stocks to show gain after two cycles will result in further harvest reduction or other measures, not expanded hatchery production. The recommended role for hatcheries in rebuilding wild coho salmon in the Strait of Georgia is summarized below.

1. Do not increase hatchery production solely to maintain catch levels during the wild stock rebuilding period.
2. Consider hatcheries to augment stocks that are not rebuilding after two cycles at average exploitation rates in the 65–70% range.
3. Augment stocks in a manner designed to maintain genetic diversity: (a) augmentation should not be necessary for most stocks; (b) native stock should be used; (c) escapement of hatchery-origin adults will not exceed 50% of the total spawning target (50% rule); and (d) hatchery operations will be done in accordance with guidelines intended to minimize negative genetic and ecological interactions.
4. Monitor augmentation efforts including rebuilding effects, genetic composition, and compliance with the 50% rule.
5. Monitor wild and hatchery coho salmon production for evidence of negative interactions in the marine environment; if found, reduce hatchery production.

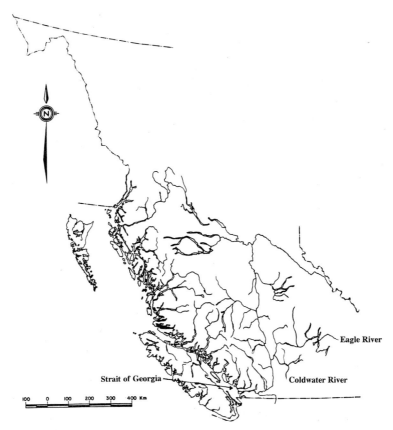

FIGURE 1.—Map of British Columbia showing the location of the Eagle and Coldwater rivers and the Strait of Georgia.

Freshwater Interactions

Hatchery augmentation of coho salmon in the Eagle and Coldwater rivers in interior British Columbia (Figure 1) was designed to use native broodstock and release fry during their first spring. The hatchery production plans were revised in the second year to add smolt releases. The fry generally rear a year in fresh water prior to migrating to sea. These augmentation programs were reviewed by Pitre and Cross (1993), who concluded that hatchery fry releases could increase stock production if habitat was underutilized, but that underestimation of wild fry abundance could lead to negative interactions between wild and hatchery fry.

Eagle River coho salmon escapement showed an initial positive response to hatchery production but has since declined. Escapement in 1991 was the lowest of the study period (Figure 2; Table 1). The ratio of return (catch plus escapement) per spawning adult for wild coho salmon and the survival to return for hatchery fry is plotted against total spring fry number in Figure 3. The hatchery survival data are based on coded wire tag (CWT) studies that involved tagging juveniles and recovering adults in the fisheries and escapement. The wild-origin escapement estimates are calculated from total escapement figures less the number of hatchery fish (from CWT data). The wild-origin catch is then estimated using the exploitation rate data for CWT hatchery coho salmon. Data sources and calculations are described by Pitre and Cross (1993). I estimated the total spring fry population in the river by assuming that each spawning adult produced 200 wild fry (this assumes fecundity is 2,000, egg-to-fry survival is 20%, and females make up 50% of escapement), and summing this estimate with the number of hatchery fry released in the spring (Table 1). The wild return per spawner and the hatchery fry survival data both indicate a decline with increasing fry abundance, suggesting that rearing capacity was exceeded.

Coldwater River data examined in the same way

TABLE 1.—Wild and hatchery coho salmon data for the Eagle and Coldwater rivers, British Columbia. Returns are equal to catch plus escapement.

Brood year	Number of wild spawners	Number of wild fry (×1,000)	Number of hatchery fry (×1,000)	Total number of fry (×1,000)	Return year	Returns Wild origin	Returns Hatchery origin	Escapement Wild origin	Escapement Hatchery origin	Return to spawner ratio for wild spawners	Hatchery fry survival (%)
					Eagle River						
1984	6,486	1,297	387	1,684	1987	13,485	7,126	6,540	3,456	2.08	1.29
1985	2,416	483	438	921	1988	19,551	7,046	6,083	2,190	8.09	1.11
1986	3,766	753	327	1,080	1989	7,486	5,132	3,279	2,248	1.99	0.57
1987	9,565	1,913	329	2,242	1990	12,332	3,897	3,315	1,056	1.29	0.63
1988	7,808	1,562	344	1,906	1991	3,871	1,176	1,475	442	0.50	0.24
					Coldwater River						
1984	600	120	61	181	1987	4,927	248	1,525	75	8.21	0.34
1985	430	86	111	197	1988	2,539	6,588	549	1,423	5.90	0.91
1986	803	161	332	493	1989	5,453	6,340	1,156	1,344	6.79	0.90
1987	1,061	212	227	439	1990	7,094	5,565	1,681	1,319	6.69	0.84
1988	1,325	265	228	493	1991	5,898	2,822	1,392	666	4.45	0.22

(Figure 4; Table 1) indicate that the hatchery releases have made a significant contribution to the rebuilding of the coho salmon stock. The wild return per spawner ratio and the survival of hatchery fry show no consistent trend relative to estimated fry abundance (Figure 5). The downturn in 1991 returns and survival at both Eagle and Coldwater are likely independent of stock-specific activities because returns were generally poor throughout southern British Columbia. The Coldwater hatchery-fry distribution differs from that at Eagle. Eagle hatchery fry are distributed throughout the available coho salmon rearing habitat. Coldwater hatchery fry are released into the upper river, which escaping adults can access only in high-water years. As a result Coldwater fry augmentation appears to be contributing to stock rebuilding with little interaction with the wild coho salmon.

Based on these studies we have discontinued fry releases into the Eagle River. The Coldwater fry program will continue as long as it is making a contribution without reducing wild production. Smolt releases are expected to continue at Eagle River as long as the additional production is needed to reach spawning targets or provide special fishery opportunities. Smolt releases are no longer considered essential for Coldwater coho salmon and have been discontinued.

We learned a number of things from this work that may be generally applicable to other species and stocks. One is that hatchery coho salmon fry may be detrimental to wild fry. Steward and Bjornn (1990) described the potential for competition and predation between wild and hatchery juveniles in streams. Release of hatchery coho salmon fry into naturally seeded areas should reflect carrying capacity estimates and wild fry abundance to avoid direct ecological and potential secondary genetic interactions.

Another lesson is that augmentation can work

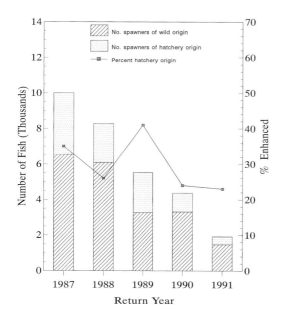

FIGURE 2.—Number of wild- and hatchery-origin spawning coho salmon returning to Eagle River and the percent of return due to hatchery-origin enhancement during 1987–1991 (from Pitre and Cross 1993).

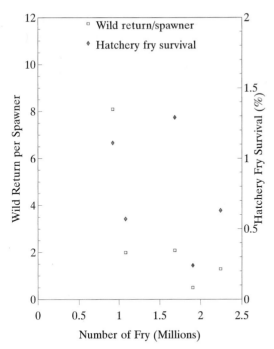

FIGURE 3.—The ratio of wild adult coho salmon return (catch plus escapement) per spawning adult and the survival to return for hatchery fry in relation to total spring fry abundance in the Eagle River during 1984–1988 brood years.

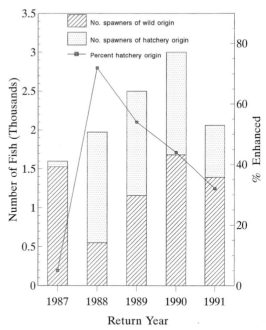

FIGURE 4.—Number of wild- and hatchery-origin spawning coho salmon returning to Coldwater River and the percent of return due to hatchery-origin enhancement during 1987–1991 (from Pitre and Cross 1993).

but populations should be monitored and hatchery release numbers should be adjusted in accordance with wild fry abundance. This is not always a simple process. Assessing natural rearing capacity and wild fry status and identifying any interactions between wild and hatchery salmon is complex and expensive work. We need simple, inexpensive methods that can be applied to a large number of streams each year. In some cases, especially for interior streams, wild adult spawner counts may be a useful index for subsequent wild fry numbers. In other cases a fry indexing system may be feasible. Simple mass-marking techniques for hatchery fish, enabling identification of the origin of returning adults, would be a major advancement.

We also learned the value of better understanding our objectives. Clearly the goal of fry augmentation programs is to rebuild the stock to the point where hatchery fry are all surplus to the natural capacity; that is, to put the hatchery out of business. There are only a few exceptions to this, due to unique habitat conditions. Both the Eagle and Coldwater hatchery coho salmon are produced in permanent facilities. It might be wiser to use portable or temporary hatchery facilities that could be moved to another problem area when the project goals are achieved. Permanent hatcheries situated to enhance several stocks from different streams offer some flexibility but may still outlive their usefulness. Along with permanent hatcheries come staff, local community, and resource user expectations, which complicate reaching a biologically based decision to terminate a hatchery program.

Ocean Interactions

Hatchery production of chinook and coho salmon smolts released from Fraser River and coastal Canadian hatcheries feeding into the Strait of Georgia (Figure 1) have been assessed using CWT marking and recovery since hatchery production of these species was started in 1971. Methodology and data sources are detailed in Cross et al. (1991).

Adult returns for chinook salmon have not increased in direct proportion to the number of smolts released (Figure 6; Table 2). A Beverton–Holt stock recruitment model was fit using log-transformed data (Hilborn and Walters 1992). The average recruitment curve is shown in Figure 6. These data are not adjusted to compensate for

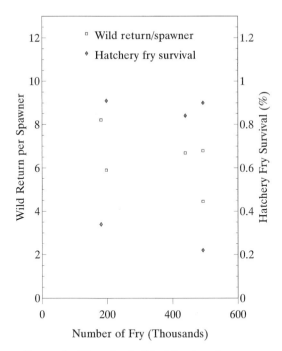

FIGURE 5.—The ratio of wild adult coho salmon return (catch plus escapement) per spawning adult and survival to adult of hatchery fry in relation to total fry abundances in the Coldwater River during 1984–1988 brood years.

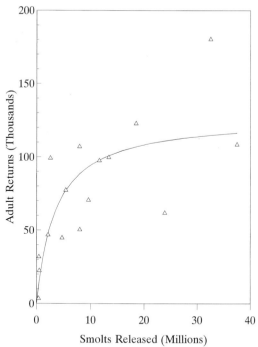

FIGURE 6.—Relationship between the number of returning adult chinook salmon and the number of smolts released from hatcheries in the Strait of Georgia during 1972–1987 brood years.

changes in fisheries regulations that have resulted in increased, unaccounted for, fishing mortality during the study period.

There is a more direct relationship between adult returns and smolt releases for coho salmon than for chinook salmon (Figure 7; Table 2). The Beverton–Holt average recruitment curve is shown for the coho salmon data; linear correlation analysis is also highly significant.

The survival data for Strait of Georgia hatchery chinook salmon suggest possible density-dependent mortality and is reminiscent of Oregon hatchery coho salmon data. The Oregon data have been examined by several authors, recently Emlen et al. (1990). The fundamental issue is whether the apparent decline in return rate (survival) is due to excessive releases of hatchery smolts resulting in density-dependent mortality. With 26 year-classes of wild and hatchery coho salmon data in Oregon the answer is still not clear. Emlen et al. (1990) caution that the conclusion of no density dependence based on earlier analyses (Nickelson 1986) is suspect and that the analyses of Emlen et al. (1990) suggest that density dependence is occurring.

Survival data for Strait of Georgia hatchery coho salmon indicate some reduction at higher smolt numbers but less reduction than that observed for chinook salmon. It is important to note that survival of both chinook and coho salmon declined dramatically in the late 1970s (Table 2). This decline is coincident with an increase in Strait of Georgia surface-water temperature during 1977 that has persisted into the 1990s (Walters 1993).

The concern for Strait of Georgia coho salmon, unlike chinook salmon, is not declining survival of hatchery fish, but the declining abundance of wild coho salmon. The estimated catch of wild Strait of Georgia coho salmon in Canadian fisheries from 1976 to 1989 decreased while the Canadian Strait of Georgia hatchery coho salmon catch increased (DFO 1992; Figure 8). Wild coho salmon escapement also declined (DFO 1992). This has led to speculation that hatchery coho are replacing wild coho salmon.

Walters (1993) described four general hypotheses that may explain the Strait of Georgia wild coho salmon decline: overfishing, freshwater habitat loss, ocean carrying-capacity limitations, and ocean conditions. As noted earlier, DFO has identified over-

TABLE 2.—Hatchery smolt release, adult return, and survival data for Strait of Georgia chinook and coho salmon.

Brood year	Coho salmon			Chinook salmon		
	Smolt release	Adult return	Survival (%)	Smolt release	Adult return	Survival (%)
1972	448,029	61,958	13.8	883,815	8,026	0.91
1973	771,037	157,399	20.4	811,065	35,596	4.39
1974	2,120,937	228,640	10.8	432,935	32,273	7.45
1975	2,160,224	369,288	17.1	2,090,609	47,271	2.26
1976	2,271,315	351,003	15.5	2,508,259	129,338	5.16
1977	2,597,727	405,551	15.6	5,382,697	78,917	1.47
1978	3,256,738	370,512	11.4	4,681,312	45,550	0.97
1979	3,809,461	353,812	9.3	8,020,445	50,856	0.63
1980	4,142,662	387,079	9.3	9,913,800	72,398	0.73
1981	3,922,413	318,400	8.1	8,141,582	108,354	1.33
1982	5,263,439	568,666	10.8	11,879,505	98,387	0.83
1983	10,789,493	718,588	6.7	14,068,365	107,436	0.76
1984	9,094,886	735,697	8.1	18,187,313	126,777	0.70
1985	6,885,616	830,536	12.1	23,521,329	64,184	0.27
1986	6,933,286	593,481	8.6	30,820,474	182,697	0.59
1987	6,681,796	571,385	8.6	36,981,046	118,531	0.32
1988	7,551,030	505,519	6.7			
1989	7,688,023	477,439	6.2			

fishing and habitat degradation as the most probable causes of the decline. Walters (1993) concludes that density-dependent competition between wild and hatchery coho salmon at sea is the most likely cause but acknowledges that oceanographic change cannot be ruled out. He discounts the overfishing and freshwater habitat hypotheses. Declines in hatchery chinook salmon survival may be due to ocean carrying-capacity limitations or oceano-

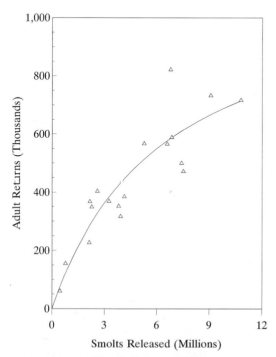

FIGURE 7.—Relationship between the number of returning adult coho salmon and the number of smolts released from Strait of Georgia hatcheries during 1972–1989 brood years.

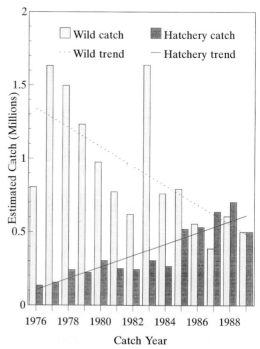

FIGURE 8.—Catch of wild-origin and hatchery-origin Strait of Georgia coho salmon in Canadian fisheries during 1976–1989.

graphic change as described in Walters' coho salmon model; estuarine capacity and smolt quality are other possible causes. Changes in the survival of hatchery chinook salmon are not the result of overfishing or freshwater habitat degradation.

It is not my intent to debate these opposing hypotheses here. Rather I have tried to describe two calamitous biological trends in Strait of Georgia salmon, the decline in survival of hatchery chinook salmon and the decline of wild coho salmon, for which there are no clear explanations. The extended debate in Oregon sends a strong message that continuing the status quo will not produce answers in the near future. Emlen et al. (1990) proposed large-scale experiments, such as described by Walters et al. (1988), to test the density dependence hypothesis for Oregon coho salmon. The DFO scientific and enhancement staff, with input from the University of British Columbia, have already invested considerable effort into design of such an experiment for Strait of Georgia chinook salmon.

A conceptual model for one experimental approach is shown in Figure 9. This experiment would require alternating hatchery production from 50% to 100% of capacity for a number of years. If density-dependent mortality is occurring, survival is expected to increase for small year-classes, resulting in relatively stable adult returns. If there is no density-dependent mortality, survival should vary independent of smolt numbers, and adult returns will be lower on average for small year-classes.

It was our intention to review the finalized design with user groups and, with their concurrence, initiate the experiment as quickly as possible. Two developments have interfered. The first is the increasing concern about density-dependent interactions between wild and hatchery coho salmon among some scientists and their suggestion that it is more important to do the experiment for coho than for chinook salmon. There was consensus that the experiment should be done for only one species at a time. It is important that all parties agree on a detailed design if we hope to agree on the interpretation of results. The second development is the output of experimental design work. While incomplete, preliminary analyses indicate a conclusive experiment for chinook salmon may require 20–30 years. A similar study for coho salmon would likely take 5–10 years. As a result, we must be prepared to make a long-term commitment to the experiment, but the species that most urgently requires study is uncertain. A final decision will depend on our analyses of the probability of success, the risks of the

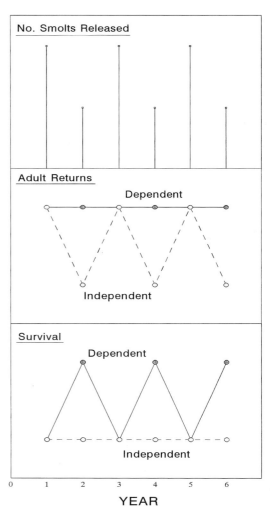

FIGURE 9.—Conceptual model of an experiment to determine if Strait of Georgia chinook or coho salmon production is limited by density-dependent interactions at sea. Number of hatchery smolts released are 100% and 50% of capacity in alternate years. If there is density dependence (solid lines) at current smolt abundance, adult returns are expected to remain relatively constant and survival rates are expected to be higher in years of low smolt numbers.

experiment, and the understanding of all interested parties.

Conclusion

Hatcheries have a demonstrated capability to produce salmon. There may be concerns over marine survival rates, but the annual harvest of over 500,000 coho salmon and 100,000 chinook salmon produced in Strait of Georgia hatcheries (Table 2) is an important contribution to our fisheries. The past 20 years have seen not only growth in fish production but also a tremendous growth in knowledge, much of which is attributable to hatchery-related research and assessment studies. We have better understanding of the life history and ocean distribution of salmon stocks and of the harvest, ecological, and genetic interactions between wild and hatchery stocks. The challenge is to apply this new knowledge when making decisions on stock and harvest management objectives and on the role of hatcheries. These decisions will often be difficult.

We must also promote research that will provide further understanding, recognizing that the most valuable research is likely going to require large-scale interventions in the production of juvenile salmon with possible effects, at least in the short term, on adult returns.

Acknowledgments

I am grateful to several Canadian Department of Fisheries and Oceans staff for their assistance with this report, including C. Cross, S. Lehmann, and K. Pitre for data compilation and discussion; D. Patterson and A. Kling for preparation of figures; and V. Palermo and B. Piper for statistical analysis. L. Blankenship and W. Smoker provided helpful review of the manuscript. Thanks also to C. Walters, University of British Columbia, for stimulating discussion on several aspects of the paper.

References

Cross, C. L., L. Lapi, and E. A. Perry. 1991. Production of chinook and coho from British Columbia hatcheries, 1971 through 1989. Canadian Technical Report of Fisheries and Aquatic Sciences 1816.

DFO (Canadian Department of Fisheries and Oceans). 1992. Strait of Georgia coho salmon planning process and recommendations: south coast coho initiative final report. DFO, Pacific Region, Vancouver.

Emlen, J. M., R. R. Reisenbichler, A. M. McGie, and T. E. Nickelson. 1990. Density-dependence at sea for coho salmon (*Oncorhynchus kisutch*). Canadian Journal of Fisheries and Aquatic Sciences 47:502–515.

Hilborn, R., and C. J. Walters. 1992. Quantitative fisheries stock assessment: choice, dynamics and uncertainty. Chapman and Hall, London.

Nickelson, T. E. 1986. Influence of upwelling, ocean temperature, and smolt abundance on marine survival of coho salmon (*Oncorhynchus kisutch*) in the Oregon production area. Canadian Journal of Fisheries and Aquatic Sciences 43:527–535.

Pitre, K. R., and C. L. Cross. 1993. Impacts of coho enhancement on three Thompson River tributaries. Pages 140–150 *in* L. Berg and P. W. Delaney, editors. Proceedings of the coho workshop. Department of Fisheries and Oceans, Vancouver.

Steward, C.R., and T.C. Bjornn. 1990. Supplementation of salmon and steelhead stocks with hatchery fish: a synthesis of published literature. U.S. Department of Energy Technical Report 90-1, Bonneville Power Administration, Portland, Oregon.

Walters, C. J. 1993. Where have all the coho gone? Pages 1–8 *in* L. Berg and P. W. Delaney, editors. Proceedings of the coho workshop. Department of Fisheries and Oceans, Vancouver.

Walters, C. J., J. S. Collie, and T. Webb. 1988. Experimental designs for estimating transient responses to management disturbances. Canadian Journal of Fisheries and Aquatic Sciences 45:530–538.

Beneficial Uses of Marine Fish Hatcheries: Enhancement of Red Drum in Texas Coastal Waters

LAWRENCE W. MCEACHRON, C. EUGENE MCCARTY, AND ROBERT R. VEGA

Texas Parks and Wildlife Department
702 Navigation Circle, Rockport, Texas 78382, USA

Abstract.—In Texas, the population of red drum *Sciaenops ocellatus* began a dramatic decline in the 1970s, prompting the Texas Parks and Wildlife Department (TPWD) to implement an assessment and recovery plan. Management approaches used were (1) initiation of an independent monitoring program to assess relative abundance; (2) implementation of restrictive regulations to reduce fishing pressure, culminating in allocation to the recreational fishery; and (3) development and implementation of a marine enhancement program based on the release of hatchery-reared fingerlings and assessment of subsequent survival. The enhancement program, begun in 1971, consisted of a systematic approach that began with determining efficient spawning and stocking techniques, continued with assessments of stocking survival and possible success, and was followed by a final phase of critically evaluating effects of enhancement on natural populations. Substantial effort to identify stocked red drum and enhancement of the population has relied on (1) fish stocking out of season; (2) tags (magnetic, jaw, and internal anchor); (3) otolith and scale analysis; (4) genetic and chemical markers; and (5) population analysis and harvest monitoring. All studies revealed stocked fish survived and stocking enhanced the natural population. Genetic and chemical marking of fish and multivariate analyses of all data are currently underway to quantify further the magnitude of enhancement. Preliminary results were encouraging. Four items were found to be critical to the assessment of stocking: (1) the smallest fish that can survive should be used to minimize hatchery and stocking costs; (2) fish should be stocked in a natural system to evaluate enhancement; (3) massive stockings may be required because small-scale experimental stockings may not be large enough to affect local variation in wild populations; and (4) benefits may be determined on a large scale but not on a small scale. Recently, the red drum population in Texas bays rebounded due to a number of factors that had a positive effect on the recovery. The TPWD's long-term management plan using hatcheries and stocking to supplement natural spawning played a role in ameliorating the decline of red drum populations.

Red drum *Sciaenops ocellatus* in Texas have been exploited by anglers and commercial fishers for over 100 years. Conflicts of allocation have occurred between these competing interests (Heffernan and Kemp 1980; Matlock 1982). Historically, management of red drum generally consisted of enacting size and bag limits and seasonal, area, and gear closures to protect the resource and to minimize conflicts between anglers and commercial fishers. Heffernan and Kemp (1980) and Matlock (1980) present detailed overviews of red drum management.

The state of Texas has experienced rapid population growth, especially in coastal areas, beginning in the 1960s and continuing to the present. With this growth has come increased conflicts over allocation of red drum, overexploitation (Matlock 1984), and possibly recruitment overfishing. In 1971, the Texas Parks and Wildlife Department (TPWD) embarked on a long-term assessment and recovery plan to increase red drum abundance: (1) a fishery-independent monitoring program was implemented in 1975 to assess relative abundance and determine reasons for trends in abundance (Matlock 1984; McEachron and Green 1987; Dailey et al. 1992; Weixelman et al. 1992); (2) red drum mortality rates were controlled through fishing regulations affecting both anglers and commercial fishers and through habitat protection to the fullest extent possible (ultimately, sale of wild red drum and use of nets in Texas marine water were prohibited, and anglers were restricted to three red drums between 508 and 711 mm in length daily); and (3) research assessing stock enhancement of the red drum population was begun.

This paper presents an overview of the TPWD's efforts at enhancing the red drum population through stocking and documents studies conducted in Texas that evaluate enhancement efforts.

Background

Based on its life history (NMFS 1986), red drum is an excellent candidate for enhancement through stocking. Red drum generally comprises two groups: adults (spawners) in offshore water and young (≤6 years old) in estuaries and nearshore water. Spawning occurs in late summer and fall (August–December) in nearshore water close to passes. Eggs, larvae, and post larvae fish are carried into estuaries on wind and tidally driven currents. Red drum remain in or near estuaries for up to 6 years, then move off shore as they grow older. Strength of a naturally spawned year-class of red drum is dependent upon larval recruitment from the Gulf of Mexico into Texas bays (Matlock 1987). If fishery managers could bypass mortality associated with larval recruitment from the Gulf, then stocking red drum into estuaries could possibly enhance the natural stock.

For stocking to proceed, mass spawning and rearing techniques had to be developed. Crucial research on physiology, maturation, and spawning was conducted both in Texas and Florida (Arnold et al. 1979; Roberts et al. 1978). Techniques for spawning red drum by use of temperature and photoperiod manipulations and rearing larvae to fingerling size (about 25 mm in total length) by use of rearing ponds were developed (Colura et al. 1976; Geiger 1983; McCarty et al. 1986; Colura et al. 1990; Geiger and Turner 1990; McCarty 1990). Once these techniques were refined, stocking of large numbers of red drum was possible. During 1983–1993, more than 2 billion eggs were spawned and over 140 million fingerlings were released into Texas bay systems.

Due to concerns over reduced heterozygosity in the natural population, the following practices were integrated into Texas marine hatchery operations from the very beginning: (1) wild broodstock were randomly collected in coastal and offshore water; (2) 140–180 broodfish were held in spawning tanks at all times, with 5–6 fish in each tank (3 female), depending on size of fish; (3) each year fish were mixed at random into new groups for next year's production; (4) at least 25% of the broodfish were rotated out of production annually and replaced with recently captured wild fish; and (5) broodfish were maintained in the program for no longer than 4 years and were then released back into the wild. Additionally, the TPWD implemented a program to catalogue the genetic phenotype of both broodfish and subsequent fingerlings stocked into the bays. These data will allow TPWD to assess the heterogeneity of fish and allow comparison of the genetic makeup of stocked fish with naturally occurring populations.

Early Research

For stocking to be successful, juveniles must survive handling stress; low survival would jeopardize success of the program. Less than 1% of fingerlings died after being transported in trailers for up to 5 h (Tomasso and Carmichael 1988). Once stocked into the wild, red drum fingerlings had an overall 24-h survival rate of over 86% (Hammerschmidt 1986).

After stocking, follow-up studies were conducted to determine if red drum could be captured. Out-of-season stocking (May–July) was used to identify stocked fish. Larvae ($N = 4.5$ million) were stocked into a small bay in 1979 but none were recaptured during subsequent sampling (Matlock 1988). Fingerlings ($N = 1.3$ million) stocked into two bays were caught in bag seines for up to 1.5 months following out-of-season stocking, after which they were not vulnerable to capture gear (Dailey and McEachron 1986). Jaw-tagged red drum ($N = 5,942$) were released into Matagorda Bay; 10 recaptures were recorded through 8 months post stocking (Matlock et al. 1984). Three red drum tagged with coded microwires were caught following a special release of 38,000 nose-tagged fingerlings in St. Charles Bay (part of Aransas Bay) during 1979 and 1980 (Matlock et al. 1986). The low recapture rate was attributed to tag loss (Gibbard and Colura 1980). Further research in Texas using cheek-tagged fish was also found to be impractical (Bumguardner et al. 1990; Bumguardner et al. 1992). Out-of-season stocked red drum ($N = 2$ million) in St. Charles Bay in 1979 and 1981 could be followed for up to 9 months by use of a combination of bag seines and 5.08-cm gill nets (Matlock et al. 1986). Analysis of these data revealed bag-seine catches in the stocked bay were significantly higher than catches in an adjacent unstocked bay. Matlock et al. (1986) concluded that stocking enhanced the population and success was based on number of fish stocked and environmental conditions immediately after stocking.

An intensive 2-year stocking program in two adjacent Texas bays was evaluated using gill nets (7.6, 10.2, 12.7, 15.2-cm stretch mesh) and angler catches during 1985–1986 (Matlock 1990). Relative abundance in gill nets was consistently higher in the stocked bay (Corpus Christi Bay) compared with the unstocked bay (upper Laguna Madre). Initially, higher catches were noted in the 7.6-cm mesh, and then in larger meshes in subsequent seasons; this pattern was not evident in the unstocked bay. Analysis of recreational data revealed angler mean land-

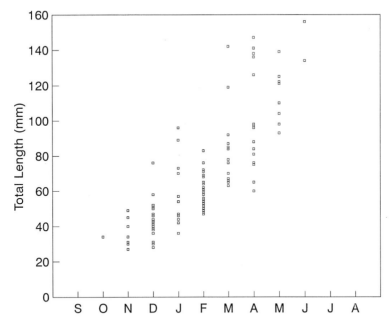

FIGURE 1.—Lengths of red drums caught in upper Laguna Madre bag-seine samples, by month, when no red drum were stocked. Data summarize catches from January 1978–July 1979; December 1981–October 1984; and November 1985–March 1987. Squares are size classes, not individual lengths.

ing rate in the stocked bay increased 150% over the historic mean (1979–1984) compared with a 50% increase in the unstocked bay during the same period (Matlock 1990).

Large-scale stocking of fishes can have deleterious effects on natural populations (Rutledge and McCarty 1989). To address this concern, studies were conducted to determine normal levels of spatial and temporal red drum genetic variability. Allozyme (Wakeman and Ramsey 1988; Gold et al. 1993a; Gold et al., in press) and mitochondrial DNA (Gold and Richardson 1991; Gold et al. 1993a, 1993b) analyses revealed little variation among samples from the Gulf of Mexico and Atlantic Ocean, an absence of spatial or temporal allelic heterogeneity among inshore and offshore red drum, and similar genetic variability (e.g., heterozygosity and haplotype frequencies) in red drum collected from North Carolina to south Texas. These results suggest that red drum in the northern Gulf of Mexico compose a randomly mating population. Similar genetic variability suggests that variability has not been reduced among Atlantic Ocean and Gulf of Mexico red drum populations as a result of the Texas supplemental stocking program.

Current Research

The TPWD implemented a fishery-independent monitoring program in 1975 to assess relative abundance of fishes. Stocked fish were captured in bag seines throughout the 1980s and 1990s in this random sampling program (TPWD, unpublished data). Independent from TPWD, National Marine Fisheries Service (NMFS) caught six known stocked red drums in Galveston Bay in 1992, 22 d poststocking (Edward F. Klima, National Marine Fisheries Service, personal communication). However, through 1989 too few fish were stocked in most bays to fully assess enhancement. Beginning in 1990, coastwide production was increased to 16–30 million fingerlings per year.

Comparing the lengths of wild red drum caught monthly in upper Laguna Madre during 1978–1987 (Figure 1) with lengths of fish caught during 1990–1993, when out-of-season red drum were intensively stocked, revealed at least 20% of the upper Laguna Madre juveniles captured during 1990–1993 routine bag-seine samples were stocked fish (Figure 2). If survival, growth, and all other effects on stocked fish are similar to wild fish, ultimately at least 20% of the fish caught by anglers could originate from

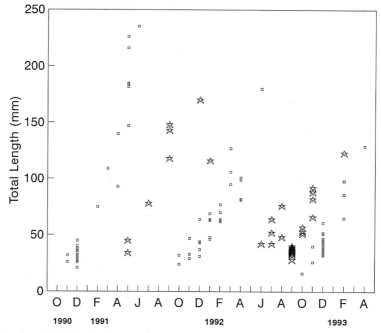

FIGURE 2.—Lengths of individual red drums caught in upper Laguna Madre bag-seine samples, by month (October 1990–April 1993), when red drum fingerlings were stocked out of season. Stars are out-of-season stocked fish caught and squares are red drums caught during normal recruitment season.

stocking. The TPWD is currently conducting multivariate statistical analyses of all data collected between 1975 and 1993 for nine bay systems to determine if coastwide enhancement can be further quantified; preliminary results reveal positive correlations between stocking and subsequent gill-net relative abundance indices for at least four Texas bays.

The Optical Pattern Recognition System (OPRS) is an image analysis software program (BioSonics 1987) being used by the TPWD to collect measurements of scale circuli patterns from hatchery and wild fish. Statistical analyses will be used to discriminate between hatchery and wild fish based on these measurements. It has been demonstrated that homogeneous conditions in a hatchery produce homogeneous circuli formations on scales. Highly variable conditions in the wild yield variable circuli patterns on wild fish. Comparison of circuli patterns from hatchery fish reared in spring and fall resulted in correct classification of 86% of spring fish and 78% of fall fish by use of OPRS (TPWD, unpublished data). Research conducted in Florida correctly classified 85% of known hatchery and 87% of known wild red drum based on daily increment measurements in the asteriscus otolith of juveniles (J. Jeffery Isely, South Carolina Cooperative Fish and Wildlife Research Unit, personal communication). Scale samples from every TPWD culture pond harvested since July 1992 and continuing into the future will provide a database to compare hatchery-reared fish from different ponds, seasons, and years. These scale samples will be used to test the feasibility of comparing wild and natural fish scales to determine percent of hatchery fish caught in sampling gear during fall-winter, when juvenile wild fish are present.

Future Research

All studies conducted revealed stocked fish were surviving and enhancement to the natural population most likely had occurred, but the magnitude of enhancement has not been determined. Research by the TPWD has identified a gene marker that can be selectively bred in red drum (King et al. 1993). Follow-up research revealed this gene marker caused no difference in mean length or weight of hatchery fish when compared with wild fish. In fall 1993, 250,000 selectively bred fingerlings that carried the gene marker were stocked into East Matagorda Bay. These fish carry the gene marker permanently and can be identified throughout their life. Collection of red drum tissue samples through independent sampling (bag seines and gill nets) and from anglers landings should allow TPWD to follow these fish while they reside in the bays.

Concurrent with stocking of selectively bred red drum, about 1,000,000 fish were marked with oxytetracycline (OTC) and stocked into upper Laguna Madre in fall 1993 to distinguish between wild and stocked fishes during the normal recruitment period (October–March); otoliths of these marked fish will fluoresce under ultraviolet light illumination (Thomas 1993). As of 30 June 1994, 15 of 65 fish caught in routine sampling carried the OTC mark. Because of this initial success, the OTC marking program will be expanded. Both the genetic and OTC marking studies provide a means of determining the proportion of hatchery to wild fish caught in routine bag-seine sampling. These studies should help determine the magnitude of enhancement and, ultimately, determine if stocked fish are being caught by anglers.

For a hatchery stocking program to be effective, there should be an acceptable cost-benefit ratio. Preliminary findings revealed a positive cost-benefit projection for red drum stocking, even at low survival rates (Rutledge 1989). However, a definitive economic analysis has not been conducted because of lack of information on the number of stocked fish being landed or entering the spawning population. Once these data have been collected, a detailed cost-benefit analyses will be conducted by the TPWD.

Conclusion

Using stocked fish following adverse climatic or anthropogenic events is an important management tool. The innovative use of stocking, in concert with traditional management practices, could provide a powerful combination to manage our natural resources wisely (Rutledge and Matlock 1986). Stocking alone is not intended to provide long-term sustainability; rather it should be used to enhance existing populations. Managers at TPWD do not intend to use the stocking program as a means to obtain long-term sustainability for the red drum population. The objective is to integrate this effective tool discerningly into TPWD fisheries management protocol.

It has taken TPWD 23 years to reach the present stage of its stocking and recovery program. To date, over 1 billion eggs, larvae, and fingerlings have been stocked in Texas marine water. Stocking hatchery fish, in addition to stringent regulations, associated research, and habitat protection, has allowed the population of red drum in Texas to recover sufficiently for the TPWD to consider liberalizing anglers' harvest. This recovery is a tribute to the foresight of Texas' fishery managers willingness to use the controversial technique of stocking to assist in restoring red drum to historical levels. Texas' long-term assessment and recovery program serves as a blueprint for managers developing future enhancement programs.

References

Arnold, C. R., W. H. Bailey, T. D. Williams, A. Johnson, and J. L. Lasswell. 1979. Laboratory spawning and larval rearing of red drum and southern flounder. Proceedings of the Annual Conference Southeastern Association of Fish and Wildlife Agencies 31(1977): 437–440.

BioSonics, Inc. 1987. Optical pattern recognition system, data acquisition program manual, version 1.08. BioSonics, Inc., Seattle.

Bumguardner, B. W., R. L. Colura, and A. F. Maciorowski. 1990. Tag retention, survival, and growth of red drum fingerlings marked with coded wire tags. American Fisheries Society Symposium 7:286–292.

Bumguardner, B. W., R. L. Colura, and G. C. Matlock. 1992. Long-term coded wire tag retention in juvenile *Sciaenops ocellatus*. U.S. National Marine Fisheries Service Fishery Bulletin 90:390–394.

Colura, R. L., B. T. Hysmith, and R. E. Stevens. 1976. Fingerling production of striped bass (*Morone saxatilis*), spotted seatrout (*Cynoscion nebulosus*), and red drum (*Sciaenops ocellatus*) in saltwater ponds. Proceedings of the World Mariculture Society 7:79–92.

Colura, R. L., B. W. Bumguardner, A. Henderson-Arzapalo, and J. D. Gray. 1990. Culture of red drum fingerlings. Texas Parks and Wildlife Department, Management Data Series 22, Fisheries Division, Austin.

Dailey, J. A., and L. W. McEachron. 1986. Survival of unmarked red drum stocked into two Texas bays. Texas Parks and Wildlife Department, Management Data Series 116, Coastal Fisheries Branch, Austin.

Dailey, J. A., J. A. Kana, and L. W. McEachron. 1992. Trends in relative abundance and size of selected finfishes and shellfishes along the Texas coast: November 1975–December 1990. Texas Parks and Wildlife Department, Management Data Series 74, Fisheries and Wildlife Division, Austin.

Geiger, J. G. 1983. Zooplankton production and manipulation in striped bass rearing ponds. Aquaculture 35:331–351.

Geiger, J. G., and C. J. Turner. 1990. Pond fertilization and zooplankton management techniques for production of fingerling striped bass and hybrid striped bass. Pages 79–98 *in* R. M. Harrell, J. H. Kerby, and R. V. Minton, editors. Culture and propagation of striped bass and its hybrids. American Fisheries Society, Southern Division, Striped Bass Committee, Bethesda, Maryland.

Gibbard, G. L., and R. L. Colura. 1980. Retention and movement of magnetic nose tags in juvenile red drum. Annual Proceedings of the Texas Chapter, American Fisheries Society 3:22–29.

Gold, J. R., T. L. King, L. Richardson, D. Bohlmeyer, and G. C. Matlock. In press. Allozyme differentiation within and between red drum (*Sciaenops ocellatus*) from the Gulf of Mexico and Atlantic Ocean. Journal of Fish Biology 44.

Gold, J. R., and L. R. Richardson. 1991. An analysis of population structure in the red drum (*Sciaenops ocellatus*) using mitochondrial DNA. Fisheries Research 12:213–241.

Gold, J. R., L. R. Richardson, T. L. King, and G. C Matlock. 1993a. Temporal stability of nuclear gene (allozyme) and mitochondrial DNA genotypes among red drum from the Gulf of Mexico. Transactions of the American Fisheries Society 122:659–668.

Gold, J. R., L. R. Richardson, C. Furman, and T. L. King. 1993b. Mitochondrial DNA differentiation and population structure in red drum (*Sciaenops ocellatus*) from the Gulf of Mexico and Atlantic Ocean. Marine Biology 116:175–185.

Hammerschmidt, P. C. 1986. Initial survival of red drum fingerlings stocked in Texas bays during 1984–1985. Texas Parks and Wildlife Department, Management Data Series 106, Coastal Fisheries Branch, Austin.

Heffernan, T. L., and R. J. Kemp. 1980. Management of the red drum resource in Texas. Pages 71–80 *in* R. O. Williams, J. E. Weaver, and F. A Kalber, co-chairmen. Proceedings: colloquium on the biology and management of red drum and seatrout. Gulf States Marine Fisheries Commission 5, Ocean Springs, Mississippi.

King, T. L., R. Ward, and I. R. Blandon. 1993. Gene marking: a viable assessment method. Fisheries (Bethesda) 18(2):4–5.

Matlock, G. C. 1980. History and management of the red drum fishery. Pages 37–53 *in* R. O. Williams, J. E. Weaver, and F. A. Kalber, co-chairmen. Proceedings: colloquium on the biology and management of red drum and seatrout. Gulf States Marine Fisheries Commission 5, Ocean Springs, Mississippi.

Matlock, G. C. 1982. The conflict between user groups of red drum and spotted seatrout in Texas. Pages 101–108 *in* R. H. Stroud, editor. Marine recreational fisheries 7: proceedings of the 7th Annual Marine Recreational Fisheries Symposium. Sport Fishing Institute, Washington, DC.

Matlock, G. C. 1984. A basis for the development of a management plan for red drum in Texas. Doctoral dissertation. Texas A&M University, College Station.

Matlock, G. C. 1987. The role of hurricanes in determining year-class strength of red drum. Contributions in Marine Science 30:39–47.

Matlock, G. C. 1988. Survival of red drum fry stocked into Christmas Bay, Texas. Texas Parks and Wildlife Department, Management Data Series 152, Coastal Fisheries Branch, Austin.

Matlock, G. C. 1990. Preliminary results of red drum stocking in Texas. NOAA (National Oceanic and Atmospheric Administration) Technical Report NMFS (National Marine Fisheries Service) 85:11–15.

Matlock, G. C., B. T. Hysmith, and R. L. Colura. 1984. Return of tagged red drum stocked into Matagorda Bay, Texas. Texas Parks and Wildlife Department, Management Data Series 63, Coastal Fisheries Branch, Austin.

Matlock G. C., R. J. Kemp, Jr., and T. L. Heffernan. 1986. Stocking as a management tool for a red drum fishery, a preliminary evaluation. Texas Parks and Wildlife Department, Management Data Series 75, Coastal Fisheries Branch, Austin.

McCarty, C. E. 1990. Design and operation of a photoperiod/temperature spawning system for red drum. Texas A&M University Sea Grant College Program TAMU-SG-90-603:44–45.

McCarty, C. E., J. G. Geiger, L. N. Sturmer, B. A. Gregg, and W. P. Rutledge. 1986. Marine finfish culture in Texas: a model for the future. Pages 249–262 *in* R. H. Stroud, editor. Fish culture in fisheries management. American Fisheries Society, Fish Culture Section and Fisheries Management Section, Bethesda, Maryland.

McEachron, L. W., and A. W. Green. 1987. Assessment of annual relative abundance and mean length of six marine fishes in Texas coastal waters. Proceedings of the Annual Conference Southeastern Association of Fish and Wildlife Agencies 38(1984):506–519.

NMFS (National Marine Fisheries Service). 1986. Secretarial fishery management plan for the red drum fishery of the Gulf of Mexico. NMFS, Washington, DC.

Roberts, D. E., Jr., B. V. Harpster, and G. E. Henderson. 1978. Conditioning and induced spawning of the red drum (*Sciaenops ocellatus*) under varied conditions of photoperiod and temperature. Proceedings of the World Mariculture Society 9:311–332.

Rutledge, W. P. 1989. The Texas marine hatchery program—it works. Pages 49–52 *in* J. Olfe, editor. California cooperative oceanic fisheries investigations. California Department of Fish and Game 30, Long Beach.

Rutledge, W. P., and G. C. Matlock. 1986. Mariculture and fisheries management—a future cooperative approach. Pages 119–127 *in* R. H. Stroud, editor. Fish culture in fisheries management. American Fisheries Society, Fish Culture Section and Fisheries Management Section, Bethesda, Maryland.

Rutledge, W. P., and C. E. McCarty. 1989. Fishery management realties versus fishery genetic fantasies. Texas Parks and Wildlife Department, Management Data Series 12, Fisheries Division, Austin.

Thomas, L. M. 1993. Chemical mark application in red drum (*Sciaenops ocellatus*). Master's thesis. Texas A&M University, College Station.

Tomasso, J. R., and G. C. Carmichael. 1988. Handling and transport-induced stress in red drum fingerlings (*Sciaenops ocellatus*). Contributions in Marine Science 30(supplement):133–137.

Wakeman, J. M., and P. R. Ramsey. 1988. Population structure and genetic variation in red drum. Contributions in Marine Science 30(supplement):49–56.

Weixelman, M., K. W. Spiller, and P. Campbell. 1992. Trends in finfish landings of sport-boat anglers in Texas waters, May 1974–May 1991. Texas Parks and Wildlife Department, Management Data Series 85, Fisheries and Wildlife Division, Austin.

A Responsible Approach to Marine Stock Enhancement

H. LEE BLANKENSHIP

Washington Department of Fish and Wildlife
600 Capitol Way North, Mail Stop 43149
Olympia, Washington 98501, USA

KENNETH M. LEBER

The Oceanic Institute, Makapuu Point
Waimanalo, Hawaii 96795, USA

Abstract.—Declining marine fish populations worldwide have rekindled an interest in marine fish enhancement. Recent technological advances in fish tagging and marine fish culture provide a basis for successful hatchery-based marine enhancement. To ensure success and avoid repeating mistakes, we must take a responsible approach to developing, evaluating, and managing marine stock enhancement programs. A responsible-approach concept with several key components is described. Each component is considered essential to control and optimize enhancement. The components include the need to (1) prioritize and select target species for enhancement; (2) develop a species management plan that identifies harvest opportunity, stock rebuilding goals, and genetic objectives; (3) define quantitative measures of success; (4) use genetic resource management to avoid deleterious genetic effects; (5) use disease and health management; (6) consider ecological, biological, and life-history patterns when forming enhancement objectives and tactics; (7) identify released hatchery fish and assess stocking effects; (8) use an empirical process for defining optimum release strategies; (9) identify economic and policy guidelines; and (10) use adaptive management. Developing case studies with Atlantic cod *Gadus morhua*, red drum *Sciaenops ocellatus*, striped mullet *Mugil cephalus*, and white seabass *Atractoscion nobilis* are used to verify that the responsible approach to marine stock enhancement is practical and can work.

Marine fish populations are declining worldwide. In the United States, current abundance trends are known for only 15 of the most important marine stocks; about half of them are declining (NOAA 1991, 1992). Current harvest rates on most declining stocks are far in excess of exploitation levels needed to maintain the high long-term average yields that could be achieved through contemporary fishery management practices. Projected increases in human population size worldwide suggest this trend will continue into the future (FAO 1991).

Three principal tactics are available to fishery managers to replenish depleted stocks and manage fishery yields: regulating fishing effort; restoring degraded nursery and spawning habitats; and increasing recruitment through propagation and release (stock enhancement). The first two methods form the basis for the current federal approach to managing marine fisheries in the United States. The potential of the third method has not been convincingly documented with marine fishes.

Marine stock enhancement is not a new concept. In fact, hatchery-based stock enhancement was the principal technique used in an attempt to restore marine fisheries during the last part of the nineteenth century and early decades of the twentieth century. However, stock enhancement fell out of favor among fishery biologists after a half century of hatchery releases produced no evidence of an increased yield. Atlantic cod *Gadus morhua*, haddock *Melanogrammus aeglefinus*, pollock *Pollachius virens*, winter flounder *Pleuronectes americanus*, and Atlantic mackerel *Scomber scombrus* were stocked. Regrettably, when the last of the early marine hatcheries in the United States closed in 1948, after 50 years of stocking marine fishes, the technology had progressed no further than the stocking of unmarked, newly hatched fry. This was partly a result of the early approach to assessment, in which the success of hatchery programs was judged by numbers of fry stocked rather than by numbers of adults surviving to enter the fishery (Richards and Edwards 1986).

A New and Responsible Approach

Two general problems have restricted development of marine stock enhancement technology this century. Lack of an evaluation capability to determine whether hatchery releases were successful has been a major obstacle. Before the development of modern marking methods, fish-tagging systems were not applicable to the small, early life history stages released by hatcheries. The other impediment to development of marine enhancement has been the inability to culture marine fishes beyond early larval stages to the juvenile stage (fingerlings and larger sizes).

A new approach to marine stock enhancement is long overdue. Faced with declining stocks and an expanding world population, managers around the globe are looking at marine enhancement with renewed interest. To develop and evaluate stock enhancement's full potential, a process is needed for designing and refining stock enhancement tactics based on the combined effects of managing the resource (i.e., the interactive effects of hatchery practices, release strategies, harvest regulations, and habitat restoration on the condition of the managed stock).

Recent advances in both tagging technology and marine fish culture provide basic tools for a new approach to marine enhancement. We now have the technology for benign tagging of fish from juvenile through adult life stages (Bergman et al. 1992). Such tagging provides the basis for a quantitative assessment of stock enhancement success. Several marine fishes can be cultured to provide a wide range of life stages for release (e.g., McVey 1991; Honma 1993). Together, these tools allow an empirical evaluation of survival of cultured fish in the wild, and feedback on hatchery-release effects can be used to refine enhancement strategies. Release effects on wild stocks, and the fisheries based on them, can be quantified and evaluated. Survival can be examined over a range of hatchery practices and release variables (such as culture practices, fish size at release, release magnitude, release site, and season) to identify optimum combinations of hatchery and release strategies.

These new tools provide the basis for significantly increasing wild stock abundances. To ensure their successful use and avoid repeating past mistakes experienced in both marine and freshwater enhancement, we must use a careful approach in developing marine stock enhancement programs. The expression "a responsible approach to marine stock enhancement" embraces a logical and conscientious strategy for applying aquaculture technology to help conserve and expand natural resources. This approach prescribes several key components as integral parts in developing, evaluating, and managing marine stock enhancement programs. Each component is considered essential to control and to optimize the results of enhancement. The components include the need to (1) prioritize and select target species for enhancement; (2) develop a species management plan that identifies harvest opportunity, stock rebuilding goals, and genetic objectives; (3) define quantitative measures of success; (4) use genetic resource management to avoid deleterious genetic effects; (5) use disease and health management; (6) consider ecological, biological, and life-history patterns when forming enhancement objectives and tactics; (7) identify released hatchery fish and assess stocking effects; (8) use an empirical process for defining optimum release strategies; (9) identify economic and policy guidelines; and (10) use adaptive management. Combining new marine fish culture and tagging technologies with these ten principles is gaining support as a responsible approach to marine stock enhancement.

Empirical data suitable for accurately assessing the effect of hatchery releases on wild populations are often lacking. Partly because of this uncertainty, there is an increasing division of conservationists into two camps—one adamantly favoring increased fishing regulations and habitat protection and restoration in preference to hatchery releases, the other supporting propagation and release as an additional tool to manage fisheries and restore declining stocks. This split must be reconciled. Is stock enhancement of marine fishes a powerful, yet undeveloped technology for rebuilding depleted wild stocks and increasing fishery yields? Or are emerging marine enhancement programs merely futile attempts at recovering precious resources, thus diverting money and attention away from habitat restoration and the regulations needed to control overfishing? To realize the full potential of marine enhancement for the conservation and rapid replenishment of declining marine stocks, we must develop the technology to supplement and replenish marine stocks responsibly and quickly.

We must act now to assess the potential of marine stock enhancement through carefully planned research programs. Using strong inference (Platt 1964), which is essentially the scientific method, and addressing all of the components of the responsible approach concept, research programs will either document the value of marine enhancement or reveal that enhancement is not a useful concept. Without determined and careful attention to the 10 points listed above, marine hatchery releases in the 1990s may serve only to fuel divisiveness between the two conservationist camps, with little or no positive effect on natural resources.

Applying the responsible approach concept to new stock enhancement initiatives is straightforward. Existing enhancement programs may find it useful to review the 10 components discussed here. Incorporating those components expanded upon below, that are not already part of ongoing enhancement programs should provide a measurable increase in the realized effectiveness of replenishment efforts.

Prioritize and Select Target Species for Enhancement

In the absence of a candid and straightforward method, targeting species for stock enhancement can become a difficult and biased process. Unless attention is focused on the full spectrum of criteria that can be used to prioritize species, consideration of an immediate need by an advocacy group or simply the availability of aquaculture technology can become the driving factors in species selection. Commercial and recreational demand are obviously important criteria, but should they take precedence over other factors?

To reduce the bias inherent in selecting species, a semiquantitative approach was developed in Hawaii to identify selection criteria and prioritize species for stock enhancement research (Leber 1994). This approach involved four phases: (1) an initial workshop, where selection criteria were defined and ranked in order of importance; (2) a community survey, which was used to solicit opinions on the selection criteria and generate a list of possible species for stock enhancement research; (3) interviews with local experts to rank each candidate species with regard to each selection criterion; and (4) a second workshop, in which the results of the quantitative species selection process were discussed and a consensus was sought. This decision-making process focused discussions, stimulated questions, and quantified participants' responses. Panelists' strong endorsement of the ranking results and selection process used in Hawaii demonstrate the potential for applying formal decision making to species selection in other regions.

A critical step in removing bias from the species selection process lies in the type of numerical analysis used. The relative importance of the various criteria can be used in the analysis by factoring the degree to which each fish meets each criterion by the criterion weight. This produces a score for each species. This same concept is used to determine dominance in ecological studies of species assemblages (i.e., relative abundance times frequency of occurrence in samples). Using a trained facilitator to conduct the workshops also reduces bias by focusing activities on achieving results and by encouraging participation by all present.

Formal decision-making tools have been used effectively to prepare comprehensive plans for fisheries research (Bain 1987). Mackett et al. (1983) discuss the interactive management system for the Southwest Fisheries Center of the National Marine Fisheries Service. Similar processes have been used for research on North Pacific pelagic fisheries, in strategic planning for Hawaii's commercial fishery for skipjack tuna *Katsuwonus pelamis* (Boggs and Pooley 1987), and for a 5-year scientific investigation of marine resources of the main Hawaiian Islands (Pooley 1988).

Develop a Species Management Plan

A management plan identifies the context into which enhancement fits into the total strategy for managing stocks. The goals and objectives of stock enhancement programs should be clearly defined and understood prior to implementation. The genetic structure of wild stocks targeted for enhancement should be identified and managed according to objectives of the enhancement program. What is the population being enhanced? Can it be geographically defined? Clearly, in the interest of both production aquaculture and conservation, effort must be made to maintain genetic diversity (Kapuscinski and Jacobson 1987; Shaklee et al. 1993a, 1993b).

Assumptions and expectations about the performance and operation of the enhancement program necessary to make it successful should be identified (such as postrelease survival, interactions with wild stocks, long-term fitness, and disease). Critical uncertainties about basic assumptions that would affect the choice of production and management strategies should likewise be identified and prioritized. Evaluation of these uncertainties should be an integral part of the species management plan, and a feedback loop to evaluate and change production and management objectives should be included.

Define Quantitative Measures of Success

Without a definition of success, how do you know if or when you have it? Explicit indicators of success are clearly needed to evaluate stock enhancement programs. The objectives of enhancement programs need to be stated in terms of testable hypotheses. To be testable, a hypothesis must be falsifiable (Popper 1965). Depending on enhancement objectives, multiple indicators of success may be needed. These could include statements such as

> Hatchery releases will provide at least a 20% increase in annual landings of *Polydactylus sexfilis* in the Kahana Bay recreational fishery by the third year of the project.

Monitoring will show less than 3% change in the frequency of rare alleles (frequency less than 0.05) after 5 years of hatchery releases (this assumes that a control for the effects of environmentally induced change in allele frequencies is possible).

Numerous indicators should be identified to track progress over time. Although simplistic, indicators like the two examples above could be linked to success and would provide a basis for evaluating enhancement efforts during the initial period of full-scale releases. Clearly, to examine such indicators requires a reliable, quantitative marking and assessment system for tracking hatchery fish.

Use Genetic Resource Management

The need for genetic resource management in stock enhancement programs is currently the subject of intense public debate, and its importance cannot be over-rated. Responsible guidelines are now becoming available to aid resource managers in revitalizing stocks without loss of genetic fitness that could follow from inbreeding in the hatchery and subsequent outbreeding depression in the wild (Kapuscinski and Jacobson 1987; Shaklee et al. 1993a, 1993b). Once the genetic status of the target stock and the genetic goals of the enhancement program are identified, the approach for managing genetic resources is similar to the approach for managing other enhancement objectives (e.g., controlling the level of impact of stocked fish on abundances of the target population). This approach includes (1) identifying the genetic risks and consequences of enhancement; (2) defining an enhancement strategy; (3) implementing genetic controls in the hatchery and a monitoring and evaluation program for wild stocks; (4) outlining research needs and objectives; and (5) developing a feedback mechanism. These points are discussed in detail by Kapuscinski and Jacobson (1987) and Shaklee et al. (1993a, 1993b).

A genetic resource management plan should encompass genetic monitoring prior to, during, and after enhancement, as well as proper use of a sufficiently large and representative broodstock population and spawning protocols, to maintain adequate effective broodstock population size. Prior to enhancement, a comprehensive genetic baseline evaluation of the wild population should be developed to describe the level and distribution of genetic diversity. This baseline evaluation should at least include the geographical range of the particular stock targeted for enhancement. The monitoring should take place over a long enough period to observe possible short-term fluctuation or long-term change. The baseline can be used as a basis to determine an effective population or broodstock size to minimize the undesirable genetic effects of inbreeding, changes in allele frequencies, and loss of alleles. Genetic monitoring of the broodstock and its released progeny should be undertaken to measure success. Long-term genetic monitoring of the wild stock after enhancement should also occur to measure possible loss of genetic diversity, which might be attributed to enhancement efforts.

Maintenance and proper use of a sufficient broodstock population may be one of the toughest and most expensive components of marine stock enhancement. It is also one of the most important. The typically high fecundity rate of marine fish provides the opportunity for a greatly reduced effective population size in a hatchery environment because relatively few adults could potentially contribute a large number of eggs. Fortunately however, marine fish are genetically more homogeneous than freshwater and anadromous species on a relative scale, and genetic studies show relatively little stock separation due to geographic, clinal, or temporal factors (Gyllensten 1985; Waples 1987; Bartley and Kent 1990; King et al. 1995, this volume). In vagile marine species gene flow is often sufficient to homogenize the genetic structures over broad areas. Regardless, sufficient numbers of broodstock must be used so that the genetic diversity (including rare alleles) of the fish being released is the same over time as their wild counterparts.

Hubbs-Sea World Research Institute (Hubbs) has been an early promoter of a responsible genetic management plan. Hubbs leads a consortium of California researchers who are evaluating the feasibility of enhancement of white seabass *Atractoscion nobilis*. Although the genetic profile of progeny from an individual spawn may differ from wild spawns, use of multiple hatchery spawns can approximate the genetic variability observed in the wild. Bartley and Kent (1990) successfully used this concept with white seabass and showed that over 98% of the genetic variability observed in the wild could be maintained with an effective population of 100 broodfish.

Texas Parks and Wildlife Department's (Texas) enhancement program for red drum *Sciaenops ocellatus* provides a good example of maintaining a large broodstock with yearly replenishment (McEachron et al. 1995, this volume). Texas has 140–170 adult broodstock for its program, with an annual replacement of at least 25%. In Norway, studies of allele frequencies are being used to compare broodstock and their progeny with wild populations of Atlantic cod (Svasand et al. 1990).

Use Disease and Health Management

Disease and health guidelines are important to both the survival of the fish being released and the wild populations of the same species or other species with which they interact. Florida Department of Environmental Protection (Florida) has developed an aggressive and responsible approach in this area in association with their red drum enhancement project (Landsberg et al. 1991). Florida's policy requires that all groups of fish pass a certified inspection for bacterial and viral infections and parasites prior to release. Maximum acceptable levels of infection and parasites in the hatchery populations are established based on the results of screening healthy wild populations.

Form Enhancement Objectives and Tactics

During the design phase of enhancement programs ecological factors that can contribute to the success or failure of hatchery releases should be considered. Predators, food availability, accessibility of critical habitat, competition over food and space, environmental carrying capacity, and abiotic factors, such as temperature and salinity, are all key variables that can affect survival, growth, dispersal, and reproduction of cultured fish in the wild. Predatory losses and food availability have long been thought to be among the principal variables that mediate recruitment success in wild populations (Lasker 1987; Houde 1987).

Habitat degradation in marine environments can also affect recruitment success. For example, seagrass meadows are important nursery habitats for fishes and crustaceans (see Kikuchi 1974). In vegetated aquatic environments, habitat availability and habitat quality (e.g., structural complexity) have been shown to mediate survival from predators (Crowder and Cooper 1982; Stoner 1982; Main 1987). In some cases, habitat degradation in marine environments may be so complete that certain habitats are unsuitable for stock enhancement (Stoner 1994). To enhance fisheries in some locales, restoration of coastal habitat may be the first priority.

The authors feel strongly that marine stock enhancement should never be used as mitigation to justify loss of habitat. However, we also feel that enhancement efforts with cultured fishes can fill a void where critically important habitats such as coastal wetlands and estuaries, which provide nurseries for early life stages, are irretrievably lost or degraded.

In addition to ecological factors, there may be physiological and behavioral deficits in hatchery-reared fish that strongly reduce survival in the wild (e.g., swimming ability, feeding behavior, predator avoidance, agonism, schooling, and habitat selection). In Japan, Tsukamoto (1993) has evaluated the effect of behavior on survival of cultured madai *Pagrus major* (called red sea bream by Tsukamoto) released into the sea. Tsukamoto's results indicate that a predator-avoidance behavior (tilting), in which wild fish lay flat against the substratum, may be reduced or absent in cultured fish during the first few days after release into the sea. Abnormal tilting behavior was directly correlated with mortality rate. For certain learned behaviors, exposure to behavioral cues and responses by wild fish in hatchery microcosms may be needed to overcome behavioral deficits (Olla and Davis 1988).

A solid understanding of the ecological and biological mechanisms mediating target species abundances can require exhaustive field studies for each species considered for enhancement. Whole careers have been dedicated to understanding mechanisms behind animal distributions and abundance; it does not seem practical to hold off on stock enhancement research until the ecological mechanisms are completely understood. However, failure to consider such factors can result in poor performance of released fish at best and at worst have negative impacts on natural stocks (Murphy and Kelso 1986).

Our viewpoint is that preliminary, pilot-scale experimental releases with subsequent monitoring of cultured fish afford a direct method for evaluating assumptions about the effects of uncontrolled environmental factors. For example, assumptions about carrying capacity in particular release habitats can and should be evaluated through pilot releases conducted prior to full-scale enhancement at those sites (Leber et al. 1995, this volume). This approach is elaborated below.

Identify Released Hatchery Fish and Assess Stocking Effects

One of the most critical components of any enhancement effort is the ability to quantify success or failure. Without some form of assessment, one has no idea to what degree the enhancement was effective or, more critically, which approaches were totally successful, partially successful, or a downright failure. Natural fluctuations in marine stock abundance can mask successes and failures. Maximization of benefits cannot be realized without the proper monitoring and evaluation system.

Tagging or marking systems that are benign and satisfy the basic assumption that identified fish are representative of untagged counterparts are essential, but weren't available until relatively recently. The detrimental effects of external tags are well documented (Isaksson and Bergman 1978; Hansen 1988; McFarlane and Beamish 1990), and few fishery managers or researchers defend their use today, especially with juvenile fish. Useful information retrieved from external tags is usually restricted to migration and growth rates of relatively large fish (Scott et al. 1990; Trumble et al. 1990).

In recent years, a few identification systems (e.g., coded wire tags, passive integrated transponder tags, genetic markers, and otolith marks) have been developed that meet the requirements that identified fish are representative of the species with regard to behavior, biological functions, and mortality factors, and thus provide unbiased data (Buckley and Blankenship 1990). The story of the development and now widespread use of the coded wire tag (Jefferts et al. 1963) is well known, and it is fair to say that it has revolutionized the approach to stock enhancement (Soloman 1990).

With an unbiased tag or mark, quantitative assessment of the effects of release is possible. In developing enhancement programs, evaluation of hatchery contributions can be partitioned into at least four distinct stages: initial survival, survival through the nursery stage, survival to adulthood (entry into the fishery), and successful contribution to the breeding pool. In Hawaii, the percent of hatchery fish in field samples taken after pilot releases of striped mullet *Mugil cephalus* has been as high as 80% in initial collections, 50% in some nursery habitats through the tenth month after release, and (in a recreational fishery in Hilo, Hawaii) as high as 20% of the catch (Leber, in press; Leber et al. 1995; Leber et al., in press). In Norway, genetic markers are beginning to show that released Atlantic cod produce viable offspring in the wild (Jorstad 1994).

Assessment of the effects of release should go further than evaluation of survival and contribution rates of hatchery fish. Evaluation of hatchery fish interactions with wild stocks is also critical. Clearly, evaluation of genetic impact is important. It is equally important to understand whether hatchery releases increase abundances in the wild or simply displace the wild stocks targeted for enhancement. At least one experimental study in Hawaii has documented that released hatchery fish can indeed increase abundances in a principal nursery habitat, without displacing wild individuals (Leber et al. 1995).

Use an Empirical Process to Define Optimum Release Strategies

Just as preliminary releases can be used to evaluate ecological assumptions, pilot release experiments afford a means of quantifying and controlling the effects of release variables and their influence on the performance of cultured fish in coastal environments (Tsukamoto et al. 1989; Svasand and Kristiansen 1990; Leber, in press; Willis et al. 1995, this volume).

Experiments to evaluate fish size at release, release season, release habitat, and release magnitude should always be conducted prior to launching full-scale enhancement programs. These experiments are a critical step in identifying enhancement capabilities and limitations and in determining release strategy. They also provide the empirical data needed to plan enhancement objectives, test assumptions about survival and cost effectiveness, and model enhancement potential. The lack of monitoring to assess survival of the fish released by marine enhancement programs early in this century (through the 1940s) was the single greatest reason for the failure of those programs to increase stock abundances and fishery yields (Richards and Edwards 1986).

Based on the results of pilot experiments by The Oceanic Institute in Hawaii, hatchery-release variables were steadily refined to maximize striped mullet enhancement potential. This resulted in an increase in recapture rates by at least 400% over a 3-year period (Leber et al. 1995; Leber et al., in press.) During the third year of pilot studies in Kaneohe Bay, hatchery fish provided at least 50% of the striped mullet in net samples during the entire 10-month collection period after releases. An understanding of how fish size at release and release habitat affected survival were the primary factors needed to increase recapture rates. However, understanding the interaction of release season with size at release and release habitat also had significant effect on refinement of release strategies. The apparent doubling effect on abundances in the third year was achieved with a release of only 80,000 juveniles into the principal striped mullet nursery habitat in Kaneohe Bay, the largest estuary in Hawaii. A subsequent study documented that mullet releases did not displace wild juveniles from that nursery habitat (Leber et al. 1995). Thus, hatchery releases in Kaneohe Bay appear to be increasing population size in the primary nursery habitat. Clearly, these pilot experiments are crucial for managing enhancement impact.

Identify Economic and Policy Objectives

Initially, costs and benefits can be estimated and economic models developed to predict the value of enhancement. This information can be used to generate funding support through reprioritization, legislation, or user fees. The information can contribute to an explicit understanding with policy makers and the general public on the time frame that is needed for components such as adaptation of culture technology and pilot-release experiments before full-scale releases can begin. The education of the public and policy makers on the need and benefits of a responsible approach is also important. In Florida, pressure is mounting to drop the responsible approach concept involving pilot-scale releases and instead plant millions of red drum fry as a neighboring state has done (Wickstrom 1993). Advocates of the latter approach simply assume that the bigger the numbers planted, the better.

Use Adaptive Management

Adaptive management is a continuing assessment process that allows improvement over time. The key to this improvement lies in having a process for changing both production and management objectives (and strategies) to control the effects of enhancement. Essentially, adaptive management is the continued use of the nine key components above to ensure an efficient and wise use of a natural resource. The use of adaptive management is central to the successful application of the approach outlined above. Some minimum level of ongoing assessment is needed, superimposed over a moderate research framework that provides a constant source of new information. New ideas for refining enhancement are thus constantly considered and integrated into the management process.

Summary

The need for marine stock enhancement has been identified, and we must learn from mistakes made in the past. The necessity and benefit of following a responsible approach in implementing enhancement cannot be overemphasized. Several organizations have started to subscribe to this new approach to marine stock enhancement. The juveniles from their pilot releases are just starting to enter the fisheries, so the results are not known.

The exception is the striped mullet enhancement program in Hawaii at The Oceanic Institute. This program has shown the benefits that can be gained from closely following the approach outlined in this paper. In addition to The Oceanic Institute's decision to develop a proper genetic management plan and to make quantitative assessments of the effects of hatchery releases on wild populations, it performed numerous pilot studies to optimize release strategies.

Without these pilot experiments, Hawaii researchers would not have increased survival rate by over 400% in Kaneohe Bay nor provided a 20% contribution to the catch in the recreational fishery in Hilo Bay. We predict that identifiable fish from each of the other programs referenced will also have a substantial effect on the catch and validate our suggested approach. What is needed now is a concerted effort by the managers of new and existing enhancement programs to use, evaluate, and refine the approach described here.

Given the worldwide decline in fisheries catch rates, bold new initiatives are needed to revitalize fisheries. We need to take care, though, to preserve existing stocks as we work to restore and increase the harvest levels of those stocks using cultured fishes.

Acknowledgments

We wish to thank Devin Bartley, Don Kent, Rich Lincoln, Stan Moberly, and Scott Willis, who have greatly contributed to the development of the ideas expressed herein. We also thank Maala Allen, Paul Bienfang, Churchill Grimes, Gary Sakagawa, Kimberly Lowe, and Dave Sterritt for insightful comments on the manuscript. Order of authorship was determined by the flip of a coin.

This paper is funded in part by a grant from the National Oceanic and Atmospheric Administration (NOAA). The views expressed herein are those of the authors and do not necessarily reflect the views of NOAA or any of its subagencies.

References

Bain, M. B. 1987. Structured decision making in fisheries management: trout fishing regulation on the Au Sable River, Michigan. North American Journal of Fisheries Management 7:475–481.

Bartley, D. M., and D. B. Kent. 1990. Genetic structure of white seabass populations from the Southern California Bight region: applications to hatchery enhancement. California Cooperative Oceanic Fisheries Investigations Report 31:97–105.

Bergman, P. K., F. Haw, H. L. Blankenship, and R. M.

Buckley. 1992. Perspectives on design, use, and misuse of fish tags. Fisheries (Bethesda) 17(4):20–24.

Boggs, C. H., and S. G. Pooley, editors. 1987. Strategic planning for Hawaii's aku industry. NOAA (National Oceanic and Atmospheric Administration) NMFS (National Marine Fisheries Service). Southwest Fisheries Center Administrative Report H-87-1:22, Honolulu, Hawaii.

Buckley, R. M., and H. L. Blankenship. 1990. Internal extrinsic identification systems: overview of implanted wire tags, otolith marks and parasites. American Fisheries Society Symposium 7:173–182.

Crowder, L. B., and W. E. Cooper. 1982. Habitat structural complexity and interaction between bluegills and their pray. Ecology 63:1802–1813.

FAO (Food and Agriculture Organization of the United Nations). 1991. Food and Agriculture Organization of the United Nations yearbook 70(1990): fishery statistics.

Gyllensten, U. 1985. The genetic structure of fish: differences in the intraspecific distribution of biochemical genetic variation between marine, anadromous and freshwater species. Journal of Fisheries Biology 26: 691–699.

Hansen, L. P. 1988. Effects of Carlin tagging and fin clipping on survival of Atlantic salmon (*Salmo salar* L.) released as smolts. Aquaculture (Netherlands) 70: 391–394.

Honma, A. 1993. Aquaculture in Japan. Japan FAO Association. Chiyoda-Ku, Tokyo.

Houde, E. D. 1987. Fish early life dynamics and recruitment variability. American Fisheries Society Symposium 2:17–29.

Isaksson, A., and P. K. Bergman. 1978. An evaluation of two tagging methods and survival rates of different age and treatment groups of hatchery-reared Atlantic salmon smolts. Journal of Agricultural Research in Iceland 10(1):74–99.

Jefferts, K. B., P. K. Bergman, and H. F. Fiscus. 1963. A coded-wire identification system for macro-organisms. Nature (London) 198:460–462.

Jorstad, K. E. 1994. Cod stock enhancement studies in Norway—genetic aspects and the use of genetic tagging. World Aquaculture Society, New Orleans, Louisiana.

Kapuscinski, A. R., and L. D. Jacobson. 1987. Genetic guidelines for fisheries management. University of Minnesota, Minnesota Sea Grant College Program, Sea Grant Research Report 17, Duluth.

Kikuchi, T. 1974. Japanese contributions on consumer ecology in eelgrass (*Zostra marina*) beds, with special reference to tropic relationships and resources in inshore fisheries. Aquaculture 4:145–160.

King, T. L., R. Ward, I. R. Blandon, R. L. Colura, and J. R. Gold. 1995. Using genetics in the design of red drum and spotted seatrout stocking programs in Texas: a review. American Fisheries Society Symposium 15:499–502.

Landsberg, J. H., G. K. Vermeer, S. A. Richards, and N. Perry. 1991. Control of the parasitic copepod *Caligus elongatus* on pond-reared red drum. Journal of Aquatic Animal Health 3:206–209.

Lasker, R. 1987. Use of fish eggs and larvae in probing some major problems in fisheries and aquaculture. American Fisheries Society Symposium 2:1–16.

Leber, K. M. 1994. Prioritization of marine fishes for stock enhancement in Hawaii. The Oceanic Institute, Honolulu, Hawaii.

Leber, K. M. In press. Significance of fish size-at-release on enhancement of striped mullet fisheries in Hawaii. Journal of the World Aquaculture Society.

Leber, K. M., D. A. Sterritt, R. N. Cantrell, and R. T. Nishimoto. In press. Contribution of hatchery-released striped mullet, *mugil cephalus*, to the recreational fishery in Hilo Bay, Hawaii. Hawaii Department of Land and Natural Resources, Division of Aquatic Resources, Technical Report 94-03, Honolulu.

Leber, K. M., N. P. Brennan, and S. M. Arce. 1995. Marine enhancement with striped mullet: are hatchery releases replenishing or displacing wild stocks? American Fisheries Society Symposium 15:376–387.

Mackett, D. J., A. N. Christakis, and M. P. Christakis. 1983. Designing and installing an interactive management system for the southwest fisheries center. Pages 518–527 in O. T. Magoon and H. Converse, editors. Coastal Zone 83. American Society of Civil Engineers, New York.

Main, K. L. 1987. Predator avoidance in seagrass meadows: prey behavior, microhabitat selection and cryptic coloration. Ecology 68(1):170–180.

McEachron, L. W., C. E. McCarty, and R. R. Vega. 1995. Beneficial uses of marine fish hatcheries: enhancement of red drum in Texas coastal waters. American Fisheries Society Symposium 15:161–166.

McFarlane, G. A., and R. J. Beamish. 1990. Effect of an external tag on growth of sablefish and consequences to mortality and age at maturity. Canadian Journal of Fisheries and Aquatic Sciences 47:1551–1557.

McVey, J. P., editor. 1991. Handbook of mariculture, volume 2: finfish aquaculture. CRC Press Inc., Boca Raton, Florida.

Murphy, B. R., and W. E. Kelso. 1986. Strategies for evaluating freshwater stocking programs: past practices and future needs. Pages 303–316 in R. H. Stroud, editor. Fish culture in fisheries management. American Fisheries Society, Fish Culture Section and Fisheries Management Section, Bethesda, Maryland.

NOAA (National Oceanic and Atmospheric Administration). 1991. Our living oceans: first annual report on the status of U.S. living marine resources. NOAA Technical Memorandum NMFS/SPO-1.

NOAA (National Oceanic and Atmospheric Administration). 1992. Our living oceans: report on the status of U.S. living marine resources. NOAA Technical Memorandum NMFS-F/SPO-2.

Olla, B. L., and M. Davis. 1988. To eat or not be eaten. Do hatchery reared salmon need to learn survival skills? Underwater Naturalist 17(3):16–18.

Platt, J. R. 1964. Strong inference. Science 146(3642): 347–353.

Pooley, S. G., editor. 1988. Recommendations for a five-year scientific investigation on the marine resources and environment of the main Hawaiian Islands.

NOAA (National Oceanic and Atmospheric Administration) NMFS (National Marine Fisheries Service). Southwest Fisheries Center Administrative Report H-88-2, Honolulu, Hawaii.

Popper, K. R. 1965. Conjectures and refutations, the growth of scientific knowledge. Harper and Row, New York.

Richards, W. J., and R. E. Edwards. 1986. Stocking to restore or enhance marine fisheries. Pages 75–80 in R. H. Stroud, editor. Fish culture in fisheries management. American Fisheries Society, Fish Culture Section and Fisheries Management Section, Bethesda, Maryland.

Scott, E. L., E. D. Prince, and C. D. Goodyear. 1990. History of the cooperative game fish tagging program in the Atlantic Ocean, Gulf of Mexico, and Caribbean Sea, 1954–1987. American Fisheries Society Symposium 7:841–853.

Shaklee, J. B., C. A. Busack, and C. W. Hopley, Jr. 1993a. Conservation genetics programs for Pacific salmon at the Washington Department of Fisheries: living with and learning from the past, looking to the future. Pages 110–141 in K. L. Main and E. Reynolds, editors. Selective breeding of fisheries in Asia and the United States. The Oceanic Institute, Honolulu, Hawaii.

Shaklee, J. B., J. Salini, and R. N. Garrett. 1993b. Electrophoretic characterization of multiple genetic stocks of barramundi perch in Queensland, Australia. Transactions of the American Fisheries Society 122: 685–701.

Soloman, D. J. 1990. Development of stocks: strategies for the rehabilitation of salmon rivers. Pages 35–44 in D. Mills, editor. Strategies for the rehabilitation of salmon rivers. Linnean Society of London, London.

Stoner, A. W. 1982. The influence of benthic macrophytes on foraging behavior of pinfish, *Lagodon rhomboides*. Journal of Experimental Marine Biology and Ecology 104:249–274.

Stoner, A. W. 1994. Significance of habitat and stock pre-testing for enhancement of natural fisheries: experimental analyses with queen conch, *Strombus gigas*. Journal of the World Aquaculture Society 25: 155–165.

Svasand, T., K. E. Jorstad, and T. S. Kristiansen. 1990. Enhancement studies of coastal cod in western Norway, part 1. Recruitment of wild and reared cod to a local spawning stock. Journal du Conseil International pour l'Exploration de la Mer 47:5–12.

Svasand, T., and T. S. Kristiansen. 1990. Enhancement studies of coastal cod in western Norway, part 4. Mortality of reared cod after release. Journal du Conseil International pour l'Exploration de la Mer 47:30–39.

Trumble, R. J., I. R. McGregor, G. St-Pierre, D. A. McCaughran, and S. H. Hoag. 1990. Sixty years of tagging Pacific halibut: a case study. American Fisheries Society Symposium 7:831–840.

Tsukamoto, K., and six coauthors. 1989. Size-dependent mortality of red sea bream, *Pagrus major*, juveniles released with fluorescent otolith-tags in News Bay, Japan. Journal of Fish Biology 35(Supplement A):59–69.

Tsukamoto, K. 1993. Marine fisheries enhancement in Japan and the quality of fish for release. European Aquaculture Society Special Publication 19:556.

Waples, R. S. 1987. Multispecies approach to the analysis of gene flow in marine shore fishes. Evolution 41:385–400.

Wickstrom, K. 1993. Biscayne redfish: yes! Florida Sportsman (March):90–91.

Willis, S. A., W. W. Falls, C. W. Dennis, D. E. Roberts, and P. G. Whitchurch. 1995. Assessment of season of release and size at release on recapture rates of hatchery-reared red drum. American Fisheries Society Symposium 15:354–365.

An Ecological Framework for Evaluating the Success and Effects of Stocked Fishes

DAVID H. WAHL

Illinois Natural History Survey, Center for Aquatic Ecology
607 East Peabody Drive, Champaign, Illinois 61820, USA

ROY A. STEIN

The Ohio State University, Aquatic Ecology Laboratory, Department of Zoology
Columbus, Ohio 43212, USA

DENNIS R. DEVRIES

Auburn University, Department of Fisheries and Allied Aquaculture
313 Swingle Hall, Auburn University, Alabama 36849, USA

Abstract.—In natural, as well as in human-manipulated communities, predation, competition, and abiotic factors interact to determine distribution and abundance of species. To develop generalities concerning the relative influence of these ecological factors in the success of stocked fishes and ultimately their effect on resident communities, we review recent research dealing with both stocked sport fishes and introduced prey species. Resident predators influence survival of stocked sport fishes, but effects vary (1) among species (e.g., esocids *Esox* spp. were more vulnerable to largemouth bass *Micropterus salmoides* than were walleyes *Stizostedion vitreum*, followed by channel catfish *Ictalurus punctatus*); (2) with size and time of stocking (larger is typically better); and (3) with the demographics of resident predator populations (species composition, size distributions, and abundance). Whereas prey preference and prey demographics also influence stocked sport fish survival and growth, the potential for competition with resident fishes has not been extensively quantified. Abiotic factors (particularly temperature) can influence stocking mortality and subsequent growth, with effects varying among species. Introduced prey can exert negative, neutral, or positive effects on resident communities. In some instances, growth of resident predators improves, but interspecific competitive interactions at early life stages can reduce recruitment of young-of-year sport fishes as well as resident prey species, producing overall negative effects. We conclude that fish stocking should be pursued within an ecological framework that integrates the relative importance of predation, competition, and abiotic factors across all life stages. This framework provides a guide for making management decisions concerning species, sizes, and timing of fish introductions into systems with specific characteristics.

Competition, predation, and abiotic factors govern community structure in natural systems. Predation can be the predominant ecological interaction (Connell 1975), with competition a temporary, unimportant process (Wiens 1977), or competition can be predominant, driving community structure (Connell 1983; Roughgardin 1983; Schoener 1983). Further, competition and predation can interact to determine species persistence (Allan 1974; Lubchenco 1978; Garvey et al. 1994). Though ecologically similar species can compete, predation often maintains populations at levels below which competition can occur. As the abiotic environment changes, the relative importance of competition and predation can vary (Goldberg 1985). Strong (1983) has emphasized the physical component of the environment and the importance of autecological factors, such as organism physiology, in determining species success in a community. Thus, within a given ecological community, three components, competition, predation, and abiotic factors, interact to determine community composition.

In aquatic communities, predation can influence community composition (Hurlbert et al. 1972; Peckarsky and Dodson 1980; Morin 1983), both by direct removal of prey and indirect effects (Kerfoot and Sih 1987; Carpenter 1988). Indirect predatory effects can lead to positive (Kerfoot 1987) or negative (Mittelbach and Chesson 1987) effects on prey and can extend beyond adjacent trophic levels (Carpenter et al. 1985).

In turn, competition can determine species composition within aquatic communities (Connell 1983; Johnson et al. 1985) via intra- and interspecific interactions (Tonn et al. 1986; Persson and Greenberg 1990). Although most research has focused on competition among adults, more recent research has focused on early life stages (DeVries et al. 1991; Welker et al. 1994). Competition may be particularly important during early life given that some larval fish are quite susceptible to starvation (Miller et al. 1988), limiting recruitment and eventual abundance within a community.

Given the importance of these ecological processes in controlling community structure, we have used them in a framework for addressing the applied management problem of developing strategies for fish introductions. Previous research using ecological approaches to solve fisheries management problems are rare (Rigler 1982), and the lack of interdisciplinary communication has been attributed to fundamental differences in approach (Werner 1980; Rigler 1982). Fishery biologists have often been concerned with autecological factors (Kerr 1980) and the relationship between fish populations and anglers, with little consideration of interactions among other biotic components. In our view, advances in fisheries management will derive from research directed at the entire aquatic community, integrating ecology and fishery science (Magnuson 1991).

Introductions of sport fish and prey fishes are important management tools. Sport-fish species are stocked to maintain fisheries in waters where habitat degradation and overexploitation have reduced existing populations (Wingate 1986) and to establish new populations in waters such as ponds and reservoirs. Upon stocking, survival and growth can be highly variable (Laarman 1978), and responsible mechanisms are often unknown. Prey fishes introduced to enhance forage availability for piscivorous sport fishes have involved a wide variety of species (see DeVries and Stein 1990) that have been introduced into diverse system types (Ney 1981; Noble 1981; Wydoski and Bennett 1981). Although successes have occurred, the ultimate goal of improved sportfishing has not always been achieved (Ney 1981; Noble 1981; Wydoski and Bennett 1981; DeVries and Stein 1990).

Herein, we review our research conducted during the past 15 years and evaluate factors influencing the survival, growth, and effect of stocked fishes on resident fish communities. Based on this work, we propose an ecological approach that integrates the relative importance of predation, competition, and abiotic factors across all life stages. Historically, fishery managers have used trial-and-error approaches when conducting species manipulations, with little consideration of those factors that ultimately influence the success and community effects of those manipulations. Our framework provides a guide for making management decisions concerning species, sizes, and timing of fish introductions into systems with specific characteristics and provides an all too often missing link between the disciplines of ecology and fisheries management.

Effects of Predation

Sport Fish Introductions

Resident predators can reduce survival of introduced fishes. Losses to predation have been quantified for several stocked sport fishes, including esocids *Esox* spp. (Stein et al. 1981; Carline et al. 1986; Wahl and Stein 1989; Wahl, in press), walleye *Stizostedion vitreum* (Santucci and Wahl 1993), and channel catfish *Ictalurus punctatus* (Santucci et al. 1994). Because predation mortality was estimated similarly among studies, we can generalize across species.

Study of the effects of predatory mortality has focused on largemouth bass *Micropterus salmoides*, which is the primary predator in midwestern reservoirs. Largemouth bass densities were estimated and combined with the number of stocked fish recovered from stomachs to estimate predatory losses (see e.g., Wahl and Stein 1989). For all three taxa, most consumption occurred during the first 7 d poststocking (predatory mortality was quantified through 30 d poststocking).

Size at stocking influenced predatory mortality for all species, with large fish less vulnerable than small ones. For muskellunge *Esox masquinongy*, northern pike *E. lucius*, and tiger muskellunge female *E. masquinongy* × male *E. lucius*, losses declined from 31% (range 8–53%) at 145 mm to 2% (range 1–2%) at 205 mm. For walleye, large (200 mm) fingerlings were less vulnerable to predation (0%) than were medium (145 mm, 17%) or small (60 mm, 6%) fingerlings. In contrast, size was not important for channel catfish (0.1% loss at 200 mm; 0% at 250 mm), but stocked sizes exceeded those for other taxa. For small fish, more predators are capable of ingesting them than their larger conspecifics. Timing of stocking can influence predatory losses as well; cool water temperatures in fall reduce food consumption rates by largemouth bass (Rice and Cochran 1984), reducing the potential for predatory impact (Stein et al. 1981).

Neither availability of alternate forage nor predator density influences predation mortality for these species. For esocids, losses to predators were unrelated to either the percentage of largemouth bass with empty stomachs (range 40–60%) or density of largemouth bass (range 1–12/ha). A similar lack of relationship was observed for walleye and channel catfish, but these conclusions are based on results from stockings of different size-groups in a single lake. In contrast, Carline et al. (1986) found positive correlations for tiger muskellunge across 14 stockings and concluded that largemouth bass were most likely to consume esocids when other prey were not abundant and predator densities were high. However, these conclusions were affected strongly by data from one pond in which largemouth bass were food limited (80% empty stomachs) and dense (75/ha). Additional stockings in lakes with differing predator populations will be required to establish relationships between losses of stocked fish and predator density and population size structure. In our view, system-specific characteristics such as these will influence predatory mortality.

Using analyses similar to Wahl (in press), we compared predatory effects from 33 reservoir stockings for esocids ($N = 20$), walleye ($N = 9$), and channel catfish ($N = 4$) based on two 30-mm size-classes (130–160 mm and 190–220 mm). Because predatory losses have not been evaluated for the small size-class of channel catfish, analyses were first completed comparing the small and large size-classes of esocids and walleye, followed by comparisons of all three groups within the large size-class. Across all stockings, esocids were more vulnerable to largemouth bass predation than were walleyes (Figure 1; analysis of variance [ANOVA], $P = 0.02$; Wahl, in press); these relationships were consistent for both size-groups. Small walleye and esocids suffered higher losses than did large ones (ANOVA, $P = 0.04$). Within the large size-class, losses of channel catfish were intermediate but did not differ from those of walleye and esocids (ANOVA, $P = 0.12$).

Differences in behavioral response to predators among introduced species may influence their relative vulnerability. Clearly, prey and predator must overlap spatially and temporally to permit predation (Noble 1986). In laboratory experiments, Wahl (in press) discovered that muskellunge were more vulnerable to largemouth bass predation than were walleyes not due to differences in antipredatory behavior but rather differences in habitat selection. Walleye spent more time associated with protective habitats such as the bottom and inshore vegetation than did muskellunge. Though channel catfish were stocked larger, dorsal spines likely reduce its losses to predation.

Several factors may influence predatory mortality including size and timing of stocking, species, and characteristics of individual impoundments, such as abundance, size distributions, and species of resident predators. Rearing technique may also influence stocking mortality. For example, experience

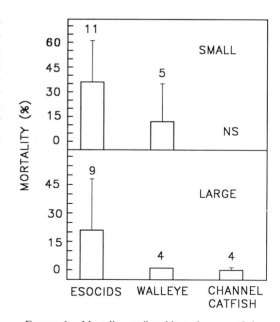

FIGURE 1.—Mortality attributable to largemouth bass predation for small (130–160 mm) and large (190–220 mm) size-classes of esocids, walleye, and channel catfish following introduction into Ohio and Illinois reservoirs. No stockings (NS) occurred for small channel catfish. Values over bars are numbers of reservoir experiments; vertical lines are the upper 95% confidence intervals.

feeding on minnows during rearing can indirectly increase survival of muskellunge by reducing its vulnerability to predation (Szendrey and Wahl, in press); this technique should be considered in the development of sport-fish stocking strategies.

Largemouth bass are the dominant littoral predators in midwestern reservoirs; few limnetic predators occur that are of sufficiently large size to consume stocked sport fishes. However, given differences in habitat selection among stocked species, predators other than largemouth bass could be important in other regions. To reduce losses to largemouth bass predation, sport-fish species should be introduced in the fall at large sizes; however, largemouth bass predation will be more likely to contribute to mortality of esocids than it will be to walleye and channel catfish.

Prey Introductions

The role of predation on success of prey fish introductions has typically been left unexamined. For example, in our review of the literature dealing with introduction of gizzard shad *Dorosoma cepedianum* and threadfin shad *D. petenense* (DeVries and Stein 1990), none of the studies that we reviewed quantified losses of stocked shad to predation, even though 72% of the studies of shad introduction eventually documented whether or not shad were present after introduction (often via the presence of young-of-year shad in predator diets). Clearly, this is an area that must be addressed in future prey fish introductions.

Resource Use and Competition

Sport Fish Introductions

Prey preference and prey demographics influence both survival and growth of stocked piscivores. Both esocids and walleye grow more slowly with centrarchid prey than with clupeid and cyprinid prey (Wahl and Stein 1988; Santucci and Wahl 1993). In laboratory experiments, centrarchids are captured less successfully by these predators compared with either clupeids or cyprinids (Mauck and Coble 1971; Moody et al. 1983; Wahl and Stein 1988). As documented in the field, these predators prefer gizzard shad over bluegill. Morphology (body depth and spines) and antipredatory behavior, unique to each prey species, contribute to differential vulnerability. In addition to prey preference, availability of appropriately sized prey will also influence growth of these stocked sport fishes (Santucci and Wahl 1993; Wahl and Stein 1993). Match to the forage base is far more important an issue for smaller stocked fishes than for larger ones. As diet breadth increases (owing to increased gape sizes), predator dependance on specific sizes and types of prey declines. With small percids (35–45 mm total length [TL]), matching stocking time with the onset of ichthyoplankton peaks improves growth (Stahl and Stein, in press) and likely survival through the first year of life. Survival of esocids and walleye is lower in centrarchid communities than in those with clupeid or cyprinid prey (Schneider 1975; Wahl and Stein 1988). Reduced survival of esocids and walleye with centrarchids may be directly related to reduced capture ability or may be indirectly related to reduced growth and the resulting increased vulnerability to predation or disease.

Sport fish are typically stocked with the expectation that they will supplement low-density resident populations; resulting populations are assumed to lie below carrying capacity. Thus, competition for resident prey resources has been assumed to be negligible. In our work, survival of stocked esocids (0–7%, Wahl and Stein 1993) and walleye (0–31%, Santucci and Wahl 1993) was usually relatively low

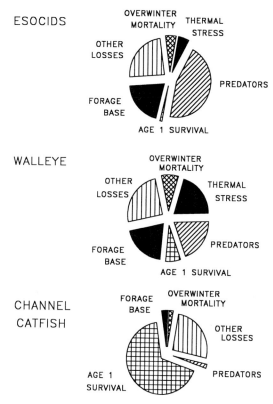

FIGURE 2.—Relative importance of stocking stress (thermal stress), resident predators, forage base, overwinter mortality, and other losses (natural mortality and outmigration) in determining age-1 survival of introduced sport fish. Values represent generalized losses for esocids (tiger muskellunge; 150 mm), walleye (150 mm), and channel catfish (200 mm). Actual losses will vary with fish size, timing of stocking, and system-specific characteristics.

(Figure 2). As a result, competition with resident fishes was also likely low. High stocking rates and high survival rates of either single year-classes, such as with channel catfish (up to 92% survival, Santucci et al. 1994), or over cumulative years of stocking can lead to greater predatory demand and perhaps substantial interspecific competition. Only with quantitative assessments of prey productivity, survival rates of stocked fishes, and total piscivore demand can accurate predictions of the effects of stocked sport fish on resident predators be determined (Johnson et al. 1988).

Prey Introductions

In response to a prey fish introduction, piscivore prey consumption should increase (Figure 3), lead-

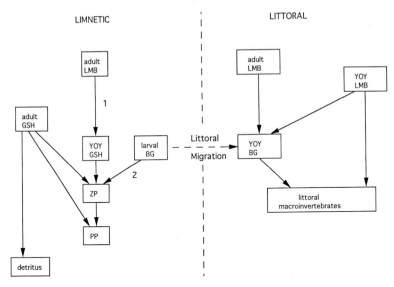

FIGURE 3.—Food web diagram incorporating spatial heterogeneity and ontogenetic niche shifts, for a system containing largemouth bass (LMB), bluegill (BG), and gizzard shad (GSH). The numbers indicate critical points relative to our suggested ecological approach: (1) the direct predator–prey interaction that leads to the positive effect of the introduced prey, and (2) competition between gizzard shad and bluegill for limnetic zooplankton that leads to the indirect negative effect of gizzard shad on largemouth bass. Abbreviations are young of year (YOY), zooplankton (ZP), and phytoplankton (PP).

ing to increased growth and recruitment to the fishery. Given appropriate prey selection studies, conducted both in the field and in the laboratory, the outcome of a prey introduction (based on direct predator–prey interactions) should be predictable. As such, given a particular system type and a target piscivore species, managers should be able to select the most appropriate prey species to be introduced to ensure an enhanced sport-fish population.

Unfortunately, how introduced prey influence a piscivore is not restricted to the direct positive effects of increased prey abundance. Rather, introduced prey can act directly as competitors with the early life stages of piscivores and indirectly via interactions with other species in the community. Because fishes exhibit ontogenetic niche shifts (Werner and Gilliam 1984), their diets and habitat use change as they grow. Early life stages of most fishes (including both piscivores and prey) are gape limited (Schael et al. 1991; Bremigan and Stein 1994), restricting their prey to small items such as zooplankton. Thus, those species that interact as predator and prey as adults may actually pass through a competitive bottleneck during their early life (Werner and Gilliam 1984). Under these circumstances, the overall outcome of a prey introduction will be some combination of the negative effects of this competitive interaction and the positive effects of the predator–prey interaction.

For example, gizzard shad is the preferred prey of many piscivores (Wahl and Stein 1988) and has often been introduced to enhance piscivore populations, including both crappie *Pomoxis* spp. and largemouth bass (DeVries and Stein 1990). Yet larvae of both gizzard shad and crappie require limnetic zooplankton, and larval and young-of-year gizzard shad can reduce zooplankton to low levels (Drenner et al. 1982; DeVries and Stein 1992; Dettmers and Stein 1992). Thus, the potential exists for gizzard shad and crappie to compete during early life, possibly explaining documented negative interactions between these two species (DeVries and Stein 1990). However, given that the ability of gizzard shad to depress zooplankton populations depends on zooplankton production (Dettmers and Stein 1992), variation in zooplankton production among systems may affect the ability of gizzard shad to influence zooplankton, and hence other planktivores (Welker et al. 1994). In addition, competition among zooplanktivorous early life stages does not occur for all piscivores. Though larval largemouth bass feed as gape-limited zooplanktivores (Chew 1974; Hirst and DeVries 1994), they do not compete with larval shad; diets and habitat use of larval shad and larval black bass *Micropterus* spp. overlapped little in two Alabama reservoirs, with no evidence for direct negative effects of larval shad on larval black bass (Hirst and DeVries 1994).

In addition to direct interactions with the target piscivore (both positive and negative), introduced prey can affect the target piscivore via indirect pathways through their interaction with other components of the aquatic community. Though the pathways by which these interactions can influence the target piscivore are complex, we suggest that they are still predictable *a priori*. For example, because young-of-year gizzard shad are typically too large for young-of-year largemouth bass to consume (Stiefvater and Malvestuto 1987; Adams and DeAngelis 1987), small largemouth bass rely on littoral young-of-year bluegill *Lepomis macrochirus* as prey when they switch from invertebrates (Chew 1974; Hirst and DeVries 1994). Because larval bluegill typically appear in the limnetic zone after larval gizzard shad, and thus after zooplankton populations have been depressed, survival of larval bluegill is compromised by the presence of gizzard shad (DeVries and Stein 1992; Stein and Garvey, The Ohio State University, unpublished data). Again, the strength of these interactions can vary among systems (Welker et al. 1994). When bluegill recruitment to the littoral zone is reduced through these interactions (Figure 3), gizzard shad exert a negative influence on littoral, young-of-year largemouth bass via indirect effects mediated through limnetic zooplankton and larval bluegill. Ultimately, because overwinter survival of young-of-year largemouth bass may be size dependent (e.g., Aggus and Elliott 1975; Isley 1979; but see Toneys and Coble 1980; Wright 1993), reduced recruitment of bluegill to the littoral zone can reduce eventual recruitment and year-class strength of largemouth bass.

As we have documented, early life stages of stocked prey species can negatively influence recruitment of sport fishes. Similarly, stocked sport fishes may do poorly if added to systems at a size at which they feed as obligate zooplanktivores. This is particularly true in systems that contain gizzard shad. In these instances, stocked sport fishes (such as hybrid striped bass *Morone* spp.) should be reared beyond zooplanktivorous life stages before stocking. In so doing, the negative interaction with resident zooplanktivorous fishes can be minimized. Alternatively, surveys can be conducted and stockings completed only in systems with adequate zooplankton populations.

Abiotic Effects

Stocking stress can cause substantial mortality in introduced fishes. Handling, confinement during transport, and thermal differences between hatchery and lake have been assessed through physiological indices (Barton et al. 1980; Carmichael et al. 1984; Mather et al. 1986) and, in the case of temperature, by determining upper lethal temperature and critical thermal maximum. Because plasma glucose concentrations and mortality are not related in esocids (Mather et al. 1986), we typically use measures of direct mortality.

In laboratory experiments with muskellunge, northern pike, and tiger muskellunge, losses due to thermal stress were more important than losses to either handling or transport, but did not differ among taxa (Mather et al. 1986; Mather and Wahl 1989). In 15°C acclimated esocids, a 10°C temperature increase killed few fish (Figure 4). A 12°C increase killed some fish across all taxa, whereas nearly all fish died in response to a 15°C temperature increase. We used similar procedures to evaluate mortality from thermal stress in walleye (D. Clapp, Y. Bhagwat, and D. Wahl, unpublished data). Again, increased temperatures caused higher mortality. An increase of 11°C resulted in no mortality, a 12°C increase caused substantial mortality, and nearly all fish died in response to an increase of 13°C (Figure 4). Direct comparisons with esocid data are difficult because acclimation temperatures were higher (20°C) and fish were smaller (100 mm). However, higher mortality for walleyes exposed to 12°C temperature increases than for esocids suggests lower thermal resistance in walleye.

We also monitored mortality during actual field stockings. Fish were transported to reservoirs in aerated tanks with 0.5% salt and were acclimated to within 5°C of lake temperature before stocking. A sample of each species was confined in predator-free cages and mortality recorded after 24 h (Mather and Wahl 1989; Santucci and Wahl 1993; Santucci et al. 1994). In the field, thermal stress translates to increased mortality. For esocids and walleye, no mortality occurred when lake temperatures were 25°C or less, but some mortality occurred when temperatures exceeded 25°C (Table 1). Discrepancies between the critical temperatures causing mortality in the laboratory and field probably resulted from different acclimation temperatures in the field for esocids (20°C) and walleye (generally 18–25°C; Table 1). Field experiments support laboratory evidence that walleye has a lower thermal tolerance than do esocids. In spite of high temper-

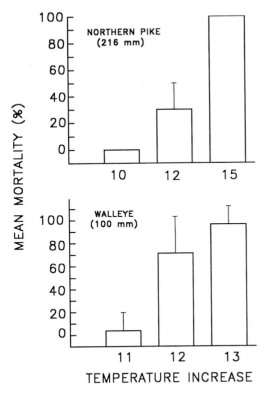

FIGURE 4.—Mean mortality at 48 hours of 15°C-acclimated northern pike and 20°C-acclimated walleye subjected to rapid temperature increase (1.5°C/min) in laboratory experiments. Vertical lines are the upper 95% confidence intervals.

atures at stocking and large temperature changes between hatchery and lake (Table 1), few channel catfish died in response to stocking stressors.

Clearly, losses of sport fish to stressors at stocking can be important and responses vary among species. Because mortality is a variable response and untested multiple stressors can reduce ability to survive high temperatures, acceptable temperature increases at stocking should be conservative. Healthy esocids acclimated to 15°C and walleyes acclimated to 20°C should suffer little stress-related mortality when reservoir temperatures are 25°C or less.

Handling and thermal effects may also increase mortality indirectly, by increasing vulnerability to predators or by reducing predatory efficacy of stocked fish. Impacts of these indirect effects are nearly impossible to quantify in the field but may lend themselves to experimental manipulation in a laboratory or hatchery-pond setting. Owing to physiological differences among species, both direct and indirect effects of abiotic factors require species-by-species assessment.

Abiotic factors, particularly temperature, also can influence growth and success of stocked fish, and the effects of abiotic factors vary among species. Variability in growth can indirectly affect other sources of mortality such as predation rates and overwinter survival. As a result, effects of introduction of fishes into systems with different thermal regimes or stratified systems with restricted habitats and access to prey (Coutant 1985; Headrick and

TABLE 1.—Percent mortality for various sizes of esocids, walleye, and channel catfish held in predator-free cages for 24 h after reservoir introductions in Ohio (OH) and Illinois (IL). Temperature change is difference between hatchery and lake temperature.

Reservoir (state)	Size (mm)	Stocking month	Temperature change (°C)	Lake temperature (°C)	Mortality (%)
			Esocids		
Kokosing (OH)[a]	145	Aug	+7 to +9	27–29	3.3
Madison (OH)[a]	145	Jul	+6 to +8	26–28	3.3
North (OH)[a]	200	Aug–Sep	+5	25	0
			Walleye		
Ridge (IL)[b]	55	Jun	+10	28 ± 2	22 ± 9
Le-Aqua-Na (IL)[c]	91	Jul	+3.5	25	0
Ridge (IL)[b]	200	Oct–Nov	−3 to +2	12 ± 6	2 ± 2
			Channel Catfish		
Ridge (IL)[d]	200	Jul–Aug	+2 to +6	24–32	0
Ridge (IL)[d]	250	Jul–Sep	+2 to +4	24–32	0

[a] Data from Mather and Wahl (1989).
[b] Data from Santucci and Wahl (1993).
[c] Data from Clapp et al. (1993).
[d] Data from Santucci et al. (1994).

Carline 1993) may be difficult to predict. Because temperature affects metabolism, food consumption, conversion efficiency, and hence growth, bioenergetics models provide a means of summarizing these physiological responses and differences among fishes (Kitchell et al. 1977; Rice and Cochran 1984; Bevelhimer et al. 1985; Wahl and Stein 1991). Though these models have been used successfully to predict growth of fish under a variety of thermal regimes, fish sizes, and stocking dates (Bevelhimer et al. 1985), they should be cautiously applied when they have not been thoroughly tested (Wahl and Stein 1991) or when physiological parameters are poorly known (e.g., with channel catfish). These simulation models can provide a powerful technique for evaluating growth of fishes and implications of this growth for survival under a variety of system-specific stocking conditions.

Overwinter Mortality

Overwinter losses of stocked fish can be important, influencing year-class strength and success of fishes. Overwinter mortality likely varies with size, energy reserves, and fish species. In our work, overwinter mortality was higher for walleye than for esocids and channel catfish (Figure 2). Other generalizations for these species are difficult given the small geographic range over which studies were conducted and the small range of fish sizes entering winter. Largemouth bass provides one example of the importance of these issues. For largemouth bass, large size translates to high lipid levels and reduced metabolic costs (per unit weight). Because largemouth bass typically do not feed at winter temperatures (i.e., below 10°C, Lemons and Crawshaw 1985) they must rely on large body size and associated energy reserves to avoid deleterious effects of starvation. Thus, large juvenile largemouth bass would be expected to have a greater probability of surviving winter starvation than would smaller conspecifics, as in young-of-year smallmouth bass *Micropterus dolomieui* (Oliver et al. 1979) and yellow perch *Perca flavescens* (Post and Evans 1989).

However, no consensus about size-related overwinter mortality in largemouth bass has been reached. In fact, the intensity of size-related overwinter mortality may be closely linked to severity of winter (likely in terms of length and average temperature). Recent studies in three distinct regions of the United States show the range of winter-related mortality. We have found strong size-dependent overwinter mortality in cohorts of age-0 stocked largemouth bass in Alabama ponds where small largemouth bass (<100 mm TL) suffered high mortality, medium-sized largemouth bass (100–149 mm TL) suffered intermediate mortality, and large largemouth bass (≥150 mm TL) had low mortality (Ludsin and DeVries, unpublished data). This difference in overwinter mortality likely occurred due to depletion of energy reserves. Conversely, other studies found weak evidence of size-dependent overwinter mortality in natural populations of largemouth bass in Illinois reservoirs and concluded this mechanism was unimportant (Kohler et al. 1993; M. Fuhr, D. Philipp, and D. Wahl, unpublished data). In northern Wisconsin, Wright (1993) found no evidence of overwinter mortality for a cohort of largemouth bass stocked into a small natural lake. Resolving this apparent discrepancy across the geographic range of this species, as well as other species, will provide useful insights into the role that overwinter conditions might play in influencing cohort success, year-class strength, and ultimately return to the angler creel of introduced sport-fish species.

Other Losses of Stocked Sport Fish

Other losses of introduced fishes during the first year of life, such as outmigration and natural mortality, will occur. Spillway losses did not influence success of walleye fingerlings (<2%; Santucci and Wahl 1993) but may be important for fry (Willis and Stephen 1987). Losses will depend on spillway design and flow rates and will likely vary among species (as per Johnson et al. 1988). The mechanisms we examined in our work did not explain all of the losses of stocked fish (Figure 2); other unidentified factors may be important. Survival through the first year of life varied with species and size at stocking. Fall survival varied among esocid taxa and was higher for large (180–205 mm, 2–43%) than for small fish (145 mm, 0%; Figure 5). Similarly, for walleye fingerlings, fall survival was highest for large (200 mm, 31%), followed by medium (140 mm, 7%) and then small (50 mm, 0%) fingerlings. After the first year of life, survival of both walleye and esocid cohorts remained relatively constant.

Interactions between abiotic and biotic factors can also affect year-class strength. Evaluations that quantify success through only the first growing season will not assess all factors that may determine cohort sizes. Because size of fish in the fall may dictate overwinter success, conclusions generated from survival rates through fall will be of limited value. Hence, we strongly recommend following stocked predators through spring and assessing

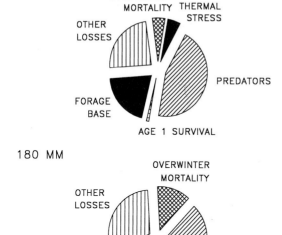

FIGURE 5.—Effect of stocking size on age-1 survival and relative losses to stocking stress (thermal stress), resident predators, forage base, overwinter mortality, and other losses (natural mortality and outmigration) for two sizes of tiger muskellunge.

population size in temperate systems in May or June. Our rationale for such a choice revolves around winter energetics and spring temperatures, predator size, and prey availability. For example, walleye must be 175 mm TL to survive a winter in northern lakes (Madenjian et al. 1991). Though individuals slightly smaller than this may have sufficient energy reserves to survive over winter, they can become energetically compromised due to rapid increases in spring water temperatures, and thus their metabolic rate, and be unable to use the forage base of large yearling and older fishes adequately. Whereas a large fish (>175 mm) likely could begin feeding on these forage fishes, a small one cannot. It may not be sufficient to estimate overwinter mortality through March or April in these populations, because total first-year mortality may have simply not as yet manifested itself by that time. For older fish, angler harvest is likely to be the most important source of mortality for introduced sport fish (Wahl and Stein 1993; Santucci and Wahl 1993; Santucci and Wahl 1994).

Conclusions and Implications for Management

Historically, fishery biologists rarely assessed the fate of stocked fishes, be they predators or prey. Through time, we as a discipline have come to realize that evaluations are necessary to justify the tremendous amount of time, money, and energy invested in hatchery-reared fishes. Unfortunately, early evaluations were haphazard and based more on trial-and-error approaches than on rigorous, methodologically well-grounded attempts. Because of the complexity and variability among aquatic ecosystems, such trial-and-error approaches have not led to broad-based generalities. Extreme variability in return to the creel across years within one system as well as within-year differences across systems indicates that approaches that provide an understanding of the underlying explanation for this variability are required. Hence, in order to frame broad stocking guidelines we advocate using an ecological approach to adding predators and prey to aquatic systems (Figure 6).

Stocking, as a management practice, must be considered within an ecosystem context, linking species management with ecological considerations. Given that fishery managers typically are attempting to enhance the population of a species of interest by means of stocking, rather than by improving the functioning of a particular community, and given differences in life-history characteristics, we must consider stocking on a species-by-species basis. Even so, we must search for commonalities across species that will generally guide stocking practices. For example, channel catfish suffer relatively low mortality from any biotic or abiotic sources whereas esocids and percids are more susceptible to thermal stress, predators, match with the forage base, and overwinter physiological stress (Figure 2). Clearly, the care with which esocids and percids must be matched to system characteristics differs greatly from that for channel catfish. Examining how other species, such as hybrid striped bass, bluegill, and largemouth bass fit into this perspective should improve our ability to draw across-species generalities.

As detailed above, using our proposed approach, we have been able to make many specific management recommendations regarding introductions of three taxonomic groups of sport fish. These recommendations regard the size, timing, and system-specific characteristics for stocking to maximize survival and growth. In spite of these advances, information gained thus far may not address all stocking scenarios, species, or aquatic ecosystems.

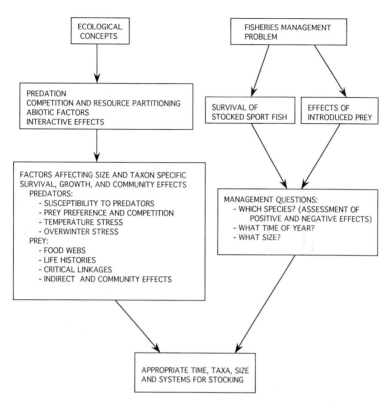

FIGURE 6.—An ecological framework for evaluating the success and effect of introduced fishes. The framework integrates the relative importance of predation, competition, and abiotic factors for all life stages.

In other situations, ecological concepts can again provide a guide to future research and management strategies.

As an example, we typically think of increasing the size of stocked fish in order to reduce vulnerability to resident predators (Figure 5). However, given what we know about optimal prey size for predators, increasing the size of stocked fishes may not always be appropriate. For percids, stocking sizes are often smaller than those sizes we have examined and, for example, decreasing the stocking size from 35 mm to 25 mm could reduce vulnerability if predators ignore smaller prey sizes. Whereas few examples of this phenomenon occur in the literature, it should be considered further. Residence time in the gut of predators will be short for small prey, and only intense sampling will yield insight into predatory impacts. When predators are important we suggest that stockings be concentrated, to saturate local resident predator populations. Conversely, if small fish are not subject to intense predation, scattered stocking may reduce intraspecific competition compared with concentrated stocking. Overall, considering the importance of predation from an ecological perspective provides fishery biology with new insights that can only be pursued by rigorously defined predator–prey sampling designs.

We suggest, as have Li and Moyle (1981, 1993) and Kohler and Stanley (1984) for exotic introductions, that, as with piscivore introductions, ecological principles must be used in making management decisions concerning the introduction of prey fishes. In general, species should be stocked only within their native range. Within these constraints, consideration must be given to not only the direct positive effects of an introduced prey species but also the indirect and complex effects that they may exert on the target species, via the rest of the aquatic community. Though it is difficult to determine all of the interactions in which an introduced prey species may be involved, use of ecological concepts provides a method for identifying critical linkages (Figure 6). Knowledge of life-history characteristics, including ontogenetic niche shifts (Werner and Gilliam 1984), of the species to be introduced, the target species, and other primary species in the recipient system is required, as is a description of the food web. Only with this type of approach will we be in a position to predict the ultimate influence of introduced prey on a target species and to anticipate the unexpected, potential interactions.

We must continue to think globally and ecologically about stocking practices. The potential genetic risks of stocking have been debated often (Philipp 1991). Once genetic issues are considered, species introduction can be a powerful management tool if it is developed within an ecological context. Using ecological, community-based perspectives that quantify important components (such as predation, competition, abiotic factors, and their interactions) provides us with a framework that serves to guide stocking evaluations for either predators or prey. Only by capitalizing on the lessons learned from basic ecology can we begin to make stocking programs meet their fullest potential in terms of return on investment, while minimizing negative and maximizing positive effects on aquatic ecosystems.

Acknowledgments

Numerous individuals associated with the Kaskaskia Biological Station, Center for Aquatic Ecology, Illinois Natural History Survey, the Aquatic Ecology Laboratory, The Ohio State University, and Auburn University assisted with this work. Special thanks are due V. Santucci, L. Einfalt, T. Szendrey, D. Clapp and S. Ludsin. We thank S. Stuewe, L. Willis, and A. Brandenburg of the Illinois Department of Conservation (IDOC), T. Nagel, P. Keyes, and J. Stafford of the Ohio Division of Wildlife (ODW), and the Alabama Game and Fish Division (AGFD) for providing fish. J. Mick and K. Cottrell coordinated activities with IDOC, D. Apgear and R. Miller with ODW, and W. Reeves with AGFD. The manuscript was improved by comments from D. Bayne, M. Freeman, B. Patter, and the Aquatic Ecology Discussion Group, Kaskaskia Biological Station. R. Wagner helped prepare the figures and text. The study was supported in part by funds from the Federal Aid in Sport Fish Restoration Act under Projects F-51-R, F-113-R, and F-118-R, administered through IDOC, Project F-57-R through ODW, and Project F-40-R through AGFD.

References

Adams, S. M., and D. L. DeAngelis. 1987. Indirect effects of early bass-shad interactions on predator population structure and food web dynamics. Pages 103–117 in W. C. Kerfoot and A. Sih, editors. Predation: direct and indirect impacts on aquatic communities. University Press of New England, Hanover, New Hampshire.

Aggus, L. R., and G. V. Elliott. 1975. Effects of cover and food on year-class strength of largemouth bass. Pages 317–322 in R. H. Stroud, editor. Black bass biology and management. Sport Fishing Institute, Washington, DC.

Allan, J. D. 1974. Balancing predation and competition in cladocerans. Ecology 55:622–629.

Barton, B. A., R. E. Peter, and C. R. Paulencu. 1980. Plasma cortisol levels of fingerling rainbow trout (Salmo gairdneri), at rest, and subjected to handling, confinement, transport, and stocking. Canadian Journal of Fisheries and Aquatic Sciences 37:805–811.

Bevelhimer, M. S., R. A. Stein, and R. F. Carline. 1985. Assessing significance of physiological differences among three esocids with a bioenergetics model. Canadian Journal of Fisheries and Aquatic Sciences 42:57–69.

Bremigan, M. T., and R. A. Stein. 1994. Gape-dependent larval foraging and zooplankton size: implications for fish recruitment across systems. Canadian Journal of Fisheries and Aquatic Sciences. 51:913–922.

Carline, R. F., R. A. Stein, and L. M. Riley. 1986. Effects of size at stocking, season, largemouth bass predation, and forage abundance on survival of tiger muskellunge. American Fisheries Society Special Publication 15:151–167.

Carmichael, G. J., J. R. Tomasso, B. A. Simco, and K. B. Davis. 1984. Confinement and water quality-induced stress in largemouth bass. Transactions of the American Fisheries Society 113:767–777.

Carpenter, S. R., J. F. Kitchell, and J. R. Hodgson. 1985. Cascading trophic interactions and lake productivity. BioScience 35:634–639.

Carpenter, S. R. 1988. Complex interactions in lake communities. Springer-Verlag, New York.

Chew, R. L. 1974. Early life history of the Florida largemouth bass. Florida Game and Fresh Water Fish Commission, Fishery Bulletin 7, Tallahassee, Florida.

Clapp, D. F., and seven coauthors. 1993. Evaluation of walleye stocking program. Illinois Natural History Survey, Aquatic Ecology Technical Report 93/1, Champaign.

Connell, J. H. 1975. Some mechanisms producing structure in natural communities: a model and evidence from field experiments. Pages 460–490 in M. Cody and J. Diamond, editors. Ecology and evolution of communities. Harvard University Press, Cambridge, Massachusetts.

Connell, J. H. 1983. On the prevalence and relative importance of interspecific competition: evidence from field experiments. American Naturalist 122:661–696.

Coutant, C. C. 1985. Striped bass, temperature, and dissolved oxygen: a speculative hypothesis for environmental risk. Transactions of the American Fisheries Society 114:31–61.

Dettmers, J. M., and R. A. Stein. 1992. Food consumption by larval gizzard shad: zooplankton effects and implications for reservoir communities. Transactions of the American Fisheries Society 121:494–507.

DeVries, D. R., and R. A. Stein. 1990. Manipulating shad to enhance sport fisheries in North America: an assessment. North American Journal of Fisheries Management 10:209–223.

DeVries, D. R., and R. A. Stein. 1992. Complex interactions between fish and zooplankton: quantifying the role of an open-water planktivore. Canadian Journal of Fisheries and Aquatic Sciences 49:1216–1227.

DeVries, D. R., R. A. Stein, J. G. Miner, and G. G. Mittelbach. 1991. Stocking threadfin shad: consequences for young-of-year fishes. Transactions of the American Fisheries Society 120:368–381.

Drenner, R. W., F. deNoyelles, Jr., and D. Kettle. 1982. Selective impact of filter-feeding gizzard shad on zooplankton community structure. Limnology and Oceanography 27:965–968.

Garvey, J. E., R. A. Stein, and H. M. Thomas. 1994. Assessing how fish predation and interspecific prey competition influence a crayfish assemblage. Ecology 75:532–547.

Goldberg, D. E. 1985. Effects of soil pH, competition, and seed predation on the distributions of two tree species. Ecology 65:503–511.

Headrick, M. R., and R. F. Carline. 1993. Restricted summer habitat and growth of northern pike in two southern Ohio impoundments. Transactions of the American Fisheries Society 122:228–263.

Hirst, S. C., and D. R. DeVries. 1994. Assessing the potential for direct feeding interactions among larval black bass and larval shad in two southeastern reservoirs. Transactions of the American Fisheries Society 123:173–181.

Hurlburt, S. H., J. Zedler, and D. Fairbanks. 1972. Ecosystem alteration by mosquitofish (*Gambusia affinis*) predation. Science 175:639–641.

Isley, J. J. 1979. Effects of water temperature and energy reserves on overwinter mortality in young-of-the-year largemouth bass (*Micropterus salmoides*). Master's thesis. Southern Illinois University, Carbondale.

Johnson, B. L., D. L. Smith, and R. F. Carline. 1988. Habitat preferences, survival, growth, foods, and harvests of walleyes and walleye × sauger hybrids. North American Journal of Fisheries Management 8:292–304.

Johnson, D. M., P. H. Crowley, R. E. Bohanan, C. N. Watson, and T. H. Martin. 1985. Competition among larval dragonflies: a field enclosure experiment. Ecology 66:119–128.

Kerfoot, W. C., and A. Sih, editors. 1987. Predation: direct and indirect impacts on aquatic communities. University Press of New England, Hanover, New Hampshire.

Kerfoot, W. C. 1987. Cascading effects and indirect pathways. Pages 57–70 *in* W. C. Kerfoot and A. Sih, editors. Predation: direct and indirect impacts on aquatic communities. University Press of New England, Hanover, New Hampshire.

Kerr, S. R. 1980. Niche theory in fisheries management. Transactions of the American Fisheries Society 109:254–257.

Kitchell, J. F., D. J. Stewart, and D. Weininger. 1977. Applications of a bioenergetics model to yellow perch (*Perca flavescens*) and walleye (*Stizostedion vitreum*). Journal of the Fisheries Research Board of Canada 34:1922–1935.

Kohler, C. C., R. J. Sheehan, and J. J. Sweatman. 1993. Largemouth bass hatching success and first-winter survival in two Illinois reservoirs. North American Journal of Fisheries Management 13:125–133.

Kohler, C. C., and J. G. Stanley. 1984. A suggested protocol for evaluating proposed exotic fish introductions in the United States. Pages 387–406 *in* W. R. Courtenay and J. R. Stauffer, editors. Distribution, biology, and management of exotic fishes. The Johns Hopkins University Press, Baltimore, Maryland.

Laarman, P. W. 1978. Case histories of stocking walleyes in inland lakes, impoundments, and the Great Lakes—100 years with walleyes. American Fisheries Society Special Publication 11:254–260.

Lemons, D. E., and L. I. Crawshaw. 1985. Behavioral and metabolic adjustments to low temperature in the largemouth bass (*Micropterus salmoides*). Physiological Zoology 58:175–180.

Li, H. W., and P. B. Moyle. 1981. Ecological analysis of species introductions into aquatic systems. Transactions of the American Fisheries Society 110:772–782.

Li, H. W., and P. B. Moyle. 1993. Management of introduced fishes. Pages 287–307 *in* C. Kohler and W. Hubert, editors. Inland fisheries management in North America. American Fisheries Society, Bethesda, Maryland.

Lubchenco, J. 1978. Plant species diversity in a marine intertidal community: importance of herbivore food preference and algal competition abilities. American Naturalist 112:23–29.

Madenjian, C. P., B. M. Johnson, and S. R. Carpenter. 1991. Stocking strategies for fingerling walleyes: an individual-based model approach. Ecological Applications 1:280–288.

Magnuson, J. J. 1991. Fish and fisheries ecology. Ecological Applications 1:13–26.

Mather, M. E., R. A. Stein, and R. F. Carline. 1986. Experimental assessment of mortality and hyperglycemia in tiger muskellunge due to stocking stressors. Transactions of the American Fisheries Society 115:762–770.

Mather, M. E., and D. H. Wahl. 1989. Comparative mortality in three esocids due to stocking stress. Canadian Journal of Fisheries and Aquatic Sciences 46:214–217.

Mauck, W. L., and D. W. Coble. 1971. Vulnerability of some fishes to northern pike (*Esox lucius*) predation. Journal of the Fisheries Research Board of Canada 28:957–969.

Miller, T. J., L. B. Crowder, J. A. Rice, and E. A. Marschall. 1988. Larval size and recruitment mechanisms in fishes: toward a conceptual framework. Canadian Journal of Fisheries and Aquatic Sciences 45:1657–1670.

Mittelbach, G. G., and P. L. Chesson. 1987. Predation

risk: indirect effects on fish populations. Pages 315–332 *in* W. C. Kerfoot and A. Sih, editors. Predation: direct and indirect impacts on aquatic communities. University Press of New England, Hanover, New Hampshire.

Moody, R. C., J. M. Helland, and R. A. Stein. 1983. Escape tactics used by bluegills and fathead minnows to avoid predation by tiger muskellunge. Environmental Biology of Fishes 8:61–65.

Morin, P. J. 1983. Predation, competition and the composition of larval anuran guilds. Ecological Monograph 53:119–138.

Ney, J. J. 1981. Evolution of forage-fish management in lakes and reservoirs. Transactions of the American Fisheries Society 110:725–728.

Noble, R. L. 1981. Management of forage fishes in impoundments of the southern United States. Transactions of the American Fisheries Society 110:738–750.

Noble, R. L. 1986. Predator–prey interactions in reservoir communities. Pages 137–143 *in* G. E. Hall and M. J. Van Den Avyle, editors. Reservoir fisheries management: strategies for the 80's. American Fisheries Society, Southern Division, Reservoir Committee, Bethesda, Maryland.

Oliver, J. D., G. F. Holton, and K. E. Chua. 1979. Overwinter mortality of fingerling smallmouth bass in relation to size, relative energy stores, and environmental temperatures. Transactions of the American Fisheries Society 108:130–136.

Peckarsky, D. L., and S. I. Dodson. 1980. Do stonefly predators influence benthic distributions in streams? Ecology 61:1275–1282.

Persson, L., and L. A. Greenberg. 1990. Juvenile competitive bottle-necks: the perch (*Perca fluviatilis*)–roach (*Rutilus rutilus*) interaction. Ecology 71:44–56.

Philipp, D. P. 1991. Genetic implications of introducing Florida largemouth bass, *Micropterus salmonides floridanus*. Canadian Journal of Fisheries and Aquatic Sciences 48:58–65.

Post, J. R., and D. O. Evans. 1989. Size-dependent overwinter mortality of young-of-the-year yellow perch (*Perca flavescens*): laboratory, in situ enclosure, and field experiments. Canadian Journal of Fisheries and Aquatic Sciences 46:1958–1968.

Rice, J. A., and P. A. Cochran. 1984. Independent evaluation of a bioenergetics model for largemouth bass. Ecology 65:732–739.

Rigler, R. F. 1982. The relation between fisheries management and limnology. Transactions of the American Fisheries Society 111:121–132.

Roughgardin, J. 1983. Competition and theory in community ecology. American Naturalist 122:583–601.

Santucci, V. J., Jr., and D. H. Wahl. 1993. Survival and growth of walleye stocked in a centrarchid-dominated impoundment. Canadian Journal of Fisheries and Aquatic Sciences 50:1548–1558.

Santucci, V. J., Jr., D. H. Wahl, and T. Storck. 1994. Growth, mortality, harvest, and cost effectiveness of stocked channel catfish in a small impoundment. North American Journal of Fisheries Management 14:781–789.

Schael, D. M., L. G. Rudstam, and J. R. Post. 1991. Gape limitation and prey selection in larval yellow perch (*Perca flavescens*), freshwater drum (*Aplodinotus grunniens*), and black crappie (*Pomoxis nigromaculatus*). Canadian Journal of Fisheries and Aquatic Sciences 48:1919–1925.

Schneider, J. C. 1975. Survival, growth and food of 4-inch walleyes in ponds with invertebrates, sunfish or minnows. Michigan Department of Natural Resources Fisheries Research Report 1833:18.

Schoener, T. W. 1983. Field experiments on interspecific competition. American Naturalist 122:240–285.

Stahl, T. P., and R. A. Stein. In press. Exploring how larval gizzard shad density influences piscivory and growth of young-of-year saugeye. Canadian Journal of Fisheries and Aquatic Sciences.

Stein, R. A., R. F. Carline, and R. S. Hayward. 1981. Largemouth bass predation on stocked tiger muskellunge. Transactions of the American Fisheries Society 110:604–612.

Stiefvater, R. J., and S. P. Malvestuto. 1987. Seasonal aspects of predator–prey size relationships in West Point Lake, Alabama–Georgia. Proceedings of the Annual Conference Southeastern Association of Fish and Wildlife Agencies 39(1985):19–27.

Strong, D. R., Jr. 1983. Natural variability and the manifold mechanisms of ecological communities. The American Naturalist 122:636–660.

Szendrey, T. A., and D. H. Wahl. In press. Size-specific growth and survival of stocked muskellunge. North American Journal of Fisheries Management.

Toneys, M. L., and D. W. Coble. 1980. Mortality, hematocrit, osmolality, electrolyte regulation, and fat deposition of young-of-the-year freshwater fishes under simulated winter conditions. Canadian Journal of Fisheries and Aquatic Sciences 37:225–232.

Tonn, W. T., C. A. Paszkowski, and T. C. Moermond. 1986. Competition in *Umbra perca* fish assemblages: experimental and field evidence. Oecologia 69:126–133.

Wahl, D. H. In press. Effect of habitat selection and behavior on vulnerability to predation of introduced fish. Canadian Journal of Fisheries and Aquatic Sciences.

Wahl, D. H., and R. A. Stein. 1988. Selective predation by three esocids: the role of prey behavior and morphology. Transactions of the American Fisheries Society 117:142–151.

Wahl, D. H., and R. A. Stein. 1989. Comparative vulnerability of three esocids to largemouth bass (*Micropterus salmoides*) predation. Canadian Journal of Fisheries and Aquatic Sciences 46:2095–2103.

Wahl, D. H., and R. A. Stein. 1991. Food consumption and growth among three esocids: field tests of a bioenergetics model. Transactions of the American Fisheries Society 120:230–246.

Wahl, D. H., and R. A. Stein. 1993. Comparative population characteristics of muskellunge, northern pike, and their hybrid. Canadian Journal of Fisheries and Aquatic Sciences 50:1961–1968.

Welker, M. T., C. L. Pierce, and D. H. Wahl. 1994. Growth and survival of larval fishes: roles of competition and zooplankton abundance. Transactions of the American Fisheries Society 123:703–717.

Werner, E. E. 1980. Niche theory in fisheries management. Transactions of the American Fisheries Society 107:257–260.

Werner, E. E., and J. F. Gilliam. 1984. The ontogenetic niche and species interactions in size-structured populations. Annual Review of Ecology and Systematics 15:393–426.

Wiens, J. A. 1977. On competition and variable environments. American Scientist 65:590–597.

Willis, D. W., and J. L. Stephen. 1987. Relationship between storage ratio and population density, natural recruitment, and stocking success of walleye in Kansas reservoirs. North American Journal of Fisheries Management 7:279–282.

Wingate, P. J. 1986. Philosophy of muskellunge management. American Fisheries Society Special Publication 15:199–202.

Wright, R. A. 1993. Size-dependence of growth, consumption, and survival in juvenile largemouth bass: combining experimental manipulations and individual-based simulations. Doctoral dissertation. University of Wisconsin, Madison.

Wydoski, R. S., and D. H. Bennett. 1981. Forage species in lakes and reservoirs of the western United States. Transactions of the American Fisheries Society 110:764–771.

Fisheries Restoration and Enhancement: Genetic Criteria and Goals

An Evaluation of Inbreeding and Effective Population Size in Salmonid Broodstocks in Federal and State Hatcheries

HAROLD L. KINCAID

*National Biological Service, Research and Development Laboratory,
Rural Delivery 4, Box 63, Wellsboro, Pennsylvania 16901, USA*

Abstract.—Federal, state, and private fisheries agencies were surveyed to identify and characterize salmonid broodstocks used in fisheries management for the National Trout Strain Registry and Database. Mating system, number of male and female parents by year-class, and generation interval were used to calculate effective population size and inbreeding rate for each broodstock. In total, 221 broodstocks of five species were identified, of which 100 (45.2%) had breeding history information adequate to calculate effective population size. Broodstock management responsibility was divided among 50 federal, 165 state, and 6 private agencies. The proportion of broodstocks with adequate breeding histories was not significantly different among agencies. The number of broodstocks, mean effective population size, and mean rate of inbreeding by species were as follows: brook trout *Salvelinus fontinalis*—16, 455, 0.97%; brown trout *Salmo trutta*—19, 476, 0.45%; cutthroat trout *Oncorhynchus clarki*—14, 1,283, 0.16%; lake trout *Salvelinus namaycush*—10, 230, 0.37%; and rainbow trout *Oncorhynchus mykiss*—41, 661, 0.28%. Inbreeding rate ranged from 0 to 10.4% among reported broodstocks; 79.2% of broodstocks had inbreeding rates less than 0.5% per generation and 6.9% had inbreeding rates greater that 1.0% per generation. Four categories of breeding systems were used to maintain broodstocks: random mating (37.6%), mating selected parents (1.0%), family mating (9.9%), and other mating systems (51.5%). Many factors influencing rate of inbreeding in broodstocks can be controlled with increased attention to the breeding and cultural techniques used in broodstock management operations.

The development of salmonid broodstocks in federal and state hatchery systems began in the 1870s with the culture of brook trout *Salvelinus fontinalis* in the eastern United States and rainbow trout *Oncorhynchus mykiss* in the West (Bowen 1970). Initially, eggs were collected from wild stocks for shipment to hatcheries across the country. Thereafter, individual hatcheries developed their own broodstocks to meet increased demand for eggs, to reduce shipping costs, and to reduce the need for repeated egg collections from natural stocks. Little is known of the original broodstocks or their performance characteristics. Since those first efforts, fish culture technology has expanded significantly. Today, hatchery personnel can culture most fish species used in sport-fish management.

Since the 1870s, salmonid broodstocks have been developed by a variety of methods including gamete and fish transfers between hatcheries, introduced strains, crosses of hatchery and natural broodstocks, and selection programs to enhance specific traits in cultured broodstocks. Most current broodstocks were developed since 1960, either from established broodstocks or periodic collections of gametes from natural broodstocks.

When a broodstock is isolated for a number of generations, selection and drift act to produce changes in population traits (physiological, morphological, behavioral, or cultural performance traits) that are reproducible and significantly different from other fish populations. This process results in strain development (Kincaid 1981). The Trout Strain Registry (Kincaid 1981) recognized strains of five trout species: rainbow trout, brook trout, brown trout *Salmo trutta*, cutthroat trout *Oncorhynchus clarki*, and lake trout *Salvelinus namaycush*. Further reports expanded the list of recognized strains (Claggett and Dehring 1984; Kincaid and Berry 1986; Kincaid et al. 1994). Preservation of genetic variability within the existing strains of each species is important because this variability defines the adaptive potential of the species for fisheries management (Krueger et al. 1981).

As resource managers are called to manage declining fish populations, they are forced to address many of the problems associated with intensive management of small populations. Progeny produced by a limited number of parents, whether in a hatchery or natural fishery, may lead to constriction of the gene pool and hence increased inbreeding. These constrictions occur as a result of breeding practices such as (1) selection of a small number of "superior" individuals as broodstock, (2) use of fish from a small segment of the spawning season, (3) use of a limited number of survivors following a major collapse of the fishery, or (4) use of a small number of broodfish because they provide adequate numbers of eggs or progeny to meet management needs.

Inbreeding is the mating of related individuals. The primary effects of inbreeding are to decrease heterozygosity and produce genotypic frequencies that depart from Hardy-Weinburg equilibrium (Gall 1987). The degree of inbreeding in finite populations, such as broodstocks, is based on the number of parents contributing progeny to each succeeding generation. Inbreeding is measured by the inbreeding coefficient F (Fisher 1965). Values of F define the probability that any two alleles at a given locus are identical and descended from a common ancestor.

The effect of inbreeding is to change genotype frequencies in a population. When selection is acting, this can lead to changes in performance traits relative to a noninbred control population. For example, inbreeding has been reported to decrease growth, survival, and feed-conversion efficiency and to increase fry abnormality rate in salmonids (Cooper 1961; Ryman 1970; Aulstad and Kittlesen 1971; Kincaid 1976a, 1976b, 1983; Gjerde et al. 1983). Decreased genetic variation associated with increased inbreeding was reported in allozyme studies of hatchery and natural populations of cutthroat trout (Allendorf and Phelps 1980) and Atlantic salmon *Salmo salar* (Stahl 1983).

The purpose of this study is to identify the trout broodstocks currently used in federal and state fisheries management programs and to estimate the inbreeding rate of each, based on calculated effective population size. Mating systems used to maintain broodstocks were classified and the frequency of each type determined. The final objective of the study was to identify hatcheries with substandard performance and to make recommendations on ways to maximize genetic diversity in trout broodstocks. Data were taken from a 1991 national survey of federal, state, and private fisheries agencies.

Methods

Identification of the existing salmonid broodstocks used in fisheries management and the breeding methods used to maintain them were the first steps in estimating effective population size and inbreeding rate. During 1991, a broodstock questionnaire was developed and distributed to fisheries resource agencies throughout the United States to develop a standard data set on salmonid broodstocks. The information collected included broodstock origin, breeding methods, generation interval, management uses of the broodstock, and performance measures for a variety of captive and non-captive traits. Parental information was recorded on the number of male and female parents used to produce each broodstock year-class from 1986 to 1991. Data from questionnaires were compiled in the National Trout Strain Registry and Database (Kincaid et al. 1994). Data printouts were returned to the originating agencies for verification and to permit reporting of updated information. Similar surveys of salmonid broodstocks managed by fisheries resource agencies were conducted in 1980 and 1983 (Kincaid and Berry 1986). Broodstock size, generation interval, and breeding methods information from the 1991 survey are summarized in this report. Broodstock managers were asked to classify the mating system used into five types: random mating of all fish mature within a single year-class, random mating of all fish mature from all year-classes of the broodstock, random mating of mature fish from a group selected for some trait or traits, and mating of single males and females to produce families that were reared and evaluated separately. The fifth category was a composite of the remaining mating systems not classified into the other four types.

Effective population size (N_e) was used to standardize the size of populations managed with different parental sex ratios. Effective population size is a generation concept used to determine the number of individuals in an idealized, panmictic population that would yield the same sampling variance or rate of inbreeding as found in the population under consideration (Falconer 1981). In effect, N_e is a calculation to adjust broodstocks maintained with nonequal parental sex ratios to a standard equal sex ratio as found in the ideal, random mating population. Effective population size was calculated for each broodstock using parental information on the 1986 to 1991 broodstock year-classes. Broodstocks that did not have breeding information were dropped from the data set. Because N_e is the number of parents contributing genes to the next generation in an idealized, panmictic population, when multiple year-classes were produced during the 6-year monitoring period, broodstock generation interval (GI) was used to determine the number of year-classes to be combined to calculate the generation N_e. Generation interval is the mean age in years at which progeny are produced for replacement broodstock (Falconer 1981). Generation interval may be the same or later than the average age of sexual maturity of females in the broodstock. In this data set, generation interval was the mean age of female sexual maturity reported in the survey (Table 1). Values of year-class effective population size ($N_{e,year}$) were calculated independently for each year-class of a broodstock by use of male and female parent numbers in the formula (modified from Falconer 1981),

$$(N_{e,\text{year}}) = 4(N_m \cdot N_f)/(N_m + N_f);$$

($N_{e,\text{year}}$) = calculated year-class effective population size,
N_m = number of male parents, and
N_f = number of female parents.

Estimates of N_e were calculated by summation of $N_{e,\text{year}}$ values for the number of years in the GI of each broodstock, starting with the oldest reported year-class based on the formula,

$$N_e = \Sigma N_{e,\text{year}};$$

N_e = calculated generation effective population size for the broodstock,
$N_{e,\text{year}}$ = calculated year-class effective population size for the broodstock,
year = 1, GI, and
GI = mean age (years) of females in the broodstock at first sexual maturity.

The relation of generation N_e and change in inbreeding level per generation (ΔF) was calculated with the formula (Falconer 1981),

$$\Delta F = 1/2 N_e;$$

ΔF = calculated increase in inbreeding level per generation, and
N_e = generation effective population size.

The relation of N_e and ΔF for N_e from 2 to 500 are shown in Figure 1.

Sex ratio was calculated as the ratio of number of male parents divided by the number of female parents for each year-class. Generation sex ratio was calculated by division of the total number of male parents divided by the total number of female parents across year-classes of each generation.

Results

The 1991 national trout broodstock survey successfully obtained responses from 100% of state fisheries agencies and national fish hatcheries. Survey response from commercial sector hatcheries was less successful with only six broodstocks reported. In total, 221 broodstocks were identified from all sources, and 100 (45.2%) had sufficient breeding history information to permit the calculation of N_e values (Table 1). The number of broodstocks and the number with breeding history information by species were rainbow trout 98, 41; brook trout 31, 16; brown trout 21, 19; cutthroat trout 24, 14; and lake trout 35, 10 (Figure 2). The number of broodstocks cultured by management agencies were 50 federal (22.6%); 165 state (74.7%); and 6 other (2.7%). The proportion of broodstocks with adequate breeding history information for N_e calculation was not significantly different among federal (50.0%) and state (44.6%) managed broodstocks. (chi-square test, $P < 0.05$; Zar 1984).

The parental sex ratio, N_e, and ΔF were calculated for each broodstock and generation (Table 1). The N_e values varied widely among broodstocks and ranged from 5 to 4,000 in the five species (Table 2). Mean N_e values for broodstocks within species ranged from 230 to 1,283. The ΔF values ranged from near 0 to 10.4% among broodstocks. Mean sex ratio values were 1.0 or higher for all species except lake trout, which included broodstocks produced from natural populations. The range in sex ratio values (0.50 to 3.33, excluding the extreme value of 9.00) was similar across species. Broodstocks reared in federal and state hatcheries did not differ in N_e, ΔF, and sex ratios among species (Table 3).

Partitioning the distribution of the 100 broodstock N_e values showed 21% of the broodstocks were maintained with less than 100 parents and 7% of the broodstocks (two brook, one brown, one cutthroat, and three rainbow trouts) with less that 50 parents (Figure 3). Forty-one percent of the broodstocks were maintained with more than 500 parents per generation. Lake trout was the only species for which an N_e value greater than 100 was maintained in all broodstocks. The proportion of broodstocks with N_e greater than 500 was highest in cutthroat trout (85.7%) and ranged from 10.0 to 41.2% in the other four species.

Partitioning the distribution of sex ratio values showed 14% of the broodstocks were maintained with an excess of males, 34% with an excess of females, and 52% with equal numbers of males and females (Figure 4). Thirty-four percent of the broodstocks (6 brook, 9 brown, 2 cutthroat, 4 lake, and 13 rainbow trouts) were maintained with sex ratios less than 0.75 or greater than 1.25 (i.e., with a surplus of one sex greater than 25%) (Table 1).

Mating systems used to maintain broodstocks fell into four types: random mating within and between year-classes (37.6%), mating selected parents (1.0%), mating as separate families without selection (9.9%), and other mating systems (51.5%; Figure 5). The proportion of broodstocks maintained

TABLE 1.—Effective population size (N_e), generation interval (GI), change in inbreeding level (ΔF), and parental sex ratio of five species of salmonid broodstocks maintained in federal (F), state (S), private (P), and tribal (T) hatchery systems from 1986 to 1991.

Broodstock (hatchery system)	Year-class N_e, year[a]						Total parents	GI[b]	Generation 1			Generation 2		
	1991	1990	1989	1988	1987	1986			N_e	ΔF (%)	Sex ratio[c]	N_e	ΔF (%)	Sex ratio[c]
Brook trout														
Henrys Lake (P)	0	0	0	300	100	20	420	2	120	0.417	1.00	300	0.167	1.00
Oswayo (S)	120	120	120	120	0	0	480	2	240	0.208	1.00	240	0.208	1.00
Reynoldsdale-1991 (S)	0	746	0	0	0	0	746	2	746	0.067	1.99			
Ford Eastern Brook (S)	0	0	436	0	0	0	436	2	436	0.115	0.83			
Tomah Lake (S)	0	9	0	0	22	0	31	3	22	2.273	1.00	9	5.417	3.33
Bellefonte Open (S)	501	460	0	0	0	0	961	2	961	0.052	0.70			
Pisgah Forest (S)	0	556	0	0	0	0	556	3	556	0.090	1.00			
Paint Bank (S)	0	0	72	0	0	0	72	2	72	0.696	1.79			
Fernwood domestic (S)	494	646	0	0	0	0	1,140	2	1,141	0.044	3.01			
Assinica (S)	3	34	0	0	5	0	42	3	5	10.420	1.50	37	1.336	1.50
Rome Lab (S)	600	600	600	0	0	0	1,800	2	1,200	0.042	1.00	600	0.083	1.00
Armstrong (S)	600	600	0	500	0	0	1,700	2	500	0.100	1.00	1,200	0.042	1.00
Benner Spring (S)	24	24	24	24	0	0	96	3	72	0.694	1.00	24	2.083	1.00
Benner Spring Select (S)	120	120	120	0	0	0	360	2	240	0.208	1.00	120	0.417	1.00
Maine Hatchery (S)	98	69	17	13	21	0	218	3	51	0.982	1.85	166	0.301	2.53
Edray (T)	766	0	600	0	0	0	1366	4	1366	0.037	1.00			
Brown trout														
New Gloucester (F)	0	19	0	0	16	0	35	3	16	3.125	2.00	19	2.679	2.00
Bitterroot (F)	0	278	113	132	0	0	523	3	522	0.096	1.56			
Rome-1 (F)	104	60	60	0	0	0	224	2	120	0.417	3.00	104	0.482	2.86
Rome-2 (F)	60	30	30	0	0	0	120	2	60	0.833	3.00	60	0.833	3.00
Rome Lab (S)	0	0	1,800	800	300	0	2,900	2	1,100	0.046	1.00	1,800	0.028	1.00
Pisgah (S)	48	0	200	0	0	200	448	3	200	0.250	1.00	248	0.202	1.00
Rome-3 (S)	0	36	24	52	0	0	112	2	76	0.658	1.00	36	1.389	1.00
Armstrong (S)	438	500	400	0	0	0	1338	2	900	0.056	1.00	438	0.114	1.00
Huntsdale/Rome (S)	50	34	40	0	0	0	124	2	74	0.676	1.00	50	1.000	1.00
Wild Rose (S)	326	0	0	0	0	0	326	2	326	0.154	2.50			
Paint Bank/Nashua (S)	0	0	0	60	0	0	60	3	60	0.833	1.67			
Walhalla-1 (S)	1,333	1,333	0	0	0	0	2,666	2	2,667	0.019	0.50			
Ford (S)	0	0	249	208	0	0	457	2	457	0.109	0.81			
Bnt-Pro-87-Wss (S)	0	0	0	0	160	0	160	2	160	0.313	1.00			
Reynoldsdale (S)	0	0	120	0	0	0	120	2	120	0.417	1.00			
Paint Bank/Crawford (S)	0	0	164	0	0	0	164	2	164	0.305	0.54			
Plymouth Rock (S)	0	0	0	1,147	0	0	1,147	2	1,147	0.044	2.00			
Walhalla-2 (S)	272	0	0	0	0	0	272	3	272	0.184	0.85			
Cortland (S)	203	192	199	219	0	0	813	3	610	0.082	1.49	203	0.247	1.73
Cutthroat trout														
Westslope (F)	1,000	1,000	1,000	1,000	1,000	1,000	6,000	4	4,000	0.013	1.00	2,000	0.025	1.00
Fish Lake Westslope (F)	0	0	0	600	800	300	1,700	3	1,700	0.029	0.86			
Greenback (F)	0	0	216	471	0	0	687	3	687	0.073	1.00			
McBride (F)	0	422	736	888	274	0	2,320	3	1,898	0.026	1.10	422	0.119	1.00
Colorado River (S)	520	420	300	400	300	0	1,940	2	700	0.071	1.00	720	0.069	1.00
Auburn Snake River (S)	338	388	262	393	384	352	2,117	4	1,391	0.036	0.84	726	0.069	1.00
Yellowstone (S)	300	300	300	300	152	0	1,352	3	752	0.067	1.07	600	0.083	1.00
Bear River (S)	430	380	400	420	440	0	2,070	3	1,260	0.040	1.00	810	0.062	1.00
Henrys Lake (S)	0	0	0	2,000	600	200	2,800	3	2800	0.018	1.00			
Snake River/Bar BC1 (S)	154	0	82	58	60	0	354	3	200	0.250	0.98	154	0.325	1.00
Snake River/Bar BC2 (S)	142	515	60	0	0	0	717	3	717	0.070	1.45			
Little Snake (S)	0	8	14	14	0	0	36	4	36	1.390	1.00			
King's Lake (S)	0	0	0	0	880	52	932	3	932	0.054	1.49			
Trapper Lake (S)	0	0	892	0	0	0	892	2	892	0.056	1.00			
Lake trout														
Seneca-1 (F)	0	0	0	126	0	0	126	6	126	0.397	0.56			
Seneca-2 (F)	0	0	0	0	126	0	126	6	126	0.397	0.56			
Isle Royale (F)	0	343	0	0	0	686	1,029	6	1,029	0.049	1.10			
Lewis Lake-1 (F)	104	136	0	0	0	0	240	6	240	0.208	1.00			
Lake Trout Iry (F)	186	0	0	0	0	65	251	6	251	0.199	0.90			
Gillis Lake (F)	0	72	0	0	0	0	72	6	72	0.694	1.00			

(*Continued on next page*)

TABLE 1.—Continued.

Broodstock (hatchery system)	Year-class N_e, year[a]						Total parents	GI[b]	Generation 1			Generation 2		
	1991	1990	1989	1988	1987	1986			N_e	ΔF (%)	Sex ratio[c]	N_e	ΔF (%)	Sex ratio[c]
Lake trout														
Seneca-3 (F)	0	0	0	126	0	0	126	6	126	0.397	0.56			
Superior (F)	0	0	0	88	0	0	88	6	88	0.568	1.00			
Seneca-4 (F)	0	0	0	0	126	0	126	5	126	0.397	1.80			
Lewis Lake-2 (S)	0	116	0	0	0	0	116	3	116	0.430	0.79			
Rainbow trout														
London (F)	100	100	100	100	100	0	500	2	200	0.250	1.00	200	0.250	1.00
Semi Annual (F)	28	28	28	0	0	0	84	2	56	0.893	1.00	28	1.786	1.00
Shasta-1 (F)	480	640	640	720	0	0	2,480	3	2,000	0.025	1.00	480	0.104	1.00
Shasta-RTS (F)	40	40	40	40	0	0	160	2	80	0.625	1.00	80	0.625	1.00
Swanson River (F)	267	267	267	267	0	0	1,068	3	800	0.063	2.00	267	0.188	2.00
Caledonia (F)	487	267	1,149	0	0	0	1,903	2	1,416	0.035	1.46	487	0.103	1.50
Arlee-1 (F)	480	640	940	0	0	0	2,060	2	1,580	0.032	1.00	480	0.104	1.00
Hot Creek-RTH (S)	0	20	20	20	0	0	60	2	40	1.250	1.00	20	2.500	1.00
Kamloops-Duncan River (F)	634	480	444	396	0	0	1,954	2	840	0.060	1.00	1,114	0.045	1.00
Hot Creek (S)	0	406	78	0	0	0	484	2	484	0.103	1.01			
Pisgah Wytheville (S)	58	100	100	0	0	0	258	2	200	0.250	1.00	58	0.862	1.00
Paint Bank (S)	0	0	132	203	0	0	335	2	335	0.149	1.58			
Eagle Lake-1 (S)	0	0	300	0	300	0	600	2	300	0.167	1.00	300	0.167	1.00
Bellefonte (S)	120	200	200	0	0	0	520	2	400	0.125	1.00	120	0.417	1.00
Hayspur (S)	0	0	0	200	100	0	300	2	300	0.167	1.00			
Avington (S)	28	12	0	0	0	0	40	2	40	1.250	1.00			
Harrison Lake (S)	242	300	0	0	0	0	542	2	542	0.092	1.00			
Skamania (S)	180	213	0	0	0	0	393	2	392	0.127	1.47			
Utah (S)	193	139	0	0	0	0	332	3	331	0.151	1.45			
Pisgah Trout Lodge (S)	110	200	0	0	0	0	310	2	310	0.161	1.00			
White Sulfur (S)	0	126	125	0	0	0	251	2	251	0.199	1.69			
Nashua (S)	54	0	117	0	0	0	171	2	117	0.429	1.40	54	0.922	1.43
Benner Spring Synthetic (S)	600	320	0	0	0	0	920	2	920	0.054	3.33			
London-1992 (S)	0	0	1,500	436	0	0	1,936	2	1,936	0.026	2.92			
Winthrop (S)	0	0	500	0	0	0	500	6	500	0.100	1.00			
Shasta-2 (S)	0	0	480	0	0	0	480	2	480	0.104	1.00			
Reynoldsdale-1990 (S)	0	0	120	0	0	0	120	2	120	0.417	1.00			
Arlee-2 (S)	850	850	850	850	0	0	3,400	2	1,700	0.029	1.00	1,700	0.029	1.00
Big Lake (S)	267	267	267	267	0	0	1,068	2	533	0.094	2.00	533	0.094	2.00
Erwin (S)	700	980	1,000	0	0	0	2,680	4	2,680	0.019	1.00			
Big Spring-1 (S)	0	40	40	40	0	0	120	2	80	0.625	1.00	40	1.250	1.00
South Tacoma (S)	0	140	124	246	153	0	663	2	399	0.125	1.11	264	0.189	0.68
Wytheville (S)	480	480	480	480	0	0	1,920	3	1,440	0.035	1.50	480	0.104	1.50
RTW (S)	1,517	40	40	40	0	0	1,637	2	80	0.625	1.00	1,557	0.032	0.76
Kamloops-1 (S)	0	0	36	0	0	0	36	1	36	1.389	9.00			
McConaughy (S)	400	480	720	720	0	0	2,320	2	1,440	0.035	1.00	880	0.057	1.00
Spokane (S)	228	211	191	311	0	0	941	2	503	0.100	2.49	439	0.114	2.48
Eagle Lake-2 (S)	422	514	596	600	0	0	2,132	2	1,196	0.042	1.00	936	0.053	1.00
Skanes Kamloops (S)	0	0	0	250	0	0	250	3	250	0.200	1.00			
Big Spring-2 (S)	48	28	0	0	0	0	76	2	76	0.658	1.00			
Kamloops-2 (S)	0	472	480	334	396	1,310	2,992	2	1,706	0.029	1.00	814	0.061	1.00

[a] Effective population size (N_e) for each broodstock year-class produced from 1986 to 1991.
[b] Average age (in years) of female parents at first maturity.
[c] Sex ratio of parents used to produce each generation calculated as total number of male parents divided by total number of female parents.

with random mating systems was lowest in the lake trout (10.0%) and highest in the brown trout (57.9%).

Discussion

Conformance to Published Guidelines

The distribution of calculated N_e values for all salmonid broodstocks in the 1991 survey ranged from 5 to 4,000 and was similar in each species. Breeding history information on salmonid broodstocks cultured in federal and state fish hatcheries showed similar ranges in N_e values for each species with means for brown trout higher in state broodstocks and lower for cutthroat, lake, and rainbow trouts in state broodstocks (Table 2). Differences in mean and range could be explained by one to three abnormally large or small broodstocks (Table 1).

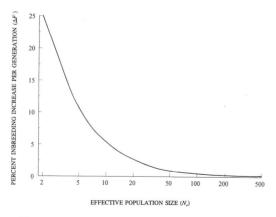

FIGURE 1.—Relationship between effective population size (N_e) and rate of inbreeding accumulation per generation (ΔF).

Most broodstocks met the general genetic guidelines for minimum N_e per generation for captive broodstocks (Kincaid 1976a, 1976b, 1983; Kapuscinski and Jacobson 1987; Allendorf and Ryman 1987). Kincaid (1976a, 1983) recommended a minimum of 100 parents, 50 males and 50 females ($N_e = 100$) be used each generation; this was achieved in 93% of the broodstocks. Allendorf and Ryman (1987) recommended a minimum of 200 parents, 100 males and 100 females ($N_e = 200$); this was achieved in 68% of the broodstocks.

Ideally, the size of any broodstock should be as large as possible to minimize the rate of inbreeding accumulation and to reduce the probability that rare alleles (gene frequency less than 0.01) will be lost from the population. In practice, however, population size limits are imposed by the management situation (e.g, limited water supply, water quality, rearing facilities, and available broodstock). The guideline of an N_e of 100 sets the expected rate of inbreeding accumulation at 0.5% per generation. The guideline of an N_e of 200 sets the expected rate of inbreeding accumulation at 0.25% per generation and provides a higher probability that rare alleles will be retained in the population. When maintenance of genetic variability in the broodstocks is a goal, a good guideline is to keep the N_e greater than 100 and as large as is possible and practical. When a broodstock situation limits the calculated N_e to less than 100, every effort should be made to expand the broodstock over the next one to two generations.

The guideline to maintain an equal sex ratio was generally followed with a parental sex ratio of 1.0 in 52% of the broodstocks and within the range of 0.75 to 1.25 in 67% of the broodstocks. The general genetic guidelines (Kincaid 1976a, 1976b, 1983) were based on the assumption that fish broodstocks approximate the ideal random mating population (i.e., one with no mutation, migration, or selection). Further, these guidelines assumed distinct generations, uniform population size from generation to generation, equal sex ratios, and a family size distribution with a mean and standard deviation of 2. In salmonid broodstocks, domestic or natural, these assumptions are not necessarily achieved. Among surveyed stocks, sex ratios ranged from 0.50 to 3.33.

The sex ratio should be kept as close to 1.0 as possible. Because each sex contributes half the gametes to the next generation, irrespective of number of individuals of each sex in the broodstock, the limiting sex will contribute disproportionately to the progeny generation resulting in a reduced N_e. The number of mature individuals of each sex available in a given year-class can be limited for a variety of reasons, especially when the broodstock must be captured from natural populations. If sex ratio is out of balance in the parental population, it should be returned to equality when the broodstock reaches maturity and before the next progeny generation is produced.

Procedures for Optimizing N_e

Family size variability can have a major effect either to increase or decrease the expected effective breeding size, but in most broodstocks that effect is unknown because individual family survival is un-

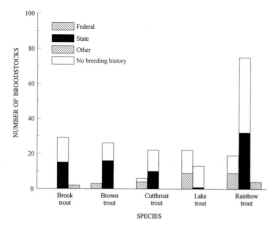

FIGURE 2.—Frequency of salmonid broodstocks with and without usable breeding history information by species and management agency.

TABLE 2.—Generation interval (GI), effective population size (N_e), estimated inbreeding rate (ΔF), and sex ratio in hatchery broodstocks of five salmonid species over two generations.

Species and generation	Number of broodstocks	GI[a] Mean	GI[a] Range	N_e Mean	N_e Range	ΔF[b] Mean	ΔF[b] Range	Sex ratio[c] Mean	Sex ratio[c] Range
Brook trout		2.41	2–4						
1	16			455	5–1,800	0.97	0.00–10.42	1.27	0.70–3.01
2	9			300	9–1,200	1.12	0.04–5.42	1.49	1.00–3.33
Brown trout		2.32	2–3						
1	19			476	17–1,333	0.45	0.02–3.13	1.42	0.54–3.00
2	9			329	19–1,800	0.77	0.03–2.68	1.62	1.00–3.00
Cutthroat trout		3.07	2–4						
1	14			1283	36–4,000	0.16	0.01–1.39	1.06	0.84–1.49
2	7			776	154–2,000	0.11	0.03–0.32	1.00	1.00–1.00
Lake trout		5.60	3–6						
1	10			230	72–1,029	0.37	0.05–0.69	0.92	0.56–1.80
2									
Rainbow trout		2.24	1–6						
1	41			661	36–2,608	0.28	0.02–1.39	1.49	1.00–9.00
2	23			493	20–1,700	0.44	0.03–2.50	1.19	0.68–2.48

[a] Generation interval of each species measured in years.
[b] Percent increase in inbreeding per generation.
[c] Calculated as number of male parents divided by number of female parents.

known. When broodstocks are managed as composite pools, the number of parents contributing gametes to the pool is known, but survival of each family from fertilized eggs to maturity is unknown. In this situation, managers typically assume that survival is random and estimate broodstock N_e based on the number of parents used to produce the pool. This overestimates N_e when survival among families is nonrandom due to factors such as differential egg quality and fecundity of females, fertility of males, stress and disease tolerance, and domestication. In the ideal, stable self-sustaining population, family size approximates a Poisson distribution with a mean and variance of 2 (Falconer 1981). The effect of variance in family size on N_e can be predicted by the formula,

TABLE 3.—Comparison of effective population size (N_e), estimated inbreeding rate (ΔF), and sex ratio in hatchery broodstocks of five salmonid species reared in federal versus state hatchery systems.

Species and source	Number of broodstocks	N_e Mean	N_e Range	ΔF[a] Mean	ΔF[a] Range	Sex ratio[b] Mean	Sex ratio[b] Range
Brook trout							
Federal	0						
State	15	1,096	5–5,100	1.07	0.01–10.42	1.32	0.70–3.01
Brown trout							
Federal	3	219	16–522	1.21	0.42–3.13	2.17	1.56–2.93
State	16	525	60–2,667	0.31	0.02–0.83	1.28	0.81–3.00
Cutthroat trout							
Federal	4	2,071	687–4,000	0.04	0.01–0.07	0.99	0.86–1.08
State	10	968	36–2,800	0.21	0.02–1.39	1.09	0.89–1.49
Lake trout							
Federal	9	243	72–1,029	0.37	0.05–0.69	0.94	0.56–1.80
State	1	116		0.43		0.79	
Rainbow trout							
Federal	9	779	40–2,000	0.36	0.03–1.25	1.16	1.00–2.00
State	32	627	36–2,680	0.25	0.02–1.39	1.55	0.77–9.00

[a] Percent increase in inbreeding per generation.
[b] Calculated as number of male parents divided by number of female parents.

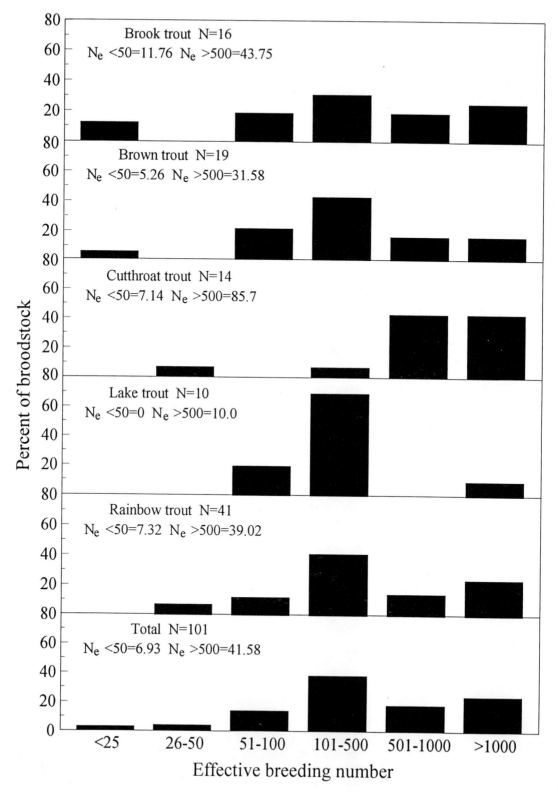

FIGURE 3.—Distribution of effective breeding number (N_e) among cultured broodstocks (N) of five salmonid species.

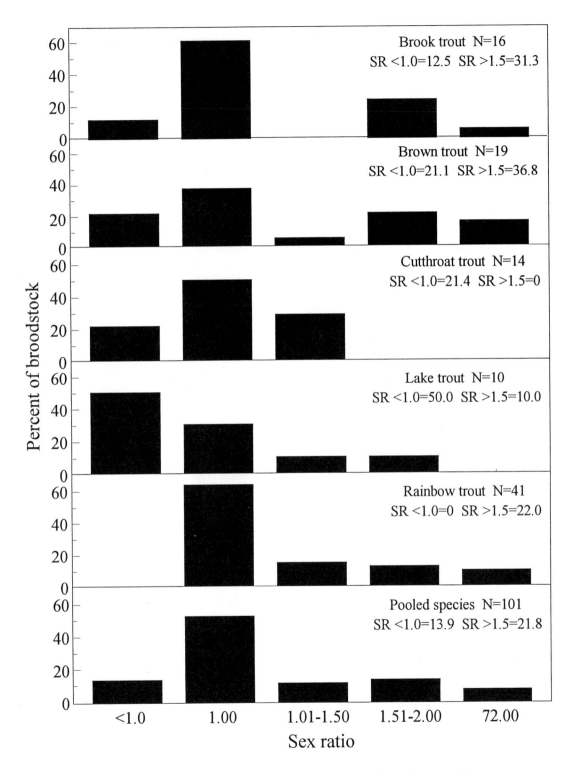

FIGURE 4.—Sex ratio (SR) distribution among cultured broodstocks (N) of five salmonid species.

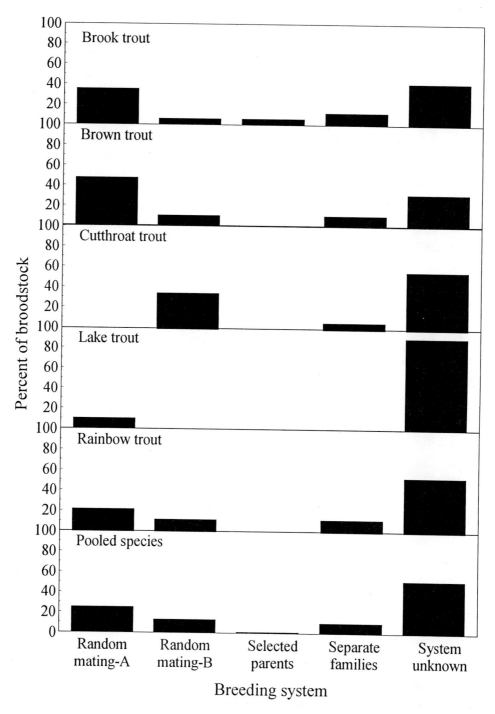

FIGURE 5.—Frequency of use of mating systems to maintain broodstocks in federal and state hatchery programs. Mating systems are random mating of all fish mature within a single year-class (random mating-A); random mating of all fish mature from all year-classes of the broodstock (random mating-B); random mating of mature fish from a group selected for some trait(s) (selected parents); mating of single males and females to produce families that were reared and evaluated separately (separate families); and broodstock for which the mating system was not reported (system unknown).

$$N_e = 4N/(2 + V_K),$$

where N is equal to the number of parents and V_K is the variance in family size (Crow and Kimura 1970). In the ideal population, N_e is equal to N when V_K is 2. Gall (1987) estimated the variance of family size in hatchery rainbow trout to be approximately 6, or,

$$N_e = 4N/(2 + 6) = 1/2\,N.$$

The effect of increasing V_K was to reduce the calculated N_e. Falconer (1981) summarized reports comparing N_e estimates in other species with variable family size and found they ranged from 0.48 to 0.94 N. This suggests that the actual N_e of hatchery salmonid broodstocks may be much lower than currently believed in many broodstocks. Until marking systems (such as DNA and allozyme markers) and culture techniques capable of identifying families throughout the life cycle are adopted, it will not be possible to completely know the effect of family size variation on the N_e of succeeding generations.

One approach to reduce family size variation in hatchery broodstocks is to hold the eggs of each family separate through egg incubation. The new broodstock lot can be established by combining an equal number of fry (5 to 25) from each family and culturing them to maturity separately. This approach equalizes family contribution to the new broodstock after differential mortality associated with embryo development occurs and equalizes family differences due to differential fecundity and egg quality of females. Family size variability will continue to be affected by inherent survival differences among families and environmental factors in the rearing environment during grow-out to maturity. If the lot size starts at two to five times the desired number of broodstock at maturity to allow for natural mortality during the rearing period, random lot reduction, which can increase family size variation, can be reduced.

Another way to increase N_e is to extend the generation interval. For example, in rainbow trout the GI could be increased from 2 years to either 3 or 4 years. This permits the addition of extra year-classes each generation to increase N_e. If N_e remained constant, the generation ΔF would be unchanged, but the effective inbreeding rate measured on a temporal basis would be reduced by 33% for a 3-year generation interval and 50% for a 4-year generation interval.

Overlapping generations can create a serious complication in measuring N_e when year-class and family size are variable and generation intervals are long (Falconer 1981; Hill 1972, 1979; Felsenstein 1971). When year-class and family size are kept uniform, overlapping generations have minimal effect. Generations typically are not discrete in natural populations and may not be in domestic broodstocks. This means that individuals present at a given time are of different ages and at different stages of the life cycle. As a result, it is difficult to determine the number of individuals in the breeding population and the assumption of random breeding is not met. When year-class size and family size variability are uniform, the number of individuals in the breeding age groups can be determined and effect of overlapping generations is small (Hill 1979; Felsenstein 1971).

The Need for More Detailed Records

A special concern made apparent by the data base information examined in this study was the proportion of broodstocks in the hatchery system that did not have information on year-class parent number, the breeding system used to maintain the broodstock, and generation interval. Whereas some broodstocks were new and had not reached first maturity in the hatchery, the records on many broodstocks simply did not have the information. This is a symptom of a larger problem, the need to maintain detailed breeding records on all broodstocks so that changes in these populations can be detected quickly and genetic variability of the populations can be protected into the foreseeable future. Information on number of male and female parents, source (lots) of parents, mating system, culling methods, and method for choosing or excluding individual broodfish is essential to evaluate the long-term integrity of the broodstock gene pool. This information is determined by hatchery personnel during the cultural operation and should be recorded in the broodstock record for management review. Without complete broodstock history information, it is not possible for management to evaluate the current status of any broodstock program fully.

Need for Genetic Monitoring

Information from genetic monitoring programs is essential for long-term broodstock management decisions. Broodstock history records are valuable because they identify the cultural and breeding methods used to maintain the broodstock and important historical events that affect broodstock N_e and performance. However, many cultured broodstocks, especially those developed from wild populations, do not have sufficient broodstock history to identify potential genetic problems. Even with good history information, population characteristics are unknown at the level of the gene (e.g, heterozygosity level and genetic variability). An initial genetic survey of the broodstock based on allozyme or DNA techniques can provide information on allele frequency, heterozygosity, and genetic variability. Comparison of genetic information from several broodstocks can provide a basis for unique identification of each broodstock. Repeated genetic surveys to collect data over multiple year-classes and generations can provide information that would allow management to detect changes in gene frequencies and genetic variability that may indicate increased inbreeding, selection, or interbreeding with other populations. Genetic monitoring information can also be used to estimate N_e (Bartley et al. 1992; Hedgecock et al. 1992). Monitoring programs also provide management with the genetic information needed to measure changes in the gene pool and predict long-term genetic consequences of cultural methods on broodstock integrity. The combination of a strong genetic monitoring program and breeding history record also would allow management to determine causes of changing performance as they occur. A genetic monitoring program is strongly recommended for all broodstocks used in fisheries management.

References

Allendorf, F. W., and S. R. Phelps. 1980. Loss of genetic variation in a hatchery stock of cutthroat trout. Transactions of the American Fisheries Society 109:537–543.

Allendorf, F. W., and N. Ryman. 1987. Genetic management of hatchery stocks. Pages 141–159 in N. Ryman and F. Utter, editors. Population genetics and fishery management. University of Washington Press, Seattle.

Aulstad, D., and A. Kittlesen. 1971. Abnormal body curvatures of rainbow trout (*Salmo gairdneri*) inbred fry. Journal of the Fisheries Research Board of Canada 28:1918–1920.

Bartley, D., M. Bagley, G. Gall, and B. Bentley. 1992. Linkage disequilibrium data to estimate effective population size of hatchery and natural fish populations. Conservation Biology 6:365–375.

Bowen, J. T. 1970. A history of fish culture as related to the development of fisheries programs. American Fisheries Society Special Publication 7:71–93.

Claggett, L. E., and T. R. Dehring. 1984. Wisconsin salmonid strain catalogue. Wisconsin Department of Natural Resources, Administrative Report 19, Madison.

Cooper, E. 1961. Growth of wild and hatchery strains of brook trout. Transactions of the American Fisheries Society 90:424–438.

Crow, J. F., and M. Kimura. 1970. An introduction to population genetic theory. Harper and Row, New York.

Falconer, D. S. 1981. Introduction to quantitative genetics, 2nd edition. Longman Inc., New York.

Felsenstein, J. 1971. Inbreeding and variance effective numbers in populations with overlapping generations. Genetics 68:581–597.

Fisher, R. A. 1965. The theory of inbreeding. Academic Press, New York.

Gall, G. A. E. 1987. Inbreeding. Pages 47–87 in N. Ryman and F. Utter, editors. Population genetics and fisheries management. University of Washington Press, Seattle.

Gjerde, B., K. Gunne, and T. Gjerde. 1983. Effects of inbreeding on survival and growth in rainbow trout. Aquaculture 34:327–332.

Hedgecock, D., V. Chow, and R. Waples. 1992. Effective population numbers of shellfish broodstocks estimated from temporal variance in allelic frequencies. Aquaculture 108:215–232.

Hill, W. G. 1972. Effective population with overlapping generations. Theoretical Population Biology 3:278–289.

Hill, W. G. 1979. A note on effective population size with overlapping generations. Genetics 92:317–322.

Kapuscinski, A., and L. Jacobson. 1987. Genetic guidelines for fisheries management. Minnesota Sea Grant Program, Sea Grant Research Report 17, St. Paul.

Kincaid, H. L. 1976a. Effects of inbreeding on rainbow trout populations. Transactions of the American Fisheries Society 105:273–280.

Kincaid, H. L. 1976b. Effects of inbreeding on rainbow trout (*Salmo gairdneri*). Journal of the Fisheries Research Board of Canada 33:2420–2426.

Kincaid, H. L. 1981. Trout strain registry. U.S. Fish and Wildlife Service, National Fisheries Research Center, Leetown, West Virginia.

Kincaid, H. L. 1983. Inbreeding in fish populations used in aquaculture. Aquaculture 33:215–227.

Kincaid, H. L., and C. R. Berry. 1986. Trout broodstocks used in management of national fisheries. Pages 211–222 in R. H. Stroud, editor. Fish culture in fisheries management. American Fisheries Society, Fish Culture Section and Fisheries Management Section, Bethesda, Maryland.

Kincaid, H. L., S. Brimm, and V. Cross. 1994. National trout strain registry. U.S. Fish and Wildlife Service, Division of Fish Hatcheries, (electronic database) Washington, DC.

Krueger, C., A. Gharrett, T. Dehring, and F. W. Allendorf. 1981. Genetic aspects of fisheries rehabilitation programs. Canadian Journal of Fisheries and Aquatic Sciences 38:1877–1881.

Ryman, N. 1970. A genetic analysis of recapture frequencies of released young of salmon (*Salmo salar* L.). Hereditas 65:159–160.

Stahl, G. 1983. Differences in the amount and distribution of genetic variation between natural populations and hatchery stocks of Atlantic salmon. Aquaculture 33:23–32.

Zar, J. 1984. Biostatistical analysis, 2nd edition. Prentice-Hall, Inc., Englewood Cliffs, New Jersey.

Use of DNA Microsatellite Polymorphism to Analyze Genetic Correlations between Hatchery and Natural Fitness

R. W. DOYLE AND C. HERBINGER

Dalhousie University, Marine Gene Probe Laboratory
Halifax, Nova Scotia B3H 4J1, Canada

C. T. TAGGART

Department of Fisheries and Oceans, Northwest Atlantic Fisheries Center
St. John's, Newfoundland A1C 5X1, Canada

S. LOCHMANN

Dalhousie University, Department of Oceanography

Abstract.—The presence of a hatchery-rearing stage in the life cycle of a fish will inevitably select for improved hatchery performance (domestication selection) even when the hatchery broodstock is collected every generation from the wild. This phenomenon poses a difficulty for enhancement programs, because the correlation between hatchery fitness and fitness in nature is usually negative. Intensity of domestication selection, genetic variance and covariance components, and the effect of domestication on fitness in the natural environment can be estimated using DNA microsatellite polymorphism. The procedures in use at the Marine Gene Probe Laboratory are briefly described. The most effective procedures employ single-locus microsatellite repeat polymorphisms analyzed by polymerase chain reaction. The procedure is illustrated with an experiment on Atlantic cod *Gadus morhua* that reveals the intensity of domestication selection in the first laboratory generation of this fish. A newly developed, maximum-likelihood procedure for detecting sib, parent–offspring, and more distant relationships in fish populations is illustrated with a hatchery population of rainbow trout *Oncorhynchus mykiss*. When applied to microsatellite polymorphism data, the procedure generates pedigree information that allows estimation of the magnitude of domestication selection, and the predicted indirect effects on natural fitness, during routine hatchery operation. The reduction in natural fitness expected in hatchery-release programs can be mitigated by identifying, and then rejecting, wild-caught potential breeders that are recognized as having hatchery parents or grandparents. An alternative, more interventionist, strategy is to deliberately choose such animals as broodstock, with the aim of coadapting the population to both the wild and hatchery environments.

Technological development in the European and North American aquaculture industry is aimed at increasing the yield and the yield-to-cost ratios of intensively managed systems. Cultivated fish and shellfish can be expected to adapt genetically (that is, evolve) in their increasingly artificial environments, whether or not there is a deliberate program of genetic broodstock improvement. It is generally recognized that adaptation to intensive cultivation is likely to result in a decrease in performance under less intensively managed artisanal conditions, and an even larger decrease in performance under natural conditions. The evidence for deterioration of natural fitness comes mainly from salmonids (Vincent 1960; Moyle 1969; Reisenbichler and McIntyre 1977; Fraser 1981; Keller and Plosila 1981; Chilcote et al. 1986; Leider et al. 1990; Hindar et al. 1992).

This paper focuses on the genetic changes that may occur when hatcheries produce fry or fingerlings that are released into environments to augment natural populations. The genetic changes of concern are the directional effects of hatchery adaptation (domestication selection) on fitness in nature. This a different problem than that created by random changes in allele frequencies in hatchery populations with small effective population sizes (Allendorf and Phelps 1980; Danzmann et al. 1989; Hedgecock and Sly 1990; Gaffney et al. 1992).

Tools are now available to study directly the tradeoff between domestication selection in hatcheries and natural selection in the wild (Queller et al. 1993). We illustrate some of the ways in which DNA microsatellite pedigrees are beginning to be used in the Marine Gene Probe Laboratory (MGPL) to estimate variances and genetic correlations between components of fitness in natural and hatchery environments. The procedures allow estimation of the magnitude of the domestication tradeoff and the implementation of various strategies for avoiding, minimizing, or (for the adventurous) using the effects of domestication selection in hatchery release programs for augmenting natural stocks.

Negative Correlations Between Natural Fitness and Performance in Hatcheries

Genotype–environment (G–E) interaction refers to the situation in which the relative performance of genotypes, strains, stocks, or breeds varies among environments. The biological properties of wild and cultivated stocks will exhibit G–E interaction when fitness in hatcheries and fitness in nature involve traits that are different but genetically correlated and heritable.

Breeds of fish adapted to high-input aquaculture tend to do less well than do other breeds when tested in less intensively managed environments, and vice versa (Fraser 1981; Doyle and Talbot 1986; Chilcote et al. 1986; Hjort and Achreck 1982). Genotype–environment interactions involving different types of domestic environment have been demonstrated in common carp *Cyprinus carpio* (Moav et al. 1975, 1976b; Wohlfarth et al. 1986) and catfish *Ictalurus* spp. (Dunham et al. 1990). The adaptation of fish other than common carp to the diversity of artificial environments is just beginning and represents a period of rapid evolution under intense environmental stress (Kohane and Parsons 1988; Suboski and Templeton 1989).

In general, the direct response of each stock to selection in its own environment will be greater than its correlated response in other environments. Selection in each environment separately, given sufficient genetic variance, will therefore eventually produce strong G–E interaction (i.e., inferior performance of each stock in the alternative environment). This begins to happen when a population is transferred from the wild into a hatchery. The rate of divergence between the wild stock, undergoing natural selection, and the hatchery stock, undergoing domestication selection, will depend on the intensity of selection in each environment and the genetic variances and correlations of traits within and between environments (Rosielle and Hamblin 1981).

Domestication Selection in the First Hatchery Generation

The presence of a hatchery-rearing stage in the life cycle of a fish will select for improved hatchery performance, even when the hatchery broodstock is collected every year from the wild. This domestication selection occurs because the young produced from wild parents suffer some mortality before being released. (It is safe to assume that 100% survival in the hatchery is rare.) The intensity of first-generation domestication selection depends on the proportion of fertilized eggs that make it to release and on the proportion of the total hatchery mortality that is selective (i.e., a function of the animal's phenotype).

The progressive loss of natural fitness caused by domestication selection can be minimized by recruiting fresh broodstock every year, as is recommended, for example, by NASCO (1991). New broodstock will not eliminate the problem, however. If released animals constitute a significant proportion of the total breeding population then genetic changes induced by hatchery selection will accumulate through time. This hatchery contribution continues until domestication is balanced by natural counter-selection at some lower level of natural fitness, or until the augmented population becomes extinct.

The existence of nonrandom, selective mortality has not been easy to demonstrate. Selection differentials in the hatchery are not well-defined selection criteria (e.g., measurements of weight, color, or shape) that are used in artificial selection. Domestication selection is a form of uncontrolled natural selection in an artificial environment. The usual way to recognize its presence is by its result—behavioral or other phenotypic changes plus enhanced fitness in the hatchery and diminished fitness in nature. The action of domestication selection has usually been determined by inference, after several generations of domestication in the hatchery, not observed directly as a process.

Use of Microsatellite DNA Polymorphism to Demonstrate First-Generation Domestication Selection

An example of the use of microsatellites for the direct detection of selective mortality in the first domestic generation of a fish is provided by an experimental rearing program at Dalhousie University, Halifax, for Atlantic cod *Gadus morhua* (Lochmann et al., in press).

Figure 1 indicates the allelic diversity present at several loci in the parental, wild population of Atlantic cod based on a DNA microsatellite pedigree analysis developed at the MGPL. Figure 1 also shows the power that these loci provide for distinguishing populations (Bruford and Wayne 1993). The primer sequences for some of the Atlantic cod loci used in the experiment have been published (Wright 1993; Brooker et al., in press). The allelic

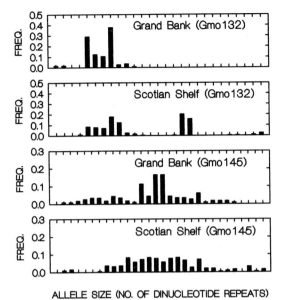

FIGURE 1.—Allelic distributions of two microsatellite loci (*Gmo132* and *Gmo145*) of Atlantic cod at two locations off the Canadian Atlantic coast. Grand Bank is located southeast of Newfoundland and Scotian Shelf east of Nova Scotia. The *x*-axis is allele size measured in number of dinucleotide repeat units (each tick represents an increase of two base pairs from an arbitrary origin). The *y*-axis is relative allele frequency.

diversity is so high in fish caught off the east coast of Canada (45 or more alleles at some microsatellite loci) that each animal is essentially unique when as few as 3 or 4 loci are scored.

In a rearing experiment, a number of wild males and females were spawned and the eggs pooled. After approximately 20 d the larvae hatched and were provided with lower-than-optimal food levels. During a 2-d period beginning about 8 d after hatching, the Atlantic cod larvae decreased in size as the yolk sac was absorbed and the larvae attempted to feed for the first time. DNA identification of larvae produced by different sires revealed that this critical 2-d period in the life of the larvae was profoundly selective: offspring of some male parents survived the transition to self feeding much better than did others. Figure 2 shows the relative survival of the offspring of seven different sires mated to the same female. Because only one female was involved, there is no between-group variance caused by genetic maternal effects and egg quality. The relative survival of offspring of the same sires mated to other females maintained the same rank order, however.

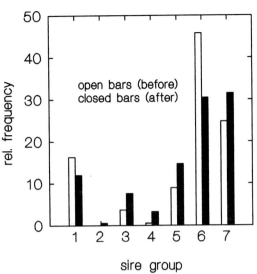

FIGURE 2.—Relative survival of seven groups of half sibs (seven males mated to one female). Mortality occurred before or after the point of "no-return," when larvae must begin to self feed.

The conclusion is that mortality was strongly selective in the laboratory, even at this very early larval stage. The coefficient of variance of sire-group survival was approximately 60%. This variance is not merely phenotypic opportunity for selection (Downhower et al. 1987), it is selection (directly observed genetic variance in survival).

Estimating the Tradeoff Between Domestication and Natural Selection

The measurement of first-generation selection gives only one-half of the answer to whether or not selective mortality in the hatchery might diminish the natural fitness of released animals. It is also necessary to estimate the correlated changes in natural fitness in the natural environment. Here, too, individual pedigrees based on microsatellite polymorphisms make the appropriate measurements possible.

Genetic correlation is calculated as the correlation between the value of a trait (e.g., growth) in one environment and the value of the same trait, observed in family members in another environment. When the trait is an all-or-nothing component of fitness, such as mating or survival, the correlation of family mean liabilities can be calculated (Falconer 1981). The procedure for observing and predicting such changes requires the identification

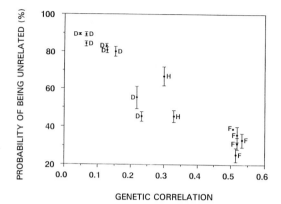

FIGURE 3.—Relationship between the mean probability of being unrelated, as calculated from microsatellite data, and the expected genetic correlations. Comparisons are between full-sib fish (F), half sibs (H), and fish more distantly related through grandparents (D).

of the family membership of individuals released from a hatchery into the wild and subsequently recaptured. It also requires the identification of individuals from the same family reared in the hatchery. Several levels of genetic relationship (e.g., full sibs, parent–offspring, and half-sibs) can be used.

Before the advent of DNA microsatellite identification the procedures for estimating genetic correlation involved rearing sib groups separately until individuals were large enough to be tagged. This procedure is operationally complex and expensive. Perhaps the greatest difficulty with this approach is that the experimental environment, with its separate rearing stage, is neither natural nor completely representative of selection under normal hatchery practice.

Mathematical Aspects of DNA Pedigrees for Estimating the Correlated Response to Domestication Selection

A maximum-likelihood procedure has been developed (C. Herbinger, unpublished) that can assign individuals to sib groups (and other groupings including grandparents, parent–offspring, and first cousins) on the basis of microsatellite information, even when the parents are not known (Figure 3).

The illustrative data in Figure 3 come from an experimental population of rainbow trout *Oncorhynchus mykiss*, in which the expected genetic correlations among the individuals are known with some confidence and can be compared with the maximum-likelihood predictions. The full-sib relationships, denoted as F in Figure 3, are known exactly from first-order pedigrees. The other groupings are corrected for suspected second-order relationships among grandparents. For example, the distantly related individuals, denoted as D, fall into three groups that are suspected to have zero, one, or three grandparents in common.

The microsatellite pedigrees greatly simplify the estimation of genetic correlations of fitness across the hatchery and natural environment. The differential survival of sib groups in both environments can be studied with little or no modification of hatchery procedures. All that is required is samples collected from hatchery-origin fish before and after release. The sampling is nondestructive; a single scale contains enough DNA for the microsatellite analyses. The calculation of direct and correlated responses to selection (hatchery adaptation and changes in natural fitness, respectively) can be complicated when there are unknown genetic correlations (family relationships) in a population. Microsatellite pedigrees can provide required information.

Molecular Aspects of Microsatellite DNA Polymorphism

Protein polymorphisms (isozymes or allozymes) have been used in the past to mark or identify genotypes in aquaculture experiments (Moav et al. 1976a). The level of isozyme polymorphism available for marking is relatively low, however, so that even with special breeding arrangements no more than 2 or 3 different genotypes (families or subpopulations) can be distinguished in pooled populations. This number is too small for measurement of genetic variances and correlations although it is useful for other purposes (e.g., for breed comparison and genetic stock identification in natural populations) (Chilcote et al. 1986; Utter et al. 1989; Shaklee et al. 1990; Brodziak et al. 1992). Allozymes have also been used very successfully to detect differences in the reproductive success of hatchery and wild fish at the population level, but not the individual or family level (Leider et al. 1990).

We have adopted single-locus microsatellite polymorphisms for genetic pedigree analysis for several reasons: (1) the level of polymorphism is very high—with heterozygosities approaching 100% at some loci—which allows the identification of all families in commercial hatcheries; (2) the DNA sampling is noninjurious, which allows identified animals to be used subsequently as broodstock; and

(3) the procedure is straightforward and cost effective.

The laboratory practice and underlying molecular biology of DNA microsatellite polymorphism in fish has recently been comprehensively reviewed by Franck et al. (1991) and Wright (1993). The aquaculture genetics projects under way at the MGPL are currently based on two of the several possible types of DNA fingerprinting; several other procedures are currently under investigation in the MGPL.

Minisatellite polymorphisms analyzed by Southern blotting.—Minisatellites consist of tandemly repeated sequences of 9–65 base pairs (e.g., Jarman and Wells 1989). These tandem arrays of core sequence repeats, which may exist at several locations in the same genome (i.e., at several loci), vary in length among individuals. Because the number of length variants (alleles) segregating at each locus may be very large, this type of polymorphism has been termed hypervariable variable number tandem repeat (VNTR) polymorphism.

When DNA extracted from the target fish is cut with restriction enzymes, subjected to Southern blotting, and probed with the core sequence, many VNTR loci can be revealed simultaneously. The result is a multilocus satellite pattern that resembles a bar code and which, because of the hypervariability of the loci, uniquely identifies individuals (Jeffreys et al. 1988). The alleles at each locus are inherited in strictly Mendelian fashion (except for a high mutation rate), so in principle the multilocus microsatellites can be used to identify the target fishes' parents (e.g., for pedigree analysis). In practice, the number of loci is often so large that alleles from different loci overlap, which introduces some ambiguities into the analysis (Lynch 1988).

The problem of overlapping alleles from different loci on a Southern blot can be overcome by using VNTR probes that hybridize to only one locus (i.e., to the unique sequence DNA that flanks or lies near the repeated core VNTR sequence) (Wong et al. 1986; Taggart and Ferguson 1990). This was accomplished in the MGPL for Atlantic salmon *Salmo salar* by Bentzen et al. (1991).

Polymerase chain reaction (PCR) procedure.—This procedure provides a simpler and less ambiguous alternative to Southern blotting, when the objective is pedigree analysis of large numbers of fish. The PCR procedure involves the synthesis of pairs of oligonucleotides that are complementary to unique sequence DNA lying adjacent to each end of the polymorphic VNTR locus. The intervening VNTR sequence is amplified by PCR and electrophoresed on acrylamide sequencing gels (Jeffreys et al. 1988). The current preference in the MGPL is to use VNTR sequences called microsatellites, which are comparatively small (1–4 base pair repeats, rather than 6–95 base pair sequence repeats that characterize minisatellites). High-resolution sequencing gels are required to avoid confounding alleles that have nearly the same length.

Discussion

One of the strategies for minimizing the effect of domestication on the natural fitness of enhanced stocks involves maintaining open, natural populations by recruiting hatchery broodstock from the wild in every generation (NASCO 1991). Unfortunately, this strategy will not completely avoid domestication selection. The microsatellite identification experiment with Atlantic cod described in this paper demonstrates that intense selective mortality can be detected in the earliest larval stages of offspring of animals from the wild. If this proves to be true for fish that are regularly stocked in enhancement programs, such as rainbow trout and Atlantic salmon, then stocking programs that spawn only wild broodstock will fail to produce offspring that are free from hatchery effects.

Fortunately, mathematical procedures and DNA pedigree techniques have been developed that can limit the accumulation of domestication selective effects in stocked populations beyond the first generation. Effecting some limit will require us to identify wild fish that have parents or grandparents of hatchery origin and then avoid spawning them. The sensitivity of the procedures that are now available seems adequate for identifying these fish.

An interesting alternative strategy for minimizing the accumulation of domestication effects is theoretically available now that DNA microsatellites have simplified the detection of pedigree relationships. Rather than trying to avoid the spawning of hatchery fish, or wild fish with recent hatchery ancestors, one could use fitness in the wild as a criterion for selection in the hatchery. There are at least two ways that this might be attempted, both based on the identification of the relatives of fish that have survived, thrived, and ultimately been captured in the wild.

In closed domestic populations, without regular addition of wild breeders, the hatchery parents or hatchery sibs of the successful wild fish can be chosen for spawning. This represents a type of progeny testing or sib selection. Because domestication selection would proceed as usual in the hatchery, the hoped for beneficial outcome would be simultaneous genetic adaptation to the hatchery and natural environments. The reduction in negative correlations between fitnesses in the two environments would be expected if the appropriate genetic variance exists in the population.

An alternative approach could be used in open populations where some or all of the broodstock is recruited from the wild every year. In this case, fish that have hatchery origins or ancestors could be deliberately chosen as broodstock rather than avoided. Such fish do carry a higher proportion of genes that have been successful in surviving (and reproducing) in both environments.

We recognize that these alternative genetic management strategies are more interventionist than those usually suggested (Moav et al. 1978). They certainly are in direct opposition to the idea that genetic changes in natural populations should be kept to a minimum (Allendorf and Ryman 1987; Nelson and Soule 1987). On the other hand, they may imply a more willing acceptance of the hatchery as an permanent, essential component of the reproductive environment of endangered fish populations.

Acknowledgments

This work was supported by the Canadian Centers of Excellence program (Ocean Production Enhancement Network), the Northern Cod Program, and a grant from the government of Nova Scotia to the Marine Gene Probe Laboratory.

References

Allendorf, F. W., and S. R. Phelps. 1980. Loss of genetic variation in a hatchery stock of cutthroat trout. Transactions of the American Fisheries Society 109:537–543.

Allendorf, F. W., and N. Ryman. 1987. Genetic management of hatchery stocks. Pages 141–159 in N. Ryman and F. Utter, editors. Population genetics and fishery management. University of Washington Press, Seattle.

Bentzen, P., A. S. Harris, and J. M. Wright. 1991. Cloning of hypervariable minisatellite and simple sequence microsatellite repeats for DNA pedigreeing of important aquacultural species of salmonids and tilapia. Pages 243–262 in T. Burke, G. Dolf, A. J. Jeffreys, and R. Wolff, editors. DNA fingerprinting: approaches and applications. Birkhaeuser Verlag, Basel, Switzerland.

Brodziak, J., B. Bentley, D. Bartley, G. A. E. Gall, R. Gomulkiewicz, and M. Mangel. 1992. Tests of genetic stock identification using coded wire tagged fish. Canadian Journal of Fisheries and Aquatic Sciences 49:1507–1517.

Brooker, A. L., D. Cook, P. Bentzen, J. M. Wright, and R. W. Doyle. In press. The organization of microsatellites differs between mammals and teleost fish. Canadian Journal of Fisheries and Aquatic Sciences.

Bruford, M. W., and R. K. Wayne. 1993. Microsatellites and their application to population genetic studies. Current Opinion in Genetics and Development 3:939–943.

Chilcote, M. W., S. A. Leider, and J. J. Loch. 1986. Differential reproductive success of hatchery and wild summer-run steelhead under natural conditions. Transactions of the American Fisheries Society 115:726–735.

Danzmann, R. G., M. M. Ferguson, and F. W. Allendorf. 1989. Genetic variability and components of fitness in hatchery strains of rainbow trout. Journal of Fish Biology 35(Supplement A):313–319.

Downhower, J. F., L. S. Blumer, and L. Brown. 1987. Opportunity for selection: an appropriate measure for evaluating variation in the potential for selection. Evolution 41:1395–1400.

Doyle, R. W., and A. J. Talbot. 1986. Artificial selection on growth and correlated selection on competitive behaviour in fish. Canadian Journal of Fisheries and Aquatic Sciences 43(5):1059–1064.

Dunham, R. A., R. E. Brummett, M. O. Ella, and R. O. Smitherman. 1990. Genotype–environment interactions for growth of blue, channel and hybrid catfish in ponds and cages at varying densities. Aquaculture 85:143–151.

Falconer, D. S. 1981. Introduction to quantitative genetics, 2nd edition. Longman, London.

Franck, J. P. C., A. P. Harris, E. M. Bentzen, E. M. Denoven-Wright, and J. Wright. 1991. Organization and evolution of satellite, minisatellite and microsatellite DNAs in teleost fishes. Oxford Surveys on Eukaryotic Genes 3:51–82.

Fraser, J. M. 1981. Comparative survival and growth of planted wild, hybrid, and domestic strains of brook trout (*Salvelinus fontinalis*) in Ontario lakes. Canadian Journal of Fisheries and Aquatic Sciences 38:1672–1684.

Gaffney, P. M., C. V. Davis, and R. O. Hawes. 1992. Assessment of drift and selection in hatchery populations of oysters (*Crassostrea virginica*). Aquaculture 105:1–20.

Hedgecock, D., and F. Sly. 1990. Genetic drift and effective population sizes of hatchery-propagated stocks of the Pacific oyster, *Crassostrea gigas*. Aquaculture 88:21–38.

Hindar, K., N. Ryman, and F. Utter. 1991. Genetic effects of cultured fish on natural fish populations. Canadian Journal of Fisheries and Aquatic Sciences 48(5):945–957.

Hjort, R. C., and C. B. Achreck. 1982. Phenotypic differences among stocks of hatchery and wild coho salmon, *Oncorhynchus kisutch*, in Oregon, Washington, and California. U.S. National Marine Fisheries Service Fishery Bulletin 80:105–119.

Jarman, A. P., and R. A. Wells. 1989. Hypervariable minisatellites: recombinators or innocent bystanders? Trends in Genetics 5:367–371.

Jeffreys, A. J., V. Wilson, R. Neumann, and J. Keyte.

1988. Amplification of human minisatellites by the polymerase chain reaction: towards DNA fingerprinting of single cells. Nucleic Acids Research 16:10953–10971.

Keller, W. T., and D. S. Plosila. 1981. Comparison of domestic, hybrid and wild strains of brook trout in a pond fishery. New York Fish and Game Journal 28(2):123–137.

Kohane, M. J., and P. A. Parsons. 1988. Domestication: evolutionary change under stress. Evolutionary Biology 23:31–48.

Leider, S. A., P. L. Hulett, J. J. Loch, and M. W. Chilcote. 1990. Electrophoretic comparison of the reproductive success of naturally spawned transplanted and wild steelhead trout through the returning adult stage. Aquaculture 88:239–252.

Lochmann, S., G. Maillet, K. T. Frank, C. T. Taggart, and S. McClatchie. In press. Lipid class composition as a measure of nutritional condition in individual Atlantic cod. Master's thesis. Dalhousie University, Halifax.

Lynch, M. 1988. Estimates of relatedness by DNA fingerprinting. Molecular Biology and Evolution 5:584–599.

Moav, R., G. Hulata, and G. Wohlfarth. 1975. Genetic differences between the Chinese and European races of the common carp, 1: analysis of genotype–environment interactions for growth rate. Heredity 34(3):323–340.

Moav, R., T. Brody, D. Wohlfarth, and G. Hulata. 1976a. Applications of electrophoretic genetic markers to fish breeding, 1: advantage and methods. Aquaculture 9:217–228.

Moav, R., M. Soller, and G. Hulata. 1976b. Genetic aspects of the transition from traditional to modern fish farming. Theoretical and Applied Genetics 47:285–290.

Moav, R., T. Brody, and G. Hulata. 1978. Genetic improvement of wild fish populations. Science 201:1090–1094.

Moyle, P. B. 1969. Comparative behaviour of young brook trout of domestic and wild origin. Progressive Fish-Culturist 31:51–59.

NASCO (North Atlantic Salmon Conservation Organization). 1991. Introductions and transfers of salmonids: their impacts on North American Atlantic salmon and recommendations to reduce such impacts. NASCO, North American Commission, Discussion Document.

Nelson, K., and M. Soule. 1987. Genetical conservation of exploited fishes. Pages 345–368 *in* N. Ryman and F. Utter, editors. Population genetics and fishery management. University of Washington Press, Seattle.

Queller, D. C., J. E. Strassmann, and C. R. Hughes. 1993. Microsatellites and kinship. Trends in Ecology and Evolution 8:285–288.

Reisenbichler, R. R., and J. D. McIntyre. 1977. Genetic differences in growth and survival of juvenile hatchery and wild steelhead trout, *Salmo gairdneri*. Journal of the Fisheries Research Board of Canada 34:123–128.

Rosielle, A. A., and J. Hamblin. 1981. Theoretical aspects of selection for yield in stress and non-stress environments. Crop Science 21:943–946.

Shaklee, J. B., C. Busack, A. Marshall, M. Miller, and S. R. Phelps. 1990. The electrophoretic analysis of mixed-stock fisheries of Pacific salmon. Pages 235–265 *in* Z-I. Ogita and C. L. Markert, editors. Isozymes: structure, function and use in biology and medicine. Progress in clinical and biological research, volume 344. Wiley, New York.

Suboski, M. D., and J. Templeton. 1989. Life skill straining for hatchery fish: social learning and survival. Fisheries Research 7:343–352.

Taggart, J. B., and A. Ferguson. 1990. Hypervariable minisatellite single locus probes for the Atlantic salmon, *Salmo salar*. Journal of Fish Biology 37:991–993.

Utter, F. M., G. B. Milner, G. Stahl, and D. J. Teel. 1989. Genetic population structure of Pacific salmon, *Oncorhynchus tshawytscha*, in the Pacific Northwest. U.S. National Marine Fisheries Service Fishery Bulletin 85:13–23.

Vincent, R. E. 1960. Some influences of domestication upon three stocks of brook trout (*Salvelinus fontinalis* Mitchill) Transactions of the American Fisheries Society 89:35–52.

Wohlfarth, G. W., R. Moav, and G. Hulata. 1986. Genetic differences between the Chinese and European races of common carp, 5. Differential adaptations to manure and artificial feeds. Theoretical and Applied Genetics 72:88–97.

Wong, Z., V. Wilson, A. J. Jeffreys, and S. L. Thein. 1986. Cloning a selected fragment from a human DNA "microsatellite": isolation of an extremely polymorphic minisatellite. Nucleic Acids Research 14:4605–4616.

Wright, J. 1993. DNA fingerprinting of fishes. Pages 57–91 *in* Hochachka and Mommsen, editors. Biochemistry and molecular biology of fishes, volume 2. Elsevier, Amsterdam.

Separating Phenotypic Variation into Genetic and Environmental Influences

Additive genetic variance and heritability.—Most life-history characters are quantitative; unlike discrete characters, they exhibit phenotypic distributions that are continuous or nearly so. Even threshold characters, such as resistance to some diseases and the incidence of many congenital malformations and (in humans) psychiatric disorders, are thought to have underlying continuous distributions (Falconer 1989). Quantitative genetic theory predicts that such distributions are an emergent property of genetic variability that is under the control of many genes. Each of these genes is generally thought to have small, independent effects on the polygenic character's expression. Quantitative genetics generally ignores individual genes underlying a quantitative character and instead concerns itself with the composite statistical behavior of the entire group of genes (Lande 1988). However, characterizing the quantitative phenotypic variation expressed by a polygenic character in a population is complicated by a number of other factors that are not at issue when dealing with single genes. Interactions between constituent genes at different loci (sites in the genome), linked inheritance of constituent genes, and sensitivity of gene expression to the environmental surroundings all act to obscure the relationship between genotypes and phenotypes. The fundamental objective of quantitative genetics is to clarify this relationship by partitioning phenotypic variation into components that relate to these factors.

In principle, the phenotypic variance (V_P) can be decomposed into two components, one genetic (V_G) and one environmental (V_E). However, the relation, $V_P = V_G + V_E$, holds only if genotypic expression is independent of environmental effects. In addition, V_G includes the contributions of dominance (i.e., interactions among alleles at the same locus) and epistasis (interactions among loci). These considerations lead to a fundamental equation of quantitative genetics that describes the phenotypic variance for a quantitative character:

$$V_P = V_A + V_D + V_I + V_E + V_{GE},$$

where V_A is the additive genetic variance (the variance due to the additive effects of the constituent loci), V_D is the variance due to dominance within loci, V_I is the variance due to epistatic interactions among loci, and V_{GE} is the variance due to the interaction between genotype and environment. It should be noted that, in addition to unlinked genes, this equation assumes that genotypes are randomly distributed with respect to environmental variation, and it does not account for particular environmental effects, such as maternal effects, shared by only some individuals (Falconer 1989).

Because quantitative geneticists are usually interested in the aspect of a character's genetic variation that is responsible for its resemblance between parents and their offspring, genetic analyses of phenotypic variance in quantitative characters typically focus on the variance of breeding values, which is also the additive genetic variance, V_A. The V_A of a character is of paramount importance in quantitative genetics because it determines the resemblance between parents and offspring and, therefore, the evolutionary properties of a population. The evolutionary significance of V_A is due to the fact that it represents the variance among gametes, which are transmitted from parents to offspring; V_D and V_I are functions of genotypes, which are not transmitted intact but instead are "reshuffled" during the meiotic processes of segregation and independent assortment (Tave 1993). Consequently, V_A makes the only direct genetic contribution to changes in phenotypic variance between generations.

The estimation of V_A and its relationship to V_P then constitutes an approach to quantifying the genetic and environmental components of phenotypic variance. The ratio V_A/V_P is referred to as the narrow-sense heritability (Falconer 1989). Narrow-sense heritability, designated by h^2, is a commonly estimated parameter that indicates the relative strength of hereditary and environmental influences on phenotypic evolution. The h^2 of a population ranges from 0 (no hereditary influence) to 1 (no environmental influence). For the purposes of supplementation, all the other components of V_P in the above equation can be considered part of the environmental variance, permitting attention to focus on V_A, V_P, and h^2.

Given the difficulties in estimating genetic parameters associated with quantitative characters, the fishery geneticist or manager must decide what information is necessary for effective genetic monitoring of these characters during supplementation. Estimates of h^2, for example, depend on gene frequencies in the population as well as on the environment in which those gene frequencies are measured. Therefore, any particular estimate is accurate only for the population and environment from which it is estimated. Estimates of h^2 for traits in laboratory populations or otherwise controlled environments are often higher than those for the same traits in populations under more natural con-

1988. Amplification of human minisatellites by the polymerase chain reaction: towards DNA fingerprinting of single cells. Nucleic Acids Research 16:10953–10971.

Keller, W. T., and D. S. Plosila. 1981. Comparison of domestic, hybrid and wild strains of brook trout in a pond fishery. New York Fish and Game Journal 28(2):123–137.

Kohane, M. J., and P. A. Parsons. 1988. Domestication: evolutionary change under stress. Evolutionary Biology 23:31–48.

Leider, S. A., P. L. Hulett, J. J. Loch, and M. W. Chilcote. 1990. Electrophoretic comparison of the reproductive success of naturally spawned transplanted and wild steelhead trout through the returning adult stage. Aquaculture 88:239–252.

Lochmann, S., G. Maillet, K. T. Frank, C. T. Taggart, and S. McClatchie. In press. Lipid class composition as a measure of nutritional condition in individual Atlantic cod. Master's thesis. Dalhousie University, Halifax.

Lynch, M. 1988. Estimates of relatedness by DNA fingerprinting. Molecular Biology and Evolution 5:584–599.

Moav, R., G. Hulata, and G. Wohlfarth. 1975. Genetic differences between the Chinese and European races of the common carp, 1: analysis of genotype–environment interactions for growth rate. Heredity 34(3):323–340.

Moav, R., T. Brody, D. Wohlfarth, and G. Hulata. 1976a. Applications of electrophoretic genetic markers to fish breeding, 1: advantage and methods. Aquaculture 9:217–228.

Moav, R., M. Soller, and G. Hulata. 1976b. Genetic aspects of the transition from traditional to modern fish farming. Theoretical and Applied Genetics 47:285–290.

Moav, R., T. Brody, and G. Hulata. 1978. Genetic improvement of wild fish populations. Science 201:1090–1094.

Moyle, P. B. 1969. Comparative behaviour of young brook trout of domestic and wild origin. Progressive Fish-Culturist 31:51–59.

NASCO (North Atlantic Salmon Conservation Organization). 1991. Introductions and transfers of salmonids: their impacts on North American Atlantic salmon and recommendations to reduce such impacts. NASCO, North American Commission, Discussion Document.

Nelson, K., and M. Soule. 1987. Genetical conservation of exploited fishes. Pages 345–368 in N. Ryman and F. Utter, editors. Population genetics and fishery management. University of Washington Press, Seattle.

Queller, D. C., J. E. Strassmann, and C. R. Hughes. 1993. Microsatellites and kinship. Trends in Ecology and Evolution 8:285–288.

Reisenbichler, R. R., and J. D. McIntyre. 1977. Genetic differences in growth and survival of juvenile hatchery and wild steelhead trout, *Salmo gairdneri*. Journal of the Fisheries Research Board of Canada 34:123–128.

Rosielle, A. A., and J. Hamblin. 1981. Theoretical aspects of selection for yield in stress and non-stress environments. Crop Science 21:943–946.

Shaklee, J. B., C. Busack, A. Marshall, M. Miller, and S. R. Phelps. 1990. The electrophoretic analysis of mixed-stock fisheries of Pacific salmon. Pages 235–265 in Z-I. Ogita and C. L. Markert, editors. Isozymes: structure, function and use in biology and medicine. Progress in clinical and biological research, volume 344. Wiley, New York.

Suboski, M. D., and J. Templeton. 1989. Life skill straining for hatchery fish: social learning and survival. Fisheries Research 7:343–352.

Taggart, J. B., and A. Ferguson. 1990. Hypervariable minisatellite single locus probes for the Atlantic salmon, *Salmo salar*. Journal of Fish Biology 37:991–993.

Utter, F. M., G. B. Milner, G. Stahl, and D. J. Teel. 1989. Genetic population structure of Pacific salmon, *Oncorhynchus tshawytscha*, in the Pacific Northwest. U.S. National Marine Fisheries Service Fishery Bulletin 85:13–23.

Vincent, R. E. 1960. Some influences of domestication upon three stocks of brook trout (*Salvelinus fontinalis* Mitchill) Transactions of the American Fisheries Society 89:35–52.

Wohlfarth, G. W., R. Moav, and G. Hulata. 1986. Genetic differences between the Chinese and European races of common carp, 5. Differential adaptations to manure and artificial feeds. Theoretical and Applied Genetics 72:88–97.

Wong, Z., V. Wilson, A. J. Jeffreys, and S. L. Thein. 1986. Cloning a selected fragment from a human DNA "microsatellite": isolation of an extremely polymorphic minisatellite. Nucleic Acids Research 14:4605–4616.

Wright, J. 1993. DNA fingerprinting of fishes. Pages 57–91 in Hochachka and Mommsen, editors. Biochemistry and molecular biology of fishes, volume 2. Elsevier, Amsterdam.

Genetic Monitoring of Life-History Characters in Salmon Supplementation: Problems and Opportunities

JEFFREY J. HARD

National Marine Fisheries Service, Northwest Fisheries Science Center
Coastal Zone and Estuarine Studies Division, 2725 Montlake Boulevard East
Seattle, Washington 98112, USA

Abstract.—Genetic monitoring is an essential element of programs designed to supplement natural salmon populations. The goal of such programs, especially when associated with genetic conservation, should be to maintain the genetic integrity of natural populations. This goal requires minimizing the phenotypic and genetic divergence of cultured and natural fish. Genetic monitoring in such programs has thus far focused primarily on genetic characters such as variability in protein and DNA markers; little attention has been devoted to life-history characters. Because population viability, adaptation, and recovery depend on life-history characters such as body size, age structure, run timing, fecundity, and some behaviors, these traits should also be monitored in supplemented populations. However, several difficulties can confound the interpretation of variation in these traits. In particular, two problems can complicate the evaluation of supplementation effectiveness. The first problem is one of genetics and arises in attempting to discriminate between genetic and environmental components of variation in life-history characters during supplementation. The second problem is one of inference and arises in equating a failure to detect changes in life-history characters that result from supplementation with the lack of such changes. This paper discusses these problems and the obstacles encountered in trying to avoid them, introduces some basic techniques that can be used to detect these problems, and suggests some tactics for effective genetic monitoring of life-history characters in salmon supplementation programs.

The use of artificial propagation to produce anadromous salmon has a long history in North America. Salmon hatcheries have increased the number of fish available for harvest in some areas (e.g., Heard 1983; Howell et al. 1985), and advances in hatchery technology have increased our understanding of several aspects of the life cycle of Atlantic salmon *Salmo salar* and Pacific salmon *Oncorhynchus* spp. Salmon hatcheries have become a mainstay of the effort to compensate for losses of salmon, especially Pacific salmon, due to overharvest, habitat degradation, and natural phenomena. Recent listings under the U.S. Endangered Species Act (ESA: 16 U.S.C.A. §§ 1531 to 1544) of Pacific salmon stocks in the Snake River basin in the Pacific Northwest have raised the question of the utility of artificial propagation for conservation and recovery of Threatened and Endangered salmon. Under the ESA, artificial propagation is one of several tools that can be considered for these purposes.

Despite the widespread use of hatchery fish for enhancement and mitigation, the use of these fish to augment natural salmon production is a controversial practice (Hard et al. 1992; Cuenco et al. 1993). Recognition that artificial propagation can pose risks to hatchery and natural salmon populations has resulted in a number of recommendations for its use. Most of these recommendations have been primarily concerned with transfers of hatchery stocks and with the loss of genetic variability within hatchery populations, but only indirectly concerned with the possible consequences for natural populations with which hatchery fish may interact (e.g., Hynes et al. 1981; Krueger et al. 1981; Allendorf and Ryman 1987). Recent declines in the abundance of natural salmon, coupled with the evident inability of many large hatchery programs to offset these declines, have led to greater attention to the consequences of interaction between hatchery and natural fish (Utter 1981; Reisenbichler and McIntyre 1986; Nelson and Soulé 1987; Hindar et al. 1991; Waples 1991; see also Campton 1995, this volume). Some recent guidelines attempt to address directly the potential adverse consequences of artificial propagation for natural populations (Hard et al. 1992; Lichatowich and Watson 1993).

Despite the concerns about genetic and ecological effects of artificial propagation on natural as well as cultured salmon populations, this technique does hold considerable potential to boost the abundance of weak, unstable, or declining populations. Consequently, artificial propagation is likely to be applied widely to the conservation and recovery of natural salmon populations. The purpose of this paper is to point out some genetic problems and opportunities associated with this practice and suggest some approaches to recognizing and dealing with them. Although the discussion has been developed with anadromous salmon populations in mind, the general considerations and recommendations should apply to any organisms propagated artificially and then released to the wild to augment natural production.

Genetic Risks in Salmon Supplementation

The artificial propagation of anadromous salmon differs from that of many other fishes, including other salmonids, because hatchery salmon are generally released to the wild for much of their life cycle. Beyond concerns about artificial selection in hatchery populations (Allendorf et al. 1987) and stock mixing (Altukhov and Salmenkova 1987), hatchery practices for enhancement and mitigation have typically shown little regard for the composition and genetic characteristics of fish produced from and returning to production facilities. Enhancement and mitigation programs have instead been concerned primarily with restoring or increasing the number of harvested fish or adults returning to spawn.

Salmon supplementation involves somewhat different concerns. Supplementation has been defined as the use of artificial propagation to reestablish, maintain, or increase natural salmon production, ideally while maintaining the long-term fitness of the target population and keeping the genetic and ecological effects on nontarget populations within specified biological limits (Cuenco et al. 1993). Supplementation differs from traditional enhancement or mitigation in that the demographic, genetic, and ecological characteristics of natural populations should be the primary considerations. Maintenance of these characteristics requires minimizing the genetic and phenotypic divergence of cultured fish from the natural fish they are supposed to augment; control of this divergence therefore should be a primary objective of supplementation.

The use of hatchery fish in supplementation can threaten the objective of supplementation in several ways. Busack (unpublished manuscript) Riggs (1990) (cited in Hard et al. 1992) identified extinction, reduction in genetic variability within populations, reduction in genetic variability among populations, and genetic change as four genetic risks that can lead to divergence between hatchery and natural fish. Salmon geneticists now regard the monitoring and evaluation of genetic characteristics in supplemented salmon populations as a promising means of reducing these risks (Lichatowich and Watson 1993). However, genetic monitoring methods to minimize these risks in salmon supplementation are not widely appreciated.

Waples et al. (1992) described a program to monitor variation in allozyme electromorphs for hatchery, unsupplemented natural, and supplemented populations of chinook salmon *O. tshawytscha* in the Snake River basin. Such a program is valuable for detecting the loss of genetic variability within and among populations. Unfortunately, the relationship between electrophoretic characters and the life-history characters that are sensitive to selection, and therefore important to local adaptation, is not at all clear (Bentsen 1991; Utter et al. 1993). Life-history characters compose the elements of fitness and therefore affect survivorship, growth and development, and reproduction (Stearns 1976; Roff 1992). Life-history characters prominent in salmon include stage-specific survival rates, sex ratio, age structure, growth rate, juvenile and adult migration timing, fecundity, egg size, stress and disease resistance, agonistic behavior, and homing ability. Few specific recommendations are available to guide the genetic monitoring of such characters (Lande and Barrowclough 1987; Cloud and Thorgaard 1993). Such guidance will have to be developed from techniques designed to detect and quantify the genetic and environmental components of observed variation (Falconer 1989; Tave 1993).

Problems in Genetic Monitoring

Two prominent problems are likely to arise during the genetic monitoring of life-history characters. The first problem is a genetic one and stems from the nature of variation in typical life-history characters; it manifests itself as difficulty in discriminating between genetic and environmental components of phenotypic variation. The second problem is one of inference and results from incorrect conclusions about the genetic effects of a supplementation program on a natural population; it occurs when a failure to detect genetic differences arising between hatchery and natural populations during supplementation leads to the conclusion that no such differences exist. The distinction between these two problems is somewhat artificial because they can be related, but this distinction is nonetheless useful in organizing methods for monitoring quantitative genetic characters during supplementation.

Separating Phenotypic Variation into Genetic and Environmental Influences

Additive genetic variance and heritability.—Most life-history characters are quantitative; unlike discrete characters, they exhibit phenotypic distributions that are continuous or nearly so. Even threshold characters, such as resistance to some diseases and the incidence of many congenital malformations and (in humans) psychiatric disorders, are thought to have underlying continuous distributions (Falconer 1989). Quantitative genetic theory predicts that such distributions are an emergent property of genetic variability that is under the control of many genes. Each of these genes is generally thought to have small, independent effects on the polygenic character's expression. Quantitative genetics generally ignores individual genes underlying a quantitative character and instead concerns itself with the composite statistical behavior of the entire group of genes (Lande 1988). However, characterizing the quantitative phenotypic variation expressed by a polygenic character in a population is complicated by a number of other factors that are not at issue when dealing with single genes. Interactions between constituent genes at different loci (sites in the genome), linked inheritance of constituent genes, and sensitivity of gene expression to the environmental surroundings all act to obscure the relationship between genotypes and phenotypes. The fundamental objective of quantitative genetics is to clarify this relationship by partitioning phenotypic variation into components that relate to these factors.

In principle, the phenotypic variance (V_P) can be decomposed into two components, one genetic (V_G) and one environmental (V_E). However, the relation, $V_P = V_G + V_E$, holds only if genotypic expression is independent of environmental effects. In addition, V_G includes the contributions of dominance (i.e., interactions among alleles at the same locus) and epistasis (interactions among loci). These considerations lead to a fundamental equation of quantitative genetics that describes the phenotypic variance for a quantitative character:

$$V_P = V_A + V_D + V_I + V_E + V_{GE},$$

where V_A is the additive genetic variance (the variance due to the additive effects of the constituent loci), V_D is the variance due to dominance within loci, V_I is the variance due to epistatic interactions among loci, and V_{GE} is the variance due to the interaction between genotype and environment. It should be noted that, in addition to unlinked genes, this equation assumes that genotypes are randomly distributed with respect to environmental variation, and it does not account for particular environmental effects, such as maternal effects, shared by only some individuals (Falconer 1989).

Because quantitative geneticists are usually interested in the aspect of a character's genetic variation that is responsible for its resemblance between parents and their offspring, genetic analyses of phenotypic variance in quantitative characters typically focus on the variance of breeding values, which is also the additive genetic variance, V_A. The V_A of a character is of paramount importance in quantitative genetics because it determines the resemblance between parents and offspring and, therefore, the evolutionary properties of a population. The evolutionary significance of V_A is due to the fact that it represents the variance among gametes, which are transmitted from parents to offspring; V_D and V_I are functions of genotypes, which are not transmitted intact but instead are "reshuffled" during the meiotic processes of segregation and independent assortment (Tave 1993). Consequently, V_A makes the only direct genetic contribution to changes in phenotypic variance between generations.

The estimation of V_A and its relationship to V_P then constitutes an approach to quantifying the genetic and environmental components of phenotypic variance. The ratio V_A/V_P is referred to as the narrow-sense heritability (Falconer 1989). Narrow-sense heritability, designated by h^2, is a commonly estimated parameter that indicates the relative strength of hereditary and environmental influences on phenotypic evolution. The h^2 of a population ranges from 0 (no hereditary influence) to 1 (no environmental influence). For the purposes of supplementation, all the other components of V_P in the above equation can be considered part of the environmental variance, permitting attention to focus on V_A, V_P, and h^2.

Given the difficulties in estimating genetic parameters associated with quantitative characters, the fishery geneticist or manager must decide what information is necessary for effective genetic monitoring of these characters during supplementation. Estimates of h^2, for example, depend on gene frequencies in the population as well as on the environment in which those gene frequencies are measured. Therefore, any particular estimate is accurate only for the population and environment from which it is estimated. Estimates of h^2 for traits in laboratory populations or otherwise controlled environments are often higher than those for the same traits in populations under more natural conditions (Prout 1958; Coyne and Beecham 1987). The sensitivity of these estimates to environmental variation and the rapid responses to selection that have been observed in several life-history characters suggest that the generally low h^2 estimates for these characters reflect large environmental variance rather than low additive genetic variance (Barton and Turelli 1989; Price and Schluter 1991). Although h^2 can sometimes forecast the actual rate of genetic change for a particular population in a particular environment over a few generations, the reliability of such estimates to predict the consequences of selection, even in controlled breeding situations, is poor (Sheridan 1988). This lack of correspondence between estimated and realized h^2 probably results from several factors, including the generation and reorganization of genetic variation during selection, genotype–environment (G–E) correlations, linkage disequilibrium, genetic drift, and genetic correlations with other selected characters. The additive genetic variance itself determines a population's potential response to selection; as such, it may be more useful to determine whether significant V_A exists for a trait than to estimate its h^2.

Quantitative genetics offers several methods to estimate V_A and h^2 for traits such as life-history characters (Falconer 1989). The two most common methods for natural populations are sib analysis and parent–offspring (P–O) regression. The specific approaches these techniques use to estimate these parameters are different, but both rely on the covariance between relatives. This paper considers only P–O regression; see Falconer (1989) or Becker (1984) for treatments of sib analysis and more comprehensive discussions of these methods. In P–O regression, offspring are produced from each of several mated pairs, and h^2 can be estimated directly from the linear regression coefficient relating the offspring and parental means for the trait in question. The phenotypic covariance between parents and their offspring provides an estimate of half the additive genetic variance of the parents, because each parent contributes half the genes to the offspring, the other half coming at random from the population through the other parent (Falconer 1989). The phenotypic variance of the parents is an estimate of V_P.

Parent–offspring regression has been used extensively to estimate heritabilities of quantitative characters in organisms where P–O identification is feasible (e.g., Boag and Grant 1978; Mitchell-Olds 1986). It is an appealing approach to estimating genetic variation: it involves the direct resemblance between parents and their offspring (these relatives are also often the easiest to identify or obtain in the wild), and linear regression is a statistical technique familiar to empirical biologists. Data from a study by Hankin et al. (1993) on age at maturity of fall-run chinook salmon in Elk River Hatchery (Oregon) can be used to illustrate how h^2 and V_A can be estimated in a life-history character from P–O regression. Age at maturity is a central feature of salmon life history because it determines age structure, which can affect long-term viability in the wild. In addition, age structure has management implications because it can be affected by size-selective fishing (Ricker 1981; Law and Grey 1989) and by hatchery practices (Rosentreter 1977).

The study by Hankin et al. (1993) is significant because, unlike other studies on age at maturity in salmonids, it examined the inheritance of this trait in an anadromous population released to the wild. Hankin et al. (1993) used response to selection to estimate h^2 of age at maturity to be 0.49–0.57 for males and 0.39–0.41 for females (no SEs available), taking into account growth at sea and ocean harvest. I used their data to demonstrate the calculation of h^2 from P–O regression, ignoring these factors. The regression data for two brood years (1974 and 1980) are shown in Figure 1. (Data for the 1979 brood were excluded from this analysis because parents were not mated in pairs.) I used data on average age structure (1968–1977 broods) and on the size of 1974 and 1980 returns given by Nicholas and Hankin (1988) to estimate the phenotypic SD of age at maturity in male and female chinook salmon to be 1.035 and 0.637, respectively. The substantial difference between the sexes in these figures requires regression on separate sexes and an adjustment in the regression coefficients that reflects the difference in SD (Falconer 1989). The resulting heritability estimates of h^2 and V_A are given in Table 1. The estimates from P–O regression did not differ significantly from the estimates given by Hankin et al. (1993); the regression estimates suggested that maternal and nonadditive genetic effects on age at maturity were not important. The substantial additive genetic variance, based on male parents, that exists for age at maturity in this hatchery population of chinook salmon suggested that this trait could respond rapidly to selection.

Several investigators have questioned the utility of estimates of genetic parameters obtained from laboratory or hatchery populations for predicting evolution in natural populations (e.g., Rose 1984; Barker and Thomas 1987). Unfortunately, estimating genetic parameters in the wild is difficult. One

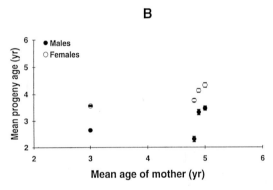

FIGURE 1.—Effect of age at maturity in fall chinook salmon from Oregon's Elk River Hatchery (1974 and 1980 broods) on mean age at maturity in their offspring (±2 SE). Data were taken from Hankin et al. (1993). (A) Correlation between fathers and their offspring. (B) Correlation between mothers and their offspring. Parent–offspring regression statistics and heritabilities generated from these data are given in Table 1.

tion should therefore prompt regular monitoring of the character in an attempt to detect change that could result from domestication in the hatchery. However, life-history characters are typically quite sensitive to environmental variation and often show evidence of phenotypic plasticity, the predictable phenotypic variation that can be expressed by a quantitative-character genotype responding to environmental variation (Via and Lande 1985; Stearns 1989). Phenotypic plasticity is important to adaptive evolution because it can help to maintain quantitative genetic variability; the existence of phenotypic plasticity can mean that no single genotype is superior to all others in all environments. As a result, genetically variable subpopulations can differentiate rapidly through local adaptation (Barton and Turelli 1989).

Phenotypic plasticity in a character can be expressed graphically (Figure 2) as a norm of reaction (Schmalhausen 1949). In a linear norm of reaction, nonzero slopes of lines relating phenotypes to environments and genotypes indicate the presence of

potential means of reducing this difficulty in estimating h^2 for a life-history character is to analyze the phenotypic regression of lab-reared offspring on their wild-caught parents. Coyne and Beecham (1987) and Riska et al. (1989) have shown that P–O regression applied in this way, when combined with an estimate of V_A, can provide an estimate of the lower bound of h^2 in the wild. For anadromous salmon, this approach could be suitable for life-history characters that can be measured on adults, such as age at maturity, fecundity, run timing, body size, and possibly some behaviors.

Phenotypic plasticity and genotype–environment interaction.—The existence of additive genetic variation in a life-history character indicates that it can respond to selection. The existence of such variation in a population considered for supplementa-

TABLE 1.—Narrow-sense heritabilities and approximate additive genetic variances for age at maturity of fall chinook salmon in Oregon's Elk River Hatchery (data from Hankin et al. 1993 and Nicholas and Hankin 1988). Heritability estimates are twice the regression coefficients (Falconer 1989). Heritability and variance estimates significantly ($P < 0.05$) greater than zero are marked with an asterisk; all other estimates are nonsignificant.

	Parent			
	Male		Female	
Offspring	Estimate	SE	Estimate	SE
Unadjusted regression coefficients				
Male	0.40	0.04	0.31	0.38
Female	0.21	0.09	0.30	0.16
Adjusted regression coefficients[a]				
Male	0.40	0.04	0.19	0.23
Female	0.34	0.15	0.30	0.16
Heritabilities				
Male	0.81*	0.08	0.39	0.46
Female	0.67*	0.30	0.59	0.32
Both[b]	0.74*	0.16	0.49	0.46
Additive genetic variances				
Male	0.86*	0.09	0.33	0.40
Female	0.27*	0.12	0.24	0.13
Both[b]	0.63*	0.14	0.42	0.40

[a] $SD_{male}/SD_{female} = 1.625$; $SD_{female}/SD_{male} = 0.616$.
[b] Male and female heritability differences nonsignificant; these values use the pooled information.

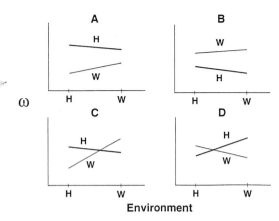

FIGURE 2.—Generalized norms of reaction (Stearns 1989) for fitness (ω) in wild (W) and hatchery (H) fish in wild and hatchery environments. If H (or W) fish in the two environments are from the same brood, the lines can be used to detect genetic variation for phenotypic plasticity (genotype–environment interaction). If H (or W) fish in the two environments are different random samples, the lines can be used to detect phenotypic plasticity but not, strictly speaking, genotype–environment interaction (Via 1993). See text for explanation. (**A**) Evidence for a general fitness advantage in hatchery fish across environments. (**B**) Evidence for a general fitness advantage in wild fish across environments. (**C**) Evidence for genotype–environment interaction that suggests local adaptation to the environment in which the fish were raised. (**D**) Evidence for genotype–environment interaction that suggests the possibility of preadaptation to foreign environments (an unlikely scenario).

plasticity, and differences in line elevations indicate the presence of genetic variation. Differences in slope magnitudes indicate variation among genotypes in phenotypic sensitivity to environmental variation (and potentially indicate G–E interaction; see Figure 2). Because they describe the correspondence between phenotype and the combination of genotype and environment—and therefore offer a more direct analysis of the causes of evolutionary change—norms of reaction may be more significant tools for investigating the evolutionary genetics of populations than are parameters such as h^2 (Lewontin 1974).

The importance of G–E in the life-history variation of salmonids is largely unknown, but evidence exists for its importance in the life histories of other organisms (van Noordwijk 1989; Via 1993). A seminal study by Reisenbichler and McIntyre (1977) on hatchery and wild summer steelhead *O. mykiss* planted as eyed eggs into several tributaries of the Deschutes River (Oregon) provides some data that suggest the importance of phenotypic plasticity to life history in salmonids in the wild. By incorporating first-generation hybrids (HW) between hatchery (HH) and wild fish (WW) in their study, Reisenbichler and McIntyre were able to provide evidence for genetic differences in survival and growth between hatchery and wild fish. They were able to identify these groups of fish by assaying electrophoretic genotypes at the *LDH-4** locus. Figure 3A expresses their findings on survival of eggs from eyeing to hatching in a form analogous to a norm of reaction. (Figures 3A–C are not, strictly speaking, norms of reaction because individuals from the same family are not exposed to all environments; see Figure 2.) The consistently higher values for survival of wild fish in Potlid, Dutchman, and Trout creeks relative to hatchery and hybrid fish indicated that genetic variation among populations exists for survival during early development. The nonparallelism of the HH and HW norms of reaction suggested that the effect of wild genetic heritage on survival may have been more important in Dutchman and Trout creeks than in Potlid Creek, but the differences in survival between hatchery and hybrid fish in each of the three environments do not appear to be significant. Thus, the results do not indicate that phenotypic plasticity contributed strongly to variation in survival.

The norms of reaction for length of HH, HW, and WW fish in the fall after planting and in the following spring indicated the existence of significant plasticity for growth in length in Deschutes River summer steelhead (Figure 3B). In the fall, HW fish were longer than either HH or WW fish in Trout and Opal creeks, but none of the three groups differed significantly in length in Dutchman or Potlid creeks. In the fall, HH fish were longest in the hatchery pond (Figure 3B). By the next spring, HW fish were longer than either WW or HH fish in Potlid Creek, but HH fish had grown faster in Trout Creek; the three groups did not differ significantly in length in Dutchman or Opal creeks (Figure 3C). The results suggest the presence of heterosis for growth in some creeks (but not others) in the fall that was not detected in spring, and relatively rapid growth of overwintering hatchery fish in one stream. Hatchery fish grew fastest in the hatchery pond.

The significant differences in survival and growth were not large. Nevertheless, these differences could give rise to local adaptations in the steelhead populations. The norms of reaction cast some doubt on the authors' conclusion that there was no evidence for a "selective advantage associated with individual LDH genotypes" (Reisenbichler and

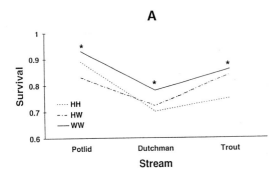

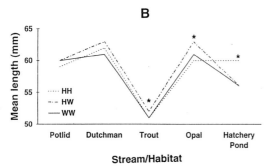

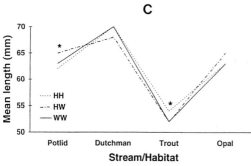

FIGURE 3.—Norms of reaction for hatchery (HH) and wild (WW) steelhead and their hybrids (HW), planted into tributaries of the Deschutes River, Oregon (data from Reisenbichler and McIntyre 1977). The wild steelhead were endemic to the Deschutes River. The hatchery steelhead had been derived from this wild stock and had been artificially propagated for two generations, and the crossbred offspring of these two groups were F_1 hybrids. (**A**) Survival from eyeing to hatching in three streams. The asterisks indicate significantly ($P < 0.05$) greater survival in WW steelhead, relative to HH and HW steelhead, in all three streams. (**B**) Length (mm) in the fall after planting in four tributaries and in a hatchery pond. (**C**) Length (mm) in the four tributaries the following spring (no data are available for the hatchery pond).

McIntyre 1977). Whether the genetically based differences observed in survival and growth would have evolutionary consequences depends on the extent to which the observed variation in growth would translate into fitness variation among the different groups of returning adults. Responses to hatchery selection may be difficult to predict if phenotypic plasticity exists for the life-history characters in question. Evidence from a few studies suggests that the genotypes that perform best in the hatchery may have low fitness in the wild (Leider et al. 1990; Fleming and Gross 1993). More evidence for phenotypic plasticity and G–E interaction in salmon life history is needed to determine the consequences of artificial propagation for supplemented natural populations as well as the genetic basis for differences in fitness of wild and hatchery fish.

Weighing the Costs of Different Errors of Inference: The Issue of Experimental Power

Minimizing the genetic and phenotypic differentiation of captive and natural fish may pose an inferential problem for the evaluation of supplementation. Consideration of this objective in light of the possibility of erroneous conclusions about the genetic effects of supplementation requires a different approach to hypothesis testing than that to which most empirical biologists are accustomed. An appropriate null hypothesis in testing supplementation is the following: hatchery and supplemented wild fish do not differ in the value of life-history character *Y*. There are two statistical errors that can result in testing this hypothesis, and only one of these has been widely appreciated by biologists.

The error most biologists are familiar with in this context is the probability of concluding that hatchery and wild fish differ when in fact they do not; this error is referred to by statisticians as a type I error (Dixon and Massey 1957). Fortunately, the acceptable level of a type I error (designated α) is under the investigator's control. The converse error is the probability of concluding that these groups do not differ when in fact they do. This error is known as a type II error (Dixon and Massey 1957), and its level (designated β) is controlled indirectly by the investigator through manipulation of α, the difference the investigator wishes to detect (the critical effect size; Cohen 1988), and sample size (Figure 4).

Biologists have frequently been led to believe that a type I error is more serious than a type II error (Toft and Shea 1983); indeed, most introductory statistics texts do not discuss type II error, much less ways to avoid it. A widely used value for β is 20%, four times the level of the conventional level for α, indicating that investigators that use this level of β consider a type I error to be four times

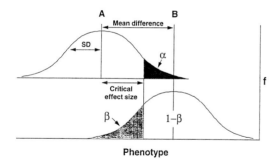

FIGURE 4.—General relationship between type I error (α), type II error (β), the mean difference (true effect size = A − B), critical effect size, sample variability (measured as phenotypic SD), and power (1−β) for comparing two normally distributed quantitative characters (f is frequency). For graphical simplicity, a one-tailed test is used to determine whether the difference between the character means is significantly different. In general, power increases with increasing mean difference or decreasing β (lower curve shifts to right), and with increasing critical effect size or decreasing α. It also increases at higher sample sizes and lower SD.

more serious than a type II error (Cohen 1988). Concern is growing in ecology (Toft and Shea 1983), conservation biology (Taylor and Gerrodette 1993), and natural resource management (Peterman 1990) over the research and management consequences of a type II error. The consequences of a type II error in conservation of Threatened and Endangered populations may be especially serious because their status "leave[s] little margin for recovery from incorrect management decisions" (Taylor and Gerrodette 1993). To use a medical analogy, a type I error would be tantamount to a physician's diagnosis that a patient has cancer when he actually does not, whereas a type II error is equivalent to the diagnosis that the patient is cancer-free, when in fact he has developed the disease. Both diagnostic errors would be cause for concern, but a type II error would ultimately be more serious for the patient.

Experimental power is the probability of detecting an effect when it actually exists, and its level is equivalent to $1 - \beta$. Thus, a type II error level of 20% in a supplementation experiment corresponds to an 80% chance (power) of detecting divergence between hatchery and natural fish. A comprehensive treatment of power and statistical error is beyond the scope of this paper. Peterman (1990) gives a concise introduction to this issue in fisheries management; texts such as Dixon and Massey (1957) and Cohen (1988) provide substantial detail on this topic, as well as a number of useful tests of power and tables of power statistics. The main point to remember for supplementation is that the anticipated costs associated with making type I versus type II errors during a monitoring program should be weighed beforehand to determine whether the desired power is achievable and, if so, the appropriate extent and frequency of monitoring. These costs include the direct costs of making the error in hypothesis testing plus the costs of remedial action. In the supplementation of natural salmon populations, a type II error is likely to pose a greater genetic threat to a natural population than is a type I error. Unlike a type I error, a type II error may not prompt the termination—or even the reevaluation—of supplementation activity until divergence is large enough to be detected.

The risk of a genetic type II error during supplementation is inversely related to the intensity of genetic monitoring (the sample size) and the direction and magnitude of genetic divergence considered acceptable (the critical effect size; see Peterman 1990). This risk, in turn, depends on character variability and how divergence is measured. For example, Cohen (Table 2.4.1: 1988) indicated that to ensure adequate safeguards against both type I and type II errors (e.g., α, β < 5%) when testing whether two character means for a normally distributed character differ, regardless of the direction of the difference (i.e., a two-tailed test), almost 300 individuals must be sampled from each group to detect a difference of 0.3 phenotypic SD. If the acceptable level of both errors can be relaxed to 10% and a one-tailed test is warranted (e.g., because previous work indicates that the mean of group A is consistently larger than that of group B), only about half this many individuals need to be sampled. Whether a divergence of 0.3 SD would constitute a threat to a natural population depends on the trait's natural variability and on its effects on fitness. Thus, information on character variation and statistical evaluation of alternative hypotheses to be tested are essential to determining the scope and intensity of quantitative genetic monitoring (Taylor and Gerrodette 1993).

A long-term study by Hershberger et al. (1990) on coho salmon *O. kisutch* reared in marine netpens in Puget Sound, Washington, shows some of the difficulties in ensuring adequate experimental power during the monitoring of change in life-history characters. This study, which extended over 10 generations, is laudable for its length and incorporation of control populations to assist in accounting for environmental components of variation affecting the intergenerational phenotypic responses to selection. The importance of control populations in measuring selection has been emphasized by Fredeen (1986). In this case, Hershberger et al. (1990) argued that the higher realized heritabilities relative to estimates obtained before selection suggested that substantial domestication selection was operating in the culture environment. However, the experiment did not provide a control for the effects of changes in the culture regime that occurred during the experiment. Consequently, some of the observed responses to natural selection during culture via "domestication selection" (Doyle 1983) could reflect environmental or G–E variation.

The change in weight among brood years observed by Hershberger et al. (1990) in two internal control populations (unselected fish raised together with fish artificially selected for high weight in the same facility) is difficult to reconcile with the authors' conclusions. These data suggest that domestication selection for weight was unimportant. It could be that the reported differences in the h^2 estimates were due to genetic effects not accounted for in an additive genetic model. Given the fact that no significant increase ($\alpha > 0.05$) in 8-month weight in seawater was detected in these lines, an analysis of trend developed by Gerrodette (1987) can be used to estimate the power of the experimental design to detect an increase arising from selection. Gerrodette's technique takes into account the strength of a trend (as measured by the rate of change in a character as a fraction of its initial value), the sample size, the coefficient of variation (SD as a percentage of the mean), and α in estimating the power of detecting a trend. He provided an inequality that can estimate the power of detecting a linear trend in the coho salmon weight data of Hershberger et al. (1990). Rearranging this inequality and solving for the level of type II error gives

$$z_\beta \leq r\, n^{3/2}/(3.464\, \mathrm{CV}) - z_{\alpha/2},$$

where z_β is the value of the standardized random normal deviate that defines the area of the normal probability curve corresponding to a level of type II error equal to β for a two-tailed test, $z_{\alpha/2}$ is the value of the deviate that corresponds to a level of type I error equal to α (two-tailed test), r is the mean rate of change in weight per generation as a fraction of the initial weight, n is the number of generations, and CV is the coefficient of variation in weight.

The coho salmon weight data (in grams, adjusted for density variation in generation 4) in the internal control lines are given in Figure 5 (estimated from

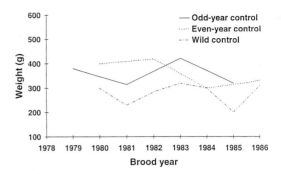

FIGURE 5.—Trends over four generations in 8-month mean weight of two unselected populations of coho salmon raised in marine net-pens in Puget Sound, Washington, and over seven generations for a wild population (from Hershberger et al. 1990). See text for analysis.

Figure 3 in Hershberger et al. 1990). The CVs of initial weight were 41 and 31% for the odd- and even-year lines, respectively (Table 2 in Hershberger et al. 1990). The slopes ($\pm$ SE) of the linear regressions on the internal control-line data in Figure 5 are 2.5 $\pm$ 39.8 g/generation (+0.7% per generation) and -26.0 ± 47.1 g/generation (-6.5% per generation) for the odd- and even-year lines, respectively; neither estimate differs significantly from zero. At a level of α of 10%, z_β is -1.61 or less for the odd-year line and z_β is -1.16 or less for the even-year line. Assuming that the weight estimates are independent and normally distributed and that the CVs are independent of the generation sampled, these values correspond to a probability of detecting (i.e., power) a trend this small of at most 12%. Thus, domestication for weight could easily have been occurring without detection in this experiment.

What are the smallest trends that could be detected with α and β both held at 10% (i.e., power = 90%)? Rearrangement of the above inequality to solve for r yields 219 g/generation for the odd-year line and -177 g/generation for the even-year line. Similarly, solving the inequality for n with a power of 90% gives a minimum number of 79 generations for the odd-year line and 15 generations for the even-year line that would be required to detect the observed trends. These results are not very encouraging; the small sample size (4 generations), the relatively large CVs, and the low estimated magnitudes of the trends all contribute to the insufficient power of this experiment to detect domestication selection.

A controlled selection experiment like that used by Hershberger et al. (1990) will not usually be possible in supplementation programs. In most cases, the best information that will be available for any given character includes estimates of the phenotypes in the hatchery and natural populations and, if the character is quantitative, sample variances around those estimates. However, without corresponding controls for environmental effects, any difference or trend that might be detected confounds environmental with genetic variation, thereby rendering a reliable genetic interpretation impossible. This will be an unavoidable problem in most supplementation applications.

Prerequisites to Effective Supplementation

This paper is based on the premise that successful supplementation of natural salmon populations requires adequate genetic monitoring of life-history characters. Very little guidance has been developed on this topic. The approach taken here has been to examine the variability of life-history characters and to ask what genetic monitoring of life-history characters is likely to entail when the objective of supplementation is to minimize the differentiation of hatchery and natural fish. In doing so, two prominent problems have been identified that should be considered before initiating supplementation: reliably discriminating between genetic and environmental components of phenotypic variation and establishing sufficient experimental power to detect population divergence. This section introduces some general recommendations intended to reduce opportunities for these problems to arise.

Discriminating Genetic and Environmental Sources of Variation

The first problem stems from the general sensitivity of life-history characters to environmental variation. Little genetic analysis has been done on salmon life-history characters, even though they are likely to directly influence supplementation success. Consequently, one recommendation is to quantify the amount of genetic variation available in life-history characters thought to be important to fitness in the wild. Ideally, this should be done for a population before supplementation is initiated, so that traits sensitive to genetic change can be identified first. For supplementation, it is not necessary to estimate additive genetic variances or heritabilities with precision (indeed, these estimates generally have such large sampling errors that the experimental requirements necessary to estimate them with reasonable precision can exceed the capacity of many rearing facilities). However, the procedures used to detect additive genetic variance should be sensitive enough to determine whether the variance is significantly nonzero.

One means of detecting significant genetic variance for a salmon life-history character is to mate several (preferably 10 or more) pairs of adults that have returned to spawn naturally and that represent as large a range of phenotypes as possible, record the parental phenotypes, mark all the offspring in each resulting full-sib family with a family-specific mark (e.g., coded microwire tags), and measure the same characters in the offspring that return (generating a mean value for each offspring family) (Falconer 1989). The utility of the method requires that several, ideally 10 or more, offspring from each family survive to return. If the slope of the regression of mean offspring on their parents is significantly different from zero (regardless of sign), and if it is assumed that (1) the genetic effects on the trait are approximately normally distributed, (2) phenotypes are scaled so that males and females have the same means and variances, and (3) maternal effects, nonadditive genetic effects, and selection can be ignored, then the additive genetic variance for that character in nature is significantly different from zero (Riska et al. 1989). Life-history characters with significant additive genetic variation in nature can have substantial effects on fitness, and care should be taken to minimize genetic change in them during hatchery supplementation. It should also be recognized that nonsignificant estimates can result from low experimental power; consequently, large nonsignificant estimates should also be treated seriously.

The characters likely to be most important to fitness in the wild and most amenable to change during hatchery culture include life-history and behavioral traits (Hynes et al. 1981; Doyle 1983). Genetic variability in behavioral traits shows some similarities to variability in life-history traits (Roff and Mousseau 1987; McIsaac and Quinn 1988; Swain and Riddell 1990). After surveying natural variation in a number of life-history characters in several populations of Pacific salmon, Lichatowich and Cramer (1979) recommended that attention be focused on growth rate, age and size at outmigration and return, migration timing, and spawn timing. All these characters are amenable to quantitative genetic analysis. Table 2 depicts life-history, physiological, and behavioral traits that have been identified as potentially important to supplementation. Maynard et al. (1995, this volume) provides experimental support for the importance of some of these

TABLE 2.—Characters that might be important for genetic monitoring during supplementation of natural salmon populations. Some of these characters have been associated with supplementation effects by Riggs (1990), Hard et al. (1992), Cuenco et al. (1993), and Lichatowich and Watson (1993).

Demographic	Phenological	Physiological	Behavioral
Survivorship	Incubation and emergence timing	Body size, morphometry, and composition	Habitat selection
Sex ratio	Juvenile out-migration timing	Egg size	Foraging ability
Growth and development rate	Adult run timing	Stress response	Agonistic behavior
Age at juvenile out-migration	Adult spawn timing	Stamina	Predator recognition
Age at maturity		Burst swimming speed	Migration
Fecundity and fertility		Disease resistance	Homing response
		Smoltification	Mating ability
		Migratory tendency	Redd construction and defense

characters for fitness during the juvenile phase of the salmonid life cycle.

Because phenotypic plasticity can obscure the responses of life-history and behavioral traits to selection, attempts to detect G–E interaction should help to identify traits that contribute to adaptation in natural habitats. Experiments like that of Reisenbichler and McIntyre (1977) may be the most effective means of characterizing differences in the selection regimes imposed by the hatchery and by the wild. These characterizations provide information essential to reducing selection differentials imposed by hatchery domestication on life-history characters. In a supplementation scheme, differentially marking the offspring of hatchery and supplemented natural fish and, where feasible, rearing each group in both environments would afford the opportunity to determine the nature of the selective differences. The most important information needed is quantification of the life-history differences between hatchery and natural fish in the wild; in combination with the determination of genetic variance, these data would identify the traits sensitive to domestication.

Evaluating Experimental Power to Detect Genetic Problems

The second major problem in the genetic monitoring of supplementation is the potential inability to detect differentiation. In principle, this problem can be controlled with adequate monitoring, primarily by maintaining large sample sizes and by considering carefully the levels of differentiation that are believed to pose threats to a natural population. It is important that monitoring be as intensive as possible during the early stages of supplementation, before divergence can exceed the acceptable level. Rates of divergence are likely to vary among characters and with the mode of supplementation. A generally acceptable level of divergence is difficult to formulate, but considerations of statistical power and variation in quantitative salmon life-history characters suggest that acceptable levels of divergence, based on examination of minimum significant differences for a variety of sample sizes and levels of type I error, are probably on the order of 0.1–0.3 SD (author, unpublished data). What does this level of detection mean for monitoring? To detect a divergence between hatchery and natural fish of 0.2 SD in the mean of a character with α and β less than 5% (power > 95%), monitoring would require the measurement of that character in about 700 fish at the appropriate stage from each of the two groups under a two-tailed test. To detect a significant trend with a correlation of 0.5 under the same levels of error, the trend must span over 40 generations. These sample sizes can be lowered—often substantially—by considering larger values for α and β or a larger effect size (i.e., SD or slope), or both (in the case of trend, a larger correlation will result in higher power, all else being equal). For example, for α and β less than 10% (i.e., power > 90%), detecting a divergence of 0.3 SD would require measuring about 240 fish from each group. If a one-tailed test is possible, the sample size could be reduced even further. Cohen (1988) provides several tables that can be used to guide choices of parameter values.

Monitoring and Evaluation in the Face of Uncertainty

Supplementation of natural salmon populations with hatchery fish should be approached as an experimental perturbation with the potential for adverse genetic and ecological effects on natural populations. A cautious approach is necessary because we do not yet understand the life-history consequences of supplementation for natural populations. These issues should be considered during the early stages of supplementation planning. If problems are detected during supplementation, the remedial actions that could be taken in response span a wide spectrum. Terminating hatchery production is an option lying at one end of this spectrum, but it is likely to be considered only as a last resort. Consequently, supplementation with hatchery fish should be conducted experimentally on relatively stable populations in systems that are conducive to intensive monitoring. Supplementation should be conducted on a scale appropriate for the natural population and its habitat and should be characterized by detailed evaluation and monitoring, so that the results of supplementation are informative enough to prompt remedial action (including immediate cessation of supplementation and recovery of artificially propagated fish prior to liberation or upon return, if necessary).

Experimental supplementation along the lines suggested here would also provide some opportunities of value to researchers and managers. Carefully regulated supplementation experiments could quickly generate information on life-history variation in hatchery and wild stocks. Such information could help to improve the management of ongoing enhancement programs. It is also likely to be useful in identifying possible causes of failures of hatchery production. Ultimately, the artificial propagation of fish that are genetically and ecologically compatible with naturally reproducing fish should reduce the threat of extinction to natural salmon populations.

References

Allendorf, F. W., and N. Ryman. 1987. Genetic management of hatchery stocks. Pages 141–159 *in* N. Ryman and F. Utter, editors. Population genetics & fishery management. University of Washington Press, Seattle.

Allendorf, F. W., N. Ryman, and F. Utter. 1987. Genetics and fishery management: past, present, and future. Pages 1–19 *in* N. Ryman and F. Utter, editors. Population genetics & fishery management. University of Washington Press, Seattle.

Altukhov, Y. P., and E. A. Salmenkova. 1987. Stock transfer relative to natural organization, management, and conservation of fish populations. Pages 333–343 *in* N. Ryman and F. Utter, editors. Population genetics & fishery management. University of Washington Press, Seattle.

Barker, J. S. F., and R. H. Thomas. 1987. A quantitative genetic perspective on adaptive evolution. Pages 3–23 *in* V. Loeschcke, editor. Genetic constraints on adaptive evolution. Springer-Verlag, Berlin.

Barton, N. H., and M. Turelli. 1989. Evolutionary quantitative genetics: how little do we know? Annual Review of Genetics 23:337–370.

Becker, W. A. 1984. Manual of quantitative genetics, 4th edition. Academic Enterprises, Pullman, Washington.

Bentsen, H. B. 1991. Quantitative genetics and management of wild populations. Aquaculture 98:263–266.

Boag, P. T., and P. R. Grant. 1978. Heritability of external morphology in Darwin's finches. Nature 274:793–794.

Campton, D. E. 1995. Genetic effects of hatchery fish on wild populations of Pacific salmon and steelhead: what do we really know? American Fisheries Society Symposium 15:337–353.

Cloud, J. G., and G. H. Thorgaard, editors. 1993. Genetic conservation of salmonid fishes. Plenum Press (in cooperation with NATO Scientific Affairs Division), New York.

Cohen, J. 1988. Statistical power analysis for the behavioral sciences, 2nd edition. Lawrence Erlbaum Associates, Hillsdale, New Jersey.

Coyne, J. A., and E. Beecham. 1987. Heritability of two morphological characters within and among natural populations of *Drosophila melanogaster*. Genetics 117:727–737.

Cuenco, M. L., T. W. H. Backman, and P. R. Mundy. 1993. The use of supplementation to aid in natural stock restoration. Pages 269–293 *in* J. G. Cloud and G. H. Thorgaard, editors. Genetic conservation of salmonid fishes. Plenum Press (in cooperation with NATO Scientific Affairs Division), New York.

Dixon, W. J., and F. J. Massey, Jr. 1957. Introduction to statistical analysis, 2nd edition. McGraw-Hill, New York.

Doyle, R. W. 1983. An approach to the quantitative analysis of domestication selection in aquaculture. Aquaculture 33:167–185.

Falconer, D. S. 1989. Introduction to quantitative genetics, 3rd edition. Longman, New York.

Fleming, I. A., and M. R. Gross. 1993. Breeding success of hatchery and wild coho salmon (*Oncorhynchus kisutch*) in competition. Ecological Applications 3:230–245.

Fredeen, H. 1986. Monitoring genetic change. Aquaculture 57:1–26.

Gerrodette, T. 1987. A power analysis for detecting trends. Ecology 68:1364–1372.

Hankin, D. G., J. W. Nicholas, and T. W. Downey. 1993. Evidence for inheritance of age of maturity in chinook salmon (*Oncorhynchus tshawytscha*). Canadian Journal of Fisheries and Aquatic Sciences 50:347–358.

Hard, J. J., R. P. Jones, Jr., M. R. Delarm, and R. S. Waples. 1992. Pacific salmon and artificial propagation under the Endangered Species Act. NOAA (National Oceanic and Atmospheric Administration) Technical Memorandum NMFS (National Marine Fisheries Service) NWFSC-2.

Heard, W. R. 1983. Chinook salmon fisheries and enhancement in Alaska: a 1982 overview. Pages 21–28 *in* C. J. Sindermann, editor. Proceedings of the eleventh U.S.–Japan meeting on aquaculture, salmon enhancement. NOAA (National Oceanic and Atmospheric Administration) Technical Report NMFS (National Marine Fisheries Service) 27.

Hershberger, W. K., J. M. Myers, R. N. Iwamoto, W. C. McAuley, and A. M. Saxton. 1990. Genetic changes in the growth of coho salmon (*Oncorhynchus kisutch*) in marine net-pens, produced by ten years of selection. Aquaculture 85:187–197.

Hindar, K., N. Ryman, and F. Utter. 1991. Genetic effects of cultured fish on natural fish populations. Canadian Journal of Fisheries and Aquatic Sciences 48:945–957.

Howell, P., K. Jones, D. Scarnecchia, L. Lavoy, W. Kendra, and D. Ortmann. 1985. Stock assessment of Columbia River anadromous salmonids, volume 1. Chinook, coho, chum, and sockeye salmon stock summaries. Report to Bonneville Power Administration (Project 83-355). (Available from Bonneville Power Administration, P.O. Box 3621, Portland, OR 97208.)

Hynes, J. D., E. H. Brown, Jr., J. H. Helle, N. Ryman, and D. A. Webster. 1981. Guidelines for the culture of fish stocks for resource management. Canadian Journal of Fisheries and Aquatic Sciences 38:1867–1876.

Krueger, C. C., A. J. Gharrett, T. R. Dehring, and F. W. Allendorf. 1981. Genetic aspects of fisheries rehabilitation programs. Canadian Journal of Fisheries and Aquatic Sciences 38:1877–1881.

Lande, R. 1988. Quantitative genetics and evolutionary theory. Pages 71–84 in B. S. Weir, E. J. Eisen, M. M. Goodman, and G. Namkoong, editors. Proceedings of the 2nd international conference on quantitative genetics. Sinauer Associates, Sunderland, Massachusetts.

Lande, R., and G. F. Barrowclough. 1987. Effective population size, genetic variation, and their use in population management. Pages 87–123 in M. E. Soulé, editor. Viable populations for conservation. Cambridge University Press, UK.

Law, R., and D. R. Grey. 1989. Evolution of yields from populations with age-specific cropping. Evolutionary Ecology 3:343–359.

Leider, S. A., P. L. Hulett, J. J. Loch, and M. W. Chilcote. 1990. Electrophoretic comparison of the reproductive success of naturally spawning transplanted and wild steelhead trout through the returning adult stage. Aquaculture 88:239–252.

Lewontin, R. C. 1974. The analysis of variance and the analysis of causes. American Journal of Human Genetics 26:400–411.

Lichatowich, J., and S. Cramer. 1979. Parameter selection and sample sizes in studies of anadromous salmonids. Oregon Department of Fish and Wildlife Information Report Series, Fisheries 80-1.

Lichatowich, J., and B. Watson. 1993. Use of artificial propagation and supplementation for rebuilding salmon stocks listed under the Endangered Species Act. Technical Report 5: Recovery issues for threatened and endangered Snake River salmon. Report to Bonneville Power Administration (Project 88-100). (Available from Bonneville Power Administration, P.O. Box 3621, Portland, OR 97208.)

Maynard, D. J., T. A. Flagg, and C. V. W. Mahnken. 1995. A review of seminatural culture strategies for enhancing the postrelease survival of anadromous salmonids. American Fisheries Society Symposium 15:307–314.

McIsaac, D. O., and T. P. Quinn. 1988. Evidence of a hereditary component in homing behavior of chinook salmon (*Oncorhynchus tshawytscha*). Canadian Journal of Fisheries and Aquatic Sciences 45:2201–2205.

Mitchell-Olds, T. 1986. Quantitative genetics of survival and growth in *Impatiens capensis*. Evolution 40:107–116.

Nelson, K., and M. Soulé. 1987. Genetical conservation of exploited fishes. Pages 345–368 in N. Ryman and F. Utter, editors. Population genetics & fishery management. University of Washington Press, Seattle.

Nicholas, J. W., and D. G. Hankin. 1988. Chinook salmon populations in Oregon coastal river basins: description of life histories and assessment of recent trends in run strengths. Oregon Department of Fish and Wildlife Information Report 88-1.

Peterman, R. M. 1990. Statistical power analysis can improve fisheries research and management. Canadian Journal of Fisheries and Aquatic Sciences 47:2–15.

Price, T., and D. Schluter. 1991. On the low heritability of life-history traits. Evolution 45:853–861.

Prout, T. 1958. A possible difference in genetic variance between wild and laboratory population. Drosophila Information Service 32:148–149.

Reisenbichler, R. R., and J. D. McIntyre. 1977. Genetic differences in growth and survival of juvenile hatchery and wild steelhead trout, *Salmo gairdneri*. Journal of the Fisheries Research Board of Canada 34:123–128.

Reisenbichler, R. R., and J. D. McIntyre. 1986. Requirements for integrating natural and artificial production of anadromous salmonids in the Pacific Northwest. Pages 365–374 in R. H. Stroud, editor. Fish culture in fisheries management. American Fisheries Society, Fish Culture Section and Fisheries Management Section, Bethesda, Maryland.

Ricker, W. E. 1981. Changes in average size and average age of Pacific salmon. Canadian Journal of Fisheries and Aquatic Sciences 38:1636–1656.

Riggs, L. A. 1990. Principles for genetic conservation and production quality: results of a scientific and technical clarification and revision. Report to the Northwest Power Planning Council (Contract Number C90-005). (Available from Genetic Resource Consultants, P. O. Box 9528, Berkeley, CA 94709.)

Riska, B., T. Prout, and M. Turelli. 1989. Laboratory estimates of heritabilities and genetic correlations in nature. Genetics 123:865–871.

Roff, D. A. 1992. The evolution of life histories. Theory and analysis. Chapman and Hall, New York.

Roff, D. A., and T. A. Mousseau. 1987. Quantitative genetics and fitness: lessons from *Drosophila*. Heredity 58:103–118.

Rose, M. R. 1984. Genetic covariation in *Drosophila* life history: untangling the data. The American Naturalist 123:565–569.

Rosentreter, N. 1977. Characteristics of hatchery fish: angling, biology, and genetics. American Fisheries Society Special Publication 10:79–83.

Schmalhausen, I. I. 1949. Factors of evolution: the theory of stabilizing selection. Blakiston, Philadelphia.

Sheridan, A. K. 1988. Agreement between estimated and realized genetic parameters. Animal Breeding Abstracts 56:877–889.

Stearns, S. C. 1976. Life history tactics: a review of the ideas. Quarterly Review of Biology 51:3–47.

Stearns, S. C. 1989. The evolutionary significance of phenotypic plasticity. BioScience 39:436–445.

Swain, D. P., and B. E. Riddell. 1990. Variation in agonistic behavior between newly emerged juveniles from hatchery and wild populations of coho salmon, Oncorhynchus kisutch. Canadian Journal of Fisheries and Aquatic Sciences 47:566–571.

Tave, D. 1993. Genetics for fish hatchery managers, 2nd edition. Van Nostrand Reinhold, New York.

Taylor, B. L., and T. Gerrodette. 1993. The uses of statistical power in conservation biology: the vaquita and northern spotted owl. Conservation Biology 7:489–500.

Toft, C. A., and P. J. Shea. 1983. Detecting community-wide patterns: estimating power strengthens statistical inference. The American Naturalist 122:618–625.

Utter, F. 1981. Biological criteria for definition of species and distinct intraspecific populations of anadromous salmonids under the U.S. Endangered Species Act of 1973. Canadian Journal of Fisheries and Aquatic Sciences 38:1626–1635.

Utter, F. M., J. E. Seeb, and L. W. Seeb. 1993. Complementary uses of ecological and biochemical genetic data in identifying and conserving salmon populations. Fisheries Research 18:59–76.

van Noordwijk, A. J. 1989. Reaction norms in genetical ecology. BioScience 39:453–458.

Via, S. 1993. Adaptive phenotypic plasticity: target or by-product of selection in a variable environment? The American Naturalist 142:352–365.

Via, S., and R. Lande. 1985. Genotype–environment interaction and the evolution of phenotypic plasticity. Evolution 39:505–522.

Waples, R. S. 1991. Genetic interactions between hatchery and wild salmonids: lessons from the Pacific Northwest. Canadian Journal of Fisheries and Aquatic Sciences 48 (Supplement 1):124–133.

Waples, R. S., and six coauthors. 1992. A genetic monitoring and evaluation program for supplemented populations of salmon and steelhead in the Snake River Basin. Report to Bonneville Power Administration (Project 89-096). (Available from Bonneville Power Administration, P. O. Box 3621, Portland, OR 97208.)

Distribution of Largemouth Bass Genotypes in South Carolina: Initial Implications

JAMES BULAK AND JEAN LEITNER

South Carolina Department of Natural Resources
1921 Van Boklen Road, Eastover, South Carolina 29044, USA

THOMAS HILBISH

University of South Carolina, Department of Biological Sciences
Columbia, South Carolina 29208, USA

REX A. DUNHAM

Auburn University, Department of Fisheries and Allied Aquacultures
Alabama Agricultural Experiment Station, Alabama 36849, USA

Abstract.—A statewide allozyme survey of largemouth bass *Micropterus salmoides* was performed in South Carolina to characterize the relative abundance of alleles characteristic of the northern and Florida subspecies. Data confirmed that South Carolina is part of a broad hybrid zone between the two subspecies. The frequency of alleles that are fixed for the Florida subspecies ranged from 98% in Lake Moultrie, a southeastern site, to 36% in Lake Wateree, a north-central site. The existence of geographically dependent allelic clines and the dependence of allele frequency on location and gene locus suggests that, in South Carolina, the hybrid zone is maintained by a balance between gene flow and selection. Analysis of the Lake Wateree population, near the center of the hybrid zone, failed to detect any linkage disequilibrium between the four loci that were tested. The absence of linkage disequilibrium and the lack of coincidence among clines at different loci suggests that this hybrid zone is not a tension zone but probably results from an environmental selective gradient across South Carolina.

The survey also provided preliminary evidence of a size-based allele selection in Lake Wateree and a suggestion that selection may modify allele frequencies within the state's hatcheries. Allele frequencies in the South Carolina hatchery system were significantly different from natural populations in the warmest and coldest areas of the state. Current hatchery practices include a semidomesticated broodstock and random selection of stocking sites for each hatchery. The regionalization of hatchery broodstocks to maximize fitness was suggested from this survey. In a vision of this strategy, broodstock with a high percentage of Florida alleles would be stocked in the warmer regions of the state. Conversely, largemouth bass with a higher percentage of northern alleles would be stocked in the cooler regions. From a genetic perspective, the decision to transfer stocks should largely rest on the fitness of a specific phenotype within a defined area.

The largemouth bass *Micropterus salmoides* is an important North American sport fish. Its original distribution in the United States included much of the Mississippi River drainage and the Atlantic slope north only to southern or central South Carolina (Lee et al. 1980). As a result of introductions, the current distribution now covers parts of all 48 continental states of the United States (Lee et al. 1980). The type locality for the species is in the vicinity of Charleston, South Carolina (Lacepède 1798). Holbrook (1860), in his *Ichthyology of South Carolina*, indicated that the largemouth bass is abundant in the waters of Florida, Georgia, and the Carolinas.

Bailey and Hubbs (1949) investigated the meristic traits of largemouth bass and concluded that two distinct subspecies occur, the northern largemouth bass *M. salmoides salmoides*, and the Florida largemouth bass *M. salmoides floridanus*. This investigation indicated that the range of the Florida subspecies was restricted to peninsular Florida, and a hybrid zone between the two subspecies occurred in northern Florida, Georgia, and the Savannah River drainage in South Carolina.

Philipp et al. (1983) used allozymes to determine the genomic contributions of each subspecies at 90 locations in the United States. Their study revealed that fixed allelic differences occurred at two loci, those which coded for the enzymes isocitrate dehydrogenase (*IDHP-1**) and aspartate aminotransferase (*sAAT-2**). Two other polymorphic loci, those which coded for the enzymes superoxide dismutase (*sSOD-1**) and malate dehydrogenase (*sMDH-B**), were also reported as helpful in distinguishing subspecies. Survey results confirmed that populations consisting of pure Florida subspecies were confined to peninsular Florida. Also, this survey indicated that the hybrid zone along the Atlantic slope extended as far north as Maryland. Philipp et al. (1985) suggested that the persistence of latitudinal clines in allele frequencies, in the face of introductions, strongly suggested that selection was acting to maintain these clines.

Hybrid Zones

The location and width of the hybrid zone between two genetically distinct populations can define the strength of natural selection in maintaining each parental genotype. A hybrid zone occurs when genetically distinct groups of individuals meet, mate, and produce at least some offspring of mixed ancestry (Harrison 1993). Within the hybrid zone, a geographic gradient in measurable characters (i.e., a cline) is observed. Clines are most likely to form in a region between genotypic optima, where net gene selection changes direction (Endler 1977).

A hybrid zone can form as a result of a geographic selection gradient (Moore and Price 1993) or as a tension zone (Barton and Hewitt 1985). In the tension zone model, hybrid genotypes (i.e., heterozygotes) will generally experience reduced fitness due to the breakdown, through recombination and random assortment, of coadapted gene complexes within their offspring (Vetukhiv 1954). Coadapted genes lead to selection acting not on individual loci but on correlated loci (Barton and Hewitt 1989). Although the general position of many hybrid zones may be determined by environmental gradients, these gradients do not seem to be the only or main factor maintaining these hybrid zones (Hewitt 1989).

A variety of events can prompt a hybrid zone to form and then change its dimensions with time. Allopatric divergence followed by secondary contact is the prominent cause of hybrid zone formation. Bermingham and Avise (1986) used mitochondrial DNA analysis to suggest that a recent (440,000 years) Pleistocene event in the Florida straits connected western with eastern populations of *M. salmoides*. Their study also suggested that recent geologic events have allowed increased dispersal of *M. s. salmoides* into coastal areas of South Carolina.

Relative to dispersal (i.e., gene flow), narrow hybrid zones require strong selection to maintain them whereas a broad hybrid zone is indicative of weak selection. Once a hybrid zone is formed, clinal variation in fitness attributes can cause it to take on a variety of shapes and widths, largely dependent on the relative fitness of the heterozygote and the intensity of gene flow (Endler 1977). The relative balance between hybrid disadvantage, or selection, and the dispersal ability of the organism thus determines the width of a hybrid zone (Barton and Gale 1993).

Coincident clines of phenotypic or genotypic traits are evidence of selective differences between populations. Coincident clines would not be expected unless there was epistasis among the alleles within each parental type (Hewitt 1989). Strong environmental gradients could also cause coincidence. It is unlikely, however, that all traits would respond at the same point along an environmental gradient.

Relative Performance of M. s. salmoides versus M. s. floridanus

The relative performance of each subspecies of largemouth bass has received considerable investigation over the last 30 years. Much of this activity was related to the still controversial fishery management strategy of introducing the Florida subspecies into waters outside its natural distribution (Philipp 1991). A substantial number of authors have compared the survival and growth of the Florida and northern subspecies in various parts of the United States (for example, Clugston 1964; Johnson 1975; and Philipp and Whitt 1991). Other studies have examined the effects of introducing the Florida subspecies into existing populations of the northern subspecies (for example, Inman et al. 1978; Bottroff and Lembeck 1978; Maceina et al. 1988). Many of these evaluations took place in geographic areas where the largemouth bass was not native. Although the genetic identity of experimental animals was not tightly controlled in some past evaluations (Philipp 1991), consistent differences in the performance characteristics of each subspecies were reported.

Data supports the conclusion of Philipp et al. (1985) that there are different degrees of fitness in different thermal environments between the two subspecies; a significant negative correlation existed between heating degree days and percent Florida alleles across the eastern United States. Carmichael et al. (1988) reported that confinement at 2°C for 5 d resulted in 48% mortality for the Florida subspecies, 5% for the Florida (female) hybrid, and 0% mortality for the northern (female) hybrid and the northern subspecies, indicating a dominant cold tolerance trait within the northern subspecies. In laboratory tests with young of the year, survival at 4°C was 85% for the northern subspecies and 0% for the

Florida subspecies (Johnson 1975). Fields et al. (1987) showed that the chronic thermal maximum was 39.2°C for the Florida subspecies and 37.3°C for the northern subspecies. After the introduction of the Florida subspecies into Oklahoma, Gilliland and Whitaker (1991) showed that significantly greater introgression of Florida alleles occurred below 35°N latitude. Gilliland (1994) recommended that introductions of the Florida subspecies into Oklahoma should not occur in areas possessing greater than 3,400 heating degree days.

Relative growth and survival appear positively correlated with the historic latitudinal range of the subspecies. In Illinois ponds, Philipp (1991) observed that the northern subspecies outgrew the Florida subspecies and the two hybrids through age 3. Florida alleles tended to decrease with time in the study populations. In a Texas reservoir, the northern subspecies grew fastest to age 1, however, females of the Florida subspecies were substantially heavier by age 3 and the incidence of Florida alleles increased with time (Maceina et al. 1988). In small ponds in Texas, Harvey (1984) observed a faster growth of the northern subspecies through age 1; F_1 hybrids demonstrated significantly faster growth after age 1. In Oklahoma, Horton and Gilliland (in press) surveyed genotype and size of 251 notable (>3.6 kg) largemouth bass that were turned into taxidermic trophies. They found that 93% were the Florida subspecies or F_1 or F_x hybrids.

The degree of mating isolation determines the potential for hybrid establishment between any two subspecies. The degree of isolation is not clearly defined for the northern and Florida subspecies. F_1 production in populations with both subspecies shows that overlap in spawning times exists (Pelzman 1980). Pond studies in both Illinois and Texas showed that juveniles were dominated by either Florida or northern alleles, possibly a function of assortative mating and environmental variability. Iseley et al. (1987) in Illinois, and Maceina et al. (1988) in Texas, collected and aged juveniles approximately 4 months after spawning. Their hatch-date data indicated that the northern subspecies spawned earlier; however, neither study considered possible differential mortality from time of spawning. Maceina et al. (1988) also noted, based on a surplus of homozygotes, that spawning was not random. Harvey (1984), in Texas, marked each subspecies, and observation of spawning pairs indicated earlier spawning by the Florida subspecies.

The goal of this study was to use allozyme analysis to characterize the relative abundance of alleles from the northern and Florida subspecies in the historic hybrid zone of South Carolina. Genetic analysis of allele frequencies was performed to understand better whether natural selection was a strong force in maintaining the hybrid zone. Two additional objectives were to make initial recommendations on efficient stocking strategies and to suggest further roles for research.

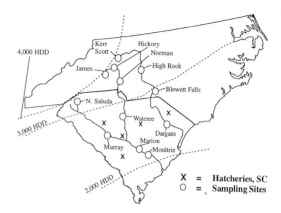

FIGURE 1.—Sampling locations, drainages, and heating degree-day (HDD) isopleths for North and South Carolina.

Methods

Largemouth bass were collected by electrofishing from six reservoirs in South Carolina (Figure 1). These reservoirs were chosen to represent the geographic and climatic diversity that exists within the state. Five reservoirs were located on the state's major drainage, the Santee, and one site, Dargan's Pond, was on the Pee Dee drainage. Additionally, samples of largemouth bass were obtained from on-site broodstock at the state's five warmwater hatcheries.

Generally, collected fish were immediately put on ice until sampling was terminated. Tissue samples were usually taken from specimens within 6 hours of collection. Collection personnel removed the liver and a sample of white skeletal muscle. These tissues were wrapped in tin foil, labeled, and stored in a freezer. Efforts were made to obtain otoliths, and the length, weight, and sex of all specimens.

Tissue samples were manually crushed in equal volumes of 0.25 M Tris(hydroxymethyl)aminomethane–HCl, pH 7.0, and resulting homogenates were used in the analysis. Allele frequency at four loci were determined for each individual with a horizontal starch gel electrophoresis system and histological stains as described and cited by Norgren (1986).

TABLE 1.—Enzyme-specific alleles that were defined as characteristic of the northern (*M. s. salmoides*) and Florida (*M. s. floridanus*) subspecies of largemouth bass.

	Alleles	
Locus	*M. s. salmoides*	*M. s. floridanus*
IDHP-1*	*IDHP-1*100*	*IDHP-1*114*
sAAT-2*	*sAAT-2*100, sAAT-2*110*	*sAAT-2*126, sAAT-2*139*
sMDH-1*	*sMDH-B*100*	*sMDH-B*140*
sSOD-1*	*sSOD-1*147*	*sSOD-1*100*

Genetic nomenclature used was that defined by Shaklee et al. (1990).

Four enzyme loci were analyzed for each largemouth bass based on the results of Philipp et al. (1983). These loci and the enzymes, including international numbers (IUBNC 1984), they code were isocitrate dehydrogenase 1.1.1.42 (*IDHP-1**), aspartate aminotransferase 2.6.1.1 (*sAAT-2**), malate dehydrogenase 1.1.1.37 (*sMDH-B**), and superoxide dismutase 1.15.1.1 (*sSOD-1**). Within each locus, an allele was designated that represented the form most characteristic of the northern and Florida subspecies (Table 1). Four alleles were expressed at *sAAT-2**. To simplify data analysis, this locus was treated as a two-allele system. Both *sAAT-2*100* and *sAAT-2*110* were defined as the northern allele, and *sAAT-2*126* and *sAAT-2*139* were defined as the Florida allele. The *sMDH-B** enzyme was obtained from white skeletal muscle; the other three enzymes were extracted from the liver.

Data defining the distribution of alleles for the same enzymes in North Carolina largemouth bass were obtained from the North Carolina Wildlife Commission. Identical collection, extraction, and analytical procedures were followed.

The frequency occurrence of alleles characteristic of the northern and Florida subspecies was defined for natural waters in South and North Carolina and warmwater hatcheries in South Carolina. Although all South Carolina largemouth bass are hybrids (i.e., F_x) when the entire genome is considered, specific allele combinations were categorized and defined for subsequent analysis. Allele frequencies at the two fixed loci, *IDPH-1** and *sAAT-2** were used to categorize largemouth bass. Individuals were classified as either heterozygous (NF) at either loci or homozygous for northern (NN) or Florida (FF) alleles at both loci.

The dependence of allele frequency on sampling location and gene locus was evaluated using the CATMOD option in SAS (SAS Institute Inc. 1987). Data from Lakes Wateree, Murray and Moultrie,

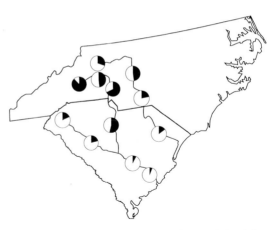

FIGURE 2.—Percentage of alleles characteristic of the northern (shaded) and Florida (clear) subspecies of largemouth bass for two enzyme-specific loci, *IDHP-1** and *sAAT-2**. Sampling locations are given in Figure 1.

South Carolina, were evaluated. The interaction between location and gene locus was also evaluated.

The possibility of size- and sex-specific selection and linkage disequilibrium were assessed in the Lake Wateree population. To examine possible linkage disequilibrium, the independence of frequencies at all two-way combinations of enzyme loci were analyzed using the *G*-test (Sokal and Rohlf 1969). The *G*-test was also used to examine the relative frequencies of northern (N) and Florida (F) alleles by sex and by a large (>375 mm) and small (<325 mm) size-class. To assess coincidence of allele frequencies, the percent of northern alleles at *IDHP-1** was plotted against the percent of northern alleles at *sAAT-2**.

All samples from hatchery populations were combined, and the resulting allele frequencies were compared with observed allele frequencies in lakes Wateree, Murray, and Moultrie. The *G*-test was used to examine the relative frequencies of northern (N) and Florida (F) alleles between the composite hatchery sample and each reservoir.

Results

Reservoir populations of largemouth bass in South Carolina were dominated by alleles that were characteristic of the Florida subspecies. The percent of Florida alleles was highest in Lake Moultrie (Figure 2; Table 2). Florida alleles were dominant in locations in the mildest climatic regions of the study. Of the six inspected reservoirs, northern alleles were dominant only in Lake Wateree.

The percentage of northern alleles tended to in-

TABLE 2.—Largemouth bass allele frequencies at four enzyme-specific loci. The allele(s) that is listed first is either fixed (IDHP-1* and sAAT-2*) or prevalent in the northern subspecies; the second allele is either fixed or prevalent in the Florida subspecies.

State and reservoir	Enzyme							
	IDHP-1*		sAAT-2*		sMDH-1*		sSOD-1*	
	IDHP-1*100	IDHP-1*114	sAAT-2*100,*114	sAAT-2*126,*139	sMDH-B*100	sMDH-B*140	sSOD-1*147	sSOD-1*100
Santee drainage								
North Carolina								
James	28	4	28	4	28	4	29	3
Hickory	19	11	10	20	16	14	20	10
Norman	9	11	18	2	13	7	19	1
South Carolina								
N. Saluda	0	34	11	15	0	34	11	23
Wateree	85	101	98	56	108	72	107	77
Murray	29	147	41	125	13	143	80	96
Marion	2	48	6	44	0	50	15	35
Moultrie	8	386	41	365	0	420	80	314
Pee Dee drainage								
North Carolina								
Kerr Scott	3	19	10	12	4	18	11	11
High Rock	7	13	12	8	11	9	13	7
Blewett Fall	6	12	2	16	1	17	5	13
South Carolina								
Dargans	0	60	19	41	0	60	20	40

crease in the most upstream areas of the Santee River drainage (Figure 2; Table 2). In Lake James, North Carolina, the most upstream reservoir on the Santee drainage, 88% of the alleles at the two diagnostic loci were characteristic of the northern subspecies. This trend toward increasing dominance of northern alleles in the North Carolina part of the drainage was not as prominent in the Pee Dee drainage (Figure 2). The number of fish inspected was less than 20 at all North Carolina reservoirs.

Allele frequencies at all four loci showed a consistent, climate-driven trend in lakes Moultrie, Murray, and Wateree, South Carolina. At each locus, the frequency of northern alleles was highest in Lake Wateree and lowest in Lake Moultrie (Figure 3).

Allele frequencies in lakes Wateree, Murray, and Moultrie were dependent upon both location ($\chi^2 = 641.1$, df = 2, $P < 0.0001$) and gene locus ($\chi^2 = 83.1$, df = 3, $P < 0.0001$). The interaction of location with gene locus was also significant ($\chi^2 = 36.6$, df = 6, $P < 0.0001$), indicating that the relationship between allele frequency and location was dependent upon which gene was examined. A plot of the two fixed alleles, IDHP-1* and sAAT-2*, demonstrated that changes in allele frequency with location were not concordant (Figure 4).

The Lake Wateree population received additional analysis due to the nearly equal frequencies of northern and Florida alleles. Analysis of allele frequencies failed to show a dependence (Table 3), indicating the absence of substantial linkage disequilibrium in the Lake Wateree population. At one of the four loci (sMDH-B*), there was a significant difference in allele frequencies between small and large individuals (Table 4). The frequency of

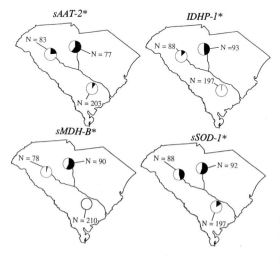

FIGURE 3.—Percentage of alleles characteristic of the northern (shaded) and Florida (clear) subspecies of largemouth bass at four enzyme loci. Number of sampled fish is denoted by N.

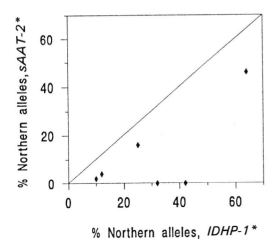

FIGURE 4.—Plot of the frequency of alleles diagnostic of the northern and Florida subspecies of largemouth bass at two fixed enzyme loci, IDHP-1* and sAAT-2*.

northern alleles at the sMDH-B* locus was 0.45 in the small and 0.66 in the large segment of the Lake Wateree population. A sex-related difference in allele frequency was not present at any of the loci.

Hatchery populations in South Carolina were dominated by Florida alleles at the two diagnostic loci (Table 5). The relative frequency of northern alleles in the composite hatchery population was significantly different at all four loci from the Lake Wateree and Lake Moultrie populations (Table 6). The composite hatchery sample most closely resembled the Lake Murray population, yet allele frequencies were significantly different at three of four loci. In all three comparisons, the sMDH-B* locus exhibited the greatest difference between the hatchery and reservoir populations.

Nine largemouth bass from South Carolina expressed a third, rare allele at the IDHP-1* locus. All specimens that expressed this uncommon allele were precluded from this investigation's data base, pending further investigation.

TABLE 4.—Comparison (G-test) of the frequency (N) of the northern and Florida alleles at four allozyme loci in small (<325 mm) and large (>375 mm) largemouth bass from Lake Wateree, South Carolina. A significance level of P greater than 0.05 is indicated by a dagger.

Allele	Small fish N	Large fish N	G-value
IDHP-1*100	15	21	0.01
IDHP-1*114	27	37	
sAAT-2*100,*110	23	26	0.27
sAAT-2*126,*139	15	12	
sMDH-B*100	19	38	4.10†
sMDH-B*140	23	20	
sSOD-1*147	26	35	0.06
sSOD-1*100	18	21	

Discussion

Allozyme survey data confirmed that South and North Carolina were part of a broad hybrid zone between the northern and Florida subspecies of largemouth bass. Allele frequencies tended to vary in response to a climatic gradient. Populations in the milder climatic regions of South Carolina were dominated by Florida alleles, whereas the Lake James, North Carolina, population in the upper reaches of the Santee drainage was dominated by northern alleles. Thus, along these drainages, the data tended to support the hypothesis of Philipp et al. (1985) of different degrees of fitness in different thermal environments.

Available data did not firmly support the existence of parallel clines in the Santee and Pee Dee drainage. Specifically, northern alleles were not dominant in Kerr Scott reservoir, the northernmost site on the Pee Dee drainage, even though this site is climatically similar to Lake James on the Santee drainage, a site where northern alleles dominate. Small sample sizes or localized collections may explain these differences between the two drainages. More intense sampling along these drainages, including riverine sections, is necessary to determine whether the clines are parallel and related to climate in a manner consistent with Philipp's hypothesis. Also, the reason for the lack of northern phenotypes at North Saluda Lake, the coldest South Carolina site, was not clear. Localized introductions and differential gene flow on the Saluda and Catawba subbasins of the Santee drainage are possible explanations. Additional sampling in the Saluda subbasin is needed.

Most hybrid zones are formed through a balance between dispersal and selection; gene flow tends to

TABLE 3.—Independence of allele frequencies on Lake Wateree, South Carolina, as measured by the G-test (Sokal and Rohlf 1969) with 4 df. A significance level of P greater than 0.05 is indicated by a dagger.

Locus 1	Locus 2	G-value
sAAT-2*	IDHP-1*	5.72†
sAAT-2*	sMDH-B*	5.80†
sAAT-2*	sSOD-1*	5.58†
IDHP-1*	sMDH-B*	8.02†
IDHP-1*	sSOD-1*	5.20†
sMDH-B*	sSOD-1*	3.44†

TABLE 5.—Allele frequency of largemouth bass in five South Carolina hatcheries at two enzyme loci that show fixed differences between the northern and Florida subspecies. The first listed allele is characteristic of the northern subspecies.

Hatchery	Enzyme			
	IDHP-1*		sAAT-2*	
	IDHP-1*100	IDHP-1*114	sAAT-2*100,*114	sAAT-2*126,*139
Barnwell	12	24	12	28
Campbell	4	16	25	15
Cheraw	11	7	9	9
Heath Springs	7	13	12	8
Shirey	0	20	1	5
Total	34	80	39	65

homogenize populations, reducing geographic variation in allele frequency whereas selection restores geographic differentiation. There are several models of this type of hybrid zone that differ in the magnitude or nature of selection operating across the zone and yield different expectations. We will discuss these models separately with reference to our results to determine whether we can infer the nature of the forces producing the hybrid zone in largemouth bass in South Carolina.

Neutral cline models posit that selection at individual loci is negligible. These models predict that, because there is no selection the rate of diffusion will be the same for all loci and allele frequency clines at all gene loci will exhibit the same pattern and be centered in the same geographic locality (i.e., they exhibit coincidence sensu Barton and Hewitt 1985). The largemouth bass hybrid zone in South Carolina does not conform to these expectations. If all loci were diffusing at the same rate we would expect a linear regression with a slope of 1 between the two fixed loci, $IDHP-1*$ and $sAAT-2*$. This was not the case (Figure 4). The frequency of northern alleles at the $sAAT-2*$ locus decreased more quickly than the frequency at the $IDHP-1*$ locus. These results indicate that the hybrid zone in South Carolina is probably influenced and perhaps maintained by a balance been dispersal and natural selection.

Neutral cline models also predict that allele-frequency clines at each locus will diffuse with time at a rate proportional to the level of gene flow. Because this study is the first to describe the presence of a hybrid zone between the northern and Florida strains of largemouth bass in South Carolina, we can not assess this prediction. This study, however, can serve as a benchmark to test this hypothesis in the future.

There are basically two types of hybrid zone models that incorporate a balance between selection and gene flow: the tension zone model and the hybrid mosaic model. In the tension zone model selection occurs against hybrid genotypes as alleles from one population disperse into the incompatible genetic background of the other population. This model predicts that clines should be coincident and that if selection is strong relative to the rate of recombination there should be nonrandom association among gene loci near the center of the clines. Barton and Hewitt (1985) reviewed the hybrid zone literature and showed that many examples of hybrid zones exhibit both coincidence and linkage disequilibrium near the center of the zone. Under the hybrid mosaic model the shape and position of allele frequency clines is a function of the manner in

TABLE 6.—Comparison (G-test) of the frequency of the northern alleles at four allozyme loci in a sample of largemouth bass from hatcheries and from three natural waters. Sample size is indicated in parentheses. All G-tests were significant at a P less than 0.01 except those noted with a dagger were significant at a P less than 0.05.

Locus	Frequency		G-value
	Hatchery	Reservoir	
Wateree			
sAAT-2*	0.38 (52)	0.64 (75)	17.48
IDHP-1*	0.29 (59)	0.47 (91)	9.76
sMDH-B*	0.24 (59)	0.61 (88)	40.58
sSOD-1*	0.39 (59)	0.58 (90)	12.67
Murray			
sAAT-2*	0.38 (52)	0.25 (81)	4.42†
IDHP-1*	0.29 (59)	0.18 (81)	4.60†
sMDH-B*	0.24 (59)	0.09 (76)	11.89
sSOD-1*	0.39 (59)	0.45 (86)	1.18†
Moultrie			
sAAT-2*	0.38 (52)	0.10 (203)	39.79
IDHP-1*	0.29 (59)	0.02 (191)	69.03
sMDH-B*	0.24 (59)	0.00 (204)	81.12
sSOD-1*	0.39 (59)	0.20 (197)	13.59

which natural selection affects different linkage groups in the genome. Smooth clines are formed under this model only when there is a uniform change in the selection gradient, and linkage disequilibria among loci may or may not occur.

As cited above, we found no evidence of coincidence among gene loci in the Santee River system. We also found no evidence of linkage disequilibria among gene loci in the Lake Wateree population. This population is near the center of the hybrid zone and is therefore the site predicted to be most likely to exhibit linkage disequilibria. Because neither linkage disequilibria nor coincidence among gene loci were evident in this study we conclude that the hybrid zone in South Carolina is probably not a tension zone. By inference then, we propose that the largemouth bass hybrid zone in South Carolina is the result of a geographic selection gradient.

This survey provided preliminary evidence of size-based selective forces operating on the Lake Wateree population. The frequency of northern alleles at the $sMDH\text{-}B^*$ locus was greater in large than in small fish. Results of size-based selection are preliminary, due to the chance of type II error, sample size, and a minimal size difference between the two classes. Future studies will obtain greater numbers of large, Lake Wateree largemouth bass to test this observation further.

The survey also provided a preliminary suggestion that selection was operating within the state's hatcheries to modify allele frequencies. Results demonstrated a significant difference between the allele frequencies at all four loci of the hatchery and Lake Wateree population, the presumptive origin of hatchery populations. The decline in the frequency of northern alleles occurred in all hatcheries at all loci, indicating this effect is not due to genetic drift. If due to selection, these results indicate that selection upon these loci may be sufficiently large to change the allele frequency of a population in only a few generations. Smith et al. (1983) showed that frequency of alleles at $MDH\text{-}A^*$ returned to natural frequencies within 10 years after a human-imposed stress was eliminated from a South Carolina largemouth bass population.

The hypothesis that selection caused these changes in hatchery populations was based on the assumption that Lake Wateree was the original source of broodstock for the hatchery program. Discussion with hatchery personnel indicated that broodstock was primarily obtained from Lake Wateree during the last decade, but documentation of broodstock origin and handling was not maintained.

Although the above conclusions are tentative, this analysis points out the need of maintaining genetic information within a hatchery program. Documentation of genetic strains would have allowed a more thorough evaluation of the effects of selection within the hatchery system. Also, selection would have to be considered if maintenance of specific genotypes was a goal of a hatchery program.

The existence of a geographically dependent allelic cline in South Carolina and the allelic differences between the composite hatchery stock and reservoir populations (Table 6) suggest the strategy of regionalizing hatchery populations to maximize fitness of stocked fish. The data suggest that fish with a high incidence of Florida alleles have greater fitness in the milder climatic areas of South Carolina. Current hatchery practices in South Carolina include emphasis on private pond stocking, a semi-domesticated broodstock, and random assignment of stocking sites for each hatchery. In a simple vision of the regionalization strategy, broodstock with a high percentage of Florida alleles would be stocked in the warmer regions of the state. Conversely, largemouth bass with a higher percentage of northern alleles would be stocked in cooler regions. Before this strategy is implemented, the state will conduct a reciprocal transplant experiment to attempt to determine the selective differences between two potential broodstock sources.

Information on the magnitude of selective forces operating on each genotype is a major data need. The broad allelic cline observed in this study could have been maintained by very small selective forces if gene flow between systems was minimal (Endler 1977). That largemouth bass have a restricted home range and South Carolina reservoirs receive minimal stockings tends to support the concept of minimal gene flow. Historic maintenance of the hybrid zone (Bailey and Hubbs 1949), differential selection on loci, and a difference between hatchery and reservoir stocks suggest that selective forces are significant. Defining the magnitude of selection will further assist management by defining the relative fitness of a specific genotype.

Information obtained in this study supported the view that populations of largemouth bass in South Carolina are hybrids (i.e., intergrades) between the northern and Florida subspecies. However, an allelic cline, apparently responding to local climatic conditions, contributed a diverse assemblage of alleles to the South Carolina populations. Determining the level of stock transfer that is genetically acceptable is an issue raised by this study.

According to theory, interstock transfers will decrease the overall fitness of the receiving population. Even without coadaptation within the population, introgression of alleles can reduce the fitness of the population (Philipp 1991). However, adding individuals from the same stock does not cause a decrease in fitness if a correct breeding protocol is followed. The potential for negative effect is minimized if the transferred population constitutes a small percent of the receiving population. Localizing broodstocks is an initial step that South Carolina hatcheries can immediately implement to minimize the risk of negative genetic interactions associated with stock transfers. Long-term performance evaluations are required to document the effect of stock transfers fully.

Acknowledgments

This study was supported by the U. S. Fish and Wildlife Service through the Sport Fish Restoration Act. Initial allozyme analysis and training in the analytical procedures was supplied by the Southeastern Cooperative Genetics Project at Auburn University. Drew Robb, Phil Kirk, Randy Geddings, Gene Hayes, Gayland Penny, Dick Christie, Robert Stroud, Dan Crochet, Scott Lamprecht, and Jim Glenn actively participated in reservoir collections in South Carolina. Mac Watson, Farrel Beck, Mitch Manis, Jimmy Singleton, Will Catoe, and Charles Gooding provided assistance obtaining specimens from warmwater hatcheries in South Carolina. We thank the North Carolina Wildlife Resources Commission for allowing us to use their allozmye survey data. Dave Philipp provided a very helpful review of the initial draft of the manuscript.

References

Avise, J. C., and M. H. Smith. 1974. Biochemical genetics of sunfish 1: geographic variation and subspecific intergradation in the bluegill (*Lepomis macrochirus*). Evolution 28:42–56.

Bailey, R. M., and C. L. Hubbs. 1949. The black basses (*Micropterus*) of Florida, with description of a new species. Occasional Papers of the Museum of Zoology, University of Michigan 516:1–40.

Barton, N. H., and K. S. Gale. 1993. Genetic analysis of hybrid zones. Pages 13–45 in R. G. Harrison, editor. Hybrid zones and the evolutionary process. Oxford University Press, New York.

Barton, N. H., and G. M. Hewitt. 1985. Analysis of hybrid zones. Annual Review of Ecology and Systematics 16:113–148.

Barton, N. H., and G. M. Hewitt. 1989. Adaptation, speciation and hybrid zones. Nature 341:497–503.

Bermingham, E., and J. C. Avise. 1986. Molecular zoogeography of freshwater fishes in the southeastern United States. Genetics 113:939–965.

Bottroff, L. J., and M. E. Lembeck. 1978. Fishery trends in reservoirs of San Diego County, California, following the introduction of Florida largemouth bass, *Micropterus salmoides floridanus*. California Fish and Game 64:4–23.

Carmichael, G. J., J. H. Williamson, C. A. Woodward, and J. R. Tomasso. 1988. Responses of northern, Florida, and hybrid largemouth bass to low temperature and low dissolved oxygen. Progressive Fish-Culturist 50:225–231.

Clugston, J. P. 1964. Growth of the Florida largemouth bass, *Micropterus salmoides floridanus*, and the northern largemouth bass, *M. s. salmoides*, in subtropical Florida. Transactions of the American Fisheries Society 93:146–154.

Endler, J. A. 1977. Geographic variation, speciation, and clines. Princeton University Press, Princeton, New Jersey.

Fields, R., S. L. Lowe, C. Kaminski, G. S. Whitt, and D. P. Philipp. 1987. Critical and chronic thermal maxima of northern and Florida largemouth bass and their reciprocal F_1 and F_2 hybrids. Transactions of the American Fisheries Society 16:856–863.

Gilliland, E. R. 1994. Experimental stocking of Florida largemouth bass into small Oklahoma reservoirs. Proceedings of the Annual Conference Southeastern Association of Fish and Wildlife Agencies 46(1992): 487–494.

Gilliland, E. R., and J. Whitaker. 1991. Introgression of Florida largemouth bass into northern largemouth bass populations in Oklahoma reservoirs. Proceedings of the Annual Conference Southeastern Association of Fish and Wildlife Agencies 43(1989):182–190.

Harrison, R. G. 1993. Hybrids and hybrid zones: historical perspective. Pages 3–12 in R. G. Harrison, editor. Hybrid zones and the evolutionary process. Oxford University Press, New York.

Harvey, W. D. 1984. An electrophoretic evaluation of bi-subspecific populations of largemouth bass in small impoundments in Texas. Doctoral dissertation. Texas A&M University, College Park.

Hewitt, G. M. 1989. The subdivision of species by hybrid zones. Pages 85–110 in D. Otte and J. Endler, editors. Speciation and its consequences. Sinauer Associates, Inc., Sunderland, Massachusetts.

Holbrook, J. E. 1860. Ichthyology of South Carolina, volume 1. Russell and Jones, Charleston, South Carolina.

Horton, R. A., and E. R. Gilliland. In press. Monitoring trophy largemouth bass in Oklahoma using a taxidermist network. Proceedings of the Annual Conference Southeastern Association of Fish and Wildlife Agencies 47(1993).

Inman, C. R., R. C. Dewey, and P. P. Durocher. 1978. Growth comparisons and catchability of three largemouth bass strains. Proceedings of the Annual Conference Southeastern Association of Fish and Wildlife Agencies 30(1976):1–17.

Iseley, J. J., R. L. Noble, J. B. Koppleman, and D. P. Philipp. 1987. Spawning period and first-year growth of northern, Florida, and intergrade stocks of largemouth bass. Transactions of the American Fisheries Society 116:757–762.

IUBNC (International Union of Biochemistry, Nomenclature Committee). 1984. Enzyme nomenclature. Academic Press, San Diego, California.

Johnson, D. L. 1975. A comparison of Florida and northern largemouth bass in Missouri. Doctoral dissertation. University of Missouri, Columbia.

Lacepède, B. G. E. 1798. Histoire naturelle des poissons. Paris.

Lee, D. L., C. R. Gilbert, C. H. Hocutt, R. E. Jenkins, D. E. McAllister, and J. R. Stauffer, Jr. 1980. Atlas of North American freshwater fishes. North Carolina State Museum of Natural History, Raleigh.

Maceina, M. J., B. R. Murphy, and J. J. Iseley. 1988. Factors regulating Florida largemouth bass stocking success and hybridization with northern largemouth bass in Aquilla Lake, Texas. Transactions of the American Fisheries Society 117:221–231.

Moore, W. S., and J. T. Price. 1993. Nature of selection in the northern flicker hybrid zone and its implications for speciation theory. Pages 196–225 in R. G. Harrison, editor. Hybrid zones and the evolutionary process. Oxford University Press, New York.

Norgren, K. G. 1986. Biochemical genetic survey of largemouth bass populations in Alabama. Master's thesis. Auburn University, Alabama.

Pelzman, R. J. 1980. Impact of Florida largemouth bass, *Micropterus salmoides floridanus*, introductions at selected northern California reservoirs with a discussion of the use of meristics for detecting introgression and for classifying individual fish of intergraded populations. California Fish and Game 66:133–162.

Philipp, D. P. 1991. Genetic implications of introducing Florida largemouth bass, *Micropterus salmoides floridanus*. Canadian Journal of Fisheries and Aquatic Sciences 48(Supplement 1):58–65.

Philipp, D. P., W. F. Childers, and G. S. Whitt. 1983. Biochemical genetic evaluation of two subspecies of largemouth bass, *Micropterus salmoides*. Transactions of the American Fisheries Society 112:1–20.

Philipp, D. P., W. F. Childers, and G. S. Whitt. 1985. Correlations of allele frequencies with physical and environmental variables for populations of largemouth bass, *Micropterus salmoides* (Lacepède). Journal of Fish Biology 27:347–365.

Philipp, D. P., and G. S. Whitt. 1991. Survival and growth of northern, Florida, and reciprocal F_1 hybrid largemouth bass in central Illinois. Transactions of the American Fisheries Society 120:58–64.

SAS Institute. 1987. SAS/STAT guide for personal computers, version 6. SAS Institute, Cary, North Carolina.

Shaklee, J. B., F. W. Allendorf, D. C. Morizot, and G. S. Whitt. 1990. Genetic nomenclature for protein-coding loci in fish. Transactions of the American Fisheries Society 119:2–15.

Smith, M. H., M. W. Smith, S. L. Scott, E. H. Liu, and J. J. Jones. 1983. Rapid evolution in a post-thermal environment. Copeia 1983:193–197.

Sokal, R. R., and F. J. Rohlf. 1969. Biometry. W. H. Freeman, San Francisco, California.

Vetukhiv, M. 1954. Integration of the genotype in local populations of three species of Drosophila. Evolution 8:241–251.

Fitness and Performance Differences between Two Stocks of Largemouth Bass from Different River Drainages within Illinois

DAVID P. PHILIPP AND JULIE E. CLAUSSEN

Illinois Natural History Survey, Center for Aquatic Ecology
607 East Peabody Drive, Champaign, Illinois 61820, USA; and
Program in Natural Resource Ecology and Conservation Biology
College of Agriculture, University of Illinois, Urbana, Illinois 61801, USA

Abstract.—Evolutionary theory predicts that through natural selection, populations of individuals within any species become genetically tailored to their local environments, attempting to maximize fitness traits through local adaptation. As a result of this selection, together with drift, populations differentiate from one another and form divergent stocks. For truly effective management it is the stock, not the species as a whole, that must be considered the operational unit for management concern. Unfortunately, many management programs fail to recognize this and, in an attempt to improve native fisheries, continue to mix stocks of fish indiscriminately. Such stock transfers negatively affect the genetic resources of a species by diminishing genetic diversity among populations and likely decreasing the fitness of resident populations through outbreeding depression. Few empirical studies have attempted to quantify fitness trait differences among stocks, and even fewer have attempted to assess the effects of stock transfers on these traits. Our long-term studies have documented the genetic and physiological differences that exist among stocks of largemouth bass *Micropterus salmoides* and have demonstrated that nonnative stocks exhibit poorer fitness and performance traits than do native stocks. To date, however, these studies have only compared the two subspecies, northern largemouth bass *M. s. salmoides* and Florida largemouth bass *M. s. floridanus*, in Illinois. We now show that there are performance and fitness differences among stocks of largemouth bass within the same subspecies that originate from two different river drainages in Illinois. These results suggest strongly that the geographic boundaries of largemouth bass stocks may be much smaller than previously believed.

It is now established that a species rarely exists as a single, randomly breeding unit, but rather as a mosaic of genetically divergent units, usually interconnected by some level of gene flow (Soulé 1986). Because of this pattern of restricted gene flow, the resulting mosaicism within species provides the opportunity for natural selection to tailor populations genetically to their environments, i.e., the process of local adaptation (Wright 1931).

In the last few years the literature devoted to describing the distribution of intraspecific genetic variation in fish has increased dramatically (see Ryman and Utter 1987). Electrophoretic analyses of proteins have established that there can be substantial levels of genetic variation among different populations of a single species (e.g., Philipp et al. 1983; Gharrett et al. 1988; Beacham et al. 1989; Krueger et al. 1989). These studies support the hypothesis that populations in different geographic locations have different genetic and physiological characteristics, each population having evolved an adaptive suite of characters for its specific environment (Philipp 1991).

It is well established that the genetic variability inherent in a species can be partitioned into two components (Wright 1978). The first component is the variation that exists among individuals within populations. Within-population variation is lost through selection or genetic drift (Allendorf et al. 1987), processes that are exacerbated when population size becomes small. The second component is the variation that exists among populations. Among-population variation is lost when previously isolated populations are mixed, resulting in the homogenization of the two previously distinct entities (Altukhov and Salmenkova 1987; Campton 1987).

Besides the loss of genetic variation, mixing two groups can also result in outbreeding depression. Outbreeding depression is a loss of fitness in the offspring produced as a result of interbreeding between two groups because the parents are too distantly related (Templeton 1986). This loss in fitness may result from a disruption in coadapted gene complexes that were derived through many years of natural selection and produced advantageous local adaptations.

The propagation of sport-fish species in hatcheries has brought about unique circumstances that permit the potential transfer and mixture of populations across large geographic areas. Jurisdiction for these activities falls to the federal, state, tribal, and local governments, as well as private citizens. Unfortunately, little coordination exists among those entities that govern quality-control aspects of fisheries management such as the genetic source of broodstocks. In fact, in the past too little concern has been given to the long-term consequences that stocking efforts may have on the genetic integrity of local populations (Philipp et al. 1993).

The Stock Concept (Kutkuhn 1981; Philipp 1991) states that because species are composed of multiple distinct units (e.g., stocks, populations, demes), for truly effective conservation it is these units, not the species as a whole, that must be considered the operational unit of concern for management programs. One of the most hotly debated and pressing questions of our day deals with defining the number and location of these units (Philipp et al. 1993). Adequate definition of units would be facilitated by knowing whether or not fitness differences exist among the proposed groupings. The purpose of this study was to provide information for one specific test case by assessing in different latitudinal regions of Illinois the fitness of two stocks of northern largemouth bass *Micropterus salmoides salmoides*, one from the Fox River drainage in northern Illinois and one from the Big Muddy River drainage in southern Illinois.

Methods

Production of Experimental Stocks

Three populations from northern Illinois and three populations from southern Illinois were selected as broodstock sources for largemouth bass (Figure 1). These populations were selected based upon their latitude of origin. The three northern lakes, Lake Catherine, Lake Marie, and Grass Lake, are all located in the extreme north of Illinois on the Fox River, a tributary to the upper Illinois River. The three southern lakes, Crab Orchard Lake, Devil's Kitchen Lake, and Washington County Lake, are all within a different major Illinois watershed, the Big Muddy River, much farther south. Both the Fox–Illinois and Big Muddy river systems are connected to the Mississippi River; the two broodstock sources are separated by approximately 380 river miles. During fall 1983 and spring 1984 broodstock from all six lakes were collected by electrofishing and transported to the production ponds at the Illinois Natural History Survey's Aquatic Research Field Laboratory. Broodstock from lakes within the same region were pooled in spring 1984 and used to produce two experimental stocks of largemouth bass, a northern Illinois stock (NILMB) and a southern Illinois stock (SILMB). In fall 1984, when fingerlings had reached sufficient size for harvesting, production ponds were drained.

FIGURE 1.—Locations of broodstock sources for northern (black diamonds) and southern (white diamonds) Illinois largemouth bass, as well as for study ponds in northern (black circles), central (gray circles), and southern (white circles) Illinois.

Individuals were weighed and measured and fins were clipped for visual identification (NILMB, right pelvic fin; SILMB, left pelvic fin). A sample ($N = 40$) of young-of-year individuals from each stock was removed to compare sizes and to determine allele frequencies at six potentially polymorphic loci.

Study Populations

Ten research ponds void of fish and ranging in size from 0.3 to 7 ha were designated by the Illinois Department of Conservation to be used as study sites in three regions of the state—north, central, and south (Figure 1). To establish the experimental populations, equal numbers of equal-size, fin-clipped individuals of each of the two stocks were introduced into each pond at a rate of 125 fingerlings/ha. In addition, adult bluegill *Lepomis macrochirus* and redear sunfish *L. microlophus* were introduced into each pond to establish a forage base. Study ponds were sampled by use of a boat-mounted electrofishing unit each summer and fall during 1985–1988. To monitor survival and growth of the fish initially stocked, collected fish were weighed, measured, and identified to stock by fin-clip. Once the originally stocked fish matured and spawning occurred, reproductive success of each stock in each pond was determined. For this, individual young-of-year (naturally spawned offspring from the test ponds) were collected through a combination of electrofishing and seining and were wrapped in foil, frozen, and stored at $-20°C$ until genetic analysis could be performed.

Electrophoretic Analysis

For electrophoretic analysis, white skeletal muscle and liver samples were prepared as described in Philipp et al. (1979). Each tissue extract was subjected to starch gel electrophoresis in conjunction with histochemical staining procedures to determine the allele frequencies at six potentially polymorphic enzyme loci: *sAAT-2** (which codes for aspartate aminotransferase, enzyme number 2.6.1.1. [IUBNC 1984]); *CK-C** (creatine kinase, 2.7.3.2); *GPI-B** (glucose-6-phosphate isomerase, 5.3.1.9); *IDHP-B** (isocitrate dehydrogenase, 1.1.1.42); *MDH-B** (malate dehydrogenase, 1.1.1.37), and *SOD-1** (superoxide dismutase, 1.15.1.1). Nomenclature used was that described by Shaklee et al. (1990). Because the NILMB and SILMB proved to have different allele frequencies at only the *GPI-B** and *MDH-B** loci, naturally spawned test pond young-of-year were analyzed for only those loci.

Statistical Analysis

The electrofishing samples for each pond provided relative abundance data for the two stocks and were used to assess relative survival after 1 year (1985) and 2 years (1986). Differences in survival of stocks were assessed two ways. First, for each collection, the distribution of NILMB and SILMB was tested against the null hypothesis of equal survivorships using a chi-square test. Second, for each year separately, the ratio of NILMB to SILMB (numbers of individuals sampled) in the ponds of each region were compared using a Kruskal–Wallis distribution-free, nonparametric one-way analysis of variance (ANOVA).

The relative growth of each stock in each pond (measured both in total length and in weight) was determined by calculating a ratio (NILMB:SILMB) of mean values. The significance of stock differences was then determined for each region independently using a Wilcoxon signed-rank test; ratios greater than 1 indicated greater growth for NILMB than for SILMB, and the reverse was indicated by ratios less than 1.

Relative fitness of the two stocks was assessed by determining the relative reproductive success of the two stocks, for which equal numbers were introduced to each pond at the beginning of the experiment. The contribution of alleles from NILMB and SILMB (average of values calculated from the observed allele frequencies at *GPI-B** and *MDH-B**) determined each year for each pond, was tested against expected equal values using a chi-square test. In addition, the ratios of NILMB:SILMB contribution of alleles in the ponds of each region were compared using a Kruskal–Wallis distribution-free, nonparametric one-way ANOVA. The relative reproductive success of the two stocks was also assessed in the same manner, but was based upon unequal numbers of parents, which was determined from the relative abundances found in the electrofishing samples.

Results

Production of Experimental Stocks

When harvested from production ponds in fall 1984, NILMB and SILMB young-of-year were of similar sizes $P > 0.05$ for total length and weight, (Scheffe's statistic) but, as expected, had quite different allele frequencies at two loci (Table 1). In fact, electrophoretic analyses of the two experimental stocks revealed that 70% of SILMB fingerlings were heterozygous at the *GPI-B** locus compared with 0% of the NILMB fingerlings, and that 80% of the SILMB fingerlings were heterozygous at the *MDH-B** locus, as opposed to only 15% of the NILMB fingerlings. Therefore, if the external marking system had failed due to regeneration of the pelvic fin clips, identification by electrophoretic analysis was still possible between the two stocks more than 90% of the time. In addition, and very importantly, this genetic difference between the two stocks also permitted an assessment of their reproductive contribution to future generations.

TABLE 1.—Total length (mean ± SD), weight (mean ± SD), and allele frequencies at *MDH-B** and *GPI-B** of northern (NILMB) and southern (SILMB) Illinois largemouth bass at harvest in 1984, as well as the allele frequencies of the mixed populations introduced to each study pond.

Stock	Total length (mm)	Weight (g)	*MDH-B**	*GPI-B**
NILMB ($N = 40$)	54.5 ± 2.5	1.71 ± 0.22	0.925 0.075	1.000 0.000
SILMB ($N = 40$)	55.9 ± 2.4	1.74 ± 0.27	0.600 0.400	0.650 0.350
Initial mixed populations			0.763 0.237	0.825 0.175

Survival of the Stocks

The relative abundances of the NILMB and SILMB collected from each of the study populations during 1985 and 1986 (Figure 2; Table 2) showed that in the northern Illinois study ponds, survival of the NILMB stock was significantly greater than that of the SILMB; the reverse was true in the southern Illinois ponds (ANOVA for 1985, $F = 7.318$, $P = 0.026$; for 1986, $F = 6.357$, $P = 0.048$; chi-square results for individual ponds are indicated in Table 2). No significant difference in survival was observed between the stocks in the central Illinois study ponds.

Growth of the Stocks

The growth of the two stocks are presented as the ratio of NILMB to SILMB (for both total length and weight) for each sampling date, for each pond, in each of the three regions (Table 3). A value greater than one indicates that the NILMB were longer or heavier than were the SILMB; the reverse is true for a ratio less than one. The results were similar to the survival data: in each of the northern study ponds during each year sampled, growth of the NILMB stock was greater than that of the SILMB; the reverse was true in the southern ponds, and the overall patterns conferred regional significance ($P < 0.05$, Wilcoxon signed-rank test). No consistent differences in growth were observed in the central study populations.

TABLE 2.—Relative abundance of two stocks of largemouth bass, northern (NILMB) and southern (SILMB) Illinois largemouth bass, among ten study ponds in northern (N), southern (S), and central (C) regions of Illinois. Values presented are actual numbers of individuals sampled during the electrofishing collections. Distributions significantly different from equal abundance ($P < 0.05$) are designated with an asterisk.

Pond	Pond size (ha)	Stock	Abundance (N) 1985	1986
		North		
N-1	7.0	NILMB	59*	35*
		SILMB	27	14
N-2	0.4	NILMB	20*	18*
		SILMB	12	11
N-3	0.9	NILMB	11	Winterkill
		SILMB	8	
		Central		
C-1	3.3	NILMB	25	16
		SILMB	29	16
C-2	1.7	NILMB	8	6
		SILMB	9	5
C-3	0.3	NILMB	13	9
		SILMB	9	11
C-4	0.8	NILMB	14	Winterkill
		SILMB	18	
		South		
S-1	1.4	NILMB	16*	23*
		SILMB	31	34
S-2	0.5	NILMB	10	12*
		SILMB	14	26
S-3	2.1	NILMB	7	Winterkill
		SILMB	10	

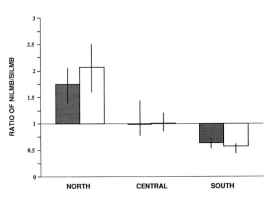

FIGURE 2.—Ratios of relative abundance of two stocks of Illinois largemouth bass, northern (NILMB) to southern (SILMB), in electrofishing collections for 1985 (shaded) and 1986 (open) in three study regions of Illinois.

TABLE 3.—Relative growth of largemouth bass among study ponds in different regions of Illinois. Total length (TL) is in millimeters and weight is in grams.

Pond and growth parameter	Year-class			3-Year average
	1985	1986	1987	
North				
N-1				
TL	1.05	1.06	1.10	
Weight	1.29	1.29	1.15	
N	80	49	21	
N-2				
TL	1.05	1.06	1.04	
Weight	1.13	1.34	1.06	
N	32	29	15	
N-3				
TL	1.15	Winterkill		
Weight	1.60			
N	19			
Central				
C-1				
TL	0.971	1.01	0.985	
Weight	0.929	1.12	0.942	
N	54	32	16	
C-2				
TL	0.984	0.948	0.957	
Weight	0.955	0.875	0.939	
N	17	11	8	
C-3				
TL	1.05	0.966	0.985	
Weight	1.02	0.940	0.937	
N	22	20	9	
C-4				
TL	1.02	Winterkill		
Weight	1.09			
N	32			
South				
S-1				
TL	0.938	0.930	0.993	
Weight	0.798	0.865	0.985	
N	47	57	47	
S-2				
TL	0.987	0.968	0.981	
Weight	0.927	0.900	0.964	
N	24	38	23	
S-3				
TL	0.941	Winterkill		
Weight	0.802			
N	17			
Summary average ratio across ponds and years				
North ponds				
TL				1.07[a]
Weight				1.27[a]
Central ponds				
TL				0.988
Weight				0.975
South ponds				
TL				0.963[a]
Weight				0.892[a]

[a] Significantly different ($P < 0.05$) as determined by Wilcoxon signed-rank test of the 1985–1987 individual pond ratios within each region; test based on 7 values for the northern, 10 values for the central, and 7 values for the southern regions.

Relative Fitness of the Stocks

Samples of young-of-year were collected (if present) from each study population during late summer in 1987 and 1988. For each pond's natural production, electrophoretic analysis was used to determine the allele frequencies at the *GPI-B** and *MDH-B** loci. Based on those values and the frequencies of the two original NILMB and SILMB stocks, we calculated the actual genetic contribution made to each year-class by each of the two stocks. Results clearly indicate (Table 4) that the fitness of the NILMB was significantly greater than that of the SILMB in northern Illinois and the reverse true in southern Illinois ($F = 9.379$, $P = 0.009$, Kruskal–Wallis one-way ANOVA of both years' values); no consistent differences were observed in central Illinois.

Furthermore, because fitness is a combination of survival and reproductive success, we have also assessed the relative reproductive success of the two stocks taking into account the relative abundances of each stock in each pond determined from the electrofishing samples (Figure 3). Based upon relative abundance values, the reproductive successes of the two stocks were not significantly different ($F = 2.538$, $P = 0.281$, Kruskal–Wallis one-way ANOVA of both years' values) although the values were still in the same direction as the first set of fitness estimates. These results indicate that survival differences played a much greater role in establishing the differential fitnesses of the two stocks in this experiment than did differences in reproductive success.

Discussion

Based upon evolutionary theory, our hypothesis has been that natural selection genetically tailors populations of largemouth bass to specific local environments by allowing them to try continually to optimize fitness traits. Furthermore, we have predicted that when comparing geographically divergent stocks that are competing in the same environment, the greatest fitness (a combination of survival and reproductive success) and likely best performance (any set of other characteristics we choose to measure, e.g., growth rate or ultimate size), should then be exhibited by the local stock. Our previous work comparing stocks of the two subspecies northern largemouth bass *M. s. salmoides* and Florida largemouth bass *M. s. floridanus* in central Illinois has confirmed those predictions at one level (Philipp 1991; Philipp and Whitt 1991). Those studies, however, used stocks that were very distant geographically. The question then becomes, how geographically close can two stocks exist and yet still show demonstrable local adaptations?

Different thermal conditions and photoperiods exist between regions of Illinois. Based upon those differences, we would predict that populations of largemouth bass from the northern extremes of the state would have different fitness and performance characteristics than do largemouth bass from the southern extremes. Our results confirm this prediction: NILMB demonstrated better survival, reproductive success, and growth than did SILMB in northern Illinois, and the reverse was true in southern Illinois. These findings demonstrate that local adaptations can and do exist among different stocks within a relatively small geographic area.

TABLE 4.—Relative fitness of two stocks of largemouth bass, northern (NILMB) and southern (SILMB) Illinois largemouth bass, among study ponds in three different regions in Illinois. Relative fitness is based upon genetic contribution to year-classes calculated from observed allele frequencies at the *MDH-B** and *GPI-B** loci. The NILMB values are normalized to 1. Values significantly different ($P < 0.05$) are designated with an asterisk.

Pond and stock	Relative fitness		
	Year 1	Year 2	Average
North			
N-1			
NILMB	1.00*	1.00*	1.00
SILMB	0.42	0.36	0.39
N	30	33	
N-2			
NILMB	NS[a]	1.00*	1.00
SILMB	NS[a]	0.46	0.46
N		120	
Central			
C-1			
NILMB	1.00	1.00	1.00
SILMB	0.91	1.03	0.97
N	27	90	
C-2			
NILMB	1.00	1.00	1.00
SILMB	0.94	1.14	1.04
N	74	44	
C-3			
NILMB	1.00	1.00	1.00
SILMB	1.06	1.04	1.05
N	90	72	
South			
S-1			
NILMB	NS[a]	1.00*	1.00
SILMB	NS[a]	1.70	1.70
N		52	
S-2			
NILMB	1.00*	1.00*	1.00
SILMB	1.92	1.70	1.81
N	35	23	
Summary average relative fitness across ponds and years			
North ponds			
NILMB			1.00
SILMB			0.41
Central ponds			
NILMB			1.00
SILMB			1.02
South ponds			
NILMB			1.00
SILMB			1.77

[a] No spawning occurred.

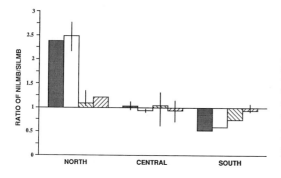

FIGURE 3.—Ratios of reproductive success by region of northern (NILMB) and southern (SILMB) Illinois largemouth bass for 1985 and 1986, assuming (1) equal numbers of parents of each stock, as determined by equal numbers originally stocked; and (2) unequal numbers of parents as determined by survival (relative abundance data) of each stock in each pond.

The genetic resources of the largemouth bass, and likely many other managed fish species as well, are eroding as a result of current management programs that inadvertently permit or even in some cases deliberately promote stock transfers. In all cases of stock transfer, these programs are causing the loss of among-population variation. In addition, it is also likely that through outbreeding depression stock transfer is having negative effects on the fitness of the native stocks involved. Natural resource management agencies need to restructure their fish stocking programs to minimize the adverse effects of stock transfers (Philipp et al. 1993). Furthermore, strict regulations to prevent similar types of stock transfers need to be developed for the private sector and strictly enforced. As stewards of our fisheries we cannot sit idle, watching the destruction of their genetic resources. On a positive note, recent activities within the American Fisheries Society have provided evidence that these concepts are becoming incorporated into at least some management agency programs. Hopefully, this trend will soon become the accepted norm.

Acknowledgments

We want to thank the many assistants who over the years helped collect fish for these experiments, particularly John Weber, John Reinke, and Mark Horner; Anna Toline, who helped with the statistical analyses; the Illinois Department of Conservation, which helped in the project's organization. This project was supported by Federal Aid in Sport Fish Restoration Act funds from the Illinois Department of Conservation, Project F-53-R.

References

Allendorf, F. W., N. Ryman, and F. M. Utter. 1987. Genetics and fishery management—past, present, and future. Pages 1–19 *in* N. Ryman and F. Utter, editors. Population genetics and fishery management. University of Washington Press, Seattle.

Altukhov, Y. P., and E. A. Salmenkova. 1987. Stock transfer relative to natural organization, management and conservation of fish populations. Pages 333–344 *in* N. Ryman and F. Utter, editors. Population genetics and fishery management. University of Washington Press, Seattle.

Beacham, T., C. B. Murray, and R. E. Withler. 1989. Age, morphology, and biochemical genetic variation of Yukon River chinook salmon. Transactions of the American Fisheries Society 118:46–63.

Campton, D. E. 1987. Natural hybridization and introgression in fishes: methods of detection and interpretation. Pages 161–192 *in* N. Ryman and F. Utter, editors. Population genetics and fishery management. University of Washington Press, Seattle.

Gharrett, A. J., C. Smoot, and A. J. McGregor. 1988. Genetic relationships of even-year northwest Alaska pink salmon. Transactions of the American Fisheries Society 117:536–545.

IUBNC (International Union of Biochemistry, Nomenclature Committee). 1984. Enzyme nomenclature. Academic Press, San Diego, California.

Krueger, C. C., E. J. Marsden, H. L. Kincaid, and B. May. 1989. Genetic differentiation among lake trout strains stocked into Lake Ontario. Transactions of the American Fisheries Society 118:317–330.

Kutkuhn, J. H. 1981. Stock definition as a necessary basis for cooperative management of Great Lakes fish resources. Canadian Journal of Fisheries and Aquatic Sciences 38:1476–78.

Philipp, D. P. 1991. Genetic implications of introducing Florida largemouth bass, *Micropterus salmoides floridanus*. Canadian Journal of Fisheries and Aquatic Sciences 48(Supplement 1):58–65.

Philipp, D. P., W. F. Childers, and G. S. Whitt. 1979. Evolution of differential patterns of gene expression: a comparison of the temporal and spatial patterns of isozyme locus expression in two closely related fish species (northern largemouth bass, *Micropterus salmoides salmoides*, and smallmouth bass, *Micropterus dolomieui*). Journal of Experimental Zoology 210: 437–488.

Philipp, D. P., W. F. Childers, and G. S. Whitt. 1983. A biochemical evaluation of the northern and florida subspecies of largemouth bass. Transactions of the American Fisheries Society 112:1–20.

Philipp, D. P., J. M. Epifanio, and M. J. Jennings. 1993. Conservation genetics and current stocking practices: are they compatible? Fisheries (Bethesda) 18(12):14–16.

Philipp, D. P., and G. S. Whitt. 1991. Survival and growth of northern, Florida, and reciprocal F_1 hybrid largemouth bass in central Illinois. Transactions of the American Fisheries Society 120:156–178.

Ryman, N., and F. Utter. 1987. Population genetics and fishery management. University of Washington Press, Seattle.

Shaklee, J. B., F. W. Allendorf, D. C. Morizot, and G. S. Whitt. 1990. Genetic nomenclature for protein-coding loci in fish. Transactions of the American Fisheries Society 119:2–15.

Soulé, M. E. 1986. Conservation biology: the science of scarcity and diversity. Sinauer Associates, Sunderland, Massachusetts.

Templeton, A. R. 1986. Coadaptation and outbreeding depression. Pages 105–116 *in* M. Soulé, editor. Conservation biology: the science of scarcity and diversity. Sinauer Associates, Sunderland, Massachusetts.

Wright, S. 1931. Evolution in Mendelian populations. Genetics 16:97–159.

Wright, S. 1978. Variability within and among populations—evolution and the genetics of natural populations, volume 4. University of Chicago Press, Chicago.

Incorporating the Stock Concept and Conservation Genetics in an Illinois Stocking Program

KIRBY D. COTTRELL

Illinois Department of Conservation, Division of Fisheries
600 North Grand Avenue West, Springfield, Illinois 62702, USA

SCOTT STUEWE

Illinois Department of Conservation, Division of Fisheries
600 North Grand Avenue West, Springfield, Illinois 62702, USA

ALAN BRANDENBURG

Illinois Department of Conservation, Little Grassy Fish Hatchery
Rural Route 1, Box 429, Carbondale, Illinois 62901, USA

Abstract.—This paper presents a discussion of stock concept theory and conservation genetics as applied to fish-stocking policy in Illinois. An historical overview of fisheries management in Illinois is given. An evaluation was conducted to identify the process by which fishery professionals would initiate and fully implement the stock concept theory and conservation genetics in the Illinois' fishery management program. The rationale and discussion regarding the assumptions, impacts, conflicts, and problems to current fish stocking policy are presented in an attempt to define the wise use of fish stocking in Illinois.

The Illinois Department of Conservation (IDOC) is legislatively delegated and mandated the responsibility to take all measures necessary for the conservation, preservation, distribution, introduction, propagation, and restoration of fish, mussels, frogs, turtles, game, wild animals, wildfowl, and birds. The Division of Fisheries (DOF) oversees 1.6-million acres of surface waters in the state. The sport fishery present in these waters currently supplies about 30-million quality angler-days per year; however it is estimated that 1.4-million sport anglers are generating a demand for over 40.3-million angler-days. Fish stocking is one of the most important tools for fisheries management to reduce this deficit and may be the only tool in certain management strategies. This latter case is particularly true in today's highly stressed waters in areas of high human populations and multirecreational use.

Fisheries' needs in the future will be far different than they were 5 to 10 years ago. Increased fishing pressure, better anglers with more efficient equipment, a decrease in the construction of new waters, pollution, and increased pressures for multiple uses of water areas are reducing the quality experience anglers expect and demand. Fisheries managers have reacted to these situations by turning to fish that will produce additional harvest and angling opportunities. The IDOC has produced fisheries for walleye *Stizostedion vitreum*, striped bass *Morone saxatilis*, hybrid striped bass *Morone chrysops* ♂ × *M. saxatilis* ♀, northern pike *Esox lucius*, muskellunge *E. masquinongy*, tiger muskellunge *E. lucius* ♂ × *E. masquinongy* ♀, coho salmon *Oncorhynchus kisutch*, chinook salmon *O. tshawytscha*, rainbow trout *O. mykiss*, brown trout *Salmo trutta*, and lake trout *Salvelinus namaycush* in waters where none of these fisheries would exist without supplemental stocking.

Stocking fish is a traditional and effective method of fish management (Radonski and Martin 1986). Traditional fish stocking has been accused of being random and destructive to the genetic integrity of native populations. Proponents of genetic conservation and the stock concept (Kutkuhn 1981; Philipp et al. 1981, 1993) have urged that these concepts be incorporated into fish management activities. They define genetic conservation as "the protection and wise use of the genetic resources of a species," and the stock concept as "the idea that a species is comprised of different stocks," and that "the stock, not the species as a whole, must be considered the operational unit of interest for fisheries management programs." The IDOC is mandated to protect the natural resources of the state, and the DOF is charged with improving fisheries resources. Fish stocking has been and will continue to be an important method for improving these resources. Genetic conservation and the stock concept have become considerations in the stocking process, and their incorporation into this process will be discussed.

History

The IDOC has a long history of fish stocking. The 1913–1914 Annual Report of the Game and Fish Conservation Commission reports that "this is the first year the state of Illinois has done anything in the way of artificially hatching black bass.... Already more than 250,000 fingerlings have been planted in 50 different places in our good bass lakes and streams." The report also discusses the practice of shipping "rescued fish" from the bottomland lakes along the Illinois and Mississippi rivers. Fish stranded in these bottomland lakes during the summer were captured and distributed to suitable waters. This fish rescue operation was carried out for decades in Illinois with fish of several species shipped to locations throughout the state by boat and rail car.

Early fisheries management efforts in Illinois consisted of little more than a mix of regulations and fish-stocking programs. But such practices did not address the problems of declining fish habitat and increasing user demand. Consequently, since the 1950s, fisheries management has been involved in sampling fish populations, conducting creel censuses and angler surveys, analyzing water quality, controlling aquatic weeds, manipulating water levels, producing the proper sizes and species of fish for stocking purposes, and when necessary, eradicating undesirable fish populations.

In 1979, hatchery system expansion was undertaken due to the inadequacy of existing facilities and declining availability of fish via trades from other states, private hatcheries, and the U.S. Fish and Wildlife Service. Plans for production were based upon projected stocking needs developed by biologists in the DOF for state, public, organizational, and private waters. An environmental assessment of the increased stocking program was completed in 1976 (IDOC 1976):

> Waters supporting threatened or endangered species will not be stocked unless it can be shown that the stocking will not further endanger the populations of threatened or endangered species.
>
> Stocking will have an impact on any and all waters receiving fish. However, the stocking program of the state of Illinois is based upon management recommendations of the various fisheries biologists and is not indiscriminate. Stocking will be used as a management tool to supplement an overall management program.
>
> The major beneficial impact to be obtained from the new stocking program is the enhancement of the existing fishery and the improvement of angling in the state. Realizing the existence of all the potential impacts and unknown variables which could accrue from a stocking program increase of the magnitude of that proposed for the waters of Illinois, the Division of Fisheries intends to take a minimum of 5 to 10 years to phase into the full production level of 50 million fish per year. Each incremental increase in fish production or new species introduction will be preceded by detailed biological systems analysis of the potential receiving waters. Before a stocking program has been initiated in a given body of water, there will be a monitoring of the dynamics operating in its aquatic ecosystem. By utilizing this approach to implement the new stocking program, the Division of Fisheries hopes to minimize the chance for adverse impacts and eco-catastrophes. The need for the new hatchery system is based upon stocking requirements necessary to establish quality fishing in the waters under the jurisdiction of state biologists. Quality fishing is a necessity because of the limited water resources available in Illinois and the large population of people using these waters. Advantage must be taken of every ecological niche, every available forage species, every fishery management tool, and any necessary regulations supplemented by the timely availability of the right sizes, species and numbers of fish stocks to produce a quality fishery.

The mission and goals of the *Strategic Plan for Illinois Fisheries Resources* (IDOC 1993a) guide the Illinois Department of Conservation Division of Fisheries in its stewardship of Illinois' fishery resources. The mission of the DOF is to demonstrate leadership through education and scientifically based management for the protection, restoration, and enhancement of fisheries and other aquatic resources and through the promotion of responsible use of Lake Michigan, reservoirs, impoundments, and streams (IDOC 1993a).

Current Practices

Illinois fish-stocking policy is driven by individual species management plans. These plans incorporate information on fish biology (life history, taxonomy, and population data), status of the fishery (quality and supply), and management (overview, regulations, stocking, and future plans). These management plans are drafted by a committee of field biologists and then reviewed and approved through the chain of command. Management plans are intended to be dynamic and to set policy and stocking priorities for the management of individual species in Illinois waters.

In an effort to make the most efficient use of annual hatchery production, the DOF developed a priority list of lakes that may be stocked in a given year. While developing priorities, consideration was given to the following factors: (1) potential adverse effects on fish species already present; (2) potential adverse effects on current lake management practice; (3) potential benefits to populations of largemouth bass *Micropterus salmoides*; (4) potential benefits to the control of panfish *Lepomis* and *Pomoxis* spp. or prey fish populations; (5) capability of the prey base to support an additional predator; (6)

anglers' desires; (7) the lake morphometry and water quality; (8) statewide distribution of coolwater or coldwater species; and (9) angling potential of the area.

Fisheries managers may propose annual changes to the statewide stocking priority list through the chain of command. Changes approved by the DOF to the statewide priority list are implemented the following calendar year. Fish stocking has remained relatively unchanged in Illinois since 1983. Species priority lists have changed very little with only a few waters added or taken off these priority lists. An average of over 52 million of some 20 species have been annually stocked in Illinois waters during the past 10 years.

Release of fish into Illinois waters is regulated by Act 5 of Illinois Conservation Law, the Fish and Aquatic Life Code (IDOC 1993b). This Act regulates the release of fish in the following manner: (1) authorizes the IDOC to issue administrative rules for carrying out provisions of this code; (2) authorizes the IDOC to take any aquatic life or their eggs and propagate these for restocking into suitable waters; (3) authorizes the IDOC to establish and maintain facilities for propagating aquatic life; (4) declares that ownership of all aquatic life within the state resides with the state (however, IDOC-permitted aquatic products propagated in permitted aquaculture facilities are exempt); (5) makes it illegal to release any fish into state waters without first securing permission from the IDOC, except for indigenous fish into waters wholly on the owners property; (6) requires that a salmonid importation permit be obtained before "live fish, viable fish eggs or viable sperm of any species or hybrid of salmon or trout may be imported into the state only by the holder of a Fish Importation Permit and other required state permits"; and (7) in reference to aquaculture permits, states "Before any person imports or receives non-indigenous aquatic life for aquaculture or stocking purposes in this state, permission must be obtained from the Department. Regulations governing non-indigenous aquatic life shall be covered by administrative rule."

Administrative rules regarding aquaculture and stocking of fish by private enterprise are covered in Part 870 of the Illinois Administrative Code: *Aquaculture, Transportation, Stocking, Importation and/or Possession of Aquatic Life*. These regulations establish an approved list of aquatic life species that may be possessed or imported and authorizes the Aquaculture Advisory Committee to review requests for species not included on this list. This is only an advisory committee, and final authority to approve possession or importation of new aquatic species resides with the Chief of the DOF. These rules also regulate the movement of approved aquatic life and transportation of restricted species, such as some carps.

Future Needs and Objectives

To implement conservation genetics and the stock concept fully would require the following actions: elimination of stock transfers, elimination of fish trades, regulation of private aquaculture, definition and assessment of stocks, institution of stocking protocols, establishment of gene refuges, development of genetic driven management plans, and education of natural resource managers and the public (Philipp et al. 1993).

Our first need was to define and assess stocks in Illinois. A review of the scientific literature was made to determine our definition of stocks. This review presented us with general options from which to define stocks (Altukhov 1981; Ihssen et al. 1981; Utter 1981), but only tended to perplex us in our quest for a definition. The greatest problem for fisheries managers is that biologists do not agree on the criteria that define a stock (Waples 1991). Rationale exists for a definition, but a truly universal definition may not be existent. We finally concluded that it is not necessary to agree on a single definition of a stock for fisheries management (Spangler et al. 1981). As technology advances the ability to detect differences among fishes, an interesting question will be whether or not the complexity will render definition of a stock impossible.

Our conclusion was that for our management purposes we lacked the necessary data (both genotypic and phenotypic) to define stocks accurately. Stocks are dynamic entities as a result of the forces exerted on them, thus it is more important that we use the stock concept rather than define Illinois stocks so precisely (Booke 1981). This is a pragmatic view, and in order to manage our fisheries, we are convinced it is also a prudent view. Our view agrees with that of other authors in that the definition of a stock is less important than adoption of the concept to provide for genetic consideration in management decisions (MacLean and Evans 1981; Spangler et al. 1981). We doubt that any natural resource agency will completely abandon their current stocking practices; rather they are more inclined to choose methods that integrate the stock concept into current practices. Management by individual species management plans incorporates the idea that stock equals species, until data proving otherwise may be obtained.

The very fact that little is known about stocks in most species limits the ability of management to develop separate management plans. However, we have instituted several actions that begin to integrate the principles of conservation genetics and the stock concept into Illinois fisheries programs. The first step was taken with the Illinois State Genetics Project, a project started in 1983 in cooperation with the Illinois Natural History Survey, and funded by the Federal Aid in Sport Fish Restoration Act (F-53-R). This 10-year study was designed to assess stocks of largemouth bass and the effects of stocked sport fish on existing populations. The final report on this project is expected later this year, and we anticipate the report may have great influences regarding our future stocking plans in Illinois. We have already begun to incorporate recommendations resulting from this project, most notably by the way we produce and stock largemouth bass. Northern Illinois and southern Illinois largemouth bass broodstocks are maintained at separate hatcheries, and the respective progeny are stocked only in the northern and southern parts of the state (Philipp et al. 1981; Philipp 1991; Philipp and Whitt 1991). We have four state hatcheries that are responsible for stocking waters in their respective area of the state and the movement of stocks between these facilities is limited. Broodstocks for the native species bluegill *Lepomis macrochirus*, white crappie *Pomoxis annularis*, black crappie *P. nigromaculatus*, largemouth bass, and smallmouth bass *Micropterus dolomieu* are captured from local populations. We also maintain a system of nursery ponds in which field biologists place local broodstocks for spawning and subsequent release of fingerlings into the adjacent waters. This practice of placing hatcheries or culture facilities near stocking areas follows recommendations of several authors (Hynes et al. 1981; Krueger et al. 1981; Spangler et al. 1981; Kapuscinski and Jacobson 1987).

The establishment of species management plans and priority lists for stocking in Illinois points to our concern over the past practice of indiscriminate stocking. Evaluation of individual stockings are being conducted under programs we refer to as the periodic index of stocking success. Sport Fish Restoration Act projects F-29-D (lake management plans) and F-69-R (creel surveys) are developing a database for Illinois waters. These assessments provide information on fish populations, lake physical and chemical characteristics, environmental conditions (e.g., size of watershed), and recreational use.

We feel that integration of these data into our fish-stocking programs is wise, although this approach is sure to draw criticism from those who feel we are taking a less than ideal position. As natural resource managers, we have to juggle the needs of society with our duty to protect and preserve the resource, as do most natural resource managers. However, we simply do not know enough about the genetic composition of Illinois' fisheries resources to apply stock concept principles strictly. Background information from work conducted by the Illinois Natural History Survey is available; this information tells us only that we have been stocking genetically pure species, and, thus far, only largemouth bass data have been incorporated into our stocking program.

There are additional considerations or conditions that exist or need to be explored. First, genetic refuges need to be identified. Although we have not formally identified any refuges, we are confident they exist. Stocking records indicate we have not stocked many streams or rivers, and many have escaped indirect stocking from spillway escapement.

Second, we are wholly dependent on other states to provide us with eggs for our Lake Michigan stocking program. Illinois may be the only state that manages an anadromous fishery without any anadromous properties. There are no streams or rivers to which anadromous species can return to complete their life cycle.

Just as the individuality of some bodies of water demand unique management, so does the individuality of documented stocks. However, we feel the riddle of stock definition must be answered before it can fit into the stocking puzzle.

We are satisfied that existing laws, regulations, and administrative orders presented in this paper are adequately addressing the movement and stocking of fish in Illinois by the private sector. The fact that other states have used our system as a model is noteworthy (R. Horner, Illinois Department of Conservation Division of Fisheries, personal communication).

References

Altukhov, Y. P. 1981. The stock concept from the viewpoint of population genetics. Canadian Journal of Fisheries and Aquatic Sciences 38:1523–1538.

Booke, H. E. 1981. The conundrum of the stock concept—are nature and nurture definable in fishery science? Canadian Journal of Fisheries and Aquatic Sciences 38:1479–1480.

Hynes, J. H., E. H. Brown Jr., J. H. Helle, N. Ryman, and D. A. Webster. 1981. Guidelines for the culture of fish stocks for resource management. Canadian Journal of Fisheries and Aquatic Sciences 38:1867–1876.

Ihssen, P. E., H. E. Booke, J. M. Casselman, J. M. Glade, N. R. Payne, and F. M. Utter. 1981. Stock identification: materials and methods. Canadian Journal of Fisheries and Aquatic Sciences 38:1838–1855.

IDOC (Illinois Department of Conservation). 1976. Environmental Assessment Report, Illinois Statewide Hatchery System, Springfield.

IDOC (Illinois Department of Conservation). 1993a. Strategic plan for Illinois fisheries resources. Springfield.

IDOC (Illinois Department of Conservation). 1993b. Illinois conservation law. Gould Publications, Inc., Longwood, Florida.

Kapuscinski, A. R., and L. D. Jacobson. 1987. Genetic guidelines for fisheries management. University of Minnesota, Minnesota Sea Grant College Program, Sea Grant Research Report 17, Duluth.

Krueger, C. C., A. J. Gharrett, T. R. Dehring, and F. W. Allendorf. 1981. Genetic aspects of fisheries rehabilitation programs. Canadian Journal of Fisheries and Aquatic Sciences 38:1877–1881.

Kutkuhn, J. H. 1981. Stock definition as a necessary basis for cooperative management of Great Lakes fish resources. Canadian Journal of Fisheries and Aquatic Sciences 38:1476–1478.

MacLean, J. A., and D. O. Evans. 1981. The stock concept, discreetness of fish stocks, and fisheries management. Canadian Journal of Fisheries and Aquatic Sciences 38:1889–1898.

Philipp, D. P. 1991. Genetic implications of introducing Florida largemouth bass, *Micropterus salmoides floridans*. Canadian Journal of Fisheries and Aquatic Sciences 48(Supplement 1):58–65.

Philipp, D. P., W. F. Childers, and G. S. Whitt. 1981. Management implications for different genetic stocks of largemouth bass (*Micropterus salmoides*) in the United States. Canadian Journal of Fisheries and Aquatic Sciences 38:1715–1723.

Philipp, D. P., J. M. Epifanio, and M. J. Jennings. 1993. Point/counterpoint: conservation genetics and current stocking practices—are they compatible? Fisheries (Bethesda) 18(12):14–16.

Philipp, D. P., and G. S. Whitt. 1991. Survival and growth of northern, Florida, and reciprocal F hybrid largemouth bass in Central Illinois. Transactions of the American Fisheries Society 120:58–64.

Radonski, G. C., and R. G. Martin. 1986. Fish culture is a tool-not a panacea. Pages 7–13 *in* R. H. Stroud, editor. Fish culture in fisheries management. American Fisheries Society, Fish Culture Section and Fisheries Management Section, Bethesda, Maryland.

Spangler, G. R., A. H. Berst, and J. F. Koonce. 1981. Perspectives and policy recommendations on the relevance of the stock concept in fishery management. Canadian Journal of Fisheries and Aquatic Sciences 38:1908–1914.

Utter, F. M. 1981. Biological criteria for definition of species and distinct intraspecific populations of anadromous salmonids under the U.S. Endangered Species Act of 1973. Canadian Journal of Fisheries and Aquatic Sciences 38:1626–1635.

Waples, R. S. 1991. Pacific salmon and the definition of a "species" under the Endangered Species Act. U.S. National Marine Fisheries Service Marine Fisheries Review 53(3):11–22.

Conservation of Genetic Diversity in a White Seabass Hatchery Enhancement Program in Southern California

DEVIN M. BARTLEY

Food and Agriculture Organization of the United Nations, Fisheries Department
Viale delle Terme di Caracalla, 00100 Rome, Italy

DONALD B. KENT AND MARK A. DRAWBRIDGE

Hubbs–Sea World Research Institute, 2595 Ingraham Street, San Diego, California 92109, USA

Abstract.—Protocols for the conservation of genetic diversity of white seabass *Atractoscion nobilis* by the Ocean Resources Enhancement and Hatchery Program (OREHP) in the Southern California Bight region are presented. Three main factors are addressed by these protocols: (1) the genetic structure of the wild population, (2) the genetic structure of the hatchery group used for enhancement, and (3) the monitoring of both enhanced and hatchery populations during the enhancement effort. Furthermore, it is necessary to define the objectives of the enhancement effort in order to adapt protocols to meet specific conservation and production goals.

Information on the genetic structure of wild and cultured white seabass has been assessed by means of starch gel electrophoresis of allozymes. Bartley and Kent (1990) found that the wild white seabass population is characterized by little population subdivision with rare alleles, those at a frequency of occurrence of approximately 2%, contributing to the differences between sampling locations. Based on this information, OREHP established a goal to preserve these rare alleles in the hatchery product. Determination of effective population size by analysis of linkage disequilibrium in hatchery progeny and binomial sampling theory indicated that approximately 150 adult wild-caught white seabass of equal sex ratio were needed to achieve this goal in the hatchery.

A broodstock management plan to maintain genetic diversity, including rotation of broodfish among spawning tanks and the infusion of new wild-caught broodfish into the hatchery population, is described. Genetic monitoring of the progeny and continued assessment of estimated effective population size is recommended because it is not known precisely how individual adults will adapt to and spawn in the hatchery environment, or how the genetic structure of the resulting progeny will change.

The use of hatcheries to enhance depleted fisheries has, for some fish species, been an important management tool since the 1800s (Netboy 1980). As aquatic husbandry and controlled breeding of aquatic species improved, hatcheries were used more frequently as a means to rehabilitate depleted commercial fisheries. This trend toward increased hatchery use was observed for salmon (Netboy 1980), red drum *Sciaenops ocellatus* (Rutlege 1989), madai *Pagrus major* (Sugama et al. 1988), Atlantic cod *Gadus morhua* (Svåsand et al. 1990), and numerous other marine species in Japan (Ikenoue and Kafuku 1992). Populations of five species of sturgeon beluga *Huso huso* and *Acipenser* spp., which support the world's richest fishery for caviar, and a species of mahi sephid *Rutilus frisii kutum* are maintained in the Caspian Sea through extensive stocking of juveniles by the Islamic Republic of Iran and the former Soviet Union (FAO 1992c). In addition, hatcheries have been utilized as a means to create new fisheries, such as the Pacific salmon fisheries of the Great Lakes of North America (Netboy 1980).

Hatcheries have also been used as a means to propagate endangered species for reintroduction into the wild and as ex-situ refugia or gene banks if reintroduction was dangerous or not possible (Johnson and Jensen 1991). For example, 24 species of endangered or threatened freshwater fish have been maintained at one time or another at the Dexter (New Mexico) National Fish Hatchery; many of these fishes have been successfully reintroduced into restored habitats. The number of whitefish *Coregonus muskun* was declining throughout its range in Finland, but hatchery enhancement has reversed this trend, at least in the Kymijobi River system (Westman and Kallio 1987).

Historically the use of hatcheries to increase marine fisheries has been questioned because of a lack of any demonstrable effect of releases on certain target populations, high costs associated with culture facilities (MacCall 1989), poor adaptability of hatchery fish to the wild (Mesa 1991) and, more recently, potential negative effects of the genetic contamination of local gene pools (Hindar et al. 1991; Philipp et al. 1993). These concerns must be addressed in any hatchery enhancement program.

Although hatcheries are often seen as appropriate mitigation for habitat destruction, very little monitoring or assessment of the effect of hatchery releases on a fishery or resource base has been conducted. In contrast, Bartley and Kent (1990) stated that, "failure to evaluate a hatchery's contribution to a target population and using hatcheries to justify habitat destruction are unacceptable"; genetic conservation of natural populations should also be a prime consideration in hatchery enhancement programs (FAO 1993).

The main objective of this paper is to describe protocols in a hatchery enhancement program that will facilitate the conservation of genetic diversity in both natural and cultured populations of white seabass *Atractoscion nobilis*.

The methods used to establish these protocols involve the application of simple population genetic principles to hatchery management of white seabass. The methods rely on clearly stated program objectives and depend on accurate descriptions of the genetic structure of the natural population, the hatchery broodstock and progeny, and the enhanced population.

Project Objectives

Program Goals

There are two primary objectives of the Ocean Resources Enhancement and Hatchery Program (OREHP). The first objective is to enhance the natural population of white seabass by releasing large numbers of juveniles into the wild, thereby increasing the catch of white seabass in both the commercial and recreational fisheries. This objective will be accomplished by producing in a hatchery viable juveniles that, after release, will be well adapted to the environment of the Southern California Bight and will interbreed with wild white seabass. The second objective is to document and conserve the genetic resources in wild and cultured stocks of white seabass. This objective will be met by analyzing isozymes of natural and cultured groups of white seabass and then establishing appropriate hatchery management protocols, based on population genetic theory.

Production Goals

Hatchery technology and basic husbandry of white seabass has progressed to a stage where large-scale releases of juvenile fish are possible (Kent et al. 1995, this volume). The hatchery production capability of the OREHP was approximately 50,000 fish annually during the pilot phase (1988–1993), but is expected to reach 400,000–500,000 fish annually in the expanded phase by 1996. The culture systems used by OREHP are described elsewhere (Kent et al. 1995, this volume).

Conservation Goals

In light of the fact that the primary goal of the OREHP is to rebuild natural populations of white seabass, it is necessary to conserve the natural genetic diversity in the wild and, to the extent possible, maintain natural levels of diversity in the cultured groups of fish.

Project Approach

Natural Population

A survey of the natural population of white seabass from the Southern California Bight (Bartley and Kent 1990) revealed little population substructuring in the area (Figure 1). Average heterozygosity values ranged from 0.033 to 0.064, genetic identity was greater than 99% in all pairwise comparisons, and only 3% of the genetic variation was attributed to between-sample differences. Gene flow was estimated to be approximately nine migrants per generation and sufficient to homogenize the genetic structure of the population. Several gene loci possessed rare alleles (frequency <2%) that contributed to genetic diversity in the region. It is recommended that a goal of the hatchery enhancement program should be conservation of this allelic diversity.

To ensure that these rare alleles are present in fish produced at the OREHP hatchery, it will be necessary to collect enough broodstock so that rare alleles will be sampled. Binomial sampling theory describes the probability of collecting an allele of frequency p as

$$N = \frac{\log_e(1 - \alpha)/\log_e(1 - p)}{2}, \quad (1)$$

where N is the number of fish required and α is the confidence level. To be 95% certain of collecting broodstock that possess rare alleles ($p = 0.02$), a minimum of 74 broodfish is needed (Figure 2).

Hatchery Population

Broodstock.—The source of broodstock is vitally important in any enhancement program. The OREHP proposes to use adult white seabass captured from wild populations off the southern California coast. These fish represent the wild genotypes and therefore represent a low genetic risk to

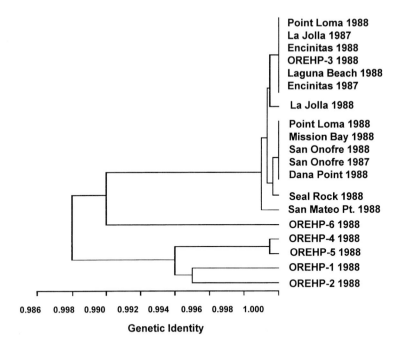

FIGURE 1.—Genetic identity among samples of white seabass (after Bartley and Kent 1990). Samples designated ORHEP-1–ORHEP-6 represent hatchery-reared progeny; other samples represent wild fish from the southern California coastal area. Year of sampling is included.

the wild population (Waples 1991). Prior to stocking the brood pools, the sex of each fish will be determined from microscopic examination of gametes obtained by catheterization or by the use of an ultrasonic imaging system. Each fish will be permanently identified using an individual passive integrated transponder tag. A biopsy will be performed on each broodfish. Tissue samples taken

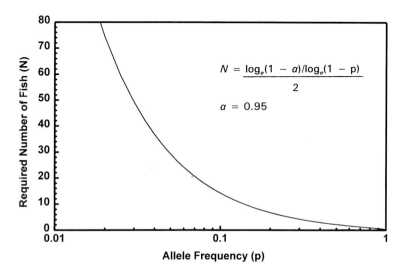

FIGURE 2.—Relationship between number of fish to be sampled (N) in order to insure, with 95% confidence level, collection of alleles of frequency p.

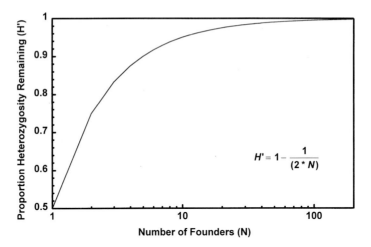

FIGURE 3.—Relationship between heterozygosity remaining (H') in the progeny generation and number of founders (N).

during the biopsy will include muscle and fin, which will be subjected to analysis of allozymes. The inadvertent use of hatchery-reared and released fish as broodstock will be strictly avoided through tagging of progeny produced in the hatchery.

Although little evidence of local population substructure was revealed in white seabass (Bartley and Kent 1990) and other commercially important sciaenids (Ramsey and Wakeman 1987; Graves et al. 1992; King and Pate 1992), it is recognized that extensive sampling of fish from offshore islands and lower Baja California will be necessary to complete the genetic description of white seabass. Until this description is completed, transfers of white seabass within their natural range will be minimized by using only broodfish collected from the southern California coast. Adults from lower Baja California or from north of Point Conception will not be used to stock southern California coastal areas.

How would the strategy of using 74 fish as broodstock affect other measures of genetic diversity? To answer this question, it is instructive to examine how a founding population of this size will affect the heterozygosity and allelic diversity of progeny generations.

The proportion of the original heterozygosity (H') of the source population represented in a founding population of size N is expressed as

$$H' = 1 - \frac{1}{(2*N)}. \qquad (2)$$

Therefore, a founding population of 74 fish will represent 99% of the heterozygosity of the source population (Figure 3). However, allelic diversity is more sensitive to small population size than is heterozygosity (Allendorf and Ryman 1987). Allelic diversity in a founding population is given by

$$n' = n - \Sigma(1 - P_j)^{2N}. \qquad (3)$$

where n' is the effective number of alleles remaining after establishing a population with N founders, n is the original number of alleles, and Pj is the allele frequency (Figure 4). For a simplified two allele model with various allele frequencies in the source or wild population, over 93% of the allelic diversity due to rare alleles (2% in this example) will be conserved if the effective size of the founding population exceeds 50 fish. Theoretically, the strategy of using 74 fish as broodstock is sound and should conserve over 90% of the natural genetic variability, measured by heterozygosity and allelic diversity.

Effective population size (N_e) is one of the primary determinants of genetic diversity. To avoid problems associated with founding hatchery populations from a restricted genetic base, as has occurred in tilapia transplanted to Asia (Eknath et al. 1993), the effective number of broodfish will be maintained at or above the estimated number needed to maintain genetic variability of the wild white seabass population. The most obvious means to accomplish this is to use large numbers of adults. However, limitations imposed by physical capabilities of the hatchery facility must also be considered. An N_e of 74 fish is required to satisfy the conservation goals of the project.

Sex ratio and variance in reproductive output

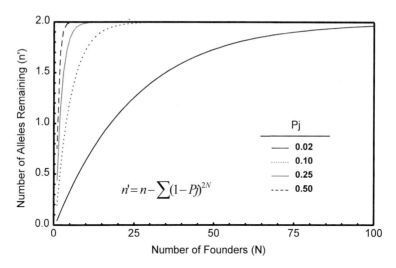

FIGURE 4.—Relationship between number of alleles remaining (n') at a two allele locus ($n = 2$) and the number of founders (N). The allele frequency is given by P_j.

influence N_e and usually result in N_e values lower than actual population size (N). Bartley et al. (1992) used linkage disequilibrium data from allozyme genotype frequencies to show that the effective population size of a mass spawning group of white seabass broodstock was about 50% of the actual population size. Therefore, to achieve the stated conservation goals, the OREHP broodstock management plan should consist of approximately 150 broodfish. The sex ratio of these broodfish should be one female to one male. Deviations will necessitate that more broodstock are maintained according to the expression

$$N_e = \frac{4N_m * N_f}{(N_m + N_f)}, \quad (4)$$

where N_m and N_f are the numbers of males and females, respectively (Figure 5). A schedule for rotating broodfish among holding pools is illustrated in Figure 6. This management strategy should maintain the diversity in progeny by increasing the number of matings with different partners per broodstock, relative to the number of different matings possible without rotation. The rotation schedule assumes that a total of 200 broodfish (1:1 sex ratio) are maintained in the hatchery with at least 5% being replaced per year. However, in light of the fact that we do not fully understand the reproductive behavior of white seabass in culture, it will be necessary to monitor the effective population size and diversity of the progeny produced by this rotation schedule. If, for example, one broodfish becomes dominant in its holding pool and makes a disproportionately high reproductive contribution to that pool's progeny, this situation should become apparent in a reduction in Ne. It is expected that the precise rotation schedule will be empirically refined.

Progeny.—Large-scale production of white seabass is possible through controlled spawning of adults and improved larval rearing and nutrition. Although the survival of juveniles in the hatchery is high, natural mortality after release into the wild is unknown. Therefore, the progeny should be physically marked and genetically characterized. The physical marker currently employed to tag fish prior to release is the binary-coded wire tag (CWT). Genetic variability of the progeny groups will be determined from allozyme analysis as described by Bartley and Kent (1990). More discriminating analyses, such as DNA fingerprinting or other molecular techniques that are capable of identifying the parents of individual offspring, are being investigated by OREHP researchers (K. Jones, California State University Northridge, personal communication). Coded wire tags will help assess the migratory and recruitment patterns of first generation hatchery fish and prevent the use of hatchery-produced fish as subsequent broodstock. The genetic characterization will serve two functions: it will allow assessment of effective population size of the broodstock and it will allow the evaluation of genetic marking programs to assess introgression of hatchery fish

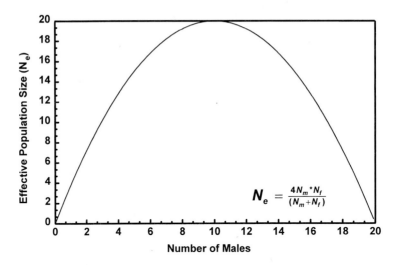

FIGURE 5.—Relationship between effective population size (N_e) and sex ratio in a population of 20 individuals.

into wild populations. Data provided by physical and genetic tags are complimentary and necessary for long-term evaluation.

To facilitate genetic description of the progeny,

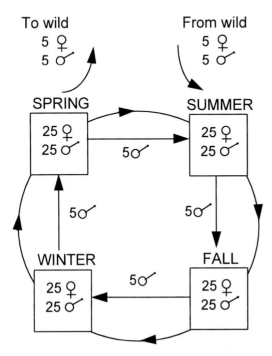

FIGURE 6.—Proposed schedule of broodstock rotation and replacement as part of overall broodstock management plan. Squares represent mass spawning tanks with 25 males and 25 females.

only loci found to be polymorphic in the broodstock will be assayed. As a consequence, only loci expressed in muscle tissue will be analyzed routinely; potentially, this will amount to over 13 polymorphic loci and should be sufficient for genetic analysis (Bartley and Kent 1990). Additional screening for new polymorphisms will be conducted on a research basis.

Currently, there is no plan to breed a genetic marker into the hatchery stock of white seabass, even though candidate allozyme markers have been identified at L-lactate dehydrogenase 1.1.1.27 (LDH), mannose-6-phosphate isomerase 5.3.1.8 (MPI), and aspartate aminotransferase 2.6.1.1 (ATT; enzyme names and international numbers follow IUBNC 1984) gene loci (Bartley and Kent 1990). These allozyme markers are presumed to be selectively neutral and should not adversely affect the viability of the progeny groups. Molecular analysis of broodstock may provide a means to identify hatchery fish without a specific breeding program (Doyle 1993).

Progeny will be collected throughout the spawning period for release and genetic description. White seabass spawn naturally between March and July with peak activity between April and June (Skogsberg 1925). In the hatchery, white seabass can be induced to spawn by manipulating temperature and photoperiod under controlled holding conditions. Differences have been reported in the genetic structure of progeny groups produced at different times of the spawning cycle (Allendorf and Ryman 1987; Sugama et al. 1988; Leary et al. 1989).

Based on genetic differences among progeny groups (i.e., occurrence of specific alleles), Bartley and Kent (1990) suggested that different adults were contributing to successive progeny groups throughout the hatchery spawning season. Therefore, to adequately describe the genetic resources of the hatchery, as well as to maximize the diversity of the genetic resources, it is imperative that progeny from the entire spawning season be utilized.

There is the possibility that certain broodstock will adapt more readily to the hatchery environment and may become over-represented in the progeny groups. Conversely, certain broodstock may become reproductively senescent over the course of the enhancement effort. Genetic monitoring of the progeny groups should help to detect these conditions and facilitate the detection and replacement of these individuals.

Enhanced Population

The effectiveness of the enhancement program can be evaluated only through systematic monitoring of the enhanced (natural and hatchery) populations. Historically, monitoring the effectiveness of hatchery programs was often neglected. Typically, hatcheries were evaluated by the numbers of fish released with little or no attempt to determine the contribution of those fish to a fishery or resource base (MacCall 1989; Hilborn 1992). Several state observer programs in the Southern California Bight region as well as local sport and commercial fishers have demonstrated their support by assisting in the sampling of white seabass (authors' observation).

Preliminary results from small-scale releases of white seabass have been encouraging. During 1992–1993 approximately 10% of fish collected from the primary release embayment were of hatchery origin. Three fish recovered from local coastal kelp beds were near the legal size limit (711 mm total length) and had survived for up to 4 years. The contribution of hatchery releases to the sport and commercial fisheries will be assessed from data on recovered CWTs. Genetic stock identification and maximum likelihood algorithms that determine the first-generation hatchery contribution to mixed stock salmon fisheries (Brodziak et al. 1992) will be evaluated in terms of their ability to assess the white seabass enhancement program. Utter and Ryman (1993) point out that, in theory, these techniques can be used with any hatchery enhancement program; the limitation is the amount and organization of the genetic variability. In this regard, white seabass have similar levels of heterozygosity and polymorphism to chinook salmon *Oncorhynchus tshawytscha* (Bartley et al. 1991), but the organization of the diversity is much different. Chinook salmon are more differentiated into subpopulations, and the genetic character of these subpopulations appears to be stable between generations; white seabass in southern California form a panmictic population with minor temporal and spatial fluctuations in genetic structure (Bartley and Kent 1990).

Discussion

Clearly defined goals are absolutely essential to design appropriate hatchery management strategies (Barber and Taylor 1990) and evaluate alternatives to hatchery enhancement. The primary goal of the OREHP is to supplement the natural population of white seabass by releasing large numbers of juveniles into the wild. Ultimately, these releases should result in increased catches of white seabass in both the commercial and recreational fisheries.

Another major goal is the conservation of genetic resources in wild and cultured stocks of white seabass. Development agencies and aquaculturists are beginning to understand that sustainable production from natural systems must involve conservation of genetic diversity (FAO 1993; UNEP 1992). This requires that the existing biological and genetic resources found in white seabass be thoroughly documented. Currently, nearly all of the genetic resources of white seabass are found in the wild populations, as practically no domestication of this fish has taken place. The existing stock structure in wild populations must be preserved, and the culture practices to expand individual populations must not adversely affect the wild stocks.

Recently, concern has been raised that these two goals are incompatible (Martin et al. 1992; Philipp et al. 1993; R. McGowan, California Coastal Commission, personal communication). That is, the interaction and interbreeding of hatchery and native fish will always adversely affect the native population. This is a valid concern. However, nearly all of the evidence for negative genetic interactions between hatchery and conspecific wild stocks comes from literature on salmonids, although the concern has been stated by researchers raising other species (Boyd 1982; Taniguchi 1986; Svåsand et al. 1990; King et al. 1993). Hindar et al. (1991) acknowledged this bias and cited the following reasons: (1) substantial data exist for salmonids; (2) salmonids generally form highly differentiated and locally adapted subpopulations; and (3) there is a long history of culture for salmonids. However, salmonid aquaculture accounts for less than 5% (593,428 of 12,200,000 metric tons) of the total production from aquaculture (FAO 1992a). Similarly, salmonid and smelt fisheries account for about only 1.5% (1,499,031 of 97,245,700 metric tons) of the total catch of capture fisheries (FAO 1992b). Although the study of genetic resources described for salmonids has greatly advanced the field of applied population genetics, using anadromous salmonids as a general model for the conservation and utilization of genetic resources of many marine and freshwater species should be done cautiously. Specific problems with using data from anadromous salmonids to set conservation guidelines for white seabass include the following.

Homing ability.—Because salmon home so precisely to spawn, subpopulations can be greatly differentiated and adapted to a local drainage or environment (Ricker 1972; Quinn 1982). Most marine fish, including white seabass, do not home as precisely and do not have as genetically differentiated subpopulations (Gyllensten 1985; Utter and Ryman 1993).

Larval and egg dispersal.—Salmon have a more limited egg and larval dispersion than most marine fish. White seabass eggs and larvae are estimated to be in the water column for 40 d and are capable of wide transport by currents along the California coast, thereby providing a mechanism to break down subpopulation structure. Waples (1987) reported that the genetic structure of populations of 10 marine fish from the Southern California Bight were correlated with egg and larval dispersal; 8 of the 10 marine inshore fish studied by Waples had very minor population differentiation.

Complex life history.—Because many salmonids have an anadromous life history and strictly defined migration patterns that appear to be genetically controlled (Ricker 1972; Bams 1976), the addition of exotic genes from conspecifics coding for other migration patterns can disrupt the fine tuning necessary for salmon to migrate to their natal streams or to the sea at the appropriate times of the year (Bams 1976). White seabass do not have as complex a life history and spawn over a longer period, April to July (Vojkovich and Reed 1983).

Another important factor in the conservation of genetic diversity is the effect of escaped fish on native stocks. Obvious differences may exist between fish that are highly selected and adapted to a culture environment and are never meant to be released into the wild, but do escape, and fish that are purposefully raised with the intention of release into the wild. These two groups may be very different, even if they are conspecific, and their effect on the environment may also be very different.

Fish in a closed-system culture facility, where progeny are used as broodstock, may become more adapted to life in a hatchery than to life in the wild (Hindar et al. 1991). In Norway, where escaped Atlantic salmon *Salmo salar* outnumber wild Atlantic salmon in some rivers, the natural resources may be seriously threatened. This is not surprising because the escaped fish represent approximately 15 years of selection, crossbreeding, and translocation efforts (Gausen and Moen 1991). There is no reason to think that introgression of these highly selected individuals would benefit the natural population. However, the primary goal of the Norwegian program was not to attempt to make its farmed Atlantic salmon like the natural population, but to increase the production of Atlantic salmon. Norway succeeded in this regard, as it is the world's leading producer of Atlantic salmon with over 157,000 metric tons produced in 1990 (FAO 1992a). Native stocks that formed the basic material for this success may now be in danger as a result of escapement of these selected stocks.

Many marine hatcheries, including the one proposed by OREHP for white seabass, are not closed systems; broodstock are taken from the wild and progeny from the hatchery are not used as broodstock or are used in limited numbers (Ikenoue and Kafuku 1992). Fish released from these hatcheries should be very similar to the wild stock because initial broodstock are wild and their progeny spend only juvenile stages in the hatchery environment. Escapes of these types of fish represent a much lower risk to wild genetic resources (Waples 1991). Here we are concerned only with genetic differences between hatchery and wild stocks. In any culture situation where animals are raised at high densities, control and prevention of disease organisms will be necessary. Obviously, if fish that are even genetically identical to wild stocks are diseased, their escape or release into the wild should be prevented.

Recently, a responsible approach to hatchery enhancement has been advocated that involves multidisciplinary analysis, action, and evaluation (Blankenship and Leber 1995, this volume). Assessment of genetic principles, along with habitat requirements, release requirements, cost analysis, modification of fishing regulations, habitat rehabilitation, and socioeconomic considerations are all components of this approach. A gathering of experts on genetic resources of aquatic animals also recommended that genetic principles be applied within broader fields of study, such as socioeconomics and marine ecology (FAO 1993). Successful hatchery enhancement in Japan has been due to combined efforts of habitat rehabilitation and fishing regulations. In the Islamic Republic of Iran and the former Soviet Union sturgeon enhancement has worked for over 40 years due to strict regulation of the fishery (Shilat, Iranian Fisheries Company, personal communication). With the breakup of the chief regulator in the Caspian Sea (i.e., the Soviet Union) the enhanced sturgeon populations are now declining (Shilat, personal communication). Adhering to genetic principles alone will not assure successful white seabass fisheries; however, ignoring genetic principles will lead to long-term problems of sustainability of aquatic resources. It is clear that the solution to fishery enhancement will be many sided.

Acknowledgments

We thank the Ocean Resources Enhancement and Hatchery Program and the California Department of Fish and Game for their support. The comments of D. Philipp and an anonymous referee are gratefully acknowledged. Publications cited in this manuscript from the United Nations and its specialized agencies are available from the first author.

References

Allendorf, F. W., and N. Ryman. 1987. Genetic management of hatchery stocks. Pages 141–160 *in* N. Ryman and F. Utter, editors. Population genetics and fishery management. University of Washington Press, Seattle.

Bams, R. A. 1976. Survival and propensity for homing as affected by presence or absence of locally adapted paternal genes in two transplanted populations of pink salmon (*Oncorhynchus gorbuscha*). Journal of the Fisheries Research Board of Canada 33:2716–2725.

Barber, W. E., and J. N. Taylor. 1990. The importance of goals, objectives, and values in the fisheries management process and organization: a review. North American Journal of Fisheries Management 10:365–373.

Bartley, D. M., and D. B. Kent. 1990. Genetic structure of white seabass population from the southern California Bight region: applications to hatchery enhancement. California Cooperative Oceanic Fisheries Investigations Report 31:97–105.

Bartley, D., B. Bentley, J. Brodziak, R. Gomulkiewicz, M. Mangel, and G. A. E. Gall. 1991. Geographic variation in population genetic structure of chinook salmon from California and Oregon. U.S. National Marine Fisheries Service Fishery Bulletin 90:77–100.

Bartley, D., M. Bagley, G. Gall, and B. Bentley. 1992. Use of linkage disequilibrium data to estimate effective population size of hatchery and natural fish populations. Conservation Biology 6:365–75.

Blankenship, H. L., and K. M. Leber. 1995. A responsible approach to marine stock enhancement. American Fisheries Society Symposium 15:167–175.

Boyd, R. O. 1982. The snapper fishery and management implications of reseeding. Pages 7–9 *in* P. J. Smith and J. L. Taylor, editors. Prospects for farming and reseeding in New Zealand. Fisheries Research Division Occasional Publication (New Zealand) 37.

Brodziak, J., B. Bentley, D. Bartley, G. A. Gall, R. Gomulkiewicz, and M. Mangel. 1992. Tests of genetic stock identification using coded wire tagged fish. Canadian Journal of Fisheries and Aquatic Sciences 49:1507–1517.

Doyle, R. W. 1993. The use of DNA fingerprint pedigrees for conserving and increasing the productivity of locally-adapted aquaculture breeds. FAO (Food and Agriculture Organization of the United Nations) Fisheries Report 491 (FIRI/R49).

Eknath, A. E., and thirteen coauthors. 1993. Genetic improvement of farmed tilapia: the growth performance of eight strains of *Oreochromis niloticus* tested in different farm environments. Aquaculture 111:171–188.

FAO (Food and Agriculture Organization of the United Nations). 1992a. Aquaculture production. 1984–1990. FAO Fisheries Circular 815, revision 4.

FAO (Food and Agriculture Organization of the United Nations). 1992b. Fishery statistics. Catches and landings. FAO yearbook, volume 72.

FAO (Food and Agriculture Organization of the United Nations). 1992c. The Islamic Republic of Iran: aquaculture sector fact finding mission. Report FI: TCP/IRA/2251 (F). FAO, Rome, Italy.

FAO (Food and Agriculture Organization of the United Nations). 1993. Report of the expert consultation on utilization and conservation of aquatic genetic resources. FAO Fisheries Report 491 (FIRI/R49).

Gausen, D., and V. Moen. 1991. Large-scale escapes of Atlantic salmon (*Salmo salar*) into Norwegian rivers threaten natural populations. Canadian Journal of Fisheries and Aquatic Sciences 48:426–428.

Graves, J. E., J. R. Mc Dowell, and M. S. Jones. 1992. A genetic analysis of weakfish *Cynoscion regalis* stock structure along the mid-Atlantic Coast. U.S. National Marine Fisheries Service Fishery Bulletin 90:469–475.

Gyllensten, U. 1985. The genetic structure of fish: differences in the intraspecific distribution of biochemical genetic variation between marine, anadromous and freshwater species. Journal of Fish Biology 26:691–699.

Hilborn, R. 1992. Hatcheries and the future of salmon in the Northwest. Fisheries (Bethesda) 17(1):5–8.

Hindar, K., N. Ryman, and F. Utter. 1991. Genetic effects of cultured fish on natural fish populations. Canadian Journal of Fisheries and Aquatic Sciences 48:945–957.

Ikenoue, H., and T. Kafuku. 1992. Modern methods of aquaculture in Japan. Elsevier, Tokyo.

IUBNC (International Union of Biochemistry, Nomenclature Committee). 1984. Enzyme nomenclature. Academic Press, San Diego, California.

Johnson, J. E., and B. L. Jensen. 1991. Hatcheries for endangered freshwater species. Pages 199–217 in W. L. Minckley and J. E. Deacon, editors. Battle against extinction. University of Arizona Press, Tucson.

Kent, D. B., M. A. Drawbridge, and R. F. Ford. 1995. Accomplishments and roadblocks of a marine stock enhancement program for white seabass in California. American Fisheries Society Symposium 15:492–498.

King, T. L., and H. O. Pate 1992. Population structure of spotted seatrout inhabiting the Texas gulf coast: an allozymic perspective. Transactions of the American Fisheries Society 121:746–756.

King, T. L., R. Ward, and I. R. Blandon. 1993. Gene marking: a viable assessment method. Fisheries (Bethesda) 18(2):4–5.

Leary, R. F., F. W. Allendorf, and K. L. Knudsen. 1989. Genetic differences among rainbow trout spawned on different days within a single season. Progressive Fish-Culturist 51:10–19.

MacCall, A. D. 1989. Against marine fish hatcheries: ironies of fishery politics in the technological era. California Cooperative Oceanic Fisheries Investigations Report 30:46–48.

Martin, J., J. Webster, and G. Edwards. 1992. Hatcheries and wild stocks: are they compatible? Fisheries (Bethesda) 17(1):4.

Mesa, M. G. 1991. Variation in feeding, aggression, and position choice between hatchery and wild cutthroat trout in an artificial stream. Transactions of the American Fisheries Society 120:723–727.

Netboy, A. 1980. Salmon: the world's most harassed fish. Andre Deutsch Limited, London.

Parrish, R. H., C. S. Nelson, and A. Bakun. 1981. Transport mechanisms and reproductive success of fishes in the California Current. Biological Oceanography 1:175–203.

Philipp, D. P., J. M. Epifano, and M. J. Jennings. 1993. Point/counterpoint: conservation genetics and current stocking practices—are they compatible? Fisheries (Bethesda) 18(12):14–16.

Quinn, T. P. 1982. Homing and straying in Pacific salmon. Pages 257–263 in J. D. McCleave, G. P. Arnold, J. J. Dodson, and W. H. Neill, editors. Mechanisms of migration in fish. Plenum Press, New York.

Ramsey, P. R., and J. M. Wakeman. 1987. Population structure of Sciaenops ocellatus and Cynoscion nebulosus (Pisces). Copeia 1987:682–695.

Ricker, W. E. 1972. Hereditary and environmental factors affecting certain salmonid populations. Pages 19–60 in R. C. Simon and P. A. Larkin, editors. The stock concept of Pacific salmon. H. R. Macmillan Lectures in Fisheries. University of British Columbia, Vancouver.

Rutlege, W. P. 1989. The Texas marine hatchery program—it works! California Cooperative Oceanic Fisheries Investigations Report 30:49–52.

Skogsberg, T. 1925. White seabass. California Department of Fish and Game Fish Bulletin 9:53–63.

Sugama, K., N. Taniguchi, and S. Umeda. 1988. An experimental study on genetic drift in hatchery populations of red seabream. Nippon Suisan Gakkaishi 54:739–744.

Svåsand, T., K. E. Jørstad, and T. S. Kristiansen. 1990. Enhancement studies of coastal cod in western Norway, part 1. Recruitment of wild and reared cod to a local spawning stock. Journal du Conseil International por l'Exploration de la Mer 47:5–12.

Taniguchi, N. 1986. Genetical problems on fish seed production. Pages 37–58 in M. Tanaka and Y. Matsumiya, editors. Sea farming technology of Red Sea bream. Koseisha Koseikaku, Tokyo.

UNEP (United Nations Environment Program). 1992. Convention on biological diversity. UNEP, Nairobi, Kenya.

Utter, F. M., and N. Ryman. 1993. Genetic markers and mixed stock fisheries. Fisheries (Bethesda) 18(8):11–21.

Vojkovich, M., and R. J. Reed. 1983. White seabass, Atractoscion nobilis, in California-Mexico waters: status of the fishery. California Cooperative Oceanic Fisheries Investigations Report 24:79–83.

Waples, R. S. 1987. Multispecies approach to the analysis of gene flow in marine shore fishes. Evolution 41:385–400.

Waples, R. S. 1991. Genetic interactions between hatchery and wild salmonids: lessons form the Pacific Northwest. Canadian Journal of Fisheries and Aquatic Sciences 48:124–133.

Westman, K., and I. Kallio. 1987. Endangered fish species and stocks in Finland and their preservation. Proceedings, World Symposium on Selection, Hybridization and Genetic Engineering in Aquaculture, volume 1:270–281.

Fish Production to Meet Needs:

Capabilities and Limitations

Adult Production of Fall Chinook Salmon Reared in Net-Pens in Backwaters of the Columbia River

JOHN W. BEEMAN

National Biological Service, Pacific Northwest Natural Science Center
Columbia River Research Laboratory, Cook, Washington 98605, USA

JERRY F. NOVOTNY

U.S. Fish and Wildlife Service, Division of Federal Aid, Regional Office
500 Northeast Multnomah Street, Suite 1692, Portland, Oregon 97232, USA

Abstract.—Adult production of fall chinook salmon *Oncorhynchus tshawytscha* reared at several densities in net-pens in two backwater areas of the Columbia River are compared with production by means of traditional hatchery methods. Rearing densities in the pens were primarily limited by water flow; densities were below recommended limits based on rearing space, but flow index recommendations were exceeded in three of four treatments. Rearing costs using net-pens were generally lower than costs in the hatchery. Lower adult contribution of fish reared in net-pen treatments compared with adult contribution of hatchery controls resulted in greater cost per adult produced in net-pens than fish reared and released using traditional methods. However, comparisons of adult return rates were confounded by release locations.

In 1967 the completion of John Day Dam, river mile 216 of the Columbia River, created a 76.4-mi-long reservoir that inundated salmon spawning and rearing habitat. To mitigate for this loss, upriver bright fall chinook salmon *Oncorhynchus tshawytscha* have been reared at Bonneville State Fish Hatchery (SFH; Oregon Department of Fish and Wildlife) and Spring Creek and Little White Salmon National Fish Hatcheries (NFH; U.S. Fish and Wildlife Service) for release above John Day Dam. In an effort to increase the return of adult chinook salmon to this area, in 1983 the U.S. Fish and Wildlife Service, with funding from the Bonneville Power Administration, began to evaluate rearing and imprinting of juvenile fall chinook salmon in temporary facilities installed in backwaters and ponds adjacent to John Day Reservoir. The goal of this research was to determine if upriver bright fall chinook salmon could be successfully reared and imprinted using temporary rearing facilities in backwaters along the Columbia River, resulting in adult contribution to various fisheries.

Methods

During 1986 and 1987 fall chinook salmon juveniles were reared in net-pens at up to four different densities as part of a larger study to investigate alternative chinook salmon rearing scenarios in the Columbia River basin. In 1983, 34 potential backwater areas were surveyed and rated according to their suitability for rearing juvenile chinook salmon. Rock Creek (river mile 228) was chosen as a study site at this time based on criteria including depth, area, accessibility, potential for water temperature fluctuations and wave action, entrance to the Columbia River, public use, and water quality (Novotny et al. 1984; Figure 1). Drano Lake (river mile 162) was added as a study site in 1987 (Novotny et al. 1987).

Fish used for this study were from upriver bright fall chinook salmon adults spawned at the Bonneville SFH. Eggs were hatched and initially reared at Spring Creek NFH. The upriver bright fall chinook salmon program was transferred from Spring Creek NFH to Little White Salmon NFH in 1987. Fish reared at Rock Creek in 1986 were from Spring Creek NFH, whereas those reared at Drano Lake in 1987 were from the Little White Salmon NFH.

The net-pens were 20 ft × 20 ft and were fitted with 0.2-in-mesh, knotless-ace nets that extended 7 ft into the water when attached to the pen frame and enclosed an area of 2,800 ft^3. The nets extended 2 ft above the water surface and were fitted with nylon covers to minimize avian predation.

Fish were reared at four densities (regular, double, triple, quadruple), ranging from 0.060 to 0.273 lb/ft^3 at release; or about 18,000 to 74,000 fish per pen. The "regular" rearing density was chosen based on available water flow and quality, maximum rearing temperatures of 61°F, and proposed release weights of 45 fish/lb (Novotny et al. 1984). Water flow was the most important criterion for determining initial densities.

Water flow through the net-pens at Rock Creek was estimated using backwater inflow and volume

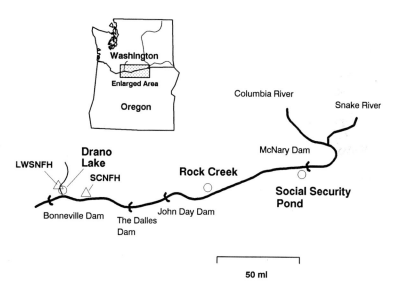

FIGURE 1.—Map of the Columbia River, from its confluence with the Snake River to Bonneville Dam, showing locations of dams as well as study sites (circles) and hatcheries (triangles) used for fish rearing during 1986–1987. Hatcheries are Spring Creek National Fish Hatchery (SCNFH) and Little White Salmon National Fish Hatchery (LWSNFH).

measurements. Flow in Drano Lake was estimated using movement of dye through a net-pen because the morphometry of Drano Lake (small inlet, backwater lake, and small outlet) made laminar flow estimates inappropriate. Water flow through the net-pens at Rock Creek was estimated as 56.5 gal/min. The estimated flow through the pens in Drano Lake ranged from 349.1 to 87.3 gal/min, based on measurements of 1–4 h for the dye to flow through a net-pen; the latter estimate was used to determine fish loading densities in Drano Lake. Carrying capacity estimates based on water flow and temperature were from Leitritz and Lewis (1980).

Each year fish were transferred from the hatcheries between 7 and 18 March and were released from the study sites between 15 and 22 May. Fish were reared until water temperatures reached and remained above 60°F, at which point fish had typically reached a size of about 90 fish/lb; enteric redmouth disease caused by the bacterium *Yersinia ruckeri* had been observed above this temperature during previous rearing trials. Fish in the net-pens were fed a commercial formulated dry feed at a rate of 3–4% body weight/d by means of automatic feeders. Fish in the hatcheries were hand-fed a moist feed at the same rate. Moist feed could not be used at the off-station sites due to the lack of refrigeration. More detailed information pertaining to rearing can be found in Novotny and Macy (1991).

In 1987, prior to the rearing trials, broodstock spawned at the Little White Salmon NFH were diagnosed with infectious hematopoietic necroses (IHN). Eggs from the entire upriver bright stock at the hatchery were subsequently exposed to IHN virus in the rearing water. Therefore, transferring fish off-station was not possible because it was contrary to disease policies of the U.S. Fish and Wildlife Service and other agencies in the Columbia River basin. To continue the study, net-pens from Rock Creek were relocated to Drano Lake, the backwater of the Little White Salmon River adjacent to the hatchery. After the juvenile fish were transferred to the net-pens in Drano Lake, those in the hatchery were diagnosed with IHN and were subsequently destroyed. Fish in the net-pens were examined for the presence of disease on eight occasions over the 9-week rearing period. Results of these examinations were negative, and fish were released from the net-pens on schedule. Fish were not reared at Rock Creek in 1987.

Adult contributions of fish reared off-station was compared with fish reared and released from the Little White Salmon NFH because all upriver bright chinook salmon reared at Spring Creek NFH were released at off-station locations. All fish were tagged with coded wire tags and had their adipose fin clipped for evaluation of adult contribution. Unique tag codes were used for each of the net-pen density treatments and hatchery control groups. Fish reared at the Little White Salmon NFH were released directly from the hatchery into Drano Lake. This site is two dams and 66 river miles downstream from the Rock Creek study area.

A combination of trap nets and weirs was used to capture adults returning to Rock Creek during 1986–1989. A Merwin trap net was the most effective capture method at this site. On-site returns from the hatchery and the 1987 rearing trials in Drano Lake were recovered from the adult collection facility at the Little White Salmon NFH.

Adult recoveries were compiled from coded-wire-tag information in the Regional Mark Information System, Pacific States Marine Fisheries Commission, Portland, Oregon. Recoveries presented in this paper are estimated numbers of adults recovered listed in the database as of June 1993.

To normalize the percent recovery data, a modification of the Freeman and Tukey arcsine transformation was used (Zar 1984). Differences between arcsine-transformed percent adult contribution of the fish reared from net-pen treatments and the hatchery control were tested for significance using analysis of variance followed by Student–Newman–Keuls' multiple-range tests when significant $P \leq 0.05$ differences existed (SAS Institute 1986). Comparisons between net-pen treatments and hatchery control were not possible for the 1987 rearing year.

Rearing costs were compared using present value theory (Senn et al. 1984). This method incorporates capital costs, project life, and operating costs of each rearing method, enabling comparisons of diverse methods on a common scale. In our estimates, hatchery efficiency ratios (HER), in U.S. dollars per pound produced, were calculated based on the costs to produce a net gain of 1,000 lb of fish by means of each rearing method. More detailed information pertaining to HER calculations for the net-pen treatments may be found in Novotny and Macy (1991).

Estimates of rearing cost per adult recovery were made for each treatment based on the HER, size at release (number per pound), number of fish released, and total number of adults recovered. The cost per adult recovery, in 1987 dollars, was calculated as

$$\frac{\text{HER} \times \left(\dfrac{\text{number of juveniles released}}{\text{number per pound at release}}\right)}{\text{number of adults recovered}}$$

Results

Rearing densities in the net-pens were approximately 3–14 times lower than the recommended maximum densities based on rearing space, and were 5–15 times lower than those in the hatchery raceways (Banks et al. 1979; Table 1). However, based on flow indices, densities in the net-pens were often greater than both the recommended maxima and those in the hatchery raceways (Wedemeyer et al. 1981). Despite the high flow indices, survival during rearing was greater than 98% in all treatments; manifestations of overcrowding were not observed in any of the net-pen or hatchery treatments in 1986, but IHN was diagnosed in control fish from the Little White Salmon NFH in 1987.

The percent adult contribution of fish from the net-pen treatments in 1986 was significantly lower than from fish released from the Little White Salmon NFH (Figure 2A). Percent contribution from the regular density treatment at Rock Creek in 1986 was significantly higher than that from the double density treatment, but the triple density treatment was not significantly different from either of these. There were no significant differences in percent recovery between the four densities tested at Drano Lake in 1987.

There was a direct relation between the number of fish reared in the pens and the total number of adult recoveries per pen. This relation was most evident in the 1987 Drano Lake trials, in which each increase in density produced a significant increase in the total number of adults recovered per pen (Figure 2B). The triple density treatment at Rock Creek in 1986 resulted in significantly more adult recoveries per pen than did the regular and double densities, which were not different from each other.

Based on cost per adult recovered, the hatchery raceway was a more economical rearing method than were the net-pen treatments at Rock Creek in 1986. Rearing cost per adult recovered for the hatchery was US$11.57 in 1986 (Table 2). The cost per adult recovered from the net-pen treatments in 1986 ranged from $25.90 (triple density) to $54.83 (double density). Cost per recovery for the Drano Lake trials in 1987 ranged from $4.06 (quadruple density) to $13.82 (regular density). These costs depended primarily on the number of adults recovered in each year, as the rearing costs (HER) varied little between years. Hatchery estimates are not available for the 1987 trials due to disease.

TABLE 1.—Summary of fall chinook salmon off-station rearing treatments tested at Rock Creek (RC) in 1986 and Drano Lake (DL) in 1987, and hatchery controls at the Little White Salmon NFH (LW) in 1986. The four treatment densities ranged from 0.06 to 0.273 lb/ft^3.

Site and enclosure type	Release information				Spatial index (lb/ft^3)[a]	Flow index[b]	Adult contribution (%)
	Treatment	Rearing survival (%)	Number fish/pen	Fish/lb			
1986							
RC, net-pen	Regular	98.4	18,030	78	0.082	1.39	0.180
	Double	99.8	37,794	70	0.193	3.34	0.080
	Triple	99.7	55,055	72	0.273	4.81	0.124
LW, raceway		99.6	201,657	108	1.250	1.22	0.408
1987							
DL, net-pen	Regular	99.7	17,873	107	0.060	0.17–0.68	0.582
	Double	99.6	36,826	101	0.130	0.36–1.44	0.495
	Triple	99.2	54,380	110	0.176	0.51–2.05	0.471
	Quadruple	99.4	74,358	105	0.253	0.72–2.91	0.523

[a] Recommended maximum 0.890 lb/ft^3 (Banks et al. 1979).
[b] Flow index equals weight fish/(fish length × gal/min); flow indices at Rock Creek based on 56.5 gal/min; those at Drano Lake based on 87.3 to 349.1 gal/min. Recommended maximum is 1.26–1.80 (Wedemeyer et al. 1981).

Discussion

The lack of differences between percent adult contribution from the different density treatments indicates that the densities tested in this study did not exceed a maximum density under the rearing conditions at the off-station sites. It may also be concluded that there was no density effect over the range of densities tested. In this study, fish reared at the highest density produced the greatest number of adults per unit of rearing space. We believe this was not due to a direct relation between increased rearing density and adult production, but to the low densities used in this study. Other investigators have found inverse relations between juvenile salmonid rearing density and adult contribution (Martin and Wertheimer 1989; Banks 1992). However, the densities tested in their studies were higher than ours, possibly due to differences in flow and temperature conditions during rearing.

The maximum practical loading densities for the rearing conditions and methods we employed are difficult to estimate, as water temperatures and flow rates in backwaters of the Columbia River can be unpredictable and are beyond the control of the fish manager. Loading densities in this study were limited not by rearing space but by water flow. We based our densities on measured water flows at the sites; however, the appropriate rearing density may have been underestimated, as it did not account for water movement caused by fish in the pen. Chacon-Torres et al. (1988) reported that water exchange through net cages can be increased by swimming behavior of fish.

The rearing densities in this study were not high enough to determine the density limits in these backwaters. However, our results indicate that a density of at least 0.273 lb/ft^3 at release can be used with success in backwaters of the Columbia River. We chose to be conservative in choosing the densities in this study, and manifestations of overcrowding in the rearing enclosures were not noted. Water temperature was the primary impetus for release, because disease problems were noted during sustained water temperatures above 60°F during prior trials.

Comparisons of adult contribution of fish reared at the off-station sites and those at the Little White Salmon NFH were confounded by differences in release locations. Juveniles reared at Rock Creek were required to migrate past two dams and 66 miles of reservoir before reaching the release site of fish from the Little White Salmon NFH. We believe much of the difference in contribution between hatchery and Rock Creek treatments was due to in-river mortality of juveniles associated with the difference in release locations. Estimates of juvenile mortality associated with dam and reservoir passage vary from 10 to 45% per project (Schoeneman at al. 1961; Sims and Ossiander 1981; McKenzie et al. 1983). This includes mortality from predation by other fish, which has been estimated as 14% in John Day Reservoir alone (Rieman et al. 1991). It is

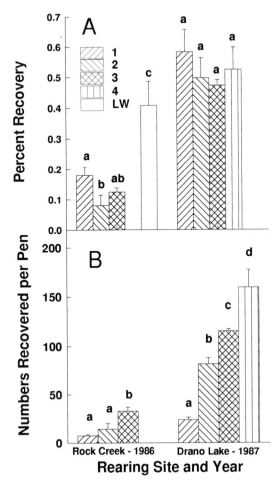

FIGURE 2.—Mean percent (A) and mean numbers (B) of adult upriver bright fall chinook salmon recovered per net-pen for regular (1), double (2), triple (3), and quadruple (4) densities of fish reared in net-pens at Rock Creek in 1986 and Drano Lake in 1987, and those reared at the Little White Salmon NFH (LW) in 1986. Vertical bars represent one standard deviation. In the same years, bars with the same letter are not significantly different from one another ($P \leq 0.05$).

TABLE 2.—Data used to calculate rearing cost per adult fish recovered. Cost per recovery is (efficiency ratio × number released ÷ number/pound) ÷ number recovered. Treatments include four rearing densities in net-pens and fish reared at the Little White Salmon NFH (Hatchery).

Year and treatment	Efficiency ratio[a]	Fish released	Fish/lb	Fish recovered	Cost per adult (US$)
1986					
Regular	7.64	205,930	78	370	54.52
Double	3.09	70,803	70	57	54.83
Triple	2.34	105,839	73	131	25.90
Hatchery	5.10	195,310	108	797	11.57
1987					
Regular	8.61	194,917	107	1,135	13.82
Double	3.96	65,880	101	326	7.92
Triple	3.15	98,005	110	462	6.07
Quadruple	2.23	121,839	105	637	4.06

[a] See Novotny and Macy (1991).

Spring Creek NFH released one reservoir upstream in the Hanford reach of the Columbia River (mean = 0.126%), and was significantly higher than that of fish released in the lower Yakima River (also one reservoir upstream; mean = 0.037%; $P = 0.0029$).

It is unfortunate that the hatchery control fish in 1987 contracted IHN and were subsequently destroyed. The comparison of adult returns from the rearing trials in Drano Lake and controls from the Little White Salmon NFH would have been a comparison of rearing methods without the confounding effects of differences in release sites. The 1987 rearing trials in Drano Lake resulted in higher adult recoveries than previous off-station trials, indicating a possible site effect. However, these differences could have also been due to changes in ocean conditions or other postrelease variables.

Fish reared in the net-pens may have been less susceptible to disease than were fish reared in the hatchery. Disease was not detected in fish transferred to the net-pens in 1987, although they were from the same bank of raceways as the fish in the hatchery that later were destroyed due to IHN. The reduced spatial densities in the net-pens is one possible reason IHN was not detected in these fish. Increased rearing density has been shown to elicit stress in juvenile salmonids, which can increase susceptibility to disease (Wedemeyer 1976; Maule et al. 1989; Salonius and Iwama 1993).

In summary, rearing fish off-station produced mixed results. Fish reared in net-pens performed well during rearing, but yielded a lower adult contribution than did fish reared and released directly at the hatchery two dams downstream. This com-

likely that much of the difference in adult recoveries between hatchery and Rock Creek treatments was due to in-river mortality of juveniles; fish were healthy during off-station rearing, grew faster, were at a more advanced stage of smoltification at release, and were larger at release compared with those reared in the hatcheries (Novotny and Beeman 1990; Novotny and Macy 1991). This premise is supported further by data indicating the contribution of fish reared at Rock Creek (mean = 0.128%) was not significantly different than that of fish from parison was confounded by differences in release sites, and disease in the hatchery limited comparisons to only one year. Based on growth and physiology during rearing and adult contribution, fish reared at the highest density (about 74,000 fish per pen) proved to be the most productive off-station method used in this study. Rearing fish in backwaters and ponds along the Columbia River may be useful as a repository for "thinning releases," as a low-cost method to hold increased production when egg take exceeds hatchery rearing capacity, or possibly as an outright addition to traditional hatchery methods.

Acknowledgments

We thank Bill Nelson, Curt Burley, and three anonymous reviewers; the staffs of the Vancouver Fisheries Research Office and Columbia River Research Laboratory who worked on this project; and the staffs of Spring Creek NFH and Little White Salmon NFH for their cooperation during this study. This project was funded by the Bonneville Power Administration (contract DE-AI79-83BP13084), U.S. Department of Energy, Portland, Oregon.

References

Banks, J. L. 1992. Effects of density and loading on coho salmon during hatchery rearing and after release. Progressive Fish-Culturist 54:137–147.

Banks, J. L., W. G. Taylor, and S. L. Leek. 1979. Carrying capacity recommendations for Olympia area national fish hatcheries. Abernathy Salmon Cultural Development Center, Longview, Washington.

Chacon-Torres, A., L. G. Ross, and M. C .M. Beveridge. 1988. The effects of fish behaviour on dye dispersion and water exchange in small net cages. Aquaculture 73:283–293.

Leitritz, E., and R. C. Lewis. 1980. Trout and salmon culture (hatchery methods). California Department of Fish and Game Fish Bulletin 164.

Martin, R. M., and A. Wertheimer. 1989. Adult production of chinook salmon reared at different densities and released as two smolt sizes. Progressive Fish-Culturist 51:194–200.

Maule, A. G., R. A. Tripp, S. L. Kaattari, and C. B. Schreck. 1989. Stress alters immune function and disease resistance in chinook salmon (*Oncorhynchus tshawytscha*). Journal of Endocrinology 120:135–142.

McKenzie, D., D. Weitkamp, T. Schadt, C. Carlie, and D. Chapman. 1983. 1982 systems mortality study. Report of Battelle Northwest Laboratories to Chelan County Public Utility District (PUD), Grant County PUD, and Douglas County PUD, Wenatchee, Washington.

Novotny, J. F., and J. W. Beeman. 1990. Use of a fish health condition profile in assessing the health and condition of juvenile chinook salmon. Progressive Fish-Culturist 52:162–170.

Novotny, J. F., and T. L. Macy. 1991. Pen rearing of juvenile fall chinook salmon in the Columbia River: alternative rearing scenarios. American Fisheries Society Symposium 10:539–547.

Novotny, J. F., T. L. Macy, and J. T. Gardenier. 1984. Pen rearing and imprinting of fall chinook salmon: annual report 1983. U.S. Fish and Wildlife Service Report (Contract DE-AI79-BP13084) to Bonneville Power Administration, Portland, Oregon.

Novotny, J. F., T. L Macy, M. P. Faler, and J. W. Beeman. 1987. Pen rearing and imprinting of fall chinook salmon: annual report 1987. U.S. Fish and Wildlife Service Report (Contract DE-AI79-BP13084) to Bonneville Power Administration, Portland, Oregon.

Rieman, B. E., R. C. Beamesderfer, S. Vigg, and T. P. Poe. 1991. Estimated loss of juvenile salmonids to predation by northern squawfish, walleyes, and smallmouth bass in John Day Reservoir, Columbia River. Transactions of the American Fisheries Society 120:448–458.

Salonius, K., and G. K. Iwama. 1993. Effects of early rearing environment on stress response, immune function, and disease resistance in juvenile coho (*Oncorhynchus kisutch*) and chinook salmon (*O. tshawytscha*). Canadian Journal of Fisheries and Aquatic Sciences 50:759–766.

SAS Institute. 1986. SAS/STAT user's guide, Version 6, 4th edition. SAS Institute Incorporated, Cary, North Carolina.

Schoeneman, D. E., R. T. Pressey, and C. O. Junge. 1961. Mortalities of downstream migrant salmon at McNary Dam. Transactions of the American Fisheries Society 90:736–742.

Senn, H., J. Mack, and L. Rothfus. 1984. Compendium of low-cost pacific salmon and steelhead trout production facilities and practices in the Pacific Northwest. Report to Bonneville Power Administration, Portland, Oregon.

Sims, C. W., and F. J. Ossiander. 1981. Migrations of juvenile chinook salmon and steelhead trout in the Snake River from 1973 to 1979, a research summary. National Marine Fisheries Service Final Report to U.S. Army Corps of Engineers, Portland, Oregon.

Wedemeyer, G. A. 1976. Physiological response of juvenile coho salmon (*Oncorhynchus kisutch*) and rainbow trout (*Salmo gairdneri*) to handling and crowding stress in intensive fish culture. Journal of the Fisheries Research Board of Canada 33:2699–2702.

Wedemeyer, G. A., R. L. Saunders, and W. C. Clarke. 1981. The hatchery environment required to optimize smoltification in the artificial propagation of anadromous salmonids. Pages 6–20 *in* L. J. Allen and E. C. Kinney, editors. Proceedings of the bio-engineering symposium for fish culture. American Fisheries Society, Fish Culture Section, Bethesda, Maryland.

Zar, J. H. 1984. Biostatistical analysis. Prentice-Hall, Englewood Cliffs, New Jersey.

Maintenance of Stock Integrity in Snake River Fall Chinook Salmon

R. M. Bugert, C. W. Hopley, C. A. Busack, and G. W. Mendel

Washington Department of Fish and Wildlife
Post Office Box 43135, Olympia, Washington 98501, USA

Abstract.—Production and survival of Snake River fall chinook salmon *Oncorhynchus tshawytscha* have progressively declined, primarily as a result of hydroelectric development in the Snake and lower Columbia rivers. In 1976 the U.S. Congress authorized the Lower Snake River Compensation Plan, under which Lyons Ferry State Fish Hatchery was built to compensate for the loss of 18,300 adult fall chinook salmon of Snake River stock. It became apparent however, that this stock was at critically low levels and could become extinct before the hatchery would become operational. An egg-bank program was established to provide interim propagation. From 1976 though 1984, adults were trapped on the Snake River and their progeny were marked and reared separately at several locations. This stock was then transferred to Lyons Ferry in 1984. Based upon electrophoretic examination of allele frequencies of 30 variable loci, we found no evidence of genetic difference between the fish maintained in the egg bank and the ancestral Snake River population. This program gave Lyons Ferry sufficient broodstock to start rebuilding this depressed run—escapement to the Snake River increased from 2,000 in 1985 to 8,400 in 1987. Concurrent with this rebuilding, however, chinook salmon released from other hatcheries strayed in increasing numbers into the Snake River and threatened the genetic integrity of this important stock. By 1989, stray chinook salmon constituted up to 43% of marked fish recovered for Lyons Ferry broodstock; presumably the same stray rate occurred to naturally spawning fish in the Snake River. A second egg-bank program was then initiated, this time with an objective to eliminate stray fish from the gene pool, both at Lyons Ferry and natural spawning grounds. From 1990 through 1993, fall chinook salmon were trapped at two dams on the Snake River and transported to Lyons Ferry and only known Lyons Ferry, Snake River-origin chinook salmon (verified by coded wire tag) were retained for production. We were unable to control escapement of unmarked stray chinook salmon into the Snake River spawning grounds, which introduced foreign genes into the natural population. Based upon analysis of temporal allele frequency changes at six loci, we suggest that Lyons Ferry stock are now genetically more similar to the original Snake River stock than are the fish spawning naturally in the river. Moreover, the natural spawners may deviate even further from their ancestral genetic composition before the straying problem can be controlled. In 1991, Snake River fall chinook salmon was listed as Threatened under the Endangered Species Act (ESA). Chinook salmon produced at Lyons Ferry are considered part of this evolutionarily significant unit, largely a result of the effort to protect genetic integrity of this population through artificial propagation. Under the ESA Recovery Plan, chinook salmon from Lyons Ferry will be used to rebuild natural production in the Snake River.

Historically, chinook salmon *Oncorhynchus tshawystcha* was abundant in the Snake River, the largest tributary to the Columbia River. Three races of chinook salmon are typically recognized in the Snake River, separated by entry time of adults to the river. Two of these races (spring and summer) exhibit a stream-type life history pattern (Healey 1983), in contrast to the ocean-type strategy of fall chinook salmon, which typically migrate to the sea as subyearlings and have a more northerly ocean distribution. Also, the primary spawning area of the fall race is distinct from the spring and summer race: fall chinook salmon use the larger, open areas of the main stem, whereas the spring and summer runs spawn and rear in the tributaries. The spring and summer races are aggregates of many distinct stocks (Ricker 1972; Thorpe et al. 1981); the extant race of fall chinook salmon in the Snake River is recognized as one interbreeding stock (Nehlson et al. 1991; WDF et al. 1993).

Prior to 1900, fall chinook salmon was widely distributed in the relatively complex Columbia River basin and contributed substantially to commercial and tribal fisheries. Within the Snake River, fall chinook salmon ranged from the confluence with the Columbia River upstream to a natural barrier at Shoshone Falls, Idaho (Fulton 1968; Howell et al. 1985), a length of more than 1,020 km. Construction of nine main-stem dams in the Snake River basin eliminated about 860 km of this habitat over a 20-year period (Irving and Bjornn 1981; Figure 1). Fall chinook salmon was particularly susceptible to effects from hydroelectric development because of inundation of its preferred spawning

FIGURE 1.—Snake River basin, showing locations of Lyons Ferry State Fish Hatchery and the five main-stem dams that inundated or blocked access to chinook salmon habitat. The reach of the Snake River between the Clearwater River confluence and Hells Canyon Dam (shown in bold) is the remaining area for fall chinook salmon natural production.

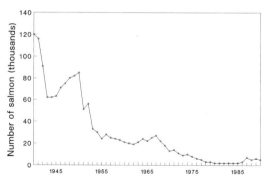

FIGURE 2.—Estimated number of fall chinook salmon entering the Snake River from 1940 to 1990 (adapted from USACE 1964, 1990; Fulton 1968).

habitat in the main stem and losses to the relatively small subyearling outmigrants during the period of the year when flow (and associated spill) levels are lowest in the Snake and Columbia rivers (Waples et al. 1991a; Connor et al. 1992). The rapid development of main-stem dams and extensive harvest in the high seas and lower Columbia River decimated this run to one-hundredth of its historical level by 1970 (USACE 1964; Fulton 1968; Figure 2).

The Lower Snake River Fish and Wildlife Compensation Plan

Federal legislation under the Water Resources Development Act of 1976 (Public Law 94-587) authorized the Lower Snake River Fish and Wildlife Compensation Plan (LSRCP). This plan provided hatchery compensation for passage mortality and loss of spawning habitat caused by construction and operation of the four lower Snake River hydropower projects (Ice Harbor, Lower Monumental, Little Goose, and Lower Granite; Figure 1). As a result of that plan, Lyons Ferry State Fish Hatchery (SFH) was designed and began operation by the Washington Department of Fisheries (WDF) in 1984. A partial objective of this hatchery was to compensate for the loss of 18,300 adult fall chinook salmon of Snake River stock (USACE 1975). Some measures to mitigate passage losses through the dams were initiated as a result of the Pacific Northwest Electric Power Planning and Conservation Act of 1980 (Public Law 96-501), but the primary means to compensate for losses was to be through the LSRCP hatcheries.

Egg-Bank Program

Purpose and Strategy

It was recognized early in the hatchery site-selection and planning stages that Snake River fall chinook salmon had a critically depressed status and was under consideration for classification as an Endangered species (Utter et al. 1982). Fishery biologists generally agreed that the run could disappear in the years between congressional appropriation and actual construction of the Snake River hatchery. The Snake River Egg-Bank Program resulted from this concern. The egg-bank concept had two components: (1) capture of natural spawners at Ice Harbor Dam for artificial propagation, and (2) release of their progeny in a lower Columbia River hatchery. Outmigrant juveniles and returning adults would therefore avoid mortality associated with passage at Columbia and Snake river dams.

The Snake River Egg-Bank Program was established in 1976 with two goals: (1) to provide an interim adult-holding and juvenile-rearing program for Snake River fall chinook salmon until Lyons Ferry SFH could be constructed; and (2) to maintain and monitor the genetic integrity of this stock during this interim period. The first effort to establish an egg bank resulted from a National Marine

TABLE 1.—Juvenile to adult survival and adult contribution to production of 1977 through 1982 broods of Snake River fall chinook salmon released from Kalama Falls State Fish Hatchery in the egg-bank program.

Brood year	Juveniles released	Adult returns	Juvenile to adult survival (%)	Eggs contributed
1977	446,889	720	0.16	1,109,400
1978	183,034	29	0.02	41,000
1979	267,813	1,546	0.58	0[a]
1980	490,782[b]	386	0.08	833,700
1981	707,723[b]	848	0.12	1,683,000
1982	451,000[b]	606	0.11	1,249,900

[a] Eggs from this brood were inadvertently mixed with Priest Rapids stock fall chinook salmon eggs. All fish were given a unique mark at release, and progeny were removed from the egg-bank program.
[b] Production at Kalama Falls in 1980 to 1981 included progeny of adults trapped at Ice Harbor Dam and returns of previous releases of Snake River fall chinook salmon from Kalama Falls.

Fisheries Service (NMFS) proposal approved by the Columbia River Fisheries Council (Council) and funded by the Pacific Northwest Regional Commission. Adults were captured at Little Goose Dam by NMFS, which incubated the eggs at Lower Granite Dam (Figure 1). Eventually, eggs were transferred to Bonneville SFH (operated by Oregon Department of Fish and Wildlife) and were hatched and reared there. After much deliberation, the fish were marked with coded wire tags (CWT) and sent to Kalama Falls SFH (operated by WDF) and released. The results of this first egg-bank effort were not rewarding, but the concept had been established in practice.

Lower Columbia River Program

The Council had discussed various problems with the trap facility at Little Goose Dam and suggested Ice Harbor Dam as a potential trap site. The Artificial Production Committee (APC) was assigned the development of an egg-bank plan for 1977. The plan was provided by WDF and approved by the Council in July 1977. The University of Idaho was to conduct the trap operation, based upon its previous experience trapping adults at hydropower projects, with funding for the adult transport and hatchery rearing supplied by NMFS. The adult-holding facility was first to be Klickitat SFH (operated by WDF), but was later changed to Tucannon SFH (Washington Department of Wildlife) because of logistical considerations. Klickitat was retained as the juvenile-rearing facility, however, because it had adequate rearing area and water supply. Kalama Falls SFH was again chosen as the release site. All chinook salmon in the program were marked (ventral-fin clip) to allow separation from the Kalama River stock as adults. Some of the marked fish were tagged with CWTs to assess survival and contribution rates (Table 1). This program continued in 1978 and 1979 (Table 2). Adults resulting from the WDF Kalama Falls releases, readily separable from the indigenous Kalama Falls returns by their ventral-fin clips, were used by WDF as egg-bank broodstock beginning in 1980.

Upper Columbia River Program

In 1978, the U.S. Fish and Wildlife Service (USFWS) proposed to the Columbia Basin Fisheries Technical Committee that Hagerman (Idaho) National Fish Hatchery (NFH) be included in the Snake River Egg-Bank Program. Plans were to rear subyearling smolts for release in the middle Snake River (upstream of Lower Granite Dam). The rationale for this decision was to maintain a second source of the Snake River stock in case a major disease outbreak, or some other factor, would decimate the primary source. The Hagerman NFH plan was approved by the Technical Committee and the APC; transfer of eyed eggs from Tucannon SFH began in 1978. At about this time, the Council stipulated that the egg-bank program would trap 400 fish or 50% of the run, whichever was lowest (Figure 3). The APC also determined that the egg take would be split evenly between the lower river program (Kalama Falls SFH) and the upper river program (Hagerman NFH). The following year, Dworshak NFH began receiving adults trapped at Ice Harbor Dam to supply Hagerman (Table 2).

Most of the 1979 through 1984 broods of fall chinook salmon reared at Hagerman NFH were released at various locations in the middle Snake River between Hells Canyon Dam and Lower Granite Dam. Several groups were also transported for release near the Columbia River Estuary (Bjornn and Ringe 1984). A portion of each brood year, except the 1982 brood, was tagged with CWTs (Roseburg et al. 1991). Adult survival and contribution rates ranged from 0.01 to 0.45% (Table 3).

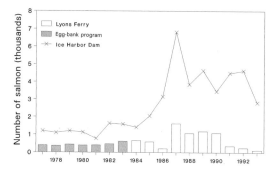

FIGURE 3.—Counts of fall chinook salmon at Ice Harbor Dam on the lower Snake River compared with collections for the Snake River Egg-Bank Program and for Lyons Ferry State Fish Hatchery.

Tags recovered from adults at Ice Harbor Dam and Dworshak NFH also indicated that hatchery fish from Hagerman releases escaped to the Snake River by 1981.

Captive Brood Program

For the 1980 through 1983 brood years, NMFS collected 15,000 eyed eggs from the egg-bank program for initial rearing in freshwater and later rearing to maturity in marine net-pens. This was an experimental attempt to provide a large number of Snake River fall chinook salmon within a generation for future recovery plans. The experimental program had difficulties with seawater tolerance of subyearlings and high losses related to bacterial kidney disease (Harrell et al. 1987) and was discontinued in 1985.

Inception of Lyons Ferry State Fish Hatchery

Beginning with the 1980 brood year, two factors occurred that improved the future of the Snake River Egg-Bank Program: (1) funding for the trap operation was provided by the USFWS under the LSRCP; and (2) construction of the fall chinook salmon hatchery in the Snake River was underway. Lyons Ferry SFH was completed in time to hold adults returning in 1984, eliminating the need to hold adults at Tucannon or Dworshak hatcheries. Trap operations continued at Ice Harbor Dam, but the adults were taken directly to Lyons Ferry (Ringe and Bugert 1989; Table 2).

Lyons Ferry SFH then began the process of broodstock development. In 1984, Lyons Ferry received all eyed eggs from the Snake River fall chinook salmon spawned at Kalama Falls SFH. This transfer would continue through 1986 (Figure 4), the last year significant numbers of chinook salmon from the egg-bank program returned to Kalama Falls. The 1982 brood was the last year the Snake River fall chinook salmon were released from Kalama Falls (Table 2). Beginning in 1986, voluntary returns of the 1983 brood released from Lyons Ferry added to the broodstock building process.

From 1984 to 1986, eyed eggs from the egg-bank program were transported from Kalama Falls SFH to Lyons Ferry SFH. In these initial years of operation, the egg-bank program provided 25 to 62% of production at Lyons Ferry (Figure 4). Some adults

TABLE 2.—Fish hatcheries used for adult holding and juvenile rearing and rivers in which Snake River fall chinook salmon were released through the egg-bank program for the 1977 to 1985 brood years.

Brood year	Adult holding	Juvenile rearing	Smolt releases
1977	Tucannon	Klickitat	Kalama River
1978	Tucannon	Klickitat	Kalama River
		Hagerman	Middle Snake River
1979	Tucannon	Klickitat	Kalama River
		Hagerman	Middle Snake River
1980	Tucannon	Klickitat	Kalama River
	Dworshak	Hagerman	Middle Snake River
1981	Tucannon	Klickitat	Kalama River
	Dworshak	Hagerman	Middle Snake River
1982	Tucannon	Klickitat	Kalama River
	Dworshak	Hagerman	Middle Snake River
1983	Tucannon	Klickitat	Lower Snake River
	Dworshak	Hagerman	Middle Snake River
1984	Lyons Ferry	Lyons Ferry	Lower Snake River
		Hagerman	Middle Snake River
1985	Lyons Ferry	Lyons Ferry	Lower Snake River

TABLE 3.—Juvenile to adult survival and contribution rate of 1978 to 1984 broods of Snake River fall chinook salmon released from Hagerman National Fish Hatchery in the egg-bank program.

Brood year	Juveniles released	Release location	Adult returns	Juvenile to adult survival (%)[a]
1978	93,000	Columbia Estuary	99	0.11
	45,361	Middle Snake River	6	0.01
1979	56,000	Columbia Estuary	24	0.04
	165,500	Middle Snake River	496	0.30
1980	61,134	Columbia Estuary	171	0.28
	120,147	Middle Snake River	377	0.31
1981	475,116	Middle Snake River	2,139	0.45
1982	78,900	Middle Snake River		[b]
1983	427,191	Middle Snake River	814	0.19
1984	128,229	Middle Snake River	549	0.43

[a] Survival rates are based upon expanded mark (coded wire tag) placement rates of juveniles released from the hatchery and mark sampling rates of adults recovered.
[b] The 1982 brood was not marked.

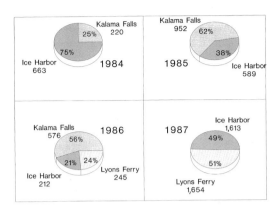

FIGURE 4.—Sources and numbers of Snake River fall chinook salmon contributed to Lyons Ferry State Fish Hatchery broodstock during its first 4 years of operation.

from the 1979 to 1985 Hagerman NFH releases were trapped at Ice Harbor Dam for Lyons Ferry broodstock. Trap operations at Ice Harbor Dam continued after 1985, but were conducted by WDF to supplement voluntary returns to Lyons Ferry. All releases of Snake River fall chinook salmon since 1985 have been from Lyons Ferry.

By 1987 the lower river program was completed. Broodstock were obtained primarily from voluntary returns to Lyons Ferry SFH and secondarily from Ice Harbor Dam. Adults from these two sources averaged about one-third of total escapement to the Snake River (Figure 3). In 1990, the NMFS upstream migrant trap at Lower Granite Dam was established as an ancillary source of broodstock (Bugert and Hopley 1991). In the initial years (1984–1989), broodstock from all sources were held and spawned together. Concerns about the high stray rate of chinook salmon from hatchery programs outside the Snake River into Lyons Ferry broodstock caused a departure from normal collection, spawning, and holding procedures in 1990.

Maintenance of Stock Integrity

Electrophoretic data dating back to 1977 exists for chinook salmon collected at Ice Harbor Dam, Lyons Ferry SFH broodstock, and adjacent Columbia River fall chinook salmon stocks. Adults collected at Ice Harbor Dam for the egg-bank program were sampled annually by NMFS from 1977 to 1982. Lyons Ferry broodstock was sampled by NMFS in 1985 and annually since then by WDF. Hatchery and natural spawners on the middle Columbia River (that reach upstream of the Snake River confluence) have also been frequently sampled since 1977. Sample sizes in general were large, usually 100 or more. The loci analyzed were those that best discriminate between Snake River and middle-Columbia River chinook salmon. In general, the loci are those used by Utter et al. (1982) to establish the genetic distinctness of middle-Columbia River and Snake River stocks, but we excluded some used by Utter et al. (1982) and added $PEPA*$ (which codes for dipeptidase, enzyme number 3.4.-.- [IUBNC 1984]).

Under contract to the USFWS, WDF geneticists collected electrophoretic samples and compared by chi-square heterogeneity test the 1986 Snake River fall chinook salmon that returned to Kalama Falls SFH and those that returned to the Snake River. Significant differences ($P = 0.03$) were found at two loci, $mAH\text{-}4*$ (aconitate hydratase, 4.2.1.3) and $MPI*$ (mannose-6-phosate isomerase, 5.3.1.8), but overall the test, based upon examination of allele frequencies of 30 variable loci, was not significant ($P = 0.32$). This result indicated a general maintenance of Snake River biochemical genetic characteristics through the egg-bank program.

The Snake River Egg-Bank Program was successful because of three factors: (1) production at Kalama Falls SFH contributed substantially to the broodstock during the first 3 years Lyons Ferry SFH was operational; (2) genetic integrity of the stock was maintained through conservative broodstock management by several hatcheries; and (3) Hagerman NFH augmented natural production in the Snake River and contributed to Lyons Ferry broodstock.

Stray Fish Management

Coded-Wire-Tag Recoveries

From 1987 to 1989, we noticed a steady increase in the percentage of stray fall chinook salmon in the Lyons Ferry SFH broodstock. Our analysis of CWTs recovered at Lyons Ferry by means of standard expansion techniques showed that strays constituted 7% of the return in 1987, 16% in 1988, and 43% in 1989 (these percentages include all age-classes). Releases of middle-Columbia River fall chinook salmon into the Umatilla River were the predominant strays to Lyons Ferry (Figure 5). Low flows and high water temperatures in lower Umatilla River probably discouraged the chinook salmon from returning to this stream. The Snake River is the next major left bank tributary to the Columbia River these chinook salmon would encounter when migrating upstream (Figure 1). At

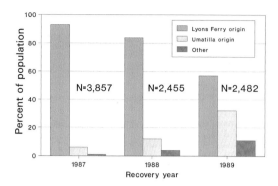

FIGURE 5.—Stock composition of fall chinook salmon at Lyons Ferry State Fish Hatchery from 1987 to 1989, based upon coded-wire-tag expansions.

TABLE 4.—Incidence of fall chinook salmon at Ice Harbor Dam, Lyons Ferry State Fish Hatchery (SFH), and Lower Granite Dam from 1990 to 1993. Values are expressed as percentages and based upon expanded mark (coded wire tag) placement and recovery rates.

Recovery site and year	Hatchery origin		
	Lyons Ferry SFH	Umatilla	Other
Ice Harbor Dam			
1990	67.8	30.0	2.2
1991	51.8	32.0	16.2
1992	69.7	27.1	3.2
1993	61.7	32.6	5.7
Lyons Ferry SFH			
1990	83.7	14.0	2.3
1991	82.1	16.1	1.8
1992	91.5	7.7	0.8
1993	88.7	9.7	1.6
Lower Granite Dam			
1990	75.0	22.9	2.1
1991	74.1	25.0	0.9
1992	83.1	8.6	8.3
1993	52.5	45.1	2.4

that time, we did not know if stray chinook salmon were trapped only at Ice Harbor Dam or if strays volunteered into Lyons Ferry and entered the middle Snake River upstream of Lower Granite Dam.

The 1989 brood was being reared at Lyons Ferry SFH when this high stray rate was detected. Under authority of a multi-agency agreement, WDF marked (CWT and blank-wire tag) and released the entire production (3,044,000) as subyearlings (Bugert et al. 1991). A strategy was then developed to ensure the entire 1989 brood chinook salmon and any stray chinook salmon would be eliminated from Lyons Ferry broodstock when they returned as adults.

Broodstock Management

In 1990, WDF and NMFS collaborated to collect wire-tagged fall chinook salmon at Lower Granite Dam annually for transport to Lyons Ferry SFH. The NMFS trap at Lower Granite Dam had the capability to detect and intercept most magnetized-wire-tagged fish. This trap could then prevent tagged strays from spawning in the middle Snake River (Figure 1). Total interception of strays was not possible, however, as the trap had no capability to detect unmarked stray chinook salmon. At that time, the average tag rate of the fall chinook salmon released in the Umatilla River was 14%. The objectives of the trap operations at Lower Granite Dam were (1) to determine the stray rate of hatchery chinook salmon entering the middle Snake River; (2) if the stray rate was significant, to prevent the tagged strays from spawning with the natural population; and (3) to supplement Lyons Ferry broodstock. Collections at Ice Harbor Dam and voluntary returns to Lyons Ferry continued. Chinook salmon collected at Ice Harbor and Lower Granite dams were placed in a holding pond separate from volunteers. All hatchery volunteers and fish collected at Ice Harbor Dam were given unique-by-week, color-coded tags in the holding ponds to allow recognition by week of arrival. Broodstock collection occurred over a 6-week period at Ice Harbor Dam and 12 weeks at Lyons Ferry, providing 18 location–week combinations. Fish collected from Lower Granite Dam were given a numbered jaw tag the day they were trapped.

Adult trapping by NMFS at Lower Granite Dam began in 1990, although some jack sampling was done in earlier years (Seidel et al. 1988). The trap was not capable of intercepting all marked fall chinook salmon; roughly 10% of the wire-tagged chinook salmon and all unmarked chinook salmon would pass the trap (Mendel et al. 1994a), allowing some gene flow between hatchery and natural chinook salmon on the spawning grounds. Based upon expansions of CWT recoveries from 1990 to 1993, fall chinook salmon from hatcheries outside the Snake River basin accounted for at least 30% of adults collected at Ice Harbor Dam, 12% of Lyons Ferry volunteers, and 32% of adults collected at Lower Granite Dam (Bugert et al. 1991; Mendel et al. 1992, 1994b; LaVoy 1994; Table 4).

Spawning Practices

During spawning, ripe fish were killed and set aside. Marked fish had CWTs removed and read, and the fish's color tag, indicating week of arrival, was recorded. The information on arrival time and CWT origin was applied to unmarked fish that arrived that same week. We could then assess the ratio of stray fish among unmarked fish on a weekly basis. Gametes from known Lyons Ferry SFH chinook salmon were fertilized and incubated separately from those of strays or fish of unknown origin. Fish without CWTs were mated together by color of external tags, hence fish of the same week and location of collection were mated together. Progeny of foreign-tagged and untagged adults were to be used for Lyons Ferry broodstock if the stray rate was below an acceptable level. We operated under the assumption that stray incidence might be less on a given week than it was on others. If the stray rate of marked fish from a given week was tolerable, we would keep progeny of unmarked fish from that same week. Stray levels did not vary by week of arrival, however, so only progeny of Lyons Ferry tagged adults were retained. The remainder were transported as eyed eggs or fry to Klickitat SFH on the lower Columbia River. Beginning with the 1990 brood, all fish released from Lyons Ferry were tagged with CWTs, greatly increasing the number of tagged adults available for future broodstock. Lyons Ferry retained only tagged Snake River origin chinook salmon from 1990 though 1993.

Stray Rates Prior to 1987

Analysis of CWT recoveries from the Snake River Egg-Bank Program showed a low incidence of strays earlier in the program. Coded-wire-tagged chinook salmon began appearing in 1981 in collections at Ice Harbor Dam. In 1984, the first year of Lyons Ferry SFH operations, tags with origins other than Snake River began to appear in the broodstock. The most common source of strays was from chinook salmon planted near the mouth of the Umatilla River, which at that time used a lower-Columbia River tule stock. (The term tule refers to an aggregate of fall chinook salmon stocks originating in the lower Columbia River; bright refers to those stocks originating in the middle Columbia and Snake rivers. In general, these stocks can be readily separated when mature based upon visual appearance.) This operation accounted for 5% of the adult (age 3 or older) recoveries in 1984, 10% in 1985, and 2% in 1986. Tules were routinely removed from the spawning population in the egg-bank program and at Lyons Ferry, thereby removing these strays from the gene pool.

Electrophoretic Analysis

To assess genetic introgression from stray chinook salmon, WDF and NMFS geneticists compared polymorphic loci common to the Snake River stock and the middle-Columbia stock (which was the donor source for the Umatilla program). Allele frequency changes at six genetic systems common to these groups demonstrated distinct and clear temporal trends (Figure 6). Two series were plotted, one for the Snake River stock (sampled at Ice Harbor Dam and Lyons Ferry SFH) and one for middle-Columbia River stock (sampled at Priest Rapids SFH and from natural spawners in the Hanford Reach). In years for which data for multiple collections are available, composite allele frequencies were calculated as the unweighted mean of the individual frequencies. In 1990 we sampled two collections at Lyons Ferry, one of known Lyons Ferry chinook salmon, determined through CWT analysis, and one of untagged fish randomly sampled during spawning (this sample would therefore include strays). We then plotted the data from untagged fish separately to evaluate effects of using only tagged fish at the hatchery.

In several series, the untagged 1990 Lyons Ferry SFH collection appears more similar to the middle-Columbia stock than it does to the tagged collection (with strays removed). Frequencies at *sAH** were too erratic to indicate trends, although it is clear that the untagged 1990 Lyons Ferry chinook salmon were more similar to middle-Columbia chinook salmon than they were to the tagged chinook salmon. The two series are nearly parallel for *MPI** frequencies. The two series of *PEPB-1** (tripeptide aminopeptidase, 3.4.-.-) frequencies appear approximately parallel, and both appear to be headed downward. The Snake River stock frequencies at *sSOD-1** (superoxide dismutase. 1.15.1.1), *sIDHP-1,2** (isocitrate dehydrogenase, 1.1.1.42), and *PEPA** exhibit unequivocal trends in the direction of the middle-Columbia series, which in each case remains stable.

At all loci where a strong trend toward middle-Columbia frequencies was observed, the trend was apparent before releases of fall chinook salmon into the Umatilla River began in 1984. This suggests introgression from the middle-Columbia stock during the egg-bank program. The possibility that some middle-Columbia chinook salmon were trapped at Ice Harbor Dam (Mendel et al. 1993) is thus compatible with the genetic data. Trap operations would have also exacerbated the collection of strays if chinook salmon from Umatilla also "dip" into

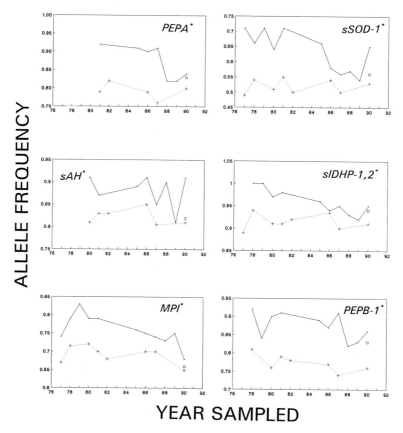

FIGURE 6.—Temporal allele frequency changes for six loci of fall chinook salmon sampled on the Snake River (solid lines) and middle Columbia River (broken lines). The 1990 Snake River value is for the known (tagged) collection at Lyons Ferry State Fish Hatchery; the open square represents the untagged collection, which includes strays.

Snake River. The hypothesis that Umatilla straying is largely an artifact of Ice Harbor Dam broodstock collection, however, is not supported by CWT data. The percentages of wire-tagged chinook salmon from Umatilla at Lower Granite Dam from 1990 to 1992 was generally the same as the percentage taken at Ice Harbor Dam (Table 4).

Despite the evidence of a genetic impact from strays, the genetic distinction between Snake River and middle-Columbia River chinook salmon remains. The difference in allele frequencies between 1990 tagged Lyons Ferry SFH and 1990 Priest Rapids SFH (middle-Columbia stock) fish by chi-square heterogeneity test is highly significant ($P = 0.00002$). A similar comparison of 1986 Lyons Ferry chinook salmon and 1990 tagged Lyons Ferry chinook salmon was not significant ($P = 0.7$), indicating the effect of strays from Umatilla was reduced by restricting broodstock to tagged Lyons Ferry salmon.

There is not much electrophoretically detectable genetic variation in late-returning stocks of chinook salmon to the middle-Columbia and Snake rivers (Schreck et al. 1986; Winans 1989; Waples et al. 1991b), yet some of the diversity is represented by the Lyons Ferry SFH population. This indicates the stock's importance in maintaining genetic diversity in upriver Columbia fall chinook salmon. The stocks that stray into the Snake River are genetically similar to that maintained at Lyons Ferry, so perhaps outbreeding depression is not a large concern. Moreover, many of the hatchery upriver stocks, because they originated from main-stem Columbia River interceptions (Chapman et al. 1991, 1994), may have Snake River ancestry.

Natural Snake River fall chinook salmon were continuously recruited into hatchery production during the egg-bank program. Likewise, the composition of Lyons Ferry SFH adults in the run over Lower Granite Dam has been continuous and stable

(Table 4), so gene flow from the hatchery to the spawning grounds occurs. From this, one would assume genetic similarity, yet the natural and hatchery spawners differ in the extent to which effects of straying can be controlled. The 1989 hatchery brood, composed of gametes from 43% strays, was marked at release for later recognition and elimination from the Lyons Ferry broodstock. In contrast, the 1989 natural production presumably has sustained the effect of this stray rate. In subsequent years, strays have been excluded from the Lyons Ferry broodstock by exclusive use of Lyons Ferry tagged fish, whereas trap operations at Lower Granite Dam could intercept most wire-tagged fish but no unmarked strays, thus allowing gene flow of stray hatchery chinook salmon into the Snake River natural spawners. The Lyons Ferry stock may now be genetically more similar to the original Snake River stock than are the fish spawning naturally above Lower Granite Dam. Moreover, the natural spawners may deviate further from their ancestral genetic composition before the straying problem can be controlled.

The Role of the Hatchery in the Endangered Species Act

In 1991, NMFS concluded that Snake River fall chinook salmon faced a high risk of extinction if factors affecting the population remain unchanged (Waples et al. 1991a). This conclusion was prompted by a substantial review of the status of this stock for inclusion as Threatened or Endangered under the U.S. Endangered Species Act of 1973 (ESA; 16 U.S.C.A. §§1531 to 1544). The genetic status of the Lyons Ferry SFH stock then became critically important to the technical decision to list the Snake River fall chinook salmon under the ESA and to use the hatchery stock in the resultant recovery plan. Chinook salmon produced at Lyons Ferry were considered part of the evolutionarily significant unit (Waples 1991), largely a result of the effort to protect genetic integrity of the Snake River population through artificial propagation. Under the ESA Recovery Plan, chinook salmon from Lyons Ferry will be used to rebuild natural production in the Snake River.

Acknowledgments

Funding for the Lyons Ferry SFH operations and evaluations and the adult trapping at Ice Harbor Dam was provided by the U.S. Fish and Wildlife Service, Boise Idaho, under the Lower Snake River Fish and Wildlife Compensation Plan. Trap operations at Lower Granite Dam were led by National Marine Fisheries Service, which also supplied electrophoretic data on early collections at Ice Harbor Dam, the middle Columbia River, and Lyons Ferry SFH.

References

Bjornn, T. C., and R. R. Ringe. 1984. Homing of hatchery salmon and steelhead allowed a short-distance voluntary migration before transport to the lower Columbia River. Idaho Cooperative Fishery Research Unit, Technical Report 84-1, Moscow.

Bugert, R., and B. Hopley. 1991. Fall chinook salmon trapping on the Snake River in 1990. Completion Report (Cooperative Agreement 14-16-0001-90525) to U.S. Fish and Wildlife Service. Washington Department of Fisheries, Olympia.

Bugert, R., and six coauthors. 1991. Lower Snake River compensation plan, Lyons Ferry fall chinook salmon hatchery evaluation program, 1990. Report (Cooperative Agreement 14-16-0001-90525) to U.S. Fish and Wildlife Service. Washington Department of Fisheries, Olympia.

Chapman, D., and ten coauthors. 1991. Status of Snake River chinook salmon. Don Chapman Consultants, Boise, Idaho.

Chapman, D., and eight coauthors. 1994. Status of summer/fall chinook salmon in the mid-Columbia region. Don Chapman Consultants, Boise, Idaho.

Connor, W. P., H. Burge, and R. Bugert. 1992. Migration timing of natural and hatchery fall chinook in the Snake River basin. Pages 46–56 *in* J. Congleton, editor. Proceedings of the workshop on passage and survival of juvenile chinook salmon migrating from the Snake River basin. University of Idaho, Moscow.

Fulton, L. A. 1968. Spawning areas and abundance of chinook salmon (*Oncorhynchus tshawytscha*) in the Columbia River basin—past and present. U.S. Fish and Wildlife Service Special Scientific Report Fisheries 571.

Healey, M. C. 1983. Coastwide distribution and ocean migration patterns of stream- and ocean-type chinook salmon, *Oncorhynchus tshawytscha*. Canadian Field-Naturalist 97:427–433.

Harrell, L. W., T. A. Flagg, and F. W. Waknitz. 1987. Snake River fall chinook salmon brood-stock program, 1981–1986. Final Report (Contract DE-A179-83-BP39642) to Bonneville Power Administration, Portland, Oregon.

Howell, P., and eight coauthors. 1985. Stock assessment of Columbia River anadromous salmonids, volume 1: chinook, coho, chum, and sockeye stock summaries. Report (Contract DE-A179-84BP12737) to Bonneville Power Administration, Portland, Oregon.

Irving, J. S., and T. C. Bjornn. 1981. A forecast of abundance of Snake River fall chinook salmon. Report (Contract 81-ABC-00042) to National Marine Fisheries Service. University of Idaho, Moscow.

IUBNC (International Union of Biochemistry, Nomenclature Committee). 1984. Enzyme nomenclature. Academic Press, San Diego, California.

LaVoy, L. 1994. Stock composition of fall chinook at

Lower Granite Dam in 1993. Columbia River Laboratory Progress Report 94-10. Washington Department of Fish and Wildlife, Battle Ground.

Mendel, G., D. Milks, M. Clizer, and R. Bugert. 1993. Upstream passage and spawning of fall chinook salmon in the Snake River. Pages 1–58 in H. L. Blankenship and G. W. Mendel, editors. Upstream passage, spawning, and stock identification of fall chinook salmon in the Snake River, 1992. Bonneville Power Administration, Project 92-046, Portland, Oregon.

Mendel, G., L. Ross, and R. Bugert. 1994a. Fall chinook salmon trapping on the Snake River in 1992. Completion Report (Cooperative Agreement 14-16-0001-92542) to U.S. Fish and Wildlife Service. Washington Department of Fisheries, Olympia.

Mendel, G., and six coauthors. 1992. Lower Snake River compensation plan, Lyons Ferry fall chinook salmon hatchery evaluation program, 1991. Report (Cooperative Agreement 14-16-0001-91534) to U. S. Fish and Wildlife Service. Washington Department of Fisheries, Olympia.

Mendel, G., and six coauthors. 1994b. Lower Snake River compensation plan, Lyons Ferry Hatchery evaluation program, fall chinook salmon, 1992. Report (Cooperative Agreement 14-16-0001-92542) to U.S. Fish and Wildlife Service. Washington Department of Fisheries, Olympia.

Nehlson, W., J. E. Williams, and J. A. Lichatowich. 1991. Pacific salmon at the crossroads: stocks at risk from California, Oregon, Idaho, and Washington. Fisheries (Bethesda) 16(2):4–21.

Ricker, W. E. 1972. Hereditary and environmental factors affecting certain salmonid populations. Pages 27–160 in R. D. Simon and P. A. Larkin, editors. The stock concept in Pacific salmon. H. R. MacMillan Lectures in Fisheries. University of British Columbia, Vancouver.

Ringe, R., and R. Bugert. 1989. Fall chinook trapping at Ice Harbor Dam in 1989. Completion Report (Cooperative Agreement 14-16-0001-87512) to U.S. Fish and Wildlife Service. Washington Department of Fisheries, Olympia.

Roseburg, R., H. Burge, B. Miller, and D. Diggs. 1991. A review of coded-wire tagged fish released from Dworshak, Kooskia, and Hagerman National Fish Hatcheries, Idaho, 1976–1990. U.S. Fish and Wildlife Service, Dworshak Fishery Resource Office, Ahsahka, Idaho.

Schreck, C. B., H. W. Li, R. C. Hjort, and C. S. Sharpe. 1986. Stock identification of Columbia River chinook salmon and steelhead. Final Report, Project 83-451, to Bonneville Power Administration, Portland, Oregon.

Seidel, P., R. Bugert, P. LaRiviere, D. Marbach, S. Martin, and L. Ross. 1988. Lower Snake River compensation plan, Lyons Ferry Hatchery evaluation program, 1987. Report (Cooperative Agreement 14-16-0001-87512) to U.S. Fish and Wildlife Service. Washington Department of Fisheries, Olympia.

Thorpe, J. E., and eight coauthors. 1981. Assessing and managing man's impact on fish genetic resources. Canadian Journal of Fisheries and Aquatic Sciences 38:1899–1907.

USACE (U.S. Army Corps of Engineers). 1964. Annual fish passage report: North Pacific Division, Bonneville, The Dalles, McNary, and Ice Harbor dams, Portland District, Oregon.

USACE (U.S. Army Corps of Engineers). 1975. Special report: lower Snake River fish and wildlife compensation plan. Walla Walla, Washington.

USACE (U.S. Army Corps of Engineers). 1990. Annual fish passage report: 1990. Columbia River Projects and Snake River Projects, Portland District, Oregon.

Utter, F. M., W. J. Ebel, G. B. Milner, and D. J. Teel. 1982. Population structures of fall chinook salmon, Oncorhynchus tshawytscha, of the mid-Columbia and Snake rivers. National Marine Fisheries Service, NWAFC Processed Report 82-10, Seattle.

Waples, R. S. 1991. Definition of "species" under the Endangered Species Act: application to Pacific salmon. NOAA (National Oceanic and Atmospheric Administration) Technical Memorandum NMFS (National Marine Fisheries Service) F/NWC-194, Seattle.

Waples, R. S., R. P. Jones, Jr., B. R. Beckman, and G. A. Swan. 1991a. Status review for Snake River fall chinook salmon. NOAA (National Oceanic and Atmospheric Administration) Technical Memorandum NMFS (National Marine Fisheries Service) F/NWC-201, Seattle.

Waples, R. S., D. J. Teel, and P. B. Aebersold. 1991b. A genetic monitoring and evaluation program for supplemented populations of salmon and steelhead in the upper Columbia River basin. Project Report (Contract DE-A179-89BP00911) to Bonneville Power Administration, Portland, Oregon.

WDF (Washington Department of Fisheries), Washington Department of Wildlife, and Western Washington Treaty Indian Tribes. 1993. 1992 Washington State salmon and steelhead stock inventory. WDF, Olympia.

Winans, G. A. 1989. Genetic variability in chinook salmon stocks from the Columbia River basin. North American Journal of Fisheries Management 9:4–52.

Supplementation: Panacea or Curse for the Recovery of Declining Fish Stocks?

EDWARD C. BOWLES

Idaho Department of Fish and Game
600 South Walnut Street, Boise, Idaho 83707, USA

Abstract.—The role of hatchery supplementation in helping recover naturally reproducing fish populations is the subject of much debate. The use of supplementation to increase natural production cannot be considered a viable alternative to solving ecosystem bottlenecks (e.g., excessive smolt mortality of anadromous salmonids during emigration through the Columbia River basin hydropower system). Supplementation can increase the number of naturally produced fish living in the wild, but not the survival of these fish. Supplementation of anadromous salmonids can potentially benefit only those populations limited by density independent or depensatory smolt-to-adult mortality. To realize this benefit, existing numbers of naturally produced smolts must be limited by adult escapement and not spawning or rearing habitat. For supplementation programs to be successful, the hatchery component must provide a net survival benefit for the target stock, as compared with the natural component. In providing this advantage, supplementation success is also dependent on circumventing some early life history mortality without compromising natural selection or incurring hatchery selection.

Using hatcheries to supplement naturally reproducing salmon *Oncorhynchus* spp. and steelhead *O. mykiss* populations is controversial. Expectations of supplementation range from a panacea for solving population declines to a curse that will accelerate declines and seal the fate of anadromous salmonid stocks. This article addresses the theoretical benefits and risks of supplementation for upper Columbia River basin stocks and puts these benefits and risks into perspective with other recovery options. Although I focus on these unique upriver stocks, numerous parallels can be inferred for other fish populations and locations. This discussion will hopefully help narrow the chasm of expectations and promote a clearer understanding of supplementation and its potential role in recovery processes.

The plight of anadromous salmonids in the Columbia River basin is a well-documented tragedy (IDFG 1992; NPPC 1993; Petrosky and Shaller 1994). This plight is particularly tragic for naturally reproducing populations, which rely heavily on upper basin tributaries for spawning and nursery areas (Figure 1). For example, Idaho contained an estimated 45 to 55% of the historical natural production potential in the entire Columbia River basin for spring and summer chinook salmon *O. tshawytscha* and summer steelhead (Bjornn 1960; Mallet 1974). Low numbers of adult returns currently limit natural production in Idaho to less than 20% of estimated capacity for spring and summer chinook salmon (Hassemer et al., in press). These declines resulted in listing several upriver stocks for protection under the U.S. Endangered Species Act of 1973 (16 U.S.C.A. §§1531 to 1544); others are likely to follow (Nehlsen et al. 1991).

Importance of Diversity

Any attempt to rebuild anadromous runs must recognize the importance of naturally reproducing populations to sustainable recovery. These populations represent a source of genetic diversity that provides the foundation for adaptation and persistence in an ever-changing environment. Recovery efforts are short-sighted if they do not maintain or enhance this natural diversity of proven performance characteristics (Riggs 1990; Currens et al. 1991; Matthews and Waples 1991; Ryman 1991). Successful recovery is thus dependent on two key elements: (1) recovery must enhance the number of naturally reproducing fish and; (2) this enhancement must not reduce the natural genetic diversity and fitness, which are assumed crucial for long-term sustainability, of fish living in the wild.

The genetic diversity of a species represents the spectrum of performance capabilities found within and among populations (a group of interbreeding individuals). This diversity represents an important hierarchy that begins at the individual population level, which has evolved through adaptation to its local environment (e.g., Schonewald-Cox et al. 1983; Ryman and Utter 1987). The diversity within this group is important for continued adaptation and resiliency to environmental fluctuations and change (Soule 1987; Waples 1990, 1991). Part of this diversity is provided by natural gene flow (straying and interbreeding) with other adjacent popula-

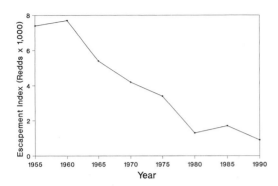

FIGURE 1.—Index of redd counts for spring and summer chinook salmon in Idaho (5-year averages).

tions. Those populations that maintain an appreciable level of gene flow combine to form a stock, whose sustainability and resiliency are a direct function of the natural diversity within and among its populations. Although additional levels in the hierarchy may be appropriate, a grouping of stocks is considered to represent the species unit, such as the Snake River spring and summer chinook salmon identified by the National Marine Fisheries Service for protection under the U.S. Endangered Species Act. Within this hierarchy, the species unit is only as strong as its stock diversity and productivity, which are only as strong as their population diversity and productivity.

Maintaining this natural diversity is crucial for sustainable recovery. Extinction theory indicates that as the number of distinct populations within a stock decline, the probability of extinction for that stock increases dramatically (Dennis et al. 1991). This concept is important because recovery options that are selective (i.e., do not benefit all natural diversity characteristics) will not promote sustainable recovery. This is true even if fish numbers have increased as a result of the management action but general diversity has declined.

Applying these concepts to fisheries management does not present the quandary that might be expected. Obviously, it is unrealistic to consider managing all anadromous fish at the individual population level, with every diversity unit within that population identified, quantified, and monitored. We need to recognize that the management unit contains a spectrum of performance capabilities whose natural diversity is critical to the health of the management unit as a whole. Management actions that do not maintain this diversity, but are selective for particular components (e.g., spawning-run timing), do not promote sustainable recovery.

Recovery Options

Management options for recovery of declining or extirpated naturally reproducing populations of anadromous salmonids in the Columbia River basin can be grouped into four main categories (CBFWA 1990; Bevan et al. 1993; NPPC 1993): (1) improve survival during migrations, particularly juvenile emigration conditions through the main-stem hydroelectric system; (2) reduce harvest impacts; (3) improve quality and quantity of spawning and freshwater rearing habitat; and (4) supplement naturally reproducing populations through artificial propagation.

To be consistent with sustainability goals, these recovery options must be assessed from the perspective of conserving natural genetic diversity and fitness as well as increasing numbers of naturally reproducing fish.

Main-stem River Migration Survival

It is generally recognized that the best hope for short- and long-term recovery of upper Columbia River basin anadromous stocks lies in improvement of juvenile migration conditions (CBFWA 1990; IDFG 1992; Bevan et al. 1993; NPPC 1993). Other options pale in comparison to this overriding factor. It is well documented that the decline and continued depression of upriver anadromous stocks results predominately from poor survival associated with dams and reservoirs (flows and fish passage problems) in the lower Snake and Columbia rivers (NPPC 1987, 1993; NMFS 1993; Petrosky and Shaller 1994). Mortality of upriver chinook salmon and steelhead stocks during migration and ocean residence is assumed to operate primarily independent of fish numbers associated with those stocks (Bjornn and Steward 1990; Bowles and Leitzinger 1991; RASP 1992).

In addition to being the best option for increasing fish numbers, main-stem river survival improvements are the best, and perhaps only, way to meet sustainability goals. This is the only recovery option that can ensure minimal selection toward particular populations or stocks, thus preserving and enhancing genetic diversity as fish numbers increase. For example, flow and passage improvements that add only 0.3% to smolt-to-adult survival could increase a target population and all other upriver anadromous stocks by 50% after one generation. Other options, such as habitat improvement or supplementation, could have that effect on only the target population.

In spite of these benefits, there is obvious reluctance to pursue this option rigorously. This is primarily because of perceived sacrifices required in the "standard operating procedures" of the Northwest—thus, the desire to find alternatives.

Harvest

Another option is to reduce harvest. In upriver areas, harvest opportunity on naturally produced anadromous fish is already severely constrained. In Idaho, for example, sport harvest of naturally produced salmon or steelhead has not occurred since 1978. With respect to downriver areas, recovery measures of any kind will be hampered without continued stringent harvest restrictions on mixed-stock fisheries in the main-stem Columbia and Snake rivers. Harvest restrictions alone, however, will not be enough to recover upriver natural populations.

Habitat

Measures to improve spawning and freshwater rearing habitat have a relatively minor but important role in the initial recovery of upriver salmon and steelhead populations. Much of Idaho's available spawning and rearing habitat is in good condition, but low numbers of returning adult chinook salmon currently limit utilization of this habitat to less than 20% of its potential (Hassemer et al., in press). With much of Idaho's quality habitat underutilized, habitat improvement will have limited benefit without concurrent improvement in adult escapements.

Habitat measures must remain an important component of any recovery package. Habitat has been severely degraded in several areas critical for naturally reproducing populations and should be enhanced to support recovery. This is particularly important when habitat degradations have adversely affected spawning or rearing success even at very low densities (i.e., density-independent mortality effects). These improvements alone, however, will only benefit the targeted population and cannot provide recovery of all components of the species unit.

Supplementation

The supplementation option uses fish hatcheries in an attempt to increase the number of naturally produced fish without eroding long-term fitness of target and nontarget natural populations (e.g., see Bowles and Leitzinger 1991; RASP 1992; Cuenco et al. 1993). Many interests consider supplementation a desirable option because it is not perceived to disrupt current management practices in the Northwest. But can supplementation provide sustainable recovery? How do we implement it? What are the risks?

Despite more than 100 years of hatchery and outplanting programs in the Northwest, existing knowledge on supplementation is quite limited (Miller et al. 1990; Steward and Bjornn 1990). We know little about supplementation mainly because true supplementation has rarely been practiced. Outplanting programs in the past typically followed traditional hatchery guidelines without the benefit of current natural production and genetic conservation theories. Performance of hatchery fish in natural habitats and the effects on existing natural populations were not major concerns (Miller et al. 1990; RASP 1991). The few programs that were monitored and adequately evaluated indicated that supplementation was rarely successful in increasing natural production, and often significant risks were incurred (Chilcote et al. 1986; Miller et al. 1990; Steward and Bjornn 1990). Outplanting programs that have successfully increased the number of naturally produced fish have typically been in areas previously extirpated or through colonization of habitats vacant of the target species (e.g., M. Cuenco, Columbia River Inter-Tribal Fisheries Commission [CRITFC], unpublished data).

Resource managers should neither fully embrace nor dismiss supplementation as a recovery tool, but reassess its role within this new context. Our challenge is to develop strategies to maximize benefits and minimize risks, and to test these strategies conservatively in low-risk areas prior to large-scale implementation. Although knowledge on supplementation is limited, we can begin this assessment by defining boundaries for its potential utility in the recovery process.

Habitat.—Supplementation of anadromous salmonids can potentially benefit only those populations where natural smolt production is limited by adult escapement, not spawning or freshwater rearing habitat. If spawning or rearing habitat is a key factor (i.e., population is at carrying capacity because of habitat degradation) then this constraint will have to be resolved before or concurrent with supplementation actions.

Survival.—Supplementation can potentially increase the number of adult progeny produced from naturally spawning adults, but supplementation is unlikely to increase the survival of these naturally produced progeny. The only potential opportunity for supplementation to increase survival of fish in natural habitats is when populations are experiencing severe inbreeding depression or exhibiting multiple stability regions (e.g., predator traps) in their stock–recruitment relationships (Peterman 1977; RASP 1991). In most cases, an optimal scenario for supplementation would be to increase the number of naturally produced fish without reducing their ability to survive in the wild.

To increase the number of naturally produced fish, artificial propagation must provide a net survival benefit for the target stock as compared with the purely natural component. In areas with existing natural populations, the combined survival (adult-to-adult) of hatchery and natural fish must exceed the natural survival occurring without supplementation. This increase in net survival (hatchery and natural combined) must not come at the expense of the natural component if sustainable recovery is desired.

To avoid reducing natural fitness, hatcheries must minimize changes in genetic, behavioral, and health characteristics of fish as they are routed through the hatchery's protective environment. As a guiding premise, hatcheries must strive to circumvent those mortality factors operating randomly in the natural environment without altering mortality factors operating selectively in the natural environment (Bowles 1994). Neither of these two requirements is easy to accomplish. Failure of either would at best negate any benefits from supplementation and at worst cause irreparable harm to the target populations we are trying to save. The overriding constraint on supplementation, however, is that because it cannot increase the survival of fish living in the wild, it cannot obviate the problems that caused the decline in the first place.

Consider a best case scenario in which supplementation increases the number of naturally produced fish without reducing natural fitness (Figure 2). Two common expectations for supplementation are to rebuild populations to near full use of available habitat and to provide self-sustaining natural production at these enhanced levels. For upper Columbia River basin anadromous stocks, realization of these goals is highly unlikely. In the above scenario, assume the hatchery increases smolt numbers to provide enough total adult returns (hatchery and natural origin) to seed the habitat fully and produce the maximum possible number of natural smolts (i.e., at carrying capacity). These natural smolts, however, cannot produce enough adult returns to seed the habitat fully, because of excessive smolt-to-adult mortality. Thus, supplementation can never provide for recovery to historical levels of naturally produced adults.

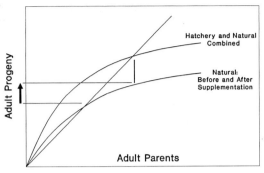

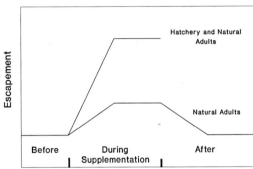

FIGURE 2.—Hypothetical stock–recruit (upper figure) and escapement time series (lower figure) curves illustrating potential benefits and risks if supplementation increases the number of naturally produced anadromous salmonids (depicted by arrow) without impairing their survival. The diagonal line in the upper figure is the replacement line. Natural variability among streams and years has been omitted for illustrative purposes.

In addition, as the number of natural adults is increased through supplementation, the survival of their progeny declines due to density-dependent natural-rearing constraints. This increased number of natural fish can be artificially maintained, but as soon as supplementation is stopped, the population will decline to presupplementation levels, assuming natural survival characteristics have not been impaired by the hatchery program. Thus even under a best case scenario, supplementation of upper Columbia River basin anadromous stocks is unable to provide self-sustaining natural production at this increased level without reductions in smolt-to-adult mortality (e.g., flow and passage constraints on the main-stem Columbia and Snake rivers).

Under this scenario, supplementation would be deemed a success because artificial propagation could be used either continually or intermittently to increase the number of naturally produced fish without significant loss in fitness. Even under this

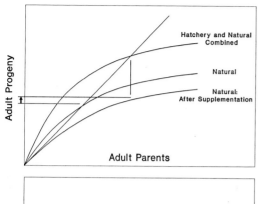

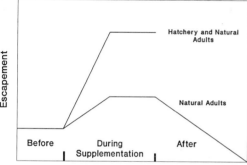

FIGURE 3.—Hypothetical stock–recruit (upper figure) and escapement time series (lower figure) curves illustrating potential benefits and risks if supplementation increases the number of naturally produced anadromous salmonids (depicted by arrow) but impairs their survival. The diagonal line in the upper figure is the replacement line. Natural variability among streams and years has been omitted for illustrative purposes.

somewhat optimal scenario, natural production cannot rebuild to historical levels, and the increased number of natural fish is not sustainable without continued hatchery support. The longer supplementation continues, the greater the risk that hatchery effects will erode natural genetic, behavioral, and health characteristics (Lichatowich and McIntyre 1987; RASP 1991; Waples 1991; Hard et al. 1992).

The consequences of eroding natural fitness is demonstrated in the next scenario, in which the number of natural fish is enhanced by supplementation; but this enhancement comes at the expense of natural survival attributes (Figure 3). Actual numbers of naturally produced adults increase during supplementation because the net adult-to-adult survival of the stock (hatchery and natural components combined) has increased from the survival advantage gained during hatchery residence. Survival of fish in the natural environment, however, has been impaired by negative hatchery influence exerted through introgression (genetics) and interaction (behavior and pathogens).

Under this scenario, the increased number of naturally produced fish is not self-sustaining, and reduced natural fitness puts the population in more serious risk of extinction.

Diversity.—Supplementation has limited potential to maintain or enhance natural genetic diversity. As discussed previously, a species is composed of a diversity of stocks, which are composed of a diversity of populations, each of which contains a diversity of locally adapted performance traits.

A major risk of supplementation is the potential loss of this among-population and within-population variability. Variability among populations can be reduced if diversity is not recognized or managed. For example, if a donor broodstock is taken from a relatively productive natural population and used to supplement other weak populations within that stock, genetic diversity of the entire stock may decline from loss of among-population variability. The genetic identity of the weak components are compromised by the excessive gene flow from the donor stock. This is one reason why many scientific forums prescribe supplementation using local populations for broodstock whenever possible (CBFWA 1990; Bowles and Leitzinger 1991; RASP 1991; Cuenco et al. 1993; Kapuscinski et al., University of Minnesota, unpublished data).

Even using local stock is not without risks to genetic diversity. Variability within the population can be lost if supplementation is genetically selective. For example, if adult collection, spawning, rearing, or release practices are selective for particular traits (e.g., age or spawning-run timing) or families within the population, natural genetic diversity may decline (Ryman and Laikre 1991; Hard et al. 1992).

Supplementation can potentially enhance only targeted populations. Obviously, independent supplementation of every population within all stocks is unrealistic. Thus supplementation can benefit only a portion of the population diversity or structure necessary for sustainability.

Managers cannot circumvent this dilemma by pooling populations so that supplementation of the entire stock becomes manageable. This misdirected approach could greatly reduce the sustainability of the stock by diluting among-population diversity, even if the number of naturally produced fish increases (Hard et al. 1992; Kapuscinski et al., University of Minnesota, unpublished data). This discussion is particularly pertinent in assessing the best recovery options for upper Columbia River basin anadromous salmonid populations. Improvements to main-stem migration conditions would benefit all upriver populations and promote sustainability with the recovery, whereas supplementation would potentially benefit only target populations.

On a positive note, supplementation may play an important interim role for populations that are on the verge of extinction. Supplementation may allow these populations to persist until adequate survival conditions are ensured. This assumes, of course, we can provide a hatchery survival advantage without adversely affecting performance in the natural environment. This is an unproven assumption that requires rigorous investigation and innovative approaches. Supplementation may also be important as a restoration tool in areas where local populations have been extirpated, assuming the problems that caused the decline are being alleviated (CRITFC 1990; RASP 1991; Cuenco et al. 1993; Kapuscinski et al., University of Minnesota, unpublished data).

Implementation.—Supplementation cannot be considered a recovery alternative to addressing the causes of decline for upper Columbia River basin anadromous stocks. Recovery must be approached as a package, with obvious emphasis placed on improving the major causes of decline—deteriorated main-stem migrational conditions as a result of hydropower development (NPPC 1987, 1993; Bevan et al. 1993; NMFS 1993).

Potential benefits of supplementation relative to risks and uncertainties do not warrant full implementation of this tool for upper Columbia River basin anadromous stocks. The potential benefits from supplementation must be kept in perspective with the overriding main-stem conditions that greatly constrain these benefits. Supplementation may help contribute to sustainable recovery, but high potential risks and large uncertainties require conservative implementation.

This is the approach taken by Idaho Supplementation Studies (Bowles and Leitzinger 1991), which was recently implemented as a cooperative effort among Idaho's resource management agencies and tribes. The purpose of this program is to help determine the utility of supplementation as a potential recovery tool for decimated stocks of spring and summer chinook salmon in Idaho. Goals include assessing the use of hatchery chinook salmon to restore or augment natural populations and to evaluate the effects of supplementation on the survival and fitness of existing natural populations.

Acknowledgments

I received valuable input for this manuscript from Eric Leitzinger, Dave Cannamela, Jim Lichatowich, Lars Mobrand, Rich Carmichael, Tom Backman, Bruce Watson, and Tom Vogel. Background work leading to this manuscript was funded in part by the Bonneville Power Administration.

References

Bevan, D. E., and six coauthors. 1993. Draft Snake River salmon recovery plan recommendations. Prepared for National Marine Fisheries Service, U.S. Department of Commerce, National Oceanic and Atmospheric Administration, Portland, Oregon.

Bjornn, T. C. 1960. Salmon and steelhead in Idaho. The Idaho Wildlife Review 13(1):6–11, Boise.

Bjornn, T. C., and C. R. Steward. 1990. Concepts for a model to evaluate supplementation of natural salmon and steelhead stocks with hatchery fish. Part 3 *in* W. H. Miller, editor. Analysis of salmon and steelhead supplementation, parts 1-3. Bonneville Power Administration, Project 88-100, Portland, Oregon.

Bowles, E. C. 1994. The hatchery challenge. Salmon management in the 21st century: recovering stocks in decline. Proceedings, 1992 Northwest Pacific chinook and coho workshop. American Fisheries Society, Idaho Chapter, Boise, Idaho.

Bowles, E. C., and E. Leitzinger. 1991. Salmon supplementation studies in Idaho rivers (Idaho supplementation studies). Idaho Department of Fish and Game, Bonneville Power Administration Project 89-089, Boise, Idaho.

CBFWA (Columbia Basin Fish and Wildlife Authority). 1990. Integrated system plan for salmon and steelhead production in the Columbia River basin. Public review draft. Prepared for Northwest Power Planning Council, Portland, Oregon.

CRITFC (Columbia River Inter-Tribal Fish Commission). 1990. Integrated tribal production plan, volume 1: production proposal for recovery of Snake River stocks. CRITFC, Portland, Oregon.

Chilcote, M. W., S. A. Leider, and J. J. Loch. 1986. Differential reproductive success of hatchery and wild summer-run steelhead under natural conditions. Transactions of the American Fisheries Society 115: 726–735.

Cuenco, M. L., T. W. H. Backman, and P. R. Mundy. 1993. The use of supplementation to aid in natural stock restoration. Pages 269–293 *in* J. G. Cloud and G. H. Thorgaard, editors. Genetic conservation of salmonid fishes. Plenum Press, New York.

Currens, K. P., and six coauthors. 1991. A hierarchical approach to conservation genetics and production of anadromous salmonids in the Columbia River basin. Product of the 1990 Sustainability Workshop, Northwest Power Planning Council, Portland, Oregon.

Dennis, B., P. L. Mulholland, and J. M. Scott. 1991. Estimation of growth and extinction parameters for endangered species. Ecological Monographs 61:115–143.

Hard, J. J., R. P. Jones, Jr., M. R. Delarm, and R. S.

Waples. 1992. Pacific salmon and artificial propagation under the Endangered Species Act. NOAA (National Oceanic and Atmospheric Administration) Technical Memorandum NMFS (National Marine Fisheries Service) NWFSC-2.

Hassemer, P. F., S. W. Kiefer, and C. E. Petrosky. In press. Idaho's salmon: can we count every last one? Proceedings, symposium on Pacific salmon and their ecosystems: status and future options. University of Washington, Center for Streamside Studies, Seattle.

IDFG (Idaho Department of Fish and Game). 1992. Anadromous fisheries management plan 1992–1996. Idaho Department of Fish and Game, Boise.

Lichatowich, J. A., and J. D. McIntyre. 1987. Use of hatcheries in the management of Pacific anadromous salmonids. American Fisheries Society Symposium 1:131–136.

Mallet, J. 1974. Inventory of salmon and steelhead resources, habitat, use and demands. Idaho Department of Fish and Game Project F-58-R-1, Boise.

Matthews, G. M., and R. S. Waples. 1991. Status review for Snake River spring and summer chinook salmon. NOAA (National Oceanic and Atmospheric Administration) Technical Memorandum NMFS (National Marine Fisheries Service) F/NWC-200.

Miller, W. H., T. C. Coley, H. L. Burge, and T. T. Kisanuki. 1990. Analysis of salmon and steelhead supplementation: emphasis on unpublished reports and present programs. Part 1 in W. H. Miller, editor. Analysis of salmon and steelhead supplementation, parts 1-3. Bonneville Power Administration, Project 88-100, Portland, Oregon.

Nehlsen, W., J. E. Williams, and J. Lichatowich. 1991. Pacific salmon at the crossroads: stocks at risk from California, Oregon, Idaho and Washington. Fisheries (Bethesda) 16(2):4–21.

NMFS (National Marine Fisheries Service). 1993. Biological opinion on 1993 operations of the federal Columbia River power system. National Oceanic and Atmospheric Administration, Washington, DC.

NPPC (Northwest Power Planning Council). 1987. Columbia River basin fish and wildlife program. Northwest Power Planning Council, Portland, Oregon.

NPPC (Northwest Power Planning Council). 1993. Columbia River basin fish and wildlife program—strategy for salmon, volumes 1 and 2. Northwest Power Planning Council Documents 92-21 and 92-21A, Portland, Oregon.

Peterman, R. M. 1977. A simple mechanism that causes collapsing stability regions in exploited salmonid populations. Journal of the Fisheries Research Board of Canada 34:1134–1142.

Petrosky, C. E., and H. A. Shaller. 1994. A comparison of the productivities for Snake River and lower Columbia River spring and summer chinook stocks. Proceedings of salmon management in the 21st century: recovering stocks in decline. 1992 Northwest Pacific chinook and coho workshop. Idaho Chapter of the American Fisheries Society, Boise.

RASP (Regional Assessment of Supplementation Project). 1991. Status report for the regional assessment of supplementation project. Prepared for the Bonneville Power Administration, Project 85-12, Portland, Oregon.

RASP (Regional Assessment of Supplementation Project). 1992. Summary report series for the regional assessment of supplementation project. Prepared for the Bonneville Power Administration, Project 85-12, Portland, Oregon.

Riggs, L.A. 1990. Principles for genetic conservation and production quality: results of a scientific and technical clarification and revision. Report (Contract C90-005) to the Northwest Power Planning Council, Portland, Oregon.

Ryman, N. 1991. Conservation genetics considerations in fishery management. Journal of Fish Biology 39(Supplement A):211–224.

Ryman, N., and F. Utter, editors. 1987. Population genetics and fishery management. University of Washington Press, Seattle.

Ryman, N., and L. Laikre. 1991. Effects of supportive breeding on the genetically effective population size. Conservation Biology 5(3):325–329.

Schonewald-Cox, C. M., S. M. Chambers, B. McBryde, and W. L. Thomas, editors. 1983. Genetics and conservation: a reference for managing wild animal and plant populations. Benjamin-Cummings, Menlo Park, California.

Soule, M. E., editor. 1987. Viable populations for conservation. Cambridge University Press, Cambridge, Massachusetts.

Steward, C. R., and T. C. Bjornn. 1990. Supplementation of salmon and steelhead stocks with hatchery fish: a synthesis of published literature. Part 2 in W. H. Miller, editor. Analysis of salmon and steelhead supplementation, parts 1-3. Bonneville Power Administration, Project 88-100, Portland, Oregon.

Waples, R. S. 1990. Conservation genetics of Pacific salmon, part 2. Effective population size and the rate of loss of genetic variability. Journal of Heredity 81:267–276.

Waples, R. S. 1991. Genetic interactions between hatchery and wild salmonids: lessons from the Pacific Northwest. Canadian Journal of Fisheries and Aquatic Sciences 48(Supplement 1):124–133.

Status of Supplementing Chinook Salmon Natural Production in the Imnaha River Basin

RICHARD W. CARMICHAEL AND RHINE T. MESSMER

Oregon Department of Fish and Wildlife
211 Inlow Hall, Eastern Oregon State College, 1410 L Avenue, La Grande, Oregon 97850, USA

Abstract.—The hatchery supplementation program for Imnaha River chinook salmon *Oncorhynchus tshawytscha* was initiated in 1982 under the Lower Snake River Compensation Plan. The program was originally conceived as a conventional hatchery mitigation program with its primary objective to provide surplus hatchery fish for harvest. However, the program has been shifted to emphasize supplementation, and its primary objectives include enhancing natural production while maintaining life history and genetic characteristics of the endemic population. Wild fish were used for broodstock from 1982 to 1985; wild and hatchery fish have been used for broodstock since 1986. Imnaha chinook salmon are supplemented by annual spring releases of yearling smolts. To determine the success of hatchery supplementation we are comparing aspects of life history and productivity of the natural and hatchery populations. High prespawning mortality, high egg loss, and poor smolt-to-adult survival rates resulted in adult progeny:parent ratios of less than 1.0 for hatchery-reared fish of the 1982–1986 brood years. Adult progeny:parent ratios for the natural spawning population were above 1.0 for the 1982–1983 broods but below 1.0 for the 1984–1986 broods. Adult progeny:parent ratios for hatchery-reared fish were equal to or greater than ratios of the natural spawning population for the 1984–1987 brood years. For the 1982–1986 broods, hatchery fish returned at a younger age and a higher percent of males returned as jacks. We have made a substantial number of hatchery production, broodstock management, and facility changes to improve program success. For example, smolt size at release has been reduced in an attempt to shift age at return to an older age, smolt production goals have been reduced, a larger proportion of the hatchery and wild fish have been passed above the weir to spawn naturally, matrix mating protocols have been implemented, and new adult capture and holding and juvenile acclimation facilities have been constructed.

Populations of summer chinook salmon *Oncorhynchus tshawytscha* in the Imnaha River basin have declined precipitously during the past 3 decades (Figure 1). Peak escapement in recent history was estimated as 3,439 in 1957 (Carmichael et al. 1990), and wild population levels have declined to below 300 individuals in recent years (Table 1). Population declines are principally attributed to reduced population productivity that has resulted from juvenile and adult mortalities which occur during migration at Snake and Columbia river dams and in the reservoirs. Historically, chinook salmon spawned in Lick Creek, Big Sheep Creek, and the main-stem Imnaha River. In recent years very few redds have been observed in Big Sheep and Lick creeks and the current spawning distribution is concentrated in about 29 km of the main stem. Four dams (Ice Harbor, Lower Monumental, Little Goose, and Lower Granite) were constructed in the lower Snake River from 1961 to 1975. It was estimated that these four dams resulted in a 48% reduction in annual production of chinook salmon in all populations above Lower Granite Dam (USACE 1975). The U.S. Congress authorized the Lower Snake River Compensation Program (LSRCP) in 1976 to mitigate for losses of salmon *Oncorhynchus* spp., steelhead *O. mykiss*, and other fishery resources that resulted from construction of the four lower Snake River dams. Mitigation goals for the Imnaha spring and summer chinook salmon were established as 3,210 adults annually. Annual hatchery production goals of 490,000 smolts at 23 g/fish (11,270 kg) were established to compensate for the loss of 3,210 adults.

The use of artificial propagation to enhance salmon abundance for the purpose of sustaining or enhancing commercial and recreational fisheries has a long history of success. These types of hatchery programs can be characterized as conventional hatcheries. However, the use of artificial propagation to conserve or enhance natural production is highly debated and has not been demonstrated to be successful in many cases (Miller et al. 1990; Hard et al. 1992; RASP 1992). Much of the evidence for failure of artificial propagation programs in supplementing natural production has come from assessment of conventional hatchery programs designed for fishery augmentation and their effects on natural production (Chilcote et al. 1986; Nickelson et al. 1986). Few hatchery programs have been designed, implemented, and managed for the primary purpose of enhancing natural production. Although the Imnaha chinook salmon program was originally

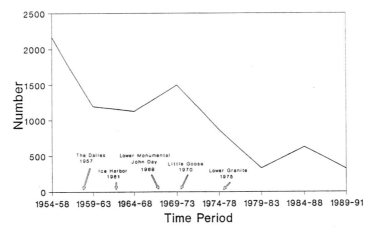

FIGURE 1.—Natural spawning escapement of chinook salmon in the Imnaha River basin. Escapements are presented as 5-year averages. The peak escapement in 1957 was 3,439 fish. The years of dam completions on the lower Snake River are also given.

conceived as a conventional type of hatchery, it has undergone considerable change to become a supplementation type hatchery.

In 1982 the Oregon Department of Fish and Wildlife initiated the hatchery program on the Imnaha River under the LSRCP. The Imnaha chinook salmon hatchery program was developed and has been managed under the guidance of the following four management objectives: restore natural populations of chinook salmon in the Imnaha River basin to historic abundance levels; reestablish traditional tribal and recreational fisheries for chinook salmon; maintain genetic and life history characteristics of the endemic wild population while pursuing mitigation goals and management objectives; and operate the hatchery program to ensure that the genetic and life history characteristics of the hatchery fish mimic the wild fish. Since 1984 we have been conducting research and monitoring and evaluating the program to document the success of achieving the management objectives. In this paper we describe the objectives for the hatchery program and highlight important aspects of the propagation program. In addition, we present the monitoring and evaluation plan that has been implemented to document the success of the program, and we describe the changes that have been implemented in the hatchery program as a result of shifting management objectives. Because we are only in the early

TABLE 1.—Total escapement, number of broodstock collected, and number and origin of natural spawners in the Imnaha River for the 1979–1990 runs. All values include adults and jacks, therefore not all broodstock removed were actually spawned. Total escapement is the sum of total natural spawners estimated from redd counts and fish retained for hatchery broodstock.

Year	Total escapement	Broodstock collected		Natural spawners		Natural spawners of hatchery origin (%)
		Wild	Hatchery	Wild	Hatchery	
1979	192	0	0	192	0	0
1980	125	0	0	125	0	0
1981	307	0	0	307	0	0
1982	419	28	0	391	0	0
1983	397	64	0	333	0	0
1984	518	36	0	482	0	0
1985	692	116	15	561	0	0
1986	798	319	21	458	0	0
1987	479	83	22	374	0	0
1988	607	140	20	435	12	3
1989	415	111	126	178	0	0
1990	566	81	153	220	112	34

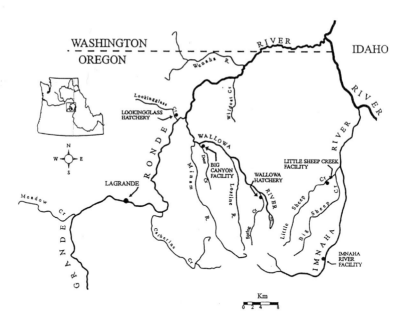

FIGURE 2.—Location of Imnaha River facility and Lookingglass State Fish Hatchery.

stages of monitoring and evaluation, a limited number of findings are presented.

Study Area

The Imnaha River is located in the northeastern corner of Oregon. The basin drains 2,461 km² of the eastern Wallowa Mountains and the plateau between the Wallowa River drainage and Hells Canyon of the Snake River. The watershed attains an elevation of 3,048 m at the headwaters in the Eagle Cap Wilderness Area. The watershed undergoes a change from alpine mountains at the headwaters to semiarid plateau in the lower main stem. Stream discharge patterns are such that maximum flows generally occur in late spring to early summer and minimum discharge occurs during the fall. The Imnaha River weir and juvenile advanced rearing pond are located at km 85.3 on the Imnaha River (Figure 2). The Imnaha River enters the Snake River at km 309.3. Eight dams reside between the Imnaha River and the ocean. Lookingglass State Fish Hatchery (SFH) is located on Lookingglass Creek, which is a tributary of the Grande Ronde River.

Description of Facilities

The Imnaha River facility is operated as a satellite of Lookingglass SFH which serves as the primary production facility. A temporary adult collection and juvenile acclimation facility was operated from 1982 to 1988. An earthen pond was used to hold adults that were collected from the river and also to acclimate smolts prior to release. A permanent facility was completed in 1989. The permanent facility includes a picket-type weir that directs fish up a stepped ladder and into an adult trap. The adult handling facility consists of an elevator, anesthetic tank, and transfer tubes that allow for transfer of adults back to the river above the weir or into an adult holding pond. The juvenile advanced rearing pond is a rectangular concrete raceway that is supplied with Imnaha River water. Lookingglass SFH serves as the incubation and rearing facility. It is equipped with indoor Canadian-style troughs for early rearing and outdoor concrete raceways for final rearing. The water source for the Canadian troughs and outdoor raceways is Lookingglass Creek.

Broodstock Development and Management

The biological uniqueness of the Imnaha River chinook salmon was recognized long before the hatchery program was initiated. This recognition led to a decision to use only the endemic stock to initiate the hatchery program and to use some natural fish for hatchery brood each year. Wild adults were collected for hatchery broodstock beginning in 1982. The weir is installed as early as is physically possible; however, in all years fish pass above the

TABLE 2.—Stock origin (W = wild, H = hatchery) and number of females spawned for Imnaha River chinook salmon hatchery program.

Brood year	Stock origin	Females spawned	Percent wild
1982	W	10	100
1983	W	31	100
1984	W	11	100
1985	W	32	100
1986	W, H	59	89.8
1987	W, H	39	97.4
1988	W, H	92	89.1
1989	W, H	83	56.6
1990	W, H	73	34.2

TABLE 3.—Summary of releases of hatchery-reared smolts of Imnaha stock chinook salmon. All smolts were reared at Lookingglass State Fish Hatchery.

Brood year	Number released	Size (g)	Date of release	Location of release
1982	24,920	14.2	Mar 22 1984	Imnaha River
1983	56,235	18.6	Sep 14 1984	Imnaha River
1983	59,595	26.1	Mar 22 1985	Imnaha River
1984	35,035	42.0	Mar 28 1986	Imnaha River
1986	101,929	41.3–45.8	Mar 21–22 1988	Imnaha River
1986	97,137	51.6	Apr 20–21 1988	Imnaha River
1987	142,320	28.4	Apr 05 1989	Imnaha River
1988	364,547	23.9–29.7	Mar 31–04 1980	Imnaha River
1988	79,953	24.7	Apr 02 1990	Big Sheep Creek
1989	267,670	20.4–28.4	Mar 22 1991	Imnaha River
1989	131,239	22.1–30.7	Apr 09 1991	Imnaha River
1990	262,548	21.0–41.3	Mar 30 1992	Imnaha River

weir prior to initiation of broodstock collection. Wild fish constituted a majority of the fish retained for broodstock and spawned for the hatchery program from 1982 to 1988 (Table 2). From 1989 to 1991 both wild and hatchery fish have been used for broodstock. In addition, in all years a significant proportion of the Imnaha population spawns below the weir.

During the early years of this program, guidelines for broodstock collection and for passage of fish above the weir for natural production were not clearly defined. In the years 1982–1986, most of the fish that were collected at the weir were retained for hatchery broodstock. Since 1987 specific guidelines for broodstock collection, retention, mating, and passage have been followed and will be discussed later in the paper.

Hatchery Production Program

The Imnaha River facility serves as an adult collection and holding facility, spawning site, advanced rearing pond for juveniles, and a release site. The facility is situated in the lower section of the river reach that is used most for natural spawning. Currently, a picket weir is installed at the site as early in June–July as is physically possible. Adults are trapped from June through early September. Some fish are retained for broodstock and others passed above the weir to spawn naturally. All fish that are trapped are anesthetized prior to being handled. Fish that are retained for hatchery broodstock are injected with antibiotics.

Spawning begins in mid-August, when the earliest fish ripen, and ends in mid-September. Eggs are transported to Lookingglass SFH for incubation. At Lookingglass SFH eggs are incubated in pathogen-free well water. All swim-up fry are ponded at standard densities into indoor Canadian-style troughs. When fry reach approximately 1–2 g they are transferred to outdoor concrete raceways. Fish are reared outside for about 1 year. Maximum density and loading factor reached just prior to release are 15 kg/m^3 and 0.6 kg/(L/min), respectively. Smolts are transferred to the advanced rearing pond at the Imnaha River facility on or near March 1 and are held for 30 d for acclimation. Fish are fed at the advanced rearing pond and are released by crowding the fish from the pond into the river. All fish released under the present program are yearling smolts.

The first release of hatchery-reared smolts under LSRCP in the Imnaha River occurred in 1984 (Table 3). Smolt production levels have varied considerably on an annual basis and have ranged from a low of 24,920 in 1984 to a high of 444,500 in 1990 (Table 3). With the exception of the 1988 and 1989 broods, released in 1990 and 1991, respectively, all hatchery smolts have been marked for identification.

Monitoring and Evaluation Program

Monitoring and evaluation was initiated in 1984, 2 years after the production program was started. The objectives of the monitoring and evaluation are to assess the effectiveness of the hatchery program in increasing adult production, adult progeny:parent ratios, and escapement to the Imnaha River; to estimate total annual adult production (catch and escapement), smolt-to-adult survival, and smolt migration success of hatchery fish; to monitor and compare life history characteristics (age composition, run timing, sex ratio, age–length relationship) of natural and hatchery fish; and to make recommendations for improving the success of achieving mitigation goals and management objectives.

Broodstock prespawning survival, egg-to-smolt survival, and number of smolts released annually are estimated by standard hatchery inventory techniques. Representative groups of production and experimental releases are marked with adipose-fin clips and coded-wire tags (CWT) during the fall prior to spring release. We determine smolt-to-adult survival of hatchery-reared smolts for each brood year based on catch and escapement of marked fish. Ocean and main-stem Columbia River catch information is obtained from the Pacific States Marine Fish Commission's CWT database.

Numbers of hatchery and naturally produced fish that escape to the Imnaha River are estimated based on adult recoveries at the Imnaha facility and on the spawning grounds. Smolt migration success and migration characteristics are determined from recovery of cold-branded smolts at Snake and Columbia rivers' dams. Adult progeny:parent ratios are estimated for the hatchery and natural populations in each brood year. Adult progeny:parent ratios are calculated as the ratio of the number of adult progeny returning to the Imnaha River to the number of adults that produced the progeny.

We have conducted spawning ground surveys throughout the entire spawning area since 1986 to assess and compare spawning distribution in years when no hatchery fish spawned naturally with years when a significant number of hatchery fish spawned naturally. We are comparing run timing of hatchery and naturally produced fish based on timing of arrival of hatchery and natural fish at the Imnaha River weir. Brood year-specific age composition at return has been determined and compared for brood years 1982–1986. Age composition of naturally produced fish was determined by scale pattern analysis, and age composition for hatchery fish was determined from known-age, marked fish.

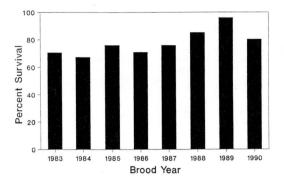

FIGURE 4.—Egg-to-smolt survival rates for hatchery-reared Imnaha River chinook salmon at Lookingglass State Fish Hatchery during brood years 1983–1990.

Results and Discussion

High prespawning mortality of wild adults collected and held for spawning occurred during the initial years of this program (Figure 3). These high mortality rates were associated with the poor holding conditions at the temporary adult facility from 1982 to 1986. Because of high mortality we transported adults to Lookingglass SFH for holding and spawning in 1987–1989. Since 1990 the adults have been held in the new facility at the Imnaha River, and prespawning mortality has been relatively low.

Egg-to-smolt survival rates have improved substantially through the years (Figure 4). For the last three complete production broods (1987–1990), egg-to-smolt survival rates have been equal to or greater than 80%. Smolt survival rates have been estimated for each brood year based on catch and escapement of marked fish. Survival rates of all age-classes have been highly variable and have ranged from 0.04 to 0.25% for completed returns of the 1982–1986 brood years (Table 4).

Because of high prespawning mortality, egg hatch failure, and poor smolt-to-adult survival, the adult progeny (excluding age-3 fish) -to-parent ratios for hatchery fish were below 1.0 for the 1982–1986

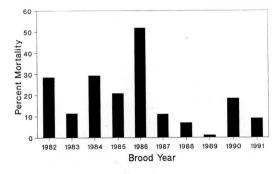

FIGURE 3.—Mortality of Imnaha stock chinook salmon adults collected and held for hatchery broodstock during 1982–1991.

TABLE 4.—Smolt-to-adult survival rate of hatchery-reared Imnaha chinook salmon smolts for the 1982–1987 brood years. Survival rates include catch and escapement of all age-classes.

Brood year	Smolt-to-adult survival (%)
1982	0.12
1983	0.04
1984	0.25
1985	0.13
1986	0.18

TABLE 5.—Adult progeny:parent ratios of hatchery and naturally produced chinook salmon in the Imnaha River for brood years 1982–1987.

Brood year	Adult progeny:parent ratio	
	Naturally produced	Hatchery produced
1982	1.3	0.5
1983	2.5	0.4
1984	0.5	0.9
1985	0.5	0.5
1986	0.6	0.7

TABLE 7.—Age-composition profiles by sex for hatchery and naturally produced chinook salmon that returned to the Imnaha River. Values are means for brood years 1982–1986. Age designations are total age.

Origin	Male age composition (%)			Female age composition (%)		
	3	4	5	3	4	5
Hatchery	60	32	8	0	72	28
Natural	15	63	22	0	39	61

brood years (Table 5). We have implemented numerous programmatic changes, and with returns completed for the 1987 brood year the hatchery adult progeny:parent ratio was 2:1. The adult progeny:parent ratio for wild fish was substantially better than that of hatchery fish for the 1982 and 1983 broods, was slightly less than that of the hatchery ratio for the 1984–1986 broods, and was substantially less than that of the hatchery ratio for the 1987 brood.

Because we release all Imnaha hatchery smolts at one location we are concerned that the natural spawner distribution may shift (concentrate near the facility) as the number of hatchery fish spawning naturally increases. We have conducted spawning ground surveys in the entire spawning area since 1986 in an effort to document shifts in spawner distribution. We saw little difference in spawning distribution in the 52 km used for spawning between 1986 and 1989, when very few, if any, hatchery fish spawned naturally, and 1990 and 1991, when a substantial number of hatchery fish spawned naturally (Table 6).

We have been monitoring and comparing aspects of the life history of hatchery and natural fish since 1986, when the first hatchery adults returned. We have seen a consistent pattern of earlier age at return for the hatchery fish (Table 7). For example, the mean age-composition profile for wild females was 39% age 4 and 61% age 5 whereas the profile for hatchery females was 72% age 4 and 28% age 5. We believe the earlier age at return is a result of releasing the hatchery smolts at a size larger than that of wild smolts. We have decreased the size at release of a proportion of the hatchery production to a size similar to wild smolts to determine if we can shift age at return to older adults.

The Imnaha chinook salmon supplementation program has been managed under the adaptive management philosophy. Extensive research, monitoring, and evaluation has been ongoing. A substantial number of programmatic changes have occurred since the initiation of this program in 1982 because of the information and knowledge gained, as well as a shift in management priorities for the hatchery. The goals and objectives that serve as the foundation of this program have shifted from emphasizing mitigation and hatchery production to emphasizing natural production enhancement. As reviewed in this text, all aspects of this hatchery program have undergone substantial changes through time (Table 8).

We have established the following broodstock management and mating protocols to guide the program in the future:

TABLE 6.—Spawner distribution of Imnaha chinook salmon in the Imnaha River for the periods 1986–1989 and 1990–1991. During the 1986–1989 period few hatchery fish spawned naturally; during 1990–1991 a substantial number of hatchery fish spawned naturally.

Location	Distance from facility (km)	Spawner distribution (%)	
		1986–1989	1990–1991
Above Imnaha River facility			
Blue Hole to Indian Crossing	19.6–22.8	19.4	19.6
Indian Crossing to Mac's Mine	7.2–19.6	55.5	48.2
Mac's Mine to adult weir	0–7.2	4.3	4.0
Below Imnaha River facility			
Adult weir to Crazyman Creek	0–5.6	13.2	24.8
Crazyman Creek to Grouse Creek	5.6–19.3	6.7	3.4
Grouse Creek to Freezeout Creek	19.3–29.0	0.9	0

TABLE 8.—Synopsis of programmatic changes made in the Imnaha chinook salmon hatchery program. Table summarizes transition of objectives, guidelines, and operations from the original program to the present program.

Program area	Original program	Present program
Production goals	490,000	Base on broodstock guidelines and research needs; well below 490,000
Management objectives	Emphasized meeting mitigation goals	Emphasize natural escapement, natural production, and genetic principles
Hatchery broodstock	Kept most fish	Keep a maximum of 30% of natural fish; no more than 50% of natural spawning population will be hatchery fish
	Collected late in the run	Attempt to collect across entire run
	Treated BKD[a] and fungus within limits	Treat BKD[a] and fungus aggressively
Natural escapement above weir	Emphasized hatchery broodstock	Pass minimum of 70% of natural fish and equal number of hatchery fish
	Had no guidelines for hatchery fish	Do not allow hatchery fish to exceed 50% of natural spawning population
Rearing and release strategies	Reared at standard densities	Rear at reduced densities
	Released large smolts	Release large smolts and smolts of natural size
Monitoring and evaluation	Focused on hatchery performance and mitigation success	Focus on hatchery and natural production performance

[a] Bacterial kidney disease.

1. All fish that return to the weir site will be captured, and we will retain no more than 30% of the natural fish by age and by sex for hatchery broodstock. The remainder will be passed above the weir to spawn naturally. This criteria is designed to prevent the removal of too many natural fish from the naturally spawning population.
2. No more than 50% of the fish passed above the weir to spawn naturally will be hatchery-origin fish. This criteria is designed to prevent overwhelming the natural fish population with hatchery fish.
3. Naturally produced fish will constitute a minimum of 30% of the fish used for hatchery broodstock. This criteria is designed to reduce the risk of domestication of hatchery-reared fish.
4. The weir will be installed and broodstock collection initiated as early each year as is physically possible to ensure all components of the run are represented in the broodstock.
5. We will use random-split cross mating; split the eggs from each female into two approximately equal groups and fertilize each group with sperm from a different male. Thus each male and female contribute to two families. This criteria is designed to increase the number of family groups.

We believe it is too early in the program to determine whether the program is a success or failure. The program is designed to minimize the genetic risks associated with a hatchery program while providing the maximum likelihood of providing a natural production enhancement benefit. Recent results indicate that we can increase total escapement by using the hatchery program. However, major uncertainties remain: How well will the hatchery fish perform in the natural system? What level of natural productivity will be achieved and sustained in the integrated population? What will be the long-term influence of the hatchery program on life history and genetic characteristics? We plan to continue this program on an experimental basis and intensively monitor and evaluate program success.

Acknowledgments

We thank D. Herrig, E. Crateau and personnel at Lookingglass Hatchery. This work is funded by the U.S. Fish and Wildlife Service, Lower Snake River Compensation Plan.

References

Carmichael, R. W., R. T. Messmer, and M. W. Flesher. 1990. Oregon's Lower Snake River compensation plan program—a status review. Pages 13–51 in Snake River Hatchery Review 1990 Workshop Summary. U.S. Fish and Wildlife Service, Boise, Idaho.

Chilcote, M. W., S. A. Leider, and J. J. Loch. 1986. Differential reproductive success of hatchery and wild summer-run steelhead under natural conditions. Transactions of the American Fisheries Society 115: 726–735.

Hard, J. J., R. P. Jones, Jr., M. R. Delarm, and R. S. Waples. 1992. Pacific salmon and artificial propagation under the endangered species act. NOAA (National Oceanic and Atmospheric Administration)

Technical Memorandum NMFS (National Marine Fisheries Service) NWFSC-2, Seattle.

Miller, W. H., T. C. Coley, H. L. Burge, and T. T. Kisanuki. 1990. Analysis of salmon and steelhead supplementation. Bonneville Power Administration, Project 88-100, Portland, Oregon.

Nickelson, T. E., M. F. Solazzi, and S. L. Johnson. 1986. Use of hatchery coho salmon (*Oncorhynchus kisutch*) presmolts to rebuild wild populations in Oregon coastal streams. Canadian Journal of Fisheries and Aquatic Sciences 43:2443–2449.

RASP (Regional Assessment of Supplementation Project Group). 1992. Supplementation in the Columbia basin. Part 1-4 RASP Summary Report series. Bonneville Power Administration, Portland, Oregon.

USACE (U.S. Army Corps of Engineers). 1975. Lower Snake River fish and wildlife compensation plan. USACE Special Report, Walla Walla, Washington.

Quality Assessment of Hatchery-Reared Spring Chinook Salmon Smolts in the Columbia River Basin

WALTON W. DICKHOFF, BRIAN R. BECKMAN, DONALD A. LARSEN,
AND CONRAD V.W. MAHNKEN

National Marine Fisheries Service, Northwest Fisheries Science Center
Coastal Zone and Estuarine Studies Division
2725 Montlake Boulevard East, Seattle, Washington 98112, USA

CARL B. SCHRECK AND CAMERON SHARPE

National Biological Service, Oregon Cooperative Fishery Research Unit, Oregon State University
Corvallis, Oregon 97331, USA

WALDO S. ZAUGG

National Marine Fisheries Service, Northwest Fisheries Science Center
Coastal Zone and Estuarine Studies Division

Abstract.—The physiological development and condition of spring chinook salmon *Oncorhynchus tshawytscha* are being studied at several hatcheries in the Columbia River basin to determine whether any or several smolt indices can be related to adult recovery and used to improve hatchery effectiveness. Hatchery fish were sampled from 1989 through 1992 at Dworshak, Leavenworth, and Warm Springs National Fish Hatcheries, and at the Oregon State Round Butte and Willamette hatcheries. Spring chinook salmon were assessed for saltwater tolerance, gill Na^+,K^+-ATPase, blood plasma concentrations of cortisol, insulin, and thyroid hormones, secondary stress tolerance, fish morphology, metabolic energy stores, immune response, blood cell counts, plasma and muscle ion concentrations, and plasma protein concentrations. Results describing differences in physiological development and condition of juvenile spring chinook salmon will be compared with adult recovery data as they become available to determine which smolt quality indices may be related to adult recovery. The completed data analysis for the first 2 years of the study indicates significant variation in smolt quality between years and between hatcheries. At most hatcheries studied during 1989, very few signs of smolt development were observed in fish prior to release. In 1990, significant variation in smolt quality indices was observed for most measures. In 1989 and 1990, major differences occurred among hatcheries in growth rates of fry and juveniles, which could account for physiological differences. A general trend in the data suggests that smolt quality is highest in fish that grow rapidly during the smolting period. If this hypothesis is confirmed, we suggest attempts should be made through regulation of temperature and feeding rate to stimulate more growth in the smaller fish, especially during the several weeks immediately prior to release. Alternatively, release timing may be delayed at some hatcheries to allow sufficient smolt development prior to release.

High-quality smolts may be operationally defined as anadromous juvenile salmonids that are physiologically ready to migrate downstream, adapt to the ocean environment, and survive to adulthood. Release of high-quality smolts from public hatcheries is an important goal, because release of poor-quality smolts may result in slow downstream migration, increased residualism and straying, precocious maturation, inability to osmoregulate, and poor survival to adulthood. Developing precise measures of smolt quality of Pacific and Atlantic salmon has been the objective of many studies during the last 15 years. Previous studies of the physiology and morphology of juvenile salmonids during the parr to smolt transformation have been summarized in the literature (Folmar and Dickhoff 1980; Wedemeyer et al. 1980; Barron 1986; Hoar 1988; Dickhoff et al. 1990; Boeuf 1993; Dickhoff 1993). Some measures of smolt development have been claimed to have either some (Ewing and Birks 1982; Wahle and Zaugg 1982; Zaugg 1989; Virtanen et al. 1991; Farmer 1994) or no (Ewing et al. 1985; Staurnes et al. 1993) correlation to adult returns. Most previous studies have evaluated a relatively small number of potential smolt quality indicators.

In 1989 we began a long-term study to examine the physiological development and condition of spring chinook salmon *Oncorhynchus tshawytscha* at several hatcheries in the Columbia River basin. We are examining 28 potential indicators of smolt quality to determine whether smolt indices could be related to adult recovery and used to improve hatchery effectiveness. Our study continues to generate a large amount of data; we report here a preliminary overview of our findings. The specific objectives of the 7-year study are: (1) to determine whether selected smolt indices differ significantly between stocks, locations, seasons, and treatments, and (2) to correlate smolt quality measurements in hatcheries with overall survival of the released groups estimated by adult recovery. The goal of objective 1 is to determine which smolt indices may reveal physiological and morphological differences in the smolts released from hatcheries. An important component of objective 1 is to determine whether variation in selected smolt indices is associated with particular fish husbandry techniques. Identification of techniques that result in better quality smolts may improve smolt survival and thereby increase the contribution of hatchery fish to fisheries or, in the case of supplementation, to natural reproduction.

The goal of objective 2 is to determine whether indices that reveal differences among groups of hatchery fish correlate with adult survival, and therefore could be used to predict adult survival. We propose that smolt indices that predict adult survival are the most appropriate indicators of smolt quality.

We have sampled juvenile fish from five hatcheries since 1989 (results of the first year of the study were summarized in Zaugg et al. [1991]). This report highlights results for the second year of the study, for which fish were released during the autumn of 1989 and spring of 1990. The tests were conducted in 1989–1990 on juvenile spring chinook salmon at Dworshak, Leavenworth, and Warm Springs National Fish Hatcheries (NFH), and the Oregon State Round Butte and Willamette hatcheries. The tests assessed saltwater tolerance, gill Na^+,K^+-ATPase (gill ATPase), cortisol, insulin, thyroid hormones, secondary stress tolerance, fish morphology, liver glycogen, liver lipid, plasma glucose, immune response, blood cell counts, plasma and muscle ion concentrations, and plasma protein concentrations.

The potential smolt indices used in the present study can be divided into two groups: indicators of development, which include measures that are expected to change during smolting, and indicators of general health, which are expected to remain relatively constant within the normal ranges for healthy fish. General health indices include blood cell counts, plasma and muscle ion concentrations, and plasma protein and glucose concentrations. For the developmental indices, data from both published and unpublished studies can be used to establish a consensus of expected patterns of change in physiology and morphology of smolts. Because most of the data came from studies of smolts produced in laboratory and hatchery studies, the expected patterns will be typical only for hatchery smolts, and not for wild or natural smolts.

An example of expected changes in gill ATPase activity, plasma concentrations of the thyroid hormone thyroxine, plasma concentrations of cortisol, and body shape during the parr–smolt transformation is shown in Figure 1. Gill ATPase activity typically increases to a peak level in smolts. Increased gill ATPase presumably reflects a heightened activity of gill tissue as a preadaptation of the fish for entering seawater, where the gill performs essential ion excretory functions. Plasma concentrations of cortisol typically increase to peak levels in smolts. Cortisol promotes general development and may specifically stimulate seawater osmoregulatory functions, including gill ATPase activity. Plasma concentrations of thyroxine begin increasing at the onset of smolting, reach peak levels during the parr–smolt transformation, and return to near baseline values when smolting is complete. Thyroid hormones have global effects on development, including increased guanine deposition in skin, which produces the silver coloration of smolts. Body shape, as measured by principal component 2 (PC 2), is essentially a measure of tail–caudal peduncle elongation in relation to total body growth in length, where a decrease in PC 2 indicates that the tail–caudal peduncle region is growing proportionally faster than the total body. Body shape changes measured by PC 2 (Winans and Nishioka 1987) typically increase slightly and then decline to a negative value when smolting is complete. The decline of PC 2 to a negative value in smolts is thought to indicate that the smolt body shape is becoming more streamlined, as a preadaptation for improved swimming efficiency of fish during the extended marine migration. It is important to note that the data shown in Figure 1 represent a consensus pattern of smolting derived from several studies. Significant variation from this pattern may be observed in individual populations of juvenile salmonids.

Data on smolt indices from particular groups of hatchery fish may depart from the expected pattern in several ways. For example, there may be little or no increase in gill ATPase activity in fish up to the point of their release from the hatchery. The lack of change in gill ATPase may or may not be associated with changes in other indices. Other differences may be manifested as quantitative differences in the maximal value, time of onset, or duration of elevation in the measured indices above some baseline

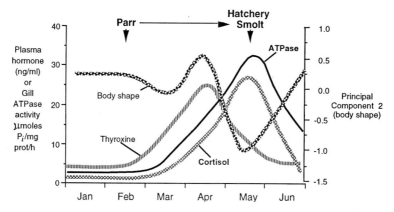

FIGURE 1.—Expected changes in selected smolt indices during the parr to smolt transformation. Plasma hormones, thyroxine and cortisol are given as ng/ml; ATPase activity is given as μmoles P_i/mg protein/h.

value. Differences in the relative timing of indices may also occur in particular groups of fish. For example, the expected pattern indicates that plasma thyroxine concentration reaches its peak value before the peak of gill ATPase. It is possible, however, that peaks in the two indices could coincide, or the peak value of gill ATPase could precede the peak value of plasma thyroxine. The differences in smolting patterns among hatchery fish are not entirely clear, but will be evaluated in the present study.

Methods

Sampled Groups

Field sampling of the 1988-brood spring chinook salmon at five hatcheries began in July 1989 and continued into April 1990. All hatcheries were located within the Columbia River basin (Figure 2). A summary of the groups sampled is shown in Table 1 and described in the following

Dworshak NFH.—Two production raceways, numbers 13 (group 1) and 14 (group 2) were sampled from 19 September 1989 through release in April 1990. Fish in both raceways had received two treatments with erythromycin in the feed. Fish in raceway 14 were from females showing infectious hematopoietic necrosis (IHN)-positive ovarian fluid, and 100% were tagged with coded wire tags (CWT) in November 1989.

Leavenworth NFH.—Fish were sampled from two production raceways, initially numbers 43 (group 1) and 44 (group 2). In November 1989, some produc-

TABLE 1.—Hatchery raceway, sampled groups, rearing density, number of fish released, and release date for fish studied.

Hatchery and raceway	Group designation	Rearing density at release (lb/ft^3)	Number of fish released	Release date
Dworshak				
13	Group 1	1.81[a]	43,755	5 April 1990
14	Group 2	1.81	42,960	5 April 1990
Leavenworth				
43	Group 1	0.85[a]		
44/54	Group 2	0.85	25,700	18 April 1990
Warm Springs				
18	Group 1	0.98	32,034	11 April 1990
19	Group 2	1.63	64,439	11 April 1990
Round Butte				
2	Group 1	1.09	28,608	19 April 1990
Pelton Ladder (lower section)	Group 1	0.30	21,328	23 April 1990
Willamette				
Asphalt pond	Group 1	0.35	28,700	7 March 1990
4	Group 2	1.93	33,183	13 November 1989

[a] Data averaged for both raceways.

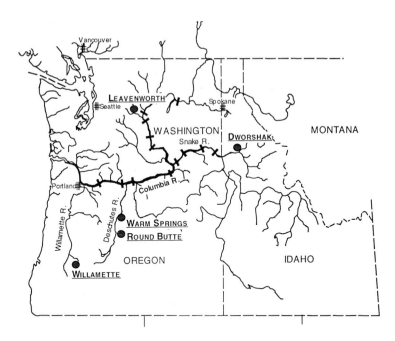

FIGURE 2.—Map of the Columbia River basin indicating the location of study hatcheries. Perpendicular bars across the Columbia and Snake rivers indicate the location of major hydroelectric dams.

tion fish were tagged with CWT and distributed to some of the ponds. On 13 December and thereafter, group 2 fish were taken from raceway 54, and one of the ponds containing tagged fish, instead of raceway 44.

Warm Springs NFH.—Fish were sampled from two production raceways (18 and 19) from 26 July to 25 September 1989. In September, fish in raceways 18 and 19 were graded into two populations, one greater than 140 mm fork length and the other less than 140 mm. Large fish were released on 27 September, and the remaining fish were established at two densities: (1) one-half normal production density (raceway 18, group 1), and (2) normal density (raceway 19, Group 2). Sampling continued in these raceways until release on 11 April 1990.

Round Butte (Oregon State) Hatchery.—Sampling began on 19 October 1989 from raceway 2 (group 1, to be reared at the hatchery) and from raceway 4 (group 2, to be transferred to Pelton Ladder). After being tagged in November, fish were transferred from raceway 4 to three sections of the Pelton Ladder (upper, middle, and lower). Further sampling of group 2 fish was from the lower section. Because of the width (3 m) and depth (>2 m) of the ladder, it was necessary to alter the normal dip-net method of fish collection. A net on a 2 m × 2 m frame was lowered to the bottom of the channel and fish were induced with food to swim over the net. The net was rapidly raised to the surface, capturing 60 to 200 fish. All fish used for tissue collection (see next section) were taken from this one seining operation. Group 2a fish (stressed) were placed in a bucket of water and, within 5 min, carried to and placed in the stress bucket. Group 2b fish (baseline cortisol) were killed as quickly as possible (generally within 1 min) in a lethal concentration of tricaine methanesulphonate (0.2 g MS-222/L). Group 2c fish (photography and tissue samples) were placed in a plastic can containing approximately 20 gallons of water and kept until the first two groups had been processed (about 1.5 h). Sampling continued from an ungraded group in raceway 2 at the hatchery.

Willamette (Oregon State) Hatchery.—Two groups of the 1988 brood were monitored at the Willamette Hatchery and the Dexter Holding Ponds, one for release in November 1989 and the other for release in March 1990. Because of extensive construction at the Willamette Hatchery, normal rearing procedures were altered. In general, fish were moved more than usual and were held at higher densities than normal. One week prior to sampling, the fish had been moved from raceways to the Burrows Ponds. Sampling began on 27 July 1989 from pond 5 (group 1, March 1990 release). Group 1 samples were later taken (beginning on 27 September) from pond 11A, which contained fish that had been taken from pond 12 and tagged with CWTs during 6–12 September. A replicate group, also taken from pond 12 and tagged, was placed in the other half of the screened Burrows Pond (11B). These fish were not sampled, however. Fish from ponds 11A and 11B were transferred to the asphalt pond at Dexter on 30 November, where they were mixed with untagged fish. Fish from Dexter pond were sampled periodically until release on 7 March 1990.

Sampling began on group 2 (November 1989 release) at Dexter on 30 August from adult holding raceway 4. Due to the size of this raceway it was necessary to net fish that had been attracted to the surface with food. Several attempts were necessary to obtain the needed number of fish. Untagged fish had been moved from the Willamette Hatchery to Dexter on 7 April 1989. Fish that were coded-wire tagged at Willamette during 13–25 July were later (1 August 1989) trucked to Dexter and mixed with the unmarked fish in raceway 4 to serve as the indicator group for survival information. Fish from raceway 4 were released on 13 November 1989.

Tissue Collection

On each sampling visit at all hatcheries, three separate 15-fish samples were taken from each population (Group); these samples were designated subgroups a, b, and c. All fish were measured and weighed to the nearest 1 mm and 0.1 g. Fish sex was also determined, including an assessment of precocious development in males. The condition factor (body weight/fork length^3) was determined from length and weight.

The 15 fish in subgroup a were obtained from the raceway by dip net and then stressed by confining for 1 h in a perforated bucket suspended in the raceway. Following the stress period, these fish were immersed in a lethal concentration of MS-222. After fish were measured and weighed, tails were severed at the caudal peduncle, and blood was collected in heparinized Natelson tubes. Plasma obtained from centrifuged samples was used to determine cortisol and glucose.

The 15 fish of subgroup b were obtained from the raceway by dip net and placed immediately in a lethal concentration of MS-222. Fish were measured and weighed. The fish tails were then severed at the caudal peduncle, and blood first collected in a heparinized microhematocrit tube for determination of hematocrits, and the remainder then collected in heparinized Natelson tubes. In order to process the fish within 10 to 15 min of death, fish were obtained and samples collected in two separate groups of 7 to 8 individuals. Plasma obtained from centrifuged samples was used to determine cortisol, glucose, blood electrolytes, and plasma total protein. Blood smears were taken from this group of fish. Anterior kidneys from these fish plus those from an additional five fish were taken for the immune competence assay.

The 15 fish of subgroup c were secured with the dip net and held alive in a bucket. One fish at a time was removed, killed by immersion in a lethal concentration of MS-222, measured, weighed, and photographed for morphometric analysis. Blood was taken and plasma obtained as indicated above. Plasma was used for thyroxine (T4), triiodothyronine (T3), and insulin determinations. A ventral incision was then made, sex noted, and a section (0.1 to 0.2 g) of the lower lobe of the liver removed and frozen immediately in liquid nitrogen; the liver tissue was used later for measuring glycogen content. A second piece of liver was excised and placed in a tube on dry ice for later liver triglyceride analysis. A section of skin (1 cm x 5 cm) in the area between the lateral line and the dorsal fin was removed, frozen on dry ice, and used later for measuring guanine content. After removal of skin, a section (0.1 to 0.2 g) of white muscle was removed and placed on dry ice and used later for measuring tissue water content. Filaments were trimmed from the lower half of two to four gill arches and placed in a tube with 1 ml of sucrose-ethylenediaminetetraacetic acid-imidazole (SEI) solution. The tube was capped, placed on dry ice, and used later for measuring adenosine triphosphatase (ATPase) activity.

Analytical Procedures

Saltwater challenge.—Groups of 20 fish each were placed for 24 h in 30 parts per thousand (‰) salt water made up with Marine Environment artificial sea salts. After the challenge period, fish were removed, killed in a lethal concentration of MS-222, weighed, and measured. Blood for plasma sodium and potassium (Clarke and Blackburn 1977) was taken using microhematocrit tubes.

Gill ATPase Activities.—Gill filaments were trimmed from arches and preserved in SEI at −80°C until analyzed for gill Na$^+$,K$^+$-ATPase activity as described in Zaugg (1982), with minor modification. Units of activity are μmoles P$_i$/mg protein/h.

Thyroxine (T4) and triiodothyronine (T3).—Blood plasma concentrations of thyroid hormones were analyzed by radioimmunoassay (RIA) according to methods described by Dickhoff et al. (1978, 1982).

Plasma insulin.—Blood plasma concentrations of insulin were analyzed by homologous RIA according to the method described by Plisetskaya et al. (1986).

Plasma cortisol, baseline and stressed.—Blood plasma concentrations of cortisol from stressed and unstressed fish were measured by RIA according to the method of Redding et al. (1984).

Stress challenge.—This stress challenge was described by Barton et al. (1985). Fish were netted and subjected to an acute handling stress by suspending them for 1 h in a perforated bucket placed in the raceways such that the backs of the median-sized fish were just under the surface of the water. The fish were then anesthetized, and a blood plasma sample was taken for later analysis for cortisol and glucose content.

Plasma glucose.—Blood glucose concentrations were determined by a colorimetric procedure (Sigma Chemical Co., St. Louis, Missouri).

Liver glycogen.—Liver glycogen was measured according to the method of Wedemeyer and Yasutake (1977). Glycogen is extracted into potassium hydroxide, precipitated, hydrolyzed to glucose, and quantified with a glucose hexokinase enzymatic determination, measured by spectrophotometry at 340 nm.

Liver triglyceride.—Liver samples were homogenized in water and centrifuged. Triglyceride concentrations were measured by the enzymatic method of Bucolo and David (1973). Glycerol is stripped by phospholipase C and then reduced with glycerol dehydrogenase. The reduced nicotine adenine dinucleotide (NADH) generated is oxidized by para-iodo-nitro-tetrazolium violet, mixed with enzymes, and measured by spectrophotometry at 500 nm.

Morphometrics.—Morphometric distances for 26 truss-network characters were calculated from each photograph and analyzed by principal component (PC) analysis (Winans 1984; Winans and Nishioka 1987).

Skin guanine.—Skin guanine content as a quantitative measure of silver coloration was determined according to the method of Staley (1984), although skin samples in our study were taken above the lateral line, not below the lateral line as described by Staley. Skin samples were extracted with 1 N HCl for 48 h at 21°C. Extracts were adjusted to pH 8.1 and treated with xanthine oxidase and guanase for 2 h at 21°C. Guanine concentration was measured by spectrophotometry at 290 nm.

Muscle water.—Water content of dorsal muscle was determined according to the method of Wedemeyer and McLeay (1981).

Blood electrolytes.—Blood plasma Na^+ and K^+ concentrations were determined by flame photometry; blood plasma Cl^- concentrations were determined using a chloridometer.

Plasma total protein.—Plasma total protein was determined using a refractometer calibrated to distilled water.

Blood white cell count and differential white cell count.—White cell counts were performed to determine the number of white cells per hundred red cells and to differentiate the number of lymphocytes, neutrophils, and monocytes in 100 white cells according to the method of Wedemeyer and Yasutake (1977).

Immune response.—The immune response of fish was determined by assessing the production of anterior kidney antibody-secreting cells (plaque-forming cells [PFC]) after in vitro exposure to a synthetic antigen (trinitrophenol-lipopolysaccharide) according to the method of Tripp et al. (1987).

Adult Contribution

Data for adult contribution were obtained from the Regional Mark Information System, Pacific States Marine Fisheries Commission, Portland, Oregon.

Results and Discussion

Comparisons were made of physiological development and general health in groups of fish within each hatchery. Measures of smolt quality included body weight, body length, condition factor, body shape (PC2), plasma levels of sodium and potassium after seawater challenge, seawater survival, gill ATPase, plasma levels of thyroxine, cortisol, insulin, glucose, sodium, chloride, potassium, and protein, skin guanine, plasma levels of cortisol and glucose after stress, liver concentrations of glycogen and triglyceride, muscle water concentration, immune response, hematocrit, white cell count, and counts of leucocytes, monocytes and neutrophils. However, due to space limitation, examples of only some data are shown in this report. Of the 28 measures of smolt quality, significant differences were seen in smolt development in all of the 17 developmental indices and 3 of the 11 general health indices. Of the 17 developmental indices, 10 showed significant variation between treatments in at least half of the treatment groups studied. These 10 measures were

TABLE 2.—Estimated adult contribution (percent of released smolts) based on recoveries of adults with coded wire tags for yearling spring chinook salmon (brood year 1988) released from some Columbia River hatcheries. Releases of fish at Willamette Hatchery included a November 1989 release (raceway 4) and a March 1990 release (asphalt pond). Releases of all other groups were in April 1990.

Hatchery (raceway)	Adult contribution (%)
Dworshak (14)	0.012
Warm Springs (19)	0.137
Leavenworth (54)	0.166
Willamette (4)	0.216
Warm Springs (18)	0.218
Willamette (asphalt pond)	0.284
Round Butte (1)	0.510
Round Butte (2)	0.659
Pelton Ladder (middle)	0.915
Pelton Ladder (lower	0.995
Pelton Ladder (upper)	1.005

body weight, body length, body shape (PC2), gill ATPase activity, plasma thyroid hormone level, liver glycogen, post-seawater-challenge survival, plasma sodium level, and plasma potassium level. Of the three general health indices that showed significant statistical variation, only blood hematocrit showed variation that was biologically significant. At Dworshak NFH, abnormally low hematocrits (below 32%) were observed in juveniles for several months before the fish were released. These low hematocrits were cause for concern because low hematocrits were also observed at Dworshak in the previous year (1989) (Zaugg et al. 1991). Subsequent analysis of spring chinook salmon at Dworshak NFH revealed the presence of erythrocytic inclusion body syndrome.

Estimated adult contribution based on recoveries of adults with coded wire tags showed an approximately 100-fold range (Table 2). The highest contributions were from the Pelton Ladder, followed by Round Butte Hatchery and the March release from the Dexter facility of Willamette Hatchery. The lowest adult contribution was from Dworshak NFH.

In general, the wide range of adult contribution coincided with the wide range in smolt quality; high quality smolts, based on similarity with the expected pattern (Figure 1), had high adult contribution. For example, fish were larger and showed high levels of gill ATPase activity at Round Butte Hatchery and Pelton Ladder (Figure 3). In contrast, fish at Dworshak, Warm Springs, and Leavenworth hatcheries were smaller and showed lower levels of gill ATPase. However, comparison of adult returns and smolt quality measures of all fish from all hatcheries is complicated because of differences in genetic stock of fish, release dates, upriver and downriver locations, river flows and estuarine conditions at the time of release, and the variable number of hydroelectric dams that fish must pass in their downstream migration. Therefore, although fish from Round Butte Hatchery and Pelton Ladder showed greater smolt quality and adult contribution than did fish from Leavenworth and Dworshak hatcheries, the Round Butte and Pelton Ladder fish had to pass only two dams whereas Leavenworth and Dworshak fish had to pass seven or eight dams, respectively (Figure 2).

A more valid comparison can be made among groups of fish released from Warm Springs and Round Butte hatcheries and Pelton Ladder, because these fish were released into the Deschutes River within a 2-week period in April 1990. As shown on Table 3, the adult contributions of fish from Warm Springs, Round Butte, and Pelton Ladder were low, intermediate, and high, respectively. Fish weight at the time of release was greatest at Round Butte, intermediate at Pelton Ladder, and smallest at Warm Springs. Gill ATPase activities were significantly lower in Warm Springs fish than in Round Butte and Pelton Ladder fish. Fish at Pelton Ladder performed significantly better in the seawater-challenge test (lower plasma sodium level) than did Round Butte and Warm Springs fish. Growth rate during the last month before release was lowest at Warm Springs, intermediate at Round Butte, and highest at Pelton Ladder. These results suggest that growth rates for the month before release show the best positive correlation with adult contribution. If greatest body size is the most important smolt quality factor, then it would be expected that Round Butte fish would have had the highest adult contribution. Gill ATPase activities differentiate the Round Butte and Pelton Ladder fish from Warm Springs fish, but are similar for Round Butte and Pelton Ladder. Seawater-challenge performance separates Pelton Ladder from Round Butte fish, but does not differ between Round Butte and Warm Springs fish.

Changes in plasma concentrations of thyroid hormones and cortisol showed clearer elevations in fish reared at Pelton Ladder compared with Round Butte Hatchery (Figure 4). Other potential smolt indicators also suggested earlier or more complete development of smolts reared in Pelton Ladder. For example, fish reared in Pelton Ladder also had a sharper and greater decline in condition factor, earlier increase in plasma insulin concentration, and a later increase in stress response.

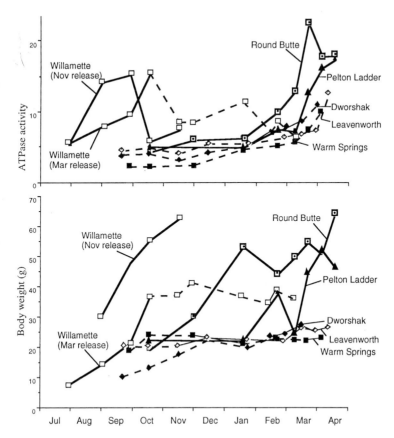

FIGURE 3.—Comparison of average body weight and gill ATPase activity of juvenile spring chinook salmon sampled during 1989–1990. Solid lines indicate groups of fish in which relatively rapid growth rate was associated with high or increasing gill ATPase activity. Dashed lines indicate groups of fish with low growth rate and low gill ATPase activity. Hatcheries are Dworshak (solid diamonds), Leavenworth (open diamonds), Round Butte (squares with dots), Pelton Ladder (solid triangles), Warm Springs (solid squares), and Willamette (open squares). The ATPase activity is given as μmoles P_i/mg protein/h.

The relationships between smolt size, smolt quality, release timing, and adult contribution have been studied (Wagner et al. 1969; Wagner 1974; Ewing et al. 1979, 1980; Bilton et al. 1982; Mahnken et al. 1982; Soivio and Virtanen 1985; Zaugg 1989; Virtanen et al. 1991; Farmer 1994). A consensus of researchers is that smolts need to attain a critical size (size threshold) at a certain date. Differences in body size are typically associated with differences in growth rate, so that larger fish are most often the fastest growing fish. Studies by Ewing et al. (1980) on the influence of growth rate on cyclic changes in gill ATPase activity of juvenile spring chinook salmon suggested that growth rates less than 0.27 mm/d suppressed development of the October peak in gill ATPase. Maximal development of ATPase activity occurred at growth rates above 0.38 mm/d.

It is interesting that growth rates of spring chinook salmon at Warm Springs NFH were below the minimum of 0.27 mm/d, at Round Butte Hatchery were between 0.27 and 0.38 mm/d, and at Pelton Ladder were above 0.38 mm/d. However, none of the published studies of the effect of smolt size have separated the effects of body size from growth rate, because the larger fish in such studies were also growing faster. Further studies on the relative contributions of body size and growth rate to smolt quality and adult survival are warranted.

The relationship between growth and smolting of Atlantic salmon *Salmo salar* has been studied most extensively by Thorpe and colleagues. By October, after their first season of growth, Atlantic salmon populations develop a bimodal size distribution (Thorpe 1977). The largest fish (upper mode) will

TABLE 3.—Body weight (g), gill ATPase activity (μmoles P_i/mg protein/h), plasma sodium after seawater challenge (mmole/L) growth rate (mm/d), and estimated adult contribution (percent of released smolts) based on recoveries of adults with coded wire tags for yearling spring chinook salmon released during 11–23 April 1990 (brood year 1988). Gill ATPase and seawater-challenge data were determined near the time of release; growth rates are from 35 to 44 d before release for the groups where data are available. Within a column, values with a common letter do not differ significantly ($P < 0.05$; analysis of variance and Fisher protected least-significant difference).

Hatchery and raceway	Body weight	Gill ATPase activity	Plasma sodium after seawater challenge	Growth rate	Adult contribution
Warm Springs					
19	22.0	9.7 ± 0.7z	228 ± 7z	0.119	0.137
18	24.0	10.7 ± 0.8z	222 ± 9z	0.095	0.218
Round Butte					
1	69.9				0.510
2	75.7	18.2 ± 2.2y	224 ± 8z	0.359	0.659
Pelton Ladder					
middle	46.8				0.915
lower	51.6	17.7 ± 1.6y	201 ± 7y	0.923	0.995
upper	42.4				1.005

become smolts the following spring, whereas the lower mode fish will remain as parr for an additional year (Thorpe 1987). However, smoltification of some lower mode fish in their first spring may be induced by elevated early winter temperature (Saunders et al. 1982; Kristinsson et al. 1985). It has been hypothesized that for Atlantic salmon critical times exist when body size or energy reserve determines smoltification (Thorpe 1987). However, the relative roles of size threshold, growth rate, and energy reserve are not fully understood (McCormick and Saunders 1987). It is not clear how models for growth and development of Atlantic salmon may apply to chinook salmon. Chinook salmon populations do not normally show bimodal length frequencies, and the potential for autumn smoltification of chinook salmon is a significant difference, among others. However, substantial evidence argues that body size and growth rate are important for juvenile development of all anadromous salmonids.

We speculate that lack of growth during the smolting period may signal a developmental stasis to suppress smolt development. Water temperature plays a key role in controlling the growth rates of fish. Feeding rates are also important. Perhaps attempts should be made through temperature, photoperiod, and feeding-rate regulation to stimulate more growth in the smaller fish, especially during the several weeks prior to release. We expect additional insight into smolt quality as we analyze data sets from broodyears 1989 and 1990 and compare these with adult returns.

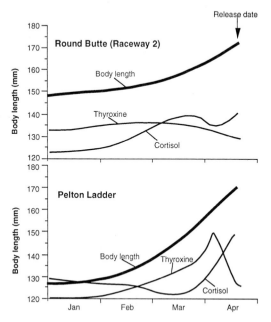

FIGURE 4.—Comparison of body length, plasma thyroxine concentration, and plasma cortisol concentration between juvenile spring chinook salmon reared at Round Butte Hatchery and at Pelton Ladder. The rapid increase in body length of fish reared at Pelton Ladder was associated with a sharp peak in plasma thyroxine and a clear elevation of plasma cortisol.

Acknowledgments

This project is a collaborative effort involving the National Marine Fisheries Service, Oregon State University, U.S. Fish and Wildlife Service, Oregon Department of Fish and Wildlife, and the Washington Department of Fisheries. The use of trade names does not imply endorsement by the National Marine Fisheries Service, NOAA.

We thank Alec Maule and Martin Fitzpatrick (Oregon State University), Aldo Palmisano, Gary Wedemeyer, and Robin Schrock (National Biological Survey); Richard Ewing (Oregon Department of Fish and Wildlife); Gary Winans (National Marine Fisheries Service); Charles Hopley (Washington Department of Fish and Wildlife); and Gerald Bouck and Robert Austin (Bonneville Power Administration) for their technical assistance in sample analysis and contributions to the design of this study. This study was supported in part by the Bonneville Power Administration (Project 89-046).

References

Barron, M. G. 1986. Endocrine control of smoltification in anadromous salmonids. Journal of Endocrinology 108:313–319.

Barton, B. B., C. B. Schreck, R. D. Ewing, A. R. Hemmingsen, and R. Patino. 1985. Changes in plasma cortisol during stress and smoltification in coho salmon *Oncorhynchus kisutch*. General and Comparative Endocrinology 59:468–471.

Bilton, H. T., D. F. Alderdice, and J. T. Schnute. 1982. Influence of time and size at release of juvenile coho salmon (*Oncorhynchus kisutch*) on returns at maturity. Canadian Journal of Fisheries and Aquatic Sciences 39:426–447.

Boeuf, G. 1993. Salmonid smolting: a pre-adaptation to the oceanic environment. Pages 105–135 *in* J. C. Rankin, and J. B. Jensen, editors. Fish ecophysiology. Chapman and Hall, London.

Bucolo, G., and H. David. 1973. Quantitative determination of serum triglycerides by the use of enzymes. Clinical Chemistry 19:475–482.

Clarke, W. C., and J. Blackburn. 1977. A seawater challenge test to measure smolting of juvenile salmon. Fisheries and Marine Service Research and Development Technical Report 705:1–11.

Dickhoff, W. W. 1993. Hormones, metamorphosis and smolting. Pages 519–540 *in* M. P. Schreibman, C. G Scanes, and P. K. T. Pang, editors. The endocrinology of growth, development, and metabolism in vertebrates. Academic Press, San Diego, California.

Dickhoff, W. W., C. L. Brown, C. V. Sullivan, and H. A. Bern. 1990. Fish and amphibian models for developmental endocrinology. Journal of Experimental Zoology 4(Supplement):90–97.

Dickhoff, W. W., L. C. Folmar, and A. Gorbman. 1978. Changes in plasma thyroxine during smoltification of coho salmon, *Oncorhynchus kisutch*. General and Comparative Endocrinology 36:229–232.

Dickhoff, W. W., L. C. Folmar, J. L. Mighell, and C. V. W. Mahnken. 1982. Plasma thyroid hormones during smoltification of yearling and underyearling coho salmon and yearling chinook salmon and steelhead trout. Aquaculture 28:39–48.

Ewing, R. D., and E. K. Birks. 1982. Criteria for parr-smolt transformation in juvenile chinook salmon (*Oncorhynchus tshawytscha*). Aquaculture 28:185–194.

Ewing, R. D., A. R. Hemmingsen, M. D. Evenson, and R. L. Lindsay. 1985. Gill (Na^+K)-ATPase activity and plasma thyroxine concentrations do not predict time of release of hatchery coho (*Oncorhynchus kisutch*) and chinook salmon (*Oncorhynchus tshawytscha*) for maximum adult returns. Aquaculture 45:359–373.

Ewing, R. D., S. L. Johnson, H. J. Pribble, and J. A. Lichatowich. 1979. Environmental effects of cyclic changes in gill (Na^+K)-ATPase activity in chinook salmon, *Oncorhynchus tshawytscha*. Journal of the Fisheries Research Board of Canada 36:1347–1353.

Ewing, R. D., H. J. Pribble, S. L. Johnson, C. A. Fustish, J. Diamond, and J. A. Lichatowich. 1980. Influence of size, growth rate, and photoperiod on cyclic changes in gill (Na^+K)–ATPase activity in chinook salmon (*Oncorhynchus tshawytscha*). Canadian Journal of Fisheries and Aquatic Sciences 37:600–605.

Farmer, G. J. 1994. Some factors which influence the survival of hatchery Atlantic salmon (*Salmo salar*) smolts utilized for enhancement purposes. Aquaculture 121:223–233.

Folmar, L. C., and W. W. Dickhoff. 1980. The parr–smolt transformation (smoltification) and seawater adaptation in salmonids. A review of selected literature. Aquaculture 21:1–37.

Hoar, W. S. 1988. The physiology of smolting salmonids. Pages 275–343 *in* W. S. Hoar and D. J. Randall, editors. Fish physiology, volume 11B. Academic Press, San Diego, California.

Kristinsson, J. B., R. L. Saunders, and A. J. Wiggs. 1985. Growth dynamics during the development of bimodal length-frequency distribution in juvenile Atlantic salmon (*Salmo salar* L.). Aquaculture 45:1–20.

Mahnken, C. V. W., E. Prentice, W. Waknitz, G. Monan, C. Sims, and J. Williams. 1982. The application of recent smoltification research to public hatchery releases: an assessment of size/time requirements for Columbia River hatchery coho salmon (*Oncorhynchus kisutch*). Aquaculture 28:251–268.

McCormick, S. D., and R. L. Saunders. 1987. Preparatory physiological adaptations for marine life of salmonids: osmoregulation, growth, and metabolism. American Fisheries Society Symposium 1:211–229.

Plisetskaya, E., W. W. Dickhoff, T. L. Paquette, and A. Gorbman. 1986. The assay of salmon insulin by homologous radioimmunoassay. Fish Physiology and Biochemistry 1:35–41.

Redding, J. M., C. B. Schreck, E. K. Birks, and R. D. Ewing. 1984. Cortisol and its effects on plasma thyroid hormone and electrolyte concentration in fresh water and during seawater acclimation in yearling coho salmon, *Oncorhynchus kisutch*. General and Comparative Endocrinology 56:146–155.

Saunders, R. L., E. B. Henderson, and B. D. Glebe. 1982. Precocious sexual maturation and smoltification in male Atlantic salmon (*Salmo salar*). Aquaculture 28:211–229.

Soivio, A., and E. Virtanen. 1985. The quality and condition of reared *Salmo salar* smolts in relation to their adult recapture rate. Aquaculture 45:335–343.

Staley, K. B. 1984. Purine deposition in the skin of juvenile coho salmon, *Oncorhynchus kisutch*. Master's thesis. Oregon State University, Corvallis.

Staurnes, M., G. Lysfjord, L. P. Hansen, and T. G. Heggberget. 1993. Recapture rates of hatchery-reared Atlantic salmon (*Salmo salar*) related to smolt development and time of release. Aquaculture 118:327–337.

Thorpe, J. E. 1977. Bimodal distribution of length of juvenile Atlantic salmon (*Salmo salar* L.) under artificial rearing conditions. Journal of Fish Biology 11:175–184.

Thorpe, J. E. 1987. Adaptive flexibility in life history tactics of mature male Baltic salmon parr in relation to body size and environment. American Fisheries Society Symposium 1:244–252.

Tripp, R. A., A. G. Maule, C. B. Schreck, and S. L. Kaattari. 1987. Cortisol mediated suppression of salmonid lymphocyte responses *in vitro*. Developmental and Comparative Immunology 11:565–576.

Virtanen, E., L. Söderholm-Tana, A. Soivio, L. Forsman, and M. Muona. 1991. Effect of physiological condition and smoltification status at smolt release on subsequent catches of adult salmon. Aquaculture 97:231–257.

Wagner, H. H. 1974. Photoperiod and temperature regulation of smolting in steelhead trout (*Salmo gairdneri*). Canadian Journal of Zoology 52:219–234.

Wagner, H. H., F. P. Conte, and J. L. Fessler. 1969. Development of osmotic and ionic regulation in two races of chinook salmon, *Oncorhynchus tshawytscha*. Comparative Biochemistry and Physiology 29:325–341.

Wahle, R. J., and W. S. Zaugg. 1982. Adult coho salmon recoveries and their Na^+–K^+–ATPase activity at release. Marine Fisheries Review 44:11–13.

Wedemeyer, G. A., and D. J. McLeay. 1981. Methods for determining the tolerance of fishes to environmental stressors. Pages 247–275 *in* A. D. Pickering, editor. Stress and fish. Academic Press, London.

Wedemeyer, G. A., R. L. Saunders, and W. C. Clarke. 1980. Environmental factors affecting smoltification and early marine survival of anadromous salmonids. Marine Fisheries Review 42:1–14.

Wedemeyer, G. A., and W. T. Yasutake. 1977. Clinical methods for the assessment of the effects of environmental stress on fish health. U.S. Fish and Wildlife Service Technical Paper 89:1–18.

Winans, G. A. 1984. Multivariate morphometric variability in Pacific salmon: technical demonstration. Canadian Journal of Fisheries and Aquatic Sciences 41:1150–1159.

Winans, G. A., and R. S. Nishioka. 1987. A multivariate description of change in body shape of coho salmon (*Oncorhynchus kisutch*) during smoltification. Aquaculture 66:235–245.

Zaugg, W. S. 1982. A simplified preparation for adenosine triphosphatase determination in gill tissue. Canadian Journal of Fisheries and Aquatic Sciences 39:215–217.

Zaugg, W. S. 1989. Migratory behavior of underyearling *Oncorhynchus tshawytscha* and survival to adulthood as related to prerelease gill (Na^+–K^+)–ATPase development. Aquaculture 82:339–353.

Zaugg, W. S., and eleven coauthors. 1991. Smolt quality assessment of spring chinook salmon. Report to Bonneville Power Administration, Project 89-046, Portland, Oregon.

Terminal Fisheries and Captive Broodstock Enhancement by Nonprofit Groups

JOHN A. SAYRE

Long Live the Kings, Inc., 19435 184th Place Northeast
Woodinville, Washington 98072, USA

Abstract.—Long Live the Kings (LLTK) is a private, nonprofit corporation that was formed in Washington State in 1986. The time seemed right for private citizens to become involved in fish restoration, and LLTK's goals are to restore wild salmon populations in selected Northwest rivers, to enhance salmonid habitat, and to rebuild regional salmon economies. Long Live the Kings works closely with 12 nonprofit regional salmon enhancement groups formed by Washington State law in 1991 and financed by a dedicated fund garnered from a surcharge on sport and commercial fishing licenses. In the last 8 years LLTK has invested over US$2.5 million of private funds into salmon restoration. This private money comes from foundations, corporations, and individuals. Additional revenue is generated by raising fish under contract for state agencies and Indian Nations.

Long Live the Kings (LLTK) is a private, nonprofit corporation that was formed in Washington State in 1986. It was formed after the state of Washington and treaty tribes ended a dozen years of litigation over salmon fishing rights and decided to work together as comanagers of Washington's salmon fisheries. In addition, the signing of the U.S.–Canada Pacific Salmon Interception Treaty gave the promise that enhancement efforts of one nation would accrue to that nation. Thus the time seemed right for private citizens to become involved in fish restoration, and LLTK was created. Long Live the Kings' goals are to restore wild salmon populations in selected Northwest rivers, to enhance salmonid habitat, and to rebuild regional salmon economies. The organization represents sport, tribal, and commercial fishers, as well as businesses, local communities, and individuals interested in restoring salmon and habitat. The organization works closely with 12 nonprofit regional salmon enhancement groups formed by Washington State law in 1991 (Title 75 RCW, Chapter 75.50) and financed by a dedicated fund garnered from a surcharge on sport and commercial fishing licenses.

In the 15 years prior to 1994, stocks of Pacific salmon *Oncorhynchus* spp. in the Northwest have been dramatically reduced. In 1994 the Northwest ocean fisheries experienced an unprecedented closure, and the economies of coastal communities collapsed. This ocean resource has supported a flourishing industry that has attracted people from around the world. Visitors annually contributed hundreds of millions of dollars to the regional economy. Sadly, the biological decline of fish stocks and the economic consequences are an increasingly common worldwide occurrence. Obviously, traditional management strategies are not working. It is the intention of LLTK to be a partner in solving the problem.

In the last 8 years LLTK has invested over US$2.5 million of private funds into salmon restoration. This private money comes from foundations, corporations, and individuals. Additional revenue is generated by raising fish under contract for state agencies and Indian Nations.

To address specific salmon stock problems and attempt to solve them, three LLTK facilities have been constructed in Washington State. These facilities serve to demonstrate low-technology enhancement techniques that can be cost effectively duplicated elsewhere by individuals, organizations, and communities. The three facilities do not produce significantly large numbers of fish—approximately 1 million smolts annually. However, the facilities are already being duplicated at several locations throughout the region. These smaller projects have the potential to benefit fisheries management significantly. All sites operate under permits from the Washington Department of Fish and Wildlife (WDFW). Each facility has a full-time manager and works in cooperation with state and federal agencies, Indian tribes, and regional salmon enhancement groups.

The original LLTK pilot project was built on the Wishkah River, which flows into Grays Harbor on the Washington coast. The river had reduced numbers of spawning chinook salmon *O. tshawytscha* for several years. With the support of local residents (many of whom were poachers), wild adult chinook were captured and spawned. Broodstock were captured with dip nets within one-fourth mile of the facility. The young were reared in earthen ponds designed to promote wild characteristics of the fish while in captivity. Trees and vegetation were allowed to grow around the pond and provided cover, shade, and natural feed. The outlet was not screened, so fish could leave volitionally. Juveniles are retained for a year; those fish that have not emigrated volitionally are forced out (usually about 25% of the original population). The original production goal was 300,000 smolts; 100,000 fish are currently released to minimize the number of adult wild fish harvested for broodstock. The goal is not to create a hatchery run of fish, but to enhance the natural spawning populations until they can sustain themselves.

In each of the last 2 years several hundred chinook salmon have returned to spawn near the Wishkah facility. In 1992 and 1993, several dozen tagged fish from our 1988 and 1990 year-classes were observed. In each of those years 100,000 age-0 smolts were released, of which one-half were tagged. Tagged juveniles are adipose fin-clipped, and a coded wire tag is placed in the nose of the fish. Starting in 1995, all smolts will be tagged and marked. Returning marked adult fish will not be used as broodstock. The results to date indicate supplementation may be a useful technique for rebuilding depleted wild stocks under certain conditions.

An added benefit of supplementation may be the maintenance of a viable population of chinook salmon in spite of severe habitat degradation. In the last 8 years, the majority of the remaining timber in the Wishkah watershed has been cut, and the valley is now essentially a totally cut watershed. Runoff regularly scours gravel beds and channelizes the river. Returning adult chinook salmon observed near the hatchery could have been largely those that were raised in its protected ponds; three storms in 1989 scoured the spawning beds after natural spawning had occurred. The survival of wild fish in 1989 was probably limited due to extremely degraded habitat on the Wishkah River. This problem also threatens other stocks of salmonids. Supplementation efforts at LLTK facilities are being initiated with naturally spawning coho salmon *O. kisutch*, chum salmon *O. keta*, and steelhead *O. mykiss*.

In areas where habitat is becoming degraded at an increasing rate and the number of spawning wild fish is declining, it is vital to provide sanctuary for wild fish gene pools while the watershed recovers, which will take at least 2 decades. Excessive harvesting during this recovery period can exacerbate the problem. The role of supplementation, therefore, may be to maintain these wild stocks until the habitat recovers sufficiently to allow population increases through natural spawning again. The habitat crisis is magnified by natural disasters (i.e., El Niño) and drought.

In addition to fish culture activities, LLTK is active in habitat restoration and recently completed a project with the U.S. Fish and Wildlife Service (USFWS) and the Chehalis Basin Fisheries Task Force. Funding came from the federally enacted Chehalis Basin Restoration Act. Five acres of overwintering and rearing ponds connected to the Wishkah River have been developed.

At first high water, native juvenile coho salmon occupied the new habitat. Several other habitat sites have been created on the Wishkah River, through efforts of sport and commercial fisheries groups. Similar projects are occurring along the Chehalis River, Washington's second largest river.

Through 7 years of operation, the Wishkah River facility has involved the local community in fish restoration by developing co-operative projects that include sport, commercial, and tribal fishers. The partnership developed between LLTK, Resource Conservation and Development Districts, the Chehalis Basin Fisheries Task Force, the Quinault Indian Nation, private landowners, and timber companies is perhaps the leading example in the region of how to begin restoring logged watersheds so the production of fish, wildlife, and timber can be maximized to the benefit of the local community.

The latest LLTK project is located on a small tributary of Hood Canal, an inlet of Puget Sound. Lilliwaup Creek is not large but has naturally spawning populations of six Pacific salmonids (chinook, coho, and chum salmon, pink salmon *O. gorbuscha*, steelhead, and cutthroat trout *O. clarki*). The site also has several springs that provide excellent fish rearing. One of LLTK's directors purchased the property and made part of the land available for the project. Working with the National Marine Fisheries Service (NMFS) staff at the Manchester (Washington) Fish Enhancement Facility, LLTK designed a captive broodstock facility to enhance wild salmon populations that are depleted to the degree that immediate restoration efforts are required.

The present operating scenario includes capturing juvenile wild salmon by electroshocking and then raising them in captivity in freshwater until they are sexually mature. Genetic guidelines require representatives of at least 50 separate families from each target stock. From a small number of captive fish (2,000), a large number of eggs can be produced that will be returned to their parents' native streams as either fry or smolts. The Lilliwaup Creek project is a cooperative endeavor of WDFW, local Indian tribes, NMFS, USFWS, and interested neighbors.

In May 1993 and 1994, WDFW and LLTK staff captured 2,000 juvenile wild coho salmon by electroshocking in 12 northern Hood Canal streams. In 2 or 3 years, these fish should reach sexual maturity and produce viable eggs. The Lilliwaup Creek facility also serves as a gene bank for the depleted stocks while the issues that caused their decline are hopefully addressed. In addition, we have stocks of chinook salmon, cutthroat trout, and steelhead that will be part of this research effort. We are also maintaining an egg bank for summer chum salmon, which spawn in Lilliwaup Creek. Only about 700 summer chum salmon returned to Hood Canal streams in 1993, and the stock is a leading candidate for Endangered Species consideration.

A spring-fed stream of 90 gal/min, which was ditched straight through the Lilliwaup Creek facility, was rerouted into a stream meander. Log weirs and spawning gravel were added, and in the fall of 1993, over 200 wild salmon were spawning in a 75-yard stretch of this 2-ft-wide creek. Provide suitable habitat, and the fish will use it.

The Lilliwaup Creek facility was constructed with $700,000 of private funds. It demonstrates what private citizens and the business community can do to help restore wild fish.

The LLTK concept originated on a 300-acre parcel of land on Orcas Island in Washington's San Juan Islands. Glenwood Springs is a small spring of 250–400 gal/min; it flows one-half mile before entering saltwater in Eastsound, a large bay of Orcas Island. Before the project, Glenwood Springs had no fish in it and was inaccessible to anadromous fish.

The project was started in the late 1970s with a small hatchery building, a large earthen pond, a concrete holding pond next to saltwater, a fish ladder, and a spawning area. Eventually three more ponds and lakes were added to the system. In 1978 the first eyed coho salmon eggs were transferred to Glenwood Springs from the WDFW Samish River Hatchery. For the next 4 years, both chinook and coho salmon eggs were donated from WDFW's Samish and Nooksack fish hatcheries. These donor stocks were incubated, reared, and released at the LLTK facility. In 1982, Glenwood Springs Hatchery reversed normal procedure and supplied an excess of 200,000 salmon eggs to WDFW's Samish hatchery, which lacked sufficient adult returns to meet its egg requirements.

By 1985 the runs were large enough that Eastsound was designated as a harvest area for gillnetters, purse seiners, and tribal fishermen, subject to the rules and regulations of the state of Washington. It was estimated that 15,000 coho salmon were caught in Eastsound, as well as several thousand chinook salmon. From the mid-1980s, there were estimated runs of 4,000 to 6,000 chinook salmon annually, with the majority being caught in the commercial fishery. These returns were from fish stockings of 300,000 age-0 smolts. Smolts were planted from a shore side holding pond, after being acclimated to saltwater pumped from a nearby cove. On release, the fish enter a remarkably healthy and productive estuary.

The adult chinook salmon, averaging 20 lb, have contributed to sport and commercial fisheries on the west coast; there have been coded-wire-tag returns from southern Oregon to Alaska.

In 1993, WDFW did not open a commercial season for chinook salmon. As a result, the largest return of chinook salmon in the history of the project occurred with over 5,000 adults returning to the pond or being caught in the sport fishery in Eastsound (some illegal gillnetting at night also took unknown numbers of fish). Mature females provided 6.3 million eggs, which were incubated at LLTK Glenwood Springs Hatchery. Most (90%) of the eyed eggs were provided to state, tribal, or educational salmon facilities that did not have adequate returns of adult fish. This return (1.6%) was from a 300,000 age-0 smolt release. Glenwood Springs adult returns (1.0 to 2.0%) over the last several years compares favorably with state and federal returns of fall chinook salmon (0.5 to 1.0%).

Long Live the Kings believes Glenwood Springs Hatchery is an example for fish hatcheries in the future. It demonstrates how to produce catchable-size fish that do not conflict with depressed stocks of wild fish. The adult fish return of over 1.5% is high for chinook salmon and operation costs are low ($40,000 per year). Long Live the Kings could produce millions of salmon, working with willing land owners. There are many landowners with good water sources on their property who want to be involved. Privatization of fish production in terminal areas also works in Alaska.

The Nooksack River flows into Puget Sound approximately 25 air-miles from Orcas Island. At present this river is managed for 95% hatchery harvest of coho and chinook salmon. This harvest results in a significant catch of wild fish in the mixed-stock fishery, and, as a result, wild fish have all but disappeared. If harvests were restricted to several terminal area sites like Eastsound, wild stocks could recover while tribal and commercial fishing continued.

Establishing a legal structure to allow private citizens to work on salmon enhancement was another goal of LLTK. In 1990–1991, working with other fishing groups, the Regional Salmon Enhancement Group Act was passed in Washington State, establishing 12 watershed-based groups. Through a dedicated fund, raised from a surcharge on sport and commercial fishing licenses, the groups receive from $500,000 to $700,000 annually for salmon enhancement. The groups are organized as nonprofit organizations, open to anyone interested in salmon, and have two full-time WDFW biologists to work with them. These groups have developed a number of projects, some focusing on habitat restoration and others on supplementation projects. These groups are an example of local people taking an interest and ownership in restoring salmon.

In summary, salmon are hardy, adaptable, and valuable animals that can recover and will greatly benefit the region's environment and economy. Hatcheries, properly used, can play a major role in recovering wild fish and, if properly located, can take pressure off wild stocks through terminal area fisheries. The challenge is coordinating private citizens, businesses, Indian tribes, and state and federal agencies to work together to reverse the Northwest salmon collapse.

A Review of Seminatural Culture Strategies for Enhancing the Postrelease Survival of Anadromous Salmonids

DESMOND J. MAYNARD, THOMAS A. FLAGG, AND CONRAD V. W. MAHNKEN

National Marine Fisheries Service
Northwest Fisheries Science Center, Coastal Zone and Estuarine Studies Division
2725 Montlake Boulevard East, Seattle, Washington 98112, USA

Abstract.—The unnatural behavioral and morphological conditioning that occurs in the fish culture environment reduces the postrelease survival of hatchery-reared salmonids compared with their wild-reared counterparts. We review innovative culture techniques that offer development of fish with more wild-like behavior and morphology and higher postrelease survival. These techniques include rearing fish over natural substrates that promote the development of proper camouflage coloration, training them to avoid predators, exercising them to enhance their ability to escape from predators, supplementing their diets with natural live foods to improve foraging ability, reducing rearing densities, and utilizing oxygen-supplementation technology. In addition to enhancing postrelease survival, these seminatural culture strategies should minimize the shift in selection pressures associated with the artificial rearing environment. We conclude that these innovative culture techniques appear effective and should be used in both enhancement and conservation hatcheries.

The success of fish culture programs for salmonids is now achieved primarily by increasing the prerelease survival of salmonid fish. Artificial propagation may increase egg-to-smolt survival by more than an order of magnitude over that experienced by wild fish. Unfortunately, the postrelease survival of these cultured salmonids is often considerably lower than that of wild-reared fish (Greene 1952; Miller 1952; Salo and Bayliff 1958; Reimers 1963). Whereas this low postrelease survival may be acceptable in put-and-take fisheries, it is intolerable in supplementation programs designed to rebuild self-sustaining natural runs and conserve genetic resources. Continued success of hatchery programs can be assured by implementing innovative fish culture techniques that increase the postrelease survival of hatchery salmonids.

Releases of hatchery strains of brook trout *Salvelinus fontinalis* failed to recolonize vacant habitats; however, releases of wild strains usually succeeded (Lachance and Magnan 1990a). Similarly, the use of hatchery coho salmon *Oncorhynchus kisutch* to supplement natural runs caused a long-term decline in production (Nickelson et al. 1986). Low postrelease survival of hatchery salmonids compared with their wild cohorts may result from the behavioral and morphological differences that develop in cultured fish. For example, the practice of feeding pellets at the surface by hand or from vehicles results in hatchery brook trout and Atlantic salmon *Salmo salar* that are more surface oriented and more likely to approach large moving objects than are wild fish (Mason et al. 1967; Sosiak 1978). This surface orientation makes these hatchery-reared salmonids more vulnerable to avian predators (e.g., herons, kingfishers, and mergansers). The conventional hatchery environment also produces brook trout, brown trout *Salmo trutta*, and coho salmon with more aggressive social behavior than is evident in wild-reared fish (Fenderson et al. 1968; Bachman 1984; Swain and Riddell 1990). After release, the heightened aggressive tendencies of these hatchery fish put them at a greater risk from predation and often result in inefficient expenditure of energy in contests over quickly abandoned feeding territories. In addition, many hatchery salmonids exhibit inept foraging behavior that results in their stomachs containing fewer digestible items than those of their wild-reared counterparts (Miller 1953; Hochackka 1961; Reimers 1963; Sosiak et al. 1979; Myers 1980; O'Grady 1983). As adults, hatchery strains of coho salmon have better developed primary sexual characteristics (egg size and number), but less well-developed secondary sexual characteristics (kype size and nuptial coloration) than do wild-reared strains (Fleming and Gross 1989). These reduced secondary sexual characteristics of hatchery strains may prohibit their ability to defend redd sites when spawning naturally. Although the effect on postrelease survival is unknown, the shape of hatchery and wild chinook salmon *O. tshawytscha* also differs at the juvenile stage (Taylor 1986).

Phenotypic differences observed between cultured and wild fish are both genetically and environmentally induced. The artificial culture environment conditions salmonids to respond to food, habitat, conspecifics, and objects in a different manner than would the natural environment. Present culture techniques also alter selection pressures, which results in cultured strains becoming innately distinct from wild strains (Flick and Webster 1964; Fraser 1981, 1989; Lachance and Magnan 1990b; Mason et al. 1967; Reisenbichler and McIntyre 1977; Swain and Riddell 1990).

Theoretically, both environmental conditioning and shifts in evolutionary selection pressure produced by the artificial culture environment can be alleviated with culture practices that simulate a more natural rearing environment. In this paper, we review fish culture methods for increasing postrelease survival. The use of antipredator conditioning, foraging training, supplemental dissolved oxygen, and reduced rearing density will be examined.

Antipredator Conditioning

Predation may be a key factor in the poor postrelease survival of cultured salmonids. The ability of an animal to avoid predation is dependent on proper cryptic coloration to avoid detection by predators, ability to recognize predators, and stamina to flee from predators. Techniques presently exist for improving each of these antipredator attributes of cultured fish.

Cryptic Coloration

Postrelease survival of cultured fish can be increased by rearing them in an environment that promotes full development of the camouflage pattern they will need after release. Both the short- and long-term camouflage coloration of salmonids is primarily affected by the background color pattern of their environment. Short-term physiological color changes are accomplished by chromatophore expansion: pigment is dispersed within the chromatophore unit and color change occurs within minutes. In contrast, morphological color changes take weeks to complete as pigments and chromatophore units are developed to match the general background coloration (Fuji 1993). The cryptic coloration ability generated by these long-term stable color adaptations provides the greatest benefit for avoiding detection by predators.

Fish culturists have long recognized that fish reared in earthen-bottom ponds have better coloration than those reared in concrete vessels (Piper et al. 1982). However, only recently has it been understood that rearing salmonids over natural substrates, similar to those over which they will be released, increases postrelease survival by enhancing cryptic coloration. Groups of brook trout reared for 11 weeks over distinct background colors were less vulnerable to predators when challenged over background colors similar to those over which they were reared (Donnelly and Whoriskey 1991).

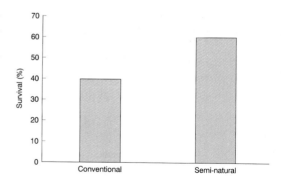

FIGURE 1.—Instream survival of fall chinook salmon released from conventional (barren; $N = 83$) and seminatural (substrate, structure, and cover; $N = 203$) raceways.

In our laboratory, fall chinook salmon reared in seminatural rectangular tanks with substrate, cover, and instream structure (plants and rootwads) had better cryptic coloration for the stream environment into which they were released than did fish reared in barren grey tanks similar to the surroundings in conventional raceways. These seminaturally reared fish had almost 50% higher postrelease survival in a coastal stream than did their conventionally reared counterparts (Figure 1). As there was no observed difference in size or disease status between the treatments, the difference in survival is probably attributable to coloration.

Similar relationships have been noted by other investigators. In one coho salmon enhancement project conducted by the Lummi Indian Nation, fish reared in dirt-bottom ponds had higher smolt-to-adult survival than those reared in concrete vessels (K. Johnson, Idaho Department of Fish and Game, personal communication). Besides having better cryptic coloration, fish reared in earthen ponds are considered to have better health, fin condition, and overall quality than those reared in concrete vessels (Piper et al. 1982). This was recently verified by Parker et al. (1990) in a study that demonstrated that coho salmon fry reared over leaf litter had higher prerelease survival than those reared in barren-bottom tanks.

Predator Avoidance

Postrelease survival of cultured salmonids can also be increased by training them to recognize and avoid predators. Thompson (1966) first determined that salmonids can learn to avoid predators in the laboratory and then demonstrated that predator avoidance training is practical in production hatcheries. He conditioned production lots of fall chinook salmon to avoid predators by moving an electrified model of a predacious trout through raceways each day for several weeks. Salmon that approached the model too closely were negatively conditioned with an electrical shock. After they were released into a coastal creek, the instream survival of the salmon trained to avoid predators was significantly higher than that of their untrained cohorts.

In the laboratory, it has been shown that coho salmon rapidly learn to recognize and avoid a predator after observing it attack conspecifics (Olla and Davis 1989). This approach to predator-avoidance training could be implemented by briefly exposing each lot of production fish to the main predators they will encounter after release. The loss of a few fish sacrificed in these training sessions should be outweighed by the larger number of trained fish that may survive later.

Swimming Performance

Swimming ability, which is critical to a fish's ability to escape from a predator, can be improved by implementing exercise programs. The swimming performance of coho salmon, Atlantic salmon, and brook trout significantly improved after they were forced to swim at higher velocities for 6 weeks or more (Besner and Smith 1983; Leon 1986; Schurov et al. 1986a). This exercise regime also enhanced their growth. The postrelease survival of exercised fish has generally (Burrows 1969; Wendt and Saunders 1972; Cresswell and Williams 1983; Leon 1986; Schurov et al. 1986b), but not always (Lagasse et al. 1980; Evenson and Ewing 1993), been higher than that of unexercised fish. The survival benefit of exercise was only realized in programs that forced salmonids to swim at high velocities for some time each day for at least 2 weeks. This exercise training may be implemented with present technology by rearing fish in either high-velocity circular or rectangular circulation ponds, or by creating high velocities in conventional raceways by temporarily drawing them down or recirculating water within them.

Foraging Training

Foraging theory suggests that supplementing standard pellet diets with live foods will profoundly

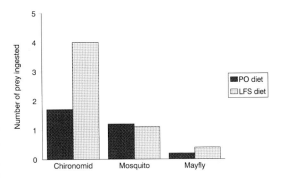

FIGURE 2.—Average number of test prey ingested by fall chinook salmon reared on pellet-only (PO; $N = 20$) or live-food-supplemented (LFS; $N = 20$) diets.

increase postrelease foraging ability of cultured fish. Gillen et al. (1981) found that previous experience in capturing live prey enhanced the foraging behavior of tiger muskellunge (F_1 hybrid of female muskellunge *Esox masquinongy* and male northern pike *E. lucius*) by decreasing the time and number of strikes required to capture natural live prey.

In our laboratory, fall chinook salmon reared on a pellet diet supplemented with live prey fed on twice as many familiar (e.g., chironmid larvae) and novel prey (e.g., mayfly larvae) as did their counterparts reared on a pellet-only diet (Figure 2). Even though food was abundantly supplied to both treatment groups, the growth of fish reared on the live-food-supplemented diet was greater than that of fish fed only pellets.

Field trials generally confirm that live-food-supplemented diets improve the postrelease foraging ability and survival of cultured fish. Tiger muskellunge reared in the hatchery on a live fish diet had higher postrelease survival than did their cohorts reared on only pellets (Johnson 1978). Similarly, brown trout reared in earthen-bottom ponds with natural food supplementation had a higher postrelease survival than did control trout reared in non-earthen-bottom tanks and fed only pellets (Hesthagen and Johnsen 1989). Live foods for salmonids can be produced by adopting techniques used in the culture of many warmwater fish species. Besides the beneficial effects on fish, live food diets have the potential both to reduce feed costs and produce less undigested waste when compared with standard diets.

Supplemental Dissolved Oxygen

The level of dissolved oxygen in the rearing environment is critical for salmonids. At rest, a fish uses up to 10% of its metabolic energy to support gill ventilation (Wooten 1990). If the oxygen content of water declines, available energy must be directed from other life functions to increase respiratory ventilation. The difference between the energy required for respiration and the total available energy is the metabolic scope for activity.

At 15°C, salmonids require 10 mg/L of dissolved oxygen to be fully active (McCauley 1991). A brook trout living in water with 7 mg/L dissolved oxygen has only three-fourths of the metabolic scope of a trout living in water with 10 mg/L dissolved oxygen (Fry 1971). Thus, although salmonids can survive and grow in a 7-mg/L-dissolved-oxygen environment, their metabolic scope is sharply curtailed.

As the metabolic scope for activity is reduced by lower levels of available dissolved oxygen, there is a commensurate decrease in activities such as sustained swimming performance. Growth and food conversion are also limited by available dissolved oxygen. In a study using coho salmon, Herman et al. (1962) showed that growth and food-conversion efficiency increased with a rise in environmental dissolved oxygen up to the highest level tested (8.3 mg/L). Theoretically, both learning ability and disease resistance of fish may similarly be limited by dissolved oxygen.

Fish culture textbooks suggest that a 7-mg/L-dissolved-oxygen environment is satisfactory for rearing salmonids and that the dissolved oxygen level should never fall below 5 mg/L (Leitritz and Lewis 1980; Piper et al. 1982; Mclarney 1984). However, these texts also indicate that higher dissolved oxygen levels are preferred for improving fish quality and reducing stress. Piper et al. (1982) indicate inflow water to ponds should be at 100% oxygen saturation and should never drop below 80% oxygen saturation anywhere in the pond. Leitritz and Lewis (1980) indicate that a 10 to 11-mg/L-dissolved-oxygen environment is best for culturing trout, which may show discomfort at a level of 7.8 mg/L. The recommended 10 to 11-mg/L-dissolved-oxygen level should provide salmonids with a full metabolic scope of activity.

A 10 mg/L-dissolved-oxygen environment can be achieved in the fish culture environment with supplementation oxygen technology. Most research on this technology has been used to increase the weight of fish that can be produced per unit volume (Dwyer et al. 1991). However, it has also been observed that in hatcheries using oxygen injection and supplemental aeration systems, disease incidence decreased and fin quality, feed conversion, and fish survival improved (Marking 1987). The cost and inconvenience of retrofitting these systems to production hatcheries is relatively low compared with the benefits in fish quality that can be achieved.

Rearing Density

Rearing density is one of the most important and well-studied factors affecting fish quality. In rainbow trout *O. mykiss*, both growth and condition factor are inversely related to rearing density (Refstie 1977). Westers and Copeland (1973) and Maheshkumar (1985) found that the fin condition of Atlantic salmon deteriorated with increasing rearing densities. However, in a study in which another strain of Atlantic salmon was reared in a different type of vessel at rearing densities of 8.5 to 68.7 kg/m^3 no relationship between rearing density and fin condition, growth, or in-culture survival was found (Soderberg and Meade 1987).

Inverse relationships between rearing density and growth, condition factor, and food conversion efficiency have been observed in coho salmon (Fagerlund et al. 1981). In addition, coho salmon reared at high densities suffered greater physiological stress as measured by body water content, fat and protein contents, interrenal cell nuclear diameter, and mortality rates. For coho salmon smolts, rearing densities as low as 16 kg/m^3 can induce physiological stress (Wedemeyer 1976), and increased rearing density reduces both gill ATPase levels (Banks 1992) and plasma thyroid hormones (Pitano et al. 1986).

In a survey of 85 variables related to strain and culture conditions, only the 5 associated with either water flow, amount of living space, or relative water level in rivers explained the postrelease survival of Atlantic salmon (Homer et al. 1979). The adult return of coho salmon also appears to be inversely related to rearing-pond density in some (Sandercock and Stone, unpublished data; Banks 1992), but not all, studies (Hopley et al. 1993).

Martin and Wertheimer (1989) examined the effect of one low, two intermediate, and one high rearing densities on the postrelease survival of chinook salmon. In the hatchery, all four rearing densities showed similar high survival (99.5% or greater), but fish reared at higher densities were smaller at release. The low-density group showed the highest adult return (1.0%), followed by the two intermediate-density groups (0.9 and 0.7%) and the high-density group (0.6%). However, the increased number of smolts produced at the two higher densities compensated for their reduced return rate and yielded a higher number of adult returns per unit volume of rearing space.

Most other chinook salmon studies have shown a consistent inverse relationship between rearing density and percentage of fish surviving to recruit to the fishery and spawning area (Hopley 1980; Fagerlund et al. 1987; Denton 1988; Downey et al. 1988; Banks 1990). However, because adult return is a function of both the number of fish released and the percentage of that number surviving to adulthood, the greatest number of fall chinook salmon adults can be produced by rearing fish at intermediate densities (Martin and Wertheimer 1989).

The relationship between rearing density and adult returns for all salmonid species indicates that a larger percentage of fish recruit to the fishery and spawning population when they are reared at a lower density. Thus, for any given number of cultured juveniles, the total adult yield will be greatest when they are reared in a large (low density) rather than a small (high density) volume vessel. Because water, not land, is the primary constraint at most fish-culture facilities, postrelease survival and total adult returns can be increased by installing larger vessels that reduce density by increasing rearing volume.

Conclusions

As demonstrated in our review, there are many culture strategies for increasing the postrelease survival of hatchery-reared salmonids. Strategies that involve rearing salmonids at low densities with naturalistic substrate, instream structure, and cover should reduce chronic stress and disease and increase survival. These strategies should also minimize potential risks from the shifts in selection pressures associated with the conventional culture environment. Strategies such as foraging training, swimming exercise, and antipredator conditioning should also behaviorally and morphologically prepare fish for survival in the postrelease environment.

Traditionally, these strategies have been rejected by hatcheries because it has been presumed that they will increase costs, maintenance, or disease. These concerns are either unfounded or can be eliminated with alternative technology. For example, salmonids can be reared at a lower density over natural substrates in large dirt-bottom raceways or ponds without increasing water consumption or incurring the higher construction costs associated with concrete ponds. Similarly, the harvest of natural feeds from on-site production facilities will enhance foraging ability and overall fish quality. Natural feeds may also reduce overall feed costs and enhance effluent water quality by reducing the generation of undigested settleable solids.

Culture strategies that increase postrelease survival can significantly reduce salmon enhancement costs. Based on several sources, we estimated the traditional cost per smolt at publicly operated facilities at about US$0.15 for coho salmon, $0.25 for spring chinook salmon, and $0.34 for steelhead (anadromous rainbow trout) (Mayo 1988; Heen 1993; R. Hager, Hatchery Consultants, Inc., personal communication). The quantity of smolts an enhancement program must produce to yield a given number of recruits is dependent on the smolt-to-adult survival. Thus, culture strategies that increase smolt-to-adult survival reduce both the total number of smolts a program must release and the cost per recruit. For example, for a spring chinook salmon smolt costing $0.25 to produce, doubling postrelease survival from 0.5 to 1.0% reduces production costs for each recruit from $50 to $25 for a net saving of $25 per recruit. For enhancement programs designed to produce half a million recruits, implementation of these culture strategies could save up to $12 million in smolt production costs each year.

There are also significant benefits to the natural spawning population that arise from increasing the postrelease survival of cultured fish. For instance, doubling postrelease survival from 0.5 to 1.0% could reduce the number of adults that culture programs must remove from wild-spawning populations by half. This reduction in the number of broodstock required is crucial for conservation and supplementation programs designed to build naturally spawning populations, as well as for enhancement facilities that are mining naturally spawning populations for broodstock. This increase in postrelease survival also halves the number of hatchery fish that must be released to produce a given number of recruits. This should reduce the postrelease competition for resources that occurs between wild and hatchery fish, thus potentially improving wild fish survival. These culture strategies may also minimize the genetic effect of cultured fish spawning with the natural population by inhibiting the development of domestic strains that are distinct from the wild strains from which they were derived. Finally, by producing fewer smolts, enhancement facilities will produce less biowaste and use less natural resources than they do with traditional fish culture practices. In summary, the reviewed innovative culture strategies could benefit wild stocks as well as target cultured salmonids by reducing broodstock collection and smolt release numbers and by lessening domestication and environmental impacts.

References

Bachman, R. A. 1984. Foraging behavior of free-ranging wild and hatchery brown trout in a stream. Transactions of the American Fisheries Society 113:1–32.

Banks, J. L. 1990. A review of rearing density experiments: can hatchery effectiveness be improved? Pages 94–102 in D. L. Park, editor. Status and future of spring chinook salmon in the Columbia River basin—conservation and enhancement. NOAA (National Oceanic and Atmospheric Administration) Technical Memorandum NMFS (National Marine Fisheries Service) F/NWC-187.

Banks, J. L. 1992. Effects of density and loading on coho salmon during hatchery rearing and after release. Progressive Fish-Culturist 54:137–147.

Besner, M., and L. S. Smith. 1983. Modification of swimming mode and stamina in two stocks of coho salmon (Oncorhynchus kisutch) by differing levels of long-term continuous exercise. Canadian Journal of Fisheries and Aquatic Sciences 40:933–939.

Burrows, R. E. 1969. The influence of fingerling quality on adult salmon survivals. Transactions of the American Fisheries Society 98:777–784.

Cresswell, R. C., and R. Williams. 1983. Post-stocking movements and recapture of hatchery-reared trout released into flowing water—effect of prior acclimation to flow. Journal of Fish Biology 23:265–276.

Denton, C. 1988. Marine survival of chinook salmon, Oncorhynchus tshawytscha, reared at three densities. Alaska Department of Fish and Game, Fisheries Rehabilitation, Enhancement, and Development Report 88, Juneau.

Donnelly, W. A., and F. G. Whoriskey, Jr. 1991. Background-color acclimation of brook trout for crypsis reduces risk of predation by hooded mergansers Lophodytes cucullatus. North American Journal of Fisheries Management 11:206–211.

Downey, T. W., G. L. Susac, and J. W. Nicholas. 1988. Research and development of Oregon's coastal chinook stocks. Oregon Department of Fish and Wildlife, Fisheries Research Project NA-87-ABD-00109, Annual Progress Report, Portland.

Dwyer, W. P., J. Colt, and D. E. Owsley. 1991. Effectiveness of injecting pure oxygen into sealed columns for improving water quality in aquaculture. Progressive Fish-Culturist 53:72–80.

Evenson, M. D., and R. D. Ewing. 1993. Effects of exercise of juvenile winter steelhead on adult returns to Cole Rivers Hatchery, Oregon. Progressive Fish-Culturist 55:180–183.

Fagerlund, U. H., J. R. McBride, B. S. Donnajh, and E. T. Stone. 1987. Culture density and size effects on performance to release of juvenile chinook salmon and subsequent ocean survival: smolt releases from Capilano Hatchery in 1980 and 1981. Canadian Technical Report of Fisheries and Aquatic Sciences 1572.

Fagerlund, U. H., J. R. McBride, and E. T. Stone. 1981. Stress-related effects of hatchery rearing density on coho salmon. Transactions of the American Fisheries Society 110:644–649.

Fenderson, O. C., W. H. Everhart, and K. M. Muth. 1968. Comparative agonistic and feeding behavior of hatchery-reared and wild salmon in aquaria. Journal of the Fisheries Research Board of Canada 25:1–14.

Fleming, I. A., and M. R. Gross. 1989. Evolution of adult female life history and morphology in a Pacific salmon (coho: Oncorhynchus kisutch). Evolution 43:141–157.

Flick, W. A., and D. A. Webster. 1964. Comparative first year survival and production in wild and domestic strains of brook trout Salvelinus fontinalis. Transactions of the American Fisheries Society 93:58–69.

Fraser, J. M. 1981. Comparative survival and growth of planted wild, hybrid and domestic strains of brook trout (Salvelinus fontinalis) in Ontario lakes. Canadian Journal of Fisheries and Aquatic Sciences 38:1672–1684.

Fraser, J. M. 1989. Establishment of reproducing populations of brook trout after stocking of interstrain hybrids in Precambrian lakes. North American Journal of Fisheries Management 9:252–363.

Fry, F. E. J. 1971. The effect of environmental factors on the physiology of fish. Pages 1–98 in W. S. Hoar and D. J. Randall, editors. Fish physiology 6. Academic Press, New York.

Fuji, R. 1993. Coloration and chromatophores. Pages 535–562 in D. H. Evans, editor. The physiology of fishes. Academic Press, New York.

Gillen, A. L., R. A. Stein, and R. F. Carline. 1981. Predation by pellet-reared tiger muskellunge on minnows and bluegills in experimental systems. Transactions of the American Fisheries Society 110:197–209.

Greene, C. W. 1952. Results from stocking brook trout of wild and hatchery strains at Stillwater Pond. Transactions of the American Fisheries Society 81:43–52.

Heen, K. 1993. Comparative analysis of the cost structure and profitability in the salmon aquaculture industry. Pages 220–238 in K. Heen, R. L. Monahan, and F. Utter, editors. Salmon aquaculture. Halsted Press, New York.

Herman, R. B., C. E. Warren, and P. Doudoroff. 1962. Influence of oxygen concentrations on the growth of juvenile coho salmon. Transactions of the American Fisheries Society 91:155–167.

Hesthagen, T., and B. O. Johnsen. 1989. Lake survival of hatchery and pre-stocked pond brown trout, Salmo trutta L. Aquaculture and Fisheries Management 20:91–95.

Hochackka, P. W. 1961. Liver glycogen reserves of interacting resident and introduced trout populations. Journal of the Fisheries Research Board of Canada 18:125–135.

Homer, M. J., J. G. Stanley, and R. W. Hatch. 1979. Effects of hatchery procedures on later return of Atlantic salmon to rivers in Maine. Progressive Fish-Culturist 41:115–119.

Hopley, C. 1980. Cowlitz spring chinook rearing density

study. Proceedings of the Annual Northwest Fish Culture Conference 31:152–159.

Hopley, C. W., S. B. Mathews, A. E. Appleby, A. Rankis, and K. L. Halliday. 1993. Effects of pond stocking rate on coho salmon survival at two lower Columbia River fish hatcheries. Progressive Fish-Culturist 55: 16–28.

Johnson, L. D. 1978. Evaluation of esocid stocking program in Wisconsin. American Fisheries Society Special Publication 11:298–301.

Lachance, S., and P. Magnan. 1990a. Performance of domestic, hybrid, and wild strains of brook trout, *Salvelinus fontinalis*, after stocking: the impact of intra- and interspecific competition. Canadian Journal of Fisheries and Aquatic Sciences 47:2278–2284.

Lachance, S., and P. Magnan. 1990b. Comparative ecology and behavior of domestic, hybrid, and wild strains of brook trout, *Salvelinus fontinalis*, after stocking. Canadian Journal of Fisheries and Aquatic Sciences 47:2285–2292.

Lagasse, J. P., D. A. Leith, D. B. Romey, and O. F. Dahrens. 1980. Stamina and survival of coho salmon reared in rectangular circulating ponds and conventional raceways. Progressive Fish-Culturist 42:153–156.

Leitritz, E., and R. C. Lewis. 1980. Trout and salmon culture (hatchery methods). University of California, Oakland.

Leon, K. A. 1986. Effect of exercise on feed consumption, growth, food conversion, and stamina of brook trout. Progressive Fish-Culturist 48:43–46.

Maheshkumar, S. 1985. The epizootiology of finrot in hatchery-reared Atlantic salmon (*Salmo salar*). Master's thesis. University of Maine, Orono.

Marking. L. L. 1987. Evaluation of gas supersaturation treatment equipment at fish hatcheries in Michigan and Wisconsin. Progressive Fish-Culturist 49:208–212.

Martin, R. M., and A. Wertheimer. 1989. Adult production of chinook salmon reared at different densities and released as two smolt sizes. Progressive Fish-Culturist 51:194–200.

Mason, J. W., O. M. Brynilson, and P. E. Degurse. 1967. Comparative survival of wild and domestic strains of brook trout in streams. Transactions of the American Fisheries Society 96(3):313–319.

Mayo, R. 1988. The California hatchery evaluation study. Report to California Department of Fish and Game, Sacramento.

Mclarney, W. 1984. The freshwater aquaculture book. Hartley and Marks, Point Roberts, Washington.

McCauley, R. 1991. Aquatic conditions. Pages 128–130 *in* J. Stolz and J. Schnell, editors. Trout. Stackpole Books, Harrisburg, Pennsylvania.

Miller, R. B. 1952. Survival of hatchery-reared cutthroat trout in an Alberta stream. Transactions of the American Fisheries Society 81:35–42.

Miller, R. B. 1953. Comparative survival of wild and hatchery-reared cutthroat trout in a stream. Transactions of the American Fisheries Society 83:120–130.

Myers, K. 1980. An investigation of the utilization of four study areas in Yaquina Bay, Oregon, by hatchery and wild juvenile salmonids. Master's thesis. Oregon State University, Corvallis.

Nickelson, T. E., M. F. Solazzi, and S. L. Johnson. 1986. Use of hatchery coho salmon (*Oncorhynchus kisutch*) presmolts to rebuild wild populations in Oregon coastal streams. Canadian Journal of Fisheries and Aquatic Sciences 43:2443–2449.

O'Grady, M. F. 1983. Observations on the dietary habits of wild and stocked brown trout, *Salmo trutta* L., in Irish lakes. Journal of Fish Biology 22:593–601.

Olla, B. L., and M. W. Davis. 1989. The role of learning and stress in predator avoidance of hatchery-reared coho salmon (*Oncorhynchus kisutch*) juveniles. Aquaculture 76:209–214.

Parker, S. J., A. G. Durbin, and J. L. Specker. 1990. Effects of leaf litter on survival and growth of juvenile coho salmon. Progressive Fish-Culturist 52:62–64.

Piper, R. G., I. B. McElwain, L. E. Orme, J. P. McCraren, L. G. Fowler, and J. R. Leonard. 1982. Fish hatchery management. U.S. Fish and Wildlife Service, Washington, DC.

Pitano, R., C. B. Schreck, J. L. Banks, and W. S. Zaugg. 1986. Effects of rearing conditions on the developmental physiology of smolting coho salmon. Transactions of the American Fisheries Society 115: 828–837.

Refstie, T. 1977. Effect of density on growth and survival of rainbow trout. Aquaculture 11:329–334.

Reimers, N. 1963. Body condition, water temperature, and over-winter survival of hatchery reared trout in Convict Creek, California. Transactions of the American Fisheries Society 92:39–46.

Reisenbichler, R. R., and J. D. McIntyre. 1977. Genetic differences in growth and survival of juvenile hatchery and wild steelhead trout, *Salmo gairdneri*. Journal of the Fisheries Research Board of Canada 34:123–128.

Salo, E. O., and W. H. Bayliff. 1958. Artificial and natural production of silver salmon, *Oncorhynchus kisutch*, at Minter Creek, Washington. Washington Department of Fisheries Research Bulletin 4.

Schurov, I. L., Y. A. Smirnov, and Y. A. Shustov. 1986a. Features of adaptation of hatchery young of Atlantic salmon, *Salmo salar*, to riverine conditions after a conditioning period before release, 1. Possibility of conditioning the young under hatchery conditions. Voprosy Ikhtiologii 26:317–320.

Schurov, I. L., Y. A. Smirnov, and Y. A. Shustov. 1986b. Peculiarities of adapting hatchery-reared juveniles of Atlantic salmon, *Salmo salar* L., to riverine conditions, 2. Behavior and feeding of trained hatchery juveniles in a river. Voprosy Ikhtiologii 26:871–874.

Soderberg, R. W., and J. W. Meade. 1987. Effects of rearing density on growth, survival, and fin condition of Atlantic salmon. Progressive Fish-Culturist 49:280–283.

Sosiak, A. J. 1978. The comparative behavior of wild and hatchery-reared juvenile Atlantic salmon (*Salmo salar* L.). Master's thesis. University of New Brunswick. Fredrickton.

Sosiak, A. J., R. G. Randall, and J. A. McKenzie. 1979. Feeding by hatchery-reared and wild Atlantic salmon

(*Salmo salar*) parr in streams. Journal of the Fisheries Research Board of Canada 36:1408–1412.

Swain, D. P., and B. E. Riddell. 1990. Variation in agonistic behavior between newly emerged juveniles from hatchery and wild populations of coho salmon, *Oncorhynchus kisutch*. Canadian Journal of Fisheries and Aquatic Sciences 47:566–571.

Taylor, E. B. 1986. Differences in morphology between wild and hatchery populations of juvenile coho salmon. Progressive Fish-Culturist 48:171–176.

Thompson, R. 1966. Effects of predator avoidance conditioning on the postrelease survival rate of artificially propagated salmon. Doctoral dissertation. University of Washington, Seattle.

Wedemeyer, G. A. 1976. Physiological response of juvenile coho salmon (*Oncorhynchus kisutch*) and rainbow trout (*Salmo gairdneri*) to handling and crowding stress in intensive fish culture. Journal of the Fisheries Research Board of Canada 33:2699–2702.

Wendt, C. A. G., and R. L. Saunders. 1972. Changes in carbohydrate metabolism in young Atlantic salmon in response to various forms of stress. International Atlantic Salmon Foundation Special Publication Series 4:55–82.

Westers, H., and J. Copeland. 1973. Atlantic salmon rearing in Michigan. Michigan Department of Natural Resources, Fisheries Division, Technical Report 73-27, Lansing.

Wooten, R. J. 1990. Ecology of teleost fishes. Chapman and Hall, London.

Evaluation of Stocked Fish:

in Hatcheries and in the Wild

Use of a National Fish Hatchery to Complement Wild Salmon and Steelhead Production in an Oregon Stream

DOUGLAS E. OLSON

U.S. Fish and Wildlife Service, Columbia River Fisheries Program Office
Vancouver, Washington 98665, USA

BRIAN C. CATES

U.S. Fish and Wildlife Service, Mid-Columbia River Fisheries Resource Office
Leavenworth, Washington 98826, USA

DANIEL H. DIGGS

U.S. Fish and Wildlife Service, Fisheries and Federal Aid
911 Northeast 11th Avenue, Portland, Oregon 97232, USA

Abstract.—Warm Springs (Oregon) National Fish Hatchery is operated by the U.S. Fish and Wildlife Service (USFWS), and is located on the Warm Springs River within the Warm Springs Indian Reservation, Oregon. The Warm Springs River, a major tributary of the Deschutes River in north-central Oregon, enters the Columbia River 330 km from the Pacific Ocean. The USFWS initiated a study of the Warm Springs River in 1975. Working closely with tribal and state comanagers, the USFWS plans to increase the fishery through a hatchery program while at the same time protecting wild fish production in the Deschutes River. The management objectives established for the hatchery are (1) optimize the natural production of anadromous and resident fish in the Warm Springs River; (2) optimize harvest opportunities for spring chinook salmon *Oncorhynchus tshawytscha* by establishing a successful hatchery program; and (3) maintain the genetic characteristics of the wild spring chinook salmon and summer steelhead *O. mykiss* in the Warm Springs River. The information collected on these stocks is currently one of the most complete databases existing in the Columbia River basin on which management decisions are made. Based on the data gathered and analyzed to date on wild and hatchery spring chinook salmon and summer steelhead, hatchery production is providing increased contribution to sport and tribal fisheries while not adversely affecting wild stock production.

Fish hatcheries for enhancement, mitigation, and outplanting have recently received some heavy criticism, especially in regards to effects on natural production (Hilborn 1992; Meffe 1992). Although some of this criticism is warranted (Lichatowich and McIntyre 1987), others have argued the beneficial use of hatcheries as a management tool (Martin et al. 1992; Nickum 1993; Radonski and Martin 1986). Hilborn (1992) noted that "instead of building more hatcheries, we need to find out how to operate hatcheries to produce fish in the catch with as little impact as possible on the wild stocks."

As originally designed, there were some concerns that fish from Warm Springs National Fish Hatchery (NFH) would negatively affect wild populations of trout, salmon *Oncorhynchus* spp., and steelhead *O. mykiss* (Fessler 1977; Wagner 1977). In consultation with the Confederated Tribes of the Warm Springs Indian Reservation (Tribes) and Oregon Department of Fish and Wildlife (ODFW), the U.S. Fish and Wildlife Service (USFWS) designed a hatchery program to minimize effects on the natural spawning populations. Also designed was a long-term monitoring and evaluation program to determine if natural and hatchery production objectives are met.

Both hatchery and wild fish production are recognized in the Hatchery Operations Plan (Plan). The Plan is policy guidance and has the concurrence of the tribes, ODFW, and USFWS. The first Plan to guide production was completed in 1977. Following the principles of adaptive management, the Plan is revised periodically to reflect changes in technology and operational experience.

The goal of Warm Springs NFH is to augment harvest while at the same time complementing wild fish populations. The management objectives identified to meet this goal are (1) optimize natural production of anadromous and resident fish in the Warm Springs River; (2) optimize harvest opportunities by establishing a successful hatchery program; and (3) maintain the biological and genetic characteristics of the Warm Springs fish populations in both the hatchery and stream environments. The respect for wild fish is an integral part of the hatchery program goal.

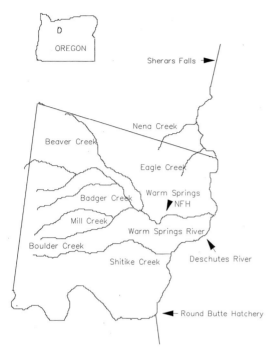

FIGURE 1.—Streams and rivers of the Warm Springs Indian Reservation of Oregon and location of Warm Springs NFH.

Study Area and Background Information

The Warm Springs River is a major tributary of the Deschutes River in north-central Oregon (Figure 1). The Warm Springs River enters the Deschutes River at river kilometer (rkm) 135, which enters the Columbia River 330 rkm from the Pacific Ocean. Headwaters of the Warm Springs River are on the eastern slope of the Cascade Mountain Range. Climate is semiarid. Warm Springs NFH is located on the Warm Springs River at rkm 16 within the Warm Springs Indian Reservation of Oregon. Water flow and temperature vary seasonally near the hatchery site; the mean annual flow is 12.5 m³/s and the minimum flow is 6.2 m³/s. River temperature at the hatchery ranges from freezing to 20°C. A detailed description of the Deschutes River subbasin can be found in the Deschutes River Subbasin Salmon and Steelhead Production Plan (Oregon Department of Fish and Wildlife and Confederated Tribes of the Warm Springs Indian Reservation 1990).

The two species that have received focus in the studies are the summer steelhead and the spring chinook salmon *O. tshawytscha*. Summer steelhead enter the Deschutes River beginning in June. They remain in the Deschutes River until entering the Warm Springs River just prior to spawning in February; the peak of the spawning run is in mid-April. Spring chinook salmon enter the Deschutes and Warm Springs rivers in April. At the hatchery two peaks are evident in run timing, with the most distinct peak in May followed by a smaller peak in late August to early September, just prior to spawning.

Because of the extended nature of summer steelhead spawning above the hatchery and high water conditions occurring at that time, little comprehensive data is available concerning summer steelhead's spawning distribution. Much more spawning information is available for spring chinook salmon. Since 1969 redd counts in the system have occurred on 35.3 km of stream. Since 1982 surveys have been expanded to include all known spawning habitat, 63.8 km in all. A range of 117 to 1,066 redds have been found, averaging 501 redds (7.85 redds/km).

Juvenile summer steelhead emigrate from the Warm Springs River during the spring and are mainly 1- and 2-year-old fish. Wild juvenile spring chinook salmon outmigration is split into fall and spring migration periods. Fall subyearling outmigrants typically overwinter in the main-stem Deschutes River and smolt the following Spring (Lindsay et al. 1989). On average 58% of the total outmigrants leave the Warm Springs River in the fall. Spring yearling migrants typically migrate through the Columbia River in May (Lindsay et al. 1989).

Materials and Methods

The data presented in this paper have been collected and compiled through cooperative efforts of the Tribes, ODFW, and USFWS. New data are continually being collected and added to the database by all three contributors. Adult return information has been collected at Warm Springs NFH trapping facilities and through creel census, spawning ground surveys, and trapping at Sherars Falls on the Deschutes River. Most juvenile fish data were obtained by sampling from a floating scoop trap in the lower Warm Springs River. Additional data were gathered via beach seining and electrofishing. Fish collected were identified to species and sampled for length, age, sex, and mark recovery. Statistical comparisons and computations are described in Ricker (1975) and Zar (1974).

The focus of this paper will be on hatchery and wild spring chinook salmon. Hatchery operations pertaining to summer steelhead and rainbow trout (nonanadromous *O. mykiss*) are briefly discussed. Procedures and protocols at the hatchery are directed by our 5-year hatchery operation plan and are discussed in the hatchery program section of this paper. A description of the facilities at Warm Springs NFH can be found in the hatchery master plan (Bureau of Sports Fisheries and Wildlife 1971). Cates (1992) describes modifications to the hatchery facilities since 1977 and also summarizes all project activities from 1975 to 1989.

Results and Discussion

The hatchery was authorized in 1966 for the purpose of increasing fish available for harvest by the tribes. Groundbreaking and intensive fish studies began in 1975, with hatchery production starting in 1978. In 1981 the hatchery program for summer steelhead was discontinued primarily due to disease problems and the apparent physical limitations of the facility in rearing 2-year-old smolts.

The hatchery program for rainbow trout was also changed in 1981. Originally Warm Springs NFH was expected to provide rainbow trout for all programs on the Warm Springs and Umatilla Indian Reservations. Due to fish disease considerations and changing USFWS priorities, the rainbow trout program has been reduced from 140,000 to 6,000 catchable-size rainbow trout.

The hatchery is currently producing catchable-size rainbow trout for release into the lower Warm Springs River near Kahneeta Resort, located downstream from the hatchery at rkm 13. The fish are *Ceratomyxa shasta*-susceptible and are obtained from various private and state hatcheries. *Ceratomyxa shasta*-susceptible fish are utilized in the Warm Springs River program to ensure that hatchery trout do not effect the wild trout populations of the Deschutes River, where *Ceratomyxa shasta* is present. It is expected that the hatchery trout will die if they migrate out of the Warm Springs River into the Deschutes River, which is managed for wild trout. The lower Warm Springs River is mainly used as a migration corridor and spawning area by wild trout; the best rearing habitat is in the main-stem Deschutes River.

Spring Chinook Hatchery Program

Broodstock collection.—As fish swim up the Warm Springs River they encounter an instream barrier dam adjacent to the hatchery. Fish are directed from the barrier dam into a fish ladder. The fish ladder is used to direct fish into holding ponds or it can be used to pass fish upstream around the barrier dam.

TABLE 1.—Spring chinook salmon broodstock (wild and hatchery stock) kept at Warm Springs NFH or passed upstream, 1978–1993.

Return year	Wild		Hatchery	
	Kept	Upstream	Kept	Upstream
1978	569	2,015	0	0
1979	416	906	0	0
1980	317	651	0	0
1981	511	1,014	0	0
1982	91	1,317	558	270
1983	442	1,081	162	170
1984	389	803	270	519
1985	322	777	532	487
1986	470	1,186	146	25
1987	147	1,550	560	0
1988	319	1,259	422	0
1989	90	1,254	849	0
1990	100	1,721	740	0
1991	0	777	576	0
1992	92	959	791	0
1993	0	528	309	0

Adult spring chinook salmon arrive at the hatchery from mid-April through September, and broodstock is collected throughout the spawning run. During the first 4 years of operation, 100% of the broodstock was of wild origin (Table 1). Initial guidelines (1978–1981) were set not to exceed one-third of the wild return, or about 400 fish, for hatchery broodstock (317 to 569 wild fish were actually retained each year). In 1978, broodstock was collected by keeping every third fish each day. In 1979 the procedure was changed to avoid size-selection bias by taking all the fish every third day and releasing all fish on other days. Based on spawning-run timing since 1975, 70% of the broodstock needed to be collected by June 1 and 90% by July 1.

Under the current operational plan, all returning fish are sorted. All fish identified as hatchery origin are retained at the hatchery. If hatchery fish return in excess of broodstock needs, then periodic decisions are made to estimate the total hatchery run and apparent excess. Any excess hatchery fish are given to the tribes for subsistence. To maintain the run-timing characteristics, excess hatchery fish are distributed before the broodstock capacity is reached. To promote retention of wild genetic traits in the hatchery, present management guidelines require that wild Warm Springs spring chinook salmon constitute 10% of the total hatchery broodstock.

Typically, only spring chinook salmon indigenous to the Warm Springs River are used for broodstock. Over the years there have been a few stray fish, based on coded-wire-tag recoveries, that have been spawned with the Warm Springs stock. The only other stock intentionally brought into the hatchery have been eggs from Round Butte State Fish Hatchery located at rkm 166 on the Deschutes River (Figure 1). Prior to hatchery operations, various stocks and life stages were planted in the Warm Springs River from 1958 to 1972 (Lindsay et al. 1989).

The current goal is to have 3,000 hatchery adult spring chinook salmon return to the mouth of the Deschutes River by use of a broodstock of 700 adults at the hatchery. This production goal, as identified in the hatchery operations plan, was determined by examining past records when better than average survival rates occurred.

Broodstock holding and spawning.—Spring chinook salmon are held for an extended period prior to spawning at Warm Springs NFH. Fish are held beginning in early May and are not ready to spawn until late August or early September. During this holding period, water temperatures in the Warm Springs River increase above 13°C. As a result, the hatchery's water chillers must cool water in the holding ponds down to a temperature of 8 to 10°C. To accomplish this, a portion of the water is recycled through the chillers and filtered. These procedures are necessary to minimize disease and other problems inherent with holding large numbers of fish.

Problems with bacterial diseases and fungus can be anticipated on a yearly basis. From 1982 to the present, all spring chinook salmon being held for broodstock have been injected with erythromycin to curtail prespawning mortalities attributable to bacterial kidney disease (BKD; causative agent *Renibacterium salmoninarum*). A second injection is given approximately 30 d after the first. For many years, fish were treated with malachite green and are now treated with formalin, chlorine, or salt to control fungus and disease.

In late August, spring chinook salmon start maturing. The hatchery crew sorts fish once or twice a week between late-August and the second week of September and spawn ripe fish. Adults are spawned on a one-to-one basis, one female to one male. The average sex ratio of adults is normally 62% female, 38% male; thus in order to accomplish the desired mating ratios, some males are used several times with different females. Gametes originating from fish with gross BKD, or enzyme-linked immunosorbent assay (ELISA) levels for *R. salmoninarum* greater than 0.5 optical density, are discarded.

Currently, fertilized eggs from each female are placed in individual incubation units to water-harden in iodophor solution. The eggs are then incubated in chilled water until river water temperatures drop to 11°C. The eggs are then incubated in river water between 1 and 11°C. After the eggs have eyed (about 6 weeks), they are shocked, sorted, and counted. They are then placed into Heath incubators with about 6,500 eggs per tray. Eggs hatch in November, and the fry are moved to inside hatchery troughs by late December or early January.

Juvenile production.—From 200,000 to 1 million juvenile spring chinook salmon have been released annually from the hatchery into the Warm Springs River. The current production goal is 750,000 juveniles. Production is typically split into fall subyearling and spring yearling release periods.

The fall subyearling releases range from 10 to 55% of total hatchery production. These fish are fast-growing, larger individuals, generally longer than 140 mm (22 fish/kg). Mortality of these fish has been high when they are reared overwinter for release the following spring. The fish not released in the fall are reared overwinter and released the following spring, typically at 26 fish/kg. Oxytetracycline has been used to mark differentially the fall and spring release groups.

Another fall and spring release strategy at Warm Springs NFH is a partial volitional release. The partial volitional releases typically start 2 to 3 weeks before the forced release dates in the fall and spring. Several ponds are not graded, and starting in early October fish are allowed to move out on their own volition through early November. Approximately 10% of the fish are estimated to exit during this fall volitional release. Past records indicate that a mixture of sizes exit the hatchery in the fall. The remaining fish are reared overwinter and then allowed to exit volitionally from late-March through April. Fish remaining at the end of April are forced out to make room for next years brood.

Some ungraded experimental groups of fish are also reared and held over until the yearling spring-release period with no fall emigration. Fish are reared at low, medium, and high densities to evaluate the best loading density. The rearing densities are similar to those described by Banks (1994).

Water temperature and rearing conditions at the hatchery are less than ideal for raising salmon because the rearing ponds are dependent upon untreated river water at the hatchery site. Daily maximum summer temperatures often hit 20°C. In winter the daily maximums are often slightly over freezing. Piper et al. (1982) indicated that between 10 and 14°C is more desirable for raising chinook salmon. The river water adjacent to the hatchery site is used only as a migration corridor by the wild fish; water temperatures are cooler in the summer much farther upstream where the natural spawning and rearing occurs. The use of untreated river water for juvenile rearing also creates fish health problems in the hatchery.

All juvenile production at Warm Springs NFH is released at the hatchery into the Warm Springs River. Our goal is to release functional smolts that emigrate quickly to the ocean. Whereas most fish released in the spring reach the Columbia River and estuary within 3 to 4 weeks, the destination of fish released in the fall is less clear. Fall-release fish have been recovered in the Columbia River in both the fall and spring periods. Scale analysis of fish released from Round Butte State Fish Hatchery indicate that over one-half of adults returning from their fall releases overwintered in the Deschutes River and migrated to the ocean in the following spring (Lindsay et al. 1989). These overwintering hatchery residuals could displace or compete with wild fish in the Deschutes River. Even though the fall release program, particularly the volitional fall release, may be a good fish cultural practice to maximize hatchery return rates, the effects on wild fish need to be assessed. The fall-release strategy is considered an experimental, limited-production program until more information is collected to evaluate its success and effects.

For the last 10 years, all juvenile fish released from Warm Springs NFH have been externally marked to identify them as hatchery fish upon return. Juvenile fish are marked in the spring of their first year of growth at approximately 350 fish/kg. Currently, all production fish are marked with an adipose fin clip plus a coded wire tag in their snout. Ventral (pelvic) fin clips and oxytetracycline have also been used to mark groups of juvenile fish. The fish are sampled for mark quality and tag retention before release. Over 90% of the fish usually retain their tags.

Various rearing and release studies to maximize adult returns are under study at the hatchery. Our goal in investigating the various release groups is to determine which of several treatments maximize adult yield back to the Deschutes River. Using coded wire tags, we can determine whether different treatments produce different outcomes by examining within-year variability (replicate tag groups) and between-year variability (at least three brood years). Based on this approach, strategies can be modified depending on the performance of the various treatment groups.

Evaluation of Hatchery and Wild Production and Life History Traits

Broodstock management.—The guidelines in the hatchery operations plan for Warm Springs NFH were designed to allow hatchery fish to remain as similar to the naturally produced wild stock as possible, given the realities of the hatchery environment. Such a similarity should benefit the hatchery fish's survival outside the hatchery environment while at the same time minimizing potential effects on the wild stocks (Reisenbichler and McIntyre 1986). It was originally anticipated that the hatchery program would be used to supplement natural production. Since 1986 the trap and weir at the hatchery have been operated to allow only wild (unmarked) spring chinook salmon and summer steelhead above the weir, keeping hatchery (marked) fish from passing upstream. From 1982 to 1985, 14 to 39% of all fish passed upstream were of known hatchery origin (Table 1). In recent years up to 10% of the hatchery fish escape upstream because of poor mark quality. Even though hatchery fish are not now intended for supplementation, efforts are still taken to minimize genetic or other effects hatchery fish may have should they come in contact with the wild fish. These efforts are part of our broodstock collection, spawning protocol, and juvenile production strategies previously described.

From 1978 to 1981, Warm Springs NFH was fully dependent on wild spring chinook salmon from the Warm Springs River for its broodstock. The adult progeny from our first egg takes began supplying most of our broodstock beginning in 1982. Although the percentage has varied, the majority of the broodstock is now of hatchery origin (Table 1). The use of wild fish in the present program continues primarily because we desire to retain the genetic characteristics of the wild fish in the hatchery stock. The purpose of this wild fish infusion into the hatchery broodstock is to reduce the effect of hatchery fish if they escape detection at the hatchery and spawn in the wild; to maintain the suitability of hatchery fish for outplanting if the wild population becomes depressed; and to reduce chances of inbreeding in the hatchery broodstock.

During 1983 to 1988, shortfalls in the hatchery broodstock were alleviated by using wild fish in excess of the number estimated for maximum wild production. Excess adult returns were estimated using spawner–recruitment data collected for this stock (Ricker 1975). A significant correlation has been observed ($r^2 = 0.896$) between the number of wild adults spawning and the eventual returns of

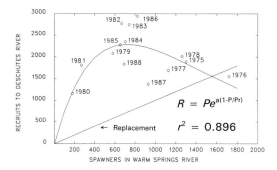

FIGURE 2.—Relationship of spawning adult salmon (spawners; P) in the Warm Springs River to recruitment (R) for wild spring chinook salmon (recruits) in the Deschutes River, 1975–1988 brood years; P_r is a replacement parameter.

their progeny (Figure 2). The number of spawning fish in the Warm Springs River was determined from the observed number of redds and the sex ratio of wild fish; each redd was assumed to represent one female. The number of recruits back to the Deschutes River was estimated from combining harvest (creel survey) and returns by age-class (scale analysis) to the hatchery. Prespawning mortality was estimated by subtracting the estimated number of spawners from the known number of fish passed upstream of the hatchery.

The term "excess wild fish" in the Warm Springs River is passé, although downstream harvest in the Deschutes River still depends on fish in excess of spawning needs. Retaining wild fish for hatchery broodstock also depends on whether spawning escapement needs have been met. Because of this, an escapement goal for wild spring chinook salmon above Warm Spring NFH has been established. From examining the spawner–recruitment relationship, 750 spawning fish in the Warm Springs River produce a maximum recruitment of approximately 2,250 adults back to the mouth of the Deschutes River (Figure 2). Considering an estimated prespawning mortality of 40%, it seems reasonable to establish an escapement goal of 1,250 wild fish passed above the hatchery. If more than 1,250 wild fish are expected, then the hatchery may take its full complement of wild fish, not to exceed 100 wild adults. The Tribes, ODFW and USFWS, will consult on the number of wild fish to use in the hatchery broodstock if less than 1,250 wild fish are anticipated to reach the hatchery. These are conservative goals that will provide for substantial wild production and greatly reduce the possibility of restricting harvests.

Although the Ricker (1975) curve is a significant relationship, there are factors other than merely number of spawners influencing adult recruitment. Peterman (1987) suggested a multivariate response. Some likely key factors affecting production are instream survival and growth, passage conditions, environment, and ocean productivity. The variation observed in spawning escapement is a natural phenomenon, and in future operations this variation will be managed for, instead of regulated.

Natural production.—The survival of spring chinook salmon from eggs to outmigrating juveniles to adults recruited back to the Deschutes River has been estimated since 1975 through the cooperative efforts of the Tribes, ODFW, and USFWS (Table 2). The data presented were derived from redd counts observed in the Warm Springs River, sex ratios of fish sampled at the hatchery, estimated fecundity of the wild fish spawned at the hatchery, population estimates from juvenile trapping in the lower Warm Springs River, estimated harvest from creel surveys in the Deschutes River, observed numbers of fish returning to the hatchery, and age composition based on scale sampling from fish returning to the hatchery. A number of relationships are worth noting. First, high egg depositions appear to result in much lower survival to migrant, whereas low egg depositions appear to result in much higher survival to migrant. A similar relationship can be seen in comparing the numbers of migrants produced per redd at varying redd counts: higher spawning densities produce fewer migrants per redd ($r^2 = 0.51$). There is also an inverse relationship in the number of migrants produced and the subsequent survival to adult return ($r^2 = 0.58$). The production data suggests survival in the Warm Springs River is density dependent (Lindsay et al. 1989).

The spawning adult–recruitment, egg–migrant, and migrant–adult relationships demonstrate the compensatory survivorship of wild spring chinook salmon in the Warm Springs River. Such an ability would help to explain how spring chinook salmon in the Warm Springs River compensate for low spawner escapements and still produce harvestable surpluses of returning adults. This stock would be considered among the more productive stocks if compared with those examined by Reisenbichler (1987).

Hatchery production.—As for hatchery production, the adult returns are above replacement, but typically well below our goal of 3,000 recruits back to the Deschutes River (Figure 3). Our goal may be unrealistic under existing conditions and constraints

TABLE 2.—Warm Springs River spawning spring chinook salmon, early life stages, and adult recruitment to the Deschutes River.

Brood year	Total redds (N)	Female (%)	Total adults (N)	Eggs deposited (× 1,000)	Juvenile migrants (N)	Adult recruits (N)
1975	808	62	1,303	2,699	69,045	1,891
1976	1,066	62	1,719	3,521	73,084	1,547
1977	699	62	1,127	2,309	50,329	1,691
1978	796	63	1,263	2,671	131,943	2,009
1979	359	62	579	1,309	50,558	2,077
1980	117	65	180	403	35,235	1,162
1981	157	58	271	539	43,885	1,807
1982	433	65	666	1,430	98,313	2,770
1983	438	59	742	1,353	120,497	2,743
1984	429	61	703	1,340	96,797	2,344
1985	398	61	652	1,315	74,326	2,274
1986	428	52	823	1,220	82,776	2,938
1987	484	52	931	1,599	86,703	1,372
1988	401	58	691	1,325	96,372	1,830
Mean (±SD)	501 (258)	0.60 (4)	832 (411)	1,645 (860)	79,276 (28,453)	2,033 (532)

at the hatchery. Limiting factors appear to be water quality and fish health. Improving water quality and fish health should improve survival. Alternative rearing strategies to improve survival should also be examined.

The effect of releases of juveniles from the hatchery on wild fish needs to be closely examined, especially in regards to the fall release program. As discussed earlier, our goal is to release from the hatchery functional smolts that emigrate quickly downstream. Steward and Bjornn (1990) report that the potential for inter- and intraspecific competition is considered minimal when migration-ready smolts are stocked. Hatchery fish released in the spring migrate quickly downstream, but less is known about the fall releases. It is estimated that 40% of the fish released in the fall which survive to adulthood migrate to the ocean during the fall-winter, whereas the other survivors overwinter in the Deschutes River and migrate the following spring. The fall release program is scheduled to continue, but on a limited scale.

The sheer numbers of smolts released from hatcheries may prematurely influence the migratory behavior of wild fish in a stream. Steward and Bjornn (1990) cite references indicating that hatchery smolts may "pull" wild fish with them during their downstream migration. At Warm Springs NFH we attempt to minimize this possibility by timing our releases to overlap with wild salmon and steelhead downstream-migration timing.

In hatchery operations, we should always be aware of the potential effects of hatchery programs on wild fish production. Potential negative effects include broodstock collection, straying, hatchery escapement upstream, disease transmission, competition for food and space, and overharvest. The goal for the hatchery program is to continue augmenting harvest while at the same time complementing wild fish.

Fisheries contribution.—Based on coded-wire-tag studies from 1977 to 1979, most Warm Springs stock spring chinook salmon were caught in the Columbia and Deschutes rivers fisheries (Figure 4). Wild fish contributed more to the ocean and Columbia River fisheries than did hatchery fish. Catch distribution was significantly different ($P < 0.05$). However, sample sizes were small for ocean and Columbia River tag recoveries, especially for wild fish. Recent tag recoveries of spring chinook salmon from Warm Springs NFH for the 1987 brood year

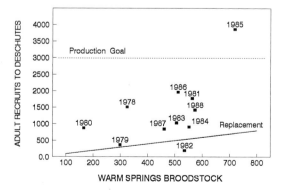

FIGURE 3.—Relationship of Warm Springs NFH spawning adult spring chinook salmon (broodstock) to adult recruitment to the Deschutes River, 1978–1988 brood years.

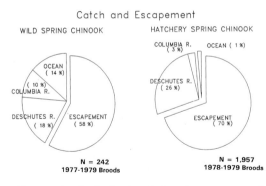

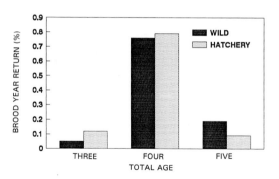

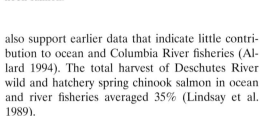

FIGURE 4.—Catch (by location) and escapement of Warm Springs River wild and hatchery stock spring chinook salmon.

FIGURE 5.—Brood-year returns (%) by age-class in years (three, four, five) of Warm Springs River wild and hatchery stock spring chinook salmon to the Deschutes River, 1978–1988.

also support earlier data that indicate little contribution to ocean and Columbia River fisheries (Allard 1994). The total harvest of Deschutes River wild and hatchery spring chinook salmon in ocean and river fisheries averaged 35% (Lindsay et al. 1989).

A significant sport and Tribal subsistence fishery for spring chinook salmon occurs in the Deschutes River at Sherars Falls (rkm 71). Sport anglers generally account for the majority of the harvest. Catch and effort have increased as a result of increased run sizes from hatchery fish from Round Butte State Fish Hatchery and Warm Springs NFH (Lindsay et al. 1989).

To reduce the fishing mortality on wild spring chinook salmon in the Deschutes River, ODFW is in the process of implementing a selective fishery on marked (hatchery) spring chinook salmon in 1995. This is similar to the selective fishery on marked summer steelhead implemented in the Deschutes River since 1979. All summer steelhead released from hatcheries throughout the Columbia River are marked with an adipose fin clip. All spring chinook salmon released from hatcheries in the Deschutes River have been marked with an adipose fin clip plus a coded wire tag since the 1991 brood year. Although selective fisheries for steelhead have been implemented for many years in Oregon, a selective fishery for hatchery spring chinook salmon will be a first. The objective of the selective fishery is to reduce the fishing mortality on wild (unmarked) fish. Bendock and Alexandersdottir (1993) found overall hooking mortality of chinook salmon released in the Kenai River, Alaska, to average 7.6%. If a similar hooking mortality is observed, a selective fishery will substantially reduce mortality due to the sport harvest of wild spring chinook salmon in the Deschutes River. From 1987 to 1993, sport harvest alone accounted for more than 25% of the total wild run returning to the Deschutes River.

Age-class strength and length frequencies.—Great care has been taken to avoid size-selection bias in the collection of broodstock for the hatchery. This care is demonstrated by the return of most wild and hatchery Warm Springs stock spring chinook salmon as 4-year-old adults (Figure 5). Still, in comparison with wild production, the hatchery does produce a higher percentage of 3-year-old fish (jacks). Gross (1991) explained that the hatchery environment of accelerated fry growth rate resulted in an increased proportion of juveniles adopting the jack tactic. Relatively few jacks are used in the Warm Springs hatchery broodstock, supporting the results presented by Gross (1991). In contrast, wild fish produce more 5-year-old adults than do hatchery fish. The higher percentage of Warm Springs wild fish maturing at 5 years may also explain why more wild fish are caught in ocean and Columbia River fisheries (Lindsay et al. 1989).

For both the 4- and 5-year-old adults, mean fork length of unmarked wild fish is greater than that of hatchery fish at return to the Warm Springs River (Figure 6). Wild and hatchery fish length frequency distributions are significantly different ($P < 0.05$). Unless there is some type of mark-induced effect, length at return to the hatchery may be a function of size at release. Juveniles produced from Warm Springs NFH are larger than are the wild downstream migrants; their length frequency distributions are also significantly different ($P < 0.05$). The rearing environment at Warm Springs NFH (like most hatcheries) facilitates faster growth rates than does the stream rearing environment (Figure 7). In

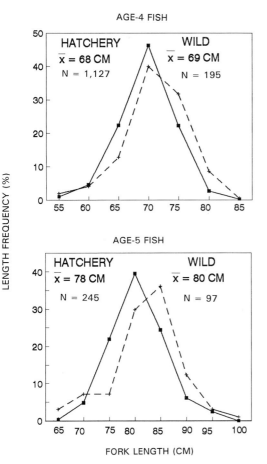

FIGURE 6.—Length frequency (%) of hatchery (solid line) and wild (dashed line) spring chinook salmon returning to the Warm Springs River during 1991–1993.

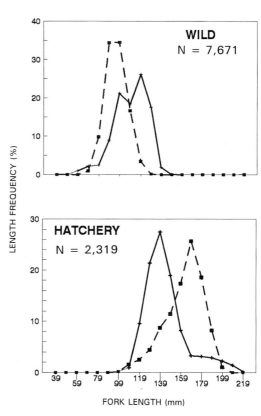

FIGURE 7.—Length frequency (%) of wild and hatchery juvenile spring chinook salmon emigrating from the Warm Springs River in the fall (dashed line) and spring (solid line), brood year 1987.

addition, studies by Dawley et al. (1986) and Cates (1992) suggest that the larger hatchery juveniles are more likely to survive the downstream migration to the estuary. As we improve our hatchery facilities and lower summer rearing temperatures, we may want to experiment with size at release to mimic the wild juvenile life history more closely in order to mimic the adult age structure and length frequency.

Cumulative spawning-run timing.—Spawning-run timing of spring chinook salmon has been closely monitored at Warm Springs NFH since the hatchery weir became operational in 1977. Once hatchery fish began returning in 1981 their timing has been compared with that of wild fish. One of the objectives of our hatchery operation plan is to maintain the natural timing of the wild run and to create a hatchery run with similar timing. The method to accomplish this goal is by collection of broodstock throughout the run in proportion to the wild returns. In actual practice, the take of broodstock for the hatchery lagged behind the proportional return of wild fish. We judged that a conservative approach to taking wild fish into the hatchery would be preferable to effecting wild production by overharvest at the hatchery. As a result the early proportion of the wild run was underrepresented in the hatchery broodstock.

The average timing for hatchery and wild stock returning from 1982 to 1989 indicates that a lag in hatchery returns is occurring and is similar to that expected from the timing of their parents. Proportional run timing of hatchery returns lags 1 to 3 weeks behind wild fish returns to the Warm Springs River (Figure 8). The current strategy in collecting broodstock is to have the hatchery manager decide how often to collect broodstock based on the manager's intuition concerning the stage and strength of

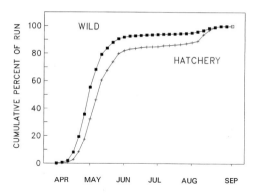

FIGURE 8.—Cumulative spawning-run timing (%) of wild and hatchery spring chinook salmon returning to the Warm Springs River, 1982–1989.

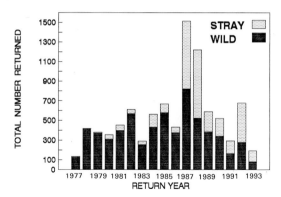

FIGURE 9.—Total number of wild and stray hatchery summer steelhead returning to the Warm Springs River, 1977–1993.

the spawning run. In refining our estimates of how strong a brood will return, we need to be more attentive in selecting broodstock in proportion to the wild run.

Judging from our experiences at Warm Springs NFH it appears that this stock of spring chinook salmon would be susceptible to altered spawning-run timing through hatchery practices. Although hatchery fish have demonstrated some changes relative to wild fish during their migrating period to Warm Springs NFH, little change in spawning time has been noted either at the hatchery or in the wild since broodstock collection began.

Summer Steelhead

The Deschutes River is one of Oregon's premier summer steelhead fishing streams (Casali and Diness 1984). Harvest rates of wild steelhead have been negligible since 1979 when sport fishing regulations began to protect the wild fish fully. Since that time, it has been illegal to harvest wild (unmarked) steelhead in the sport fishery, although Indian dipnetters continue to harvest some wild steelhead in their subsistence fishery at Sherars Falls. The return numbers of wild steelhead to Warm Springs NFH, since the restrictive fishery, probably reflect a reduction in harvest rates; however, with the exception of 1987, no dramatic trends of increase in the numbers of wild steelhead have been observed at the hatchery. If this situation continues, it may indicate that the harvest restrictions are not greatly influencing production of wild steelhead in the Warm Springs River. This could be due to a number of factors, including hook and release mortalities, possible increased harvest in the mainstem Columbia River, and rearing capacity limitations in the Warm Springs River.

As stated earlier, steelhead hatchery production at Warm Springs NFH was terminated in 1981. Since that time, we have sacrificed all known strays at the hatchery and pass only unmarked fish upstream. All sacrificed fish are distributed to the Tribes.

Starting in 1987 a high number of marked strays have been counted and sacrificed at the hatchery (Figure 9). More than 1,500 steelhead returned in 1987, including 692 stray hatchery fish. Coded-wire-tag-recovery data indicate that most strays are coming from the upper Columbia River, primarily Snake River tributaries. Large numbers of these stray fish are apparently entering the Deschutes River. Judging from the number of strays seen near spawning time at Warm Springs NFH and Round Butte State Fish Hatchery, the steelhead strays are probably entering many of the spawning streams. The effect of these fish on the native Deschutes summer steelhead stocks is unknown, but studies by Reisenbichler and McIntyre (1977) and Chilcote et al. (1986) indicate reduced performance of wild and hatchery crosses of steelhead in the natural stream environment. Data indicate that the number of hatchery strays continue to contribute a large proportion of the total return. This demonstrates the success of the Columbia River steelhead hatchery program basinwide, and the decreasing production of wild fish. It is suspected that Columbia River hydropower and fish transportation operations also influence stray rates. Straying fish, both hatchery and wild, complicate any analysis of the status of wild steelhead in this system. In a sense through operation of the hatchery weir on the lower Warm Springs River we have created a wild fish refuge in the upper Warm Springs River.

Future Operations and Research Needs

The following nine tasks are recommended in order to continue responsible management of not only Warm Springs NFH but all the fisheries resources in the Deschutes River basin.

1. Form a Deschutes River Restoration Council to address protection and restoration of anadromous and resident fish populations within the basin. This cooperative effort would combine the expertise from different agencies and maximize the benefit to the resource, tribes, and public.
2. Assess hooking mortality of unmarked (wild) fish when ODFW implements a selective sport fishery of marked spring chinook salmon in the Deschutes River.
3. Evaluate and monitor fish passage at Warm Springs NFH. We are in the design stage for a new fish passageway at Warm Springs NFH. The new fish passageway will be used to detect coded-wire-tagged hatchery fish. These fish will be shunted into the hatchery holding pond. Non-tagged (wild) fish will be monitored via video technology and then directed to the fish ladder to continue their upstream migration. Sampling protocol and operational and monitoring plans need to be developed. Our goal is to reduce the handling of wild fish, obtain a 95% detection efficiency of tagged fish, and reduce the prespawning mortality observed in the stream.
4. Continue monitoring the health of both wild and hatchery fish to document changes in fish health with changes in hatchery operations, habitat, and environmental influences.
5. Work with the ODFW and tribes to repair damaged habitat in the Deschutes River basin. "Continued survival of wild salmonid populations in the Pacific Northwest depends on reversing the degradation of freshwater habitat quality" (Lawson 1993). The USFWS needs to engage in a more active role in habitat restoration. A long-term habitat restoration and monitoring program needs to complement the long-term population monitoring program.
6. Continue to improve the water quality, fish health, and fish cultural practices at Warm Springs NFH. The complete adult to adult cycle should be investigated.
7. Quantify the fate of wild and hatchery juvenile downstream migrants, especially the fall migrants. Within the next 5 years, passive integrated transponder (PIT) monitoring stations will be installed at Columbia River dams downstream from the mouth of the Deschutes River. At that time PIT tags should be used on the Warm Springs stocks.
8. Determine the genetic variability of Deschutes River fish populations. We are currently cooperating with the National Biological Service in a genetics study examining the performance of wild and hatchery fish in both the hatchery and stream environment. With hope this will help fisheries managers make decisions in supplementation strategies and will help estimate how hatchery management actions may effect wild stocks.
9. With ODFW and Tribal comanagers, continue the population dynamics and life history monitoring program for summer steelhead and spring chinook salmon. This monitoring is necessary to estimate carrying capacity, quantify escapement needs, manage contribution to fisheries, identify limiting factors and bottlenecks, and evaluate the long-term interplay between hatchery and wild stocks in this system.

Summary

The survival of spring chinook salmon from eggs to outmigrating juveniles to adults recruited back to the Deschutes River has been estimated since 1975 through the cooperative efforts of the Tribes, ODFW, and USFWS. The information collected on this stock is currently one of the most complete databases existing in the Columbia River basin on which management decisions are made. From the data gathered and analyzed to date, production from Warm Springs NFH is providing increased contribution to sport and tribal fisheries while not adversely affecting wild stock production.

Acknowledgments

We wish to acknowledge contributions from the Confederated Tribes of the Warm Springs Indian Reservation, Department of Natural Resources; Oregon Department of Fish and Wildlife; and staff at Warm Springs NFH. Don Campton, Robert Piper, and an anonymous reviewer provided helpful comments on this manuscript.

References

Allard, D. 1994. Annual coded wire tag program: missing production groups. U.S. Fish and Wildlife Service

1993 Annual Report to Bonneville Power Administration, Project 89-065, Portland, Oregon.

Banks, J. L. 1994. Raceway density and water flow as factors affecting spring chinook salmon (*Oncorhynchus tshawytscha*) during rearing and after release. Aquaculture 119:201–217.

Bendock, T., and M. Alexandersdottir. 1993. Hooking mortality of chinook salmon released in the Kenai River, Alaska. North American Journal of Fisheries Management 13:540–549.

Bureau of Sports Fisheries and Wildlife. 1971. Warm Springs National Fish Hatchery. Master plan. U.S. Fish and Wildlife Service, Portland, Oregon.

Casali, D., and M. Diness. 1984. The new Henning's Guide to fishing in Oregon, 6th edition. Flying Pencil Publications, Portland, Oregon.

Cates, B. C. 1992. Warm Springs National Fish Hatchery evaluation and anadromous fish study on the Warm Springs Indian Reservation, 1975–1989. Progress Report. U.S. Fish and Wildlife Service, Lower Columbia River Fisheries Resource Office, Vancouver, Washington.

Chilcote, M. W., S. A. Leider, and J. J. Loch. 1986. Differential reproductive success of hatchery and wild summer-run steelhead under natural conditions. Transactions of the American Fisheries Society 115:726–735.

Dawley, E. M., and eight coauthors. 1986. Migrational characteristics, biological observations, and relative survival of juvenile salmonids entering the Columbia River estuary, 1966–1983. National Marine Fisheries Service Final Report to the Bonneville Power Administration, Project 81-102, Portland, Oregon.

Fessler, J. 1977. A case study in fish management—the Deschutes River: What we know and don't know. American Fisheries Society Symposium 10:121–126.

Gross, M. R. 1991. Salmon breeding behavior and life history evolution in changing environments. Ecology 72(4):1180–1186.

Hilborn, R. 1992. Hatcheries and the future of salmon in the Northwest. Fisheries (Bethesda) 17(1):5–8.

Lawson, P. W. 1993. Cycles in ocean productivity, trends in habitat quality, and the restoration of salmon runs in Oregon. Fisheries (Bethesda) 18(8)6–10.

Lichatowich, J. A., and J. D. McIntyre. 1987. Use of hatcheries in the management of Pacific anadromous salmonids. American Fisheries Society Symposium 1:131–136.

Lindsay, R. B., B. C. Jonasson, R. K. Schroeder, and B. C. Cates. 1989. Spring chinook salmon in the Deschutes River, Oregon. Oregon Department of Fish and Wildlife, Information Report 89-4, Portland.

Martin, J., J. Webster, and G. Edwards. 1992. Hatcheries and wild stocks: Are they compatible? Fisheries (Bethesda) 17(1):4.

Meffe, G. K. 1992. Techno-arrogance and halfway technologies, salmon hatcheries on the Pacific Coast of North America. Conservation Biology 6:350–354.

Nickum, J. 1993. The role of fish hatcheries in the 21st century. Fisheries (Bethesda) 18(2):2.

Oregon Department of Fish and Wildlife and Confederated Tribes of the Warm Springs Indian Reservation. 1990. Deschutes River subbasin salmon and steelhead production plan. Report to the Northwest Power Planning Council, Portland, Oregon.

Peterman, R. M. 1987. Review of the components of recruitment of pacific salmon. American Fisheries Society Symposium 1:417–429.

Piper, R. G., I. B. McElwain, L. E. Orme, J. P. McCraren, L. G. Fowler, and J. R. Leonard. 1982. Fish hatchery management. U.S. Fish and Wildlife Service, Washington, DC.

Radonski, G. C., and R. G. Martin. 1986. Fish culture is a tool, not a panacea. Pages 7–13 *in* R. H. Stroud, editor. Fish culture in fisheries management. American Fisheries Society, Fish Culture Section and Fisheries Management Section, Bethesda, Maryland.

Reisenbichler, R. R. 1987. Basis for managing for harvest of chinook salmon. North American Journal of Fisheries Management 7:589–591.

Reisenbichler, R. R., and J. D. McIntyre. 1977. Genetic differences in growth and survival of juvenile hatchery and wild steelhead trout. Journal of the Fisheries Research Board of Canada 34:123–128.

Reisenbichler, R. R., and J. D. McIntyre. 1986. Requirements for integrating natural and artificial production of anadromous salmonids in the Pacific Northwest. Pages 365–374 *in* R. H. Stroud, editor. Fish culture in fisheries management. American Fisheries Society, Fish Culture Section and Fisheries Management Section, Bethesda, Maryland.

Ricker, W. E. 1975. Computation and interpretation of biological statistics of fish populations. Bulletin of the Fisheries Research Board of Canada 191.

Steward, C. R., and T. C. Bjornn. 1990. Supplementation of salmon and steelhead stocks with hatchery fish: a synthesis of published literature. Technical Report 90-1 *in* W. Miller, editor. Analysis of salmon and steelhead supplementation. Report to Bonneville Power Administration, Project 88-100, Portland, Oregon.

Wagner, H. H. 1977. Options for managing the anadromous fisheries of the lower Deschutes River. American Fisheries Society Symposium 10:127–132.

Zar, J. H. 1974. Biostatistical analysis. Prentice-Hall, Inc., Englewood Cliffs, New Jersey.

The Contribution of Hatchery Fish to the Restoration of American Shad in the Susquehanna River

MICHAEL L. HENDRICKS

Pennsylvania Fish and Boat Commission, Benner Spring Fish Research Station
State College, Pennsylvania 16801, USA

Abstract.—Spawning migrations of anadromous American shad *Alosa sapidissima* in the Susquehanna River are blocked by four large hydroelectric dams. Restoration efforts include trapping prespawn adult American shad at the first barrier (Conowingo Dam) and transplanting them to upstream spawning areas and releasing hatchery-reared larvae and fingerlings. Since 1985, otoliths of all hatchery-reared American shad have been marked with tetracycline antibiotics to distinguish them from naturally produced American shad. Cohorts of American shad produced in the hatchery prior to the initiation of tetracycline marking were distinguished from naturally produced American shad based on microstructural characteristics of otoliths. By means of tetracycline marking and otolith microstructure, it was estimated that hatchery contribution to the population of juvenile American shad above Conowingo Dam ranged from 79 to 99% during 1985 to 1991. Hatchery contribution decreased to 62% in 1992 and 44% in 1993, suggesting increased spawning success by transplanted adults. Catch of adult American shad at the Conowingo Dam fish lifts averaged only 127 fish/year during 1972 to 1981, but increased to a record 27,000 in 1991. Hatchery contribution to the adult population entering the lifts ranged from 67 to 83% during 1989 to 1993. Hydroelectric companies have agreed to construct fish passage facilities at the remaining three main-stem dams by the year 2000, opening up several hundred miles of historical spawning habitat. The success of the restoration effort, driven by the hatchery component, has played a major role in achieving that agreement.

The Susquehanna River is the second largest river in eastern North America and drains over 71,225 km², including parts of New York and Maryland and nearly one-half of Pennsylvania. Combined seasonal flow averages approximately 900 m³/s (Carlson 1968). American shad *Alosa sapidissima* was abundant in the Susquehanna River until the early 1900s. The Pennsylvania portion of the river alone supported a commercial fishery of nearly 181,818 kg in 1901 (Mansueti and Kolb 1953). Intercept fisheries in Maryland undoubtedly took large numbers of American shad before they could reach Pennsylvania.

American shad were cut off from their historical spawning areas in the Susquehanna River by the construction of four large hydroelectric dams between 1904 and 1931. In the 1960s, state and federal agencies sought cooperation from hydropower developers to restore American shad and other diadromous fish to their historical range above dams. Power companies requested solid evidence that restoration was feasible before agreeing to spend millions of dollars for fish passage facilities. Feasibility of a full-scale restoration could be established only by demonstrating success with a modest-scale restoration.

Remnant populations of wild American shad continued to spawn in the lower 16 km of the Susquehanna River and upper Chesapeake Bay, but commercial catches had declined severely since 1890 (Foerster and Reagan 1977). Since few wild Susquehanna stock American shad remained, a demonstration project was initiated with out-of-basin stocks.

Between 1971 and 1976, over 216 million American shad eggs were stocked into the Susquehanna River from sources in the Columbia River and many East Coast rivers. Between 1981 and 1987, over 30,000 prespawn adult American shad from the Hudson and Connecticut rivers were released into the Susquehanna River. These efforts produced few outmigrating juvenile American shad and were ultimately discontinued (St. Pierre 1994).

In 1972, Philadelphia Electric Company built a fish lift at Conowingo Dam to collect prespawn adult American shad for transplant into upstream spawning areas. Few adult American shad were collected until 1982, when transplanting American shad began. In 1976, the Pennsylvania Fish and Boat Commission established the Van Dyke Research Station for Anadromous Fish to develop techniques for American shad culture for use in the Susquehanna River restoration effort. Since that time, the restoration effort has consisted of two primary components: (1) trapping adults at Conowingo Dam and transplanting them to upstream spawning areas, and (2) producing hatchery-reared larvae and fingerlings. The objective of the restoration plan is to produce a self-sustaining population of 3 million adult American shad. The role of the hatchery is to produce adequate numbers of juvenile American shad, imprinted to return to upstream spawning areas, to demonstrate the long-term feasibility of restoration and to convince power companies to support the project by building fish passage facilities. In effect, the hatchery is producing a founder population by use of out-of-basin stocks. This mixed stock will expand by natural reproduction, which will include some degree of crossbreeding with remnant wild upper Chesapeake Bay stocks. No definitive target has been set, but hatchery operations will cease when natural reproduction becomes dominant. The purpose of this paper is to evaluate the contribution of hatchery fish to the restoration effort.

Methods

Trap and transplant.—Prespawn adult American shad are collected in lifts at Conowingo Dam (river kilometer 15). The West Tailrace lift was built in 1972 and requires sorting and trucking to move American shad upstream. The East Tailrace lift was built in 1991 and is capable of direct release to Conowingo Reservoir in addition to sorting and trucking to upstream sites. American shad are attracted to the lifts by flow from the turbines or spill gates, then passed over a weir gate, and crowded into a steel hopper. The hopper is lifted and its contents released into sorting tanks. American shad are removed from the sorting tank by net and placed in holding tanks or directly into transport units. Prespawn American shad in good condition are transported to upstream spawning areas whenever 50 or more are collected in one day. Up to 200 adult American shad are transported in 2.4-m-diameter circular tanks mounted on trucks or trailers. A circular current is maintained in the tanks by a pump, and oxygen is bubbled into the tank to maintain desirable dissolved oxygen levels. Transport time ranges from 2 to 3 h.

American shad culture.—American shad are cultured by methods developed at the Van Dyke Research Station and are similar to those reported by Howey (1985). Ripe adult American shad are collected in rivers with abundant spawning populations. These have included the Pamunkey, James, Delaware, Hudson, Connecticut, and Columbia rivers. Eggs are fertilized and water-hardened at the collection site and delivered to the hatchery for incubation. Eggs are disinfected by 10 min immersion in 80 mg free iodine/L upon arrival at the hatchery. Average fecundity is 250,000 eggs (Leggett and Carscadden 1978), but since American shad are fractional spawners, each stripped female releases an average of only 30,000 eggs. Eggs are fertilized by sperm from males at a ratio of one male per female whenever possible. When few males are collected, one male may be used for as many as three or four females. Eggs are incubated at 15–17°C, and hatching occurs in about 7 d. Larvae are reared at 18–21°C in circular, 1,200-L tanks at densities of 100,000 to 1,000,000 larvae per tank. Freshwater flow to the tanks is 6 L/min. On day 3 or 4, American shad are fed *Artemia* sp., at a rate of 12 nauplii/fish/d and Zeigler AP-100 larval fish food (150 micron), at a rate of 64.5 g/250,000 fish/d (Wiggins et al. 1986). The diets are delivered by automatic feeders on timed circuits and adjusted to feed for 5 s every 5 min during daylight hours.

Most larvae are stocked at 18–21 d of age in the Susquehanna River or a major tributary, the Juniata River. Some larvae are stocked earlier, and some are stocked in raceways or ponds, fed on Abernathy diet, and released as fall fingerlings.

Tetracycline marking.—Since 1985, all American shad larvae reared at the Van Dyke Research Station have been immersed in tetracycline antibiotics to mark their otoliths in order to distinguish hatchery-reared American shad from naturally produced American shad spawned in the wild (Hendricks et al. 1991). In 1985 and 1986, marks were produced by immersion in 25 or 50 mg oxytetracycline/L for 12 h/d on 4 or 5 consecutive days. Double marks were produced by a second immersion series 7 d later.

Beginning in 1987, larvae were marked by a single 6-h immersion in 200 mg/L of oxytetracycline or tetracycline hydrochloride. Multiple marks were produced by subsequent immersions 4 d apart. Unique multiple immersion marks were applied to groups of larvae according to stocking site and egg source river.

Since 1987, fingerling American shad received additional marks that were produced by feeding them tetracycline-laced feed for 3 d at 88 g tetracycline/kg of food. Multiple feed marks were produced by repeating the feeding series at 5- to 7-d intervals. Combinations of immersion and feed marks were applied to fingerlings to produce unique marks according to rearing site (pond or raceway).

Mark detection.—Marked American shad were grown out in ponds or raceways and sampled for mark retention. Specimens were processed immediately after collection or frozen whole in water. Otoliths were extracted and examined according to the methods of Hendricks et al. (1991).

Otolith microstructure.—Otoliths of hatchery-reared American shad were also distinguished from those of naturally produced American shad by examination of microstructural characteristics (Hendricks et al. 1994). This methodology permitted identification of cohorts of American shad produced in the hatchery prior to the initiation of tetracycline marking. Otoliths from hatchery-reared American shad exhibited narrow increments related to slow growth in the hatchery and a "stocking check," followed by an increase in increment width after release into the wild. Otoliths from naturally produced American shad exhibited characteristically wide increments throughout. Blind trials were conducted to determine the rate of erroneous classification of known origin otoliths. A correction factor, based on these error rates, was developed to estimate hatchery contribution to the population (Hendricks et al. 1994).

Sample collection.—Juvenile American shad were collected annually from July to November using a variety of gears, including seines, trawls, electrofishers, cast nets, and lift nets. Additional samples were collected from impinged specimens at power plant cooling water intakes and strainers. Collection sites (Figure 1) ranged from the Juniata River downstream to the Susquehanna Flats area of the upper Chesapeake Bay. Below Conowingo Dam, juvenile American shad were collected during summer through early fall, prior to the outmigration, thereby targeting juveniles spawned or stocked there and excluding out-migrating juveniles of upstream origin.

Random samples of prespawn adult American shad were collected from the Conowingo Dam fish lifts beginning in 1989. In 1993, prespawn adults were also collected in commercial-type pound nets in the upper Chesapeake Bay.

Results and Discussion

The catch of adult American shad at the Conowingo Dam fish lifts was extremely low from 1972 to 1981, averaging 127 fish/year (Table 1). Lift catch increased dramatically from 1982 to 1991 when a record 27,000 adults were caught. The decline in lift catch in 1993 is troubling, but was coincident with declines in runs in other East Coast rivers in 1993 (R. St. Pierre, U.S. Fish and Wildlife Service, personal communication). Since 1982, over 91,000 prespawn adults have been transported to upstream spawning areas (Table 1).

Since 1976, over 125 million hatchery-reared American shad larvae and 1.5 million fingerlings have been

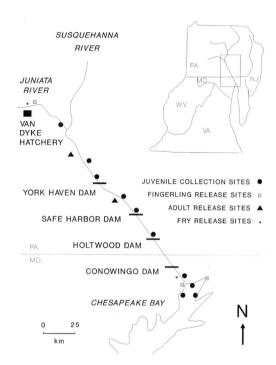

FIGURE 1.—Map of the Susquehanna River showing the major dams, stocking sites, and juvenile collection sites.

stocked in the Susquehanna River and upper Chesapeake Bay (Table 2). Since 1985, otoliths of all hatchery-reared fish (95 million larvae and 838 thousand fingerlings) were marked with tetracycline antibiotics. Tetracycline marks produced in 1985 and 1986 by immersion at 25 or 50 mg/L were faint, diffuse, and difficult to detect. Immersion in 200 mg tetracycline/L (beginning in 1987) produced intense, narrow marks with close to 100% mark retention (Hendricks et al. 1991). The presence of these marks, coupled with classification using microstructural characteristics of otoliths (Hendricks et al. 1994), was used to distinguish between naturally produced and hatchery fish and permitted evaluation of the relative contribution of the hatchery to the overall restoration effort.

Since 1985, hatchery fish have dominated the catch of juvenile American shad in the Susquehanna River above Conowingo Dam, constituting an estimated 44 to 99% of the population (Table 3). In 1985 and 1987 through 1990, naturally produced fish (progeny of transplanted adults) were scarce, accounting for less than 4% of the catch of juvenile American shad above Conowingo Dam. Naturally produced American shad constituted 16% of the population above Conowingo Dam in 1986 and

TABLE 1.—Trap and transport to upstream spawning areas of adult American shad at the Conowingo Dam fish lifts, Susquehanna River, 1972–1973.

Year	Lift catch	Number transported
1972	182	0
1973	65	0
1974	121	0
1975	87	0
1976	82	0
1977	165	0
1978	54	0
1979	50	0
1980	139	0
1981	328	0
1982	2,039	842
1983	430	36
1984	167	0
1985	1,546	967
1986	5,195	4,172
1987	7,667	7,202
1988	5,146	4,736
1989	8,311	6,469
1990	15,964	15,075
1991[a]	27,227	24,662
1992[a]	25,721	15,764
1993[a]	13,546	11,171

[a] Includes catch at East Tailrace fish lift, completed in 1991.

have increased in relative abundance in recent years: 21% in 1991, 38% in 1992, and 56% in 1993. These increases are likely due to increased numbers of adults transplanted, improvements in transplanting techniques, and favorable environmental conditions.

Below Conowingo Dam, collections of juvenile American shad have been more sporadic, ranging from 0 to 111 fish (Table 4). Collection gear and effort have varied from year to year and it is unclear whether the observed variation in catch is related to these changes or changes in abundance. Sample sizes were too low in 1985, 1987, 1988, and 1992 to draw any conclusions regarding hatchery contribution. Hatchery contribution exceeded 54% in 1986 and 1989 through 1991. Hatchery fish released above Conowingo Dam represented only 3% of the juvenile American shad collected below Conowingo Dam due to the timing of the juvenile collection efforts there. Marked hatchery fish were not stocked below Conowingo Dam in 1993. Since collections below the dam are terminated prior to outmigration from the river, all juvenile American shad collected there in 1993 were naturally produced. Since long-term restoration success depends upon development of a self-sustaining population, continued natural production below Conowingo Dam is encouraging.

The most important indicator of the success of the hatchery is the return of hatchery-reared, prespawn adult American shad to the Conowingo Dam fish lifts. Since 1989, hatchery-reared fish, as determined from otolith evaluation, have contributed an estimated 67 to 83% of the catch of adult American shad at the lifts (Table 5).

The majority of the tetracycline-marked adult American shad returning to Conowingo Dam exhibited marks indicative of upstream release sites (Table 5). At York Haven Dam, outmigrating juvenile

TABLE 2.—Stocking of hatchery-reared American shad in the Susquehanna River, during the period 1976–1993.

Year	Above Conowingo Dam		Below Conowingo Dam		Total	
	Larve	Fingerling	Larve	Fingerling	Larve	Fingerling
1976	518,000	266,000			518,000	266,000
1977	968,900	34,500			968,900	34,500
1978	2,124,000	6,400			2,124,000	6,400
1979	629,500	34,100			629,500	34,100
1980	3,526,300	5,000			3,526,300	5,000
1981	2,029,650	23,600			2,029,650	23,600
1982	5,018,800	40,700			5,018,800	40,700
1983	4,047,600	98,300			4,047,600	98,300
1984	11,995,700	30,500			11,995,700	30,500
1985[a]	6,227,590	92,000		23,200	6,227,590	115,200
1986[a]	9,899,400	61,200	5,171,200		15,070,600	61,200
1987[b]	5,179,800	81,500	4,408,700		9,588,500	81,500
1988[b]	6,450,700	74,000	3,649,500		10,100,200	74,000
1989[b]	13,464,600	60,400	7,652,500		21,117,100	60,400
1990[b]	5,619,000	90,000	3,943,200	163,000	9,562,200	253,000
1991[b]	7,218,000	54,400	4,877,000	111,500	12,095,000	165,900
1992[b]	3,039,400	21,800	1,249,800	4,800	4,289,200	26,600
1993[b]	6,541,500	79,400		100,000	6,541,500	179,400
Total	94,498,440	1,153,800	30,951,900	402,500	125,450,340	1,556,300

[a] All fish marked by immersion in 25 or 55 mg oxytetracycline/L.
[b] All fish marked by immersion in 200 mg oxytetracycline or tetracycline hydrochloride/L.

TABLE 3.—Composition of catch of juvenile American shad in the Susquehanna River above Conowingo Dam, based on microstructure classification and tetracycline marking (TC), 1985–1993. Estimates of population proportions were derived from sample classifications and then corrected based on error from a blind classification trial.

Year, sample size (N), and proportion (%)	Otolith microstructure				
	Naturally produced	Hatchery		Hatchery total	Total
		Unmarked	TC marked		
1985					
N	5	99	141	240	245
Sample (%)	2	40	58	98	
Population (%)	10	33	58	90	
1986					
N	40	9	203	212	252
Sample (%)	16	4	81	84	
Population (%)	16	3	81	84	
1987					
N	19	2	506	508	527
Sample (%)	4	0	96	96	
Population (%)	4	0	96	96	
1988					
N	4	0	284	284	288
Sample (%)	1	0	99	99	
Population (%)	1	0	99	99	
1989					
N	7	3	245	248	255
Sample (%)	3	1	96	97	
Population (%)	3	1	96	97	
1990					
N	14	4	747	751	765
Sample (%)	2	1	98	98	
Population (%)	2	0	98	98	
1991					
N	133	0	479	479	612
Sample (%)	22	0	78	78	
Population (%)	21	1	78	79	
1992					
N	153	0	238	238	391
Sample (%)	39	0	61	61	
Population (%)	38	1	61	62	
1993					
N	355	1	262	263	618
Sample (%)	57	0	42	43	
Population (%)	56	2	42	44	

American shad are being diverted around turbines by means of strobe lights and high frequency sound (Stone and Webster Environmental Technology and Services 1994). At Safe Harbor, Holtwood, and Conowingo dams, most of the juvenile outmigrants likely passed through turbines due to typically low flows during the outmigration period. Recent studies have shown that turbine passage survival is higher than expected. Forty-eight-hour survival was over 94% at Safe Harbor Dam (Heisey et al. 1992) and over 92% at Conowingo Dam (RMC Environmental Services 1994). At Holtwood Dam, 48-h survival was as low as 60% in one trial (RMC Environmental Services 1992). The potential for use of strobes lights and high frequency sound to divert outmigrants around turbines at Holtwood Dam is currently under investigation.

Uniquely marked hatchery American shad larvae and fingerlings released below Conowingo Dam have been recovered at the Conowingo Dam fish lifts in relatively low numbers (Table 5). It is not known whether this is due to low survival or the lack of migratory urge to move to areas above Conowingo Dam.

In 1993, adult American shad were also collected for otolith analysis in pound nets at two upper Chesapeake Bay locations. Otoliths from 48 fish were analyzed, and hatchery contribution was estimated to be 48%. In contrast, hatchery contribution to the adults collected at the Conowingo Dam fish

TABLE 4.—Composition of catch of juvenile American shad in the Susquehanna River below Conowingo Dam, based on microstructure classification and tetracycline (TC) marking, 1985–1993. Estimates of population proportions were derived from sample classifications and then corrected based on error from a blind classification trial.

Year, sample size (N), and proportion (%)	Otolith microstructure			Hatchery total	Total
	Naturally produced	Hatchery			
		Unmarked	TC marked		
1985					
N	0	0	0	0	0
Sample (%)	0	0	0	0	
Population (%)	0	0	0	0	
1986					
N	10	0	11	11	21
Sample (%)	48	0	52	52	
Population (%)	46	1	52	54	
1987					
N	3	0	0	0	3
Sample (%)	100	0	0	0	
Population (%)	98	2	0	2	
1988					
N	2	0	1	1	3
Sample (%)	67	0	33	33	
Population (%)	65	2	33	35	
1989					
N	5	18	88	106	111
Sample (%)	5	16	79	95	
Population (%)	7	13	79	93	
1990					
N	28	0	31	31	59
Sample (%)	47	0	53	53	
Population (%)	46	1	53	54	
1991					
N	8	1	20	21	29
Sample (%)	28	3	69	72	
Population (%)	28	3	69	72	
1992					
N	2	0	1	1	3
Sample (%)	67	0	33	33	
Population (%)	65	2	33	35	
1993					
N	67	0	0	0	67
Sample (%)	100	0	0	0	
Population (%)	98	2	0	2	

lifts was estimated to be 83%. These proportions are significantly different (chi-square = 22.8, df = 1). This difference suggests that the pound nets represent intercept fisheries and include American shad populations destined for both upper Chesapeake Bay and Susquehanna River spawning sites.

In summary, the Van Dyke Research Station has produced the vast majority of the outmigrating juvenile American shad collected above Conowingo Dam from 1985 to 1992. Transplanted adult American shad produced relatively fewer progeny until 1993, when an estimated 56% of the juveniles collected above Conowingo Dam were naturally produced. The catch of returning adult American shad at the Conowingo Dam fish lifts has increased dramatically in recent years, and, most importantly, an estimated 67 to 83% of these adults were of hatchery origin. Power companies are now convinced that restoration is feasible and on 1 June 1993, they joined with the governors of Maryland and Pennsylvania in signing an historic agreement to construct fish passage facilities at the remaining three main-stem dams by the year 2000. This agreement was a direct result of the success of the restoration effort. The high proportions of hatchery-reared fish in the juvenile collections and adult fish lift catch are evidence that the Van Dyke Research Station played the key role in achieving that agreement.

TABLE 5.—Composition of catch of adult American shad at Conowingo Dam fish lifts, based on microstructure classification and tetracycline (TC) marking, 1989–1993. Estimates of population proportions were derived from sample classifications and then corrected based on error rates from a blind classification trial.

Year, sample size (N), and proportion (%)	Otolith microstructure						Total
	Naturally produced	Unknown	Unmarked	Hatchery release site			Hatchery total	
				Below Conowingo Dam	Above Conowingo Dam		
1989							
N	29	1	94	0	36	130	160
Sample (%)	18	1	59	0	23	81	
Population (%)	29	1	48	0	23	71	
1990							
N	32	2	42	1	48	91	125
Sample (%)	26	2	34	1	38	73	
Population (%)	31	2	28	1	38	67	
1991							
N	68	0	63	8	114	185	253
Sample (%)	27	0	25	3	45	73	
Population (%)	31	0	21	3	45	69	
1992							
N	54	0	19	8	156	183	237
Sample (%)	23	0	8	3	66	77	
Population (%)	24	0	7	3	66	76	
1993							
N	21	0	4	20	79	103	124
Sample (%)	17	0	3	16	64	83	
Population (%)	17	0	3	16	64	83	

Acknowledgments

Pennsylvania Power and Light Company, Safe Harbor Water Power Corporation, and York Haven Power Company provided funding for the restoration program, including hatchery operations, through the auspices of the Susquehanna River Anadromous Fish Restoration Committee. Philadelphia Electric Company built and operated the fish lifts at Conowingo Dam. The Maryland Department of Natural Resources, Pennsylvania Fish and Boat Commission, New York Department of Environmental Resources, U. S. Fish and Wildlife Service, and the Susquehanna River Basin Commission provided guidance and direction for the restoration effort. Rickalon L. Hoopes and Richard St. Pierre reviewed the manuscript. Gratitude is also extended to the hundreds of individuals who collected and delivered eggs, cultured American shad, evaluated otoliths, collected adults and juveniles, and operated the fish lifts.

References

Carlson, F. T. 1968. Suitability of the Susquehanna River for restoration of shad. Special Report for U.S. Department of the Interior, Maryland Board of Natural Resources, New York Conservation Department, Pennsylvania Fish Commission, Washington, DC.

Foerster, J. W., and S. P. Reagan. 1977. Management of the northern Chesapeake Bay American shad fishery. Biological Conservation 12:179–201.

Heisey, P. G., D. Mathur, and T. Rineer. 1992. A reliable tag-recapture technique for estimating turbine passage survival: application to young-of-the-year American shad (*Alosa sapidissima*). Canadian Journal of Fisheries and Aquatic Sciences 49(9): 1826–1834.

Hendricks, M. L., T. R. Bender, Jr., and V. A. Mudrak. 1991. Multiple marking of American shad otoliths with tetracycline antibiotics. North American Journal of Fisheries Management 11:212–219.

Hendricks, M. L., D. L. Torsello, and T. W. H. Backman. 1994. Use of otolith microstructure to distinguish wild from hatchery-reared American shad in the Susquehanna River. North American Journal of Fisheries Management 14:151–161.

Howey, R. G. 1985. Intensive culture of juvenile American shad. Progressive Fish-Culturist 47:203–212.

Leggett, W. C., and J. E. Carscadden. 1978. Latitudinal variation in reproductive characteristics of American shad (*Alosa sapidissima*): evidence for population specific life history strategies in fish. Journal of the Fisheries Research Board of Canada 35(11):1469–1477.

Mansueti, R., and H. Kolb. 1953. A historical review of

the shad fisheries of North America. Chesapeake Biological Laboratory Publication 97, Solomons, Maryland.

RMC Environmental Services. 1992. Turbine passage survival of juvenile American shad at the Holtwood Hydroelectric Station, FERC Project 1881, Pennsylvania. RMC Environmental Services, Drumore, Pennsylvania.

RMC Environmental Services. 1994. Turbine passage survival of juvenile American shad at Conowingo Hydroelectric Station. *In* Restoration of American shad to the Susquehanna River. Annual Progress Report to Susquehanna River Anadromous Fish Restoration Committee, Harrisburg, Pennsylvania.

St. Pierre, R. 1994. American shad restoration in the Susquehanna River. Pages 81–85 *in* J. E. Cooper, R. T. Eades, R. J. Klauda, and J. G. Loesch, editors. Anadromous *Alosa* symposium, American Fisheries Society, Tidewater Chapter, Bethesda, Maryland.

Stone and Webster Environmental Technologies and Services. 1994. 1993 Evaluation of behavioral fish protection technologies at the York Haven Hydroelectric Project. *In* Restoration of American shad to the Susquehanna River. Annual Progress Report to Susquehanna River Anadromous Fish Restoration Committee, Harrisburg, Pennsylvania.

Wiggins, T. A., T. R. Bender, Jr., V. A. Mudrak, J. A. Coll, and J. A. Whittington. 1986. Effect of initial feeding rates of Artemia nauplii and dry-diet supplements on the growth and survival of American shad larvae. Progressive Fish-Culturist 48:290–293.

Genetic Effects of Hatchery Fish on Wild Populations of Pacific Salmon and Steelhead: What Do We Really Know?

DONALD E. CAMPTON

Department of Fisheries and Aquatic Sciences, University of Florida
7922 Northwest 71st Street, Gainesville, Florida 32653, USA

Abstract.—A general perception or belief has arisen in recent years that hatcheries and hatchery fish may be negatively affecting the genetic constitution of wild populations of conspecific fish. Several factors have contributed to this perception including several recent reviews in the scientific literature. However, virtually all previous assessments of this problem have not distinguished clearly the direct biological effects of hatcheries and hatchery fish from the indirect—and biologically independent—effects of stock transfers, mixed-stock fisheries on hatchery and wild fish, and other human factors related to management. Collectively, the potential genetic effects of hatcheries and hatchery fish can be grouped into three categories: (1) the genetic effects of hatcheries and artificial propagation on hatchery fish, (2) the direct genetic effects of hatchery fish on wild populations due to natural spawning and potential interbreeding, and (3) the indirect genetic effects of hatchery fish on wild populations due to ecological interactions or management decisions that affect abundance. Review of the scientific literature for Pacific salmon *Oncorhynchus* spp. and steelhead *O. mykiss* reveals that most genetic effects detected to date appear to be caused by hatchery or fishery management practices and not by biological factors intrinsic to hatcheries or hatchery fish. Although widely accepted theoretical and conceptual arguments suggest that the direct genetic effects of hatchery fish on wild populations could be substantial and potentially detrimental, the empirical data supporting those arguments are absent or largely circumstantial. This absence of direct evidence may be due, in part, to the general absence of baseline data on wild populations prior to the stocking of hatchery fish and to the relative difficulty of detecting genetic effects resulting from biological causes (e.g., natural selection) versus management practices. Nevertheless, the evaluation presented here suggests that many of the problems attributed to hatcheries and hatchery fish may be solvable (or circumvented) by implementing fishery and hatchery management practices that follow established guidelines for conserving genetic resources. Still, our science has only scratched the surface regarding our understanding of the biological interactions between hatchery and wild fish, and many critical questions remain unanswered.

> We contend that interbreedings among genetically diverged salmonid populations are generally disadvantageous to natural populations. The direct experimental evidence for this contention is admittedly sparse, basically because—to our knowledge—appropriate experiments to clarify this critical issue have not been carried out.
>
> (Utter, Hindar, and Ryman 1993)

A general perception has arisen in recent years that hatcheries and hatchery fish may negatively affect the genetic constitution of wild populations (Allendorf and Ryman 1987; Hindar et al. 1991; Waples 1991a). Several factors have contributed to this perception including a general observation that the abundance of wild fish often decreases subsequent to the initiation of a hatchery-release program (Hilborn 1992; Washington and Koziol 1993), and a general belief that hatchery fish are adapted to the hatchery environment; hence, their natural spawning and interbreeding with wild fish may reduce the fitness of natural populations (Taylor 1991). In general, the genetic effects of hatcheries and hatchery fish can be grouped into three major categories (Waples 1991a; Krueger and May 1991): (1) the genetic effects of hatcheries and artificial propagation on hatchery fish; (2) the direct genetic effects of hatchery fish on wild populations due to natural spawning and potential interbreeding; and (3) the indirect genetic effects of hatchery fish on wild populations due to competition, predation, disease transfer, and other ecological factors not associated directly with the natural spawning of hatchery fish. Indirect genetic effects also include those caused by human activities that are influenced by the presence or existence of hatchery fish (e.g., fishing mortality). In general, any factor that reduces the abundance or effective population size of a natural population can be interpreted as a negative genetic effect. Likewise, any factor which reduces the viability or reproductive capability of an individual or population can be interpreted as a negative genetic effect.

Several recent review papers and perspectives have described the potential negative genetic effects of hatcheries and hatchery fish (including cultured or farm-raised fish) on wild populations of conspecific, anadromous salmonid fish (Hindar et al. 1991; Waples 1991a; Utter et al. 1993). However, those reviews and perspectives did not clearly distinguish genetic effects caused directly by biological factors intrinsic to hatcheries or hatchery fish from genetic effects caused by extrinsic management practices (e.g., stock transfers, selective breeding, and mixed-stock fisheries on hatchery and wild fish). This confounding of the two types of causes (biological versus management) is pervasive in both the scientific and popular literature. The net result is that hatcheries and hatchery fish have become scapegoats for virtually every perceived negative biological effect associated with the artificial propagation, release, and management of anadromous salmonid fish (Hilborn 1992; Meffe 1992). Not only is the confounding of biology and management potentially misleading, it also prevents us from clearly distinguishing causes from effects, and hence, from identifying and addressing the critical, unanswered, scientific questions. Obtaining answers to such questions has reached a critical stage in the management of anadromous salmonid fish in the Pacific Northwest (Nehlsen et al. 1991) because the very existence of many populations, or evolutionary significant units (Waples 1991b), may depend upon having a clear understanding of both causes and effects, especially as those factors relate to artificial propagation, restoration, and supplementation of threatened or endangered populations. In this context, using hatcheries and hatchery fish as retrospective scapegoats for ill-advised management practices overlooks the potential benefits of hatcheries in supplementation and recovery programs, or to serve as living repositories for conserving genetic resources.

Here, I review the scientific literature regarding what is known and, more importantly, what is not known about the genetic effects of hatcheries and hatchery fish on wild populations of Pacific salmon *Oncorhynchus* spp. and steelhead *O. mykiss*. My goal is to provide an objective, comprehensive evaluation of our current state of knowledge regarding this problem. My objectives are to (1) identify and clarify potential genetic effects, (2) distinguish biological causes from management causes of those perceived genetic effects, (3) distinguish fact from speculation regarding our understanding of those effects, and (4) identify critical unanswered questions in order to stimulate future research. My perspective is genetical from the standpoint of conserving genetic resources, whether in natural populations or through hatchery propagation, as opposed to a fisheries management perspective of enhancing or augmenting exploitive fisheries. I focus on Pacific salmon and steelhead along the west coasts of the United States and Canada because (1) artificial propagation and management goals for these fish relate directly to the conservation of wild populations (in contrast to farm-raised Atlantic salmon *Salmo salar* in Norway, for example); (2) a relatively large body of empirical data exists for these fish; and (3) these fish are generating the greatest controversies regarding the hatchery versus wild fish issue (Hilborn 1992; Martin et al. 1992; Meffe 1992; Daley 1993). My primary source of information is the peer-reviewed, scientific literature. I have referenced only a limited number of state and federal agency reports because, in most instances, such information has not been completely analyzed or stood the test of peer review. However, I have consulted the reports of Miller et al. (1990) and Steward and Bjornn (1990) for their many references and insightful conclusions. For the presentation here, offspring of naturally spawning hatchery fish are treated as wild fish, and hatchery-produced fish that are the offspring of wild fish are treated as hatchery fish. I focus on intraspecific interactions and effects.

Genetic Effects of Hatcheries and Artificial Propagation on Hatchery Fish

At least four direct genetic effects can occur in hatchery populations as a result of artificial propagation (Table 1): (1) loss of genetic variation due to genetic drift and small effective population size (N_e) or effective number of breeders (N_b; Waples 1990); (2) introgression of exogenous genes via importation of fish or gametes; (3) artificial selection or selective breeding by hatchery personnel; and (4) domestication selection or natural selection due to the hatchery environment. Of these four potential effects, all but domestication selection are caused or influenced primarily by management practices.

Loss of Genetic Variation

Several studies have reported significant losses of genetic variation or changes in allozyme frequencies in hatchery populations of salmonid fish (Allendorf and Phelps 1980; Ryman and Stahl 1980; Cross and King 1983; Stahl 1983; Vuorinen 1984). These losses and changes were attributed to genetic drift or founder effects due to small effective population sizes. Until recently, however, similar reports for Pacific salmon or steelhead were lacking. For example, Steward and Bjornn (1990) stated that they "found few cases of reduced levels of genetic

TABLE 1.—Potential genetic effects of hatcheries and artificial propagation on hatchery fish, and the biological and management causes of those potential effects. A plus (+) or minus (−) in parentheses implies the existence or absence, respectively, of strong empirical evidence for the indicated cause or effect in populations of Pacific salmon or steelhead.

Genetic effect	Biological causes	Management causes
Loss of genetic variation (?)	Genetic drift (+)	Small effective number of breeders (+)
Introgression of exogenous genes (+)	Natural straying (−)	Stock transfers (+)
Phenotypic change in life history or other quantitative characters (+)	Domestication selection (+)	Artificial selection (+)

variability among hatchery stocks of Pacific salmon and steelhead."

Considerable evidence now exists that salmon hatcheries in Oregon have historically suffered from a relatively small effective number of breeders per year (Simon et al. 1986; Waples and Smouse 1990; Waples and Teel 1990). In most instances, estimated N_b was substantially less than the total number of spawners or the total number of adults available for spawning. Several factors appear to have contributed to those low effective numbers including unequal sex ratio (1 male:≥2 females), unequal family size, and the wholesale exportation of entire families (all of the fertilized eggs from many females) to other hatcheries. These practices may also have contributed to a genetic and phenotypic similarity among hatchery stocks and a dissimilarity to wild stocks (Hjort and Schreck 1982). Furthermore, the practice of mixing eggs and sperm from several males and females simultaneously (Simon et al. 1986) may have contributed significantly to substantial reductions in the effective number of breeders because of variation among males in relative potency due to sperm competition (Gharrett and Shirley 1985; Withler 1988).

Despite strong supportive evidence that hatchery practices (in Oregon) have promoted substantial reductions in the effective number of breeders, and potentially in the genetic variation and stability of those hatchery populations, we really have no evidence at this time that drift effects have reduced genetic variation or affected the fitnesses of those populations, either in the hatchery or in the wild. Computer simulations suggest that low frequency alleles are subject to rapid extinction in populations of Pacific salmon with an N_b less than 100 per year (Waples 1990). This number is approximately twice the estimated N_b for many hatchery populations in Oregon (Waples and Teel 1990). Computer simulations also suggest that genetic drift is the overriding factor controlling the loss of genetic variation in captively bred populations (Lacy 1987). However, the extent to which genetic variation for fitness traits has been lost in hatchery populations of anadromous salmonid fish is unknown. Furthermore, Steward and Bjornn (1990) state that "there is little evidence of extensive inbreeding depression among hatchery stocks of Pacific salmon used for supplementation."

The genetic effects associated with genetic drift and small effective population size can be minimized by appropriate spawning protocols and hatchery practices. By spawning equal numbers of males and females in a strictly pairwise manner, or alternatively in a replicated factorial design (e.g., 2 × 2 or 3 × 3), N_b can be maximized. Theoretically, N_b can be nearly twice ($2N - 1$) the actual number of spawners, if the variance in family size is zero; that is, if each male and female parent contribute an equal number of progeny to the spawning population of the next generation (Falconer 1981). The equalization of family size can also help to counteract the potential genetic effects of domestication selection (Allendorf 1993).

Introgression of Exogenous Genes

Stock or fish transfers among hatcheries or watersheds are well documented in the fisheries literature (Billington and Hebert 1991). This is especially true for salmon and steelhead hatcheries in the Pacific Northwest where artificial gene flow and mixing of previously isolated gene pools have historically been standard hatchery practices (Simon et al. 1986). For example, the potential population genetic effects of mixing were recently revealed in the lower Columbia River where similar allozyme frequencies characterized several hatchery populations of chinook salmon *O. tshawytscha* (Utter et al. 1989). However, the extent to which such mixing and potential introgression has affected the fitnesses of the recipient populations is unknown (Emlen 1991; Waples 1991a). One possible exception to this latter generalization are those instances, not documented in the scientific literature, involving differences in disease susceptibility between the donor and recipient stocks (Beacham and Evelyn 1992; Fjalestad et al. 1993). Nevertheless, any genetic effect caused by the importation of exogenous fish or gametes cannot be considered an effect caused by hatcheries or artificial propagation per se, but rather must be considered an effect caused by humans and the management process.

Artificial Selection

A clear distinction must be made between artificial selection, or selective breeding by hatchery personnel (Donaldson and Menasveta 1961), and domestication (or natural) selection imposed by the hatchery environment (Doyle 1983). Clearly, selective breeding or artificial selection by hatchery personnel, whether deliberate or inadvertent, is a potential source of genetic change that can be rectified or minimized by appropriate hatchery practices. On the other hand, some form of domestication selection is probably an unavoidable outcome of artificial propagation and natural selection in the hatchery environment.

Anyone who has worked in a production salmon hatchery in the Pacific Northwest is well aware of the degree of selective breeding that can occur as part of routine operations. These operations include selection for early-run timing by excluding late-spawning fish, exclusion of precocious males (i.e., males that return to spawn after spending less than 1 year in saltwater), selective breeding for large males, and a variety of other deliberate or inadvertent selective pressures. Even the standard practice of grading fish to prevent cannibalism and reduce competition can impose an unknown selective pressure if the slower growing fish are eventually discarded as surplus production. Ultimately, however, a response to selection will be obtained only if the heritability of the trait is greater than zero.

In general, production hatcheries have made little attempt to monitor quantitative traits or other characters (e.g., allozyme frequencies) genetically despite the millions of adult salmon and steelhead that have been spawned artificially over the past 30 years. Relatively simple statistics such as heritabilities and genetic correlations for life history or other quantitative characters have rarely been calculated as part of standard hatchery operations but have, instead, been the domain of academic and research institutions. Such statistics provide basic biological information regarding the genetic and environmental contributions to traits directly under the influence of natural or artificial selection and thus allow one to predict and assess the genetic and phenotypic consequences of altered selection regimes (e.g., directional selection versus stabilizing selection). This absence of genetic data for salmon and steelhead, and fish in general, contrasts sharply with the situation for other types of animals that are bred artificially or in captivity. The problem with fish is that maintaining pedigrees and family identities for such large numbers of individuals would be a significant logistic and economic problem, even for a small hatchery.

Despite the virtual absence of any pedigreed databases on quantitative characters for hatchery populations of salmon and steelhead, circumstantial and experimental data do suggest that some selection responses have occurred in spawning-run timing and age at sexual maturity (Ayerst 1977; Rosentreter 1977; Leider et al. 1984). For rivers with long-established hatchery operations, the return of adults is typically bimodal with hatchery and wild fish constituting the early and late groups, respectively. By themselves, those data are circumstantial, because one cannot exclude differences in environmental conditions experienced by these two groups during the freshwater phase for the observed phenotypic differences in time of return. However, experimental studies have revealed that spawning-run timing, within-season spawn date, and within-season age at sexual maturity generally have very high heritabilities (>0.5) in salmonid fish (Crandell and Gall 1993; Gharrett and Smoker 1993; Hankin et al. 1993). The suggested implication of bimodal run timing, therefore, is that hatchery populations have been selectively bred for early maturity and have either responded directly to selection or have replaced wild fish that historically constituted the early part of the runs. Again, circumstantial evidence suggests that both of these postulated effects may have occurred; however, the absence of baseline data for most wild populations and pedigree data for hatchery populations precludes being able to unequivocally draw those conclusions.

The available data also support earlier interpretations that size-selective commercial fisheries have had a major genetic effect on both hatchery and wild populations of Pacific salmon by reducing the mean size of adults and age at sexual maturity (Ricker 1981). For example, in pink salmon *O. gorbuscha*, where age and size at sexual maturity are biologically uncoupled because of the strict 2-year life cycle, size at maturity appears to have a very high heritability ($h^2 = 0.3 - 0.8$; Smoker et al., in press b), thus suggesting the capacity for rapid responses to selection (at least in hatchery fish), although interannual variation in environmental variance is high.

Clearly, all of these perceived genetic effects (loss of genetic variation, importation of exogenous genes, and artificial selection) are the direct result of fishery and hatchery management practices and are not caused by the hatchery environment itself. Although these distinctions are generally recognized by salmon biologists in the Pacific Northwest, this message has not been transmitted clearly to the general public, politicians, or fishery biologists in other parts of the country.

Domestication Selection

Domestication selection is simply natural selection in an artificial or domestic environment. Behavioral or physiological traits are the ones most often affected. In general, domestication results in increased fitness under culture conditions but decreased fitness under natural or feral conditions (Price 1984; Kohane and Parsons 1988).

Effects.—The behavioral and physiological effects of domestication in salmonid fish have been documented best in captive strains of trout (Vincent 1960; Green 1964; Moyle 1969; Hynes et al. 1981; Bachman 1984). Under controlled laboratory or hatchery conditions, domesticated trout, compared with wild trout, are more active physically, swim more often near the water surface and in open water, show decreased swimming stamina and temperature tolerance in challenge tests, and exhibit faster growth rates on hatchery diets. In contrast, under natural conditions, domestic trout show decreased growth, decreased survival, increased swimming activity, and in some instances, increased aggression. A common observation is that domesticated or hatchery trout often exhibit excess swimming activity and aggression, and this behavior may contribute to increased mortality in the wild (Mesa 1991; Ruzzante 1994). This more active behavior may be a selection response to high rearing densities and pulsed surface feeding in hatchery environments; however, to my knowledge, this latter hypothesis has never been tested experimentally.

Fish geneticists and salmon biologists have wondered for many years whether domestication selection can occur to any detectable or significant extent in hatchery populations of anadromous salmonid fish (Hynes et al. 1981). Such domestication might be associated with the relaxation of natural selection for spawning behavior and adaptation of presmolt juveniles to hatchery conditions. A number of recent studies have revealed that, indeed, hatchery populations of anadromous salmonid fish may be undergoing some degree of domestication. Reisenbichler and McIntyre (1977) first demonstrated growth and survival differences between hatchery and wild steelhead in two environments, natural streams and hatchery ponds. Their results revealed a strain × environment interaction, suggesting increased adaptation of hatchery fish to the hatchery environment but decreased adaptation to the stream environment. More recently, hatchery juveniles of coho salmon *O. kisutch* and anadromous cutthroat trout *O. clarki* in laboratory and field experiments, respectively, were shown to be behaviorally more aggressive than were their wild counterparts (Swain and Riddell 1990, 1991; Mesa 1991; Ruzzante 1991, 1992, 1994; Holtby and Swain 1992). These latter results parallel those obtained previously for domesticated brook trout *Salvelinus fontinalis* (Vincent 1960; Moyle 1969), suggesting that the same mode of selection is operating for both groups of fish under hatchery conditions. An earlier study had also revealed a genetic basis for variation in agonistic behavior between two wild populations of coho salmon (Rosenau and McPhail 1987; see also Riddell and Swain 1991). One implication of these results is that wild offspring produced by natural hatchery × wild spawnings may show increased vulnerability to predators under natural conditions (Johnsson and Abrahams 1991).

Recent studies have also revealed differences in spawning behavior between hatchery and wild fish that are just the opposite of the differences reported for juvenile behavior (Fleming and Gross 1992, 1993). In a series of experiments involving spawning behavior of adult coho salmon, hatchery males were less aggressive, less active, more submissive, and competitively inferior to wild males. Behavioral differences between hatchery and wild females were not as great as those between males, but hatchery females, compared with wild females, did suffer greater egg losses due to temporal delays in spawning and redd superposition. Overall, the breeding successes of hatchery males and females were estimated to be 62% and 82%, respectively, of that of wild males and females. In a separate study, differences in body morphology of juveniles between hatchery and wild populations of coho salmon were found to be due primarily to environmental effects (Swain et al. 1991). On the other hand, significant differences in secondary sex characteristics of adult females (kype length and egg size) were detected in accordance with evolutionary and life history predictions (Fleming and Gross 1989), although the

TABLE 2.—Potential direct genetic effects of hatchery fish on wild populations due to natural spawning and potential interbreeding, and the biological and management causes of those potential effects. A plus (+) or minus (−) in parentheses implies the existence or absence, respectively, of strong empirical evidence for the indicated cause or effect in populations of Pacific salmon or steelhead.

Genetic effect	Biological causes	Management causes
Decreases in between-population genetic variation (?)	Natural straying (−)	Time and location of releases (+); Stock transfers (+)
Decreases in within-population genetic variation (−)	Genetic swamping (−)	Low effective population size of hatchery fish (+)
Decreases in fitness (?)	Introgressive hybridization (+)	Genetic changes in hatchery stocks (+); Stock importation (+)
Increases in between- or within-population genetic variation and fitness (−)[a]	Introgressive hybridization (+)	Stock importation coupled with environmental change (+)

[a] One possible example of these effects was cited (see text).

genetic basis for those differences was not tested explicitly.

In summary, experimental results obtained to date confirm earlier suggestions that domestication selection could be a significant source of genetic change in hatchery populations of anadromous salmonid fish. Such changes have to be interpreted as a direct genetic effect of the hatchery environment. How such changes are affecting wild populations, either as a direct or indirect genetic effect, is unknown at this time.

Solutions.—The genetic effects of domestication selection could be minimized by modifying the hatchery environment to reduce the presumed selective pressure (e.g., artificial spawning channels) and by implementing hatchery practices that would help to counteract the effects of natural selection in the hatchery. Hatcheries could potentially reduce or retard substantially the effects of domestication selection by adopting a husbandry practice that is well known to population and quantitative geneticists: equalize family size by culling surplus individuals within families prior to release (Allendorf 1993). Maintaining family identities on all fish in a hatchery prior to release could be logistically unfeasible. However, as the breeding goals of production hatcheries change from subsidizing commercial and recreational fishing to conserving genetic resources, the logistic and financial problems of maintaining family identities may seem a relatively small investment with respect to long-term maintenance of genetic variation.

Direct Genetic Effects on Wild Populations

The natural spawning of hatchery fish in the habitat of a self-sustaining wild population can potentially lead to one or more of several possible outcomes (Table 2). These potential effects include (1) decreases in between-population genetic variation; (2) decreases in within-population genetic variation; and (3) decreases in fitness of the wild population (outbreeding depression). Alternatively, the natural spawning and genic introgression of hatchery fish could also have the opposite effects; that is, introgression could cause increases in between- or within-population genetic variation and, under certain circumstances, increases in population fitness.

Decreases in Between-Population Genetic Variation

This direct genetic effect is expected to occur if (1) hatchery fish stray, and subsequently reproduce, at a rate that is greater than that of wild fish; (2) fish from a single hatchery population are transported and stocked into rivers and streams other than the one to which their parents returned as maturing adults; (3) exogenous fish are imported to the hatchery and spawned with indigenous fish; or (4) all of the above.

Quinn (1993) has recently reviewed the evidence for the common belief that hatchery fish stray more than do wild fish. Some of his conclusions are (1) "straying is an integral component of salmon behavior and population biology"; (2) "introduced (nonnative) populations and salmon displaced from their rearing site for release stray more than native salmon and those reared and released on-site"; (3) "evidence that standard hatchery practices increase the tendency of salmon to stray is equivocal, but releases of salmon at a different season from the normal migration period can increase straying"; and (4) "at present, fundamental gaps in our understanding of the genetic and environmental factors that influence straying hinder accurate prediction of the levels and consequences of straying." Quinn (1993) cites studies showing both increased

(McIsaac 1990) and decreased (Labelle 1992) straying by hatchery fish. The existing evidence is further equivocal because virtually all of the data on straying in hatchery fish comes from studies in which a nonnative stock was evaluated, or the fish were transported away from the hatchery rearing site prior to release. Hence, one cannot distinguish the biological effect of straying from the management effect of transporting or releasing a nonnative stock. Furthermore, as pointed out by Quinn (1993), hatcheries often use multiple water sources (well water versus surface water) or rearing locations (upstream hatcheries versus downstream homing and release ponds), and the effect of these practices on the imprinting process and subsequent homing of returning adults is unknown. In addition, a common practice in the Pacific Northwest is for smolts from a single hatchery population to be transported and released throughout several geographic regions (Leider et al. 1990). This practice can obviously reduce between-population genetic variation irrespective of any straying by hatchery fish.

Despite many arguments that straying or stock transfers could be reducing genetic variation among wild populations of salmon and steelhead, little direct evidence exists that such reductions have indeed occurred (Waples 1991a). Two potential reasons for this lack of evidence are (1) baseline genetic data for most populations prior to the introduction of hatchery fish do not exist; and (2) hatchery fish may not be making significant genetic contributions to wild populations. For example, in a study of genetic diversity among major population groups of chinook salmon in the Pacific Northwest, Utter et. al. (1989) concluded that "the persistence of these geographic patterns in the face of natural opportunities for introgression, and sometimes massive transplantations, suggests that genetically adapted groups within regions have resisted large-scale introgression from other regions." These results for wild fish differed dramatically from the homogeneous allele frequencies observed among hatchery populations (Utter et al. 1989). However, populations of wild steelhead on the northwest coast of Washington State, an area where nonnative hatchery steelhead had been stocked extensively since the 1940s, were genetically more homogenous than unstocked, native populations in British Columbia (Parkinson 1984; Reisenbichler and Phelps 1989). These results suggest a potential genetic effect by hatchery fish in the former populations. However, this interpretation must be considered conjectural because one cannot exclude other possible causes (e.g., different colonization histories) for those results.

An accumulating body of evidence indicates that nonnative fish and transported fish stray at a much higher rate, or return at a much lower rate, than do indigenous fish that are reared and released in their streams of origin (Ricker 1972; Bams 1976; Reisenbichler 1988; Altukhov and Salmenkova 1990; Quinn 1993). Such increased rates of straying cannot be considered an effect of hatcheries or artificial propagation but rather an effect caused by management by using a nonnative stock. Furthermore, the preciseness of homing and the viability of hatchery fish following their release appear to be negatively correlated with distance transported from the donor hatchery (Ricker 1972; Reisenbichler 1988). Such results are usually interpreted as reflecting some form of local adaptation and the inability of transplanted hatchery fish to home properly, although the exact nature and mechanism responsible for the adaptation has, to my knowledge, never been investigated. The work of Bams (1976), however, suggests the possibility that local adaptation may involve the evolution of increased olfactory sensitivity to the specific chemical characteristics of the home stream in response to natural selection for homing ability following initial colonization. To my knowledge, experiments addressing this possibility have not been conducted. Furthermore, natural straying rates of wild fish are, in most instances, unknown. Nevertheless, the data do suggest, at least for anadromous salmonids, that transplantation of fish stocks between hatcheries and watersheds does indeed increase the rate of straying, and that there is a genetic component associated with homing ability and natal stream detection (McIsaac and Quinn 1988). Collectively, these data and the summarizations of Quinn (1993) suggest that the increased straying associated with stock transfers, as well as the stock transfers themselves, may have a much greater potential of reducing between-population genetic variation than does any direct biological effect on innate homing of the hatchery environment itself. However, studies to address many of these key questions have not been performed.

Decreases in Within-Population Genetic Variation

A general axiom in theoretical population genetics is that the interbreeding of two populations with different allele frequencies will result in an immediate increase in within-population genetic variation and average heterozygosities (the Wahlund effect). Interspecific hybridization is one extreme example of this phenomenon. However, this aspect of the theory is based strictly on parametric gene frequencies for infinitely large populations and thus ignores the genetic properties of finite populations. In finite populations, any significant reduction in effective population size can accelerate the effects of genetic drift and thereby cause a loss of genetic variation over time, particularly for rare or low frequency alleles (Allendorf 1985).

Ryman and Laikre (1991) have outlined the conservation biology consequences of a population admixture derived from the random mating of two populations with different effective population sizes. Their general equation for two populations is

$$\frac{1}{N_e} = \frac{x^2}{N_c} + \frac{(1-x)^2}{N_w},$$

where N_w and N_c are the effective population sizes of the wild (recipient) and cultured (donor) populations, respectively, and N_e is the effective population size of the genetic admixture (introgressed population) derived from a fraction x of the donor population and a fraction $1 - x$ of the recipient population. Some general conclusions of this random interbreeding are readily apparent in Figure 1. First, if N_c is greater than $0.5N_w$, interbreeding of hatchery and wild fish can increase the effective population size of the wild population over a relatively wide range of values for x ($0 < x < 0.6$). However, one normally does not expect a hatchery population to have a greater effective population size than does the wild population except, perhaps, in cases dealing with threatened or endangered populations (Bartley et al. 1992). On the other hand, if the effective size of the hatchery population is substantially less than that of the wild population ($N_c < 0.1N_w$), then a substantial reduction in the effective size of the wild population will result from interbreeding over a very wide range of admixture proportions ($x > 0.2$). The reason for this reduction is commonly called genetic swamping; that is, a large number of hatchery fish, representing the offspring or descendents of a relatively small number of parents, interbreed with wild fish, thereby swamping the indigenous gene pool with genetic material from the hatchery population. The potential genetic implications of this swamping are obvious. The extreme case is a value of x equal to 1 which is equivalent to complete replacement of the wild population by the hatchery population (see Hutchings 1991; Byrne et al. 1992). Note, however, that these arguments implicitly assume equal fitness

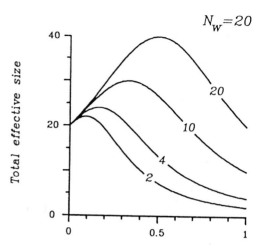

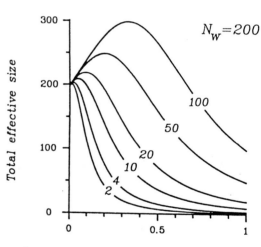

FIGURE 1.—Total effective population size (wild + captive) when a natural population of (**a**) 20 or (**b**) 200 effective parents (N_w) is supplemented by offspring from varying effective numbers of captive parents (represented by the different curves). The x-axis represents the proportion of offspring contributed by captive breeding. (Figure from Ryman and Laikre 1991; reprinted by permission of the authors, the Society for Conservation Biology, and Blackwell Scientific Publications, Inc.)

of hatchery and wild fish, and their recombinant genotypes, in the natural environment.

To my knowledge, no study has ever documented decreases in effective population size or loss of genetic variation in a wild population of Pacific salmon or steelhead due primarily to genetic swamping or natural spawning by hatchery fish. However, this does not mean that this effect has not occurred. Computer simulations suggest that sample sizes substantially larger than the effective population size are required to detect changes in the latter parameter with any statistical precision using biochemical genetic markers (C. Busack, Washington State Department of Fish and Wildlife, personal communication). On the other hand, this direct genetic effect may be negligible if hatchery fish or their wild offspring are maladapted to the natural environment or do not contribute genetically to wild populations.

Decreases in Fitness

Decreases in fitness can occur when two genetically diverged or reproductively isolated populations interbreed. This decrease in fitness is commonly called outbreeding depression. Two potential causes of outbreeding depression are genetic adaptation to the local environment (i.e., local adaptation), and genomic coadaptation (Templeton 1986). The former occurs when an introduced population interbreeds with the indigenous population, and the introduced population is maladapted to the local environment. Genomic coadaptation reflects epistatic or chromosomal interactions among alleles at different loci that have a positive effect on organismic fitness. However, these allelic combinations can be broken up or disrupted in the F_2 generation due to meiotic recombination in the F_1 generation following interpopulation matings. These two potential consequences of interpopulation hybridization are the basis for the commonly held belief that the fitness of a wild population will decrease following natural spawning of hatchery fish.

Extensive arguments have been made in the scientific literature regarding the potential for outbreeding depression in Pacific salmon (Emlen 1991; Waples 1991a; Gharrett and Smoker 1993). These arguments are based largely on experimental studies of *Drosophila*, theoretical considerations, and our current state of knowledge regarding life history adaptations in anadromous salmonid fish (Taylor 1991). For example, local adaptations in homing ability (Quinn 1993), rheotactic swimming behavior of emerging fry (Brannon 1972), outmigration timing of smolts (Gharrett and Smoker 1993), and within-season timing of returning adults (Smoker et al., in press a) all appear to have a genetic basis. However, virtually no empirical data documenting outbreeding depression for conspecific or local populations of Pacific salmon or steelhead actually exist. For example, regarding interbreeding between hatchery and wild fish, Steward and Bjornn (1990) stated that they were "unable to locate any published studies in which the fitness of progeny of hatchery × wild matings were measured over multiple generations and compared with the fitness of the original hatchery and wild parental stocks." Similarly, there is a general lack of data demonstrating outbreeding depression at the local population level in other groups of fish and ectothermic vertebrates (Waldman and McKinnon 1993). Most of the available empirical evidence for outbreeding depression comes from studies of *Drosophila* (Dobzhansky 1970) and isolated examples with other organisms (Endler 1977; Templeton 1986; Barrett and Kohn 1991).

Recently, outbreeding depression was suggested among F_2 generation progeny of inter-race crosses of even- and odd-year pink salmon (Gharrett and Smoker 1991). The F_2 hybrid adults showed decreased adult return survival and increased bilateral asymmetry relative to F_1 controls (F_2 controls were not produced). Bilateral or fluctuating asymmetry is generally believed to be one manifestation of developmental instability resulting from a dysgenic genome (Leary et al. 1985). The results of the above experiment with pink salmon implicate disruption of genomic coadaptation (as opposed to local adaptation) as the cause of the suspected outbreeding depression, because the two year-classes are only temporally isolated. The environments to which the two year-classes are presumably adapted are one and the same, although one cannot rule out the possibility of substantially different migration routes in the marine environment.

The most extensive study of natural spawning and potential interbreeding of hatchery and wild fish to date has been conducted in the Kalama River, western Washington State, where summer-run and winter-run steelhead populations have been studied extensively since 1975 (Leider et al. 1990). The general conclusions for summer-run fish were (1) returning hatchery adults outnumbered wild adults by a ratio of 4.5 to 1; (2) hatchery adults genetically produced more than 50% of the outmigrating wild smolts and more than 40% of the returning wild adults; and (3) each hatchery adult produced, on average, only 10–20% as many surviving smolt and adult progeny as did each wild adult (Chilcote et al. 1986; Leider et al. 1990; Campton et al. 1991). The reasons for these differences in relative reproductive success are unknown; however, neither biological factors nor management practices can be excluded as potential causes (Campton et al. 1991). In addition, there are no obvious indications to date that hatchery fish introgression has significantly affected the fitness of the wild population, although this latter question has not yet been specifically evaluated. Future evaluations of hatchery–wild fish interactions in the Kalama River system will, perhaps, provide insights regarding many of these key questions.

Increases in Within- and Between-Population Genetic Variation and Fitness

If hatchery fish represent a nonnative or genetically diverged stock relative to a wild population, then the natural interbreeding of the two groups would be expected to initially increase the level of within-population genetic variation. This, in turn, would most likely cause an increase in genetic variation between wild populations if hatchery fish were not straying to other systems. For this increased level of variation to persist, however, the introgression of hatchery fish genes must be either selectively neutral or selectively positive. A positive response may occur in threatened or endangered populations with very low effective population sizes. However, as pointed out in the Kalama River study and elsewhere (Taylor 1991; Gharrett and Smoker 1993), the general expectation is that such interbreeding would be expected to result in a decrease in fitness. Consequently, if one accepts the concept of local adaptation in Pacific salmon and steelhead, then it is difficult to conceive of a situation in which introgressive hybridization with hatchery fish could result in an increase in fitness of a wild population.

The one situation in which interbreeding with hatchery fish may actually increase the fitness of an indigenous wild population is one in which the habitat has been environmentally perturbed, and the native fish are no longer locally adapted. I am aware of one example where such an effect may have occurred. Rainbow trout (nonanadromous *O. mykiss*) currently inhabiting the upper Yakima River of Washington State appear to represent a genetic admixture of native and nonnative fish (Campton and Johnston 1985). Abundant runs of steelhead historically ascended and spawned in the upper Yakima River, but an irrigation diversion dam (Roza Dam) built in 1939 effectively blocked all upstream migration because of a nonfunctional fish ladder. During that time and continuing for several decades later, nonanadromous rainbow trout from domesticated hatchery strains were continuously stocked in the Yakima River for the sole purpose of providing a put-and-take sport fishery. Genetic and other data suggest strongly that those domesticated rainbow trout interbred with native fish in the upper Yakima River to generate the present nonanadromous hybrid population. The net result is that this population is currently doing very well; the population is self-sustaining, fish are abundant, and a trophy rainbow trout fishery exists because of catch-and-release regulations (S. Leider, Washington State Department of Fish and Wildlife, personal communication). Recent studies with more extensive sampling and loci have confirmed our earlier (Campton and Johnston 1985) interpretations (S. Phelps, Washington State Department of Fish and Wildlife, personal communication). Obviously, for steelhead populations in the upper Yakima River, anadromous behavior was at a selective disadvantage after the construction of Roza Dam; however, the interbreeding of native fish and introduced, nonanadromous fish may have allowed a wild population to persist and subsequently flourish because the perturbed environment selected positively for an altered or different phenotype. The implicit assumption here, of course, is that migratory behavior has a genetic component and that the indigenous population of *O. mykiss* may have been severely depressed or gone extinct without an infusion of hatchery fish genes. In this context, O'Brien and Mayr (1991) have argued that preserving genes through hybridization is more important biologically than is loss of indigenous gene pools or alleles due to extinction or small effective population sizes.

Indirect Genetic Effects of Hatchery Fish on Wild Populations

Indirect genetic effects are those that affect the abundance of wild populations but are not associated specifically with the natural reproduction of hatchery fish (Table 3). Causes of indirect genetic effects include a variety of ecological factors such as predation, competition (Nickelson et al. 1986), and attraction of predators (Beamish et al. 1992). In this context, humans are also predators that are clearly attracted to large numbers of hatchery fish. Management factors causing indirect genetic effects on wild populations include overharvest of wild fish in mixed-stock fisheries (Evans and Willox 1991; Kope 1992), disease introductions via stocking of infected hatchery fish, broodstock extraction, and environmental manipulations associated with the management of hatchery fish in the wild. Broodstock exploitation (mining) occurs when overzealous

TABLE 3.—Potential indirect genetic effects[a] of hatchery fish on wild populations due to ecological interactions or management decisions that affect abundance, and the biological and management causes of those potential effects. A plus (+) or minus (−) in parentheses implies the existence or absence, respectively, of strong empirical evidence for the indicated cause or effect in populations of Pacific salmon or steelhead.

Genetic effect	Biological causes	Management cause
Decreases in abundance (+)	Predation (+); Competition (?); Attraction of predators (?)	Overharvest due to mixed hatchery–wild fisheries (?); Disease transfers (−); Broodstock extraction or mining (?); Environmental degradation due to mitigation hatchery (+)
Increases in abundance (?)	Natural selection (?)	Fisheries targeted on hatchery fish (+); Successful rebuilding of wild populations through hatchery supplementation (?)

[a] Indirect genetic effects were not specifically reviewed here. Evidence for these causes and effects was obtained primarily from Steward and Bjornn (1990).

hatchery programs remove more adult spawning fish from the wild population than can be replaced by adult recruits from the hatchery population or by natural propagation. Environmental manipulations include a broad array of factors including the situation in which politicians, responding to the economic demands of society, use mitigation by hatcheries to allow the wholesale elimination of spawning or rearing habitat of wild populations. Construction of dams on the Columbia River is one example of this type of impact on wild populations, a process Meffe (1992) refers to as techno-arrogance. Also, environmental regulations or decisions that are based primarily on information available on hatchery fish, in ignorance or absence of information on wild fish, can also have an indirect genetic effect on wild populations (Waples 1991a). For instance, Waples (1991a) points out that fishery management decisions can have long-term effects on wild populations by creating altered selection regimes; for example, size-selective fisheries that target large fish (Ricker 1981; McAllister and Peterman 1992; McAllister et al. 1992). On the other hand, fishery management programs that specifically target hatchery fish could, in theory, have a positive genetic effect on wild populations if the habitat allows those populations to rebound or propagate naturally. However, I am not aware of a single example where this latter effect has been documented in the scientific literature.

From the review of Steward and Bjornn (1990), there appear to be few generalizations one can draw regarding ecological interactions and potential indirect genetic effects. Such effects appear to be a function of the specific populations studied, the number and size of hatchery fish released, and the associated environmental conditions encountered by the two groups. For example, competition between hatchery and wild fish in freshwater may be minimal if the hatchery fish are released as smolts in main-stem rivers; however, Nickelson et al. (1986) demonstrated that stocking of fry or presmolt coho salmon in small streams where adults typically spawn can potentially cause a significant decrease in abundance of the wild populations due to presumed competition and the apparent lower reproductive success of returning hatchery adults (versus wild adults).

Similarly, the effects of disease transfer from hatchery stocks to wild populations of Pacific salmon and steelhead are also equivocal. Steward and Bjornn (1990) concluded that, "in spite of the comparatively high incidence of disease among some hatchery fish stocks, there is little evidence to suggest that diseases or parasites are routinely transmitted from hatchery to wild fish." However, Steward and Bjornn (1990) also noted elsewhere that "there has not been much research on this subject," and that "the full impact of disease of supplemented stocks [of wild fish] is probably underestimated." In contrast, disease transfer from cultured stocks to wild populations of Atlantic salmon in Norway appear to be well documented (Heggberget et al. 1993). However, in this latter situation, the disease transfer was not a result of artificial propagation per se, but rather the result of exogenous stock transfers. The parasite *Gyrodactylus salaris* was imported to Norway from Swedish fish farms in Baltic Sea drainages. Likewise, the bacterium *Aeromonas salmonicida* (causative agent of furunculosis) was imported to Norway with rainbow trout from Denmark and with Atlantic salmon from Scotland. These disease transfers were thus caused by human factors related to management and not by biological factors intrinsic to the operation of fish farms in Norway. Presumably, infections by *Aeromonas* sp. and *Gyrodactylus* sp. would not be a problem today in Norway (Heggberget et al. 1993) if the stock transfers had not been performed.

Finally, overharvesting of wild fish in mixed-stock fisheries on hatchery and wild fish has been commonly cited or suggested as a major cause of decreases in the abundance of salmon and steelhead in the Pacific Northwest (Utter et al. 1993; Wright 1993). However, I am unaware of any study in the scientific literature in which this factor has been documented as the primary cause of decline in abundance of a wild population of Pacific salmon or steelhead; virtually all of the data related to this question appear to be conjectural, theoretical, or circumstantial. For example, Wright (1993) implied that overexploitation due to managing the salmon fishery for hatchery fish was the cause of extinction of wild, native coho salmon in the lower Columbia River. Although this interpretation is probably true in the strictest sense, these populations were already severely depressed due to overfishing when the hatchery program was initiated over 30 years ago. Indeed, the specific goal of the hatchery program was to restore the fishery. That goal was clearly achieved, but apparently at the expense of the remaining wild populations (Johnson et al. 1991). Also, it may be very difficult to distinguish, scientifically, the overharvesting of wild populations from the degradation of freshwater environments or from competition in the marine environment as potential causes of the apparent inverse relationship over time between the abundance of hatchery and wild salmon in the Pacific Northwest (Hilborn 1992; Washington and Koziol 1993). For example, regarding density-dependent factors in the marine environment, Steward and Bjornn (1990) concluded that "competition between hatchery and wild salmonids in the ocean has not been unequivocally demonstrated, because there is little or no competition, or perhaps because of the complexity of factors involved, a paucity of experimental data, and natural variability in the occurrence of competition and its effects."

My description of indirect genetic effects of hatchery fish on wild populations of Pacific salmon and steelhead is purposefully superficial. These effects relate, in one way or another, to the ecological well being of wild populations independent of any direct genetic effect. My general conclusions regarding indirect genetic effects caused by ecological factors, obtained primarily from the review of Steward and Bjornn (1990) and the references cited therein, are that (1) ecological effects of hatchery fish are very difficult to detect without controlled studies (Nickelson et al. 1986); (2) very few key questions have been investigated experimentally; (3) interspecific interactions may be more detrimental than intraspecific interactions; and (4) ecological factors affecting the abundance of wild populations may prove to be the ones most difficult to control. On the other hand, indirect genetic effects on wild populations caused by management decisions or practices can potentially be eliminated by simply changing the philosophy and priorities of fisheries resource management. Blaming hatcheries for negative effects associated with stock transfers or overharvest of wild populations diverts the public's attention from the real problems and only exemplifies how hatcheries have been used as scapegoats for errors (or ignorance) in human judgement.

Conclusions

Examination of Tables 1–3 reveals that a significant proportion of the perceived or potential genetic effects associated with hatcheries and hatchery fish are caused directly by fishery resource management, either as management relates to fish in the hatchery or fish in the wild. Furthermore, there appears to be substantially more scientific documentation associated with management causes than with biological causes; this difference may be due, in part, to the relative difficulties of detecting effects associated with the two types of causes. For example, it is much easier to document artificial selection in the hatchery than natural selection in the wild. Furthermore, baseline genetic data prior to the introduction of hatchery fish do not exist for most wild populations. Despite these limitations, many of the negative genetic effects attributed to hatcheries or hatchery fish are not caused by hatcheries or artificial propagation, per se, but rather by humans and their mismanagement of hatchery and wild populations. Perhaps many of these perceived or anticipated effects can be reduced, circumvented, or eliminated by simply changing the way we manage hatchery fish and their fisheries (Brannon 1993).

In this paper, I have attempted to present an objective, comprehensive review of our current state of knowledge regarding the genetic effects of hatcheries on wild populations of Pacific salmon and steelhead. I have also tried to distinguish fact and scientific data from untested dogma or speculation and separate my personal values from my perspective as a scientist. In my evaluation of this problem of hatchery versus wild fish, I have often seen, in both the scientific literature and in the popular press, a general blurring between fact and speculation, between data and interpretation, and between science and values. If this blurring continues, then we may undermine our ability to evaluate and understand the perceived problems clearly, objectively, and scientifically.

In the long run, questions regarding the use of hatcheries as a management tool for Pacific salmon and steelhead may have to be resolved in terms of our values (Hilborn 1992) irrespective of the scientific data that we may or may not have to justify management decisions. As Dennis Scarnecchia (1988) stated,

> Salmon resources have been aided and harmed by technology, and managers must carefully assess how current and future technologies will be used to manage salmon. Effective managers must be knowledgeable of fishery science and human values. The science in fishery management is the objective, logical, and systematic method of obtaining reliable knowledge about fishery resources. The art in fishery management involves our values, that is, what we judge to be good, desirable, and important in the long run.

Given our current state of knowledge, perhaps the vast majority of our management decisions will thus be based on what we judge to be good, desirable, and important in the long run.

Waples (1991a) has suggested that a guiding principle of hatchery fish supplementation should be, "First, do no harm." I would further suggest that such a principle should be, in a more general sense, the first goal or prime directive of fisheries management. Such a goal, from the perspective of conserving genetic resources, can perhaps be achieved by simply following the recommendations of the International Symposium on Fish Gene Pools, Preservation of Genetic Resources in Relation to Wild Fish Stocks, which was held in Stockholm, Sweden, in January 1980 (Ryman 1981). Those recommendations, as summarized by Hindar et al. (1991) and Utter et al. (1993), are (1) identify genetic resources, (2) maintain natural ecosystems, (3) avoid selective harvest of natural populations, (4) preserve genetic variability within cultured populations, (5) release fish into natural environments with great care, (6) provide adequate funding for basic and applied research, and (7) inform those responsible for management of existing knowledge. To those recommendations, I would add an eighth recommendation: (8) understand, scientifically, the biological consequences of management decisions.

Achieving the objective of the eighth recommendation will obviously require a closer working relationship between the scientific community at large and the agencies entrusted with monitoring and conserving genetic resources. In the past, hatcheries were used primarily as production factories for subsidizing or enhancing commercial or recreational fisheries with little or no interest by the public—or the agencies themselves—to evaluate, scientifically, the biological consequences of those factories on natural populations. Clearly, the tide has changed as evidenced by the proceedings of this symposium. However, our science has only scratched the surface regarding our understanding of the biological interactions between hatchery and wild fish, and many critical questions remain unanswered. Consequently, I suggest that one of the most useful goals of hatcheries should be to integrate scientific research and management into a mutually beneficial relationship in order to learn as much as possible about the biology of the fish we are propagating and the effects of those fish on natural populations. Then, perhaps, controversies regarding the pros and cons of hatcheries will be understood scientifically before they become more polarizing.

Acknowledgments

I thank E. L. Brannon, J. J. Hard, W. W. Smoker, and F. M. Utter for their many helpful comments and suggestions on an earlier version of this manuscript. This contribution is Florida Agricultural Experiment Station Journal Series No. R-04354.

References

Allendorf, F. W. 1985. Genetic drift and the loss of rare alleles versus heterozygosity. Zoo Biology 5:181–190.

Allendorf, F. W. 1993. Delay of adaptation to captive breeding by equalizing family size. Conservation Biology 7:416–419.

Allendorf, F. W., and S. R. Phelps. 1980. Loss of genetic variation in a hatchery stock of cutthroat trout. Transactions of the American Fisheries Society 109:537–543.

Allendorf, F. W., and N. Ryman. 1987. Genetic management of hatchery stocks. Pages 141–159 in N. Ryman and F. Utter, editors. Population genetics & fishery management. University of Washington Press, Seattle.

Altukhov, Y. P., and E. A. Salmenkova. 1990. Introductions of distinct stocks of chum salmon, *Oncorhynchus keta* (Walbaum), into natural populations of the species. Journal of Fish Biology 37(Supplement A):25–33.

Ayerst, J. D. 1977. The role of hatcheries in rebuilding steelhead runs of the Columbia River system. American Fisheries Society Special Publication 10:84–88.

Bachman, R. A. 1984. Foraging behavior of free-ranging wild and hatchery brown trout in a stream. Transactions of the American Fisheries Society 113:1–32.

Bams, R. A. 1976. Survival and propensity for homing as affected by presence or absence of locally adapted paternal genes in two transplanted populations of pink salmon (*Oncorhynchus gorbuscha*). Journal of the Fisheries Research Board of Canada 33:2716–2725.

Barrett, S. C. H., and J. R. Kohn. 1991. Genetic and evolutionary consequences of small population size in plants: implications for conservation. Pages 3–30 *in* D. A. Falk and K. E. Holsinger, editors. Genetics and conservation of rare plants. Oxford University Press, New York.

Bartley, D., M. Bagley, G. Gall, and B. Bentley. 1992. Use of linkage disequilibrium data to estimate effective size of hatchery and natural fish populations. Conservation Biology 6:365–375.

Beacham, T. D., and T. P. T. Evelyn. 1992. Genetic variation in disease resistance and growth of chinook, coho, and chum salmon with respect to vibriosis, furunculosis, and bacterial kidney disease. Transactions of the American Fisheries Society 121:456–485.

Beamish, R. J., B. L. Thomson, and G. A. McFarlane. 1992. Spiny dogfish predation on chinook and coho salmon and the potential effects on hatchery-produced salmon. Transactions of the American Fisheries Society 121:444–455.

Billington, N., and P. D. N. Hebert, editors. 1991. International symposium on the ecological and genetic implications of fish introductions (FIN). Canadian Journal of Fisheries and Aquatic Sciences 48(Supplement 1).

Brannon, E. L. 1972. Mechanisms controlling migration of sockeye salmon fry. Bulletin of the International Pacific Salmon Commission 21:1–86.

Brannon, E. L. 1993. The perpetual oversight of hatchery programs. Fisheries Research 18:19–27.

Byrne, A., T. C. Bjornn, and J. D. McIntyre. 1992. Modelling the response of native steelhead to hatchery supplementation programs in an Idaho river. North American Journal of Fisheries Management 12:62–78.

Campton, D. E., and J. M. Johnston. 1985. Electrophoretic evidence for a genetic admixture of native and nonnative rainbow trout in the Yakima River, Washington. Transactions of the American Fisheries Society 114:782–793.

Campton, D. E., and six coauthors. 1991. Reproductive success of hatchery and wild steelhead. Transactions of the American Fisheries Society 120:816–827.

Chilcote, M. W., S. A. Leider, and J. J. Loch. 1986. Differential reproductive success of hatchery and wild summer-run steelhead under natural conditions. Transactions of the American Fisheries Society 115:726–735.

Cross, T. F., and J. King. 1983. Genetic effects of hatchery rearing in Atlantic salmon. Aquaculture 33:33–40.

Crandell, P. A., and G. A. E. Gall. 1993. The genetics of age and weight at sexual maturity based on individually tagged rainbow trout (*Oncorhynchus mykiss*). Aquaculture 117:95–105.

Daley, W. J. 1993. The use of fish hatcheries—polarizing the issue. Fisheries (Bethesda) 18(3):4–5.

Dobzhansky, T. 1970. Genetics of the evolutionary process. Columbia University Press, New York.

Donaldson, L. R., and D. Menasveta. 1961. Selective breeding of chinook salmon. Transactions of the American Fisheries Society 90:160–164.

Doyle, R. W. 1983. An approach to the quantitative analysis of domestication selection in aquaculture. Aquaculture 33:167–185.

Emlen, J. M. 1991. Heterosis and outbreeding depression: a multi-locus model and an application to salmon production. Fisheries Research 12:187–212.

Endler, J. A. 1977. Geographic variation, speciation, and clines. Monographs in population biology 10. Princeton University Press, Princeton, New Jersey.

Evans, D. O., and C. C. Willox. 1991. Loss of exploited indigenous populations of lake trout, *Salvelinus namaycush*, by stocking of non-native stocks. Canadian Journal of Fisheries and Aquatic Sciences 48:134–147.

Falconer, D. S. 1981. Introduction to quantitative genetics, 2nd edition. Longman, London.

Fjalestad, K. T., T. Gjedrem, and B. Gjerde. 1993. Genetic improvement of disease resistance in fish—an overview. Aquaculture 111:65–74.

Fleming, I. A., and M. R. Gross. 1989. Evolution of adult female life history and morphology in a Pacific salmon (coho: *Oncorhynchus kisutch*). Evolution 43:141–157.

Fleming, I. A., and M. R. Gross. 1992. Reproductive behavior of hatchery and wild coho salmon (*Oncorhynchus kisutch*)—does it differ? Aquaculture 103:101–121.

Fleming, I. A., and M. R. Gross. 1993. Breeding success of hatchery and wild coho salmon (*Oncorhynchus kisutch*) in competition. Ecological Applications 3:230–245.

Gharrett, A.J., and S.M. Shirley. 1985. A genetic examination of spawning methodology in a salmon hatchery. Aquaculture 47:245–256.

Gharrett, A. J., and W. W. Smoker. 1991. Two generations of hybrids between even- and odd-year pink salmon (*Oncorhynchus gorbuscha*): a test for outbreeding depression? Canadian Journal of Fisheries and Aquatic Sciences 48:1744–1749.

Gharrett, A. J., and W. W. Smoker. 1993. A perspective on the adaptive importance of genetic infrastructure in salmon populations to ocean ranching in Alaska. Fisheries Research 18:45–58.

Green, D. M., Jr. 1964. A comparison of stamina of brook trout from wild and domestic parents. Transactions of the American Fisheries Society 93:96–100.

Hankin, D. G., J. W. Nicholas, and T. W. Downey. 1993. Evidence for inheritance of age of maturity in chinook salmon (*Oncorhynchus tshawytscha*). Canadian Journal of Fisheries and Aquatic Sciences 50:347–358.

Heggberget, T. G., and six coauthors. 1993. Interactions between wild and cultured Atlantic salmon: a review of the Norwegian experience. Fisheries Research 18:123–146.

Hilborn, R. 1992. Hatcheries and the future of salmon in the northwest. Fisheries (Bethesda) 17(1):5–8.

Hindar, K., N. Ryman, and F. Utter. 1991. Genetic effects of cultured fish on natural fish populations.

Canadian Journal of Fisheries and Aquatic Sciences 48:945–957.

Hjort, R. C., and C. B. Schreck. 1982. Phenotypic differences among stocks of hatchery and wild coho salmon, *Oncorhynchus kisutch*, in Oregon, Washington, and California. U.S. National Marine Fisheries Service Fishery Bulletin 80:105–119.

Holtby, L. B., and D. P. Swain. 1992. Through a glass, darkly: a response to Ruzzante's reappraisal of mirror image stimulation studies. Canadian Journal of Fisheries and Aquatic Sciences 49:1968–1969.

Hutchings, J. A. 1991. The threat of extinction to native populations experiencing spawning intrusions by cultured Atlantic salmon. Aquaculture 98:119–132.

Hynes, J. D., E. H. Brown, Jr., J. H. Helle, N. Ryman, and D. A. Webster. 1981. Guidelines for the culture of fish stocks for resource management. Canadian Journal of Fisheries and Aquatic Sciences 38:1867–1876.

Johnson, O. W., T. A. Flagg, D. J. Maynard, G. B. Milner, and F. W. Waknitz. 1991. Status review for lower Columbia River coho salmon. NOAA (National Oceanic and Atmospheric Administration) Technical Memorandum NMFS (National Marine Fisheries Service) F/NWC-202, Seattle.

Johnsson, J. I., and M. V. Abrahams. 1991. Interbreeding with domestic strain increases foraging under threat of predation in juvenile steelhead trout (*Oncorhynchus mykiss*): an experimental study. Canadian Journal of Fisheries and Aquatic Sciences 48:243–247.

Kohane, M. J., and P. A. Parsons. 1988. Domestication. Evolutionary change under stress. Evolutionary Biology 23:31–38.

Kope, R. G. 1992. Optimal harvest rates for mixed stocks of natural and hatchery fish. Canadian Journal of Fisheries and Aquatic Sciences 49:931–938.

Krueger, C. C., and B. May. 1991. Ecological and genetic effects of salmonid introductions in North America. Canadian Journal of Fisheries and Aquatic Sciences 48(Supplement 1):66–77.

Labelle, M. 1992. Straying patterns of coho salmon (*Oncorhynchus kisutch*) stocks from southeast Vancouver Island, British Columbia. Canadian Journal of Fisheries and Aquatic Sciences 49:1843–1855.

Lacy, R. C. 1987. Loss of genetic diversity from managed populations: interacting effects of drift, mutation, immigration, selection, and population subdivision. Conservation Biology 1:143–158.

Leary, R. F., F. W. Allendorf, and K. L. Knudsen. 1985. Developmental instability as an indicator of the loss of genetic variation in hatchery trout. Transactions of the American Fisheries Society 114:230–235.

Leider, S. A., M. W. Chilcote, and J. J. Loch. 1984. Spawning characteristics of sympatric populations of steelhead trout (*Salmo gairdneri*): evidence for partial reproductive isolation. Canadian Journal of Fisheries and Aquatic Sciences 41:1454–1462.

Leider, S. A., P. L. Hulett, J. J. Loch, and M. W. Chilcote. 1990. Electrophoretic comparison of the reproductive success of naturally spawning transplanted and wild steelhead trout through the returning adult stage. Aquaculture 88:239–252.

Martin, J., J. Webster, and G. Edwards. 1992. Hatcheries and wild stocks: are they compatible? Fisheries (Bethesda) 17(1):4.

McAllister, M. K., and R. M. Peterman. 1992. Decision analysis of a large-scale fishing experiment designed to test for a genetic effect of size-selective fishing on British Columbia pink salmon (*Oncorhynchus gorbuscha*). Canadian Journal of Fisheries and Aquatic Sciences 49:1305–1314.

McAllister, M. K., R. M. Peterman, and D. M. Gillis. 1992. Statistical evaluation of a large-scale fishing experiment designed to test for a genetic effect of size-selective fishing on British Columbia pink salmon (*Oncorhynchus gorbuscha*). Canadian Journal of Fisheries and Aquatic Sciences 49:1294–1304.

McIsaac, D. O. 1990. Factors affecting the abundance of 1977-1979 brood wild fall chinook salmon (*Oncorhynchus tshawytscha*) in the Lewis River, Washington. Doctoral dissertation. University of Washington, Seattle. (Not seen; cited in Quinn 1993.)

McIsaac, D. O., and T. P. Quinn. 1988. Evidence for a hereditary component in homing behavior of chinook salmon (*Oncorhynchus tshawytscha*). Canadian Journal of Fisheries and Aquatic Sciences 45:2201–2205.

Meffe, G. K. 1992. Techno-arrogance and halfway technologies—salmon hatcheries on the Pacific coast of North America. Conservation Biology 6:350–354.

Mesa, M. G. 1991. Variation in feeding, aggression, and position choice between hatchery and wild cutthroat trout in an artificial stream. Transactions of the American Fisheries Society 120:723–727.

Miller, W. H., T. C. Coley, H. L. Burge, and T. T. Kisanuki. 1990. Analysis of salmon and steelhead supplementation: emphasis on unpublished reports and present programs. Bonneville Power Administration, Project 88-100, part 1. Portland, Oregon.

Moyle, P. B. 1969. Comparative behavior of young brook trout of wild and hatchery origin. Progressive Fish-Culturist 31:51–56.

Nehlsen, W., J. E. Williams, and J. A. Lichatowich. 1991. Pacific salmon at the crossroads: stocks at risk from California, Oregon, Idaho, and Washington. Fisheries (Bethesda) 16(2):4–21.

Nickelson, T. E., M. F. Solazzi, and S. L. Johnson. 1986. Use of hatchery coho salmon (*Oncorhynchus Kisutch*) to rebuild wild populations in Oregon coastal streams. Canadian Journal of Fisheries and Aquatic Sciences 43:2443–2449.

O'Brien, S. J., and E. Mayr. 1991. Bureaucratic mischief: recognizing endangered species and subspecies. Science 251:1187–1188.

Parkinson, E.A. 1984. Genetic variation in populations of steelhead trout (*Salmo gairdneri*) in British Columbia. Canadian Journal of Fisheries and Aquatic Sciences 41:1412–1420.

Price, E. O. 1984. Behavioral aspects of animal domestication. Quarterly Review of Biology 59:1–32.

Quinn, T. P. 1993. A review of homing and straying of wild and hatchery-produced salmon. Fisheries Research 18:29–44.

Reisenbichler, R. R. 1988. Relation between distance transferred from natal stream and recovery rate for hatchery coho salmon. North American Journal of Fisheries Management 8:172–174.

Reisenbichler, R. R., and J. D. McIntyre. 1977. Genetic differences in growth and survival of juvenile hatchery and wild steelhead trout. Journal of the Fisheries Research Board of Canada 34:123–128.

Reisenbichler, R. R., and S. R. Phelps. 1989. Genetic variation in steelhead (*Salmo gairdneri*) from the north coast of Washington. Canadian Journal of Fisheries and Aquatic Sciences 46:66–73.

Ricker, W. E. 1972. Hereditary and environmental factors affecting certain salmonid populations. Pages 19–160 *in* R. C. Simon and P. A. Larkin, editors. The stock concept in Pacific salmon. H. R. McMillan Lectures in Fisheries, University of British Columbia, Vancouver.

Ricker, W. E. 1981. Changes in the average size and average age of Pacific salmon. Canadian Journal of Fisheries and Aquatic Sciences 38:1636–1656.

Riddell, B. E., and D. P. Swain. 1991. Competition between hatchery and wild coho salmon (*Oncorhynchus kisutch*): genetic variation for agonistic behavior in newly-emerged wild fry. Aquaculture 98:161–172.

Rosenau, M. L., and J. D. McPhail. 1987. Inherited differences in agonistic behavior between two populations of coho salmon. Transactions of the American Fisheries Society 116:646–654.

Rosentreter, N. 1977. Characteristics of hatchery fish: angling, biology, and genetics. American Fisheries Society Special Publication 10:79–83.

Ruzzante, D. E. 1991. Variation in agonistic behavior between hatchery and wild populations of fish: a comment on Swain and Riddell (1990). Canadian Journal of Fisheries and Aquatic Sciences 48:519–520.

Ruzzante, D. E. 1992. Mirror image stimulation, social hierarchies, and population differences in agonistic behavior—a reappraisal. Canadian Journal of Fisheries and Aquatic Sciences 49:1966–1968.

Ruzzante, D. E. 1994. Domestication effects on aggressive and schooling behavior in fish. Aquaculture 120:1–24.

Ryman, N., editor. 1981. Fish gene pools. Ecological Bulletin 34.

Ryman, N., and L. Laikre. 1991. Effects of supportive breeding on the genetically effective population size. Conservation Biology 5:325–329.

Ryman, N., and G. Stahl. 1980. Genetic changes in hatchery stocks of brown trout (*Salmo trutta*). Canadian Journal of Fisheries and Aquatic Sciences 37:82–87.

Scarnecchia, D. L. 1988. Salmon management and the search for values. Canadian Journal of Fisheries and Aquatic Sciences 45:2042–2050.

Simon, R. C., J. D. McIntyre, and A. R. Hemingsen. 1986. Family size and effective population size in a hatchery stock of coho salmon (*Oncorhynchus kisutch*). Canadian Journal of Fisheries and Aquatic Sciences 43:2434–2442.

Smoker, W. W., A. J. Gharrett, and M. S. Stekoll. In press a. Genetic variation in seasonal timing of anadromous migration in a population of pink salmon (*Oncorhynchus gorbuscha*). Canadian Special Publications in Fisheries and Aquatic Sciences, Ottawa, Ontario.

Smoker, W. W., A. J. Gharrett, M. S. Stekoll, and J. E. Joyce. In press b. Genetic analysis of size in an anadromous population of pink salmon. Canadian Journal of Fisheries and Aquatic Sciences 51(Supplement).

Stahl, G. 1983. Differences in the amount and distribution of genetic variation between natural populations and hatchery stocks of Atlantic salmon. Aquaculture 33:23–32.

Steward, C. R., and T. C. Bjornn. 1990. Supplementation of salmon and steelhead stocks with hatchery fish: a synthesis of published literature. Bonneville Power Administration, Project 88-100, part 2. Portland, Oregon.

Swain, D. P., and B. E. Riddell. 1990. Variation in agonistic behavior between newly emerged juveniles from hatchery and wild populations of coho salmon, *Oncorhynchus kisutch*. Canadian Journal of Fisheries and Aquatic Sciences 47:566–577.

Swain, D. P., and B. E. Riddell. 1991. Domestication and agonistic behavior in coho salmon: reply to Ruzzante. Canadian Journal of Fisheries and Aquatic Sciences 48:520–522.

Swain, D. P., B. E. Riddell, and C. B. Murray. 1991. Morphological differences between hatchery and wild populations of coho salmon (*Oncorhynchus kisutch*): environmental versus genetic origin. Canadian Journal of Fisheries and Aquatic Sciences 48:1783–1791.

Taylor, E. B. 1991. A review of local adaptation in Salmonidae, with particular reference to Pacific and Atlantic salmon. Aquaculture 98:185–207.

Templeton, A. R. 1986. Coadaptation and outbreeding depression. Pages 105–116 *in* M. E. Soule, editor. Conservation biology: the science of scarcity and diversity. Sinauer Associates, Sunderland, Massachusetts.

Utter, F., K. Hindar, and N. Ryman. 1993. Genetic effects of aquaculture on natural salmonid populations. Pages 144–165 *in* K. Heen, R. L. Monahan, and F. M. Utter, editors. Salmon aquaculture. Wiley, New York.

Utter, F., G. Milner, G. Stahl, and D. Teel. 1989. Genetic population structure of chinook salmon, *Oncorhynchus tshawytscha*, in the Pacific Northwest. U.S. National Marine Fisheries Service Fishery Bulletin 87:239–264.

Vincent, R. E. 1960. Some influences of domestication upon three stocks of brook trout (*Salvelinus fontinalis* Mitchell). Transactions of the American Fisheries Society 89:35–52.

Vuorinen, J. 1984. Reduction of genetic variability in a hatchery stock of brown trout, *Salmo trutta* L. Journal of Fish Biology 24:339–348.

Waldman, B., and J. S. McKinnon. 1993. Inbreeding and outbreeding in fishes, amphibians, and reptiles. Pages 250–282 *in* N. W. Thornhill, editor. The natural history of inbreeding and outbreeding. University of Chicago Press, Chicago, Illinois.

Waples, R. S. 1990. Conservation genetics of Pacific

salmon, part 2. Estimating effective population size and the rate of loss of genetic variability. Journal of Heredity 81:267–276.

Waples, R. S. 1991a. Genetic interactions between hatchery and wild salmonids: lessons from the Pacific Northwest. Canadian Journal of Fisheries and Aquatic Sciences 48:124–133.

Waples, R. S. 1991b. Pacific salmon, *Oncorhynchus* spp., and the definition of "species" under the Endangered Species Act. Marine Fisheries Review 53(3):11–22.

Waples, R. S., and P. E. Smouse. 1990. Gametic disequilibrium analysis as a means of identifying mixtures of salmon populations. American Fisheries Society Symposium 7:439–458.

Waples, R. S., and D. J. Teel. 1990. Conservation genetics of Pacific salmon, part 1. Temporal changes in allele frequency. Conservation Biology 4:144–156.

Washington, P. M., and A. M. Koziol. 1993. Overview of the interactions and environmental impacts of hatchery practices on natural and artificial stocks of salmonids. Fisheries Research 18:105–122.

Withler, R. E. 1988. Genetic consequences of fertilizing chinook salmon (*Oncorhynchus tshawytscha*) eggs with pooled milt. Aquaculture 68:15–25.

Wright, S. 1993. Fishery management of wild Pacific salmon stocks to prevent extinctions. Fisheries (Bethesda) 18(5):3–4.

Assessment of Season of Release and Size at Release on Recapture Rates of Hatchery-Reared Red Drum

SCOTT A. WILLIS, WILLIAM W. FALLS, CLYDE W. DENNIS, DANIEL E. ROBERTS, AND PETER G. WHITCHURCH

Florida Marine Research Institute, Florida Department of Environmental Protection
14495 Harllee Road, Port Manatee, Florida 34221, USA

Abstract.—Stock enhancement is a management tool that may be used to augment depleted wild populations of fish. Red drum *Sciaenops ocellatus* was used as a test species to evaluate the efficacy of marine stock enhancement in Florida. Fingerlings were marked with either coded wire tags or internal anchor tags. Three experiments were conducted to compare tag return rates from fingerlings released during different seasons and at different sizes, to develop guidelines for using internal anchor tags on larger fingerlings, and to assess the entry of hatchery-reared fish into the subadult population.

In the red drum size-at-release experiment, approximately 60,000 fingerlings were graded equally into three size-classes: 60, 90, and 120 mm mean total length (TL). The experiment consisted of two phases: (1) using fingerlings produced during the natural fall spawning period (in season); and (2) using fingerlings produced approximately 6 months later (out of season) by photothermal manipulation of broodstock to induce spawning. Two sampling designs were used at the study site: seine sampling at randomly selected stations and at fixed stations. Red drum recaptures from the released group that had been produced in season totaled 821; there was only one return from the released group that had been produced out of season. In random-station sampling there was no significant difference in recapture rates between size-classes. In fixed-station sampling there was a significant positive correlation between recapture rate and increasing fish size at release. The correlation was already established by day six, and the rates of recapture for the three size-classes remained unchanged during 6 months of poststocking sampling; large fish were 3.7 times more likely to be recaptured than were small fish.

In the second experiment, fingerlings were graded into two size-classes (157.1 and 201.7 mm mean TL), tagged with internal anchor tags, and released. Total returns for the two size-classes after 767 d were 52 (1.4%) and 189 (5.4%), respectively. Fish in the larger size-class were 3.9 times more likely to be returned than were smaller fish, and there was no significant difference between size-classes in either growth rate or net distance traveled postrelease.

In the third experiment, we used trammel nets to sample throughout Volusia County to determine the rate of entry of hatchery-reared fish into the local stocks. A total of 2% of the 395 red drums caught during trammel-net sampling (218–828 mm TL) were hatchery reared. Recapture rates localized in the area where fish were released were higher (4.2%; $N = 168$).

Historically, many people thought that the marine environment contained a limitless supply of marine organisms. In Florida, however, a number of marine fisheries stocks have been recognized as being in numerical decline during the past decade; these declining stocks include common snook *Centropomus undecimalis* (Bruger and Haddad 1986); spotted seatrout *Cynoscion nebulosus* (Laguna 1994); and red drum *Sciaenops ocellatus* (Goodyear 1991). Overfishing has been identified as the most likely factor leading to the red drum stock decline in the mid-1980s (Goodyear 1991). This decline prompted the 1987 closure of the commercial red drum fishery in federally regulated coastal waters and resulted in stricter sport-harvest rules in Florida (Johnson and Funicelli 1991).

Marine stock enhancement is a management tool that may have the potential to augment depleted wild populations of some fish species by supplementing or restoring the stocks. Using hatchery-propagated fish in early life stages to remedy declines in fish abundances is particularly attractive because many marine species produce very large numbers of offspring from a single mating. In nature, these fish generally experience high mortality in the early developmental stages. Hatchery culture, however, can mitigate larval mortality factors and increase early life stage survival dramatically. Technological advances now make it possible to produce large numbers of young animals in marine hatcheries and release them into the sea (Matsuoka 1989). Striped mullet *Mugil cephalus* (Leber et al., 1995, this volume; in press); Atlantic cod *Gadus morhua* (Svasand et al. 1990); striped bass *Morone saxatilis* (Dorazio et al. 1991); madai *Pagrus major* (Tsukamoto et al. 1989); and red drum (Rutledge 1989) are marine species that have been shown to have the potential for successful stock enhancement. To evaluate the efficacy of using marine stock enhancement as a tool to augment natural populations, carefully planned assessment studies must follow hatchery releases. Key components, identified by Blankenship and Leber (1995, this volume), that are needed to develop, evaluate, and manage marine enhancement programs have been incorporated into Florida's marine stock enhancement assessment program.

Marine stock enhancement is not a new concept. As early as the late-1800s, large numbers of Atlantic cod fry were released into Norwegian waters (Mac-Call 1989). Over a period of 90 years, more than 70 billion Atlantic cod yolk-sac larvae were released into the ocean, but no perceptible evidence of benefit to the population was ever discovered (Shelbourne 1965). Development of new sampling techniques, however, has created renewed interest in Atlantic cod stock enhancement (Svasand et al. 1990). Further interest in marine fish stock enhancement is being spurred by the emergence of new technologies for spawning and rearing fish and by the development of release and monitoring techniques that allow the effects of enhancement to be assessed. Florida, as well as other states (Blankenship and Leber 1995), has adopted a conservative and responsible approach to stock enhancement. To ensure that wild stocks are not adversely affected by the release of cultured animals, Florida Department of Environmental Protection (FDEP) biologists formulated a marine stocking policy that is in draft form (Futch and Willis 1992). All releases have been in accord with the policy requirements for broodstock genetic identification, independent fish health certification, and marking. In addition, FDEP biologists have developed fish health protocols to assess health during culture, before release, and after release. Recently, a 5-year genetics program was initiated to assess the effects that released fish have upon the genetic identities of wild stocks.

Red drum was selected to be a model species for stock enhancement assessment research in Florida for three principal reasons: the technology for culturing red drum had reached a point at which large numbers of fingerlings could be produced routinely in the hatchery (Roberts et al. 1978); there were perceived declines in the wild stock (Goodyear 1991); and the early life history of red drum, knowledge of which can aid in determining the time and place of release, was well documented (Yokel 1966; Peters and McMichael 1987; Daniel 1988). We determined the short-term goals for assessing stock enhancement by conducting a series of pilot studies as well as structured experiments. These pilot studies and experiments, begun in 1988 and still in progress, were designed to (1) identify and characterize suitable release sites; (2) standardize experimental designs, sampling gear, and sampling procedures; (3) evaluate and refine tagging, transport, and assessment technology; (4) determine the residency time of hatchery-reared juveniles at the release site, their dispersal rates, and their short-term survival; and (5) assess temporal and spatial relationships between hatchery-reared and wild red drum.

Three experiments that were conducted by FDEP biologists are included in this paper. A size-at-release experiment in which the coded wire tag (CWT) was used as the marking technique was designed to yield information on the optimum release size and release season for small fingerlings. A size-at-release experiment in which the internal anchor tag (IAT) was used as the marking technique for larger-sized fingerlings was designed to establish fish growth and dispersal rates and to determine the optimum fish size for use with this tag type. In the third experiment, we used trammel nets to sample subadults and thus assess the entry of hatchery-reared fish into the local wild stocks.

All of the fish released in the three experiments were marked with either CWTs or IATs so that we could evaluate their growth, survival, migration, and subsequent contribution to the wild population.

Tag-return studies can be classified into two main categories: fishery independent and fishery dependent. Each type of study has its own constraints and advantages, uses different methodologies, and gathers different types of information (Wydoski and Emery 1983). In fishery-independent studies, the tagged fish are recovered by biologists, data can be directly correlated to structured assessment experiments, analysis of tag return information can yield very accurate results, and tags do not have to be visible. The tag that was selected for use in the fishery-independent portion of the project was the coded wire tag (Jefferts et al. 1963). The CWT is a stainless steel wire, approximately 1 mm long and 0.254 mm in diameter. The wire contains four rows of notches in a binary code that identifies the batch of fish being marked. The wire is cut from a roll, magnetized, and injected into the fish. The tags are detected during field sampling by passing the captured fish in front of a detector, which identifies the magnetic field associated with the tag.

Coded wire tags, placed into the nasal cartilage of salmonids, have been successfully used for several decades as a mark to differentiate hatchery-reared fish from wild stock (Bergman et al. 1968; Buckley and Blankenship 1990). Many nonsalmonid fish, including red drum, have shown poor CWT retention when the tag was placed into nasal cartilage (Gibbard and Colura 1980). Selection of the fish's cheek musculature as an alternate tag site has been attempted with generally good results in several other species (Klar and Parker 1986; Dorazio et al. 1991; Leber, in press). Bumgardner et al. (1990) reported low rates of CWT retention in red drums; however, they tagged very small fish, and Bumgardner et al. (1992) speculated that horizontal placement of the tag into the cheek musculature of red drums may have contributed to the low rates of retention. We have conducted short-term retention studies (approximately 4 h post-tagging) on subsamples of all red drums tagged at our facility, and the mean tag retention rate has been 97.3% for the more than 14,600 fish examined between 1990 and 1993. In addition, the results of a 168-d, CWT retention experiment (S. A. Willis et al., FDEP, unpublished data) yielded a tag retention rate of 95.3%.

Fishery-dependent studies depend on the recovery of tags by the fishing public, whose large numbers potentially increase the sampling effort. Tags used in these studies should be highly visible, the return mechanism must be uncomplicated, and a reward system should be used to assure a high return rate (Nielsen 1992). The tag that was selected for use by FDEP staff was an internal anchor tag. The IAT consists of a flat, anchoring base plate and an external streamer. To implant the tag, an incision is made in the lower wall of the fish abdomen. The base plate is inserted into the body cavity, and the streamer extends through the body wall and is externally visible. The combination of an internal base plate and an external streamer prevents the tag from being rejected or migrating to other areas of the body. Fish must be of sufficient size to carry this relatively large tag satisfactorily.

Methods

Experiment 1: CWT Size at Release

The primary release site was Murray Creek, a tidally influenced creek near Ponce Inlet in central Volusia County (Figure 1). This site was selected based upon habitat suitability indices previously established for juvenile red drum (Buckley 1984; Peters and McMichael 1987). This site has consistently yielded young-of-the-year red drums during pilot studies and contains three distinct areas: the creek proper, which is about 2 ha; an upstream area that is separated from the creek itself by a culvert and is tidally influenced only during storm surges; and a large, shallow borrow pit, about 10 ha in size, that is connected to Murray Creek by a narrow (about 2-m-wide), tidally influenced stream. Murray Creek was the first release site selected in Volusia County, and numerous red drums had already been released there, principally in pilot studies that helped us refine field sampling techniques.

In the size-at-release experiment, we released two separate groups of fish: fingerlings that were produced during the natural fall spawning period (in season) and those that were produced during an artificially induced spawning period approximately 6 months later (out of season). Fish were mechanically graded into three size-classes with mean total length (TL) of approximately 60, 90, and 120 mm. Each size-class consisted of approximately 20,000 fish, and size-classes were differentiated by CWTs injected into the left cheek musculature of each fish (Table 1). All fish were released into the stream that joins the borrow pit with Murray Creek proper.

We sampled at both fixed and randomly selected stations during this experiment. The five fixed stations were selected because they had consistently provided catches of both wild and hatchery-reared red drums during the pilot studies. Two sites (CR1 and CR2) are located in the lower reaches of Murray Creek proper, one site (BP1) is directly adjacent to the borrow pit, and two sites (FW1 and FW2) are in the upper reaches of Murray Creek (Figure 1). Sites FW1 and FW2 are located upstream of a road culvert and are tidally influenced only during flood tides and storm surges. Therefore, this area, which receives freshwater from overland drainage, tends to remain relatively lower in salinity than all other sampling sites.

The randomly selected sampling stations were delineated by dividing Murray Creek, the borrow pit, and the adjacent area to the north of the creek into four sampling zones (Figure 1). The zones are somewhat natural divisions based upon physical parameters, but each area contains a mixture of habitat types. Each zone was then subdivided into 100 ft × 100 ft grids. Zone 1 (88 grids) contains the borrow pit area. Zone 2 (89 grids) consists of Murray Creek proper and includes the low-salinity area in its upper reaches. Zone 3 (100 grids) includes the area northwest of Murray Creek and extends to the junction of the south fork of Spruce Creek. Zone 4 (108 grids) is the area to the northeast of Murray Creek. In each of the four zones, we sampled three randomly selected grids with a beach seine. The seine was deployed in a circle, and the entire grid was encompassed by the net. Sampling was generally

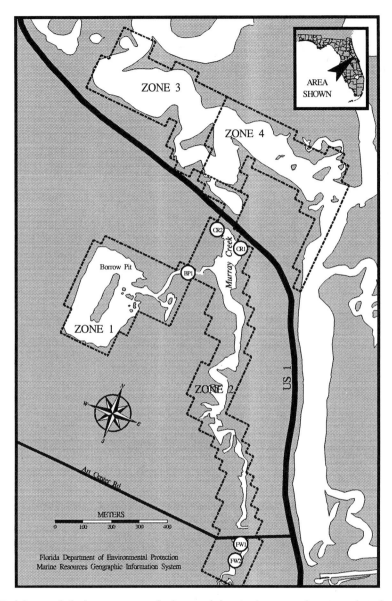

FIGURE 1.—Red drum coded-wire-tag, season-of-release and size-at-release experiments conducted at Murray Creek, Volusia County, Florida. Fixed-station sites are designated by symbols CR1, CR2, BP1, FW1, and FW2. Random-station sites were within zones 1, 2, 3, and 4.

completed in 2 d, took place between 0800 and 1800 hours, and was accomplished biweekly on full and new moon phases each month. To minimize diurnal bias due to sampling time and changing tidal stage, we randomly reordered the sampling sequence of both the random and fixed stations. All captured red drums were enumerated and measured, and all tagged red drums were sacrificed for identification. We also measured dissolved oxygen, pH, temperature, salinity, and conductivity at each site.

Experiment 2: IAT Size at Release

Fish were hand graded into size-classes smaller and larger than 175 mm TL and tagged with IATs. Subsamples of 225 small and 285 large fish were

TABLE 1.—Size (mean ± SD), number measured, and total number of red drums released in the coded-wire-tag, size-at-release experiment. Fish were produced during two seasons, and all fish were released at Site BP1 (see Figure 1). Size-classes are described in text.

Fish size-class	Total length (mm)	Number measured	Total released
Fall 1992[a]			
Small	62.4 (±7.5)	175	21,857
Medium	83.6 (±10.1)	200	18,231
Large	118.6 (±12.6)	251	21,971
Spring 1993[b]			
Small	65.9 (±8.4)	300	15,785
Medium	83.3 (±11.3)	333	21,297
Large	116.0 (±11.6)	333	14,490

[a] In-season release dates were 8–18 March 1993.
[b] Out-of-season release dates were 20 August–2 September 1993.

randomly selected from each size-class to determine the mean fish length for each size-class. The total numbers of fish released in each of the two size-classes were 3,780 fish, with a mean TL of 157.1 ± 12.0 mm (range 131–183 mm), and 3,519 fish, with a mean TL of 201.7 ± 24.6 mm (range 166–375 mm; only 3 measured fish were longer than 283 mm). The fish were transported approximately 250 km to Wilbur Bay in Volusia County, acclimated to ambient conditions, and released after dark. Wilbur Bay is a small embayment adjacent to the Halifax River, is surrounded by numerous small islets, and is located about 6 km north of Murray Creek (Figure 2).

Anglers would be returning the tags from these hatchery-reared fish, making the results fishery dependent. To encourage anglers to return tags, we distributed more than 250 permanent reward signs outlining project goals and the procedure for reporting data on recaptured fish. The reward for reporting tag-related fishery information is a t-shirt that has a replica of the program's reward sign on the back. Requested recapture information includes capture location, fish size and weight, and tag number. Information reported by anglers has provided data on estimates of fish growth and distances traveled as well as on recapture rates for the two sizes of fish released.

Experiment 3: Subadult Survey

Trammel-net sampling of subadult red drums was used to determine if hatchery fish are entering the local fishery. The trammel net that we used has an inner panel of 7.6-cm stretch mesh and an outer panel of 12.7-cm stretch mesh. It is 600 m long by 2.4 m high. Monthly trammel-net sampling (3 d/month) began in November 1992 and continues. We made sets in open-water areas, generally deeper than 1.8 m, from Tomoka Basin in the northern portion of the county to Mosquito Lagoon in the south, a distance of approximately 70 km. All red drums captured were measured (TL) and checked for tags (CWT or IAT). Trammel-net results are reported through 10 February 1994.

Results

Experiment 1: CWT Size at Release

During the first 125 d following the release of the group produced in season (fall 1992), a total of 821 hatchery-reared red drums were recaptured during random- and fixed-station sampling. Of the 821 fish recovered, 426 were recaptured during random-station sampling from 120 seine hauls, and 395 fish were recaptured during fixed-station sampling from 50 seine hauls. Analyses do not include adjustments for estimated tag loss (approximately 3%) or fish loss due to handling.

During the 6 months we sampled after the release of the group produced out of season (spring 1993) only one fish was recaptured. The recovered fish was from the large size-class; however, because it represented less than 0.01% of the fish released, this fish was not included in further analyses.

In random-station sampling of the fish that had been produced in season, significantly fewer fish were recaptured in zones 3 and 4 (7 fish) than were captured in zones 1 and 2 (419 fish); therefore, zones 3 and 4 were not included in the analyses. Zones 3 and 4 are downstream extensions of the original nursery area sampled during pilot studies and were added in an attempt to determine fish distribution beyond the creek proper. Random-station sampling (in zones 1 and 2) data show that there was little difference in the number of fish recovered from each of the three different size-classes of fish released (Table 2). The slope of a regression of the log of the number of fish recaptured on the days postrelease is the average daily mortality rate. This mortality rate, however, includes not only mortality but also emigration of fish from the assessment sampling area and losses due to nonreplacement sampling. An analysis of covariance (ANCOVA; Zar 1984) for random-station sampling in which sampled fish were not replaced in zones 1 and 2 compared regressions by the numbers of fish recaptured in the three size-classes on the number of days the fish had been at large when recaptured. The ANCOVA indicated that there was no significant difference in the mortality rates

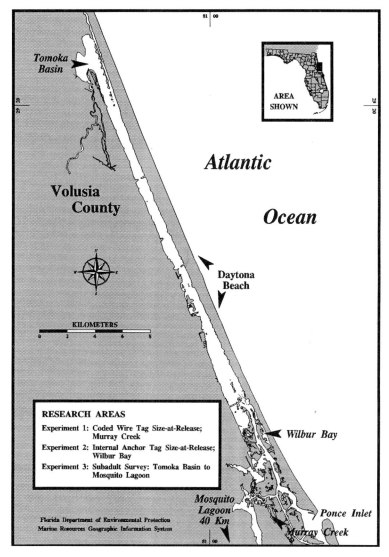

FIGURE 2.—Red drum stock enhancement assessment research areas in Volusia County, Florida.

among size-classes ($F = 0.06$; df = 2,24; $P = 0.05$; Figure 3).

The total number of fish recaptured during fixed-station sampling was 63, 117, and 215 for the small, medium, and large size-classes, respectively. Analysis of covariance for fixed stations also showed that there were no significant differences among the mortality rates for the three size-classes of fish ($F = 0.26$; df = 2,16; $P = 0.05$). There was, however, a significant difference in the elevation between each of the regression lines ($F = 15.09$; df = 2,18; $P < 0.05$), indicating that more larger fish were recaptured than were smaller fish (Figure 4). This difference between size-classes in the number of fish recaptured in fixed-station sampling was established by the first day of postrelease sampling (day 6) and continued to be evident during most of the biweekly sampling periods for the balance of the experiment (Table 2).

Comparisons of the mean lengths of wild red drums that were captured during both fixed- and random-station sampling also show that more large fish were collected at fixed stations than at random stations during the same sampling periods (Table 3). Several factors may be contributing to higher numbers of fish collected at fixed sites: fixed sites

TABLE 2.—Number of coded-wire-tagged red drums recaptured at fixed (5 samples) and random (12 samples) stations during biweekly sampling in the Murray Creek area. Size-classes are small (S), medium (M), and large (L) as described in text.

Days at large	Fixed stations				Random stations			
	S	M	L	Total	S	M	L	Total
6	23	33	66	122	38	85	84	207
14	25	49	85	159	3	8	12	23
28	7	21	33	61	90	28	11	129
41	4	7	14	25	3	0	1	4
53	3	1	5	9	8	9	8	25
67	1	4	6	11	4	5	4	13
83	0	2	4	6	4	3	5	12
98	0	0	1	1	1	0	10	11
112	0	0	1	1	0	1	0	1
125	0	0	0	0	1	0	0	1
Total	63	117	215	395	152	139	135	426

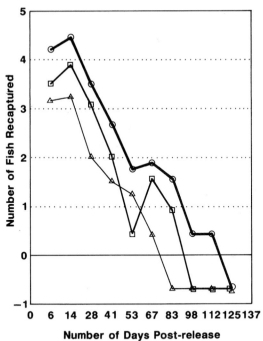

FIGURE 4.—Red drum recapture data for the coded-wire-tag, size-at-release experiment (fall 1992 production season release). The graph shows the number of fish recaptured at fixed stations on days postrelease (0.5 was added to each fish recapture number to allow zero recaptures to be plotted). Size-classes are small, medium, and large as described in text.

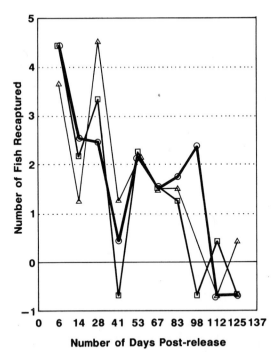

FIGURE 3.—Red drum recapture data for the coded-wire-tag, size-at-release experiment (fall 1992 production season release). The graph shows the number of fish recaptured at random stations on days postrelease (0.5 was added to each fish recapture number to allow zero recaptures to be plotted). Size-classes are small, medium, and large as described in text.

are optimal red drum habitat and larger fish preferentially selected these areas; once established in the preferred habitat, the larger fish may exclude the smaller fish from these areas through intraspecific competition; and fish in an optimal habitat grow faster and larger than do cohorts in other areas.

Hatchery-released red drums resided in Murray Creek in relatively high numbers through about days 67–83 postrelease, after which the recovery of hatchery-reared and wild red drums during random- and fixed-station sampling decreased dramatically (Tables 2 and 3). On day 67 after their release, hatchery-reared red drums captured at random and fixed stations had mean TL of 118mm ($N = 13$) and 145mm ($N = 11$), respectively, and the wild red drums captured at both types of stations a mean TL of 138mm ($N = 5$). The total lengths of these fish are consistent with the lengths at which wild red

TABLE 3.—Size and number of wild red drums captured at fixed (5 samples) and random (12 samples) stations during biweekly sampling in the Murray Creek area.

Day at large	Fixed stations		Random stations	
	Mean total length (mm)	N	Mean total length (mm)	N
6	102.8	15	78.5	25
14	126.7	27	123.5	2
28	132.0	9	90.1	11
41	144.0	4	90.0	4
53	155.8	4	136.8	14
67	140.3	3	134.0	2
83	182.0	2	185.5	2
98	206.0	1	0	0
112	0	0	0	0
125	0	0	0	0

drum have been shown likely to move to a different habitat (Peters and McMichael 1987; Daniel 1988).

There were also certain stations, both fixed and random, that offered preferred habitat. At fixed-station CR2, a large number of fish were recaptured during a single sample (75 fish on day 6), and hatchery fish were recaptured there during every biweekly sampling period throughout the experiment. Many fish were also recaptured during single samples at two random stations: 142 fish at station 139 (day 6) and 63 fish at station 142 (day 28). These stations are located nearly adjacent to each other in the middle of Murray Creek proper. The size structure of the fish sampled at these two sites might be expected to be similar because these sets were sampled only 22 d apart, but it was in fact quite different. The ratio of fish in small:medium:large size-at-release classes that were recaptured at station 139 (6 d) was 1:4:3, and at station 142 (28 d) was 25:8:1. Such a change in size-class composition is difficult to explain. These differences in stock size structure as well as differences in results between random- and fixed-station data may be related to the differential use of habitat over time.

Although the fish recaptured during random-station sampling showed no difference between size-classes, 3.7 times as many fish in the large size-class as in the small size-class were recaptured during the fixed-station sampling. Preliminary data gathered from the trammel-net survey of subadults (Table 4) showed that of the seven CWT fish that were recaptured during sampling, six (86%) were from the larger size-classes which were released during pilot studies. Although this is a very small data set, it supports the fixed-station, size-at-release data indicating that more fish from the larger size-classes may survive than do smaller fish.

Experiment 2: IAT Size at Release

As of 1 December 1993, 241 of the red drums released 15–25 October 1991 (a period of 767 d), had been recaptured (Table 5). Fish from the larger size-class were 3.9 times more likely to be returned as were those from the smaller size-class.

All fish recaptures were classified by time period blocks of 100 d at large, and the two size-classes were compared for mean growth rate (Table 6) and mean distance traveled (Table 7). Analyses of variance indicated that there were no significant differences in either mean growth rate ($F = 0.167$; df = 1,7; $P = 0.05$) or mean distance traveled ($F = 0.008$; df = 1,7; $P = 0.05$) between the fish in the two size-classes.

The first-year growth of the IAT hatchery-reared fish (0.93 mm/d) was comparable to that reported by Simmons and Breuer (1962) and Peters and McMichael (1987) for wild red drum growth (0.83 mm/d). The first-year growth rate of 0.93 mm/d was calculated from lengths of IAT fish reported by anglers (0–200 d post release; Table 6) and lengths at the hatchery prior to release (165 d).

Experiment 3: Subadult Sampling

A total of 395 red drums ranging in size from 218 mm to 828 mm TL were captured in 110 trammel-net sets throughout Volusia County, from Tomoka Basin in the north to Mosquito Lagoon in the south (Table 4). The eight tagged fish recaptured represent 2% of the total number of red drums captured.

TABLE 4.—Results of subadult red drum trammel-net survey for Volusia County, Florida. Survey conducted from 17 November 1992 through 10 February 1994.

	Total fish captured			Tagged fish recaptured			
		Total length				Total length	
Area of county	N	Mean (mm)	Range (mm)	N	Mean (mm)	Range (mm)	Percent tagged fish
Murray Creek	168	403.0	280–758	7	354.0	280–549	4.2
North and south	227	384.7	218–828	1	586		0.4
Total	395	392.5	218–828	8	383.0	280–586	2.0

TABLE 5.—Release and recapture data for red drums released in the internal-anchor-tag, size-at-release experiment. Release was during 15–25 October 1991, and recaptures were reported for the period 25 October 1991 through 1 December 1993.

Mean total length at release (mm)	Fish released (N)	Fish recaptured (N)	Fish recaptured (%)	Mean distance traveled (km)	Mean growth (mm/d)	Mean days at large
157.1	3,780	52	1.4	26.1	0.62	348
201.7	3,519	189	5.4	25.4	0.71	291

Trammel-net sets in the area of Murray Creek, where most of the hatchery fish had been released, yielded a total of 168 red drums, 7 of which (4.2%) were tagged with CWTs. The presence of hatchery-reared fish should continue to increase because none of the approximately 120,000 CWT red drums from the size-at-release experiment should have grown sufficiently since their release to have entered this fishery.

Discussion

Although we saw no difference between the three size-classes in the total number of fish recaptured in the random-station sampling conducted during the CWT size-at-release experiment, we did see significant differences between size-classes in total number of fish recaptured in the fixed-station sampling. These differences occurred after release but prior to day 6 and remained consistent throughout the experiment. Because a difference in apparent success was established so soon after stocking and the degree of difference between size-classes did not change through time, fishery managers could use this information to predict long-term success of hatchery releases of red drum by concentrating assessment sampling efforts soon after release. A size-at-release effect has also been demonstrated by Leber (in press) who worked with hatchery-reared striped mullet (45–120 mm TL). His analyses suggested that the success of different size-classes is directly related to season and that hatchery contributions can result in as much as a four fold increase in hatchery fish contribution if releases are properly timed.

Less than 0.01% of the red drum fingerlings released in summer (produced out of season in the spring) were recaptured. Several factors could have contributed to the poor recapture rate for these fish. It is possible that some of the fish rapidly emigrated soon after they were released and moved to another habitat out of the sampling area. These fish may yet be recovered in future subadult sampling. Fewer fish of other species were captured during the out-of-season study period than were captured during other portions of the year. The lower numbers of fish may be related to high summer temperatures, which can cause daily temperature fluctuations in shallow backwater areas and therefore make these areas less viable as fish habitat.

In pilot studies, red drums that were produced in the spring (out of season), held in ponds at the

TABLE 6.—Mean growth rate (mm/d) of red drums in two size-classes used in the internal-anchor-tag size-at-release experiment. Numbers in parentheses are numbers of fish recaptured during 100-d time periods. Mean total lengths of fish in the small and large size-classes were 157.1 ± 12 mm ($N = 225$) and 201.7 ± 24.6 mm ($N = 285$), respectively.

Time periods (d) from release to recapture	Growth rate (mm/d)	
	Small	Large
0–100	0.91 (4)	1.18 (30)
101–200	0.70 (5)	0.66 (23)
201–300	0.66 (14)	0.72 (51)
301–400	0.56 (9)	0.57 (42)
401–500	0.56 (8)	0.51 (16)
501–600	0.53 (7)	0.44 (6)
601–700	0.42 (2)	0.46 (12)
701–800	0.55 (2)	0.52 (5)

TABLE 7.—Mean net distance traveled by red drums in the two size-classes used in the internal-anchor-tag size-at-release experiment. Numbers in parentheses are the number of fish recaptured during the 100-d time periods. Mean total lengths of fish in the small and large size-classes were 157.1 ± 12.0 mm ($N = 225$) and 201.7 ± 24.6 mm ($N = 285$), respectively.

Time periods (d) from release to recapture	Net distance traveled (km)	
	Small	Large
0–100	20.2 (4)	18.0 (31)
101–200	9.1 (5)	24.8 (25)
201–300	26.9 (15)	26.7 (52)
301–400	25.4 (9)	24.1 (42)
401–500	21.3 (8)	25.2 (16)
501–600	30.0 (7)	32.8 (6)
601–700	64.5 (2)	25.0 (12)
701–800	46.5 (2)	64.7 (5)

hatchery through the summer, and then harvested, tagged, and released in the fall were recaptured at considerably higher rates than were fish produced similarly but released in the summer (S. A. Willis, FDEP, unpublished data). Although this strategy of holding fish produced out of season and delaying their release until the fall results in higher red drum recapture rates, the fish are being released approximately one-half year out of phase with wild cohorts. Because these delayed-release fish are considerably larger at the time of release than are the wild young-of-the-year red drums that are normally recruited in the fall, intraspecific competition may occur. Releasing hatchery-reared fish out of phase with wild stocks may also result in interspecific competition because red drums released out of season may interact with all other species that are normally using the nursery areas and compete for limited space, food, and territory.

The decline in the number of red drums recaptured within the study site during the CWT size at release experiment was due to a combination of factors, including mortality, emigration, and loss due to nonreplacement sampling. Emigration may be a major contributing factor because the number of red drums recaptured in the study area rapidly decreased just as the fish attained the size at which wild red drums have been shown to move normally from nursery areas. Daniel (1988) found that in coastal South Carolina smaller juveniles apparently remain in marsh areas until they are about 150 mm TL, and then they move into river channels and lower harbors.

Hatchery-reared red drums mixed with similarly sized wild red drums after release. Patchy distribution and schooling behavior were compared between random stations in zones 1 and 2 and fixed stations (all fixed stations lie within the bounds of zones 1 and 2). Patchy red drum distribution was evident: only 27 of the 60 random stations (45%) and 29 of the 50 fixed stations (58%) sampled contained hatchery fish. Wild red drums were captured in only 33 of the 110 random and fixed stations sampled (30%). The association of hatchery fish with wild fish was very high: of the 33 stations where wild red drums were caught, 29 (88%) also contained hatchery fish. Further analysis of the data by identification of microhabitat within zones in relation to red drum recapture rates may yield more complete answers concerning emigration, schooling, and association of hatchery-reared with wild red drums.

The IAT size-at-release experiment illustrates that release of large, hatchery-produced juveniles (201.7 mm TL) contributes to the public fishery. A total of 5.4% of the large size-class of IAT fish were recaptured by the public (Table 5). Yeager (1988) also demonstrated a size-related benefit in northern Florida: striped bass that were released at 150–250 mm TL were returned 100 times more often than were striped bass that were released at 30–45 mm TL. Analyses also indicated that there was no significant difference in growth rate between the two size-classes of fish released and that the overall growth rate of hatchery-reared fish was similar to that reported for wild red drums.

During the IAT size-at-release experiment, only 18 of the 241 IAT returns were from areas to the south of the release site, and all southerly returns were from areas close to the release site (within 10–12 km). Several interrelated factors may be contributing to the distribution of IAT red drums in the general area to the north of the release site (Figure 2). One factor may be that the direction of water flow in the Halifax River is to the north, although water movement is slow and strongly influenced by tides. Another factor may be that the only input of oceanic water into the Halifax River system in this area is through Ponce Inlet, located about 8 km to the south of the release site. Tomoka Basin, approximately 25 km north of the release site, contributed 96 recaptured fish (40%) and also has a large freshwater input, with salinity generally around 15–20 ppt. Therefore, lower salinity, which has been shown to be important for red drum movement (Yokel 1966), is probably an important factor in the location of recapture. Another factor may simply be that fishing pressure is greater in the more northern portions of the county. The city of Daytona Beach is located on the barrier island near the site where 29 fish were recaptured (Figure 2). These 29 recaptures were made near a bridge that connects the city with the mainland.

Even with several factors influencing movement of fish, hatchery-reared red drums remained within 25 km of the release site during the 767 d postrelease. Osburn et al. (1982) and Murphy and Taylor (1990) also reported low emigration rates from estuarine to nearshore areas for subadult red drums.

Detecting the entry of hatchery-released fish into the public fishery is one of the main goals of this project. Subadult sampling shows that some of the hatchery fish that were released in the early years of the program are now reaching legal size and are near the reproductive size described by Murphy and Taylor (1990). Preliminary analysis of subadult, trammel-net sampling data from November 1992 through February 1994 indicates that hatchery contributions to one local stock of subadult red drums are at least 4.2% and have the potential to be much higher. The criteria for long-term success are (1) entry of hatchery fish into subadult stocks, (2) attainment of reproductive status by these same fish, and (3) contribution of offspring by these hatchery fish to the wild population. The first criterion has been met, the second should soon be met, and the confirmation of the third is the goal of a recently initiated 5-year genetic project that will provide allozyme and DNA analyses to show if hatchery fish are contributing offspring.

Acknowledgments

The authors want to acknowledge the hard work and dedication of the entire hatchery crew, without whom this project would not have been possible. We also want to thank the field crew, Greg White, Dale Cronin, and Kent Benken, for their dedication during difficult and often uncomfortable work conditions. Lee Blankenship of the Washington Department of Fisheries and Frank Haw and Peter Bergman of Northwest Marine Technology, Inc., provided valuable information on coded wire tag protocol. Lee Blankenship and Ken Leber provided critical project review and valuable direction of research efforts. Robert Muller provided guidance for, and review of, statistical analyses. Cartography was provided by Henry Norris. Excellent editorial review was provided by Ruth Reese, James Quinn, and Judy Leiby of the FDEP, Cynthia Dohner of the U. S. Fish and Wildlife Service, and Richard Eades of the Virginia Department of Game and Inland Fisheries.

References

Bergman, P. K., K. B. Jefferts, H. F. Fiscus, and R. L. Hager. 1968. A preliminary evaluation of an implanted coded wire fish tag. Washington Department of Fisheries Research Paper 3:63–84.

Blankenship, H. L., and K. M. Leber. 1995. A responsible approach to marine stock enhancement. American Fisheries Society Symposium 15:167–175.

Bruger, G. E., and K. D. Haddad. 1986. Management of tarpon, bonefish, and snook in Florida. Proceedings of a Symposium on Marine Recreational Fisheries 11:53–57. Sport Fishing Institute, Washington, DC.

Buckley, J. 1984. Habitat suitability index models: larval and juvenile red drum. U.S. Fish and Wildlife Service FWS/OBS-82/10.74.

Buckley, R. M., and H. L. Blankenship. 1990. Internal extrinsic identification systems: overview of implanted wire tags, otolith marks, and parasites. American Fisheries Society Symposium 7:173–182.

Bumgardner, B. W., R. L. Colura, and A. F. Maciorowski. 1990. Tag retention, survival, and growth of red drum fingerlings marked with coded wire tags. American Fisheries Society Symposium 7:286–292.

Bumgardner, B. W., R. L. Colura, and G. C. Matlock. 1992. Long-term coded wire tag retention in juvenile *Sciaenops ocellatus*. U.S. National Marine Fisheries Service Fishery Bulletin 90:390–394.

Daniel, L. B. III. 1988. Aspects of the biology of juvenile red drum, *Sciaenops ocellatus*, and spotted seatrout, *Cynoscion nebulosus* (Pisces: Sciaenidae), in South Carolina. Master's thesis. College of Charleston, Charleston, South Carolina.

Dorazio R. M., B. M. Florence, and C. M. Wooley. 1991. Stocking of hatchery-reared striped bass in the Patuxent River, Maryland: survival, relative abundance, and cost-effectiveness. North American Journal of Fisheries Management 11:435–442.

Futch C. R., and S. A. Willis. 1992. Characteristics of the procedures for marine species introductions in Florida. Pages 59–60 *in* M. R. DeVoe, editor. Introductions and transfers of marine species. South Carolina Sea Grant Consortium, Charleston, South Carolina.

Gibbard, G. L., and R. L. Colura. 1980. Retention and movement of magnetic nose tags in juvenile red drum. Annual Proceedings of the Texas Chapter, American Fisheries Society 3:22–29.

Goodyear, P. G. 1991. Status of red drum stocks in the Gulf of Mexico. National Marine Fisheries Service, Southeast Fisheries Center, Coastal Resources Division Contribution MIA-90/91-87, Miami, Florida.

Jefferts, K. B., P. K. Bergman, and H. F. Fiscus. 1963. A coded wire identification system for macro-organisms. Nature (London) 198:460–462.

Johnson, D. R., and N. A. Funicelli. 1991. Spawning of the red drum in Mosquito Lagoon, east-central Florida. Estuaries 14:74–79.

Klar, G. T., and N. C. Parker. 1986. Marking fingerling striped bass and blue tilapia with coded wire tags and microtaggants. North American Journal of Fisheries Management 6:439–444.

Laguna, J. 1994. Spotted seatrout (*Cynoscion nebulosus*): background report and review of management alternatives. Florida Marine Fisheries Commission Report. March 4, 1994.

Leber, K. M. In press. Significance of fish size-at-release on enhancement of striped mullet fisheries in Hawaii. Journal of the World Aquaculture Society.

Leber, K. M., N. P. Brennan, and S. M. Arce. 1995. Marine enhancement with striped mullet: are hatchery releases replenishing or displacing wild stocks? American Fisheries Society Symposium 15:376–387.

Leber, K. M., D. A. Sterritt, R. N. Cantrell, and R. T. Nishimoto. In press. Contribution of hatchery released striped mullet, *Mugil cephalus*, to the recreational fishery in Hilo Bay, Hawaii. Hawaii Department of Land and Natural Resources, Division of Aquatic Resources-Technical Report 94-03.

MacCall, A. D. 1989. Against marine fish hatcheries: ironies of fishery politics in the technological era. California Cooperative Fishery Institute Report 30:46–48.

Matsuoka, T. 1989. Japan sea-farming association (JASFA).

International Journal of Aquaculture and Fisheries Technology 1:90–95.

Murphy M. D., and R. G. Taylor. 1990. Reproduction, growth, and mortality of red drum (*Sciaenops ocellatus*) in Florida waters. U.S. National Marine Fisheries Service Fishery Bulletin 88:531–542.

Nielsen, L. A. 1992. Methods of marking fish and shellfish. American Fisheries Society Special Publication 23.

Osburn, H. R., G. C. Matlock, and A. W. Green. 1982. Red drum (*Sciaenops ocellatus*) movements in Texas bays. Contributions in Marine Science 25:85–97.

Peters, K. M., and R. H. McMichael, Jr. 1987. Early life history of the red drum, *Sciaenops ocellatus* (Pisces: Sciaenidae), in Tampa Bay, Florida. Estuaries 10(2): 92–107.

Roberts, D. E. Jr., B. V. Harpster, and G. E. Henderson. 1978. Conditioning and induced spawning of the red drum (*Sciaenops ocellatus*) under varied conditions of photoperiod and temperature. Proceedings of the World Aquaculture Society 9:311–332.

Rutledge, W. P. 1989. The Texas marine hatchery program—it works! California Cooperative Fishery Institute Report 30:49–52.

Shelbourne, J. E. 1965. Rearing marine fish for commercial purposes. California Cooperative Fishery Institute Report 10:53–63.

Simmons E. G., and J. P. Breuer. 1962. A study of redfish, *Sciaenops ocellata* Linnaeus, and black drum, *Pogonias cromis* Linnaeus. Publications of the Institute of Marine Science University of Texas 8:184–211.

Svasand, T., K. E. Jorstad, and T. S. Kristiansen. 1990. Enhancement studies of coastal cod in western Norway, part 1. Recruitment of wild and reared cod to a local spawning stock. Journal du Conseil International pour l'Exploration de la Mer 47:5–12.

Tsukamoto, K., and six coauthors. 1989. Size-dependent mortality of red sea bream, *Pagrus major*, juveniles released with fluorescent otolith-tags in News Bay, Japan. Journal of Fishery Biology 35(Supplement A): 59–69.

Wydoski, R., and L. Emery. 1983. Tagging and marking. Pages 215–237 *in* L. A. Nielsen and D. L. Johnson, editors. Fisheries techniques. American Fisheries Society, Bethesda, Maryland.

Yeager, D. M. 1988. Evaluation of Phase I and Phase II hybrid striped bass in the Escambia River, Florida. Proceedings of the Annual Conference Southeastern Association of Fish and Wildlife Agencies 41(1987): 41–47.

Yokel, B. J. 1966. A contribution to the biology and distribution of the red drum, *Sciaenops ocellata*. Master's thesis. University of Miami, Miami, Florida.

Zar, J. H. 1984. Biostatistical analysis, 2nd edition. Prentice-Hall, Englewood Cliffs, New Jersey.

The Effect of Hatcheries on Native Coho Salmon Populations in the Lower Columbia River

THOMAS A. FLAGG, F. WILLIAM WAKNITZ, DESMOND J. MAYNARD,
GEORGE B. MILNER, AND CONRAD V. W. MAHNKEN

National Marine Fisheries Service, Northwest Fisheries Science Center
2725 Montlake Boulevard East, Seattle, Washington 98112, USA

Abstract.—In May 1990, the National Marine Fisheries Service (NMFS) was petitioned to list lower Columbia River (LCR) coho salmon *Oncorhynchus kisutch* under the U.S. Endangered Species Act (ESA). The NMFS' biological status review of this petition determined that out-of-basin transplants, releases of an average 40,000,000 hatchery fish per year and harvest rates often exceeding 90% had changed the native LCR coho salmon runs to a genetic mixture of native and hatchery stocks. The NMFS Biological Review Team concluded that the available data failed to identify an existing evolutionarily significant unit of coho salmon in the LCR, and NMFS recommended that the fish not be listed under ESA. This paper reviews the basis for the NMFS' decision. In addition, new data is presented to suggest causal relationships between hatchery practices and the over 10-fold reduction in wild spawner densities observed over the last 30 years. These hatchery practices include (1) hatchery selection for early spawn timing, rendering the fish maladapted for establishing self-reproducing populations; (2) pervasive fry stockings of up to seven times carrying capacity, displacing wild fish; and (3) planting of fry over two times the size of wild counterparts, placing wild fish at a competitive disadvantage. Efforts to restore viable self-sustaining LCR coho salmon stocks should include reductions in harvest to a level that will allow naturally spawning populations to rebuild, restrictions on interhatchery transfer of fish to allow stocks to develop watershed-specific characteristics, and termination of planting nonmigratory juveniles. Establishment of conservation hatcheries that produce smolts with behavior, physiology, and genetic diversity similar to those of their wild counterparts is recommended where the conservation of depleted gene pools is of concern.

The status of the population of coho salmon *Oncorhynchus kisutch* in the lower Columbia River (LCR) below Bonneville Dam provides an excellent example of the effects of large-scale-production hatchery operations on native populations of Pacific salmon. In May 1990, Oregon Trout, Oregon Natural Resources Council, Northwest Environmental Defense Center, American Rivers, and the Oregon and Idaho chapters of the American Fisheries Society petitioned the National Marine Fisheries Service (NMFS) to list indigenous populations of LCR coho salmon as Threatened or Endangered under the Endangered Species Act of 1973 (ESA; 16 U.S.C.A. §§ 1531 to 1544). In response, NMFS conducted a formal biological status review for LCR coho salmon (Johnson et al. 1991). The NMFS Biological Review Team (BRT) concluded that years of intense hatchery operation had changed the native coho salmon runs in the LCR to a genetic mixture of native and hatchery stocks. The BRT concluded that the available data failed to identify an existing evolutionarily significant unit of coho salmon in the LCR. As defined by NMFS, a population (or group of populations) will be considered distinct (and hence a species) for purposes of the ESA if it represents an evolutionarily significant unit (ESU) of the biological species. A population must satisfy two criteria to be considered an ESU: (1) it must be reproductively isolated from other conspecific population units, and (2) it must represent an important component in the evolutionary legacy of the species (Waples 1991a). The BRT recommended that LCR coho salmon not be listed under ESA (Johnson et al. 1991).

The NMFS biological status review for LCR coho salmon dealt with the basic question of whether these coho salmon comprised a distinct species or ESU, and therefore qualified for protection under ESA. The use of the term species in the context of ESA can refer to true taxonomic species, subspecies, and distinct population segments. The definition of what constitutes a species under the ESA is addressed by Waples (1991a, 1991b). The reader should note that the use of the term species in the context of ESA should be interpreted in the legal sense.

We review data from the biological status review and include new data derived from state and federal hatchery records to suggest causal relationships between hatchery practices (broodstock selection and release strategies) and harvest rates and alterations in the population structure and abundance of wild

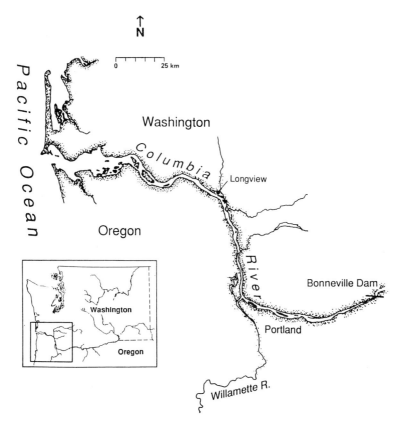

FIGURE 1.—Lower Columbia River below Bonneville Dam. Lower Columbia River basin shown in inset.

LCR coho salmon (as used here, wild fish are progeny of naturally spawning, but not necessarily indigenous, fish). Where appropriate we include data from other West Coast stocks to establish correlations. This information is presented in the context of examination of historic management practices in an effort to develop recommendations for the operation of hatcheries where the conservation of depleted gene pools is of concern.

Population Status

Coho salmon once occurred throughout the Columbia River basin (Figure 1). The population abundance of Columbia River basin coho salmon experienced an almost continuous decline from 1920 to the 1960s (Figure 2) (Beiningen 1976; Mullan 1984; PFMC 1991). This decline in production was widespread throughout the river system and was attributed to a combination of overharvest and habitat loss (McKernan et al. 1950). Prior to 1941, all middle and upper Columbia River stocks of coho salmon were drastically reduced by overharvest, habitat loss, construction of impassable tributary dams, and unscreened irrigation diversions (Mullan 1984). All coho salmon stocks spawning above Grand Coulee Dam were eliminated with the completion of the dam in 1941, because no facilities were provided for fish passage. The abundance of coho salmon in the Snake River was drastically reduced (to about 3,000 fish) by 1962 (COE 1990). The last known coho salmon stocks in the Snake River system were lost by 1986 (COE 1990).

Mullan (1984) indicated that two-thirds of the historical Columbia River coho salmon production may have originated in the LCR; an estimated peak abundance of over 400,000 adult coho salmon returned to the LCR prior to the 1920s. By the late 1950s, fewer than 25,000 coho salmon were returning annually to the LCR (Wahle and Pearson 1987; CBFWA 1990), a number representing less than 5% of historic population levels. This drastic decline in abundance of LCR coho salmon precipitated a

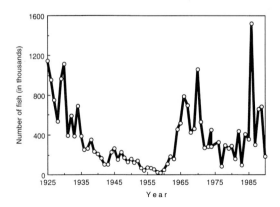

FIGURE 2.—Estimate of population abundance of Columbia River basin coho salmon from 1925 to 1990 (data from Beiningen 1976; Mullan 1984; PFMC 1991).

widespread hatchery enhancement program after 1960. This program included increased production of coho salmon at existing state and federal fish hatcheries and construction of new hatchery facilities (ODFW 1990a, 1990b; WDF 1991a). Between the 1960s and 1990, 16 hatcheries released an average of 40,000,000 juvenile coho salmon per year into the LCR (Johnson et al. 1991).

The hatchery enhancement program for coho salmon in the LCR built the population back to historic levels. Estimates of LCR coho salmon abundance often exceeded 400,000 fish between the mid-1960s and 1990 and reached a historic high of over 1.5 million fish in 1986 (Figure 2) (Howell et al. 1985; CBFWA 1990; PFMC 1991). Between the 1960s and 1990, the LCR coho salmon population was managed almost exclusively for hatchery fish (Cramer et al. 1991). Most of these fish were harvested in the sport and commercial fisheries. However, at the time of the biological status review in 1990, it was estimated that up to 100,000 adult coho salmon were returning to hatchery ponds annually and another 20,000 hatchery fish were spawning naturally in streams associated with hatcheries (Johnson et al. 1991).

Even though the hatchery population of LCR coho salmon was large and stable after 1960, the wild population was severely depressed (Wahle and Pearson 1987; CBFWA 1990; Johnson et al. 1991). Surveys indicated that coho salmon spawning from October through December in LCR tributaries away from hatcheries had been further reduced from about 62 fish/km of stream in the 1950s to about 6 fish/km by the late 1980s (WDF 1991b). During the same period the December-to-March run component, representing putative native strains, had been reduced from up to 50 fish/km to under 0.6 fish/km (WDF 1991b). At the time of the biological status review in 1990, it appeared that there were less than 6,000 wild coho salmon spawning annually in LCR streams (Johnson et al. 1991).

Relationship Between Hatchery Management Practices and Decline of Wild LCR Coho Salmon

The coho salmon hatchery program stabilized LCR coho salmon abundance and increased the opportunity for commercial and sport harvest. Nevertheless, production in wild LCR coho salmon stocks continued to drop. It was hypothesized that these further declines in natural production were the result of excessively high harvest rates on the abundant hatchery stocks (Cramer and Chapman 1990; ODFW 1990a, 1991b). Harvest rates of 80 to 90% were maintained from the 1960s through 1980s, even though theoretical optimum sustained harvest was only about 75% for LCR coho salmon stocks (Chapman 1986). This overexploitation was a primary factor in the decline of wild stocks. However, we believe genetic and ecological interactions between hatchery and wild fish were also substantial factors in the decline of nonhatchery coho salmon in the LCR.

Genetic Changes

Although many protein electrophoretic studies have demonstrated distinct population structure between some coho salmon populations (Utter et al. 1970, 1973, 1980; May 1975; Allendorf and Utter 1979; Winter et al. 1980), genetic data examined during the NMFS biological status review were inconclusive in determining whether distinct, native coho salmon populations existed in the LCR (Johnson et al. 1991, 1993). Although minimizing the loss of genetic variability in wild and hatchery stocks of salmonids is crucial to maintaining the flexibility to respond to natural perturbations (Hard et al. 1992), the practice of maintaining stock integrity did not appear to be included among hatchery management priorities in the LCR. The hatchery-based LCR coho salmon population had been established from combinations of Columbia River, coastal, and Puget Sound stocks rather than from expressly indigenous LCR stocks, and these nonindigenous stocks of fish had subsequently been introduced throughout the LCR (Johnson et al. 1991).

For example, the coho salmon population in Still Creek, Oregon, designated as wild by Oregon De-

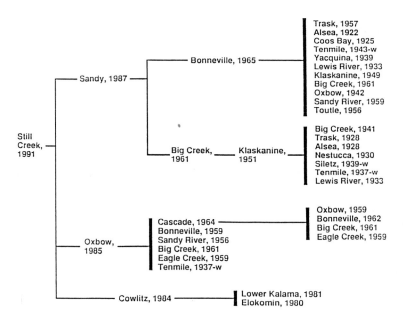

FIGURE 3.—Influence of hatchery coho salmon in streams managed for wild fish in the lower Columbia River. All stocks in the pedigree for Still Creek were of hatchery origin, except those marked with the letter "w," which were taken from a naturally spawning population (data from ODFW 1991a).

partment of Fish and Wildlife (ODFW; ODFW 1991c), was planted with Cowlitz Hatchery (Washington Department of Fisheries; WDF) fish in 1984, Oxbow Hatchery (ODFW) fish in 1985, and Sandy Hatchery (ODFW) fish in 1987. All three of these hatchery stocks were the products of various transfers of coho salmon from other river systems, some of which were outside the Columbia River basin (Figure 3). It is highly probable that similar repeated plantings of mixed-stock hatchery fish have confounded the genetic structure of native LCR coho salmon through introgression with these stocks (Johnson et al. 1991, 1993). Since 1960, most naturally spawning LCR coho salmon are likely feral, first-generation, hatchery fish of mixed origin rather than wild fish (Johnson et al. 1991; ODFW 1991a; WDF 1991a).

Furthermore, the stock integrity of LCR hatchery populations has been altered through domestication and selective breeding and by the use of universal donor stocks. For example, in 1984, ODFW hatcheries apparently did not have sufficient coho salmon juveniles for their nonsmolt planting program. This deficit was rectified by massive transfers of eggs or juveniles from the WDF Cowlitz Hatchery: of the 30 separate releases that year of unsmolted coho salmon into Oregon LCR streams, 27 were Cowlitz River hatchery fish (1,256,000 fish) and 3 were hatchery stocks from the Oregon side of the Columbia River (57,000 fish) (ODFW 1991a).

Life History Alterations

In addition to introgression with nonindigenous hatchery stocks, the life history characteristics of LCR coho salmon were altered by selective breeding. After 1960, the increased survival of hatchery coho salmon that resulted from improvements in diet and disease control increased juvenile output and produced a large surplus of adults returning to LCR hatcheries (McGie 1980). Hatchery personnel could then choose which fish to spawn, as the egg supply frequently exceeded the capacity of the incubation facilities. It was known that larger hatchery coho salmon survived better after release than did smaller salmon (Hager and Noble 1976; Bilton et al. 1982; Mahnken et al. 1982), so early-returning fish were spawned so that the eggs would hatch early to provide maximum growth of juveniles prior to liberation (WDF 1991a). Thus, hatchery stocks were selected for a very narrow range of spawner timing.

The exclusive spawning of early-run coho salmon had a rapid and dramatic effect on the return time of hatchery coho salmon. For example, prior to the 1960s, coho salmon returned to five WDF hatcher-

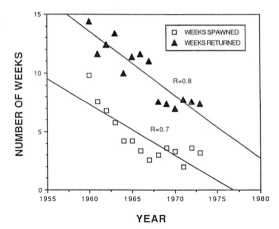

FIGURE 4.—Changes in time of spawning and time of adult return for coho salmon at five Washington Department of Fisheries hatcheries (data from Rasch and Foster 1978a, 1978b, 1978c).

ies in western Washington, including two from the LCR, over a range of about 14 weeks and were spawned over a period of almost 10 weeks (Rasch and Foster 1978a, 1978b, 1978c). However, by 1973, the period for egg taking at all five hatcheries had decreased to about 3 weeks (Figure 4). In response to this greatly condensed spawning time, the period of return for adult fish from the ocean also diminished rapidly. From 1960 to 1973, or in about four generations, adult return time was reduced by about one-half (Figure 4). A reduced period of egg taking on the Oregon side of the LCR was similar to the Washington situation (Table 1; ODFW 1991a).

The alteration and compression of spawn timing appears to have had little effect on hatchery operations; hatchery fish seem to be well suited for hatchery and harvest requirements. However, hatchery-altered early spawning stocks of coho salmon may no longer be well adapted for establishing permanent self-reproducing populations. The BRT concluded that LCR hatchery spawning regimes produced feral LCR hatchery coho salmon that spawned before native fish (Johnson et al. 1991). Scouring floods in late fall and early winter have become a regular phenomenon in the degraded salmonid habitat of the LCR (Chamberlin et al. 1991). As a result, eggs of feral hatchery fish often have fallen victim to scouring floods, whereas later spawning wild fish have not been as affected by the fall storms on the Pacific coast (Nickelson et al. 1986). In addition, fry from earlier spawning feral hatchery fish usually have emerged before natural food supplies were abundant, further reducing their survivability (Nickelson et al. 1986).

Hatchery Outplants

Although the life history patterns of hatchery coho salmon were drastically altered, the fish were still expected to adapt successfully to natural rearing environments after outplanting. The LCR streams were extensively stocked with up to 10,000,000 presmolts/year (ODFW 1991a; WDF 1991c) even though returns to streams outplanted with presmolts were known to be uniformly poor compared with returns from streams outplanted with smolts (Willis 1962; Wallis 1968; Johnson 1970; Hager and Noble 1976; Johnson 1982; ODFW 1991a; WDF 1991a). Although large numbers of fish were outplanted in streams each year, returns were poor and many LCR streams remained below juvenile carrying capacity for naturally spawning coho salmon (Nickelson et al. 1993).

Nonetheless, presmolt releases appear to have exceeded the carrying capacities of tributaries by hundreds of thousands of fish in many cases. For example, ODFW (1991a) monitored several unplanted Oregon tributaries in the LCR as index sites and estimated carrying capacities of 100 to 2,100 coho salmon smolts per year for these streams

TABLE 1.—Range of spawning dates for coho salmon at Oregon Department of Fish and Wildlife's lower Columbia River hatcheries, 1984–1990. Data is taken from ODFW 1991a.

Hatchery	Year			Average duration (d)
	1984	1987	1990	
Klaskanine	Oct 23–Nov 01	Nov 03	Oct 29–Nov 19	10.3
Big Creek	Oct 25–Nov 15	Oct 22–Nov 16	Nov 01–Nov 26	23.7
Eagle Creek	Oct 17–Nov 1	Oct 16–Nov 16	Oct 12–Nov 08	24.3
Sandy	Oct 29–Nov 14	Nov 09–Nov 30	Nov 05–Nov 27	19.7
Bonneville		Nov 05–Nov 17	Nov 19–Nov 09	16.5
Cascade	Oct 29–Nov 16	Nov 03–Nov 17	Oct 24–Nov 26	21.7
Mean				19.4

TABLE 2.—Estimated number of emergent coho salmon fry per lineal meter in surveyed sections of coho salmon index streams, lower Columbia River, Oregon. Data taken from ODFW 1982, 1991.

Index streams	Estimated number		Survey length (km)	Estimated fry/m
	Smolts	Fry		
Little Creek	1,700	17,000	4.3	3.9
Carcus Creek	2,100	21,000	5.1	4.2
Raymond Creek	500	5,000	3.5	1.4
Hortill Creek	200	2,000	5.8	3.5
Klickitat Creek	100	1,000	3.3	3.0
Speelyai Creek	200	2,000	0.9	2.3
Stavebolt Creek	300	3,000	1.3	2.4
Mean				3.0

TABLE 3.—Estimated number of coho salmon fry planted per lineal meter in selected nonindex streams in the lower Columbia River, Oregon, 1981–1988.

Nonindex stream and year	Number of fry planted[a]	Stream length (km)[b]	Estimated fry/m
Mill Creek			
1985	5,000	2.7	1.9
1983	29,500		10.9
Mary's Creek			
1983	40,000	4.0	10.0
1982	10,000		2.5
Bear Creek			
1987	42,500	7.9	5.4
1981	142,500		18.0
Beaver Creek			
1988	65,500	9.1	7.2
1987	151,000		16.6
Goble Creek			
1987	30,000	6.3	4.8
1981	236,500		37.5
Tide Creek			
1987	102,000	5.8	17.6
1985	19,000		3.3
1981	274,000		47.2
Milton Creek			
1987	49,500	4.2	11.8
1983	131,000		31.2
Multnoma Creek			
1987	43,000	1.1	39.1
Mean			16.6

[a] Data taken from ODFW 1991a.
[b] Data taken from CBIAC 1966; DeLorme 1991.

(Table 2). Survival from the emergent fry to smolt stage of naturally reared coho salmon in nearby Oregon coastal streams had been estimated at about 10% (ODFW 1982). Therefore, the estimated carrying capacity of emergent fry in the index streams would be approximately 1,000 to 21,000 fry, or about 3.0 fry/m (Table 2). Yet, similar-sized LCR streams in Oregon have been planted with unsmolted coho salmon at a rate of about 16.6 fry/m (Table 3). Washington LCR streams, when planted, have received about 22.2 coho salmon fry/m (Table 4).

Overseeding of natural habitat may be related to pressure on hatcheries to reduce densities. The large variation in the number of unsmolted coho salmon planted each year (Tables 3 and 4) suggested that outplanting is the means by which excess production is thinned to avoid exceeding hatchery space limitations. The need to remove excess production apparently has left managers with few options other than massive releases of nonmigratory coho salmon juveniles in nearby watersheds.

Overstocking of streams may have adversely affected wild populations in several ways. At a minimum, the wild fish population would have experienced the mortality required by the carrying capacity of the stream. However, several studies have shown that the first coho salmon juveniles to occupy rearing habitat enjoy ecological and competitive advantages over subsequent inhabitants (Mason 1966), sometimes forcing the later fish out of the stream altogether (Chapman 1962). In addition, planted fish are usually larger than the wild fish in the stream, due to growth advantages experienced in the hatchery environment (Solazzi et al. 1983; Nickleson et al. 1986). For example, between 1985 and 1987, unsmolted coho salmon planted into natural habitats on the Washington side of the LCR averaged about 1.8 g (Table 5). Many of these fish were planted before mid-April, when wild coho salmon fry weighed less than 0.7 g, or had not yet emerged from the gravel (Tagart 1976). In addition, wild fish prefer lower rearing densities than do their hatchery-reared counterparts (Allee 1974). Wild juveniles also have a propensity to outmigrate prematurely from areas that have been planted with hatchery presmolts (Solazzi et al. 1983; Nickelson et al. 1986). Therefore, hatchery fish may have displaced wild fish each time an area was overstocked.

Hatchery records also reveal a lack of consistency in planting a given stream over time, which may be due to yearly variations in the number of unsmolted fish removed from hatchery populations. Since 1980, some Washington LCR streams received multiple plants, yet others were planted only once (Coleman and Rasch 1981; Castoldi and Rasch 1982; Hill 1984; Kirby 1985; Abrahamson 1986, 1987, 1988; WDF 1991a; Table 4). For example,

TABLE 4.—Estimated number of coho salmon fry planted per lineal meter in selected streams in the lower Columbia River, Washington, 1980–1987.

Nonindex stream and year	Number of fry planted[a]	Stream length (km)[b]	Estimated fry/m
Brim Creek			
1987	127,000	5.0	25.4
Olequa Creek			
1985	811,000	27.0	30.0
1981	1,400,000		51.9
Lacamas Creek			
1987	556,000	31.0	17.9
Unnamed creek			
1986	37,000	4.6	8.0
Ostrander Creek			
1987	274,200	17.0	16.1
1984	95,000		5.6
1981	390,000		22.9
Mean			22.2

[a] Data taken from Coleman and Rasch 1981; Castoldi and Rasch 1982; Hill 1984; Kirby 1985; Abrahamson 1986, 1987, 1988.
[b] Data taken from DeLorme 1988.

Olequa Creek, a small tributary to the Cowlitz River, was not planted before 1980. However, between 1981 and 1988, this stream received as many as 1.4 million unsmolted coho salmon annually (Table 4). Other Washington LCR streams also received massive plants of coho salmon for the first time in the 1980s (Table 4).

Conclusions and Recommendations

It appeared that the introduction of stocks from outside the LCR, release of large numbers of hatchery fish, and extreme overharvest combined to change the native coho salmon populations in the LCR to a genetic mixture of indigenous and mixed-hatchery stocks. Although hatchery fish performed well in hatchery environments and supported a large, stable ocean fishery, they do not appear ecologically well suited for recolonizing vacant habitat in the LCR. As Wagner (1967) cautioned, natural habitat should not be used as a postliberation rearing area for the hatchery product, but only as a highway to the sea. Nevertheless, outplants of unsmolted coho salmon in LCR streams have been both massive and pervasive. Outplanting of presmolts may have been practiced as a convenient means of disposing of hatchery overproduction rather than an effective mechanism for enhancement.

We believe that for LCR coho salmon, efforts should be undertaken to (1) reduce harvest to a level that will allow naturally spawning populations to rebuild (it is likely that such a harvest rate will have to be less than 50% for stocks to be restored); (2) restrict interhatchery transfer of fish to allow stocks to develop watershed-specific characteristics; and (3) terminate the planting of nonmigratory coho salmon juveniles.

Most Pacific Northwest salmon hatcheries were established to mitigate for runs that were depressed due to lost habitat. However, in most cases this mitigation has worked to the detriment of native populations. Because hatchery populations have been viewed as an interchangeable substitute for native populations, reliance on hatcheries has inadvertently contributed to a lack of determination to protect and repair natural habitats. Circumventing habitat problems with hatcheries has been the focus of fisheries legislation for more than a century (Johnson 1984). As a result, critical salmonid habitats, including winter pools and instream structures such as rootwads and large woody debris (McMahon and Hartman 1989; Hicks et al. 1991), have disappeared to a significant extent in the watersheds of the Pacific Northwest (Sedell and Luchessa 1982).

We are now entering an era where, even though salmon hatcheries can produce hundreds of thousands of salmon, fish will not be harvested due to restrictions necessary to protect wild fish. We believe that where hatchery operation and management policies conflict with the welfare of coexisting wild populations, conservation hatchery strategies must replace outdated production practices. New policies must be adopted so that hatcheries become part of comprehensive ecosystem restoration plans. Conservation hatcheries should provide fish for supplementation of natural spawners rather than simply providing fish for harvest.

TABLE 5.—Average presmolt coho salmon weight and spring release timing from some Washington Department of Fisheries lower Columbia River hatcheries, 1985–1987.[a]

Hatchery	First release	Last release	Weight (g)
Grays	Mar 17	Mar 27	1.1
Elokomin	Apr 21	May 12	2.7
Cowlitz	Apr 13	July 10	3.7
Kalama Falls	Apr 21	Apr 30	0.7
Lewis	Apr 18	June 8	0.8
Speelyia	Apr 11	Apr 11	2.2
Washougal	Apr 18	June 5	1.5
Mean			1.8

[a] Data taken from Delarm and Smith 1990.

To accomplish this goal, conservation hatcheries must rear local stocks under conditions that ensure production of smolts with similar behavior, physiology, and genetic diversity as their wild counterparts. There appears to be little hope of significantly increasing production in natural habitats without directed efforts both to restore and protect watersheds and riparian habitat and to ensure that released hatchery fish do not exceed the carrying capacity of natural habitats.

References

Abrahamson, P. 1986. A detailed listing of the liberations of salmon into the open waters of the state of Washington during 1985. Washington Department of Fisheries Report 243, Olympia.

Abrahamson, P. 1987. A detailed listing of the liberations of salmon into the open waters of the state of Washington during 1986. Washington Department of Fisheries Report 259, Olympia.

Abrahamson, P. 1988. A detailed listing of the liberations of salmon into the open waters of the state of Washington during 1987. Washington Department of Fisheries Report 267, Olympia.

Allee, B. J. 1974. Spatial requirements and behavioral interaction of juvenile coho salmon (*Oncorhynchus kisutch*) and steelhead trout (*Salmo gairdneri*). Doctoral dissertation. University of Washington, Seattle.

Allendorf, F. W., and F. M. Utter. 1979. Population genetics. Pages 407–454 *in* W. S. Hoar, D. J. Randall, and J. R. Brett, editors. Fish physiology, volume 8. Academic Press, New York.

Beiningen, K. T. 1976. Fish runs. Investigative Reports of Columbia River Fisheries Project. Pacific Northwest Regional Commission, Vancouver, Washington.

Bilton, H. T., D. F. Alderdice, and J. T. Schnute. 1982. Influence of time and size at releases of juvenile coho salmon (*Oncorhynchus kisutch*) on returns at maturity. Canadian Journal of Fisheries and Aquatic Sciences 39:426–447.

Castoldi, P., and T. Rasch. 1982. A detailed listing of the liberations of salmon into the open waters of the state of Washington during 1981. Washington Department of Fisheries Report 160, Olympia.

Chamberlin, T. W., R. D. Herr, and F. H. Everest. 1991. Timber harvesting, silviculture, and watershed processes. American Fisheries Society Symposium 19:181–205.

Chapman, D. W. 1962. Aggressive behavior in juvenile coho salmon as a cause of emigration. Journal of the Fisheries Research Board of Canada 19:1047–1080.

Chapman, D. W. 1986. Salmon and steelhead abundance in the Columbia River in the nineteenth century. Transactions of the American Fisheries Society 115:662–679.

Coleman, P., and T. Rasch. 1981. A detailed listing of the liberations of salmon into the open waters of the state of Washington during 1980. Washington Department of Fisheries Report 132, Olympia.

CBFWA (Columbia Basin Fish and Wildlife Authority). 1990. Integrated system plan for salmon and steelhead production in the Columbia River basin. Northwest Power Planning Council, Portland, Oregon.

CBIAC (Columbia Basin Inter-Agency Committee). 1966. River mile index: Klaskanine, Sandy, Hood, Umatilla, Walla Walla rivers and minor left bank Columbia River tributaries. CBIAC, Portland, Oregon.

COE (Corps of Engineers). 1990. Fish Passage Report, 1989: Columbia and Snake rivers, for salmon, steelhead, and shad. U.S. Army Corps of Engineers, Portland, Oregon.

Cramer, S., and D. Chapman. 1990. Prefatory review of status of coho salmon in the lower Columbia River. Report to the ESA administrative record for coho salmon, November 1990. Pacific Northwest Utilities Conference Committee, Portland, Oregon.

Cramer, S., A. Maule, and D. Chapman. 1991. The status of coho salmon in the lower Columbia River. Pacific Northwest Utilities Conference Committee, Portland, Oregon.

Delarm, M. R., and R. Z. Smith. 1990. Assessment of present anadromous fish production facilities in the Columbia River basin, volume 4: Washington Department of Fisheries. Final Report to Bonneville Power Administration, Project Number 89-045, Portland, Oregon.

DeLorme Mapping. 1988. Washington atlas and gazetteer. DeLorme Mapping Company, Freeport, Maine.

DeLorme Mapping. 1991. Oregon atlas and gazetteer. DeLorme Mapping Company, Freeport, Maine.

Hager, R. C., and R. E. Noble. 1976. Relation of size at release of hatchery-reared coho salmon to age, size and sex composition of returning adults. Progressive Fish-Culturist 38:144–147.

Hard, J. J., R. P. Jones Jr., M. R. Delarm, and R. S. Waples. 1992. Pacific salmon and artificial propagation under the Endangered Species Act. NOAA (National Oceanic and Atmospheric Administration) Technical Memorandum NMFS (National Marine Fisheries Service) NWFSC-2.

Hicks, B. J., J. D. Hall, P. A. Bisson, and J. R. Sedell. 1991. Responses of salmonids to habitat change. American Fisheries Society Special Publication 19:483–518.

Hill, P. M.. 1984. A detailed listing of the liberations of salmon into the open waters of the state of Washington during 1983. Washington Department of Fisheries Report 210, Olympia.

Howell, P., and eight coauthors. 1985. Stock assessment of Columbia River anadromous salmonids, volume 1. Chinook, coho chum, and sockeye salmon stock summaries. Oregon Department of Fish and Wildlife Report to the Bonneville Power Administration, Project 83-335, Portland.

Johnson, K. A. 1970. The effect of size at release on the contribution of 1964 brood Big Creek Hatchery coho salmon to the Pacific coast sport and commercial fisheries. Oregon Fish Commission Research Report 2, Portland.

Johnson, O. W., T. A. Flagg, D. J. Maynard, G. B. Milner, and F. W. Waknitz. 1991. Status review for lower

Columbia River coho salmon. NOAA (National Oceanic and Atmospheric Administration) Technical Memorandum NMFS (National Marine Fisheries Service) F/NWC-202.

Johnson, O. W., D. J. Maynard, G. B. Milner, and F. W. Waknitz. 1993. Genetic considerations in the status review of lower Columbia River coho salmon. Pages 161–181 in L. Berg and P. W. Delaney, editors. Proceedings, 1993 northeast Pacific chinook and coho salmon workshop. American Fisheries Society, North Pacific International Chapter, Nanaimo, British Columbia.

Johnson, S. L. 1982. A review and evaluation of release strategies for hatchery reared coho salmon. Oregon Department of Fish and Wildlife Report 82-5, Portland.

Johnson, S. L. 1984. Freshwater environmental problems and coho production in Oregon. Oregon Department of Fish and Wildlife Report 84-11, Portland.

Kirby, L. L. 1985. A detailed listing of the liberations of salmon into the open waters of the state of Washington during 1984. Washington Department of Fisheries Report 231, Olympia.

Mahnken, C., E. Prentice, W. Waknitz, G. Monan, C. Sims, and J. Williams. 1982. The application of recent smoltification research to public hatchery releases: an assessment of size/time requirements for Columbia River hatchery coho salmon (*Oncorhynchus kisutch*). Aquaculture 28:251–268.

Mason, J. C. 1966. Behavioral ecology of juvenile coho salmon (*Oncorhynchus kisutch*) in stream aquaria with particular reference to competition and aggressive behavior. Doctoral dissertation. Oregon State University, Corvallis.

May, B. 1975. Electrophoretic variation in the genus *Oncorhynchus*: the methodology, genetic basis, and practical application to fisheries research and management. Master's thesis. University of Washington, Seattle.

McGie, A. M. 1980. Analysis of relationships between hatchery coho salmon transplants and adult escapements in Oregon coastal watersheds. Oregon Department of Fish and Wildlife Report 80-6, Portland.

McKernan, D. L., D. R. Johnson, and J. L. Hodges. 1950. Some factors influencing the trends of salmon populations in Oregon. Transactions of the North American Wildlife Conference 15:427–449.

McMahon, T. E., and G. F. Hartman. 1989. Influence of cover complexity and current velocity on winter habitat use by juvenile coho salmon (*Oncorhynchus kisutch*). Canadian Journal of Fisheries and Aquatic Sciences 46:1551–1557.

Mullan, J. W. 1984. Overview of artificial and natural propagation of coho salmon (*Oncorhynchus kisutch*) on the mid-Columbia River. U.S. Fish and Wildlife Service Biological Report 84(4):1–37.

Nickelson, T. E., M. F. Solazzi, and S. L. Johnson. 1986. Use of hatchery coho salmon (*Oncorhynchus kisutch*) presmolts to rebuild wild populations in Oregon coastal streams. Canadian Journal of Fisheries and Aquatic Sciences 43:2443–2449.

Nickelson, T. E., M. F. Solazzi, S. L. Johnson, and J. D. Rodgers. 1993. An approach to determining stream carrying capacity and limiting habitat for coho salmon (*Oncorhynchus kisutch*). Pages 251–260 in L. Berg and P. W. Delaney, editors. Proceedings, 1993 northeast Pacific chinook and coho salmon workshop. American Fisheries Society, North Pacific International Chapter, Nanaimo, British Columbia.

ODFW (Oregon Department of Fish and Wildlife). 1982. Comprehensive plan for production and management of Oregon's anadromous salmon and trout, part 2. Coho salmon plan considerations. Oregon Department of Fish and Wildlife, Portland.

ODFW (Oregon Department of Fish and Wildlife). 1990a. History of Oregon's lower Columbia River coho and Snake River chinook hatchery populations. Report to ESA administrative record for coho salmon, December 1990. Oregon Department of Fish and Wildlife, Portland.

ODFW (Oregon Department of Fish and Wildlife). 1990b. Lower Columbia River coho salmon, evaluation of stock status, part 1. Report to ESA administrative record for coho salmon, December, 1990. Oregon Department of Fish and Wildlife, Portland.

ODFW (Oregon Department of Fish and Wildlife). 1991a. Lower Columbia River coho salmon, evaluation of stock status, causes of decline, and critical habitat, part 2. Report to ESA administrative record for coho salmon, February 1991. Oregon Department of Fish and Wildlife, Portland.

ODFW (Oregon Department of Fish and Wildlife). 1991b. Numbers of late spawning coho salmon in the lower Columbia River. Report to ESA administrative record for coho salmon, February, 1991. Oregon Department of Fish and Wildlife, Portland.

ODFW (Oregon Department of Fish and Wildlife). 1991c. Wild coho salmon surveys in the lower Columbia River. Report to ESA administrative record for coho salmon, September, 1991. Oregon Department of Fish and Wildlife, Portland.

PFMC (Pacific Fishery Management Council). 1991. Review of 1990 ocean salmon fisheries. PFMC, Portland, Oregon.

Rasch, T., and R. Foster. 1978a. Hatchery returns and spawning data for Columbia River, 1960–1976. Washington Department of Fisheries Report 61, Olympia.

Rasch, T., and R. Foster. 1978b. Hatchery returns and spawning data for Puget Sound, 1960–1976. Washington Department of Fisheries Report 59, Olympia.

Rasch, T., and R. Foster. 1978c. Hatchery returns and spawning data for the Straits and coast, 1960–1976. Washington Department of Fisheries Report 60, Olympia.

Sedell, J. R., and K. J. Luchessa. 1982. Using the historical record as an aid to salmonid habitat enhancement. Pages 210–223 in N. B. Armantrout, editor. Acquisition and utilization of aquatic habitat inventory information symposium. American Fisheries Society, Western Division, Bethesda, Maryland.

Solazzi, M. F., S. L. Johnson, and T. E. Nickelson. 1983. The effectiveness of stocking hatchery coho presmolts to increase the rearing density of coho salmon in

Oregon coastal streams. Oregon Department of Fish and Wildlife Report 83-1, Portland.

Tagart, J. V. 1976. The survival from egg deposition to emergence of coho salmon in the Clearwater River, Jefferson County, Washington. Master's thesis. University of Washington, Seattle.

Utter, F. M, W. E. Ames, and H. O. Hodgins. 1970. Transferrin polymorphism in coho salmon (*Oncorhynchus kisutch*). Journal of the Fisheries Research Board of Canada 27:2371–2373.

Utter, F. M., D. Campton, S. Grant, G. Milner, J. Seeb, and L. Wishard. 1980. Population structures of indigenous salmonid species of the Pacific Northwest, 1. A within and between species examination of natural populations based on genetic variation of proteins. Pages 285–304 *in* W. J. McNeil and D. C. Himsworth, editors. Proceedings of a symposium on salmonid ecosystems of the North Pacific. Oregon State University, Oregon Sea Grant College Program, Corvallis.

Utter, F. M, H. O. Hodgins, F. W. Allendorf, A. G. Johnson, and J. L. Mighell. 1973. Biochemical variation in Pacific salmon and rainbow trout: their inheritance and application in population studies. Pages 329–339 *in* J. H. Schroder, editor. Genetics and mutagenesis of fish. Springer-Verlag, Berlin, Germany.

Wagner, H. H. 1967. A summary of investigations of the use of hatchery reared steelhead in the management of a sport fishery. Oregon State Fish Commission, Fisheries Report 5, Portland.

Wahle, R. J., and R. E. Pearson. 1987. A listing of Pacific coast spawning streams and hatcheries producing chinook and coho salmon. NOAA (National Oceanic and Atmospheric Administration) Technical Memorandum NMFS (National Marine Fisheries Service) F/NWC-122.

Wallis, J. 1968. Recommended time, size and age for release of hatchery-reared salmon and steelhead trout. Fish Commission of Oregon Processed Report, Portland.

Waples, R. S. 1991a. Definition of "species" under the Endangered Species Act: application to Pacific salmon. NOAA (National Oceanic and Atmospheric Administration) Technical Memorandum NMFS (National Marine Fisheries Service) NWFSC-194.

Waples, R. S. 1991b. Pacific salmon, *Oncorhynchus* spp., and the definition of "species" under the Endangered Species Act. Marine Fisheries Review 53:11–22.

WDF (Washington Department of Fisheries). 1991a. Historical production of coho salmon from Washington state hatcheries on the lower Columbia River. Report to ESA administrative record for coho salmon, April, 1991. Washington Department of Fisheries, Olympia.

WDF (Washington Department of Fisheries). 1991b. Washington Department of Fisheries historical lower Columbia River stream surveys. Report to ESA administrative record for coho salmon, February, 1991. Washington Department of Fisheries, Olympia.

WDF (Washington Department of Fisheries). 1991c. Overview of Washington Department of Fisheries Columbia River hatchery coho production. Report to ESA administrative record for coho salmon, February, 1991. Washington Department of Fisheries, Olympia.

Willis, R. A. 1962. Gnat Creek weir studies. Final Report to the Fish Commission of Oregon. Oregon Department of Fish and Wildlife, Portland.

Winter, G. W., C. B. Schreck, and J. D. McIntyre. 1980. Resistance of different stocks and transferrin genotypes of coho salmon (*Oncorhynchus kisutch*) and steelhead trout (*Salmo gairdneri*) to bacterial kidney disease and vibriosis. U.S. National Marine Fisheries Service Fishery Bulletin 77:795–802.

Marine Enhancement with Striped Mullet: Are Hatchery Releases Replenishing or Displacing Wild Stocks?

KENNETH M. LEBER, NATHAN P. BRENNAN, AND STEVE M. ARCE

The Oceanic Institute, Makapuu Point
Waimanalo, Hawaii 96795, USA

Abstract.—The hypothesis that marine hatchery releases can increase fish abundances has at least two corollaries that need to be tested: (1) cultured fish can survive and grow when released into coastal environments; and (2) cultured fish do not displace wild individuals. Both are being tested in Hawaii. The present study was conducted to evaluate whether hatchery releases of striped mullet *Mugil cephalus* actually increase abundances or displace wild stock. In summer 1993, 5,811 wild striped mullet were captured, tagged, and released in three lots back into two primary nursery habitats in Kaneohe Bay. Three weeks later, quantitative sampling with cast nets was conducted in several striped mullet nursery habitats within the bay to evaluate pretreatment dispersal of wild fish. Following those initial collections, cultured striped mullet were released to establish the primary treatment condition, a hatchery release. A total of 29,354 cultured striped mullet were tagged and released, but at only one of the nursery sites (treatment site). Monthly monitoring was conducted over an 8-month period to determine if there was greater dispersal of wild fish at the treatment site. There was no significant difference in the dispersal rates of wild fish from the treatment site compared with the control (no hatchery release) site. As expected, based on earlier pilot hatchery releases, a majority of tagged and released cultured and wild striped mullet remained within those nursery habitats where they were released.

Hatchery releases in this study did not result in displacement of wild individuals from the principal nursery habitat in Kaneohe Bay. The cultured fish released there increased abundances of striped mullet at the treatment site by around 33%. Thus it appears that even small-scale releases could help replenish the depleted striped mullet fishery in Kaneohe Bay; conducting small-scale hatchery releases in several nursery habitats in Kaneohe Bay should increase overall striped mullet abundances in this estuary. This study also corroborated earlier experiments in Hawaii showing a direct relationship between fish size at release and recapture rate.

These results indicate hatchery releases can increase abundances of targeted inshore fish populations in Hawaii. If a careful approach is used, marine stock enhancement appears to have considerable potential as an additional fishery management tool.

World fishery resources currently face enormous risk of severe depletion. As capture fisheries continue to be exploited at a nonsustainable pace (NOAA 1991, 1992) and as human population size doubles by the middle of the next century (FAO 1992), what will become of today's fisheries? Will commercial fishing virtually pass from existence in the next 50 years? Will recreational angling be an option for coastal residents? Hardin's (1968) "tragedy of the commons" holds that our public resources will eventually be squandered if their fate is left in the hands of user groups.

Current fisheries management practices have been generally ineffective at stemming overexploitation. Better control over fishing effort is clearly needed (Coutin and Payne 1989; Ross and Nelson 1992). It is also clear that habitat degradation is a major factor in the decline of coastal organisms (Moyle et al. 1992; Bryan et al. 1992).

Lacking a direct responsibility for protecting fisheries resources, public response will likely continue to be apathetic towards the hard choices that are needed to reduce fishing effort and restore critical habitat. Limited entry to fisheries and subsequent ownership of those fisheries by a select few has been considered as a management alternative (Fox 1992). Political reality, however, has prevented widespread use of this concept. Containing development in coastal habitats could become more acceptable if the steady loss of wilderness and recreation areas leads to public discontent.

The two current methods of managing coastal fisheries, control of fishing effort and habitat protection (including pollution control), might be augmented by an additional management strategy to help prevent further depletions and extinctions. Recent research suggests that propagation and release (stock enhancement) of marine organisms has potential to increase population abundances in coastal environments (Tsukamoto et al. 1989; Svasand et al. 1990; Honma 1993; Kent et al. 1995, this volume; Willis et al. 1995, this volume; Leber, in press;

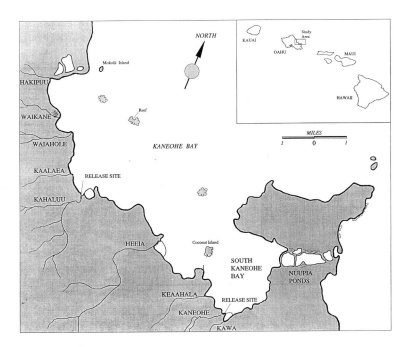

FIGURE 1.—Map of the study area in Kaneohe Bay. Control fish (wild striped mullet) were released near the mouth of Kaneohe stream, and treatment fish (wild striped mullet at treatment site) and cultured fish were released near the mouth of Kahaluu stream. Recapture collections were conducted at release sites and neighboring streams in Kaneohe Bay (Kaneohe, Keaahala, Kahaluu, Kaalaea, and Waiahole streams).

Leber et al., in press). Despite over 50 years of marine hatchery releases, serious effort at evaluating the potential of coastal stock enhancement has come about during only the past 10 years (see Stroud 1986). To realize the full potential of marine enhancement to help conserve and replenish rapidly declining stocks, we must develop this technology carefully and quickly (Blankenship and Leber 1995, this volume).

The central issue that needs resolution in marine stock enhancement is the hypothesis that hatchery releases can actually increase population size of a variety of coastal species. Two important and largely untested predictions underlie this hypothesis—the first is that significant numbers of hatchery fish can survive in the wild; the second is that released hatchery fish actually increase abundances rather than displace wild stocks.

Both of these predictions are being tested in Hawaii with two inshore fishes, striped mullet *Mugil cephalus* and *Polydactylus sexfilis*. Previous experiments have documented that cultured striped mullet can survive in the wild. Juvenile cultured striped mullet regularly exceed 20% of the striped mullet collected in quantitative samples in striped mullet nursery habitats 6 to 8 months after small-scale pilot releases (Leber, in press; Leber et al., in press). Pilot releases of juveniles in Hilo Bay, Hawaii, also contribute around 20% of the catch in the recreational pole and line fishery (Leber et al., in press). The results of an experiment to evaluate whether such hatchery releases supplement or displace wild striped mullet in Kaneohe Bay, the largest estuary in Hawaii, are presented here.

Methods

A factorial-design, release–recapture experiment was performed to test the effects of the release of juvenile cultured striped mullet on the dispersal of existing wild striped mullet juveniles in the estuaries of Kaneohe Bay, Hawaii. In June–July 1993, 5,811 wild striped mullet fingerlings were captured, tagged, and released at two of the most productive nursery habitats in Kaneohe Bay—Kaneohe stream and Kahaluu stream (Figure 1). The release of tagged wild fish provided baseline data on the dispersal patterns of wild striped mullet before and after releasing cultured striped mullet. In August 1993, approximately 4 weeks after the wild releases,

29,354 hatchery fish were tagged and released into Kahaluu stream (treatment site) to establish the primary treatment condition. No cultured fish were released at Kaneohe stream (control site).

Wild striped mullet juveniles were collected during a peak recruitment period, from 1 June through 21 June, at Kaneohe stream and Kahaluu lagoon. To increase the total number of wild striped mullet fingerlings tagged and released in this study, additional collections were conducted at neighboring striped mullet nursery habitats within Kaneohe Bay.

Beach seines were used to capture wild striped mullet. Fish were caught in seines, transferred into 1,000-L hauling tanks, and transported by truck to The Oceanic Institute. Upon arrival, the wild striped mullet were transferred to 40,000-L quarantine tanks where they recovered from handling stress.

After quarantine procedures were completed, the wild striped mullet fingerlings were graded into four size-classes ranging in total length (TL) from 45 mm to 110 mm (45 to 60 mm; 60 to 70 mm; 70 to 85 mm; and 85 to 110 mm). The striped mullet were then tagged with binary coded wire tags (CWT; Northwest Marine Technology, Inc., Olympia, Washington) to identify fish origin (wild), initial capture site, size at release, release site (Kaneohe stream or Kahaluu stream), and release lot (date). Buckley and Blankenship (1990) and Bergman et al. (1992) discuss the utility of the CWT for marking fishes. The tags were implanted in cartilaginous tissue in the snout by use of an automatic tag injector equipped with head molds designed specifically for striped mullet by researchers from the State of Washington Department of Fish and Wildlife.

A total of 5,811 wild striped mullet were released in 3 replicate lots at Kahaluu stream and Kaneohe stream on 25 June, 30 June, and 2 July 1993. Numbers released per lot were kept as similar as possible. Releases at the control site and treatment site were conducted on the same day, 3–4 hours apart, for each lot. The wild fish were released about 4 weeks prior to releases of cultured fish.

Cultured striped mullet were reared at The Oceanic Institute. Striped mullet broodstock were spawned throughout the winter of 1993. Hatched striped mullet were reared through mid-August to fingerlings for tagging and release. Larval striped mullet were hatched and cultured in 5,000-L, round, conical-bottom tanks for 45 d. Stage-one juveniles, 45 d old and 20 mm TL, were transferred to 8,000-L round tanks and nursed for an additional 40 d until stage-two juveniles (85 d old and 40 mm TL). Stage-two juveniles were transferred to 30,000-L round tanks and nursed to the sizes released in this study (60 to 130 mm TL). In culture tanks, juvenile striped mullet growth rates averaged 0.5 mm/d.

Cultured striped mullet were graded into four size-classes ranging from 60 mm to 130 mm TL (60 to 70 mm; 70 to 85 mm; 85 to 110 mm; and 110 to 130 mm). The fish were provided a 2-d period to recover from grading stress, then tagged with CWTs. Batch codes were used to identify fish type (cultured), size at release, release site (Kahaluu stream), and release lot.

A CWT retention rate of 97% has been documented over a 12-month period for striped mullet (Oceanic Institute 1991). To verify tag retention rates in this study, at least 5% of the fish tagged for each release site were randomly subsampled from each release lot. The subsamples totaled 307 tagged wild fish and 1,533 tagged cultured fish, which were retained in tanks for up to 6 months for periodic tag retention checks. The subsampled fish were not released.

Previous studies have shown a direct relationship between striped mullet size at release and survival following summer releases in Kaneohe Bay (Leber, in press). Those data were used to target sizes for release of cultured fish during this study. Mostly large stage-two juveniles (70 to 130 mm TL) were released at the treatment site, Kahaluu stream.

A total of 29,354 cultured striped mullet juveniles were tagged and released. The replicate lots were released over a 3 week period, on 30 July, 6 August, and 13 August 1993. Due to a shortage of cultured striped mullet ranging in size from 70 to 110 mm TL for the final release lot, that lot was supplemented with 3,852 tagged striped mullet from a smaller size-class (60 to 70mm TL).

All releases were conducted during midday or early afternoon. Successive release lots spanned tide ranges from low to high. At the release sites, salinities at the top of the water column ranged from 0 to 5‰, whereas salinities at the bottom ranged from 20 to 32‰ (Table 1). Releases were conducted near the shoreline in water from 0.5 to 1.5 m deep.

Releases of wild striped mullet at the control site, Kaneohe stream, were conducted approximately 200 m upstream from the stream mouth. At the treatment site, Kahaluu stream, all releases were conducted within a lagoon, approximately 300 m upstream from the stream mouth. Both of the release sites are primary striped mullet nursery habitats in Kaneohe Bay (Oceanic Institute 1991). Kahaluu stream is located on the north end of Kaneohe Bay (Figure 1). This tributary is fed by

TABLE 1.—Environmental conditions at release sites and in the transport tanks during each release.

Variable	Kaneohe stream (wild striped mullet release)			Kahaluu stream (wild striped mullet release)			Kahaluu stream (cultured striped mullet release)		
	Lot 1[a]	Lot 2[a]	Lot 3[a]	Lot 1[a]	Lot 2[a]	Lot 3[a]	Lot 1[b]	Lot 2[b]	Lot 3[b]
Field conditions									
Salinity (‰)									
Top	2	5	5	5	5	5	0	0	4
Bottom	32	32	32	29	29	26	20	30	23
Water temperature (°C)									
Top	26	27	26	29	32	30	31	27	25
Bottom	28	28	27	31	30	29	27	26	26
Dissolved oxygen (mg/L)									
Top		7.5	7.8		8.2	8.5	6.8	7.2	8.8
Bottom		5.6	5.5		3.8	6.0	6.2	3.2	3.9
Secchi disk depth (cm)									
Top	65	80	65	25	80	65	70	70	90
Bottom	70	80	85	75	105	105	115	75	115
Tide (cm)									
Stage	Low	Incoming	Incoming	Incoming	Outgoing	High	Incoming	Low	High
Height	18.3	61	36.6	21.4	51.9	67.1	64.1	24.4	64.1
Weather									
Cloud cover	60%	20%	70%	100%	30%	10%	70%	100%	100%
Condition		Trades	Trades	Breezy	Calm	Windy		Rain	Rain
Loading time	1100	1015	1020	1100	1445	1405	1100	1030	1100
Release time	1152	1120	1122	1445	1550	1510	1240	1230	1220
Transfer tank conditions									
Salinity (‰)	23	20	19	23	20	16	23	21	18.5
Dissolved oxygen (mg/L)									
Beginning	6.5	7.6	8.2	6.5	7.5	7.9	7.6	10.3	7.7
Ending	12.8	8.3	10.6	8.4	7.9	8.1	5.6	8.7	8.2
Water temperature (°C)									
Beginning	25.5	26.0	27.0	26.0	26.0	25.0	25.8	25.0	24.5
Ending	26.0	26.0	27.0	26.0	26.0	26.0	25.0	26.7	25.0

[a] Release dates were lot 1, 25 June; lot 2, 30 June; and lot 3, 2 July 1993.
[b] Release dates were lot 1, 30 July; lot 2, 6 August; and lot 3, 13 August 1993.

several stream systems that originate in the Koolau mountain range. The mouth of Kaneohe stream lies 11.6 km south of the Kahaluu stream mouth in South Kaneohe Bay.

To evaluate the effect of hatchery-released fish and to compare growth and survival of tagged wild fish with cultured fish, monthly collections were made with cast nets in several Kaneohe Bay nursery habitats. Monitoring began on 19 July 1993, 2 1/2 weeks after the last of the wild striped mullet releases, to establish initial dispersal patterns of wild fish before cultured striped mullet were released. Sampling during the post-treatment period began on 23 August 1993, 10 d after releasing the last lot of cultured fish. Each field collection was done over a 5-d period. There were 8 monthly collections.

Sampling design included collections at five nursery sites within Kaneohe Bay. The sites were Kaneohe stream (control site), Keaahala stream (1.1 km north of Kaneohe stream), Kahaluu stream (treatment site; 11.6 km north of Kaneohe stream), Kaalaea stream (12.6 km north of Kaneohe stream), and Waiahole stream (15 km north of Kaneohe stream). Collections were made during the day over an 8-h period at each sampling station. All collection sites were established in the vicinity of documented primary nursery habitats for striped mullet (Oceanic Institute 1991).

To standardize sampling effort, two substations were established at each station—one upstream, the other near the stream mouth. Within substations, 15 cast-net throws were made. Thus, a total of 150 cast-net samples were taken each month. To broaden the range of microhabitats and fish size

TABLE 2.—Summary statistics for 3,619 wild stripped mullet released into Kaneohe stream (control site) and 2,192 wild stripped mullet released into Kahaluu stream (treatment site). Beginning one month later, 29,354 cultured striped mullet were released at the treatment site to evaluate the effect of hatchery releases on dispersal of wild juveniles. All individuals were identified using coded wire tags.

Size (mm)	Wild fish				Cultured fish			
	Lot[a]				Lot[b]			
	1	2	3	Total (N)	1	2	3	Total (N)
Kaneohe stream								
45–60	519	512	821	1,852				
60–70	356	340	380	1,076				
70–85	204	193	197	594				
85–110	11	47	39	97				
110–130								
Total (N)	1,090	1,092	1,437	3,619				
Kahaluu stream								
45–60	468	227	0	695				
60–70	218	216	166	600			3,852	3,852
70–85	312	312	171	795	4,977	5,015	1,386	11,378
85–110	38	36	28	102	3,014	3,019	2,158	8,191
110–130					2,012	1,982	1,939	5,933
Total (N)	1,036	791	365	2,192	10,003	10,016	9,335	29,354

[a] Release dates were lot 1, 25 June; lot 2, 30 June; and lot 3, 2 July 1993.
[b] Release dates were lot 1, 30 July; lot 2, 6 August; and lot 3, 13 August 1993.

ranges sampled, two different size cast nets were used. Of the 15 casts per substation, 10 were conducted with a 5-m-diameter, 10-mm-mesh net, and 5 casts were made with a 3-m-diameter, 6.5-mm-mesh net. The smaller net was more effective in narrow upstream habitats.

Placement of cast nets was stratified over schools of striped mullet juveniles, rather than completely random. Random sampling yielded few wild striped mullet and very few tagged individuals. Striped mullet schooled in fairly low densities within the clearwater nursery habitats, and collections targeted these schools. Nevertheless, the data used to determine recapture rates, hatchery contribution rates, and proportions of tagged wild or cultured striped mullet in samples were randomly distributed, because we had no indication that schools, once sighted, contained tagged individuals.

Striped mullet sampled in these collections were measured and checked for tags by means of a portable tag detector (Northwest Marine Technology, Inc.). Tagged striped mullet were placed on ice and returned to the laboratory at The Oceanic Institute to be weighed and measured. Tags were extracted using a binary search to locate them within the snout region. Tag codes were read using a binocular microscope (40 × magnification). Less than 1% of the tags from recaptured fish were lost during extraction. To verify tag codes, each was read twice (once each by two different research assistants).

Data were analyzed using Systat (Wilkinson 1990). A randomized-block-design analysis of variance (ANOVA) was used to compare means. Treatments (fish type, release site, and fish size at release) were blocked over time (3 release lots). Proportions were arcsine transformed. Systat Basic was used to write tag decoding algorithms. For each recaptured fish, the algorithms identified fish type, place of origin, size at release, release site, release date, release lot, and number of fish released per treatment–lot combination, based on the binary tag codes. An error check algorithm was also used. Variance estimates are expressed throughout as standard errors.

Results

Release Statistics

After releasing 5,811 tagged wild striped mullet by early July at the control and treatment sites (Kaneohe and Kahaluu streams), 29,354 tagged cultured juvenile striped mullet were released at the treatment site in August to establish the primary treatment effect (hatchery release). Numbers of tagged and released fish varied among size groups. An attempt was made to keep numbers similar (within experimental groups) from lot to lot (Table 2).

To establish and acclimate identifiable wild fish at both the treatment and control sites, plans called for releasing at least 1,000 tagged wild fish in each

TABLE 3.—Total numbers of wild and cultured striped mullet recovered in cast-net samples at the release sites following releases at Kahaluu stream (treatment site) and Kaneohe stream (control site). Wild fish were tagged and released at Kahaluu and Kaneohe streams in July. Cultured fish were released at Kahaluu stream in August. The percentages of total fish recovered that were cultured fish are given.

Fish recovered	1993						1994		Total
	Jul	Aug	Sep	Oct	Nov	Dec	Jan	Feb	
Kahaluu stream									
Wild fish recovered									
Total (N)	440	372	192	323	107	167	360	77	2,038
Tagged (N)	23	15	9	17	14	11	16	1	106
Cultured fish recovered									
Total (N)		224	49	64	12	37	75	62	523
Percent		37.6	20.3	16.5	10.1	18.1	17.2	44.6	
Kaneohe stream									
Wild fish recovered									
Total (N)	232	53	58	110	13	142	96	102	806
Tagged (N)	62	5	5	14	1	21	11	10	129
Cultured fish recovered									
Total (N)				2					2
Percent				1.8					

of three release lots approximately 1 month prior to the release of cultured fish. However, because of difficulties inherent in capturing and holding wild fish, that target was achieved in only four of the six release lots. Only one-third of the 1,000 wild individuals targeted were available for the 2 July release at the treatment site.

All wild fish released at the treatment site were initially collected at that site. However, because of difficulties in collecting sufficient numbers of wild juveniles at the control site, 86% of the fish released at the control site were initially captured either at Waiahole stream or outside Kaneohe Bay. To evaluate any effect this may have had on dispersal from the control site, all tagged wild fish were coded to identify initial capture site in addition to the primary coding variables. Capture site had an insignificant effect on dispersal (G-test = 1.01, df = 1, $P = 0.31$).

Tag retention for cultured fish ($N = 1,533$) and wild fish ($N = 307$) subsampled across release lots averaged 98.1% ($\pm 0.9\%$) and 99.3% ($\pm 0.7\%$), respectively, after 75 d. With one exception (93.9%), tag-retention rates within all release lots exceeded 96%.

Recapture Summary

At least 596 tagged cultured fish were recaptured in monthly cast-net collections made in striped mullet nursery habitats over the 7-month period following their release. Based on the average 98.4% tag retention rate, the number of cultured fish sampled can be extrapolated as 607, or 2.1% of the cultured fish released. Seven tags (0.8%) were lost during the retrieval and reading process and thus are not included in this analysis.

Of the wild striped mullet tagged and released at the control site (control fish), 145 (4.0%) were recaptured in cast-net samples over the 8-month study period following their release. Of the wild fish released at the treatment site (treatment fish), 108 (4.9%) were recaptured during the same (8-month) period.

Numbers of tagged fish retrieved at the release sites decreased after 1 month postrelease (Table 3), but were fairly constant during the remainder of the study. Wild striped mullet were clearly more abundant at the treatment site. The total number of tagged wild fish recaptured was greatest at the control site (Kaneohe stream), which was expected because more wild fish were tagged and released there. However, this pattern varied considerably from month to month. Nearly one-half of the tagged wild fish recaptured at the control site were sampled within 30 d after their release; experimental wild fish were more abundant at the treatment site in five of the seven remaining collections.

Recaptured cultured fish constituted up to 44% of the total striped mullet in monthly samples at the treatment site (mean = $25.7 \pm 4.8\%$, $N = 7$ collection dates). Only 2 cultured fish were recaptured at the control site. After an initial decline in abundance, cultured fish clearly became established at the treatment site, where they persisted in samples through the end of the study (Table 3).

Effect of Hatchery Releases on Wild Striped Mullet

Pretreatment comparison of dispersal patterns.—Collections made in five nursery habitats in July 1993 revealed initial dispersal patterns of the wild fish released a month earlier. Because those collections were conducted prior to the primary treatment effect, the hatchery release in August, they established a control condition for examining hatchery-release effect.

There was little movement away from release sites prior to the release of hatchery fish. There was no significant difference in mean percent dispersal of treatment fish, compared with control fish, out of release habitats and into adjacent streams in July (Figure 2; ANOVA, $P > 0.18$, df = 2, $N = 3$ release lots). Data in Figure 2 are mean percent per lot of total fish recaptured within experimental treatment groups in July. Proportions of treatment and control fish recaptured at the sites where they were released also were not significantly different (ANOVA, $P > 0.49$, df = 2, $N = 3$). No treatment fish were collected outside the treatment site in July. However, five of the control fish collected in July had dispersed outside the control site; three of those had moved 12 km and were caught at the treatment site. Although this shows a trend towards slight movement out of the control site, this difference between treatment and control fish was trivial. Combining data from the 3 lots within treatments, 100% of the treatment fish and 92.5% of the control fish recaptured in July were collected at the site where they were released.

Post-treatment comparison of dispersal patterns.—Following the release of nearly 30,000 cultured fish at the treatment site in August, there was no trend towards greater dispersal of wild fish from the treatment site than from the control site (Figure 3; ANOVA, $P > 0.80$, df = 2, $N = 3$). The data plotted in Figure 3 are mean percent per lot ($N = 3$) of the total fish recaptured within treatment groups after the hatchery release. Summing across lots within treatments, 97.7% of the 85 treatment fish sampled from August 1993 through February 1994, and 85.9% of the 78 control fish sampled, were collected at the site where they were released.

Cultured striped mullet also showed a tendency to remain in the vicinity of the release site (Fig-

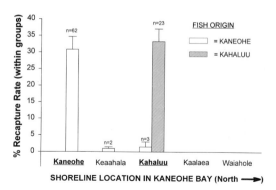

FIGURE 2.—Pretreatment dispersal patterns of tagged wild striped mullet by release location. Mean percent recaptured (per lot, within treatment groups) is shown with standard errors for wild fish released at the control site, Kaneohe stream (control fish) and wild fish released at the treatment site, Kahaluu stream (treatment fish). Wild fish were initially recaptured 2.5 weeks following the collection, tagging, and release of 5,811 wild juveniles at the control and treatment sites in July 1993.

ure 3). Only 12.2% of cultured fish recaptured had dispersed away from the treatment site. A total of 523 cultured fish were recaptured at the treatment site, an average of 25.7% of the total striped mullet collected there.

There was a clear trend towards less movement of the wild treatment fish than of the control fish from their respective release sites through the end

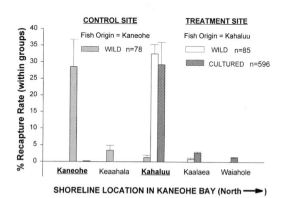

FIGURE 3.—Post-treatment dispersal patterns of wild and cultured striped mullet from release sites. Mean percent recaptured (per lot, within treatment groups) is shown with standard errors for control fish, wild fish released at the treatment site (treatment fish), and cultured fish released at the treatment site. Post-treatment collections began about 3 weeks after tagging and release of 29,354 cultured fish into Kahaluu stream.

of 1993 (Table 4). There was also clear movement of a portion of the cultured fish out of the treatment site and into two streams to the north. Cultured fish were consistently recaptured both at and outside the release site. With the exception of the November samples from Waiahole stream, cultured fish were always collected at Kaalaea and Waiahole streams as well as at the treatment site, Kahaluu stream.

A comparison of the relative contribution of monthly collections to total tagged wild fish caught within treatments reveals another similarity. Monthly fluctuations in recapture rates were nearly identical for treatment and control fish collected at release sites from month 2 on (Figure 4). After the first month in the wild, proportions were evenly distributed in collections over time. For wild treatment fish, no single collection dominated. Cultured fish were more similar to control fish, for which total recaptures were dominated by the number sampled during the first month after release (Table 4).

After analyzing these data, we noted that recapture rates of the treatment fish approached zero in February 1994 (month 8, Figure 4). To evaluate whether this was a sampling artifact, or a permanent decline in wild fish at the treatment site, we conducted an additional standard cast-net collection at Kahaluu stream in April 1994. The April collection yielded 155 striped mullet, of which 64 were tagged. Of the tagged fish, there were 8 wild and 56 cultured striped mullet. Thus, the low frequency of treatment fish in month 8 samples reflected natural sampling variability, as seen with control fish in month 5 (Figure 4).

Size-Specific Survival

The effect of fish size at release (SAR) on recapture rates was examined to evaluate whether hatchery releases at the treatment site affected size-specific survival rates of wild fish. Fish SAR had a clear effect on survival of cultured fish (Figure 5). Although four size-classes of cultured striped mullet were tagged and released (Table 2), the smallest size range released (SAR group 2, 60 to 70 mm TL) were prevalent in samples only during initial collections. By the last 2 months of the study, recapture rates were clearly a direct function of SAR, and proportions of fish that ranged from 60 to 85 mm TL at release (SAR groups 2 and 3) were underrepresented in samples.

Comparison of size-specific recapture rates of treatment fish, before versus after release of hatchery fish, revealed no significant differences in sur-

TABLE 4.—Dispersal patterns of 5,811 wild striped mullet released into Kaneohe Bay and then sampled in cast-net collections, both before and after releases of 29,354 cultured striped mullet into Kahaluu stream (treatment site) in August 1993. Dispersal of Kaneohe stream (control site) was used in both pretreatment and post-treatment collections to establish a control pattern. Data are mean (SE) number of individuals per release lot ($N = 3$) retrieved in 30 cast-net samples, July 1993–February 1994. Streams are sorted north to south.

Collection period and recapture site	Number of tagged wild fish recaptured		Number of tagged cultured fish recaptured
	Control fish	Treatment fish	
Pretreatment			
19–23 Jul			
Waiahole	0 (0)	0 (0)	
Kaalaea	0 (0)	0 (0)	
Kahaluu	1.0 (1.0)	7.7 (.9)	
Keaahala	0.7 (0.3)	0 (0)	
Kaneohe	20.7 (2.6)	0 (0)	
Post-treatment			
23–27 Aug			
Waiahole	0 (0)	0 (0)	1.7 (.9)
Kaalaea	0 (0)	0 (0)	4.7 (1.8)
Kahaluu	0.3 (.3)	5.0 (1.5)	74.7 (5.4)
Keaahala	0.7 (0.7)	0 (0)	0 (0)
Kaneohe	1.7 (0.3)	0 (0)	0 (0)
20–24 Sep			
Waiahole	0 (0)	0 (0)	0.7 (0.3)
Kaalaea	0 (0)	0.3 (.3)	0.7 (0.3)
Kahaluu	0.3 (0.3)	3.0 (1.0)	16.3 (3.9)
Keaahala	0.7 (0.3)	0 (0)	0 (0)
Kaneohe	1.7 (0.3)	0 (0)	0 (0)
18–22 Oct			
Waiahole	0 (0)	0 (0)	0.7 (0.3)
Kaalaea	0 (0)	0 (0)	1.0 (0.6)
Kahaluu	0.3 (0.3)	5.7 (1.5)	21.3 (11.1)
Keaahala	0.0 (0.0)	0 (0)	0 (0)
Kaneohe	4.7 (1.9)	0 (0)	0.7 (0.7)
18–24 Nov			
Waiahole	0 (0)	0 (0)	0 (0)
Kaalaea	0 (0)	0 (0)	0.7 (0.3)
Kahaluu	0 (0)	4.7 (2.2)	4.0 (0.6)
Keaahala	0.7 (0.3)	0 (0)	0 (0)
Kaneohe	0.3 (0.3)	0 (0)	0 (0)
13–17 Dec			
Waiahole	0 (0)	0 (0)	1.0 (1.0)
Kaalaea	0 (0)	0.3 (0.3)	3.7 (1.9)
Kahaluu	0 (0)	3.7 (0.7)	12.3 (5.2)
Keaahala	0.7 (0.3)	0 (0)	0 (0)
Kaneohe	7.0 (1.0)	0 (0)	0 (0)
10–14 Jan			
Waiahole	0 (0)	0 (0)	3.3 (0.3)
Kaalaea	0 (0)	0 (0)	1.0 (0.6)
Kahaluu	0 (0)	5.3 (.9)	25.0 (9.6)
Keaahala	0 (0)	0 (0)	0 (0)
Kaneohe	3.7 (3.2)	0 (0)	0 (0)
7–11 Feb			
Waiahole	0 (0)	0 (0)	0.3 (0.3)
Kaalaea	0 (0)	0 (0)	4.3 (1.3)
Kahaluu	0 (0)	0.3 (0.3)	20.7 (7.8)
Keaahala	0 (0)	0 (0)	0 (0)
Kaneohe	3.3 (1.2)	0 (0)	0 (0)

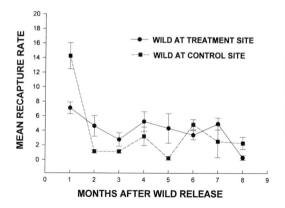

FIGURE 4.—The mean percent recaptured (within treatment groups) with standard errors for wild striped mullet released at the treatment site (Kahaluu stream) and wild fish released at the control site (Kaneohe stream) over the 8 months following the release of tagged wild fish.

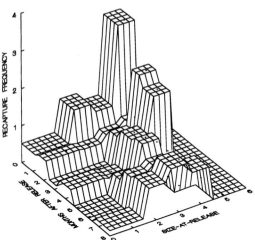

FIGURE 6.—Size-at-release effect on recapture rates of tagged wild striped mullet recaptured in cast-net samples at the treatment site, Kahaluu stream. Data are given for each of five size-classes (1 = 45–60 mm total length; 2 = 60–70 mm; 3 = 70–85 mm; 4 = 85–110 mm; and 5 = 110–130 mm). Data are percent recaptured of the total fish released per size interval. No wild striped mullet larger than 110 mm were released.

vival patterns. Four size-classes of wild fish were released (SAR groups 1 through 4). Note that no wild fish in the 110 to 130 mm bracket, the largest group of cultured fish released, were captured and released. Treatment and control fish from the smallest size groups released (SAR groups 1 and 2, 45 to 70 mm TL) were frequently encountered in samples throughout the study (Figures 6 and 7). Wild fish were less affected by SAR than were cultured fish. The relationship between SAR and recapture rates for treatment fish was similar to that for control fish.

Discussion

Dispersal of wild striped mullet at the treatment site (treatment fish) was clearly unaffected by the release of cultured striped mullet. There was slight movement of treatment fish after the hatchery release in August, but there was also even greater movement of control fish out of the control site during that period.

The data within individual collections reveal that there was no hatchery release effect on recapture rates of wild fish. Had the hatchery fish displaced wild fish, we would have expected higher proportions of treatment fish outside the release habitat. This was not the case. Only two (2.4%) of the treatment fish that were recaptured after the hatchery release in August were collected outside the release site. In contrast, 15% of the control fish and 12.2% of the cultured fish sampled after the hatchery release were collected outside their respective release sites. Had the treatment fish moved from Kahaluu stream, they would have been picked up

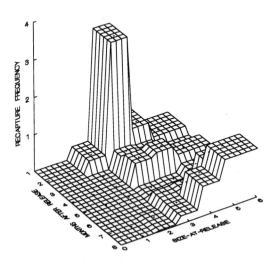

FIGURE 5.—Size-at-release effect on recapture rates of cultured striped mullet collected in cast-net samples at the treatment site, Kahaluu stream. Data are given for each of five size-classes (1 = 45–60 mm total length; 2 = 60–70 mm; 3 = 70–85 mm; 4 = 85–110 mm; and 5 = 110–130 mm). Data are percent recaptured of the total fish released per size interval. No cultured fish smaller than 60 mm were released.

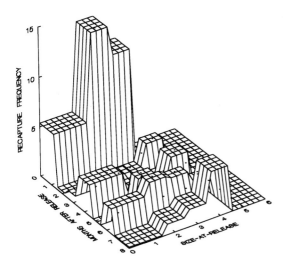

FIGURE 7.—Size-at-release effect on recapture rates of tagged wild striped mullet recaptured in cast-net samples at the control site, Kaneohe stream. Data are given for each of five size-classes (1 = 45–60 mm total length; 2 = 60–70 mm; 3 = 70–85 mm; 4 = 85–110 mm; and 5 = 110–130 mm). Data are percent recaptured of the total fish released per treatment group. No wild striped mullet larger than 110 mm were released.

elsewhere. This is supported by the fact that cultured fish moved out of Kahaluu stream and were subsequently sampled in adjacent streams during every collection period (Table 4). This same pattern was also evident in previous releases of cultured fish in Kaneohe Bay in 1990, 1991, and 1992 (Leber, in press; Leber, unpublished data). This study documents that Kahaluu stream, an important striped mullet nursery in Kaneohe Bay, could support a moderate hatchery release of around 30,000 striped mullet in addition to the wild juveniles already present in that nursery system at the time of the hatchery release.

The greater dispersal of wild fish from the control site may be related to the fact that wild fish were brought into the control site from other collection sites. Although most of the released fish that dispersed from the control site were originally captured elsewhere, G-test results indicated that differences in dispersal between striped mullet introduced from other streams and resident striped mullet could be explained by chance. Also, if the fish brought into the control site were dropped from this analysis, dispersal patterns of the treatment and control fish would be virtually identical.

As expected, based on previous studies in Hawaii (Leber, in press; Leber et al., in press) fish SAR had a marked effect on recapture rates of cultured striped mullet in this study. Survival of cultured striped mullet in the previous studies was directly related to SAR and was particularly poor for the smallest individuals released (45 to 60 mm TL). A similar, but less pronounced, trend was evident in this study for tagged wild fish. This trend was more pronounced at the treatment site. All sizes of wild fish released were represented in samples from both release sites throughout the study. Thus, cultured striped mullet did not appear to affect size-selective survival of wild fish.

Size at release is likely a principal mechanism controlling survival of cultured marine fish released into the wild. Although this relationship has been recognized with salmonids and other freshwater fishes, it has only recently been investigated with marine organisms. In addition to the Hawaiian studies, SAR effect has been shown for other marine fishes including madai *Pagrus major* in Japan (Tsukamoto et al. 1989; referred to as red sea bream by Tsukamoto), Atlantic cod *Gadus morhua* in Norway (Svasand and Kristiansen 1990), white seabass *Atractoscion nobilis* in California (Drawbridge et al., this volume), and red drum *Sciaenops ocellatus* in Florida (Willis et al., this volume).

Cultured fish averaged 25% of the total striped mullet collected monthly. This hatchery release increased striped mullet abundance at Kahaluu by one-third (i.e., from 3:0 to 3:1, wild:cultured fish). Thus, small-scale hatchery releases have the potential to make a substantial contribution to striped mullet production at this nursery site.

A key question remaining is what level of striped mullet abundances can the Kahaluu nursery support without adversely affecting wild stocks? How far below carrying capacity is this nursery now, and how near to that level should one attempt to get?

A large proportion of cultured fish in this study remained at the treatment site, yet they had a higher dispersal rate than did their wild counterparts at that site. The greater dispersal of hatchery fish suggests that cultured striped mullet might provide a useful tool for evaluating carrying capacity in these nursery habitats. One way to examine carrying capacity in Kaneohe Bay might be to use releases of cultured striped mullet to identify release magnitudes at which cultured fish show accelerated movement out of the habitat. In this study, about 12 to 15% appeared to be the dispersal rate for striped mullet in an unfamiliar nursery habitat.

The quantified results of hatchery contributions to wild stocks in Hawaii, Norway (Svasand et al. 1990), Japan (Honma 1993), California (Drawbridge et al., this volume), and Florida (Willis et al., this volume) indicate that marine enhancement has potential to increase abundances of targeted marine organisms. Given this potential, a responsible approach is needed to develop its use as an additional fishery management tool in coastal ecosystems (Blankenship and Leber, this volume). The results of this study document that small-scale releases of juveniles can increase striped mullet abundances in nursery habitats in Kaneohe Bay without displacing wild stocks. These results, combined with data showing that release season and release habitat appear to interact with fish SAR in controlling survival of cultured striped mullet in the wild (Leber, in press; Leber et al., in press; Leber, unpublished data), provide the minimum information needed to control hatchery-release effect on striped mullet population size.

To ensure that stocks are actually enhanced by hatchery-release activities, this sort of information from pilot studies needs to be coupled with additional management considerations to provide a controlled approach to marine enhancement—development of species management plans, well-defined indicators of success, prevention of genetic inbreeding and outbreeding depression, disease and health management, consideration of ecological interactions, identification of socioeconomic realities, and use of an adaptive management strategy (Blankenship and Leber, this volume).

In summary, there was no treatment effect in this experiment; hatchery releases did not displace wild individuals from the principal nursery habitat in Kaneohe Bay. The cultured fish released in this study increased abundances of striped mullet at the treatment site by around 33%. If these results were repeated in releases over several time periods, then even small-scale releases could help replenish the depleted striped mullet fishery in Kaneohe Bay. Conducting small-scale releases of cultured striped mullet in several nursery habitats in Kaneohe Bay should increase overall striped mullet abundances in this estuary.

This study corroborates earlier experiments that indicate hatchery releases can increase inshore fish abundances in Hawaii. If a careful approach is used, marine stock enhancement has considerable potential as an additional fishery management tool.

Acknowledgments

The authors wish to acknowledge several colleagues who participated in this study. We thank Lee Blankenship and Dan Thompson, of the Washington Department of Fish and Wildlife (WDFW), and Anthony Morano, Glenn Karimoto, Marcus Boland, Ryan Takushi, and all of the researchers in The Oceanic Institute Finfish Program for their tireless dedication in the tagging, field collection, laboratory work, and aquaculture work involved in this project. We thank Lee Blankenship, Rich Lincoln, Jim Shaklee, Ray Buckley, Dave Sterritt, and Jim West at WDFW for many fruitful discussions about the need to evaluate interactions of cultured fish with wild populations. We also thank Laurie Peterson for editing assistance and help with the graphics and Dave Sterritt for help with the database and graphics. This paper is funded by a grant from the National Oceanic and Atmospheric Administration (NOAA). The views expressed herein are those of the authors and do not necessarily reflect the views of NOAA or any of its subagencies.

References

Bergman, P. K., F. Haw, H. L. Blankenship, and R. M. Buckley. 1992. Perspectives on design, use, and misuse of fish tags. Fisheries (Bethesda) 17(4):20–24.

Blankenship, H. L., and K. M. Leber. 1995. A responsible approach to marine stock enhancement. American Fisheries Society Symposium 15:167–175.

Buckley, R. M., and H. L. Blankenship. 1990. Internal extrinsic identification systems: overview of implanted wire tags, otolith marks and parasites. American Fisheries Society Symposium 7:173–182.

Bryan, D. F., D. A. Rutherford, and B. Walker-Bryan. 1992. Acidification of the lower Mississippi River. Transactions of the American Fisheries Society 121(3):369–377.

Coutin, P. C., and I. A. Payne. 1989. The effects of long-term exploitation of demersal fish populations off the coast of Sierra Leone, West Africa. Journal of Fish Biology 35:163–167.

Drawbridge, M. A., D. B. Kent, M. A. Shane, and R. F. Ford. 1995. The assessment of marine stock enhancement in southern California: a case study involving the white seabass. American Fisheries Society Symposium 15:568–569.

FAO (Food and Agriculture Organization of the United Nations). 1992. FAO Yearbook: fishery statistics 70 (1990).

Fox, W. 1992. Testimony at Congressman G. Studds' Congressional Committee Hearing. U.S. Congressional Record HMM2750221992, Washington, DC.

Hardin, G. 1968. The tragedy of the commons. Science 162:1243–1248.

Honma, A. 1993. Aquaculture in Japan. Japan FAO Association, Chiyoda-Ku, Tokyo.

Kent, D. B., M. A. Drawbridge, and R. F. Ford. 1995. Accomplishments and roadblocks of a marine stock enhancement program for white seabass in California. American Fisheries Society Symposium 15:492–498.

Leber, K. M. In press. Significance of fish size-at-release

on enhancement of striped mullet fisheries in Hawaii. Journal of the World Aquaculture Society.

Leber, K. M., D. A. Sterritt, R. N. Cantrell, and R. T. Nishimoto. In press. Contribution of hatchery-released striped mullet, *Mugil cephalus*, to the recreational fishery in Hilo Bay, Hawaii. *In* Proceedings of the First Biennial Symposium for the Main Hawaiian Islands Marine Resources Investigation. Hawaii Department of Land and Natural Resources, Division of Aquatic Resources, Technical Report 94-03, Honolulu.

Moyle, P. B., B. Herbold, D. E. Stevens, and L. W. Miller. 1992. Life history and status of Delta Smelt in the Sacramento–San Joaquin Estuary, California. Transactions of the American Fisheries Society 121(1):67–77.

NOAA (National Oceanic and Atmospheric Administration). 1991. Our living oceans: first annual report on the status of U.S. living marine resources. NOAA Technical Memorandum NMFS (National Marine Fisheries Service) F/SPO-1.

NOAA (National Oceanic and Atmospheric Administration). 1992. Our living oceans: report on the status of U.S. living marine resources 1992. NOAA Technical Memorandum NMFS (National Marine Fisheries Service) F/SPO-2.

Oceanic Institute. 1991. Stock enhancement of marine fish in the state of Hawaii (SEMFISH) phase 4. Annual Report to National Marine Fisheries Service, Honolulu.

Ross, M. R., and G. A. Nelson. 1992. Influences of stock abundance and bottom-water temperature on growth dynamics of haddock and yellowtail flounder on Georges Bank. Transactions of the American Fisheries Society 121(5):578–587.

Stroud, R. H., editor. 1986. Fish culture in fisheries management. American Fisheries Society, Fish Culture Section and Fisheries Management Section, Bethesda, Maryland.

Svasand, T., K. E. Jorstad, and T. S. Kristiansen. 1990. Enhancement studies of coastal cod in western Norway, part 1. Recruitment of wild and reared cod to a local spawning stock. Journal du Conseil International pour l'Exploration de la Mer 47:5–12.

Svasand, T., and T. S. Kristiansen. 1990. Enhancement studies of coastal cod in western Norway, part 4. Mortality of reared cod after release. Journal du Conseil International pour l'Exploration de la Mer 47:30–39.

Tsukamoto, K., and six coauthors. 1989. Size-dependent mortality of red sea bream, *Pagrus major*, juveniles released with fluorescent otolith-tags in News Bay, Japan. Journal of Fish Biology 35(Supplement A):59–69.

Wilkinson, L. 1990. SYSTAT: the system for statistics. Systat, Evanston, Illinois.

Willis, S. A., W. W. Falls, C. W. Dennis, D. E. Roberts, and P. G. Whitchurch. 1995. Assessment of season of release and size at release on recapture rates of hatchery-reared red drum. American Fisheries Society Symposium 15:354–365.

Role of Exotic Species:

Past and Future Uses

in Fisheries Management

Introduced Species as a Factor in Extinction and Endangerment of Native Fish Species

DENNIS R. LASSUY

209 Poplar Road, Riva, Maryland 21140, USA

Abstract.—In a previous analysis of extinctions of 40 North American native fishes, habitat alteration was cited as a factor in 29 cases (73%), introduced species in 27 cases (68%), and contaminants in 15 cases (38%). The present analysis of factors cited in Endangered Species Act (ESA) fish listings revealed a similar pattern. Of the 92 species listed through 1991, 69 final listing notices provided sufficient information about factors contributing to endangerment to allow analysis in a manner similar to the extinctions analysis. In these 69 cases, habitat degradation was again the most commonly cited factor (63 listings, 91%); contaminants were cited in 28 listings (41%); and introduced species were cited in 48 listings (70%)—in 40 (58%) as a factor in species decline and in 8 others (12%) as a continuing threat. Of these 48 listings, 35 introductions related to sportfishing (i.e., introduced as game, forage, or bait species). As with extinctions, most ESA listings cited more than one factor, and most cases in which introduced species were cited appeared to have been the consequence of intentional introductions. A recently completed study by the U.S. Congressional Office of Technology Assessment concluded that a pattern of intentional introductions of fish and other species causing harm as often as do unintentional introductions reflects "a history of poor species choices and complacency regarding their potential harm." These patterns suggest the need for greatly improved decision making in species introductions if we are to reduce threats to native fish fauna and avoid this impression of complacency.

The U.S. Congressional Office of Technology Assessment (OTA) concluded that nonnative species "are here to stay and many of them are welcome" (OTA 1993). However, in the same document OTA concluded that other nonnative species "have had profound environmental consequences, exacting a significant toll on U.S. ecosystems." In reviewing the data for this paper, the validity of both of these seemingly contradictory statements was apparent. In presenting the results in various forums it has also been apparent that there is strong resistance, both by those who would introduce and those who generally oppose introductions, to accepting this validity.

The importance and value of nonnative species, at least in the United States, is exemplified by their extensive use in research, biological control, the aquarium industry, aquaculture, and fisheries management. While the benefits of using nonnative species are recognized, in this paper, I will focus on problems associated with fisheries management uses of nonnative species.

Nonnative species have become a component of current sportfishing programs in most states. The brown trout *Salmo trutta*, for example, is native to Europe and western Asia but was introduced across the United States and is now a popular recreational species. Large recreational fisheries have developed for Pacific salmon *Oncorhynchus* spp. introduced into the Great Lakes, although it should be noted that hatchery production is used to maintain yields for these fisheries. Anglers who fish for rainbow trout *O. mykiss* in Virginia, Colorado, and Pennsylvania, and the millions of anglers who fish for largemouth bass *Micropterus salmoides* outside its native range (e.g., Oregon, California, and Arizona) clearly depend on a nonnative species for their angling enjoyment. The introduction of several nonnative species, for all their sportfishing value, has also been the source of substantial problems for native species. In testimony before the U.S. Congress, for example, OTA noted that biodiversity has declined both by the loss of native species and the addition of nonnative species.

Fisheries managers have facilitated the demand for and expectation that nonnative fishes will continue to be available for recreational fishing. However, we have done a less thorough job of anticipating or understanding the potential consequences of introductions (e.g., predation, competition, habitat alteration, hybridization, and disease transfer) and their outcomes for native species. There is a growing literature on the uses of introduced species and their effects on native species to which readers may turn for additional information (see Rosenthal 1980; Garman and Nielsen 1982; Crossman 1991; DeVoe 1992; Rosenfield and Mann 1992). This paper provides an analysis of some of the most severe results of managing with nonnative fishes.

Methods

Miller et al. (1989) compiled information from a number of sources to provide a review of factors associated with the extinctions of 27 species and 13 subspecies of North American fishes over the past 100 years. They also generally assessed whether cited factors played a major role in the demise of each taxon. All analyses of extinctions in this paper are based on information presented by Miller et al. (1989).

Endangered Species Act (ESA; 16 U.S.C.A. §§1531 to 1544) listings (i.e., determinations of Threatened or Endangered status) are published in the *Federal Register* and are now required to include a description of factors that led to the listing. Information on factors cited in the listing of fish species under the ESA was derived from the files of the U.S. Fish and Wildlife Service for all fish listings through 1991. Many early listings provided only the name of the species with no specific information on the causes of decline or continuing threat and, therefore, were not included in my analysis. Adequate information existed for 69 of 92 U.S. species listed at the time of the analysis. Unfortunately, ESA listings did not consistently state the relative importance of the cited listing factors as Miller et al. (1989) did for extinctions.

Results

Extinctions

More than one factor was cited for most of the 40 extinct taxa analyzed by Miller et al. (1989). Habitat alteration was cited as a factor in 29 of 40 cases (73%), introduced species in 27 cases (68%), and contaminants in 15 cases (38%). Though the importance of factors other than introduced species is recognized and is presented here to clarify the relative frequency with which they have been cited by the reviewed sources, the remainder of this analysis is limited to introduced species effects.

Among the 27 cases in which introduced species were cited as a factor in extinction, over two-thirds (19 cases) were apparently the consequence of intentional introductions. Intentional introduction is used here to refer to purposefully bringing a species into an ecosystem, including containment facilities within them, to which that species is not native. The term intentional introduction therefore includes those taxa that are introduced directly (though not always legally) or indirectly into an aquatic habitat beyond their natural range through such actions as stocking game or forage species, or releasing bait or aquarium species. Species that have escaped from containment (e.g., aquacultural and aquarium production, rearing, or holding facilities) are thus also intentional introductions because escape from such facilities is a consequence of the initial introduction.

Introduced species were cited by Miller et al. (1989) as a "major" or "primary" factor in extinction of the native species in 10 of the 27 introduced species cases examined (37%). All 10 of these cases appear to have been the result of intentional introductions, with 7 of the 10 involving sportfishing introductions (i.e., introductions of game fish, forage for game fish, or bait species likely used in sportfishing). According to the information presented by Miller at al. (1989), habitat alteration was not cited as a factor in the extinctions of 6 of these 10 species. While I recognize that species decline in many cases is likely due to a combination of factors, Miller et al.'s (1989) findings indicate that habitat alteration is not, as is often suggested, always a necessary precursor to severe impacts by introduced species.

ESA Listings

The OTA (1993) noted that "biological communities can be radically and permanently altered without extinctions occurring." The first stated purpose of the ESA (Section 2(b)) is to provide a means of protecting the ecosystems upon which Threatened and Endangered species depend. There is perhaps no clearer signal, short of extinction, of the disruption of ecosystem integrity than the listing of one of its component species under the ESA.

In the 69 fish listings analyzed (Table 1), habitat alteration was the most commonly cited factor (63 species, 91%). Contaminants were cited in 28 species listings (41%), and introduced species were cited in 48 cases (70%)—in 40 (58%) as a factor in the decline of and in 8 others (12%) as a continuing threat to native species. Often several introduced species were cited and most ESA listings cited more than one factor. As with extinctions, most cases that cited introduced species appeared to have been the consequence of intentional introductions.

Among the 48 cases that cited introduced species as a listing factor, 7 involved ornamental species, 7 involved aquacultural species (other than ornamentals), and 6 related to pest control. Of the 48 cases, 35 involved sportfishing introductions; centrarchids were the most frequently cited taxon of sport fish.

The largemouth bass was the most frequently cited individual species (21 cases). Green sunfish *Lepomis cyanellus* was cited in nine cases. Other centrarchids included bluegill *L. macrochirus*, crap-

TABLE 1.—Analysis of factors cited in listing of fish species under the Endangered Species Act.[a] Names follow Robins et al. (1991).

Common name	Scientific name	Listing factor(s)			Purpose of introduction			
		Habitat alteration	Pollution	Introduced species	Sportfishing	Pest control	Ornamental	Aquaculture
Catfish, Yaqui	Ictalurus pricei	X		X	X			
Cavefish, Alabama	Speoplatyrhinus poulsoni		X					
Cavefish, Ozark	Amblyopsis rosae		X					
Chub, bonytail	Gila elegans	X		X	X			X
Chub, Borax Lake	Gila boraxobius	X		X[b]				
Chub, Chihuahua	Gila nigrescens	X	X	X	X	X		
Chub, humpback	Gila cypha	X		X	X			X
Chub, Hutton tui	Gila bicolor spp.	X	X[b]	X[b]				
Chub, Owens tui	Gila bicolor snyderi	X		X	X			
Chub, slender	Erimystax cahni	X	X					
Chub, Sonora	Gila ditaenia			X	X			
Chub, spotfin	Cyprinella monacha	X	X					
Chub, Virgin River	Gila robusta seminuda	X		X	X			
Chub, Yaqui	Gila purpurea	X		X	X			
Dace, Ash Meadows speckled	Rhinichthys osculus nevadensis	X		X	X	X		
Dace, blackside	Phoxinus cumberlandensis	X		X	X			
Dace, Clover Valley speckled	Rhinichthys osculus oligoporus	X		X	X			
Dace, desert	Eremichthys acros	X		X[b]				
Dace, Foskett speckled	Rhinichthys osculus spp.	X		X[b]				
Dace, Independence Valley speckled	Rhinichthys osculus lethoporus	X		X	X			
Dace, Moapa	Moapa coriacea	X		X			X	
Darter, amber	Percina antesella	X	X	X[b]				
Darter, bayou	Etheostoma rubrum	X	X					
Darter, Elk River	Etheostoma wapiti	X						
Darter, goldline	Percina aurolineata	X	X					
Darter, leopard	Percina pantherina	X	X					
Darter, Niangua	Etheostoma nianguae	X		X	X			
Darter, slackwater	Etheostoma boschungi	X						
Darter, snail	Percina tanasi	X						
Logperch, Conasauga	Percina jenkinsi	X	X	X[b]				
Logperch, Roanoke	Percina rex	X	X					
Madtom, Neosho	Noturus placidus	X	X					
Madtom, Scioto	Noturus trautmani	X						
Madtom, Smokey	Noturus baileyi	X	X					
Madtom, yellowfin	Noturus flavipinnis	X	X					
Minnow, loach	Rhinichthys cobitis	X		X	X			
Pupfish, Ash Meadows Amargosa	Cyprinodon nevadensis mionectes	X		X	X	X	X	
Pupfish, desert	Cyprinodon macularius	X	X	X	X		X	X
Pupfish, Devils Hole	Cyprinodon diabolis	X						
Pupfish, Leon Springs	Cyprinodon bovinus	X		X	unclear			
Sculpin, pygmy	Cottus pygmaeus	X	X					
Shiner, beautiful	Cyprinella formosa	X		X	X			
Shiner, blue	Cyprinella caerulea	X	X					
Shiner, Cahaba	Notropis cahabae		X					
Shiner, Cape Fear	Notropis mekistocholas	X	X					
Shiner, Pecos bluntnose	Notropis simus pecosensis	X	X	X	unclear			
Silverside, Waccamaw	Menidia extensa		X	X[b]	X[b]			
Spikedace	Meda fulgida	X		X	X			
Spinedace, Big Spring	Lepidomeda mollispinis pratensis	X		X			X	
Spinedace, Little Colorado	Lepidomeda vittata	X	X	X	X			
Spinedace, White River	Lepidomeda albivallis	X	X	X		X	X	
Springfish, Hiko White River	Crenichthys baileyi grandis	X		X	X		X	
Springfish, Railroad Valley	Crenichthys nevadae	X		X			X	X
Springfish, White River	Crenichthys baileyi	X		X	X		X	
Squawfish, Colorado	Ptychocheilus lucius	X		X	X			
Sturgeon, pallid	Scaphirhynchus albus	X						
Sturgeon, Gulf	Acipenser oxyrhynchus desotoi	X	X	X[b]				X[b]

(Continued on next page)

TABLE 1.—Continued.

Common name	Scientific name	Habitat alteration	Pollution	Introduced species	Sportfishing	Pest control	Ornamental	Aquaculture
Sucker, June	Chasmistes liorus	X	X	X	X			X
Sucker, Lost River	Deltistes luxatus	X	X	X	X			
Sucker, Modoc	Catostomus microps	X		X	X			
Sucker, razorback	Xyrauchen texanus	X		X	X			X
Sucker, shortnose	Chasmistes brevirostris	X	X	X	X			
Sucker, Warner	Catostomus warnerensis	X		X	X			
Topminnow, Gila	Poeciliopsis occidentalis	X		X	X	X		
Trout, Apache	Oncorhynchus apache	X		X	X			
Trout, greenback cutthroat	Oncorhynchus clarki stomias	X		X	X			
Trout, Lahontan cutthroat	Oncorhynchus clarki henshawi	X		X	X			
Trout, Little Kern golden	Oncorhynchus aguabonita whitei	X		X	X			
Trout, Paiute cutthroat	Oncorhynchus clarki seleniris	X		X	X			

^aAnalysis limited to those species for which information in Fish and Wildlife Service ESA final rule file included the five ESA listing factors.
^bCited as continuing threat rather than cause of decline.

pie *Pomoxis* spp., smallmouth bass *Micropterus dolomieu*, and other sunfish. Ictalurids were also commonly cited: channel catfish *Ictalurus punctatus* was cited in 7 cases and a variety of bullhead species *Ameiurus* spp. were cited in 11 cases. Cited baitfish species included red shiner *Cyprinella lutrensis* and the fathead minnow *Pimephales promelas* and "other baitfish." Rainbow trout and brown trout were cited in seven and seven cases, respectively, primarily for having caused problems through hybridization with native trout species or as predators of smaller species. In most cases, the listing information did not indicate whether the introduction was sanctioned by a public agency.

Discussion

Primack (1993) pointed out that whereas patterns of evolution have proceeded largely as a result of geographic isolation, "humans have radically altered this pattern by transporting species throughout the world." Any introduced species that survives the transfer necessarily affects the receiving ecosystem. In a recent text on biological pollution, Courtenay (1993) summarized that "every introduction will result in impacts to native biota, which may range from almost nil to major, including extinction, with time." Nonnative species can affect native species through a number of mechanisms including hybridization, competition, predation, pathogen transfer, and habitat alteration.

As noted earlier in this paper, prior habitat degradation is not a necessary precursor to severe impacts from introduced species. However, habitat degradation clearly can make a species and its supporting ecosystem more vulnerable to the effects of a nonnative species. This is apparently the case in the Colorado River where the combination of dams and introduced species has led to the endangerment of four native fish species adapted to large, flowing river systems (Minckley 1991). Moyle and Williams (1990) determined that large water projects, in concert with introductions of fish species better able to cope in altered habitats, were largely responsible for the decline of California's native fish fauna. In particular, the presence of introduced species was a "very important factor" or "principal factor" (Moyle and Williams 1990) in the status of 49% of species described as extinct, endangered, or in need of special protection.

Whether habitat has been altered or not, the decision to introduce must be made with great care. Unfortunately, the results of a recent investigation of a group of aquatic taxa (OTA 1993) suggests this may not have always been the case. Whereas the view is often expressed that unintentional introductions constitute the major source of problems to natural ecosystems, OTA (1993) found that intentional introductions, even using a narrower interpretation of intentional (*viz*, deliberate releases) than is used in this paper, are as likely to cause problems as are unintentional introductions. The OTA concluded that this pattern reflects "a history of poor species choices and complacency regarding their potential harm." Whereas the results of this analysis may support OTA's conclusion about fisheries management choices, I am less convinced that the source of our mistakes is complacency.

I suggest that the record of "poor species choices" is one of false assumptions and unrealistic expectations. For example, in situations where human activity has so altered ecosystems that native species have been lost or severely reduced, nonnative species or specific different life stages of native species have been used in efforts to restore some perceived ecosystem function. When an altered environment cannot support a particular life stage of a native species, culture techniques may serve a useful purpose in bridging the gap until the native species is again able to persist on its own. An example of this type is the reintroduction of cordgrass *Spartina alterniflora* for shoreline stabilization along the U.S. Atlantic coast. However, this same species was then used outside its native range and is now the source of increasingly severe problems in the Pacific Northwest. Though the same introduction decision was made, the outcome when the species was used outside its native range was very different.

Use of nonnative species to maintain ecosystem function must rely on solid understanding and realistic expectations. To some extent, expectation and prediction can be improved by gathering information on both the species being considered for introduction and the receiving environment. However, I believe OTA (1993) identified a particularly important basis for false assumptions when it singled out fisheries managers for continuing to use the "erroneous concept" of the vacant niche (i.e., "filling" a perceived void in an ecosystem with an introduced species).

For example, the waters behind a new dam may concentrate detritus and silt-dwelling invertebrates where a previously abundant stream-dwelling native salmonid now survives only in low numbers. The ecosystem continues to function in some manner; we simply don't care for what the altered energy and nutrient use pattern is now producing as a result of the manipulation. Because we do not see the outputs of the altered system as anything of immediate use, some refer to the new pattern as having "voids."

In many past cases, species chosen to "fill the void" appear to have been selected without considering potential effects on the receiving ecosystem because those species were deemed to be of more immediate benefit to humans than what persisted of the native community in the altered ecosystem. Perhaps it was seen as simpler to look for ways to channel the altered resource use pattern into a product of more immediate human benefit than to address alternatives to the proposed manipulation seriously or even to look for ways to minimize its consequences. Often the choice has been instead to manipulate the system further by introducing new species to fill these illusory empty niches.

In the waters behind the new dam cited above, one biologist may see just a single "empty niche" and introduce carp to convert the detritus and invertebrate biomass into fish flesh. Another biologist (or creative but misguided angler) may imagine any number of "empty niches" to fill and decide, for example, to introduce a crayfish and a small catostomid to feed on the detritus now concentrated in that portion of the watershed, plus maybe a small centrarchid to prey on the newly abundant benthic invertebrates. Then a large predatory centrarchid or two may be introduced to feed on this prey base and create a new fishery. Some refer to this approach of filling imaginary empty niches by introducing a whole suite of species as "ecosystem management," though most often it involves only a portion of the ecosystem. I believe it is more akin to ecosystem recreation, with all of the attendant evolutionary ramifications for native species throughout that and any interconnected ecosystems. Because of the uncertainties of predicting a particular result in such cases, OTA appropriately warns that "application of this approach to natural communities is inappropriate."

One other issue that must be addressed to understand the record on introductions, one that clearly links introductions to activities that alter habitats, is that introductions have often been driven by required mitigation for federal activities (e.g., dams). The species chosen to meet mitigation demands could have been native but often has been a nonnative species. Often nonnative species are the simplest alternative because culture techniques for a few commonly used species are well understood. Because of their prior use in other environments, these commonly used nonnatives are also species the public has become accustomed to seeing portrayed as the preferred species.

Conclusions

Nonnative aquatic species have been and continue to be both a source of economic benefits and costs to many sectors of society and a major factor in the loss of biological diversity. Despite this importance, the implications of nonnative species introductions have in general been underrecognized. This may be changing.

Recent headlines have included such items as the following: "Exotic Plants, Animals Imperil U.S. Ecosystems" (*Los Angeles Times*); "Court Action is Studied to Shut CAP" (*Arizona Republic*, 21 January 1994—referring to the potential for nonnative species transported by the Central Arizona Project, to harm native species); "Biology That's Alien and Expensive" (*Washington Post*, 7 October 1993); and "Introduction of Nonnative Fish is Devastating Many Local Rivers and Lakes" (*Oregonian*, 28 November 1990). The articles have not projected positive images of fisheries management decisions, but I believe the increasing awareness of this issue within and outside the fisheries profession suggests the need to improve upon our record.

Though calls by Congress for further research have been used as a delaying or obstructionist tactic, additional research can help clarify the risks of nonnative species introductions, prioritize actions intended to minimize such risks, and enable and promote the use of native species. However, I believe the greatest need is a change in attitude from one dominated by value judgements based on immediate human benefit to one that values the integrity of native ecosystems and all of its component species, a huge long-term benefit to our children's children, indeed to the human species. Soulé (1986) warned that "dithering and endangering are often linked;" let us not dither any longer.

Acknowledgments

This paper is in part the result of work done while the author chaired the Intentional Introductions Policy Review (review) Committee of the federal interagency Aquatic Nuisance Species (ANS) Task Force established under the Nonindigenous Aquatic Nuisance Prevention and Control Act of 1990. Though much of the information and language used in this paper are directly adapted from the review, this paper is presented as the position of the author only, not the ANS Task Force or any of its member agencies.

References

Courtenay, W. R., Jr. 1993. Biological pollution through fish introductions. Pages 35–61 *in* B. N. McKnight, editor. Biological pollution: the control and impact of invasive exotic species. Indiana Academy of Science, Indianapolis.

Courtenay, W. R., Jr., and J. R. Stauffer, Jr., editors. 1984. Distribution, biology, and management of exotic fishes. Johns Hopkins University Press, Baltimore, Maryland.

Crossman, E. J. 1991. Introduced freshwater fishes: a review of the North American perspective with emphasis on Canada. Canadian Journal of Fisheries and Aquatic Sciences 48(supplement):46–57.

DeVoe, M. R., editor. 1992. Introductions and transfers of marine species: achieving a balance between economic development and resource protection. South Carolina Sea Grant Consortium, Charleston.

Garman, G. C., and L. A. Nielsen. 1982. Piscivory by stocked brown trout (*Salmo trutta*) and its impact on the nongame fish community of Bottom Creek, Virginia. Canadian Journal of Fisheries and Aquatic Sciences 39:862–869.

Miller, R. R., J. D. Williams, and J. E. Williams. 1989. Extinctions of North American fishes during the last century. Fisheries (Bethesda) 14(6):22–38.

Minckley, W. L. 1991. Native fishes of the Grand Canyon region: an obituary? Pages 124–177 *in* Colorado River ecology and management. National Academy Press, Washington, DC.

Moyle, P. B., and J. E. Williams. 1990. Biodiversity loss in the temperate zone: decline of the native fish fauna of California. Conservation Biology 4(3):275–284.

OTA (Office of Technology Assessment). 1993. Harmful non-indigenous species in the United States. U.S. Government Printing Office, Washington, DC.

Primack, R. B. 1993. Essentials of conservation biology. Sinauer Associates, Sunderland, Massachusetts.

Robins, C. R., and six coauthors. 1991. Common and scientific names of fishes from the United States and Canada. American Fisheries Society Special Publication 20.

Rosenfield, A., and R. Mann, editors. 1992. Dispersal of living organisms into aquatic ecosystems. University of Maryland Press, College Park.

Rosenthal, H. 1980. Implication of transplantations to aquaculture and ecosystems. Marine Fisheries Review 42:1–14.

Soulé, M. E. 1986. Conservation biology: the science of scarcity and diversity. Sinauer Associates, Sunderland, Massachusetts.

Coldwater Fish Stocking and Native Fishes in Arizona: Past, Present, and Future

JOHN N. RINNE

U.S. Forest Service, Rocky Mountain Forest and Range Experiment Station
Southwest Forest Science Complex, 2500 South Pineknoll Drive
Flagstaff, Arizona 86001, USA

JOE JANISCH

Arizona Game and Fish Department, 2221 West Greenway Road
Phoenix, Arizona 85023, USA

Abstract.—Since the 1930s almost 70 million fish representing 17 nonnative species have been introduced into lakes and streams of the Little Colorado and Black river drainages in the White Mountains of east-central Arizona. The two drainages historically contained populations of native Apache trout *Oncorhynchus apache* and a native cyprinid species, Little Colorado spinedace *Lepidomeda vittata*. Both are classified as Threatened species. The declines of these fishes have resulted from stocking of nonnative species, principally rainbow trout *Oncorhynchus mykiss*, brown trout *Salmo trutta*, and brook trout *Salvelinus fontinalis*. Establishment of nonnative salmonids was facilitated in many cases by stream renovations with fish toxicants and baitfish introductions. Habitat alterations related to land management activities, principally timber harvest and livestock grazing, further affected the Apache trout and Little Colorado spinedace. Proposed changes in stocking strategies and innovative management activities will be instrumental in sustaining these two species and other native fishes of the state.

As Europeans moved west across the continental United States they brought not only personal possessions but their habits, traditions, hobbies, pets, and desire to fish for recreation and food. Water in the southwestern United States was more limited than the colonists were accustomed to, and stream discharges commonly fluctuated from flood to drought. The native fish fauna of the region was depauperate and largely comprised of smaller minnows and killifishes (Minckley 1973; Rinne and Minckley 1991).

Alteration of natural flow patterns of the arid intermountain southwestern United States began with completion of the first U. S. Bureau of Reclamation dam in Arizona in 1911 (Figure 1). Roosevelt Dam and the dams that followed in the next half century markedly altered the hydrology of this arid region (Rinne 1990a, 1994); rivers and streams that once flowed seasonally or annually in wet cycles were now either altered in quantity and quality of water or completely dried.

Water supply reservoirs provided relatively stable water bodies and regulated downstream river flows, thereby providing habitats conducive to establishment and proliferation of nonnative sport-fish species. Stocking common carp *Cyprinis carpio*, channel catfish *Ictalurus punctatus*, and black bass *Micropterus* spp. began about 1900 and has continued to the present (U.S. Fish and Wildlife Service 1979; Stephenson 1985; Rinne, 1995). Presently, Arizona's fish fauna, numbering about two dozen native species, has been doubled by nonnative fish introductions (Figure 2). Although these introductions were largely of sport species, many nonnative minnow species were introduced as forage for sport fishes and as bait for the sport-fishing industry (Miller 1952). Several species were also introduced for biological control.

In addition to warmwater species introductions to lower elevation impoundments, nonnative salmonids, principally rainbow trout *Oncorhynchus mykiss*, brown trout *Salmo trutta*, and brook trout *Salvelinus fontinalis* were introduced into upper elevation streams and reservoirs throughout the southwestern United States. Stocking activities for sport fishes in Arizona were largely under the jurisdiction and carried out by the U. S. Fish and Wildlife Service and the Arizona Game and Fish Department (Stephenson 1985). Streams in the Rocky Mountain region already were inhabited by native cutthroat trout *Oncorhynchus clarki* and subspecies (Gresswell 1988), Gila trout *O. gilae*, and Apache trout *O. apache* (Miller 1950, 1972; Rinne 1988, 1991a, 1991b). Because of spawning habits (Rinne 1985; Rinne and Minckley 1985; Rinne et al. 1986), increased competition, or predation (Rinne et al. 1981), the native trouts declined after introductions of nonnative trouts. Furthermore, native minnows and suckers also inhabiting these upper elevation montane

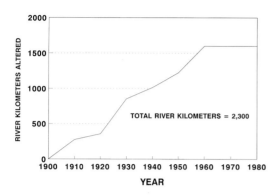

FIGURE 1.—Cumulative modification of mainstream rivers by dams in Arizona, 1900–1980.

streams (Minckley 1973; Rinne and Minckley 1991; Rinne and Medina, in press) were affected through predation by the nonnative salmonids (Blinn et al. 1993). In addition to the widespread introductions of nonnative salmonids stocked from hatcheries, management activities including fish eradications were common practice to enhance survival of nonnative salmonids (Rinne and Turner 1991).

In this paper we discuss (1) coldwater stockings of principally salmonid species in the White Mountains of east-central Arizona, (2) management activities and regulations attendant to sport-fishery management, and (3) changes in agency philosophy to restore native fishes.

Stocking History

Since the 1930s over 61 million nonnative sport fishes have been introduced into lakes in the Little Colorado and Black river drainages (Figure 3; Ta-

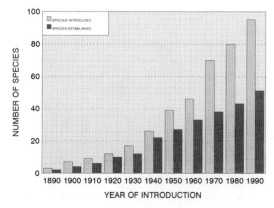

FIGURE 2.—Nonnative fish species introduced and established in Arizona, 1890–1990.

FIGURE 3.—The Little Colorado and Black river drainages in the White Mountains, Arizona. Lakes and reservoirs are (1) Big Lake, (2) Crescent Lake, (3) Lee Valley Reservoir, (4) Lyman Lake, (5) Nelson Reservoir, (6) Becker Lake, and (7) Colter Reservoir.

ble 1). In addition, about 8 million nonnative sport fish were introduced to these two rivers and their tributaries (Tables 2 and 3). Combining lake and stream introductions, almost 41 million hatchery-cultured nonnatives were stocked into the Black River system and 28 million individuals were stocked into all water bodies in the Little Colorado River system. In addition to sport fishes, fathead minnows, tadpoles, and crayfish were introduced as bait in lake environments. Stockings consisted predominantly of rainbow trout (60%), followed by cutthroat trout (12%), brook trout (7%), and brown trout (3%).

Effects of Introductions

The Little Colorado and Black river drainages are part of the historic range of Apache trout and Little Colorado spinedace *Lepidomeda vittata*. These two native species are currently listed as Threatened under the Endangered Species Act of 1973 (16 U.S.C.A. §§ 1531 to 1544). The Apache trout ranged throughout the headwaters of the Black, White, and Little Colorado rivers (Minckley 1973; Rinne 1985). Rinne and Minckley (1985), based on morphometric evidence, documented that pure populations of the native Apache trout occurred in streams where rainbow trout were not

TABLE 1.—Nonnative fishes stocked into White Mountain lakes, Arizona, 1930–1991.

Lake and year	Species	Number of stocking events	Number stocked
Becker			
1937	Largemouth bass[a]	1	510
1979–1982	Fathead minnow[b]	2	750
1985	Brook trout	1	10,000
1978–1988	Brown trout	14	121,700
1979–1987	Cutthroat trout	6	135,100
Big			
1940–1970	Arctic grayling[c]	33	3,941,000
1936–1991	Brook trout	41	1,532,000
1942	Brown trout	1	400
1940–1991	Cutthroat trout	61	6,604,000
1936–1939	Native cutthroat trout[d]	7	178,000
1936–1991	Rainbow trout	234	11,434,356
1968	Tadpole[e]	1	575
Colter			
1973–1982	Brook trout	2	15,000
1952–1975	Brown trout	4	16,600
1953–1978	Rainbow trout	12	87,560
Concho			
1977–1991	Brook trout	22	619,000
1971	Crayfish	1	325
1980–1991	Cutthroat trout	7	245,300
1957	Rainbow trout	123	4,213,000
Crescent			
1973–1991	Brook trout	31	1,008,000
1945–1953	Cutthroat trout	7	482,000
1940–1991	Rainbow trout	141	3,837,000
Lee Valley			
1965–1991	Arctic grayling[c]	11	107,500
1953–1985	Brook trout	28	356,300
1955–1991	Rainbow trout	4	50,800
Luna			
1943–1991	Brook trout	19	502,200
1935–1979	Brown trout	5	126,900
1952–1990	Cutthroat trout	10	316,400
1958–1964	Kokanee[f]	2	156,100
1937–1938	Native cutthroat trout[d]	4	154,100
1937–1991	Rainbow trout	222	7,084,251
1986	Tadpole[e]	1	2,700
Lyman			
1978	Black crappie[g]	1	20,000
1935–1976	Bluegill[h]	4	6,993
1953–1968	Brown trout	8	219,000
1965–1991	Channel catfish	24	1,678,000
1952–1966	Cutthroat trout	4	369,000
1978	Fathead minnow[b]	1	40,000
1935–1991	Largemouth bass[a]	8	40,500
1967–1973	Northern pike[i]	8	303,100
1952–1991	Rainbow trout	89	3,184,000
1979	Redear sunfish[j]	1	15,000
1967	Tadpole[e]	1	5,000
1959	Threadfin shad[k]	1	750
1973–1984	Walleye[l]	6	3,250,000
1979	Yellow perch[m]	1	100,000
Nelson			
1977–1991	Brook trout	18	450,400
1978–1991	Brown trout	10	73,100
1975	Channel catfish	1	10,000
1987–1991	Cutthroat trout	10	208,300
1958–1991	Rainbow trout	106	3,312,000
1968–1969	Tadpole[e]	2	9,500

(*Continued on next page*)

TABLE 1.—Continued.

Lake and year	Species	Number of stocking events	Number stocked
Showlow			
1969	Arctic grayling[c]	1	5,000
1982–1989	Brook trout	5	138,700
1986	Bluegill[h]	1	3,000
1977–1991	Brown trout	21	391,100
1976–1991	Channel catfish	19	149,700
1972	Cutthroat trout	1	15,000
1960	Kokanee[f]	1	20,000
1969–1970	Northern pike[i]	3	125,000
1954–1991	Rainbow trout	386	3,918,000
Total			61,399,570

[a]*Micropterus salmoides*; [b]*Pimephales promelas*; [c]*Thymallus arcticus*; [d]black spotted mountain cutthroat trout and Wyoming native cutthroat trout *Oncorhynchus clarki* subsp.; [e]*Rana* spp.; [f]lacustrine sockeye salmon *Oncorhynchus nerka*; [g]*Pomoxis nigromaculatus*; [h]*Lepomis macrochirus*; [i]*Esox lucius*; [j]*Lepomis microlophus*; [k]*Dorosoma petenense*; [l]*Stizostedion vitreum*; [m]*Perca flavescens*.

stocked on a sustained basis. Rainbow trout introduced into waters of the Black River drainage (Tables 1 and 2) have negatively affected the Apache trout through hybridization (Miller 1972; Rinne 1985; Rinne et al. 1986). In addition, the extensive introductions of brook and brown trouts reduced Apache trout distribution and abundance through competition for food and habitat (Rinne et al. 1981).

The Little Colorado spinedace occurred throughout the Little Colorado River drainage (Miller 1963; Minckley and Carufel 1967). Distribution and abundance of this species are now greatly reduced (Miller 1963; Minckley and Carufel 1967; Minckley 1984). Rainbow trout prey on Little Colorado spinedace, and Little Colorado spinedace abundance is inversely related to rainbow and brook trout abundance (Blinn et al. 1993).

The effects of other introduced sport and bait fishes that are abundant in waters now occupied by the Little Colorado spinedace in streams in the White Mountains are unknown. Interbasin transfers of bait species and a lack of consideration for the recipient drainage and its fauna is a national and international problem (Litvak and Mandrak 1993). The baitfish industry has had considerable influence in the lower Colorado River (Miller 1952). Bullfrog tadpoles, for example, are opportunistic in feeding habits (Clarkson and Devos 1986) and could negatively affect fry and larvae of native and juvenile fishes. The movement of introduced minnows, crayfish, and frogs from one habitat to another by anglers, although illegal, continues today.

Fishery management activities to ensure the success of nonnative sport fishes also impacted the native ichthyofauna. In the 1950s and 1960s management activities for salmonid sport fisheries in upper elevation montane streams involved extensive removal of native fish with toxicants. In most cases, these fishes were minnows and suckers, some of which are now Rare, Threatened, or Endangered species. Earliest records are from the 1950s when toxaphene was used to eradicate fish from Lyman and Becker lakes (Hemphill 1954). Effects of the toxicant downstream in the Little Colorado River were not measured. However, the Little Colorado spinedace was noted as part of the kill in that eradication activity. Because 23 km of stream treated on the Little Colorado River between Springerville and Lyman Lake were prime Little Colorado spinedace habitat, the effect on Little Colorado spinedace probably was significant.

Reports in the 1960s referred to most stream renovation activities as "minor stream habitat manipulations of a developmental nature" (Bruce 1961). Stream renovations with rotenone were extensive in the state of Arizona in the early 1960s; Rinne and Turner (1991) reported 135 km treated with fish toxicants during 1958–1968. More recent efforts have been designed to remove nonnative sport salmonids and reintroduce the Apache trout—a native, and potential, sport species (Table 4). Often, however, other species such as speckled dace *Rhinichthys osculus* and desert sucker *Catostomus clarki* were natural components of the native fish community in these montane streams and were adversely affected. Restoration of natural fish communities was incomplete at best.

Land Management

By the 1930s, as sport-fish management was being initiated in the White Mountains, these lands already had sustained extensive timber harvesting and livestock grazing for a quarter of a century or more. Maximum timber harvest was the primary objective of multiple use management. Effects of logging on

TABLE 2.—Nonnative fishes stocked in streams in the historic range of Apache trout in the Apache Sitgreaves National Forest, White Mountains, Arizona, 1920–1991.

Stream and year	Species	Number of stocking events	Number stocked
Bear Wallow Creek			
1920–1933	Native cutthroat trout[a]	4	9,400
1933	Brook trout	1	6,400
1935–1954	Rainbow trout	2	2,300
Beaver Creek			
1939–1953	Rainbow trout	2	6,700
1933–1977	Brown trout	2	6,500
Black River			
1959–1981	Brown trout	3	6,981
1938–1981	Rainbow trout	50	45,430
East Fork Black River			
1969	Arctic grayling	1	10,000
1933–1940	Brook trout	6	48,620
1935–1977	Brown trout	50	286,267
1933–1937	Native cutthroat trout[a]	5	57,760
1933–1991	Rainbow trout	692	1,427,350
North Fork Black River			
1969	Arctic grayling	2	10,000
1933–1963	Brook trout	2	30,500
1938–1959	Brown trout	11	48,515
1936–1986	Rainbow trout	174	143,541
West Fork Black River			
1970	Arctic grayling	1	5,100
1935–1937	Brook trout	3	50,400
1938–1981	Brown trout	20	76,149
1938	Native cutthroat trout[a]	2	6,100
1934–1991	Rainbow trout	705	573,130
Boggy and Centerfire Creeks			
1941–1975	Brown trout	3	5,000
1937	Native cutthroat trout[a]	2	2,688
1938–1948	Rainbow trout	3	4,838
Conklin Creek			
1933	Brook trout	1	5,400
1933	Native cutthroat trout[a]	1	5,400
Fish Creek			
1933	Brook and cutthroat trout	1	900
1954	Brown trout	1	1,000
1935–1952	Rainbow trout	6	26,620
Wildcat Creek			
1966–1975	Rainbow trout	2	2,600
Willow Creek			
1933	Brook trout	1	6,400
1933	Native cutthroat trout[a]	1	6,400
Total			2,924,389

[a] Black spotted mountain cutthroat trout and Wyoming native cutthroat trout.

riparian vegetation and instream habitat for the Apache trout and Little Colorado spinedace were not considered. Trees were cut adjacent to stream channels, and logging roads became as numerous as the streams themselves. These roads frequently crossed and recrossed the streams and increased sediment in stream channels. These effects are typical of how forest management practices affect the aquatic environment (Lynch et al. 1977; Rinne 1990b; Rinne and Medina, in press).

An essential component of sustaining and enhancing native fishes is ownership of stream habitat (Rinne 1994, 1995). Most of the watersheds of the two river drainages in the White Mountains is under

TABLE 3.—Nonnative fishes stocked into streams historically containing Little Colorado spinedace in the White Mountains, Arizona, 1930–1991.

Stream and year	Species	Number of stocking events	Number stocked
Colter Creek			
1933	Brook trout	1	3,800
1933	Native cutthroat trout[a]	1	3,800
1936–1963	Rainbow trout	5	19,500
1936	Brown trout	1	300
Little Colorado River			
1975	Brook trout	1	500
1975	Channel catfish	1	10,000
1969–1975	Rainbow trout	3	2,550
Upper Little Colorado River			
1968–1975	Brook trout	2	6,000
1933–1966	Brown trout	63	593,200
1973	Coho salmon[b]	1	1,800
1937	Native cutthroat trout[a]	1	11,200
1933–1991	Rainbow trout	1,100	1,906,700
East Fork Little Colorado River			
1935	Brook trout	1	2,500
1940–1958	Brown trout	5	21,850
1933	Native cutthroat trout[a]	1	14,000
1937–1989	Rainbow trout	96	111,013
South Fork Little Colorado River			
1938–1975	Brown trout	5	33,200
1938	Native cutthroat trout[a]	1	2,500
1937–1991	Rainbow trout	510	266,750
West Fork Little Colorado River			
1935–1937	Brook trout	2	29,264
1946–1973	Brown trout	17	59,394
1973	Coho salmon[b]	1	480
1933	Native cutthroat trout[a]	1	44,000
1936–1991	Rainbow trout	879	975,428
Nutrioso Creek			
1937	Rainbow trout	1	2,400
1935	Brown trout	1	6,100
Paddy Creek			
1935	Brook trout	1	500
1933	Native cutthroat trout[a]	1	7,000
1936–1947	Rainbow trout	4	20,060
Silver Creek			
1983	Brook trout	1	1,000
1934–1963	Brown trout	18	194,830
1969–1983	Channel catfish	4	10,600
1937–1991	Rainbow trout	297	533,123
Total			4,895,342

[a] Black spotted mountain cutthroat trout and Wyoming native cutthroat trout.
[b] *Oncorhynchus kisutch*.

U.S. Forest Service jurisdiction. However, native fish species such as Apache trout and Little Colorado spinedace have fragmented distributions. Frequently, critical stream reaches are necessary for their survival. For this reason, a 2.4-km reach of Nutrioso Creek immediately upstream of Nelson Reservoir (Figure 3), critical habitat for Little Colorado spinedace, was acquired by the U.S. Forest Service in a land exchange in 1993. Also in 1993, the Arizona Game and Fish Department (Department) acquired 3.2 km of the Little Colorado River immediately north of Springerville, Arizona, and the watershed for Rudd Creek, a 3-km tributary of Nutrioso Creek (Figure 3). The latter acquisition will secure habitat for both the Apache trout and Little Colorado spinedace.

TABLE 4.—Stream renovations conducted and proposed in the Black and Little Colorado river drainages, Apache Sitgreaves National Forest, Arizona, to reestablish Apache trout.

Stream	Year	Stream length treated (km)	Target species removed
Completed to Date			
Bear Wallow Creek	1981	18.0	Rainbow and hybrid trouts
Lee Valley Creek	1982	4.8	Brook trout
	1987	4.8	Brook trout
Home Creek	1987	11.3	Rainbow trout
Wildcat Creek	1988	6.5	Rainbow trout
Hayground Creek	1989	6.5	Brown trout
Total		47.1[a]	
Proposed			
West Fork Black River		8.0	Brown trout
Thompson Creek		3.2	Brown trout
Burro Creek		9.6	Brown trout
Stinky Creek		3.8	Brown and hybrid trouts
Snake Creek		7.2	Brown and hybrid trouts
Bear Wallow Creek		18.0	Rainbow and hybrid trouts
Total		49.8	

[a] Represents net kilometers; Lee Valley was treated twice.

Domestic livestock grazing had its inception even earlier than the timber industry. In the late 1800s cattle grazing was unrestricted on the White Mountain landscape prior to its designation as U.S. Forest Service land. Lack of regulation led to overstocking of the rangeland, which resulted in degradation of ground cover across watersheds and increased sediment delivery to stream channels. In addition, livestock congregated in stream bottoms during hot, dry periods and negatively affected riparian vegetation.

By the time of the Taylor Grazing Act of 1934 (43 U.S.C.A. §§ 315 et. seq.) in the 1930s, the Multiple Use and Sustained Yield Act of 1960 (16 U.S.C.A. §§ 528 to 531), and the environmental legislation in the late 1960s and early 1970s, much of the impact to these two native fishes and their habitats had occurred (Rinne and Medina, in press). Almost a century of land use and more than half a century of extensive sport-fish management had brought these two fishes to the brink of extinction.

In summary, the combination of extensive nonnative sport-fish introductions into waters of the Black and Little Colorado river drainages accompanied by stream renovations with fish toxicants, baitfish introductions, and devastating land management activities has negatively affected the Apache trout and Little Colorado spinedace and their habitats.

Management Strategies

Current Activities

Trout angling is a large component of the overall sportfishing industry in Arizona (Persons 1990). Over 130,000 anglers prefer to fish for trouts and most of this activity is on mountain lakes and reservoirs (Persons 1990). Today, trout fishing in Arizona is the result of more than 60 years of hatchery development and introduction of trouts into streams and lakes in the wild (Stephenson 1985). In 1988, over 5 million trout were stocked into Arizona waters by state and federal hatcheries (Arizona Game and Fish Department 1990) at a cost of US$1.2 million. Any significant changes in management that will reduce trout fishing opportunities must be approached with caution. Changes will have economical, political, and social implications for both the Department and the state of Arizona. Currently, 380,000 anglers spend $57–79 on an average fishing trip and contribute $355–561 million into the Arizona economy annually (Anonymous 1992).

Indeed, from the 1930s to the early 1980s the fisheries program in Arizona was insensitive to native fish management. The Department now recognizes the need to manage for both sport and native fishes in Arizona. The Department must be innovative in managing both coldwater sport fisheries and native, nongame species (Cain 1993).

Currently, fisheries managers are at the crossroads of opportunity to commence positive, innovative, and aggressive management activities to sustain native fishes in Arizona. Socially and politically, the climate is more favorable than in the past. Department fishing license sales have decreased gradually since 1986. Almost two-thirds of Arizona residents believe that every effort should be made to manage native fishes—even if it means restricting stocking of sport fishes such as trout (Behavior Research Center, Inc., Phoenix, Arizona, 1992). In addition, 26% of Arizona anglers think we should do everything we can to preserve native fish (Anonymous 1992). The Department has recognized not only the responsibility but the opportunity to effect important management measures that will enhance both native fish ranges and populations.

Some changes have already taken place. By Department policy enacted in 1982, no additional nonnative species will be stocked into the waters of the state. Furthermore, no additional species (native or nonnative) will be transplanted into a watershed or range where it currently does not exist. Also, baitfish regulations keep anglers from transporting baitfish between lakes and from one area of the state to another, and baitfish are not allowed as a legal method of take in over half of the state.

Native fishes were included in the Department's sport-fish management program beginning in 1984. However, to date, only the Apache trout has received considerable attention. Plans are now underway to establish and maintain lake fisheries for the Apache trout in addition to those already in existence in Bear Canyon Lake, Lee Valley Reservoir, and Becker Lake.

In contrast to the 1960s, the current objective of stream renovations is to remove nonnative salmonids and reestablish or enhance native trouts (Minckley and Brooks 1985; Rinne and Turner 1991). Within the historic range of Apache trout, for example, stream renovation projects to remove rainbow, brown, and hybrid trouts and reintroduce Apache trout are ongoing (Table 4). Apache trout are raised in hatcheries or salvaged from wild populations for reintroduction. All stocked fish are first generation progeny of cultured stock of pure strain Apache trout from the type locality, East Fork of the White River. In addition, nonnative salmonids are not stocked into streams containing Apache trout populations. However, the mainstream Black and Little Colorado rivers and Silver Creek were stocked with nonnative trouts as recently as 1991 (Tables 2 and 3).

Future Opportunities

Put-and-take fisheries for Apache trout will be developed on the mainstream Little Colorado and Black rivers where rainbow trout are currently stocked. The Department, the U.S. Forest Service, and Trout Unlimited have initiated a program to enhance the Apache trout on the West Fork of the Black River (U.S. Forest Service 1993). This project encompasses about 29 km of the mainstream West Fork of the Black River and an additional 30 km of tributary streams; management activities include special fishing regulations, habitat improvement including instream and streambank structures, willow plantings, road closures, and vegetation management via fencing and grazing strategy alteration. Stream renovations involving habitat enhancement and removal of nonnative species with fish toxicants will be major components of this project (Table 4). This project alone will increase the stream kilometers reclaimed for Apache trout on the Black River drainage by almost 50%. With additional renovations of Snake and Bear Wallow creeks, current stream kilometers reclaimed for Apache trout will be doubled.

It has been relatively easy to conduct stream renovations that replace nonnative trouts with Apache trout. This native trout has sportfishing potential, and efforts to restore it are accepted by the public. However, implementing future stream renovations and removing nonnative sport species (trouts) and reestablishing a native, nonsport species will be both more controversial and less easily implemented. Adoption of the same course of action to reestablish the Little Colorado spinedace, which has no recreational value, will be the true test of public support for native fish management in these two montane river systems and in Arizona in general.

Native fish management is long overdue in Arizona. New approaches will be phased into the current coldwater sport-fish management program throughout the state. Several different strategies will be used to introduce native fishes and yet maintain sportfishing opportunities for nonnative trouts. Key watersheds and drainages will be designated specifically for the management of native fishes. In addition, native fish management will be integrated into some existing sport-fish populations to provide sustainability of both. Often, this will involve placement of natives upstream from habitats occupied by nonnative fishes and separation by either natural or artificial barriers (Rinne and Turner 1991). Land acquisitions will more readily facilitate designation of these waters as special management areas for the Apache trout, the Little Colorado spinedace, and other native species.

Many montane streams have been impounded for irrigation or recreation purposes. An alternative to spatial separation of nonnative and native species may be strain or special species management. Here, strains or species that will not have the same competitive effect, will not reproduce, or that are less piscivorous will be used for sport-fish management. Certain strains of rainbow trout or genetically altered rainbow trout, Arctic grayling, and kokanee will be utilized in the new management strategy.

The basic philosophy for stocking coldwater species for sportfishing will no longer be dictated by angler demands (e.g., size and diversity of stock) but rather by the probable effect on the native species. To accomplish this, we need to realign the thinking of our many constituents through education. Both Department and private avenues of educating the public will be needed to attain success in this area. The Apache trout, and perhaps the Gila trout, will be used in the future for coldwater sportfishing. Rinne (1988) suggested that promotion of the Apache and Gila trouts for sportfishing will aid in their long-term sustainability through angler affiliation with the species.

Timely and effective management of native species in the state will reduce the probability of loss of native fishes and will prevent listing as Endangered species. Endangered status can limit management alternatives and curtail or eliminate recreational sportfishing.

Conclusions

Management of native fishes and sport fishes in Arizona is at the crossroads. The challenge to institutionalize management and conservation activities for native fishes is great. Changes in the Department in both the philosophy of management for native fishes and actual activities on the ground have occurred. An ecosystem or watershed approach to managing riparian-stream areas that provide fish habitat will be equally important and essential in sustaining changes in fisheries management.

It took many years to bring about the decline of native fishes because of past management philosophy; it will also require some time to reverse this trend and reach the goal of native fish sustainability. However, time is a precious commodity and may be short for species such as the Little Colorado spinedace. Achieving the goal of a compatible nonnative and native fisheries program will require both immediate and novel sport-fish-native-fish management and proper stewardship of the habitats these native species will require to perpetuate themselves.

References

Anonymous. 1990. Arizona coldwater sport fisheries strategic plan, 1991–1995. Arizona Game and Fish Department, Phoenix.
Anonymous. 1992. Arizona angler survey: preliminary results. Arizona Game and Fish Department, Phoenix.
Blinn, D. W., C. Runck, A. Clark, and J. N. Rinne. 1993. Effects of rainbow trout predation on Little Colorado River spinedace. Transactions of the American Fisheries Society 122:139–143.
Bruce, J. 1961. Statewide fishery investigations. Postdevelopment evaluation (investigations). Arizona Game and Fish Department, Federal Aid in Sport Fish Restoration, Project F-7-R-7, Jobs 1 and 2, Completion Report, Phoenix.
Cain, T. 1993. Beyond dollars and sense: debating the value of nongame fish. Fisheries (Bethesda) 18(7):20–21.
Clarkson, R. W., and J. C. Devos, Jr. 1986. The bullfrog, *Rana catesbeiana* Shaw in the lower Colorado River, Arizona–Colorado. Journal of Herpetology 20:42–49.
Gresswell, R., editor. 1988. Status and management of interior stocks of cutthroat trout. American Fisheries Society Special Publication 4.
Hemphill, J. E. 1954. Toxaphene as a fish toxin. Progressive Fish-Culturist 16:41–42.
Litvak, M. K., and N. E. Mandrak. 1993. Ecology of freshwater baitfish use in Canada and the United States. Fisheries (Bethesda) 18(12):6–13.
Lynch, J. A., E. S. Corbett, and R. Hoopes. 1977. Implications of forest management practices on the aquatic environment. Fisheries (Bethesda) 2(2):16–22.
Miller, R. R. 1950. Notes on the cutthroat and rainbow trouts with the description of a new species from the Gila River, New Mexico. Occasional Papers of the Museum of Zoology University of Michigan 529:1–42.
Miller, R. R. 1952. Bait fishes of the lower Colorado River, from Lake Mead, Nevada, to Yuma, Arizona, with a key for their identification. California Fish and Game 38:7–42.
Miller, R. R. 1963. Distribution, variation, and ecology of *Lepidomeda vittata*, a rare cyprinid fish endemic to eastern Arizona. Copeia 1963:1–5.
Miller, R. R. 1972. Classification of the native trouts of Arizona with the description of a new species, *Salmo apache*. Copeia 1972(3):401–422.
Minckley, C. O. 1984. Current distribution and status of *Lepidomeda vittata* (the Little Colorado spinedace) in Arizona. Report to Arizona Game and Fish Department, Phoenix.
Minckley, W. L. 1973. Fishes of Arizona. Arizona Game and Fish Department, Phoenix.
Minckley, W. L., and J. E. Brooks. 1985. Transplantations of native Arizona fishes: records through 1980. Journal of the Arizona–Nevada Academy Sciences 20:73–89.
Minckley, W. L., and L. H. Carufel. 1967. The Little Colorado River spinedace, *Lepidomeda vittata*, in Arizona. Southwest Naturalist 13:291–302.
Persons, B. 1990. Statewide survey of 1986 and 1989 Arizona anglers. Arizona Game and Fish Department, Phoenix.
Persons, B. 1992. Historical stocking summary. Arizona Game and Fish Department, Phoenix.
Rinne, J. N. 1985. Variation in Apache trout populations in the White Mountains, Arizona. North American Journal of Fisheries Management 5:146–158.
Rinne, J. N. 1988. Native southwestern (USA) trouts: status, taxonomy, ecology and conservation. Polskie Archiwum Hydrobiology 35(3-4):305–320.
Rinne, J. N. 1990a. An approach to management and conservation of a declining regional native fish fauna: southwestern United States. Pages 74–78 *in* Proceed-

ings of the Fifth International Congress of Zoology. Yokohama, Japan.

Rinne, J. N. 1990b. The utility of stream habitat and biota for identifying potential conflicting forest land uses: montane riparian areas. Forest Ecology and Management 33/34:363–383.

Rinne, J. N. 1991a. Apache trout. Pages 178–183 in J. Stoltz and J. Schnell, editors. Trout. Stackpole Books, Harrisburg, Pennsylvania.

Rinne, J. N. 1991b. Gila trout. Pages 274–279 in J. Stoltz and J. Schnell, editors. Trout. Stackpole Books, Harrisburg, Pennsylvania.

Rinne, J. N. 1994. Declining southwestern aquatic habitats and fishes: are they sustainable? Pages 256–265 in Proceedings of the sustainability symposium, Flagstaff, Arizona.

Rinne, J. N. 1995. The effects of introduced fishes on native fishes: Arizona, southwestern United States. Pages 148–158 in Proceedings of the World Fisheries Conference. Oxford & IBH Publishing Co., Pvt. Ltd., New Delhi.

Rinne, J. N., and W. L. Minckley. 1985. Patterns of variation and distribution in Apache trout (*Salmo apache*) relative to co-occurrence with introduced salmonids. Copeia 1985:285–292.

Rinne, J. N., and W. L. Minckley. 1991. Native fishes in arid lands: a dwindling resource of the desert Southwest. U.S. Forest Service General Technical Report RM-206:1–45.

Rinne, J. N., and P. R. Turner. 1991. Reclamation and alteration as management techniques, and a review of methodology in stream renovation. Pages 219–244 in W. L. Minckley and J. E. Deacon, editors. Battle against extinction: native fish management in the American West. University of Arizona Press, Tucson.

Rinne, J. N., and A. L. Medina. In press. Implications of multiple use management strategies on native Southwestern (USA) fishes. In Proceedings of the World Fisheries Congress. Oxford & IBH Publishing Co. Pvt. Ltd., New Delhi.

Rinne, J. N., W. L. Minckley, and J. N. Hanson. 1981. Chemical treatment of Ord Creek, Apache County, Arizona, to re-establish Arizona trout. Arizona–New Mexico Academy of Sciences 16(3):74–78.

Rinne, J. N., R. Sorenson, and S. C. Belfit. 1986. An analysis of F_1 hybrids between Apache (*Salmo apache*) and rainbow trout (*Salmo gairdneri*). Journal of the Arizona–Nevada Academy of Sciences 20(2):63–69.

Stephenson, R. L. 1985. Arizona cold water fisheries strategic plan, 1985–90. Arizona Game and Fish Department, Phoenix.

U. S. Fish and Wildlife Service. 1979. The fish car era. U.S. Fish and Wildlife Service, Washington, DC.

U. S. Forest Service. 1993. West Fork of the Black River watershed and fisheries restoration project: implementation plan. U.S. Forest Service, Springerville Ranger District, Springerville, Arizona.

Problems and Prospects for Grass Carp as a Management Tool

JOHN R. CASSANI

Lee County Hyacinth Control District, Post Office Box 60005
Fort Myers, Florida 33906, USA

Abstract.—Grass carp *Ctenopharyngodon idella* have been used extensively for vegetation control in the United States since the early 1970s. Based on a review of the literature, the most common result of stocking grass carp to manage aquatic vegetation is elimination of macrophytes and, subsequently, changes in water quality, primary and secondary productivity, structural habitat complexity, and changes in the fish community. Despite the extensive use of grass carp, the probability of predicting these changes is often low due to differences in such dynamic processes as nutrient loading, macrophyte seasonality, and climate among lakes. Variations in consumption rates and preference for certain plant species can further complicate vegetation management with grass carp, especially where mixed-species communities occur. Use of stocking rates that do not account for these differences are largely responsible for the unpredictable impacts of stocking grass carp. Other factors that contribute to the unpredictability of using grass carp for vegetation control are inconsistent survival rates associated with size at stocking, poor water quality, and variable predation pressure. Stocking rates for macrophyte suppression, rather than elimination, are in the developmental stages and preliminary results indicate potential application. These findings, in conjunction with more sophisticated simulation models, refined removal and containment techniques, and integration with other control methods, will likely improve vegetation management with grass carp in the future. However, resource managers will need to place more value on macrophyte suppression rather than elimination so that the low cost of using grass carp is balanced with potentially long-term impacts.

Grass carp *Ctenopharyngodon idella* were introduced into the United States in 1963 by Auburn University and the U.S. Fish and Wildlife Service (Guillory and Gasaway 1978). The primary intent of this introduction was to assess the potential of grass carp for controlling aquatic plants. The Arkansas Game and Fish Commission was the first agency to use grass carp on an operational basis in 1970 when Lake Greenlee was stocked with 2,100 diploid grass carp (Bailey and Boyd 1972). Grass carp were rapidly spread to other areas of the United States, and by 1972 the fish had been introduced to 40 states (Pflieger 1978). Many of the original stockings in Arkansas were in lakes or reservoirs open to stream systems, and by 1974 there were numerous reports of grass carp captured in the Missouri and Mississippi rivers (Pflieger 1978).

The collection of eggs or larvae documented that grass carp can reproduce in U.S. rivers (Conner et al. 1980; Zimpher et al. 1987; Brown and Coon 1991; Anonymous 1993). However, impacts of these naturally reproducing populations have not been documented.

During the late 1970s, the controversy over natural reproduction of grass carp resulted in the production of an intergeneric triploid hybrid. However, the hybrid did not effectively control aquatic weeds (Osborne 1982). Further refinements of triploid induction techniques led to the development of a nonhybrid triploid grass carp (Malone 1984; Cassani and Caton 1986; Thompson et al. 1987). Functional sterility of the triploid grass carp further liberalized its use (Allen et al. 1986; Van Eenennaam et al. 1990; Sanders et al. 1991). As of 1994, 12 states prohibit grass carp introductions, eight have no restrictions, and the remainder allow some form of application, usually by permit, for triploid grass carp (R. J. Wattendorf, Florida Game and Fresh Water Fish Commission, personal communication).

Management Dilemmas

The relatively low costs of using grass carp for aquatic plant control have made it attractive to many user groups. Chemical control of submersed macrophytes in small ponds is 2.6 to 5.4 times more expensive than the equivalent level of control with grass carp (Shireman et al. 1985). Mechanical control is even more expensive, ranging from 2 to 28 times the cost of using grass carp (Cooke et al. 1993). The temptation to rapidly reduce and often eliminate submersed macrophytes with grass carp is always a factor when evaluating management objectives.

Choosing grass carp as a management tool for economic reasons alone reflects a short-sighted approach to managing aquatic plants. Rapid elimination of submersed macrophytes in single-use systems such as irrigation canals and potable water reservoirs—especially if chemical and mechanical plant control methods are incompatible with water use—is a different situation from large multi-use systems with a sport fishery or other biota dependent on some level of macrophyte abundance.

A review of grass carp impacts indicates that submersed macrophytes are usually eliminated from the target area. Macrophyte eradication can have varying effects, some of which result in undesirable impacts to fish community structure, invertebrate diversity and abundance, waterfowl food sources, and water quality (Gasaway and Drda 1976; Bailey 1978; Leslie et al. 1983; Cassani and Caton 1985; Maceina et al. 1992; Bettoli et al. 1993). Other studies have shown that extremes in macrophyte cover, either from some method of control or lack thereof, can negatively affect sport-fish population size structure and abundance, and that intermediate levels are generally desirable (Colle and Shireman 1980; Savino and Stein 1982; Durocher et al. 1984; Engel 1987).

Lake Conroe, a large Texas reservoir, provides a well-documented example of grass carp eradicating submersed macrophytes and the effects on the sport fishery. Grass carp stocked at 74/ha of vegetation eliminated submersed vegetation, primarily *Hydrilla verticillata*, in 2 years. A 7-year poststocking study revealed increases in mean annual chlorophyll-*a* levels with a concomitant decrease in water transparency; blue-green algal density relative to phytoplankton abundance increased and total zooplankton decreased 1.5 years after macrophyte removal (Maceina et al. 1992). Effects on the fishery in Lake Conroe were also documented over the same 7-year period (Bettoli et al. 1993) (Table 1). Species declining after macrophyte removal included phytophilic *Lepomis* spp. (*L. punctatus*, *L. marginatus*, and *L. gulosus*), bluegill *Lepomis macrochirus*, and brook silverside *Labidesthes sicculus*. The density of threadfin shad *Dorosoma petenense*, a pelagic planktivore, increased dramatically after vegetation removal and may have accounted for subsequent large year-classes of yellow bass *Morone mississippiensis* and white bass *M. chrysops*. Biomass and density of channel catfish *Ictalurus punctatus* also increased. The authors concluded that the sport fishery could not be fully assessed but that a change in the structure of the sport-fish community did occur. The original largemouth bass–crappie–hybrid striped bass fishery was replaced by a channel catfish–white bass–hybrid striped bass–largemouth bass–black crappie fishery after macrophyte removal. Limited creel data indicated a decline in largemouth bass catch rates but an increase in average size.

TABLE 1.—Common and scientific names of fish species (Robins et al. 1991) impacted by stocking grass carp into Lake Conroe, Texas.

Common name	Scientific name
Bluegill	*Lepomis macrochirus*
Brook silverside	*Labidesthes sicculus*
Threadfin shad	*Dorosoma petenense*
Yellow bass	*Morone mississippiensis*
White bass	*Morone chrysops*
Channel catfish	*Ictalurus punctatus*
Largemouth bass	*Micropterus salmoides*
Crappie	*Pomoxis* sp.
Hybrid striped bass	*Morone chrysops* × *M. saxatilis*
Black crappie	*Pomoxis nigromaculatus*

Evaluating the significance of changes to a fishery can be highly subjective and may vary according to the evaluator's concept of a desirable fishery. The importance of submersed vegetation to fish populations and lake systems remains controversial, especially when management priorities are focused on one or two species (e.g., largemouth bass and bluegill). The success or failure of vegetation decline from grass carp is thus measured, in many instances, by the way the target fishery responds.

Macrophyte elimination with grass carp is rarely the intended objective, but there are several reasons why intermediate control is rarely achieved. A primary reason is that each lake system has unique conditions that will affect grass carp stocking rates. When these conditions or variables are not measured or accounted for, it invalidates comparisons of stocking rates among studies. Factors that should be considered include climate, nutrient loading, levels of predation affecting grass carp survival, target plant phenology, and different target weed species that affect the rate of consumption by grass carp. These factors are not routinely assessed, because they are often difficult to measure and are highly dynamic on a temporal basis.

Another reason why intermediate control is rarely achieved is that researchers have not taken a uniform approach to quantifying and evaluating important stocking rate considerations. Use of different sizes of grass carp, which have different survival and plant consumption rates, and noncomparable approaches to quantifying macrophyte abundance weaken attempts to build a knowledge base resulting from repetitive experiences. The end result has been a relatively large amount of information, but little that can be used to determine stocking rates.

It is necessary to contain grass carp in the target area to maintain effective stocking levels. Containment is generally practical and inexpensive in relatively small isolated systems, but difficult in large lakes or impounded rivers. Dams, locks, and spillways may restrict grass carp movement temporarily but periodic floods and sporadic escapement can dilute the original stocking density and potentially impact nontarget areas. In Guntersville Reservoir, a large open mainstream reservoir of the Tennessee River, juvenile triploid grass carp remained close to food sources (*Hydrilla verticillata*). Grass carp dispersion increased as the fish matured, resulting in movement well beyond the target area (Bain et al. 1990).

Of even greater concern is the potential natural reproduction of escaped diploid grass carp and the impact of the resulting population. Confirmed spawning of grass carp in the Trinity River, Texas, may have been the result of grass carp that escaped from Lake Conroe (Anonymous 1993). The value of using a sterile triploid grass carp becomes obvious when containment is a problem. Triploid grass carp are certainly as capable of impacting nontarget areas, but the effects are generally limited to the life of the fish.

A consideration of stocking any cultured fish is the possibility of transferring diseases or parasites to wild stocks. Grass carp carry several diseases and parasites known to be transmissible or potentially transmissible to native North American fishes (Riley 1978; Shireman and Smith 1983). However, limiting the use of grass carp on the premise of controlling the spread of new diseases or parasites is somewhat moot because the fish has already been widely distributed in the United States for over 20 years. In addition, triploid grass carp, all of which are produced in hatcheries where it is advantageous to maintain disease- and parasite-free stock, are the only source of grass carp in almost all states where stocking is permitted. Bain et al. (1990) suggested that native fish in open systems stocked with grass carp should have limited vulnerability to new diseases and parasites because those systems may be interconnected to a wide range of aquatic habitats and pre-existing sources of disease and parasites. However, restricting grass carp from pristine systems and connecting water bodies where endangered native cyprinids or other endangered fish species occur would seem appropriate. Much of the concern about parasite dispersal from grass carp stems from the assumption that the Asian tapeworm *Bothriocephalus opsarichthydis* was first introduced into the United States when grass carp were imported in 1963 (Hoffman and Schubert 1984), and that grass carp will aid dispersal of this parasite.

Management Prospects

Adjusting grass carp stocking rates for the diverse and dynamic set of variables affecting consumption rates and plant phenology is a difficult task. As mentioned earlier, the methodological approach to this problem has not been consistent, resulting in little comparability between studies. However, there has been more emphasis in recent years on refining stocking rates and attempting to avoid eradication of submersed macrophytes (Kirk 1992; Bonar et al. 1993a). Consideration for site- or region-specific variables is increasing, and concentrated research on stocking-rate effects in certain regions will allow resource managers to estimate grass carp stocking rates more accurately. In Illinois, Wiley et al. (1987) developed a model that provides an alternative to a single stocking approach. Here, stocking recommendations are based on a serial stocking approach where grass carp are stocked a second time, usually 5 to 7 years after the initial stocking. Serial stocking allows for more management interaction and flexibility and an overall goal of leaving some residual plant population. The stocking recommendations are further qualified by the type and areal coverage of various plant groups based on their preference by triploid grass carp.

In Washington, Bonar (1990) evaluated stocking rates based on estimates of the quantity (metric tons) of vegetation present. Vegetation was quantified by underwater collection of all plant material within randomly located 0.25 m^2 quadrats and expressed as g/m^2 wet weight. In addition, the rates were adjusted for altitude and associated differences in climate. Research on small warmwater impoundments in southwest Florida using stocking rates of three 25–30-cm-total-length triploid grass carp per metric ton of vegetation (preferred plant species) resulted in macrophyte suppression without eradication up to 5 years after stocking at two sites (Cassani et al., in press) (Figure 1).

Grass carp consumption rates and plant growth are highly dynamic variables and are related to temperature, grass carp size, nutrient loads, and grass carp survival. Computer models may be the best approach to dealing with complex and interactive variables associated with determining grass carp stocking rates. Several models are now available, most of which have been developed for application in certain regions of the United States with similarities in climate or target plants. Models already developed include the AMUR/STOCK model (Boyd and Stewart 1991), the Illinois model

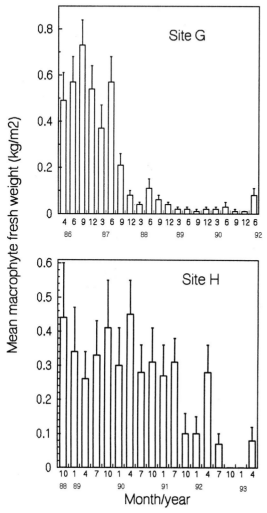

FIGURE 1.—Mean macrophyte density per sampling period at two impoundments in Lee County, Florida, stocked with three 25–30-cm triploid grass carp per metric ton of vegetation. Site G was stocked in July of 1986 and Site H was stocked in January of 1989. Dominant macrophytes at Site G were *Chara* sp. and *Najas guadalupensis*. At Site H the dominant macrophyte was *Najas guadalupensis*. Error bars are 95% confidence intervals. From Cassani et al., in press.

(Wiley et al. 1987), the CONTROL and BIOMASS models developed for areas in Washington by Bonar (1990), and a coldwater model described by Swanson and Bergerson (1988). These tools hold considerable potential for refining and interpreting stocking rates for macrophyte suppression. However, the validity of the simulations is only as accurate as the information supplied. Entering a range of variables (e.g., stocking rates, percent grass carp mortality, etc.) will result in multiple simulations (vegetation decline scenarios), allowing the user to select a stocking rate that will most likely meet the management objective.

Increasingly, stocking grass carp is integrated with other methods of plant control for more precise management, but few if any published studies are available. The value of this approach is that plant density is reduced prior to stocking so that fewer grass carp are used, therefore reducing their impact on water quality and other system components if plant eradication occurs. Also, intentionally understocking grass carp and selectively treating plants with herbicides to maintain desirable levels is an example of the flexibility possible with integrated control. Sutton and Vandiver (1986) found that only 13 triploid grass carp per hectare were able to prevent the regrowth of *Hydrilla verticillata* after initially eliminating it with herbicides, and suggested that three to eight grass carp per hectare may be a low enough rate to allow beneficial macrophytes to reestablish. Costs associated with integrated control will be higher than using grass carp as the only method of control and will have to be prioritized according to the management objectives and anticipated results.

Another approach toward mitigating the long-term effects of overstocking grass carp is to reduce the grass carp population with some type of capture method after the desired level of control has been achieved. Recapture techniques have been evaluated extensively in Florida (Rue Hestand, Florida Game and Fresh Water Fish Commission, personal communication) but nothing has been published to date on the results. The earliest attempts to reduce grass carp populations were with rotenone. Selective removal of grass carp from 80-ha Lake Baldwin (Florida) was attempted by applying rotenone at 0.1 mg/L (Colle et al. 1978). Over half of the grass carp were removed and the authors concluded that this method was practical for obtaining population estimates with potential broader applications.

Bonar et al. (1993b) evaluated seven capture methods in five Washington lakes and concluded that herding was the most efficient method and that it may be effective in small (less than 10 ha), overstocked lakes. Herding involves progressively driving fish with noisemakers (boat motor, plungers, scarelines) from one side of the lake, and preventing their return with a gill net until the entire lake has been traversed. Up to 8.2% of the original population was captured in one lake by herding and seining (Bonar et al. 1993b).

The recent development of a feed pellet containing rotenone has demonstrated relatively good removal rates in small lakes (Rue Hestand, Florida Game and Fresh Water Fish Commission, personal communication). Testing this removal method involved manipulating bait components to increase its palatability to grass carp while reducing its attractiveness to other species. Applications that increased effectiveness of the bait were also refined. Preliminary results indicate that 20–40% of the grass carp in a lake could be removed in a year with this method.

In conclusion, many of the problems associated with the use of grass carp for aquatic weed control have not been adequately addressed. Grass carp are a cost-effective tool for eradicating submersed macrophytes, but their application in large multi-use systems will demand more careful consideration in the future. Progress toward more sophisticated stocking rate models, refined removal and containment techniques, and integration with other plant control methods is slowly taking place and will likely improve the applicability of grass carp for aquatic plant management.

Future research to estimate grass carp stocking rates should address reasons for discrepancies among past studies and avoid inconsistent approaches. The following criteria should be incorporated in new research on grass carp efficacy:

1. stock triploid grass carp that are 250–300 mm total length to minimize predation losses;
2. estimate area of macrophyte cover in hectares and abundance as metric tons wet weight;
3. report stocking rates as number of grass carp per vegetated hectare and number per metric ton of vegetation;
4. determine the lake trophic state as a minimal measure of the system productivity (Carlson 1977; Canfield et al. 1983) that may relate to macrophyte growth potential;
5. stock grass carp during the spring months or before the season of maximum macrophyte growth;
6. assure that transportation and on-site introduction stress are minimized to reduce uncertainties about poststocking survival; and
7. use an appropriate model for developing stocking rates that may more accurately achieve the vegetation management objective.

References

Allen, S. K., Jr., R. G. Thiery, and N. T. Hagstrom. 1986. Cytological evaluation of the likelihood that triploid grass carp will reproduce. Transactions of the American Fisheries Society 115:841–848.

Anonymous. 1993. Confirmed grass carp spawning. Fisheries (Bethesda) 18(3):36.

Bailey, W. M. 1978. A comparison of fish populations before and after extensive grass carp stocking. Transactions of the American Fisheries Society 107:181–206.

Bailey, W. M., and R. L. Boyd. 1972. Some observations on the white amur in Arkansas. Hyacinth Control Journal 10:20–22.

Bain, M. B., D. H. Webb, M. D. Tangedal, and L. N. Mangum. 1990. Movements and habitat use by grass carp in a large mainstream reservoir. Transactions of the American Fisheries Society 119:553–561.

Bettoli, P. W., M. J. Maceina, R. L. Noble, and R. K. Betsill. 1993. Response of a reservoir fish community to aquatic vegetation removal. North American Journal of Fisheries Management 13:110–124.

Bonar, S. A. 1990. Efficacy of sterile grass carp (*Ctenopharyngodon idella*) for aquatic plant control in the Pacific Northwest. Doctoral dissertation. University of Washington, Pullman.

Bonar, S. A., G. L. Thomas, S. L. Thiesfeld, G. B. Pauley, and T. B. Stables. 1993a. Effect of triploid grass carp on the aquatic macrophyte community of Devils Lake, Oregon. North American Journal of Fisheries Management 13:757–765.

Bonar, S. A., S. A. Vecht, C. R. Bennett, G. B. Pauley, and G. L. Thomas. 1993b. Capture of grass carp from vegetated lakes. Journal of Aquatic Plant Management 31:168–174.

Boyd, W. A., and R. M. Stewart. 1991. Biocontrol simulations. U.S. Army Corps of Engineers, Waterways Experiment Station, Miscellaneous Paper A-91-3, Vicksburg, Mississippi.

Brown, D. J., and T. G. Coon. 1991. Grass carp larvae in the lower Missouri River and its tributaries. North American Journal of Fisheries Management 11:62–66.

Canfield, D. E., Jr., K. A. Langeland, M. J. Maceina, W. T. Haller, and J. V. Shireman. 1983. Trophic state classification of lakes with aquatic macrophytes. Canadian Journal of Fisheries and Aquatic Sciences 40:1713–1718.

Carlson, R. E. 1977. A trophic state index for lakes. Limnology and Oceanography 22:361–369.

Cassani, J. R., and W. E. Caton. 1985. Effects of chemical and biological weed control on the ecology of a South Florida pond. Journal of Aquatic Plant Management 23:51–58.

Cassani, J. R., and W. E. Caton. 1986. Efficient production of triploid grass carp (*Ctenopharyngodon idella*) utilizing hydrostatic pressure. Aquaculture 55:43–50.

Cassani, J. R., E. Lasso de la Vega, and H. A. Allaire. In press. An assessment of triploid grass carp stocking rates in small warmwater impoundments. North American Journal of Fisheries Management.

Colle, D. E., J. V. Shireman, R. D. Gasaway, R. L. Stetler, and W. T. Haller. 1978. Utilization of selective removal of grass carp (*Ctenopharyngodon idella*) from an 80-hectare Florida lake to obtain a population estimate. Transactions of the American Fisheries Society 107:724–729.

Colle, D. E., and J. V. Shireman. 1980. Coefficients of condition for largemouth bass, bluegill, and redear sunfish in hydrilla-infested lakes. Transactions of the American Fisheries Society 109:521–531.

Cooke, G. D., E. B. Welch, S. A. Peterson, and P. R. Newroth. 1993. Restoration and management of lakes and reservoirs, 2nd edition. Lewis Publishers, Boca Raton, Florida.

Conner, J. V., R. P. Gallagher, and M. F. Chatry. 1980. Larval evidence for natural reproduction of the grass carp (Ctenopharyngodon idella) in the lower Mississippi River. Pages 1–19 in L. E. Fuiman, editor. Proceedings of the Fourth Annual Larval Fish Conference. U.S. Fish and Wildlife Service, Biological Services Program FWS-OBS-80/43:1–19.

Durocher, P. P., W. C. Provine, and J. E. Kraai. 1984. Relationship between abundance of largemouth bass and submerged vegetation in Texas reservoirs. North American Journal of Fisheries Management 4:84–88.

Engel, S. 1987. The impact of submerged macrophytes on largemouth bass and bluegills. Lake and Reservoir Management 3:227–234.

Gasaway, R. D., and T. F. Drda. 1976. Effects of grass carp introduction on waterfowl habitat. North American Wildlife and Natural Resources Conference 42:73–85.

Guillory, V., and R. D. Gasaway. 1978. Zoogeography of the grass carp in the United States. Transactions of the American Fisheries Society 107:105–112.

Hoffman, G. L., and G. Schubert. 1984. Some parasites of exotic fishes. Pages 233–261 in W. R. Courtney and J. R. Stauffer, Jr., editors. Distribution, biology, and management of exotic fishes. The Johns Hopkins University Press, Baltimore, Maryland.

Kirk, J. P. 1992. Efficacy of triploid grass carp in controlling nuisance aquatic vegetation in South Carolina farm ponds. North American Journal of Fisheries Management 12:581–584.

Leslie, A. J., Jr., L. E. Nall, and J. M. Van Dyke. 1983. Effects of vegetation control by grass carp on selected water-quality variables in four Florida lakes. Transactions of the American Fisheries Society 112:777–787.

Maceina, M. J., M. F. Cichra, R. K. Betsill, and P. W. Bettoli. 1992. Limnological changes in a large reservoir following vegetation removal by grass carp. Journal of Freshwater Ecology 7:81–95.

Malone, J. M. 1984. Triploid white amur. Fisheries (Bethesda) 9(2):36.

Osborne, J. A. 1982. The potential of the hybrid grass carp as a weed control agent. Journal of Freshwater Ecology 1:353–360.

Pflieger, W. L. 1978. Distribution and status of the grass carp (Ctenopharyngodon idella) in Missouri streams. Transactions of the American Fisheries Society 107(1):113–118.

Riley, D. M. 1978. Parasites of grass carp and native fishes in Florida. Transactions of the American Fisheries Society 107:207–212.

Robins, C. R. and six coauthors. 1991. Common and scientific names of fishes from the United States and Canada, 5th edition. American Fisheries Society Special Publication 20.

Sanders, L., J. J. Hoover, and K. J. Killgore. 1991. Triploid grass carp as a biological control of aquatic vegetation. U.S. Army Corps of Engineers, Waterways Experiment Station, Aquatic Plant Control Research Program Report A-91-2, Vicksburg, Mississippi.

Savino, J. F., and R. A. Stein. 1982. Predator-prey interactions between largemouth bass and bluegills as influenced by simulated submersed vegetation. Transactions of the American Fisheries Society 111:255–266.

Shireman, J. V., D. E. Colle, and D. E. Canfield. 1985. Efficiency and cost of aquatic plant control in small ponds. Aquatics 6:14–20.

Shireman, J. V., and C. R. Smith. 1983. Synopsis of biological data on the grass carp Ctenopharyngodon idella (Cuvier and Valenciennes, 1844). FAO (Food and Agriculture Organization of the United Nations) Fisheries Synopsis 135.

Sutton, D. L., and V. V. Vandiver, Jr. 1986. Grass carp—a fish for biological management of hydrilla and other aquatic weeds. University of Florida, Institute of Food and Agricultural Sciences Bulletin 867, Gainesville.

Swanson, E. D., and E. P. Bergerson. 1988. Grass carp stocking model for coldwater lakes. North American Journal of Fisheries Management 8:284–291.

Thompson, B. Z., R. J. Wattendorf, R. S. Hestand, and J. L. Underwood. 1987. Triploid grass carp production. Progressive Fish-Culturist 49:213–217.

Van Eenennaam, J. P., R. K. Stocker, R. G. Thiery, N. T. Hagstrom, and S. I. Doroshov. 1990. Egg fertility, early development and survival from crosses of diploid female × triploid male grass carp (Ctenopharyngodon idella). Aquaculture 86:111–125.

Wiley, M. J., P. P. Tazik, and S. T. Sobaski. 1987. Controlling aquatic vegetation with triploid grass carp. Illinois Natural History Survey, Circular 57, Champaign.

Zimpher S. P., C. F. Bryan, and C. H. Pennington. 1987. Factors associated with the dynamics of grass carp larvae in the lower Mississippi River Valley. American Fisheries Society Symposium 2:102–108.

The Case for Caution with Fish Introductions

WALTER R. COURTENAY, JR.

*Florida Atlantic University, Department of Biological Sciences
Boca Raton, Florida 33431, USA*

Abstract.—Intentional and accidental introductions of fishes to and within North America, particularly over the past century, have resulted in successful establishment of a large number of species in new habitats. Every American and most Mexican states and Canadian provinces now contain nonnative fishes. Whereas a majority of introductions was made without declared intent, purposeful releases occurred for many reasons including sport and commercial fishing and biological control. Because introductions affect ecological niches (including ontogenetic niches) of resident species, responsible biological stewardship and management demand that potentially detrimental effects to habitats and ecosystems from intentional releases receive equal attention to arguable perceived social, economic, and political benefits. The growing problem of accidental introductions by hobbyists, ballast water discharges, escapes from aquacultural facilities, and releases of baitfishes by anglers requires more attention than it has received and far more concern and vigilance on the part of management agencies.

Fish culture and introductions into new habitats have been entwined in history for so long that it is impossible to separate them. Culture practices appear to have begun in China about three millennia ago (Nichols 1943; Hickling 1968; Bardach et al. 1972; Balon 1974; Ling 1977). Thus, fish culture closely parallels the history of plant and animal culture in developing agricultural societies, as does the history of species transfers from native ranges into distant locales.

Successful transfers require efficient transportation. Using early species movements as indicators of future success with introductions, humans drastically rearranged the flora and fauna of much of the world, particularly over the past century as transportation mechanisms improved. Unlike other species, we humans have the ability to modify environments to suit our perceived needs. We have done so with agricultural production and technology and with culture and introductions of fishes.

In this paper I examine biological concerns and management attitudes regarding species transfers. The concern shared by biologists and managers is conservation and, for many, production; nevertheless, we may differ on conservation or production of what and why. The one factor that unites us is fishes, their perpetuation and preservation, and improvement of their habitats, especially in waters degraded by human activities to the detriment of aquatic communities. That we are forced to try to improve on nature or correct mistakes through introductions is where we fall short of fulfilling our responsibilities for fishes and, in fact, magnify problems created by those who have shown little or no concern for fishery resources.

Extent of Species Introductions

Movement of fish species has been global, especially over the past century. Welcomme (1988) listed 237 inland fish species of foreign origin as introduced into 140 nations by 1,354 reported introductions. Marine species and transfers of fishes within borders of those nations were not included. Although many introductions were made a century or more ago as Europeans colonized many parts of the world, the majority occurred within the past 40 years following the advent of commercial, intercontinental jet cargo aircraft (Courtenay 1993). Much of this increase in fish movement following World War II correlates with growth of the aquarium fish industry in developed nations and expansion of aquaculture in both developed and developing countries (Courtenay and Stauffer 1990; Courtenay and Williams 1992).

Although importation is not necessarily synonymous with introduction of exotic species, it presents opportunities for release or escape. Welcomme (1988) observed that allowing any species to be imported is tantamount to approving its introduction. Releases, from whatever source, may or may not result in establishment of reproducing populations. The United States now hosts established populations of at least 70 exotic fishes (Courtenay et al. 1991; Courtenay 1993), most of which are warmwater species. As of a decade ago, Canada contained at least 10 exotic fishes (Crossman 1984) and Mexico had 26 (Contreras and Escalante 1984), some of which emigrated from U.S. introductions. Whereas Canada has few habitats hospitable to warmwater species, Mexico has not been exposed to extensive culture activities that otherwise would have increased its number of established exotic fishes; I expect this to change with passage of the North American Free Trade Agreement and almost certain development of aquarium and food fish culture in Mexico.

Within North America north of the Mexican plateau, almost 200 fishes occupy waters beyond their known native ranges (Lee et al. 1980; Courtenay and Taylor 1984; Hocutt and Wiley 1986; Courtenay 1993). Of these transplants, 177 species are confirmed introductions, 20 have populations that are strongly suspected of having been transplanted, and 12 (perhaps more) were moved to safer waters to prevent their extirpation in portions or all of their known ranges (Table 1). The 179 known transplants represent nearly 23% of the native fish fauna. Most (about 75%) relate to recreational fishing activities, particularly releases of bait fishes (Courtenay 1991, 1993). This estimate of recreational fishing releases was derived by combining those obviously moved as sport species (several ictalurids, some catostomids, all esocids, most salmonids, percichthyids, and centrarchids, one gadid, one haemulid, one sciaenid, and three percids) with those transplanted as forage (two clupeids and one osmerid) and those most likely released as unwanted live bait (small cyprinids, catostomids, one umbrid, one cyprinodontid, and small percids). Nowhere else on this planet have there been more fish introductions than in the United States, not necessarily an accomplishment in which we can take pride.

Why Introductions Occur

Benefits have accrued from introductions of exotic species and transplants used for agricultural purposes. An interdependence between humans and food plants and many domesticated animals developed through culture and husbandry, effected in large part by generations of genetic manipulation of those organisms through hybridization. This interdependence and these "improvements" on nature have prevented most cultivated and several domesticated animals from becoming feral pests that invaded and altered nonagricultural habitats.

Natural fish distributions are the result of evolution of both species and habitats. Humans have created new distributional ranges by moving fishes or presenting them with opportunities to do so on their own. Contemporary recreational fishing often is based on introductions and stocking to provide the public with fishes with which they are familiar or with new angling opportunities, a practice begun this past century. Among the most successful intro-

TABLE 1.—Native fishes known or suspected to have been introduced beyond their native ranges in the United States. No symbol indicates known transfer; ? indicates suspected introduction; * indicates transplant to prevent extirpation in native habitat. Table is mostly from Lee et al. (1980) and Hocutt and Wiley (1986).

Scientific name	Common name
Petromyzontidae—lampreys	
Petromyzon marinus	sea lamprey
Acipenseridae—sturgeons	
Acipenser transmontanus	white sturgeon
Amiidae—bowfins	
Amia calva	bowfin
Clupeidae—herrings	
Alosa pseudoharengus	alewife
Alosa sapidissima	American shad
Dorosoma cepedianum	gizzard shad
Dorosoma petenense	threadfin shad
Cyprinidae—carps and minnows	
Agosia chrysogaster	longfin dace
Campostoma anomalum	central stoneroller
Campostoma oligolepis	largescale stoneroller
Cyprinella galactura	whitetail shiner
Cyprinella lutrensis	red shiner
Cyprinella spiloptera	spotfin shiner
Cyprinella venusta	blacktail shiner
Dionda episcopa ?	roundnose minnow
Exoglossum maxillingua ?	cutlips minnow
Gila atraria	Utah chub
Gila bicolor	tui chub
Gila coerulea	blue chub
Gila copei	leatherside chub
Gila orcutti	arroyo chub
Gila pandora	Rio Grande chub
*Gila purpurea**	Yaqui chub
Hesperoleucus symmetricus	California roach
Hybognathus placitus	plains minnow
Luxilus cerasinus ?	crescent shiner
Luxilus chrysocephalus	striped shiner
Luxilus coccogenis ?	warpaint shiner
Luxilus zonistius	bandfin shiner
Lythrurus ardens ?	rosefin shiner
Lythrurus atrapiculus	blacktip shiner
Macrhybopsis storeriana	silver chub
Nocomis biguttatus	hornyhead chub
Nocomis leptocephalus ?	bluehead chub
Nocomis micropogon	river chub
Nocomis raneyi ?	bull chub
Notemigonus crysoleucas	golden shiner
Notropis amoenus ?	comely shiner
Notropis atherinoides	emerald shiner
Notropis baileyi	rough shiner
Notropis bairdi	Red River shiner
Notropis bifrenatus ?	bridle shiner
Notropis buccula	smalleye shiner
Notropis buchanani	ghost shiner
Notropis chiliticus ?	redlip shiner
Notropis dorsalis	bigmouth shiner
Notropis girardi	Arkansas River shiner
Notropis hudsonius	spottail shiner
Notropis leuciodus ?	Tennessee shiner
Notropis lutipinnis	yellowfin shiner
Notropis oxyrhynchus	sharpnose shiner

(Continued on next page)

TABLE 1.—Continued.

Scientific name	Common name
Cyprinidae—carps and minnows	
Notropis ozarcanus	Ozark shiner
Notropis procne ?	swallowtail shiner
Notropis rubellus	rosyface shiner
Notropis rubricroceus ?	saffron shiner
Notropis schumardi	silverband shiner
Notropis spectrunculus ?	mirror shiner
Notropis telescopus	telescope shiner
Notropis volucellus	mimic shiner
Notropis winchelli	clear chub
Orthodon microlepidotus	Sacramento blackfish
Phenacobius mirabilis	suckermouth minnow
Phoxinus oceas	mountain redbelly dace
Pimephales notatus	bluntnose minnow
Pimephales promelas	fathead minnow
Pimephales vigilax	bullhead minnow
Ptychocheilus grandis	Sacramento squawfish
Rhinichthys cataractae	longnose dace
Rhinichthys osculus	speckled dace
Richardsonius balteatus	redside shiner
Richardsonius egregius	Lahontan redside
Semotilus atromaculatus	creek chub
Semotilus corporalis	fallfish
Catostomidae—suckers	
Carpiodes carpio	river carpsucker
Catostomus commersoni	white sucker
Catostomus fumeiventris	Owens sucker
Catostomus platyrhynchus	mountain sucker
Catostomus plebeius	Rio Grande sucker
Catostomus santaanae	Santa Ana sucker
Catostomus tahoensis ?	Tahoe sucker
Erimyzon sucetta ?	lake chubsucker
Hypentelium etowanum	Alabama hog sucker
Hypentelium nigricans	northern hog
Ictiobus bubalus	smallmouth buffalo
Ictiobus cyprinellus	bigmouth buffalo
Minytrema melanops	spotted sucker
Moxostoma duquesnei ?	black redhorse
Moxostoma erythrurum	golden redhorse
Moxostoma macrolepidotum	shorthead redhorse
Moxostoma rhothoecum ?	torrent sucker
Moxostoma rupiscartes ?	striped jumprock
Characidae—characins	
Astyanax mexicanus	Mexican tetra
Ictaluridae—bullhead catfishes	
Ameirus catus	white catfish
Ameirus melas	black bullhead
Ameirus natalis	yellow bullhead
Ameirus nebulosus	brown bullhead
Ictalurus furcatus	blue catfish
Ictalurus punctatus	channel catfish
Noturus gyrinus	tadpole madtom
Noturus insignis	margined madtom
Pylodictus olivaris	flathead catfish
Esocidae—pikes	
Esox americanus americanus	redfin pickerel
Esox americanus vermiculatus	grass pickerel
Esox lucius	northern pike
Esox masquinongy	muskellunge
Esox niger	chain pickerel
Umbridae—mudminnows	
Dallia pectoralis	Alaska blackfish
Umbra limi	central mudminnow

TABLE 1.—Continued.

Scientific name	Common name
Osmeridae—smelts	
Osmerus mordax	rainbow smelt
Salmonidae—trouts	
Coregonus artedi	cisco or lake herring
Coregonus clupeaformis	lake whitefish
Oncorhynchus aguabonita	golden trout
Oncorhynchus apache*	Apache trout
Oncorhynchus clarki	cutthroat trout
Oncorhynchus gilae*	Gila trout
Oncorhynchus gorbuscha	pink salmon
Onchorhynchus keta	chum salmon
Oncorhynchus kisutch	coho salmon
Oncorhynchus mykiss	rainbow trout
Oncorhynchus nerka	sockeye salmon
Oncorhynchus tshawytscha	chinook salmon
Salmo salar	Atlantic salmon
Salvelinus alpinus	Arctic char
Salvelinus fontinalis	brook trout
Thymallus arcticus	Arctic grayling
Amblyopsidae—cavefishes	
Chologaster agassizi	spring cavefish
Gadidae—cods	
Lota lota	burbot
Cyprinodontidae—killifishes	
Crenichthys nevadae*	Railroad Valley springfish
Cyprinodon bovinus*	Leon Springs pupfish
Cyprinodon diabolis*	Devils Hole pupfish
Cyprinodon elegans*	Comanche Springs pupfish
Cyprinodon macularius*	desert pupfish
Cyprinodon rubrofluviatilis	Red River pupfish
Empetrichthys latos*	Pahrump poolfish
Fundulus diaphanus	banded killifish
Fundulus grandis	gulf killifish
Fundulus sciadicus	plains topminnow
Fundulus seminolis	Seminole killifish
Fundulus stellifer	southern studfish
Fundulus waccamensis	Waccamaw killifish
Fundulus zebrinus	plains killifish
Leptolucania ommata	pygmy killifish
Lucania goodei	bluefin killifish
Lucania parva	rainwater killifish
Poeciliidae—livebearers	
Gambusia affinis	western mosquitofish
Gambusia gaigei*	Big Bend gambusia
Gambusia geiseri	largespring gambusia
Gambusia holbrooki	eastern mosquitofish
Gambusia nobilis*	Pecos gambusia
Heterandria formosa	least killifish
Poecilia formosa	Amazon molly
Poecilia latipinna	sailfin molly
Poeciliopsis occidentalis*	Gila topminnow
Atherinidae—silversides	
Labidesthes sicculus	brook silverside
Menidia beryllina	inland silverside
Gasterosteidae—sticklebacks	
Culaea inconstans	brook stickleback
Gasterosteus aculeatus	threespine stickleback
Pungitius pungitius	ninespine stickleback

(*Continued on next page*)

TABLE 1.—Continued.

Scientific name	Common name
Cottidae—sculpins	
Cottus bairdi	mottled sculpin
Percichthyidae—temperate basses	
Morone americana	white perch
Morone chrysops	white bass
Morone mississippiensis	yellow bass
Morone saxatilis	striped bass
Centrarchidae—sunfishes	
Ambloplites cavifrons	Roanoke bass
Ambloplites constellatus	Ozark bass
Ambloplites rupestris	rock bass
Archoplites interruptus	Sacramento perch
Enneacanthus gloriosus	bluespotted sunfish
Lepomis auritus	redbreast sunfish
Lepomis gibbosus	pumpkinseed
Lepomis gulosus	warmouth
Lepomis humilis	orangespotted sunfish
Lepomis macrochirus	bluegill
Lepomis megalotis	longear sunfish
Lepomis microlophus	redear sunfish
Lepomis punctatus	spotted sunfish
Micropterus coosae	redeye bass
Micropterus dolomieui	smallmouth bass
Micropterus salmoides	largemouth bass
Micropterus treculi	Guadalupe bass
Pomoxis annularis	white crappie
Pomoxis nigromaculatus	black crappie
Percidae—perches	
Etheostoma blennioides ?	greenside darter
Etheostoma caeruleum	rainbow darter
Etheostoma chlorosomum	bluntnose darter
Etheostoma edwini	brown darter
Etheostoma fusiforme	swamp darter
Etheostoma olmstedi	tessellated darter
Etheostoma zonale	banded darter
Perca flavescens	yellow perch
Percina caprodes	logperch
Percina macrolepida	bigscale logperch
Percina roanoka	Roanoke darter
Stizostedion canadense	sauger
Stizostedion vitreum	walleye
Haemulidae—grunts	
Anisotremus davidsoni	sargo
Sciaenidae—drums	
Aplodinotus grunniens	freshwater drum
Cichlidae—cichlids	
Cichlasoma cyanoguttatum	Rio Grande cichlid
Gobiidae—gobies	
Gillichthys mirabilis	longjaw mudsucker

ductions have been brown trout *Salmo trutta*, bairdiella *Bairdiella icistia*, and orangemouth corvina *Cynoscion xanthulus* in the Salton Sea and striped bass along the Pacific coast. It can be argued that transfers of many salmonids, centrarchids, and percids, among others, have been beneficial, but often not without costs to resident nonpredatory species (Taylor et al. 1984; Moyle et al. 1986; He and Kitchell 1990; Minckley 1991).

There are important contrasts here between using nonnative, often genetically modified, species to feed humans and nondomesticated species about which we can make few predictions and over which we have minimal, if any, control once they are introduced. Furthermore, managing things we live with in our terrestrial environment, and observe on a daily basis, is quite different than making introductions and hoping they function as planned in aquatic environments, where we do not live. We quickly see changes in terrestrial systems when imbalances occur or when some pest threatens our food supplies, forests, or other factors we need for survival. We are far less aware of what goes on under water.

The first exotic fish introduction in North America was not of a food species. Goldfish *Carassius auratus* were released in the 1680s (DeKay 1842). For the next 150 years, no additions of exotic fishes were made, but canals were dug that began interdrainage transfers of native fishes (Smith 1985). One of these canals set the stage for the invasion by sea lamprey *Petromyzon marinus* of the western Great Lakes a century later (Ashworth 1986). Common carp *Cyprinus carpio* arrived without government-sponsored importation and were released by private citizens into the Hudson River in 1831 (DeKay 1842) and into California in 1872 (Moyle 1976). Mexico obtained common carp from Haiti in 1872 or 1873 (Contreras and Escalante 1984). The U.S. government did not get involved in fish transfers until the 1870s, when the "introduction paradigm" was dominant (Courtenay and Moyle, in press).

These early introductions only hint at the myriad ways nonnative species enter new habitats. Fishes have been intentionally released to enhance or improve food resources, to provide recreational opportunities, to control pest organisms (some of which were introduced earlier), to establish species for questionable aesthetic reasons (Li and Moyle 1993), to free unwanted bait or pets, and to reestablish Threatened and Endangered species in habitats from which they might be, or previously had been, extirpated (Johnson and Hubbs 1989; Carlson and Muth 1993). There have been instances of experimental baitfish releases resulting in established populations in marine waters (Lobel 1980; Maciolek 1984; Randall 1987). Unintentional introductions have resulted from escape from aquacultural facilities, release of ballast water, interdrainage canal construction, and nontarget species accompanying intentionally released target species (Courtenay and Robins 1989; Courtenay and Stauffer 1990; Courtenay 1991, 1993; Courtenay and Williams 1992; Carlton 1985, 1987, 1989, 1992; Carlton and Geller 1993; Li and Moyle 1993). Many fishes introduced with or without intent were cultured in North America before releases were made (Courtenay and Kohler 1986).

Habitat Modification

Human need for water brought about damming of tributaries and major rivers. By 1980, almost 1,600 large reservoirs greater than 200 ha had been created (Hayes et al. 1993). Included are extensive impoundments in arid regions that had no large lakes since the last postglacial pluvial period (10,000 or more years ago). As human populations settled and expanded, water demands increased. Deforestation around natural lakes and riverine riparian zones raised water temperatures (Ashworth 1986). Significant portions of river systems were converted into lakes, others were modified by diversion, channelization, dams, and locks. Some natural lakes of the Intermountain West have been almost drained for municipal water supplies. Wetlands have been emptied for agricultural, industrial, and real estate development, some under the guise of reducing problems caused by insects. Postglacial aquifers in the American Southwest have been and are being mined at an alarming rate, dropping levels in some springs and destroying others. The net result has been immense modification and loss of aquatic ecosystems. Few undisturbed freshwater areas still exist, and water flowing out of those areas into estuaries and oceans is often of reduced quality (Nielsen 1993). None of these activities was undertaken to protect, enhance, or improve fish resources, a practice that has begun to change only in recent years (Nielsen 1993).

When habitat is modified, our "opportunity" to manage begins, often with species transfers. Consequences of such actions require critical examination.

Effects of Introductions and the Vacant Niche Myth

Because introducing fishes has a long history and is recognized as an important management tool, continuation is expected. Nevertheless, those who select this option should be aware of the possible consequences. To base an introduction solely on what it might provide in terms of socioeconomic benefits or enhancing recreational opportunities ignores the biological implications that are part of every introduction and may well create or intensify previous biological disturbance. As noted earlier, Welcomme (1988) observed that allowing importation of a species is equivalent to sanctioning its introduction; this applies equally to culture of fishes beyond historical ranges where escape is virtually certain (Shelton and Smitherman 1984) and to unauthorized releases of fishes by the public.

Frequently, rationale for intentional introductions is based on a so-called "vacant" niche, usually meaning abundant food availability, in a particular habitat, that could be converted into usable fish flesh (Li and Moyle 1993). Hutchinson (1957) defined niche as, in effect, where and how an organisms fits into its abiotic and biotic surroundings. Thus, a niche is multidimensional and takes into account, for aquatic organisms, such factors as kinds and size of prey, water depth, velocity, and quality, substrate type, and relationships with other biota within a habitat. It defines the role of a species within a multispecies community. Therefore, to predict effects of a species introduction into one or more habitats within an ecosystem based mainly on food availability completely ignores what the new niche of the introduced species will be (regardless of what it may have been in its native range) and how it will affect niches of resident biota.

Price (1975) noted that two organisms having the same niche cannot coexist in a steady state, resulting in local extinction or displacement for one. Within a habitat, coexistence depends on the range of resources and places used by each species and the number of species exploiting certain resources or places. Species packing (MacArthur and Levins 1967; Abrams 1983) occurs when a large number of species uses a particular resource (or series of resources) and parts of a habitat. Therefore, where species packing occurs, niche breadths of each species will be compressed (Price 1975; de Moor 1993).

In aquatic communities, abiotic factors as well as seasonal changes in resource availability continually alter the niche of each species. Thus, through evolution and the genetic makeup of each species, population sizes will typically be in a state of flux as niches expand or contract. Habitat alterations, either natural (e.g., earthquakes, devastating floods, sudden or gradual temperature changes over evolutionary time) or human induced (e.g., pollution, water diversion or withdrawal, reservoir construction), modify niche dimensions for species within a community. When niche dimensions become too confining, one or more species loses the ability to survive, resulting in extinction. Where habitat alteration is less severe, niche dimensions will be altered to favor some species while compressing niches of others, perhaps into a threatened or endangered status. If a species has the genetic makeup to diversify its resource utilization, it has a survival advantage over those that do not by reducing niche gradients (Whittaker 1975; de Moor 1993); we typically select such species for culture and introductions.

With intentional introductions, positive and negative effects are typically focused on adults, ignoring the fact that fishes have different niches during their ontogeny. An introduced species may affect life history stages of one or more resident species differently during stages of its own ontogeny. Therefore, ontogenetic niche shifts of fishes (Werner and Hall 1988) must become a concern with fish introductions, recognizing that the degree of niche shift within a particular habitat is often influenced by local conditions (Werner and Hall 1988; Schramm and Jirka 1989; Belk and Lydeard 1994; Schaefer et al. 1994).

Rarely does an introduced species not have niche overlap with one or more resident species. If, however, the introduced species cannot fit into a niche in its new environment, even one that overlaps with those of resident species, it will not become established. If a niche is realized, (1) niche compression of residents with similar niches may result, (2) no niche compression may occur because there are no residents with similar niches, (3) extinction may eliminate one or more resident forms, or (4) habitat alteration can affect abiotic or biotic habitat, which in turn affects food webs and can lead to multiple resident species becoming extinct (Li and Moyle 1993). If any of these four alternatives occur, the introduced species is likely to establish and, where opportunities present themselves, become invasive by expanding into new habitats, sometimes with effects that may not have occurred at the site of introduction. This is why it is difficult to predict what an introduced species will do in a new environment (which normally includes many different habitats), why considering only a vacant or unoccupied trophic level cannot predict such outcomes, and why risks are involved with every introduction, regardless of geographic origin. There are no vacant niches in nature (Contreras and Escalante 1984; Li and Moyle 1993).

Magnuson (1976) likened lakes to islands, isolated from other such bodies of water by land, and suggested that island biogeography theory (MacArthur and Wilson 1967) applied to lakes. If land barriers isolate lakes, they also isolate drainage systems. Crossing land barriers by directly moving aquatic organisms, connecting lakes and drainages by canals, or pumping foreign ballast water that contains live organisms sets the stage for colonization (establishment and perhaps invasions) in a manner similar to natural mechanisms that permitted colonizations of island ecosystems. The difference is that island colonizations involved relatively long periods of geologic time, whereas anthropogenic movements of organisms occur very quickly. Colonizations of islands, although slow, were accompanied over time by species packing and niche compressions from earlier colonists. Thus, a colonization saturation point is reached where immigration and extinction rates determine the number of species inhabiting an island. As Magnuson (1976) and Li and Moyle (1993) observed, because freshwater habitats are "islands of water surrounded by a sea of land," it follows that multiple introductions will increase extinction rates by compressing or overloading niches, resulting in unstable communities. Because most inland habitats are disturbed from other activities, further destabilization of communities only makes management more complex and difficult.

Finally, because some ecosystems are richer in number of species than are others, it is likely that speciose ecosystems would be less hospitable to establishment and invasion by nonnative species through species packing over time. This may explain why fish introductions into reservoirs of the Southeast and Midwest have not been accompanied by dramatic biotic alterations as have occurred in many parts of the Intermountain West (Courtenay 1990; Douglas et al. 1994). This is not to say, however, that species-rich ecosystems always resist introduced biota, Lake Victoria in eastern Africa being a case in point (Barel et al. 1985; Hughes 1986; Goldschmidt and Witte 1990; Goldschmidt et al. 1990). The Nile perch *Lates niloticus* introduced into a rapidly eutrophicating lake created multiple extinctions of native haplochromine cichlids and contributed to dramatic alteration of food webs. As cichlid forage dominoes tumbled, so did the fishery productivity of Lake Victoria. The cichlid fishes of Lake Victoria had never experienced intense predation; the same applies to many fishes of the American West (Minckley et al. 1986; Minckley and Douglas 1991; Minckley et al. 1991; Douglas et al. 1994).

Biological Implications of Introductions

As Regier (1968) observed, we introduce fishes on the basis of what they can do for us, not what they can do for the environment. Based on what I presented above, the most responsible option from a biological standpoint would be not to make any more introductions and try, to the best of our abilities, to reduce impacts made with previous introductions. Nevertheless, this is not a practical or realizable option.

What should be clear is that there will be impacts. What should be equally clear is that we know so little of the inner workings of aquatic ecosystems, particularly effects of introductions on different life history stages of fishes, that we are incapable of accurately predicting or measuring even a modicum of the full range of effects from introductions. Furthermore, negative effects on receiving habitats and their biota does not require introduction of a predator. The realized niche of an introduced species in a new system, not necessarily its trophic habits, determines both short- and long-term effects of an introduction to resident habitat and biota. Multiple introductions will have multiple effects that are different with each receiving system and each species released. These effects contribute to further biological instability of the receiving system, often resulting in imagined needs for yet more introductions to muddle management efforts and goals further. Managing aquatic resources to meet human desires with introductions is risky at best, and can become a dangerous situation to biological resources, underscoring the need for caution.

In the Southeast and to a lesser extent the Midwest, where few natural lakes existed and most reservoirs have been created, introductions and, when necessary, continued stocking, have provided the angling public with a good variety of target species. Black bass *Micropterus* spp. and other centrarchids, trouts, walleye, and striped bass are among those existing in new habitats. As western reservoirs were built, many of the same species were introduced far beyond their native ranges. Most reservoirs were initially stocked with centrarchid and trout complexes, with later additions of walleye and striped bass. Often, what was introduced into western reservoirs was based on successes in the Southeast and Midwest, areas that had a greater fish diversity. These copycat introductions (Courtenay and Robins 1989) sometimes resulted in predator overload, creating more management problems than they were supposed to have solved. Additionally, many of these introductions resulted in endangering native fishes by affecting them far more than the perturbations caused by dam and reservoir construction (Minckley 1991). Walleye were illegally introduced to the Columbia River basin, adding to previously introduced fishes and creating problems with managing native salmonids already depressed by other habitat alterations (Li et al. 1987).

Other Players in Making Introductions

In nearly every state, it is illegal for any person to introduce a nonnative fish without a permit from the state fishery agency. Canada essentially prohibits introductions (Crossman 1984); Mexico, however, has yet to develop similar regulations (Contreras Balderas 1991). Much to the frustration of fishery managers, escapes from culture facilities and irresponsible actions by individuals in making their own introductions have resulted in establishment of many unwanted species. Whereas states usually have management plans for fishes they introduce or stock, those plans do not normally include contingencies for biologically irresponsible activities.

The aquarium fish industry and hobbyists are implicated in over half of the exotic fish introductions that have become established in the United States (Courtenay and Stauffer 1990). In Florida, where aquarium fish culture is a major industry, escapes have been a major source of introductions into natural habitats (Shafland 1991; Courtenay 1993). Similar situations exist in southern California (Courtenay et al. 1984, 1986). Most other aquarium fish introductions appear to have been made intentionally by hobbyists, dumping unwanted pets or trying to effect establishment in thermal springs (Courtenay et al. 1988).

Releasing unwanted bait fishes, either seined and moved to other drainage basins or cultured, transported, and sold well beyond their native ranges, has resulted in most establishments of transplanted fishes (Courtenay 1993). Anglers often believe such actions will only provide forage for sport fish. Anglers also have illegally, in many cases successfully, introduced sport fishes that were not included in management plans of receiving states or drainages; these introductions in many cases have confounded management of other species.

There is a segment of the public that is either ignorant of the consequences of their actions, does not care as long as their selfish needs are met, or maybe both. This topic must be addressed and serious attempts made to change prevailing public perception that releasing nonnative fishes will cause no harm.

The Costs of Introductions

Economic values can be placed on costs versus earnings to arrive at a profit or loss in business, including commercial fish culture. Similar values can be developed for commercial fishing. The facts speak for themselves. It is more difficult to arrive at values for recreational fishing because they require estimates. Costs of raising, stocking, and, for exotic species, importing fishes; funds generated from fishing license sales; and taxes levied on fishing tackle, bait, and related factors (e.g., boats, motors, trailers, rentals of such equipment) can be accurately measured. From there on, things become more nebulous. Only estimates can be made of what anglers actually generate to local economies from fishing activities. Nevertheless, values can be estimated that generally describe economic contributions from fishing activities.

In only one case do we know what the costs of an introduction have been, and we have those figures only because the introduction (the sea lamprey) had enormous negative impact on commercial and recreational fishing (Pister, in press). Annual control efforts cost about US$10 million, and a similar amount is spent each year for fish stocking in the Great Lakes (U.S. Congress General Accounting Office 1992). If control efforts were terminated, estimated lost value in terms of fishing resources and related costs could top $500 million annually (Spaulding and McPhee 1989). The Great Lakes Fishery Commission (1992) suggested that costs from accidentally introduced ruffe *Gymnocephalus cernuus* could exceed $90 million annually, based on its economic effects on fishing and related activities.

There is no compilation on yearly costs of common carp control, largely because so many different agencies (federal, state, and local) are involved. Total funds expended to eradicate brown trout to protect native golden trout in the Kern River, California, or to eliminate rainbow trout populations threatening brook trout in Great Smoky Mountain National Park, pale in comparison with sea lamprey and ruffe control (Courtenay 1991). In most waters where nonnative fishes have become established and have been implicated in declines of native species, they have spread so far that attempts at eradication would be both fiscally and physically impossible.

What is the value of some small fish in an Alabama stream or an equally small one in a remote thermal spring in Nevada? Would elimination of any threatened and endangered large river fish in the Colorado or Green rivers result in negative economic effects to human communities in states through which these rivers flow? Clearly we have a dichotomy in defining value. What is important in aquatic systems (the niche of a species or subspecies) has meaning within the biotic communities of those systems and to the process of evolution of the entire community but is unmeasurable in economic terms. Does introduction of a species capable of eliminating one or more resident fishes constitute a value gain or a value loss? Herein lies an important difference between managing for fishing versus managing for fishes.

Where Do We Go from Here?

Fisheries management originated for the purpose of increasing availability of selected species, and fish culture was a means of achieving that purpose. Management also involves manipulation of wild fish populations (Nielsen 1993), primarily for production of food and sport species; minimal resources have been dedicated to conservation of nongame species (Deacon and Minckley 1991; Carlson and Muth 1993; Warren and Burr 1994). Heidinger (1993) reviewed current philosophies of sport-fish management, noting that these range from purist (preservationist) to pragmatist (management for production) attitudes, with most biologists falling somewhere in the middle. Passage of the Endangered Species Act of 1973 (U.S.C.A. §§ 1531 to 1544) added a focus on nongame species. Since 1973, attention to nongame species, which outnumber game species 9:1 in North America (Warren and Burr 1994), seems to have further divided the purist-pragmatist spectrum described by Pister (1992) and Heidinger (1993). This schism is most unfortunate and counterproductive to what we should all be concerned with—fishes and the well-being of their habitats and ecosystems. We all abhor degraded water quality, drained wetlands, or similar kinds of insults to fish habitat that mean reduced fish biodiversity and fishing opportunities. Further, misunderstanding of the word biodiversity has struck unnecessary fear and panic in the minds of pragmatists, causing them to lose sight of the fact that natural systems, already disturbed, do not require additional species to make them function (Courtenay and Moyle 1992).

Management of U.S. inland and nearshore marine fisheries resources is largely in the hands of state agencies. Therefore, those resources are usually managed on a state-by-state basis, making surveyed state boundaries the limits and ignoring the fact that those boundaries are not biological. Although cooperation between states sharing drainages and other ecosystems is clearly the biologically responsible approach to aquatic resource management (Wilcove et al. 1992; Warren and Burr 1994), such an approach, unfortunately, is unlikely to receive much intrastate political support.

With regard to introductions, it should be understood that any addition to the biota of a given body of water will affect niches of resident species, whether native or results of earlier releases. How heavily those niches (with fishes, their ontogenetic niches) can be overloaded and their dimensions reduced before some, or perhaps, many of them collapse is the unknown factor. To know these factors and to be able to predict what might happen is presently beyond our abilities. Our past knowledge of what occurred following introductions, either anecdotal or with data, gives us some indication of what can happen. The question is, do we want to make introductions compatible with biological systems as well as with us?

Pragmatists may argue that introductions have had nothing to do with the collapse of native populations, blaming such alterations on actions of others affecting aquatic resources, such as habitat change through dam construction. Purists and those in between will point to a growing body of evidence that introductions have worked, by themselves or in synergy with other perturbations, to bring about such alterations (Minckley 1991; Wilcove et al. 1992).

Li and Moyle (1993) presented a series of options to managing fisheries resources without introductions or, at least, managing with introductions over which we have some control. The political realities of today suggest that many of these options will be ignored in favor of status quo management, which is easier, perhaps presently less expensive, and requires less challenge. From an ecological niche perspective, introductions from any source are risky and accurate prediction of the range of postintroduction effects is impossible.

At a minimum, attitudes toward introductions should be directed to (1) working toward interstate cooperation—preferably on an ecosystem basis—in fisheries management, including open discussion of proposed intentional introductions; (2) focusing on the entire ecosystem in management, not just its segments (Wilcove et al. 1992; Warren and Burr 1994); (3) following professional guidelines and policies for aquatic species introductions (Kohler and Courtenay 1986a, 1986b) and urging their adoption as agency policies; (4) reducing or eliminating escapes from culture facilities; and (5) involving educational staff in instructing the public that releasing bait and pets or trying to create a new fishery that is biologically unsound and unmanageable is a dangerous and illegal practice. Protecting biodiversity and providing a wealth of fishing values are not mutually exclusive when properly handled, and the sooner that is recognized, the easier it will be to accomplish both goals.

Acknowledgments

I thank Mark M. Konikoff and the organizers of this symposium for inviting my participation and allowing me to share my views with symposium participants and those who read this contribution. Special thanks go to B. M. Burr and E. P. Pister for providing comments on an earlier draft, to three anonymous reviewers for input on a later version, and R. E. Jenkins, W. L. Minckley, and J. D. Williams for their critique of parts or all of Table 1.

References

Abrams, P. 1983. The theory of limiting similarity. Annual Review of Ecology and Systematics 14:359–376.

Ashworth, W. 1986. The late, Great Lakes: an environmental history. Wayne State University Press, Detroit, Michigan.

Balon, E. K. 1974. Domestication of the carp *Cyprinus carpio* L. Miscellaneous Publications of the Royal Ontario Museum, Life Sciences: 1–37, Toronto, Ontario.

Bardach, J. E., J. H. Ryther, and W. O. McLarney. 1972. Aquaculture, the farming and husbandry of freshwater and marine organisms. Wiley-Interscience, New York.

Barel, C. D. N., and twelve coauthors. 1985. Destruction of fisheries in Africa's lakes. Nature (London) 315: 19–20.

Belk, M. C., and C. Lydeard. 1994. Effect of *Gambusia holbrooki* on a similar-sized, syntopic poeciliid, *Heterandria formosa*: competitor or predator? Copeia 1994:296–302.

Carlson, C. A, and R. T. Muth. 1993. Endangered species management. Pages 355–381 *in* C. C. Kohler and W. A. Hubert, editors. Inland fisheries management in North America. American Fisheries Society, Bethesda, Maryland.

Carlton, J. T. 1985. Transoceanic and interoceanic dispersal of coastal marine organisms: the biology of ballast water. Oceanography and Marine Biology Annual Review 23:313–371.

Carlton, J. T. 1987. Patterns of transoceanic marine biological invasions in the Pacific Ocean. Bulletin of Marine Science 41:452–465.

Carlton, J. T. 1989. Man's role in changing the face of the ocean: biological invasions and implications for conservation of near-shore environments. Conservation Biology 3:265–273.

Carlton, J. T. 1992. Dispersal of living organisms into aquatic ecosystems as mediated by aquaculture and fisheries activities. Pages 13–46 in A. Rosenfield and R. Mann, editors. Dispersal of living organisms into aquatic ecosystems. Maryland Sea Grant College, College Park.

Carlton, J. T., and J. B. Geller. 1993. Ecological roulette: the global transport of nonindigenous marine organisms. Science 261:78–82.

Contreras Balderas, S. 1991. Conservation of Mexican freshwater fishes: some protected sites and species, and recent federal legislation. Pages 191–197 in W. L. Minckley and J. E. Deacon, editors. Battle against extinction: native fish management in the American west. University of Arizona Press, Tucson.

Contreras-B., S., and M. A. Escalante-C. 1984. Distribution of exotic fishes in Mexico. Pages 102–130 in W. R. Courtenay, Jr., and J. R. Stauffer, Jr., editors. Distribution, biology, and management of exotic fishes. Johns Hopkins University Press, Baltimore, Maryland.

Courtenay, W. R., Jr. 1990. Fish conservation and the enigma of introduced species. Pages 11–20 in D. A. Pollard, editor. Introduced and translocated fishes and their ecological effects. Proceedings of the Bureau of Rural Resources 8, Canberra, Australia.

Courtenay, W. R., Jr. 1991. Pathways and consequences of the introduction of non-indigenous fishes in the United States. Report to the Office of Technology Assessment, Congress of the United States, Washington, DC.

Courtenay, W. R., Jr. 1993. Biological pollution through fish introductions. Pages 35–61 in B. N. McKnight, editor. Biological pollution: the control and impact of invasive exotic species. Indiana Academy of Science, Indianapolis.

Courtenay, W. R., Jr., D. A. Hensley, J. N. Taylor, and J. A. McCann. 1984. Distribution of exotic fishes in the continental United States. Pages 41–77 in W. R. Courtenay, Jr., and J. R. Stauffer, Jr., editors. Distribution, biology, and management of exotic fishes. Johns Hopkins University Press, Baltimore, Maryland.

Courtenay, W. R., Jr., D. A. Hensley, J. N. Taylor, and J. A. McCann. 1986. Distribution of exotic fishes in North America. Pages 675–698 in C. H. Hocutt and E. O. Wiley, editors. Zoogeography of North American freshwater fishes. John Wiley & Sons, New York.

Courtenay, W. R., Jr., D. P. Jennings, and J. D. Williams. 1991. Exotic fishes of the United States and Canada. American Fisheries Society Special Publication 20: 97–110.

Courtenay, W. R., Jr., and C. C. Kohler. 1986. The role of exotic fishes in North American fisheries management. Pages 401–413 in R. H. Stroud, editor. Fish culture in fisheries management. American Fisheries Society, Fish Culture Section and Fisheries Management Section, Bethesda, Maryland.

Courtenay, W. R., Jr., and P. B. Moyle. 1992. Crimes against biodiversity: the lasting legacy of non-native fish introductions. Transactions of the North American Wildlife and Natural Resources Conference 57: 365–372.

Courtenay, W. R., Jr., and P. B. Moyle. In press. Biodiversity, fishes, and the introduction paradigm. In R. C. Szaro, editor. Biodiversity in managed landscapes. Oxford University Press, New York.

Courtenay, W. R., Jr., and C. R. Robins. 1989. Fish introductions: good management, mismanagement, or no management? Reviews in Aquatic Sciences 1:159–172.

Courtenay, W. R., Jr., C. R. Robins, R. M. Bailey, and J. E. Deacon. 1988. Records of exotic fishes from Idaho and Wyoming. Great Basin Naturalist 47:523–526.

Courtenay, W. R., Jr., and J. R. Stauffer, Jr. 1990. The introduced fish problem and the aquarium fish industry. Journal of the World Aquaculture Society 21: 145–159.

Courtenay, W. R., Jr., and J. N. Taylor. 1984. The exotic ichthyofauna of the contiguous United States with preliminary observations on intranational transplants. European Inland Fisheries Advisory Committee Technical Paper 42:466–487.

Courtenay, W. R., Jr., and J. D. Williams. 1992. Dispersal of exotic species from aquaculture sources, with emphasis on freshwater fishes. Pages 49–81 in A. Rosenfield and R. Mann, editors. Dispersal of living organisms into aquatic ecosystems. Maryland Sea Grant College, College Park.

Crossman, E. J. 1984. Introduction of exotic fishes into Canada. Pages 78–101 in W. R. Courtenay, Jr., and J.R. Stauffer, Jr., editors. Distribution, biology, and management of exotic fishes. Johns Hopkins University Press, Baltimore, Maryland.

Deacon, J. E., and W. L. Minckley. 1991. Western fishes and the real world: the enigma of "endangered fishes" revisited. Pages 405–413 in W. L. Minckley and J. E. Deacon, editors. Battle against extinction: native fish management in the American west. University of Arizona Press, Tucson.

DeKay, J. E. 1842. Zoology of New York IV: fishes. W. & A. White and J. Visscher, Albany, New York.

de Moor, I. J. 1993. Methods for assessing the susceptibility of freshwater ecosystems in southern Africa to invasion by alien aquatic animals. Master's thesis. Rhodes University, Grahamstown, South Africa.

Douglas, M. E., P. C. Marsh, and W. L. Minckley. 1994. Indigenous fishes of western North America and the hypothesis of competitive displacement: *Meda fulgida* (Cyprinidae) as a case study. Copeia 1994:9–19.

Goldschmidt. T., and F. Witte. 1990. Reproductive strategies of zooplanktivorous haplochromine cichlids (Pisces) from Lake Victoria before the Nile perch boom. Oikos 58:356–368.

Goldschmidt, T., F. Witte, and J. de Visser. 1990. Ecological segregation in zooplanktivorous haplochromine species (Pisces: Cichlidae) from Lake Victoria. Oikos 58:343–355.

Great Lakes Fishery Commission. 1992. Ruffe in the Great Lakes: a threat to North American Fisheries.

Great Lakes Fishery Commission, Ruffe Task Force, Ann Arbor, Michigan.

Hayes, D. B., W. W. Taylor, and E. L. Mills. 1993. Natural lakes and large impoundments. Pages 493–515 *in* C. C. Kohler and W. A. Hubert, editors. Inland fisheries management in North America. American Fisheries Society, Bethesda, Maryland.

He, X., and J. F. Kitchell. 1990. Direct and indirect effects of predation on a fish community: a whole lake experiment. Transactions of the American Fisheries Society 119:825–835.

Heidinger, R. C. 1993. Stocking for sport fisheries enhancement. Pages 309–333 *in* C. C. Kohler and W. A. Hubert, editors. Inland fishes management in North America. American Fisheries Society, Bethesda, Maryland.

Hickling, C. F. 1968. The farming of fish. Pergamon Press, Oxford, England.

Hocutt, C. H., and E. O. Wiley, editors. 1986. Zoogeography of North American freshwater fishes. Wiley, New York.

Hughes, N. F. 1986. Changes in the feeding biology of the Nile perch *Lates nilotica* (L.) (Pisces: Centropomidae) in Lake Victoria, East Africa since its introduction in 1960 and its impact on the native fish community of the Nyanza Gulf. Journal of Fish Biology 29:541–548.

Hutchinson, G. E. 1957. Concluding remarks. Cold Spring Harbor Symposia on Quantitative Biology 22:415–427.

Johnson, J. E., and C. Hubbs. 1989. Status and conservation of poeciliid fishes in the United States. Pages 301–317 *in* G. K. Meffe and F. N. Snelson, Jr., editors. Ecology and evolution of livebearing fishes (Poeciliidae). Prentice-Hall, New York.

Kohler, C. C., and W. R. Courtenay, Jr. 1986a. Regulating introduced aquatic species: a review of past initiatives. Fisheries (Bethesda) 11(2):34–38.

Kohler, C. C., and W. R. Courtenay, Jr. 1986b. American Fisheries Society position on introductions of aquatic species. Fisheries (Bethesda) 11(2): 39–42.

Lee, D. S., C. R. Gilbert, C. H. Hocutt, R. E. Jenkins, D. E. McAllister, and J. R. Stauffer, Jr. 1980. Atlas of North American freshwater fishes. North Carolina State Museum, Raleigh.

Li, H., and P. B. Moyle. 1993. Management of introduced fishes. Pages 287–307 *in* C. C. Kohler and W. A. Hubert, editors. Inland fisheries management in North America. American Fisheries Society, Bethesda, Maryland.

Li, H., C. B. Schreck, C. E. Bond, and E. Rexstad. 1987. Factors influencing changes in fish assemblages of Pacific Northwest streams. Pages 193–202 *in* W. J. Matthews and D. C. Heins, editors. Community and evolutionary ecology of North American stream fishes. University of Oklahoma Press, Norman.

Ling, S-W. 1977. Aquaculture in southeast Asia, a historical overview. University of Washington Press, Seattle.

Lobel, P. S. 1980. Invasion by the Mozambique tilapia (*Sarotherodon mossambicus*; Pisces: Cichlidae) of a Pacific atoll marine ecosystem. Micronesica 16:349–355.

MacArthur, R. H., and R. Levins. 1967. The limiting similarity, convergence and divergence of coexisting species. American Naturalist 101:377–385.

MacArthur, R. H., and E. O. Wilson. 1967. The theory of island biogeography. Princeton University Press, Princeton, New Jersey.

Maciolek, J. A. 1984. Exotic fishes in Hawaii and other waters of Oceania. Pages 131–161 *in* W. R. Courtenay, Jr., and J. R. Stauffer, Jr., editors. Distribution, biology, and management of exotic fishes. Johns Hopkins University Press, Baltimore, Maryland.

Magnuson, J. J. 1976. Managing with exotics—a game of chance. Transactions of the American Fisheries Society 105:1–10.

Minckley, W. L. 1991. Native fishes of the Grand Canyon region: an obituary? Pages 124–177 *in* Colorado River ecology and dam management. National Academy Press, Washington, DC.

Minckley, W. L., and M. L. Douglas. 1991. Discovery and extinction of western fishes: a blink in the eye of geologic time. Pages 7–17 *in* W. L. Minckley and J. E. Deacon, editors. Battle against extinction: native fish management in the American west. University of Arizona Press, Tucson.

Minckley, W. L., D. A. Hendrickson, and C. A. Bond. 1986. Geography of western North American freshwater fishes: description and relationships to intracontinental tectonism. Pages 519–613 *in* C. H. Hocutt and E. O. Wiley, editors. Zoogeography of North American freshwater fishes. Wiley, New York.

Minckley, W. L., P. C. Marsh, J. E. Brooks, J. E. Johnson, and B. L. Jensen. 1991. Management toward recovery of the razorback sucker. Pages 303–357 *in* W. L. Minckley and J. E. Deacon, editors. The battle against extinction: native fish management in the American west. University of Arizona Press, Tucson.

Moyle, P. B. 1976. Inland fishes of California. University of California Press, Berkeley.

Moyle, P. B., H. W. Li, and B. A. Bean. 1986. The Frankenstein effect: impact of introduced fishes on native fishes in North America. Pages 415–426 *in* R. H. Stroud, editor. Fish culture in fisheries management. American Fisheries Society Fish Culture Section and Fisheries Management Section, Bethesda, Maryland.

Nichols, J. T. 1943. The fresh-water fishes of China. American Museum of Natural History, New York.

Nielsen, L. A. 1993. History of inland fisheries management in North America. Pages 3–31 *in* C. C. Kohler and W. A. Hubert, editors. Inland fisheries management in North America. American Fisheries Society, Bethesda, Maryland.

Pister, E. P. 1992. Ethical considerations in conservation of biodiversity. Transactions of the North American Wildlife and Natural Resources Conference 57:355–364.

Pister, E. P. In press. Ethics of native species restoration: the Great Lakes. Bulletin of the Great Lakes Fishery Commission.

Price, P. W. 1975. Insect ecology. Wiley, New York.

Randall, J. E. 1987. Introductions of marine fishes to the

Hawaiian Islands. Bulletin of Marine Science 41:490–502.

Regier, H. A. 1968. The potential misuse of exotic fish as introductions. Ontario Department of Lands and Forests Research Report 82.

Schaefer, J. F., S. T. Heulett, and T. M. Farrell. 1994. Interactions between two poeciliid fishes (*Gambusia holbrooki* and *Heterandria formosa*) and their prey in a Florida marsh. Copeia 1994:516–520.

Schramm, H. L., Jr., and K. J. Jirka. 1989. Epiphytic macroinvertebrates as a food resource for bluegills in Florida lakes. Transactions of the American Fisheries Society 118:416–426.

Shafland, P. L. 1991. Management of introduced freshwater fishes in Florida. Pages 214–137 *in* New directions in research, management and conservation of Hawaiian stream ecosystems. Hawaii Department of Natural Resources, Division of Aquatic Resources, Honolulu.

Shelton, W. L., and R. O. Smitherman. 1984. Exotic fishes in warmwater aquaculture. Pages 262–301 *in* W. R. Courtenay, Jr. and J. R. Stauffer, Jr., editors. Distribution, biology, and management of exotic fishes. Johns Hopkins University Press, Baltimore, Maryland.

Smith, C. L. 1985. The inland fishes of New York state. New York Department of Conservation, Albany.

Spaulding, W. M., Jr., and R. J. McPhee. 1989. The report of the evaluation of the Great Lakes Fishery Commission by the Bi-National Evaluation Team, volume 2. An analysis of the economic contribution of the Great Lakes Sea Lamprey Program. U.S. Fish and Wildlife Service, Great Lakes Region, Twin Cities, Minnesota.

Taylor, J. N., W. R. Courtenay, Jr., and J. A. McCann. 1984. Known impacts of exotic fishes in the continental United States. Pages 322–373 *in* W. R. Courtenay, Jr., and J. R. Stauffer, Jr., editors. Distribution, biology, and management of exotic fishes. Johns Hopkins University Press, Baltimore, Maryland.

U.S. Congress General Accounting Office. 1992. Great Lakes Fishery Commission—actions needed to support an expanded program. U.S. General Accounting Office, NSAID-92-108, Gaithersburg, Maryland.

Warren, M. L., and B. M. Burr. 1994. Status of freshwater fishes of the United States: overview of an imperiled fauna. Fisheries (Bethesda) 19(1):6–18.

Welcomme, R. L. 1988. International introductions of inland aquatic species. FAO (Food and Agriculture Organization of the United Nations) Fisheries Technical Paper 213.

Werner, E. F., and D. J. Hall. 1988. Ontogenetic habitat shifts in bluegill: the foraging rate-predation risk trade-off. Ecology 69:1352–1366.

Whittaker, R. H. 1975. Communities and ecosystems. Macmillan, New York.

Wilcove, D., M. Bean, and P. C. Lee. 1992. Fisheries management and biological diversity: problems and opportunities. Transactions of North American Wildlife and Natural Resources Conference 57:373–383.

From Sportfishing Bust to Commercial Fishing Boon: A History of the Blue Tilapia in Florida

MARTY M. HALE, JOE E. CRUMPTON, AND RALPH J. SCHULER, JR.

Florida Game and Fresh Water Fish Commission
Eustis Fisheries Research Laboratory
Eustis, Florida 32727, USA

Abstract.—The blue tilapia *Tilapia aurea* is an important food fish as exemplified by worldwide aquacultural operations and extensive commercial fisheries. This fish was brought to the southeastern United States in 1957 by Auburn University for evaluation as a potential sport and culture fish. In 1961, the Florida Game and Fresh Water Fish Commission (GFC) obtained blue tilapia from Auburn University to evaluate its potential (1) to replace gizzard shad *Dorosoma cepedianum* as the dominant fish in eutrophic waters with a fish higher in human food value; (2) to become an additional sport-fish species; (3) to control nuisance aquatic vegetation such as duckweed *Lemna* spp. and the filamentous algae *Pithophora* spp.; and (4) to become an additional forage fish species. After completing the studies, the GFC decided blue tilapia was undesirable for stocking in Florida's public waters given its low catchability, its ability to overwinter even in north Florida, and the lack of a need for additional forage species. Before blue tilapia could be eradicated from the study site, fish were removed by the public and stocked into several rivers and public lakes. Once blue tilapia were established in public waters, attempts to eradicate them permanently were unsuccessful. In some highly eutrophic lakes, blue tilapia did replace gizzard shad as the dominant species in biomass. However, age-0 blue tilapia had similar food habits to age-0 centrarchids, and adult blue tilapia almost completely displaced largemouth bass *Micropterus salmoides* in some spawning areas. By 1968, substantial snag-hook and cast-net fisheries existed in several lakes. The commercial sale of blue tilapia was legalized in February 1973 after the GFC determined the blue tilapia range was expanding and several lakes contained high biomass. It was also hoped that native fish populations would benefit from intensive commercial harvest of blue tilapia. In 1992–1993, 2,701,165 kg of blue tilapia worth US$1.6 million (wholesale) were reported harvested. This commercial fishery currently offers economic opportunities to several hundred Floridians and a low-cost food fish to consumers. Given the limited number of management options available to control blue tilapia, it appears that commercial harvest is making the best of the situation by using this abundant resource.

The Florida Game and Fresh Water Fish Commission (GFC) imported blue tilapia *Tilapia aurea* in 1961 for evaluation as a food and sport fish (Crittenden 1962). By 1984, blue tilapia was the most widely distributed exotic fish species in Florida (Courtenay et al. 1984). Ten years later, the blue tilapia still holds that distinction. Approximately 10 years after their introduction into public waters by the public, a commercial fishery for blue tilapia developed. The objectives of this paper are to describe the status of blue tilapia in Florida from its introduction to its commercialization and to document the importance of the commercial fishery for blue tilapia that has developed.

Introduction and Subsequent Escapement of Tilapia in Florida

In August 1961, a shipment of 3,000 blue tilapias was delivered to the Pleasant Grove Research Station 32 km southeast of Tampa, Florida, from Auburn University. The fish sent to Florida were initially identified as Nile tilapia *Tilapia nilotica* but were correctly identified as blue tilapia in 1966 at Auburn University. In pond studies at Auburn University, Swingle (1958) was encouraged by the tilapia's apparent ability to control the filamentous algae *Pithophora* spp. Swingle (1960) was also encouraged by angling catch rates of tilapias and stated that tilapias may prove useful as sport fish under a variety of conditions. As proposed by Benson et al. (1961), Crittenden (1962) wanted to assess the blue tilapia's potential as a food and sport fish and its productivity in the highly eutrophic water inhabited primarily by low-value fishes in Florida such as gizzard shad *Dorosoma cepedianum*. Blue tilapias (2.5–12.5 cm long) were stocked at 2,470/ha; reproduction occurred 4–5 weeks after stocking; and blue tilapias overwintered with no observed mortality in 1961–1962 (Crittenden 1962).

A pond study in south Florida suggested that largemouth bass *Micropterus salmoides* actively preyed on young blue tilapia (Crittenden 1962). Based on these results, Crittenden (1962) was encouraged at the apparent ability of largemouth bass to reduce the blue tilapia population and stated that "bass predation might prevent tilapia from becoming the nuisance fish carp and other carelessly introduced exotics have sometimes turned out to be."

Swingle (1960) reported angling harvests in ponds of 667 kg/ha in 1957 and 664 kg/ha in 1958 for Mozambique tilapia *Tilapia mossambica,* but only 144 kg/ha for blue tilapia. In later studies, Swingle et al. (1965) reported that when Mozambique and blue tilapia were stocked together, blue tilapia was more readily caught. When stocked separately, Mozambique tilapia was more readily caught. He offered no explanation for these unusual differences in catch rates.

To compare the sportfishing values of blue tilapia and bluegill *Lepomis macrochirus,* Crittenden (1962) tagged and released 100 bluegills and 100 blue tilapias into a 6.9-ha pond with an existing sport-fish population. Rewards ranging from US$3–500 were offered by the Joseph Schlitz Brewing Company to motivate tag returns. Because blue tilapia and bluegill were both caught with worms, fishing effort was believed to have been equal. Over a 5-month period, 21% of the tagged bluegills were captured and 23% of the tagged blue tilapias were captured. Apart from tagging returns, Crittenden (1962) reported that blue tilapia appeared to be gaining favorable recognition among Florida's bream anglers.

In a preliminary report, Swingle (1958) described the use of Mozambique tilapia to control the filamentous algae *Pithophora spp.* McBay (1961) reported that adult blue tilapia consumed relatively large quantities of *Pithophora spp.* In Florida, Crittenden (1962) thought that when stocked at a rate of 2,470 or more per hectare, blue tilapia might control duckweed *Lemna spp.* in ponds. The brood pond at Pleasant Grove was two-thirds covered by duckweed and within 4 months of stocking blue tilapia, the duckweed was reduced to a narrow, 30-cm margin along the shore.

Crittenden then planned to conduct pond studies to evaluate the effect of blue tilapia on gizzard shad populations. However, in 1964 the blue tilapia's future in Florida changed. Barkuloo (1964) stated blue tilapia was "not a desirable sport fish" and he recommended that "no further stocking of this fish be done in Florida." Barkuloo gave the following reasons for his recommendation: (1) sportfishing success for blue tilapia was low in all waters regardless of population density or species composition; (2) blue tilapia could be caught by hook and line only during short periods of the year (Barkuloo believed that a fish that did not furnish year-round sportfishing should not be considered a desirable sport species); (3) blue tilapia was found to be capable of surviving winters even in north Florida, which could result in a serious control problem; and (4) because forage species cause most of the fishery problems in Florida, another forage species would likely serve no useful purpose and may cause additional problems.

Unfortunately, before the negative aspects of blue tilapia could be publicized, several GFC employees gave selected public individuals blue tilapia for stocking (Buntz and Manooch 1968). Some anglers who believed the GFC was proceeding too slowly also stocked them in public water bodies. Buntz and Manooch (1968) attributed much of the excitement about blue tilapia to a very supportive media. Articles extolling the virtues of blue tilapia appeared in newspapers, national outdoors magazines, and even in the GFC publication *Florida Wildlife*. Buntz and Manooch (1968) wisely recommended that better control of the media (by not releasing preliminary data) and experimental species was needed to avoid further releases like blue tilapia in Florida.

Many authors have warned against the importation of nonnative fish species. Hubbs (1968) strongly recommended that each blue tilapia release be carefully evaluated; he feared that scientists might create a second "carp problem" from these introductions. After discovering that redbelly tilapia *Tilapia zilli* could overwinter in southern and central California, Pelzman (1973) recommended that they be designated a prohibited species in northern and central California. After observing blue tilapia introduction in Florida, Courtenay and Robins (1973) concluded that "availability of such experimental stocks almost always leads to introduction into open waters." Unfortunately, most of these warnings came after tilapia became established in many states.

By 1968, blue tilapia had spread to the Peace and Alafia rivers, four creeks, six lakes, and at least 10 private ponds in south-central Florida (Buntz and Manooch 1968). The presence of blue tilapia in Lake Parker (a 928-ha hypereutrophic lake near Lakeland, Florida) was verified in 1966, and within 3 years Lake Parker was credited with supporting the largest population of blue tilapia in the United States (Buntz and Manooch 1970). By 1977, the blue tilapia range in Florida had expanded to 21 counties and the Alafia, Hillsboro, Kissimmee, Manatee, Myakka, Peace, St. Johns, and Withlacoochee river basins (Foote 1977). Buntz and Manooch (1968) correctly ascertained that "time will be the major factor limiting spread from central Florida south." Courtenay et al. (1984) described blue tilapia as the most widely distributed exotic fish species in Florida in 1984, a distinction it still holds.

Early attempts to eradicate blue tilapia before they became established were unsuccessful. One of the more notable efforts occurred in Lake Morton, a 16-ha mesotrophic lake in downtown Lakeland that has connections to other lakes (Buntz and Manooch 1968). Lake Morton also contained an abundant largemouth bass population. Before renovation in October 1966, 66 kg of blue tilapia were collected from a 0.4-ha block-net sample in Lake Morton. Blue tilapia again became abundant in Lake Morton shortly after renovation and continue to thrive there (T. Champeau, GFC, personal communication). The Lake Morton experience indicated largemouth bass may not be able to prevent blue tilapia from becoming a nuisance fish.

Blue Tilapia's Transition to a Commercial Fish in Florida

Conclusive studies that describe actual effects of blue tilapia on sport-fish populations in Florida have not been published. Several studies in Florida and Texas have documented potential negative effects. On Lake Parker, Chapman et al. (1975) observed the displacement of largemouth bass by blue tilapia and concluded that "competition for spawning space was inevitable in this littoral area." Horel (1951) counted 28 largemouth bass nests in Silver Glen Springs Run, a short spring-fed creek (3.2 m^3/s) that flows into Lake George on the St. Johns River. In 1983, Zale (1984) observed only 2 largemouth bass nests and 2,962 blue tilapia nests in Silver Glen Springs Run. Noble et al. (1976) reported the population of northern largemouth bass *Micropterus salmoides salmoides* in Trinidad Lake, Texas, declined to one-third its abundance in 3 years in the presence of a dense population of blue tilapia. In a pond experiment, Noble et al. (1976) also determined that Florida largemouth bass *M. s. floridanus* failed to reproduce successfully at a blue tilapia density of 2,245 kg/ha, but at 1,123 kg/ha some largemouth bass spawning occurred. Gonosomatic indices suggested that eggs were retained by largemouth bass at high blue tilapia densities. Instead of largemouth bass controlling blue tilapia densities as Crittenden (1962) hoped, blue tilapia effected largemouth bass populations.

One of the original objectives for stocking blue tilapia in Florida was to replace gizzard shad in eutrophic lakes. Gizzard shad populations have declined in the presence of dense blue tilapia populations (Babcock 1972; Hendricks and Noble 1980). Two years after blue tilapia introduction into Lake Parker, Florida, the biomass was 5% blue tilapia and 60% gizzard shad (Horel 1969). Four years later, biomass was 68% blue tilapia and 5% gizzard shad (Babcock 1972).

After observing similarities in food habits of blue tilapia, gizzard shad, and threadfin shad *Dorosoma petenense* in Trinidad Lake, Texas, Hendricks and Noble (1980) reported a corresponding decline in shad biomass as blue tilapia biomass increased to 2,000 kg/ha. Germany (1977) also observed the shad population decline in Trinidad Lake when the blue tilapia population increased. The shad population increased in Lake Trinidad after a complete die-off of blue tilapia due to cold weather (Germany 1977).

Shafland and Pestrak (1984) suggested displacement of gizzard shad by blue tilapia may partially be the result of selective largemouth bass predation on gizzard shad. Given the blue tilapia's general body shape, spiny-rayed fins, and intermediate length–depth relation, its forage value may lie between that of gizzard shad and bluegill. Shafland and Pestrak (1984) concluded blue tilapia introductions may cause changes in largemouth bass predator–prey relations, and the effect of blue tilapia in Florida may ultimately be judged on its value as largemouth bass prey.

Shortly following the introduction of blue tilapia in Florida, Buntz and Manooch (1968, 1970) described a unique sport fishery for it on Lake Parker. Blue tilapia composed over 50–87% of the total harvest by number in April, May, November, and December 1968. Anglers actively pursued this new fish by use of cast nets (legalized in 1967), baited hooks, and snag hooks. Baited hooks accounted for only 7.3% of the 5,934 tilapias enumerated from a creel survey between 1 August 1968 and 31 July 1969 (Buntz and Manooch 1970). Most of these were caught incidentally with the exception of young-of-year catches in late summer. This finding concurred with McBay's (1961) findings that 7.5–12.5-cm blue tilapias feed on insects. Snag hooking was very successful at the warmwater outflow of the power plant on Lake Parker during cold months. The warm water attracted blue tilapia, making them vulnerable to snag hooking. Catch rates for snag hooking peaked at 2.6 blue tilapias/h. Cast netting was the most successful harvest technique. Blue tilapia was susceptible to harvest by cast nets year-round, but peak harvest periods were observed during spawning activities. Catch rates for cast netting peaked at 89 fish/h.

The presence of blue tilapia increased fishing activity in Lake Parker. Two local tackle shops collectively sold more than 27,000 treble hooks, and repair work and sales of rods and reels increased five times. From 1 July 1968 through 30 June 1969, 1,264 cast-net permits were issued. Sales of cast nets increased from 6 to 36 annually at one tackle shop. Net repairs increased four times from the previous year.

A haul-seine program to remove blue tilapia was initiated in the Winter Haven Chain of Lakes, Lake Hancock, and Lake Parker to enhance their game-fish populations. Ware (1971) observed greater numbers, size, and spawning success of largemouth bass in Lake Hollingsworth, Florida, following a haul-seine program there. Potential benefits in addition to enhancing game-fish populations were nutrient removal and utilization of gizzard shad and blue tilapia. In 46 hauls in Lake Parker between late-May and mid-September 1972, 101,461 kg of fish were harvested; 84.1% were blue tilapia. In nine hauls in Lake Hancock, a 1,830-ha hypereutrophic lake with a poor sport-fish population, 7,173 kg of gizzard shad and 136 kg of blue tilapia were harvested. The Lake Hancock haul-seine commercial fisher claimed he was unable to make a profit selling only gizzard shad but would continue if the sale of whole blue tilapia was legalized (only filleted blue tilapia could be sold in 1972).

Whereas the 1972 haul-seine program did little to enhance game-fish populations, it did bring the GFC to an important crossroad. Because of the higher cost of filleting blue tilapia, two of the three commercial fishers quit the haul-seine program because they could not make enough money selling gizzard shad or filleted blue tilapia. The GFC then had to decide how it would regard commercialization of blue tilapia and the sale of whole fish. To maintain the haul-seine program, Babcock (1972) recommended that the GFC legalize the sale of whole blue tilapia in or out of the state of Florida as long as the fish were eviscerated, headed, or frozen but otherwise left intact before leaving Polk County. This recommendation ensured that fish would be dead and identifiable.

Commercial Use of Tilapia in Florida

As early as 1968, Buntz and Manooch (1968) suggested the possibility of commercial exploitation of blue tilapia but believed Florida law regarding fishing gear and sale of only filleted fish inhibited this opportunity. Buntz and Manooch (1970) mentioned the possibility of a commercial fishery for blue tilapia because "it is palatable and can be harvested in large numbers." In keeping with the recommendation of Babcock (1972) to authorize the sale of whole, dead fish, the GFC legalized the sale of whole blue tilapia in February 1973.

Blue tilapia are harvested commercially in Florida with haul seines and cast nets. The haul-seine fishery is regulated by the GFC through a permit system. Of the five haul-seine permits available, four are currently actively used. Haul-seine length and mesh size is regulated by the GFC and allows for optimal blue tilapia harvest. Most haul seines are about 1,000-m-long and 2.1–3.7-m-deep nylon webbing with 7.5-cm-stretch mesh wings and a 5.5-cm-stretch mesh pocket. When making a deepwater set, commercial fishers pull both ends of the seine until it is tightly stretched. One end is anchored and a boat pulls the other end to form a large circle and returns to the anchored end. The net is further restricted until fish are herded into the pocket. After fish are concentrated in the pocket, commercial fishers use dip nets to remove all fish and keep those legal to harvest (all nongame fish and turtles). When making a shoreline set, commercial fishers anchor one end of the seine at the shoreline and pull the seine by boat to form a large circle. Because one haul seine set takes about 5 h to complete, most commercial fishers make only one set per day. Haul seines can be fished year-round, but commercial fishers prefer the warmer months when catch rates are higher.

Commercial cast nets used to harvest blue tilapia are constructed of monofilament webbing. When fishing in shallow water for spawning fish, most cast netters use 5.5-cm-stretch mesh. When fishing deeper water, commercial fishers switch to a larger mesh net (6.9–8.1-cm-stretch mesh) to allow for faster sinking. Most nets are 7.3 m in diameter. Cast nets are used year-round but are more productive in spring when blue tilapia spawn and in winter near warmwater attractants.

Many large blue tilapia catches were recorded shortly after commercialization began. One seine operator harvested 112 kg/ha of blue tilapia in 3 months from Lake Parker. Another operator harvested 393 kg/ha of blue tilapia from Banana Lake in a similar period. During 1974, haul seines alone harvested 601 kg/ha of blue tilapia from Banana Lake (Chapman et al. 1975).

In a creel survey on Lake Parker from May 1973 through May 1974, Babcock (1974) estimated that 312 kg/ha of blue tilapia were caught by snag hooks and cast nets (primarily cast nets). The total harvest

TABLE 1.—Harvest (total gutted weight), wholesale value, and average price per kilogram of blue tilapia caught in Florida from 1 July 1991 through 31 December 1993.

Harvest period	Gutted weight (kg)	Wholesale value ($US)	Average price ($/kg)
1 Jul 1991–30 Jun 1992	2,486,743	1,516,913	0.61
1 Jul 1992–30 Jun 1993	2,701,165	1,674,722	0.62
1 Jul 1993–31 Dec 1993	1,052,633	736,843	0.70

for Lake Parker approached 490 kg/ha for that time period when fish caught in haul seines were added.

Blue tilapia harvest data since 1991 have revealed a valuable commercial fishery (Table 1). The 2.7 million kilograms harvested in 1992–1993 made blue tilapia the most important freshwater commercial fish by weight and the second most important fish in monetary value ($1.6 million wholesale) behind ictalurid catfishes (Hale et al. 1993). The 1993–1994 blue tilapia harvest looks promising. Approximately 1.0 million kilograms were reported for the first half of the year with the more productive winter and spring season remaining.

Commercial fishers were paid $0.26–0.33/kg for gutted blue tilapia in 1973–1974; these fish retailed for $0.99–1.65/kg in Florida. By 1977, the wholesale price averaged $0.33/kg (Langford et al. 1978). From 1991 through 1993, commercial fishers received $0.64/kg for gutted blue tilapia. Whereas a doubling of the wholesale value in 20 years may not appear high, the retail value has not increased very much either. In February 1994, gutted blue tilapia retailed for $2.18–2.75/kg.

Langford et al. (1978) described the outlook for commercial blue tilapia harvest as good, considering the large number of highly eutrophic lakes in Florida. Langford et al. (1978) stated that markets already were well developed and at least 100 people were employed in the harvesting, processing, and marketing of the species. According to a survey conducted in 1993 (Hale et al. 1993), at least 252 people are currently employed in the south Florida blue tilapia fishery. Seven commercial fish-processing houses employed approximately 56 people to handle and process commercial blue tilapia catches. These processors purchased fish from about 176 cast-net fishers who contributed more than 90% of the commercial harvest. The four active blue tilapia haul-seine permittees employed about 20 crew members. The number of cast-net fishers is a conservative one. This number included only south Florida cast netters who brought fish to fish houses; it did not include those selling directly to the public or recreational cast netters from central Florida counties like Lake and Marion.

In addition to the wild-caught blue tilapia fishery, blue tilapia is also cultured in ponds and tanks in Florida. Fifty-five permits were issued in 1993 for aquaculturists to raise blue tilapia. These fish were raised in 235 ponds (55.7 ha) and 412 tanks (4.4 million liters). Most of the culturists are not in direct competition with the of wild-caught fishery because most cultured fish are sold to the out-of-state live fish market. Only 15,227 kg of blue tilapia (whole weight, not gutted) were reported sold by this industry in 1987–1988. Compared with the wild-caught fishery, the cultured fishery was very small and constituted less than 1% of all blue tilapia marketed in 1987–1988.

In conclusion, we can confidently say that blue tilapia is permanently established in Florida. We also are confident that most biologists, ecologists, and conservationists are opposed to the escapement and spread of blue tilapia in Florida. However, the commercial fishery that has developed currently offers economic opportunities to several hundred Floridians and a low-cost food fish to consumers. Given the limited number of management options available to control blue tilapia, commercial harvest is appropriate. The current commercial fishery appears to be making the best of the situation and should continue under present restrictions.

References

Babcock, S. C. 1972. South Florida region haul seine program: end of 1972 calendar year review and recommendations. Eustis Fisheries Research Laboratory Technical Publication 71, Eustis, Florida.

Babcock, S. C. 1974. A creel survey of blue tilapia *Tilapia aurea* fishermen on Lake Parker, Florida. Eustis Fisheries Research Laboratory Technical Publication 72, Eustis, Florida.

Barkuloo, J. M. 1964. Experimental stocking of *Tilapia* sp. Florida Game and Fresh Water Fish Commission, Federal Aid in Sport Fish Restoration, Project F-14-R-4, Completion Report, Tallahassee.

Benson, N. G., J. R. Greeley, M. T. Huish, and J. H. Kuehn. 1961. Status of management of natural lakes. Transactions of the American Fisheries Society 90:218–224.

Buntz, J., and C. S. Manooch, III. 1969. *Tilapia aurea* (Steindachner), a rapidly spreading exotic in south central Florida. Proceedings of the Annual Conference Southeastern Association of Game and Fish Commissioners 22(1968):495–501.

Buntz, J., and C. S. Manooch, III. 1970. Fisherman utilization of *Tilapia aurea* (Steindachner) in Lake Parker, Lakeland, Florida. Proceedings of the Annual Conference Southeastern Association of Game and Fish Commissioners 23(1969):312–319.

Chapman, P., F. H. Langford, D. Reece, and C. C. Starling. 1975. South Region Annual Report 1974–75. Florida Game and Fresh Water Fish Commission, Tallahassee.

Courtenay, W. R., Jr., and C. R. Robins. 1973. Exotic aquatic organisms in Florida with emphasis on fishes: a review and recommendation. Transactions of the American Fisheries Society 102:1–12.

Courtenay, W. R., Jr., D. A. Hensley, J. N. Taylor, and J. A. McCann. 1984. Distribution of exotic fishes in the continental United States. Pages 41–77 in W. R. Courtenay, Jr. and J. R. Stauffer, Jr., editors. Distribution, biology and management of exotic fishes. Johns Hopkins University Press, Baltimore, Maryland.

Crittenden, E. 1965. Status of Tilapia nilotica (Linnaeus) in Florida. Proceedings of the Annual Conference Southeastern Association of Game and Fish Commissioners 16(1962):257–262.

Foote, K. J. 1977. Blue tilapia investigations. Annual Report 1976–77. Florida Game and Fresh Water Fish Commission, Tallahassee.

Germany, R. D. 1977. Population dynamics of the blue tilapia and its effects on the fish populations of Trinidad Lake, Texas. Doctoral dissertation. Texas A&M University, College Station.

Hale, M. M., J. E. Crumpton, and R. J. Schuler, Jr. 1993. Commercial Fisheries Investigations. Annual Report 1992–93. Florida Game and Fresh Water Fish Commission, Tallahassee.

Hendricks, M. K., and R. L. Noble. 1980. Feeding interactions of three planktivorous fishes in Trinidad Lake, Texas. Proceedings of the Annual Conference Southeastern Association of Fish and Wildlife Agencies 33(1979):324–330.

Horel, G. 1951. The major bedding areas of largemouth bass in Lake George, Florida. Florida Game and Fresh Water Fish Commission Technical Report, Tallahassee, Florida.

Horel, G. 1969. Fish population studies. Florida Game and Fresh Water Fish Commission, Federal Aid in Sport Fish Restoration, Project F-12-10, Tallahassee.

Hubbs, C. 1968. An opinion on the effects of cichlid releases in North America. Transactions of the American Fisheries Society 97:197–198.

Langford, F. H., F. J. Ware, and R. D. Gasaway. 1978. Status and harvest of introduced Tilapia aurea in Florida lakes. Pages 102–108 in R. O. Smitherman, W. L. Shelton, and J. H. Grover, editors. Symposium on the culture of exotic fishes. American Fisheries Society, Fish Culture Section, Auburn University, Alabama.

McBay, L. G. 1961. The biology of Tilapia nilotica Linneaus. Proceedings of the Annual Conference Southeastern Association of Game and Fish Commissioners 15(1961):208–218.

Noble, R. L., R. D. Germany, and C. R. Hall. 1976. Interactions of blue tilapia and largemouth bass in a power plant cooling reservoir. Proceedings of the Annual Conference Southeastern Association of Game and Fish Commissioners 29(1975):247–251.

Pelzman, R. J. 1973. A review of the life history of Tilapia zillii with a reassessment of its desirability in California. California Department of Fish and Game, Inland Fisheries Administrative Report 74-1, Sacramento.

Shafland, P. L., and J. M. Pestrak. 1984. Predation on blue tilapia by largemouth bass, in experimental ponds. Proceedings of the Annual Conference Southeastern Association of Fish and Wildlife Agencies 35(1981):443–448.

Swingle, H. S. 1958. Further experiments with Tilapia mossambica as a pondfish. Proceedings of the Annual Conference Southeastern Association of Game and Fish Commissioners 11(1957):152–154.

Swingle, H. S. 1960. Comparative evaluation of two tilapias as pondfishes in Alabama. Transactions of the American Fisheries Society 89:142–148.

Swingle, H. S., E. E. Prather, G. N. Greene, J. W. Avault, H. A. Swingle, and T. Scott. 1965. Management techniques for public fishing waters. Alabama Department of Conservation, Federal Aid in Sport Fish Restoration, Project F-10-R-6, Montgomery.

Ware, F. J. 1971. Lake management and research, haul seine study. Florida Game and Fresh Water Fish Commission, Federal Aid in Sport Fish Restoration, Project F-12-13, Tallahassee.

Zale, A. V. 1984. Applied aspects of the thermal biology, ecology and life history of the blue tilapia, Tilapia aurea (Pisces: Cichlidae). Florida Cooperative Fish and Wildlife Research Unit, Technical Report 12, Gainesville.

Splake as a Control Agent for Brook Trout in Small Impoundments

JAMES R. SATTERFIELD, JR.

Colorado Division of Wildlife
6060 Broadway, Denver, Colorado 80216, USA

KEITH D. KOUPAL

Colorado State University, Department of Fishery and Wildlife Biology
Fort Collins, Colorado 80523, USA

Abstract.—During 1988–1992, fingerlings (70–120 mm total length) of splake, the hybrid of male brook trout *Salvelinus fontinalis* and female lake trout *S. namaycush*, were stocked annually into three, 3.5–40.5-ha Colorado impoundments at a density of 100 per hectare to improve population structure of existing brook trout populations. After 5 years of splake stocking, brook trout in all three impoundments exhibited significant improvements in population structure. Density of stock-length (≥125 mm total length) brook trout, as measured by gill-net catch per unit effort (CPUE), declined in all waters from a pre-splake-introduction average of 42 fish per net-night to 23 fish. Conversely, brook trout proportional stock density (PSD) increased over the 5 years of splake stocking from an average of 4 (range: 1–7) to an average of 69 (range: 44–89). After 5 years of stocking, gill-net CPUE for stock-length (≥150 mm total length) splake averaged 21 fish per net-night in the three impoundments. Splake fingerlings reached 300 mm in three growing seasons (1991) and 350 mm in four growing seasons (1992) in all waters. Brook trout PSD and splake PSD were positively related, suggesting that splake populations with a substantial proportion of quality-length fish (≥250 mm total length) can affect brook trout population structure. Thus, besides providing angling opportunities for an additional game fish, splake stocking may also be an effective means of reducing brook trout density and improving average size of brook trout. However, if splake populations are established in small waters containing brook trout, anglers may have to accept lower brook trout catch rates.

Brook trout *Salvelinus fontinalis* were first stocked into Colorado waters in the 1870s (Wiltzius 1985). Over the following 50 years, brook trout became widely distributed into drainages throughout the state (U.S. Fish and Wildlife Service 1993). This widespread introduction of brook trout led to displacement of many subspecies of cutthroat trout *Oncorhynchus clarki*, as well as other native fishes, and the creation of thousands of naturally recruiting brook trout populations (Behnke and Zarn 1976; Behnke 1979; Woodling 1984).

Today, these nonnative brook trout populations in Colorado and other western states are often characterized by excessive reproduction and recruitment, leading to high population densities of small fish in poor condition (Rabe 1970; Donald and Alger 1989; Johnson et al. 1992), i.e., stunted populations. Such populations often provide high catch rates of small individuals, but opportunities to catch larger fish may be limited or nonexistent (Johnson et al. 1992). This constitutes low-quality sportfishing for many anglers and poses a difficult challenge to fishery managers.

Most approaches to rehabilitating populations of stunted brook trout have involved some means of reducing population density to reduce intraspecific competition and provide more food for remaining fish. Donald and Alger (1989) improved population structure by employing gill nets to simulate 20% annual fishing mortality for 3 years. Klein (1961) successfully used rotenone to reduce brook trout population density and improve growth. Reducing population density by blocking access to spawning sites has been proposed as a means of improving brook trout population structure (Donald and Alger 1989).

Although these approaches have produced some degree of success, there are also drawbacks and limitations. Removal of surplus fish by netting, for example, is labor intensive and expensive. Likewise, chemical treatment is also expensive and laborious. Furthermore, relying on angler harvest implies the presence of substantial angling effort that may not be available, particularly in remote waters. Most importantly, all of these approaches are relatively short-term remedies that do not provide a long-term solution to brook trout overpopulation.

One potential long-term remedy for improving brook trout population structure in closed systems is stocking a piscivorous species to achieve a balance between predator and prey, similar to management of largemouth bass *Micropterus salmoides* and

TABLE 1.—Physicochemical characteristics of Colorado study impoundments. Secchi disk transparency, alkalinity, hardness, pH, and conductivity were measured during August 1993.

Impoundment	Elevation (m)	Surface area (ha)	Maximum depth (m)	Secchi disk transparency (cm)	Alkalinity (mg/L)	Hardness (mg/L)	pH	Conductivity (μmhos/cm)
Chinn's Lake	3,354	4.0	4.0	300	52.3	2.0	7.0	3.0
Lefthand Creek	3,232	40.5	10.4	160	17.1	2.0	6.8	1.6
Sherwin Lake	3,381	3.5	6.4	430	34.2	3.0	7.0	4.3

bluegill *Lepomis macrochirus* in small impoundments (Swingle 1950; Anderson 1980). We are aware of no literature regarding the use of a piscivorous species to improve brook trout population structure in small waters. This study was undertaken to evaluate predator stocking to rehabilitate brook trout populations.

Splake, the hybrid of male brook trout *Salvelinus fontinalis* and female lake trout *S. namaycush*, was chosen as a predator because it has fast growth and introductory stockings often exhibit high survival, particularly in small lakes (Eipper 1964; Fraser 1980; Ihssen et al. 1982). Adult splake are piscivorous and can reach sufficient sizes to prey on most individuals in a population of stunted brook trout (Burkhard 1962; Klein 1972; Fraser 1980). Further, splake are probably more suited than are lake trout for use in small waters without extensive deep water areas (Spangler and Berst 1978).

This paper presents the results of a management evaluation of introducing splake in small waters containing populations of stunted brook trout. Our specific objectives were to evaluate the feasibility of establishing splake populations by stocking fingerlings in waters containing populations of stunted brook trout and to determine the effects of splake on brook trout populations.

Study Sites

Three small impoundments located west of Denver, Colorado, in the front range of the Rocky Mountains were selected for this study. Lefthand Creek Reservoir is located within the Boulder Ranger District of the Roosevelt National Forest. Chinn's Lake and Sherwin Lake are located within the Clear Creek Ranger District of the Arapaho National Forest. Physicochemical characteristics of the three impoundments are presented in Table 1. Chinn's Lake and Sherwin Lake are connected by a small stream that has little or no flow during summer months. Lefthand Creek Reservoir occasionally releases surface water for irrigation use during summer months, but typically remains nearly full throughout the summer. Access to all three waters is by only unimproved, four-wheel-driveable roads. Although no creel surveys were conducted on these waters during the study, fishing effort was estimated to be 400 hours per hectare per year, the fishing effort on similar lakes surveyed in central Colorado (Colorado Division of Wildlife 1993).

Naturally recruiting brook trout populations occurred in all three impoundments. No stocking records are available for these waters, but brook trout have been sampled in all three impoundments since the 1970s. The study lakes were chosen because their brook trout populations all exhibited excessive recruitment and were dominated by small (<200 mm) individuals.

The only other fish species present in any of the waters was rainbow trout *Oncorhynchus mykiss*, which were stocked as fingerlings in Chinn's Lake and Sherwin Lake for a brief period in the late 1980s; very few of these fish remained.

Methods

During June and July of 1988–1992, splake fingerlings (70–120 mm total length, mean = 100 mm) were stocked annually into all three study impoundments at a density of 100 fish per hectare. Splake were the F_1 generation produced at the Story Hatchery, Wyoming, by fertilization of Jenny Lake lake trout eggs with Soda Lake brook trout sperm. Splake were shipped as eyed eggs and reared in Colorado at the Bellvue-Watson Hatchery. Fish were aerially stocked from fixed-wing aircraft.

Between 1988 and 1993, fish populations in study impoundments were sampled according to standardized Colorado Division of Wildlife sampling procedures (Powell 1980; Satterfield 1986). This consisted of overnight sets of gill nets in July and August. Nets were 45.7 m long and 1.8 m deep and had six 7.6-m-long panels of 1.3, 1.9, 2.5, 3.2, 3.8, and 5.1-cm mesh (bar measure). Gill nets were bottom set perpendicular to shore, starting at a depth of approximately 1 m, and gill net locations were standardized. In each water, gill nets were alternately set with small and large mesh placed next to shore. Each overnight set was considered one unit of effort. Annual netting effort was four sets in Chinn's Lake and Sherwin Lake and six sets in Lefthand Creek Reservoir. All captured fish were individually measured for total length (to the nearest millimeter) and weighed (to the nearest gram).

Brook trout and splake length-frequency data were summarized by calculation of proportional stock density (PSD; Anderson 1976) and relative stock density of preferred-length fish (RSD-P; Gabelhouse 1984). Minimum stock, quality, and preferred lengths were defined, respectively, as 125, 200, and 250 mm for brook trout, and 150, 250, and 350 for splake (Gabelhouse 1984; Dumont 1991; Johnson et al. 1992).

We used relative weight (W_r; Wege and Anderson 1978) to evaluate condition of brook trout; no standard weight equation exists for splake. The standard weight equation used for brook trout was from Burton et al. (1984):

$$\log_{10}(W_s) = -5.085 + 3.043 \log_{10}(L);$$

where W_s is weight in grams and L is total length (TL) in millimeters.

Gill-net catch per unit effort, defined as the number of stock-length fish caught per net-night, was used as an index of relative abundance for brook trout and splake. Differences in CPUE between 1988 and 1993 were tested by paired Student's t-tests. Because PSD, RSD-P, and W_r values are not normally distributed, the Wilcoxon matched-pairs, signed-ranks test was used to test for differences in these structural indices between 1988 and 1993. All statistical analyses were performed with the Statistical Analysis System (SAS; SAS Institute 1985). Significance was determined at a P of 0.05 or less in all cases.

Results

Initial sampling in 1988, prior to splake introduction, revealed brook trout populations in all study lakes had relatively high densities and few large individuals. Brook trout CPUE ranged from 23 to 61, PSD ranged from 1 to 7, and RSD-P was 0 in all lakes (Table 2). Despite this high number of relatively small fish, mean W_r values were over 100 for most length ranges.

Between 1988 and 1993, during the period of annual splake stocking, brook trout populations in all waters exhibited substantial changes in population structure (Figure 1). Catch rate of stock-length brook trout gradually declined in all waters (Figure 2) and was significantly lower in 1993 than it was in 1988 (Table 2). Brook trout PSD was significantly higher in all lakes in 1993 (Table 2). Brook trout RSD-P increased significantly in Chinn's Lake. No significant changes in W_r between 1988 and 1993 were found for any of the brook trout populations.

Splake catch rate of stock-size fish increased from near 0 in all waters in 1988 to 3.5–47.0 in 1993 (Figure 3; Table 3). After 5 years of stocking, splake PSD ranged from 31 to 81 in the study impoundments. Stocked splake first reached quality length (250 mm TL) in 1991 after three growing seasons (Figure 4). Furthermore, splake reached preferred

TABLE 2.—Proportional stock density (PSD), preferred-length relative stock density (RSD-P), catch per unit effort (CPUE) of stock-size fish,[a] and mean relative weight (W_r) by size category[b] of brook trout sampled in Colorado study impoundments in 1988, prior to introduction of splake, and in 1993, after 5 years of splake stocking.

Impoundment	PSD	RSD-P	CPUE	W_r S–Q	Q–P	≥P
1988						
Chinn's Lake	7	0	42.7	119	112	
Lefthand Creek	1	0	60.5	106	99	
Sherwin Lake	4	0	23.0	100	96	
1993						
Chinn's Lake	75	9	19.5	125	121	118
Lefthand Creek	89	1	39.7	98	94	94
Sherwin Lake	44	0	9.8	104	105	

[a] The number of fish 125 mm or greater caught per net-night.
[b] Size ranges are stock–quality (S–Q, 125–199 mm); quality–preferred (Q–P, 200–299 mm); and greater than preferred (≥P, ≥300 mm).

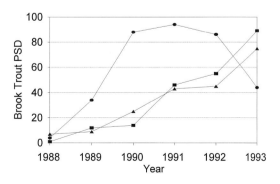

FIGURE 1.—Brook trout proportional stock density (PSD) in Colorado study impoundments, 1988–1993.

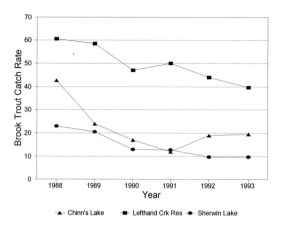

FIGURE 2.—Number of stock-length brook trout captured per net-night in Colorado study impoundments, 1988–1993.

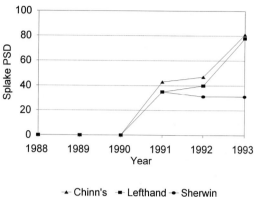

FIGURE 4.—Splake proportional stock density (PSD) in Colorado study impoundments, 1988–1993.

length (350 mm TL) in two of three waters after four growing seasons (Table 3).

Discussion

After 5 years of splake stocking, populations of stunted brook trout were converted to populations containing substantial numbers of quality-length fish; several fish were in excess of 300 mm TL. Although length structure of the brook trout populations improved, W_r did not change, suggesting that factors limiting brook trout condition, and possibly growth rate, had not changed. We used the only brook trout standard weight equation available; this equation was formulated using the 50th percentile and included lentic and lotic populations. The presence of brook trout with W_r greater than 100 in populations considered to be stunted suggests that refinement of this equation is needed; however, statistical comparisons of values of W_r calculated from this equation are valid.

Annual stocking of fingerling splake succeeded in creating populations with several year-classes in the presence of brook trout. Despite uniform stocking densities, population densities of splake ranged widely among the lakes. Splake were most numerous in the lake that contained the highest abundance of brook trout, Lefthand Creek Reservoir, and least numerous in the lake with the lowest abundance of brook trout, Sherwin Lake. Our results suggest that splake standing crop may be limited by the availability of brook trout. Water quality data, particularly transparency, also suggest that there are probably significant differences in the basic productivity of the study waters. Despite these differences, growth of splake was similar in all waters. Quality-length fish (≥250 mm TL) were produced in three growing seasons in all waters. Given the high elevation and relatively low fertility of the study lakes, splake growth as documented in this

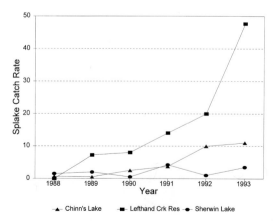

FIGURE 3.—Number of stock-length splake captured per net-night in Colorado study impoundments, 1988–1993.

TABLE 3.—Proportional stock density (PSD), preferred-length (≥350 mm) relative stock density (RSD-P), and catch per unit effort (CPUE) of splake (>150 mm) sampled in Colorado study impoundments in 1993, after 5 years of splake stocking.

Impoundment	PSD	RSD-P	CPUE
Chinn's Lake	81	30	11.0
Lefthand Creek	78	1	47.0
Sherwin Lake	31	0	3.5

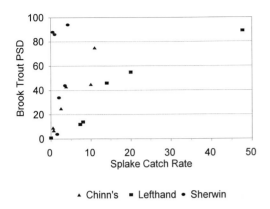
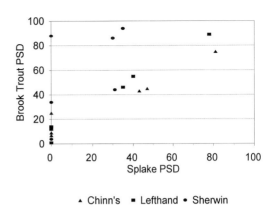

FIGURE 5.—Relationship between brook trout proportional stock density (PSD) and splake captured per net-night in Colorado study impoundments, 1988–1993.

FIGURE 6.—Relationship between brook trout proportional stock density (PSD) and splake PSD in Colorado study impoundments, 1988–1993.

study is probably a conservative indication of their growth elsewhere.

Another management objective of this study was to determine the effect of splake on brook trout populations. Although there was only a relatively weak relationship between brook trout PSD and splake CPUE (Figure 5), brook trout PSD and splake PSD were positively related (Figure 6). This may indicate that stock-length to quality-length splake (150–249 mm TL) alone do not exert significant predation on brook trout; splake of less than 250 mm may not be sufficiently large enough to prey on brook trout effectively, or their food habits may not include a high percentage of fish. Conversely, splake populations with a greater proportion of quality-size fish ($\geq$250 mm TL) appear to have a much greater effect on brook trout population structures and dynamics.

Results of this study suggest splake stocking may provide a means of improving brook trout populations and providing angling opportunities for an additional sport fish. Development of higher quality angling for larger brook trout may be expedited by stocking larger splake to reduce lag time between initial stocking and establishment of a reasonable number of quality-length splake. Another approach to minimize this time may be to reduce splake stocking densities, particularly in relatively infertile waters with short growing seasons. Only one stocking rate was used in this study, but a range of rates should be evaluated in future research to optimize stocking procedures.

Angler reaction to splake-brook trout management in small impoundments should be evaluated. Splake management in brook trout waters appears to reduce the number of brook trout. Presumably this will reduce angler catch rate for brook trout. Splake and larger brook trout will be available to the angler, but perception of anglers towards this management approach should be assessed to determine whether further splake introductions in small impoundments are justified. Angler willingness and ability to comply with splake minimum length limits should also be evaluated because splake may require protection in waters receiving moderate to heavy fishing effort. A 350-mm-minimum-length limit was placed on splake at the conclusion of this project in 1993 in Chinn's Lake and Sherwin Lake, and the efficacy of this regulation on these fisheries will be evaluated in future research.

Many approaches have been taken to manage nonnative brook trout populations in western states, ranging from various attempts to manipulate population densities to complete eradication. Given the widespread distribution of naturally recruiting brook trout in this region, however, it seems obvious populations of stunted fish will occur for some time. An important fisheries management challenge is to improve recreational opportunities provided by these brook trout populations. Introduction of splake appears to be one approach to improving such fisheries through improvement of brook trout population structure and the creation of an additional sport-fish population.

Acknowledgments

We thank J. Ramsey, J. Edwards, R. Antonio, J. Staley, and W. Hundley for their assistance in collection of field data. G. Satterfield assisted in statistical analyses and preparation of graphics. S. Flickinger, D. Willis, R. Knox, and T. Nesler reviewed the manuscript. The Wyoming Game and Fish Department provided eyed splake eggs for this research. Funding for this project was provided by Central Region Fisheries Management, Colorado Division of Wildlife.

References

Anderson, R. O. 1976. Management of small warmwater impoundments. Fisheries (Bethesda) 1(6):5–7, 26–28.

Anderson, R. O. 1980. Proportional stock density (PSD) and relative weight (W_r): interpretive indices for fish populations and communities. Pages 27–33 in S. Gloss and B. Shupp, editors. Practical fisheries management: more with less in the 1980's. American Fisheries Society, New York Chapter, Bethesda, Maryland.

Behnke, R. J. 1979. Monograph of the native trouts of the genus *Salmo* of western North America. U.S. Fish and Wildlife Service, Denver, Colorado.

Behnke, R. J., and M. Zarn. 1976. Biology and management of threatened and endangered western trout. U.S. Forest Service General Technical Report RM-28.

Burkhard, W. T. 1962. A study of the splake trout in Parvin Lake, Colorado. Master's thesis. Colorado State University, Fort Collins.

Burton, T. M., R. W. Merritt, R. J. Stout, and W. W. Taylor. 1984. Compilation of 1983 annual report of the Navy ELF communications system-aquatic ecosystems. Illinois Institute of Technology Technical Report E06549-8, Chicago.

Colorado Division of Wildlife. 1993. Lake categorization system: administrative report. Colorado Division of Wildlife, Denver.

Donald, D. B., and D. J. Alger. 1989. Evaluation of exploitation as a means of improving growth in a stunted population of brook trout. North American Journal of Fisheries Management 9:177–183.

Dumont, S. C. 1991. Evaluation of the Colorado Division of Wildlife's fish sampling procedures in coldwater impoundments. Master's thesis. Colorado State University, Fort Collins.

Eipper, A. W. 1964. Growth, mortality rates, and standing crops of trout in New York farm ponds. Cornell University Memoir 388, Ithaca, New York.

Fraser, J. M. 1980. Survival, growth, and food habits of brook trout and F_1 splake planted in precambrian shield lakes. Transactions of the American Fisheries Society 109:491–501.

Gabelhouse, D. W., Jr. 1984. A length-categorization system to assess fish stocks. North American Journal of Fisheries Management 4:273–285.

Ihssen, P. E., M. J. Powell, and M. Miller. 1982. Survival and growth of matched plantings of lake trout (*Salvelinus namaycush*), brook trout (*Salvelinus fontinalis*), and lake × brook F_1 splake hybrids and backcrosses in northeastern Ontario lakes. Ontario Ministry of Natural Resources, Ontario Fisheries Technical Report Series 6, Toronto.

Johnson, S. L., F. J. Rahel, and W. A. Hubert. 1992. Factors influencing the size structure of brook trout populations in beaver ponds in Wyoming. North American Journal of Fisheries Management 12:118–124.

Klein, W. D. 1961. Removal of brook trout with derris root from a section of the South Fork of the Frying Pan River. Colorado Division of Wildlife Special Report 66, Denver.

Klein, W. D. 1972. Splake trout. Colorado Division of Wildlife Fishery Information Leaflet 18, Fort Collins.

Powell, T. G. 1980. Standardized fish population sampling procedures for lakes and reservoirs. Colorado Division of Wildlife Administrative Report, Fort Collins.

Rabe, F. W. 1970. Brook trout populations in Colorado beaver ponds. Hydrobiologia 35:431–448.

SAS Institute. 1985. SAS/STAT guide for personal computers, version 6. SAS Institute, Cary, North Carolina.

Satterfield, J. R., Jr. 1986. Small impoundment sampling guidelines and procedures. Colorado Division of Wildlife Technical Bulletin, Denver.

Spangler, G. R., and A. H. Berst. 1978. Questions and answers on splake. Ontario Fish and Wildlife Review 17(2):3–8.

Swingle, H. S. 1950. Relationships and dynamics of balanced and unbalanced fish populations. Alabama Polytechnic Institute, Agriculture Experiment Station Bulletin 274.

U.S. Fish and Wildlife Service. 1993. Greenback cutthroat trout recovery plan. U.S. Fish and Wildlife Service, Denver, Colorado.

Wege, G. J., and R. O. Anderson. 1978. Relative weight (W_r): a new index of condition for largemouth bass. Pages 79–91 in G. Novinger and J. Dillard, editors. New approaches to the management of small impoundments. American Fisheries Society, North Central Division, Special Publication 5, Bethesda, Maryland.

Wiltzius, W. J. 1985. Fish culture and stocking in Colorado, 1872–1978. Colorado Division of Wildlife Report 12, Fort Collins.

Woodling, J. 1984. Game fish of Colorado. Colorado Division of Wildlife Publication M-1-25-84, Denver.

An Evaluation of Aquatic Vegetation Control by Triploid Grass Carp in Virginia Ponds

RICHARD EADES

Virginia Department of Game and Inland Fisheries
6530 Indian River Road, Virginia Beach, Virginia 23464, USA

ED STEINKOENIG

Virginia Department of Game and Inland Fisheries
1320 Belman Road, Fredericksburg, Virginia 22401, USA

Abstract.—The effectiveness of triploid grass carp *Ctenopharyngodon idella* in reducing aquatic vegetation in small impoundments in Virginia was evaluated through mail surveys to pond owners who had applied for stocking permits from the Virginia Department of Game and Inland Fisheries (VDGIF). Seventy-eight percent of surveyed pond owners who had stocked triploid grass carp were satisfied with their effectiveness and would recommend their use to others. In ponds where aquatic vegetation was reduced, the reduction was evident within 1 year in 82% of the ponds. In ponds where triploid grass carp were not effective, lack of control was attributed to several factors. Most pond owners did not follow VDGIF stocking recommendations, so in some ponds insufficient numbers of triploid grass carp were stocked to achieve desired results. Losses of grass carp to predation by largemouth bass *Micropterus salmoides* and other fishes, as well as river otters *Lutra canadensis,* osprey *Pandion haliaetus,* and bald eagles *Haliaeetus leucocephalus,* were reported by some pond owners and believed to be common because stocked fish averaged only 8–10 in. Some pond owners attempted to control plants not preferred by grass carp. A majority (69%) of pond owners did not know what species of aquatic vegetation was growing in their ponds. Many pond owners did not know the exact acreage of their pond or vegetation coverage. If pond owners accurately determine acreage of plant coverage, identify the vegetation growing in their pond, and follow VDGIF stocking recommendations, we expect triploid grass carp to control aquatic vegetation effectively. In ponds less than 5 acres in size, less than 5 ft in depth, with high predator densities, or having vegetation not preferred by triploid grass carp, chemical control would yield more reliable results.

Grass carp *Ctenopharyngodon idella* were first brought to the United States in 1963 by Auburn University and the U. S. Fish and Wildlife Service (USFWS) (Stanley and Lewis 1978; Sutton and Vandiver 1986) for aquatic vegetation control research (Guillory and Gasaway 1978). Grass carp are voracious consumers of aquatic plants, they are hardy, and they grow quickly (Baker et al. 1993), which makes them an attractive alternative to more expensive and potentially hazardous chemical control methods (Allen and Wattendorf 1987). Grass carp are native to large Asian rivers that flow into the Pacific Ocean, such as the Amur River in Russia and the West River in China (Chilton and Muoneke 1992).

Grass carp are not a plant control panacea (Baker et al. 1993), and problems associated with their use (Forester and Lawrence 1978) led many states to prohibit their use for vegetation control (Sport Fishing Institute 1977; Allen and Wattendorf 1987). Large numbers of grass carp could potentially eliminate all aquatic plants, resulting in lost habitat for fish and aquatic invertebrates, decreased food for waterfowl, and increased shoreline erosion, turbidity, nutrient accumulation, and algal growth (Leslie et al. 1983; Hardin et al. 1987; Wattendorf and Anderson 1987).

With the development of the functionally sterile triploid grass carp (Malone 1984; Cassani and Caton 1986), reproduction of grass carp in the wild was no longer a major concern and the potential for uncontrolled spread was minimized (Allen and Wattendorf 1987; Clugston and Shireman 1987). Several states began allowing the use of triploid grass carp for aquatic vegetation control. Currently, 37 states allow either diploid or triploid grass carp or both to be stocked under various conditions (USFWS, unpublished list).

Since 1973, the Virginia Department of Game and Inland Fisheries (VDGIF) has regulated the importation of certain nonnative fish species. Grass carp was added to the regulation in 1979. In spite of the regulation, unknown numbers of grass carp were imported into Virginia prior to 1979 by private pond owners. During that time the use of diploid grass carp for aquatic vegetation control was recommended by the Soil Conservation Service (SCS).

In 1984, VDGIF started permitting the use of

triploid grass carp. Only grass carp certified as triploid by the USFWS were allowed to be imported. At that time, only two distributors were approved to deliver triploid grass carp to Virginia. VDGIF currently recognizes 10 triploid grass carp distributors.

The stocking rate recommended by VDGIF in 1984 was 16 triploid grass carp per acre. In 1987, the recommended stocking rate was revised to 16 fish per vegetated acre. In 1989, the recommended stocking rate was revised again. Based on staff experience and available literature, two rates were established: one for controlling vegetation and one for eradicating vegetation. Grass carp do not consume all types of plants equally (Shireman and Smith 1983; Sutton and Vandiver 1986; Baker et al. 1993). For plants preferred by grass carp, VDGIF recommended stocking 8 fish per vegetated acre for control and 16 for eradication. For less palatable plant species, the rates are doubled.

Between 1990 and 1993, VDGIF annually approved an average of 547 permit applications requesting approximately 9,000 triploid grass carp. Approximately 75% of the permit requests were for 15 or fewer fish and 90% were for 29 or less. Some pond owners receiving permits did not stock any triploid grass carp, and the actual total number of fish stocked was approximately 6,750 per year. Triploid grass carp were stocked in private ponds in all but five counties in Virginia during that period. Stocked fish generally ranged from 8 to 10 in.

Fisheries biologists from VDGIF make site visits to privately owned ponds to provide technical assistance on fish population management, water quality problems, fish kills, aquatic vegetation control, and other subjects. Through these visits, biologists get a general idea of whether stocked triploid grass carp are successfully controlling vegetation. However, since the biologists did not have records of when a pond was stocked, how many triploid grass carp were stocked, or what types of aquatic weeds may have been in the pond in the past, they were lacking necessary information to make adjustments to the existing vegetation management programs. Furthermore, the effectiveness of recommended stocking rates of triploid grass carp in ponds over a range of different conditions throughout Virginia had not been evaluated. To ensure VDGIF biologists would be able to provide pond owners with accurate information and make sound management recommendations about grass carp use, we evaluated successes and failures of triploid grass carp stocked in private ponds.

Methods

A survey was developed for pond owners who had been issued permits to stock triploid grass carp into their ponds. Several VDGIF fisheries biologists were asked to randomly select 50 permit holders to be surveyed in the biologists' district or region. All physiographic regions of Virginia were covered by the survey. Permittees for 1993 were not included so that those surveyed had triploid grass carp in their pond for at least 1 year.

In 1993, 330 surveys were mailed out to permit holders across the state. Because the pond owners to be surveyed were randomly selected within each region, the number of pond owners surveyed within a given county was assumed to be proportional to the number of permits issued in that county. No effort was made to include every county in the survey. A pre-addressed, stamped envelope was included with each questionnaire. No follow-up mailings were made to nonrespondents.

Each permit holder surveyed was asked if they had actually stocked triploid grass carp in their pond after receiving a permit. Those who had not were asked to provide a brief reason. Only those who had actually stocked their pond were asked to respond to the remaining questions (Table 1).

Results

Of the 330 surveys mailed out, 211 (64%) were returned completed. Approximately 20 surveys were returned by the Post Office as undeliverable; another 20 were not completed because the permit holder no longer managed the pond, had moved, or had died. Of the 211 respondents, 184 (87%) had stocked triploid grass carp in their pond.

Of the 27 respondents who had not stocked any triploid grass carp, only four stated the fish were too expensive and three said they were unable to obtain the fish from a supplier. The remaining 20 chose not to reduce the aquatic vegetation in their ponds or chose another control method.

Only 31% of the respondents stated they identified the species of aquatic vegetation growing in their pond. Many thought they knew what the plants were, but it was apparent from their responses and VDGIF site visits that some respondents were not identifying the vegetation correctly. Survey responses indicated that triploid grass carp rarely displayed a preference for a particular aquatic plant. In ponds containing more than one species of aquatic vegetation, 89% of the respondents stated the triploid grass carp ate some of each plant, whereas 11% said the triploid grass carp ate

TABLE 1.—Survey sent to pond owners who had been issued permits to stock triploid grass carp in Virginia.

1. Did you stock triploid grass carp in your pond?
   A) If no, why not?
   B) If yes, number stocked   date stocked
2. Did someone positively identify the vegetation in your pond before stocking?
   A) No   B) Yes   If yes, who?
   What type(s) of vegetation?
3. If you had more than one type of vegetation in your pond, did the triploid grass carp:
   A) Eat some of each type.
   B) Prefer some type(s). Which ones?
   C) Eat only one type. Which one?
4. If the triploid grass carp were ineffective, why? (In your opinion)
5. Were spillway barriers installed to prevent escapement of the triploid grass carp from the pond?
6. If the triploid grass carp were effective, how soon after stocking was a reduction in vegetation evident?
7. Do you think the stocking rates recommended were:
   A) Too high   B) About right   C) Too low
8. Were other control measures ever used in the pond to kill the vegetation?   When?   What type?
   (Herbicide, water level drawdown, mechanical, other)
9. Since you stocked the triploid grass carp, would you say fishing in your pond has:
   A) Improved greatly   B) Improved somewhat   C) Not changed
   D) Declined somewhat   E) Declined greatly   F) Don't know or don't fish
10. Overall how satisfied are you with the triploid grass carp?
    A) Very satisfied   B) Somewhat satisfied   C) Somewhat dissatisfied   D) Very dissatisfied
11. Would you recommend triploid grass carp stocking to others?

Please provide the following information about your pond:
Size in acres   Average depth   Maximum depth
Location (County)
Percent of pond having vegetation growth: before stocking   now
Normal water clarity: muddy   tea colored   greenish   clear

more of some plants than others. No respondent stated that the triploid grass carp fed on one plant exclusively.

Only 28% of the respondents had previously tried other methods of vegetation control prior to stocking triploid grass carp. Of these respondents, 76% had used herbicides, 14% had used mechanical control methods, 6% had used water-level manipulation techniques, and 4% had used both herbicides and mechanical control methods.

About half (51%) of the respondents said they were very satisfied with the triploid grass carp and another 27% were somewhat satisfied. Of the 22% who expressed dissatisfaction with the triploid grass carp, 11% were somewhat dissatisfied and 11% were very dissatisfied. Three-fourths of the respondents would recommend stocking triploid grass carp to other pond owners. Another 19% said they might recommend their use depending on pond conditions; 6% stated they definitely would not recommend triploid grass carp stocking to other pond owners.

When respondents reported reductions in aquatic vegetation, 82% reported reductions were evident within the first year, 13% said reductions were not evident until the second year after stocking, and 5% said it was the third year before a reduction was evident. One-fourth of the respondents reported a noticeable decrease in vegetation within 3 months after stocking.

In total, 121 respondents reported vegetation coverage at the time of stocking as well as the current coverage. Thirty-eight respondents reported complete eradication of aquatic vegetation, 70 reported some degree of reduction, and 13 reported no reduction in vegetation coverage.

Poor survival and inadequate stocking were the most common reasons postulated for triploid grass carp having little or no effect. One-fifth of the respondents thought the recommended stocking rates were too low. Few pond owners thought escapement occurred; only five respondents reported triploid grass carp leaving their pond. Spillway barriers had been installed by 61% of the respondents prior to stocking.

Pond owners reported observing triploid grass carp being eaten by river otters *Lutra canadensis*, osprey *Pandion haliaetus*, and bald eagles *Haliaeetus leucocephalus*. Predation by largemouth bass *Micropterus salmoides* and other fishes was believed to be an important factor in some ponds.

Many pond owner responses conveyed the attitude that all aquatic plants are basically the same, and owners stocked triploid grass carp to control probably every species of aquatic vegetation found in Virginia. Those respondents who had their vegetation identified by a VDGIF biologist, county extension agent, or other knowledgeable person reported successfully controlling common waterweed *Elodea canadensis*, coontail *Ceratophyllum demersum*, pondweeds *Potamogeton* spp., milfoil *Myriophyllum* spp., and hydrilla *Hydrilla verticillata*. Success rates were also very high for controlling muskgrass *Chara* spp. and *Nitella* spp. Although water lilies *Nymphaea* spp. are not a preferred food, some pond owners reported triploid grass carp eating them after eating the more palatable species.

One aquatic plant that was not controlled in most ponds was duckweed *Lemna* spp. Even though some pond owners had limited success using triploid grass carp, most reported the fish were unable to control this plant, particularly in those ponds with heavy infestations. Visits by VDGIF biologists to stocked ponds confirmed that triploid grass carp were generally unable to control duckweed. However, ponds with existing triploid grass carp populations did not develop duckweed densities as thick as nonstocked ponds. Triploid grass carp were observed feeding on duckweed during several site visits, but it was apparent that very high stocking rates would be needed to control the plant. In Virginia, the use of herbicides is normally recommended for duckweed control rather than stocking triploid grass carp.

Pond size and depth were apparent factors in triploid grass carp stocking success. Triploid grass carp were generally more effective in larger and deeper ponds. All respondents who were very dissatisfied with the triploid grass carp's performance had stocked ponds less than 5 acres in size. Overall, 23% of the pond owners who had stocked ponds less than 5 acres were dissatisfied and 77% were satisfied (48% very satisfied). Of the pond owners who stocked triploid grass carp in ponds larger than 20 acres, 92% were satisfied (84% very satisfied) with the reductions in vegetation. Pond owners reported poor results in ponds less than 5 ft average depth and better results in deeper ponds.

Water clarity was not a major factor in triploid grass carp effectiveness. Pond owner satisfaction was about equal for ponds described as clear, muddy, or tea colored. Satisfaction was lower in ponds having "green" water. Based on their comments, many pond owners expected triploid grass carp to control planktonic algae as well as plants.

Because many pond owners did not follow VDGIF stocking recommendations, we were able to examine the effectiveness of stocking rates of 2–200 triploid grass carp per acre. High stocking rates did not always lead to large reductions in vegetation, suggesting that plant palatability affects the amount of aquatic vegetation a triploid grass carp will consume. Reductions in aquatic vegetation were achieved in ponds stocked with as few as two fish per vegetated acre. Stocking rates of 11 to 16 fish per vegetated acre gave the best overall results. At stocking rates greater than 25 fish per vegetated acre, reductions in vegetation varied greatly.

The practice of preventive stocking was apparent. Some pond owners stocked low numbers of triploid grass carp in ponds with little vegetation in an effort to prevent infestations. Several respondents reported no vegetation was present in their pond when triploid grass carp were stocked.

Fifty-two respondents said fishing in their pond improved after stocking triploid grass carp whereas 13 said fishing declined. Sixty-nine respondents reported no change in the fishing and 31 did not use their pond for fishing. All of the pond owners reporting improved fishing had large decreases in weed coverage; 85% had weed coverage decrease by more than half after stocking. Those respondents that said fishing had declined had no change in weed coverage after stocking so it is unlikely that triploid grass carp were a factor in the decline.

Discussion

We expected the percentage of survey respondents reporting they had stocked triploid grass carp to be higher than the true percentage of all permit holders that had stocked, because those who did not stock would be less likely to respond to the survey. However, Perry's Minnow Farm, one of the major triploid grass carp distributors in Virginia, reported that from 1991 to 1993, 543 (86%) of the 634 pond owners with permits to purchase triploid grass carp from them had purchased the fish (S. Perry, personal communication). Therefore we believe the survey results are representative of all permit holders.

Numerous studies have been conducted to assess the use of grass carp for vegetation control. Many of these have been short-term studies in small ponds to determine feeding habits (Terrell and Fox 1975; Colle et al. 1978; Wattendorf and Shafland 1985), appropriate stocking rates (Fowler and Robson 1978), and effects on other fishes (Forester and Lawrence 1978; Kilgen 1978; Baur et al. 1979). The results of some of these studies have indicated that grass carp are not always effective at controlling aquatic vegetation. Kirk (1992) found that triploid grass carp were generally ineffective in controlling aquatic vegetation in 33 ponds ranging from 0.25 to

14 acres and concluded that triploid grass carp were an unsatisfactory management tool for controlling unwanted vegetation in South Carolina farm ponds. Our study showed that triploid grass carp were less consistently effective in small (<5 acres) ponds than in large ponds, particularly those greater than 20 acres. Other studies in large ponds and reservoirs have also shown grass carp to be effective for controlling aquatic vegetation (Mitzner 1978; Bain et al. 1990). Fisheries researchers should be cautious about making statements about potential grass carp effectiveness based solely on studies in small experimental ponds. Studies should also be conducted in larger impoundments to evaluate grass carp more fully for a given area.

Caution should also be used in vegetation preference studies. Several lists of preferred and nonpreferred plant species have been published (Wiley et al. 1987; Chilton and Muoneke 1992); there is general agreement for most aquatic plant species, but some show up on both lists in different publications. Feeding preferences change as grass carp grow larger (J. Malone, personal communication). Grass carp feed heavily on duckweed in their first year but not in later years. Short-term studies using small grass carp might indicate that duckweed was a preferred species, but as shown in our survey, triploid grass carp are not an effective long-term control for duckweed.

The survey responses and VDGIF site visits suggest that triploid grass carp stocking by private pond owners in Virginia has been generally haphazard. Less than one-third of the pond owners knew what kind of vegetation was growing in their ponds before stocking triploid grass carp. Many pond owners did not know the acreage of their pond, and therefore few pond owners accurately followed the recommended stocking rates of 8, 16, or 32 fish per vegetated acre. Because some pond owners stocked triploid grass carp based on the acreage of the pond rather than the vegetation coverage, their stocking rates were higher than recommended. However, generally satisfactory results were achieved from stocking 11–16 fish per vegetated acre, evidence that VDGIF's current recommended stocking rates should produce positive results if followed correctly. Nationwide, rates of 10–25 fish per acre are generally used (Allen and Wattendorf 1987).

Although a majority of respondents were satisfied with triploid grass carp effectiveness, it is noteworthy that more pond owners who had used other vegetation control methods were dissatisfied with triploid grass carp (37%) than those using triploid grass carp only (18%). This suggests that whereas triploid grass carp generally yielded good results, they were not always considered the best control methods.

Most pond owners apparently relied on information provided by other pond owners rather than VDGIF biologists or literature when determining the number of triploid grass carp to stock. Knowing that triploid grass carp were working in other ponds was apparently all most pond owners needed to know. Under VDGIF's triploid grass carp permitting system, pond owners were sent an informational brochure along with a permit application. Pond owners could contact a district fisheries biologist for more information or specific recommendations if they wished to. However, very few did so. Completed applications were sent to VDGIF headquarters for approval. The district biologist in the applicant's area received copies of issued permits and could contact the pond owners if necessary.

A few modifications have been made to the VDGIF triploid grass carp permitting program since this study was completed. Pond owners are now required to provide more information on their permit application and include a map showing the exact location of their pond. Permit applications are sent to the appropriate district fisheries biologist for review and approval or denial. If the number of fish requested is too high, the biologist can approve the permit but for fewer fish. The biologist can deny the permit application if escapement into an open system is probable. Due to potential liability for subsequent spillway or dam failure, VDGIF does not require spillway barriers to be installed. By gathering more information and carefully reviewing permit applications, VDGIF will hopefully ensure triploid grass carp are more effectively and wisely used in Virginia.

Whereas herbicides yield almost immediate results, more time is required to see results from triploid grass carp. For pond owners wanting immediate results, herbicides are the recommended method. For pond owners looking for an alternative to repeated chemical treatments, triploid grass carp are an attractive option. Pond owners should have good results if they make the effort to have the vegetation in their ponds correctly identified, determine the extent of vegetation coverage accurately, and follow the stocking recommendations provided with the permit application. For ponds less than 5 acres in area or less than 5 ft in depth, herbicide usage would be more reliable. Herbicides may also be a better choice in ponds where losses to predation may be high.

Acknowledgments

We thank Tom Gunter, John Kauffman, and Ron Southwick for their assistance in designing and conducting the pond owner surveys and Rebecca Wajda for providing permit data.

References

Allen, S. A., and R. J. Wattendorf. 1987. Triploid grass carp: status and management implications. Fisheries (Bethesda) 12(4):20–24.

Bain, M. B., D. H. Webb, M. D. Tangedal, and L. N. Mangum. 1990. Movements and habitat use by grass carp in a large mainstream reservoir. Transactions of the American Fisheries Society 119:553–561.

Baker, J. P., H. Olem, C. S. Crager, M. D. Marcus, and B. R. Parkhurst. 1993. Fish and fisheries management in lakes and reservoirs. U.S. Environmental Protection Agency, Clean Lakes Program, Washington, DC.

Baur, R. J., D. H. Buck, and C. R. Rose. 1979. Production of age-0 largemouth bass, smallmouth bass, and bluegills in ponds stocked with grass carp. Transactions of the American Fisheries Society 108:496–498.

Cassani, J. R., and W. E. Caton. 1986. Efficient production of triploid grass carp (*Ctenopharyngodon idella*) utilizing hydrostatic pressure. Aquaculture 55:43–50.

Chilton, E. W., and M. I. Muoneke. 1992. Biology and management of grass carp (*Ctenopharyngodon idella*, Cyprinidae) for vegetation control: a North American perspective. Reviews in Fish Biology and Fisheries 2:283–320.

Clugston, J. P., and J. V. Shireman. 1987. Triploid grass carp for aquatic plant control. U. S. Fish and Wildlife Service Fish and Wildlife Leaflet 8.

Colle, D. E., J. V. Shireman, and R. W. Rottmann. 1978. Food selection by grass carp fingerlings in a vegetated pond. Transactions of the American Fisheries Society 107:149–152.

Forester, T. S., and J. M. Lawrence. 1978. Effects of grass carp populations on bluegill and largemouth bass in ponds. Transactions of the American Fisheries Society 107(1):172–175.

Fowler, M. C., and T. O. Robson. 1978. The effects of food preferences and stocking rates of grass carp on mixed plant communities. Aquatic Botany 5:261–276.

Guillory, V., and R. D. Gasaway. 1978. Zoogeography of the grass carp in the United States. Transactions of the American Fisheries Society 107:105–112.

Hardin, S., R. Land, M. Spelman, and G. Morse. 1987. Food items of grass carp, American coots, and ring-necked ducks from a central Florida lake. Proceedings of the Annual Conference Southeastern Association of Fish and Wildlife Agencies 38(1984):313–18.

Kilgen, R. H. 1978. Growth of channel catfish and striped bass in small ponds stocked with grass carp and water hyacinths. Transactions of the American Fisheries Society 107:176–180.

Kirk, J. P. 1992. Efficacy of triploid grass carp in controlling nuisance aquatic vegetation in South Carolina farm ponds. North American Journal of Fisheries Management 12:581–584.

Leslie, A. J., Jr., L. E. Nall, and J. M. VanDyke. 1983. Effects of vegetation control by grass carp on selected water quality variables in four Florida lakes. Transactions of the American Fisheries Society 112:777–787.

Malone, J. M. 1984. Triploid white amur. Fisheries (Bethesda) 9(2):36.

Mitzner, L. 1978. Evaluation of biological control of nuisance aquatic vegetation by grass carp. Transactions of the American Fisheries Society 107:135–145.

Shireman, J. V., and C. R. Smith. 1983. Synopsis of biological data on the grass carp *Ctenopharyngodon idella* (Cuvier and Valenciennes, 1844) FAO (Food and Agriculture Organization of the United Nations) Fisheries Synopsis 135.

Sport Fishing Institute. 1977. Grass carp banned in 33 states. SFI Bulletin 281:5.

Stanley, J. G., and W. M. Lewis. 1978. Special section: grass carp in the United States. Transactions of the American Fisheries Society 197:104.

Sutton, D. L., and V. V. Vandiver, Jr. 1986. Grass carp. A fish for biological management of hydrilla and other aquatic weeds in Florida. Florida Agricultural Experiment Station, Institute of Food and Agricultural Sciences, Bulletin 867. University of Florida, Gainesville.

Terrell, J. W., and A. C. Fox. 1975. Food habits, growth, and catchability of grass carp in the absence of aquatic vegetation. Proceedings of the Annual Conference Southeastern Association of Game and Fish Commissioners 28(1974):251–259.

Wattendorf, R. J., and P. L. Shafland. 1985. Hydrilla consumption by triploid grass carp in aquaria. Proceedings of the Annual Conference Southeastern Association of Fish and Wildlife Agencies 37(1983):319–326.

Wattendorf, R. J., and R. S. Anderson. 1987. Hydrilla consumption by triploid grass carp. Proceedings of the Annual Conference Southeastern Association of Fish and Wildlife Agencies 38(1984):319–326.

Wiley, M. J., P. P. Tazik, and S. T. Sobaski. 1987. Controlling aquatic vegetation with triploid grass carp. Illinois Natural History Survey Circular 57, Champaign.

Introduction and Establishment of a Successful Butterfly Peacock Fishery in Southeast Florida Canals

PAUL L. SHAFLAND

Florida Game and Fresh Water Fish Commission, Non-Native Fish Research Laboratory
801 Northwest 40th Street, Boca Raton, Florida 33431, USA

Abstract.—After reviewing the literature, water temperature records, resident fish population data, previous reports of introductions of peacock bass *Cichla* spp., and many personal communications, it was concluded that the introduction of the butterfly peacock *Cichla ocellaris* into southeast Florida coastal canals posed no serious environmental risks and offered several biological and socioeconomic benefits. The purpose of this paper is to summarize the evaluation process and criteria presented in the 1984 proposal that led to the introduction and establishment of the butterfly peacock in Florida. Between 1984 and 1987 approximately 20,000 pond-raised fingerlings were introduced into 11 coastal canals, where they are now considered established (i.e., self-sustaining populations). In a 1989 sample, the exotic spotted tilapia *Tilapia mariae* made up 75% of the identifiable stomach contents of canal-caught butterfly peacocks. Positive correlations between density estimates of butterfly peacocks and redear sunfish *Lepomis microlophus*, bluegill *L. macrochirus*, and largemouth bass *Micropterus salmoides* support the hypothesis that the butterfly peacock has not detrimentally affected these fishes and may even be benefiting them. The butterfly peacock sport fishery has an estimated direct economic value exceeding US$1 million annually. The introduction of the butterfly peacock has been socioeconomically and biologically successful.

Introduction of the butterfly peacock *Cichla ocellaris* (see endnote) into the canals of southeast Florida was authorized in 1984 to increase predation on abundant nonnative forage species and to enhance freshwater sportfishing opportunities in the metropolitan Miami–Fort Lauderdale area. The purpose of this paper is to review the information and rationale that led to the first authorized introduction of an exotic fish expected to become established in Florida. This manuscript is largely historical, as it is based on the proposal that led to the introduction of the butterfly peacock (Shafland 1984). A preliminary analysis of postintroduction data and observations is included to complement the historical objective of this manuscript.

Introductions of exotic fishes should be approached cautiously because their long-term effects cannot be predicted precisely and some have had serious environmental effects. Important questions to answer prior to approving an introduction include (1) is there a justifiable need, (2) has the receiving community been thoroughly evaluated, (3) are there reasonable predictions of beneficial effects and an absence of unacceptable adverse effects, (4) is the species selected uniquely suited to fit the pre-identified vacant or underutilized niche (e.g., will the introduced species survive, reproduce, and become established in its new environment), (5) could a native species be expected to have similar beneficial effects and, if so, why was it not selected for introduction, (6) is there an effective means for limiting or controlling the exotic should it become too abundant or have unexpected detrimental effects, and (7) have all other reasonable options been considered?

After each of the above and any other necessary considerations have been fully evaluated, a formal proposal should be prepared and distributed to a group of professionals, including some who are likely to disagree with it. If the proposal passes the peer-review process without substantive objections or problems, it should then be submitted to the appropriate governmental agency for the agency's consideration. If substantial objections are encountered, the proposal should be rejected or rewritten to address these objections before proceeding. This is the general process that led to the introduction of the butterfly peacock in Florida, and it required about 4 years to complete.

Every introduction involves a unique set of circumstances, and no single set of considerations is appropriate in all situations. Therefore, each introduction should be evaluated using species-, community-, and habitat-specific criteria. However, some generalities do exist and, as one of few such accounts, the information summarized here should be beneficial to others considering or opposing purposeful introductions.

Peacock Bass Biology

Peacock bass *Cichla* spp. are tropical, freshwater, opportunistic piscivores with a body shape similar to that of black bass *Micropterus* spp. They are the largest South American member of the family Cichlidae and attain weights in excess of 12 kg but more commonly weigh less than 2 kg (Ogilvie 1966a; Lowe-McConnell 1969; Zaret 1980). Three species of peacock bass are widely recognized (Machado-Allison 1971), but, as with many other tropical groups of fishes, taxonomic problems exist. The most common and widespread species of peacock bass is the butterfly peacock, which attains a maximum size of about 5 kg; the larger speckled peacock *C. temensis* grows to more than 12 kg (Ogilvie 1966a; Machado-Allison 1971; Zaret 1980).

Peacock bass are substrate spawners capable of reproducing more than once a year. Both parents guard broods of 2,000–3,000 young for extended periods of time (>2 months) (Ogilvie 1966b; Zaret 1980). Given sufficient food, young butterfly peacocks grow to sexually mature sizes of 25–30 cm total length (TL) in less than 12 months (Fontenele 1950; Devick 1972a). Growth is rapid to sizes of 1–1.5 kg, after which it may slow. Like many other cichlids, male peacock bass grow larger than do females (Zaret 1980).

The presence of only one largemouth bass *Micropterus salmoides* in the stomachs of 1,459 Hawaiian butterfly peacocks suggested they would not prey significantly on largemouth bass in Florida (Devick 1972b, 1972c). Because peacock bass were known to feed on tilapias *Tilapia* spp. (McGinty 1983; Verani et al. 1983) and Hawaii's experience showed butterfly peacocks did not prey upon largemouth bass (Devick 1972b, 1972c), increased predation resulting from the butterfly peacock's introduction into south Florida canals was expected to be directed primarily at the spotted tilapia *Tilapia mariae* established in these canals, rather than at less abundant native fishes.

Due to their potentially similar forage preferences, peacock bass could compete with largemouth bass if the combination of these predators eliminated forage populations. This seemed unlikely due to the abundance, fecundity, and competitive nature of the exotic forage species present (especially the spotted tilapia). Moreover, one study concluded butterfly peacocks were less efficient predators than were largemouth bass (Swingle 1967). If peacock bass were to compete for food unexpectedly to the detriment of largemouth bass, the competition would be caused by a major reduction in the abundance of exotic forage species that contribute most to the diet of largemouth bass in southeast Florida canals. For example, 93% of the identifiable fish in the stomachs of largemouth bass from the Black Creek Canal were exotics (Shafland et al. 1985).

Preintroduction field observations, personal communications, and published information indicated substantial differences between peacock bass and largemouth bass feeding behaviors. Peacock bass primarily use their exceptional speed to run down forage fishes rather than relying on the typical ambush tactics of largemouth bass (Erdman 1969; G. P. Garrett, Texas Parks and Wildlife Department, personal communication). Unlike largemouth bass, peacock bass feed only during daylight hours because they are inactive at night, as are most other cichlids (Lowe-McConnell 1969; Devick 1972b; Zaret 1980). Furthermore, peacock bass often feed in shallower water than is characteristic of largemouth bass (Devick 1972b). Therefore, it would be erroneous to consider the butterfly peacock and largemouth bass as ecological homologues or as mutually exclusive competitors.

Prior Introductions of the Butterfly Peacock

Florida's interest in peacock bass dates back to the early 1960s. Studies conducted then documented excellent growth and successful reproduction of the butterfly peacock in Florida (Ogilvie 1966a, 1966b). These studies were discontinued because butterfly peacocks could not consistently overwinter even in south Florida ponds (McClane 1971).

The butterfly peacock was introduced to Hawaii in 1957 (Kanayama 1968) and continues to support a popular sport fishery there (Devick 1988). Following a comprehensive evaluation of the butterfly peacock and largemouth bass in Wahiawa Reservoir, Hawaii, Devick (1976, 1988) concluded the two species had different reproductive strategies and different spawning seasons, plus distinct although partially overlapping food and habitat preferences. Moreover, Devick (1988) concluded the adult butterfly peacock and largemouth bass in Wahiawa Reservoir were complementary predators (i.e., two or more predator species that use forage more efficiently together than when they occur alone).

In 1974 the Texas Parks and Wildlife Department began evaluating the use of peacock bass in manmade lakes with heated waters from electrical power generating plants (Lyons 1980). The excellent sportfishing qualities of peacock bass were again confirmed; however, periodic power plant shutdowns resulted in complete winterkills (e.g., Rutledge and Lyons 1978; Lyons 1980; Garrett 1982). As in the earlier Florida studies, intolerance to low water temperatures was considered an insurmountable problem, and studies involving peacock bass in Texas were discontinued.

The only published detrimental effects associated with any peacock bass introduction was a report of the butterfly peacock in Lake Gatun, Panama, by Zaret and Paine (1973). These effects included dramatic reductions in native fishes; however, a subsequent report suggested these effects were exaggerated (Welcomme 1988). In contrast to Zaret and Paine (1973), Fontenele and Peixoto (1979) reported the butterfly peacock demonstrated no tendency to eliminate other fishes 32 years after being introduced into northeastern Brazil reservoirs. Exaggerated results or not, Zaret and Paine's (1973) work provides good reason to approach all introductions from a conservative environmental and philosophical perspective.

Introductions of the butterfly peacock in Panama, Puerto Rico, and Hawaii all resulted in successful sport fisheries being established (Chew 1975; Fontenele and Peixoto 1979; Welcomme 1988; Erdman 1984; Devick 1988). Even in Lake Gatun, the butterfly peacock had important socioeconomic benefits: "So far, *Cichla* has completely lived up to all expectations; its capture has provided entertainment for fishermen, and its taste has pleased many palates. Further, it is the only freshwater fish sold for consumption in this area" (Zaret and Paine 1973). Fisheries scientists investigating peacock bass have not reported any serious environmental problems associated with their introduction, with the exception of Lake Gatun (Zaret and Paine 1973).

Need, Niche, and Justification for Introduction

Human alteration of south Florida's natural water-flow pattern, caused largely by the construction of an extensive interconnecting canal system, has created an environment conducive to the establishment of many unwanted exotic tropical fishes. Currently, 14 illegally introduced exotic fishes are permanent residents of these canals.

Extensive fish sampling prior to the introduction of the butterfly peacock estimated the average standing crop of fishes in coastal Dade County canals at 229 kg/ha and 17,790 fish/ha (Metzger and Shafland 1985). Average standing crop estimates of exotic fishes in these canals were 87 kg/ha (37%) and 4,210 fish/ha (24%). The exotic spotted tilapia was the most abundant species collected by weight (30%), and second-most abundant by total number (22%).

Butterfly peacock occupies canal-type habitats in its native range (Lowe-McConnell 1969), which suggested the canals of south Florida would also be suitable for it. The box-cut urban canals of southeast Florida are unique and particularly vulnerable to being invaded by undesirable exotic freshwater tropical fishes. Shafland et al. (1985) predicted peacock bass would become established and abundant if introduced into these canals. Moreover, the butterfly peacock was expected to establish permanent populations in few, if any, other habitats due to its intolerance of low temperatures and salt water (Shafland 1984).

Biological benefits.—Quantitative fish population surveys in south Florida canals during the late 1970s and early 1980s confirmed native predators were using only a small fraction of the abundant exotic forage fishes (forage to carnivore biomass ratios averaged greater than 12:1), and that native fishes might benefit from a management plan that would exploit these exotic species (Metzger and Shafland 1985; Shafland et al. 1985). It was also apparent that largemouth bass, together with other native predators, had been unable to prevent the rapid range extension and population explosion of spotted tilapia (Courtenay and Hensley 1979). Devick (1988) suggested that in Wahiawa Reservoir, the combination of adult butterfly peacock and largemouth bass was more efficient at utilizing forage fish than either species alone. Hence, the introduction of the butterfly peacock in Florida was expected to increase predation primarily on exotic fishes, which might indirectly benefit native fishes.

Urban freshwater fishing opportunities in southeast Florida are largely limited to canals, and largemouth bass is the premier sport fish and primary predator in these communities. Because fishing pressure on canal largemouth bass is high (Shafland 1990, 1993b), the presence of peacock bass might also benefit largemouth bass by reducing the amount of fishing effort directed towards it.

Socioeconomic benefits.—Peacock bass are a highly sought sport fish that aggressively attack artificial and live baits, and they are known for their powerful runs and occasional aerial acrobatics (McClane 1964, 1971). They are also an excellent food fish, placing high on the preference lists of those who have eaten them.

Butterfly peacocks can be caught from shore using a variety of sportfishing equipment, which makes the fish popular with nearly every type of freshwater angler. If the butterfly peacock became established in metropolitan Miami–Fort Lauderdale canals, many anglers were expected to fish for it. The sociological importance of providing quality outdoor recreational opportunities in urban areas is considered important by most fisheries resource managers.

The economic contribution of fishery resources to local and state economies has often been overlooked or ignored. If successful, establishment of a butterfly peacock fishery in southeast Florida canals was expected to contribute millions of dollars to the economy.

Limiting Factors

Temperature.—Low water temperature is the most important range-limiting factor for introduced tropical fishes in Florida (Shafland and Pestrak 1982). Early attempts to evaluate the butterfly peacock as a purposeful introduction failed even in south Florida ponds due to its inability to survive low temperatures (V. E. Ogilvie, Florida Game and Fresh Water Fish Commission, personal communication). Controlled temperature studies indicate peacock bass cannot survive water temperatures less than 15°C (Swingle 1967; Guest et al. 1980; Guest and Lyons 1980).

In the early 1980s it was discovered that the canal systems of coastal southeast Florida had warmer than expected winter temperatures (Shafland et al. 1985). The main reason for this is that the canals were built into the Biscayne Aquifer, which is extremely transmissive (VanArman et al. 1984); groundwater intrusion from this shallow aquifer typically warms these canals during the winter. Also contributing to warmer winter water temperatures in canals than in natural aquatic habitats in south Florida (e.g., the Everglades) are a low surface area to water volume ratio due to the canals' deep and narrow morphometry, shoreline windbreaks of trees and buildings, and multidirectional orientation that reduces the cooling effect of directional winds associated with cold fronts. Warmer minimum winter water temperature is the major environmental reason why so many exotic tropical fishes are established in these canals.

Incidental winter temperatures recorded during 1982–1983 indicated the minimum water temperature in the main Black Creek Canal was about 20°C. Use of a continuous recording thermograph during the winter of 1983–1984 documented a minimum main canal temperature of 19.6°C (Shafland 1984). These data indicated peacock bass could overwinter in Black Creek Canal and other area canals. The intolerance of peacock bass to water temperatures below 15°C prohibits their widespread expansion into Florida's natural freshwater habitats (Shafland 1982).

Peacock bass are less tolerant of low water temperatures than are other tropical fish currently established in Florida (Shafland and Pestrak 1982). Minimum January water temperatures in the Miami Canal near Miami International Airport average about 17°C and reportedly fall below 16°C in parts of the North New River Canal near Fort Lauderdale (Anderson 1975). Thus even in coastal south Florida canals, butterfly peacocks might occasionally experience partial winterkills (Shafland 1982).

Salinity.—Freshwater canals in south Florida flow directly into saltwater habitats. Distributional records of peacock bass in their native (South America) and introduced (Hawaii, Puerto Rico, and Panama) ranges indicate they live in only fresh water (Lowe-McConnell 1969; Zaret 1980; W. S. Devick, Hawaii Department of Land and Natural Resources, and D. S. Erdman, Commonwealth of Puerto Rico, personal communications). In order to substantiate this observation, butterfly and speckled peacocks' salinity tolerances were tested in aquaria; both species died at 18 ppt when increased 2 ppt/d (Shafland and Hilton 1985, 1986). For comparison, simultaneously tested largemouth bass died at 20 ppt. These data duplicated the salinity tolerance reported by Guest et al. (1980) and supported field observations that both the butterfly and speckled peacocks will be restricted to fresh water with salinity similar to habitats inhabited by largemouth bass.

Regulations.—Butterfly peacocks are susceptible to being harvested by sportfishing techniques. Shortly after their introduction, a regulation was put into place that prohibited their legal harvest. Five years later, the bag limit was set at two butterfly peacocks per day, only one of which could be greater than 43.2 cm (17 inches) TL. This is one of the strictest harvest regulations for freshwater game fish in Florida and was specifically designed to minimize the harvest of butterfly peacocks. If butterfly peacocks should unexpectedly have detrimental consequences, this regulation could be relaxed to encourage their harvest and presumably reduce such effects.

Preintroduction Discussion and Conclusions

During the 20 years prior to 1984, Florida and Texas biologists had conducted studies on peacock bass in North and South America (Ogilvie 1966a, 1966b; Rutledge and Lyons 1978). All of these studies had the common objective of introducing these fish into the southern United States to convert overly abundant forage fish into quality sport fisheries. Without exception, fisheries managers who had worked with the butterfly peacock in Hawaii, Texas, and Florida believed the butterfly peacock could be a valuable fisheries management tool in artificial habitats overcrowded with forage fishes. Furthermore, no significant adverse effects on native fishes were identified in these or other published studies, with the exception of Zaret and Paine (1973).

The warm stenothermic nature of peacock bass prohibited their successful introduction in North America prior to their 1984 introduction in Florida. Ironically, the intolerance of peacock bass to low temperatures is one of the key criteria that make them uniquely suited for introduction into the canal habitats of south Florida (i.e., intolerance to low temperatures largely restricts them to artificial habitats in Florida).

When the idea of introducing peacock bass was being seriously considered, no alternative management recommendations were proposed that could accomplish the identified needs better than would the introduction of peacock bass (Shafland et al. 1985). Several native predators (common snook *Centropomus undecimalis*, largemouth bass [20–24 cm TL], and *Morone* hybrids) were considered for introduction but rejected due to the absence or cost of hatchery production technology, limited duration of benefits, or behavior indicating these species would leave the targeted area.

Based on this review of the literature, personal communications, water temperature records, extensive assessments of canal fish communities, and previous reports of peacock bass evaluations in Brazil, Panama, Puerto Rico, Hawaii, Texas, and Florida, it was concluded in 1984 that the introduction of peacock bass in south Florida canals would have significant socioeconomic benefits, some important biological benefits, and no serious environmental consequences. It was also concluded that initial efforts to introduce peacock bass should focus on the butterfly peacock because it matures in less than 12 months. The speckled peacock requires 3 years to mature (V. E. Ogilvie, Florida Game and Fresh Water Fish Commission, personal communication), is somewhat less tolerant of low water temperatures (Guest et al. 1980; Guest and Lyons 1980), and is less abundant than is the butterfly peacock in most areas where they coexist (Machado-Allison 1971).

Source of Fish Stocked in Florida

Butterfly peacock fingerlings were obtained from commercial fish dealers who imported them from Brazil, Guyana, and Peru. Progeny from butterfly peacocks originally imported from Guyana in the mid-1970s and subsequently used in Texas studies were also used in this introduction. These different stocks of fingerling butterfly peacocks were grown to adults and then spawned in ponds at the Florida Game and Fresh Water Fish Commission's Non-Native Fish Research Laboratory. Pond-produced fingerlings from all stock origins were released to maximize genetic variability, which ultimately was expected to produce a stock better adapted to these canals than if fishes from one randomly selected origin were used (D. P. Philipp, Illinois Natural History Survey, personal communication).

About 20,000 fingerling butterfly peacocks were stocked into southeast Florida canals beginning in October 1984 and ending in August 1987. These fishes were stocked into 11 major canal systems including and between the Mowry Canal in southern Dade County and the Pompano Canal in northern Broward County. Butterfly peacocks were expected to expand from these canals to all hospitable canals without additional stockings. Prior to introduction, all stocks of butterfly peacocks were screened and found free of known harmful parasites and diseases by U.S. Fish and Wildlife Service personnel (C. P. Carlson, U.S. Fish and Wildlife Service, personal communication). No butterfly peacocks have been stocked into Florida canals since 1987.

Preliminary Postintroduction Analysis

Butterfly peacocks have overwintered and reproduced successfully in south Florida canals every year since their first introduction in 1984 (Shafland 1993a). Annual fish community biomass estimates in the primary study canal, Black Creek Canal, ranged from 95 to 403 kg/ha and averaged 232 kg/ha between 1984 and 1993, but no obvious trends in these estimates were observed.

A direct comparison of the average spotted tilapia standing crop estimates for the first and second 5-year segments of this study indicated they declined 36% by weight and 55% by number (Shafland 1993b). Spearman rank correlation coefficients (r_s) associating annual population changes between spotted tilapia and butterfly peacock populations were low ($r_s = -0.21$ by weight and 0.00 by number). Stronger and positive correlations were found between biomass estimates of butterfly peacocks and redear sunfish *Lepomis microlophus* ($r_s = 0.66$) and bluegill *L. macrochirus* ($r_s = 0.59$). Similarly, numerical estimates of butterfly peacocks were positively related to those of largemouth bass ($r_s = 0.60$), redear sunfish ($r_s = 0.64$), and bluegill ($r_s = 0.55$; Shafland 1993b). No detrimental effects on native fishes have been associated with the presence of butterfly peacocks. These results support the hypothesis that the butterfly peacock may be benefitting some native fishes (Shafland 1993b).

Stomachs from 378 canal butterfly peacocks were examined, and 127 (34%) contained identifiable fish remains; one had a shrimp and another had fish eggs in its stomach. Over 80% of the stomachs containing identifiable fish remains (104 of 127) contained at least one spotted tilapia. Furthermore, 255 of the total 339 (75%) identified fishes found in butterfly peacock stomachs were spotted tilapia (Shafland 1989).

Two 12-month roving creel surveys using nonuniform probability yielded excellent angler catch-rate estimates of 0.64 and 0.85 fish/h for butterfly peacocks and average to above average largemouth bass catch rates of 0.30 and 0.41 fish/h (Shafland 1990, 1993b). Total estimated fishing effort ranged from 228 to 792 h/ha in canals, of which approximately one-third was directed at the butterfly peacock and one-fifth at largemouth bass. The butterfly peacock was the most sought after species by anglers surveyed in both studies.

There are now approximately 530 km of major canals and many kilometers of secondary canals and small lakes with fishable populations of butterfly peacocks in them. Using the above creel results and average expenditures associated with fishing in Florida, the butterfly peacock sport fishery in the major canals alone is conservatively contributing a direct economic benefit of more than US$1 million annually to south Florida's economy (Shafland 1990).

1994 Discussion and Conclusions

The Florida Game and Fresh Water Fish Commission has a comprehensive program dealing with introduced fishes that is divided into three major components: prevention, assessment, and management (Shafland 1986). Purposeful introduction is one of the four defined objectives within the assessment component of this program; hence, purposeful introductions are a small though important aspect of a much larger program. This comprehensive program and the overlapping nature of its major components enable each aspect of the program to be addressed thoroughly. The purposeful introduction of the butterfly peacock could not have been planned, implemented, and assessed as well as it was without the support of this program.

The specific need for introducing an exotic species should be identified *before* any consideration is given to proposing an introduction. One important reason the introduction of butterfly peacocks was successful is that specific data-justified needs were identified prior to the development of a proposal to introduce the species. In fact, the comprehensive long-term studies conducted in Florida (Shafland et al. 1985; Metzger and Shafland 1985) that led directly to this introduction were philosophically conceived on the premise that all introductions were deleterious and that purposeful introductions should be avoided.

The 1984 proposal for introducing peacock bass predicted they would be restricted to freshwater canals of south Florida by their physiological tolerances to temperature and salinity. South Florida has experienced warmer than usual winters since 1984, except in 1989 when a minor butterfly peacock winterkill occurred. Recently, the butterfly peacock has been reported from a canal about 20 km west of the predicted western limit for its establishment (Shafland and Hilton 1985). There are no physical barriers preventing butterfly peacocks from leaving the designated area nor are there any reasonable means of prohibiting its unauthorized transport by individuals wishing to do so. Since 1984 the butterfly peacock has been limited to the geographic range originally predicted with the exception noted.

Since the introduction of the butterfly peacock, urban freshwater anglers in southeast Florida canals have gained access to a sport fish previously accessible only to anglers traveling outside the continental United States. The butterfly peacock has created or renewed the freshwater fishing interests of many urban anglers (Lampton 1994) and contributed millions of dollars to both the state and local economies. Moreover, the presence of the butterfly peacock has increased predation on the abundant exotic forage fishes without detrimentally affecting native fish populations. Public interest in urban fishing has increased considerably as evidenced by numerous articles in newspapers and magazines that feature the butterfly peacock fishery. Increased public recognition of the importance of these resources will create more public support for programs that protect and enhance these and other fisheries resources.

There are philosophical and biological trade-offs and drawbacks to nearly every management recommendation, and the introduction of a "good" exotic species is no exception. Although no detrimental effects have been identified during the 10 years since the butterfly peacock was introduced, its introduction is not expected to be without some consequences. Given the 1970 description of tropical Florida being a "biological cesspool of introduced life" (Lachner et al. 1970), it seems at the very least the introduction of the butterfly peacock has made a bad situation better.

The major biological and socioeconomic objectives for introducing the butterfly peacock into south Florida canals have been accomplished. These objectives were to convert abundant and underutilized exotic forage fishes, especially the spotted tilapia, into a socioeconomically important urban recreational resource. Clearly, the intentional introduction of the butterfly peacock into the canals of southeast Florida has been successful.

Acknowledgments

Numerous people contributed directly or indirectly to the development of the proposal to introduce peacock bass in Florida; for these contributions I am grateful to N. W. Carter, P. G. Chapman, J. P. Clugston, W. R. Courtenay, Jr., W. S. Devick, D. S. Erdman, O. Fontenele, A. Forshage, G. P. Garrett, W. C. Guest, R. C. Heidinger, A. Ramos Henao, D. E. Holcomb, O. G. Kelley, W. M. Lewis, Sr., W. F. Loftus, R. W. Luebke, J. A. McCann, A. S. McGinty, R. J. Metzger, V. E. Ogilvie, D. P. Philipp, J. N. Taylor, R. L. Welcomme, and F. J. Ware. I also thank H. Rambarron, A. I. Schwartz, and the Texas Parks and Wildlife Department for donating peacock bass fingerlings; R. J. Metzger, B. D Hilton, and M. S. Stanford, who assisted with all the field work; and, H. Cruz-Lopez, N. J. Musselman, C. Prosperi, and M. S. Stanford for critically reviewing drafts of this manuscript.

Endnote

Several common names are used for *Cichla* spp., which has created confusion in some discussions involving these fishes. In the English language the earliest and most consistently applied common name given to *Cichla* was peacock bass. This name was selected by North Americans in South America because *Cichla*'s body shape is similar to that of black bass *Micropterus* spp., and the Spanish common name "pavon" means peacock (McClane 1964). The principal source of confusion with *Cichla* common names is the inconsistent use of the name peacock bass, which sometimes refers to a particular species and at other times to all members of the genus *Cichla*. All peacock bass are brightly colored and have on the caudal fin an obvious ocellated eye-spot that resembles markings on the tail feathers of the male peacock *Pavo cristatus*. Although other names are occasionally used (e.g., lukanani), the only other common names used for these fishes in English-speaking areas is "tucunaré" or "tuc," which is used for *C. ocellaris* in Hawaii. (Tucunaré is the Portuguese name used for *Cichla* spp. in Brazil.) In 1991 the American Fisheries Society's Committee on Names of Fishes added yet another name, peacock cichlid, for *C. ocellaris* (Robins et al. 1991). My review of *Cichla* literature and communications with other scientists working with fishes in this genus had not previously encountered the common name peacock cichlid.

Based on the common names recommended by Ogilvie (1966a), the principles set forth by the American Fisheries Society (Robins et al. 1991), and the common names used by others studying the genus *Cichla*, the following common names have worked well throughout the course of this study and are used in this paper: (1) peacock bass refers collectively to all members of the genus *Cichla*; (2) butterfly peacock for *C. ocellaris*, which is English for its Spanish name "pavon mariposa"; (3) speckled peacock for *C. temensis*, which is a rough translation of its Spanish name "pavon pinta de lapa"; and (4) blackstriped peacock for *C. intermedia* (*C. nigrolineatus* of Ogilvie 1966a), which refers to the seven to eight vertical bars of this species versus the three typical bars of other *Cichla* spp.

References

Anderson, W. 1975. Temperature of Florida streams. Map series 43 (revised). United States Geological Survey, and Bureau of Geology, Florida Department of Natural Resources, Tallahassee.

Chew, R. L. 1975. Peacock bass—biologist goes to South America for possible new Texas game fish. Texas Parks and Wildlife Magazine 33(8):24–28.

Courtenay, W. R., Jr., and D. A. Hensley. 1979. Range extension in southern Florida of the introduced spotted tilapia, with comments on their environmental impress. Environmental Conservation 6(2):149–151.

Devick, W. S. 1972a. Life history of the tucunare, *Cichla ocellaris*. Hawaii Department of Land and Natural Resources, Federal Aid in Sport Fish Restoration, Project F-4-R-17, Job Completion Report, Honolulu.

Devick, W. S. 1972b. Life history study of the tucunare, *Cichla ocellaris*. Hawaii Department of Land and Natural Resources, Federal Aid in Sport Fish Restoration, Project F-9-1, Job Completion Report, Honolulu.

Devick, W. S. 1972c. Studies on food habits of the tucunare. Hawaii Department of Land and Natural Resources, Federal Aid in Sport Fish Restoration, Project F-9-2, Job Completion Report, Honolulu.

Devick, W. S. 1976. Studies on the limnology of tucunare habitats. Hawaii Department of Land and Natural Resources, Federal Aid in Sport Fish Restoration, Project F-9-5, Job Progress Report, Honolulu.

Devick, W. S. 1988. Studies on the relationship between tucunare and largemouth bass. Hawaii Department of Land and Natural Resources, Federal Aid in Sport Fish Restoration, Project F-14-R-12, Job Progress Report, Honolulu.

Erdman, D. S. 1969. Culture and stocking of peacock bass. Puerto Rico Department of Natural Resources, Federal Aid in Sport Fish Restoration, Project F-1-17, Job 16, San Juan.

Erdman, D. S. 1984. Exotic fishes in Puerto Rico. Pages 162–176 in W. R. Courtenay, Jr. and J. R. Stauffer, Jr., editors. Distribution, biology and management of exotic fishes. The Johns Hopkins University Press, Baltimore, Maryland.

Fontenele, O. 1950. Contribuição para o conhecimento da biologia dos tucunarés (Actinopterygii, Cichlidae), em captiveiro. Aparelho de reprodução. Hábitos de desova e incubação. Revista Brasileira de Biologia 10(4):503–519 (English translation from the library of R. L. Chew, Texas Parks and Wildlife Department, Austin).

Fontenele, O., and J. T. Peixoto. 1979. Apreciação sobre os resultados da introdução do tucunaré comum, *Cichla ocellaris* (Bloch and Schneider, 1801), nos açudes do nordeste Brasileiro, através da pesca comercial. B. Téc. DNOCS, Fortaleza, Brazil 37(2):109–134. (In Portuguese, translation by Dr. M. Sovereign, Pompano Beach, Florida.)

Garrett, G. P. 1982. Status report on peacock bass (*Cichla* spp.) in Texas. Annual Proceedings of the Texas Chapter, American Fisheries Society 5:20–28.

Guest, W. C., and B. W. Lyons. 1980. Temperature tolerance of peacock bass (*Cichla temensis*). Annual Proceedings of the Texas Chapter, American Fisheries Society 3:42–48.

Guest, W. C., B. W. Lyons, and G. Garza. 1980. Effects of temperature on survival of peacock bass fingerlings. Proceedings of the Annual Conference Southeastern Association of Fish and Wildlife Agencies 33(1979): 620–627.

Kanayama, R. K. 1968. Hawaii's aquatic animal introductions. Proceedings of the Annual Conference Western Association of State Game and Fish Commissioners 47:123–131.

Lachner, E. A., C. R. Robins, and W. R. Courtenay, Jr. 1970. Exotic fishes and other aquatic organisms introduced into North America. Smithsonian Contributions to Zoology 59.

Lampton, B. 1994. Pennies for the peacocks—adding these fish to the water has created jobs and enhanced the south Florida lifestyle. Florida Sportsman 26(5): 198–205.

Lowe-McConnell, R. H. 1969. The cichlid fishes of Guyana, South America, with notes on their ecology and breeding behaviour. Zoological Journal of the Linnean Society 48:255–302.

Lyons, B. W. 1980. Peacock bass. Texas Parks and Wildlife Department, Federal Aid in Sport Fish Restoration, Project F-31-R-1, Final Report, Austin.

Machado-Allison, A. 1971. Contribucion al conocimiento de la taxonomia del genero *Cichla* (Perciformes: Cichlidae) en Venezuela, parte 1. Acta Biologica Venezuela 7(4):459–497 (English translation from the library of R. L. Chew, Texas Parks and Wildlife Department, Austin).

McClane, A. J. 1964. McClane's standard fishing encyclopedia and international angling guide. Holt, Rinehart and Winston Publishers, New York.

McClane, A. J. 1971. Pavón paradise. Field and Stream Magazine (February):49–51, 88, 90.

McGinty, A. S. 1983. Population dynamics of peacock bass, *Cichla ocellaris* and *Tilapia nilotica* in fertilized ponds. Pages 86–94 in L. Fishelson and Z. Yaron, compilers. Proceedings of the International Symposium on Tilapia in Aquaculture. Tel Aviv University, Tel Aviv, Israel.

Metzger, R. J., and P. L. Shafland. 1985. Fish populations of south Florida canals. Completion report for Study X: 1980–84. Florida Game and Fresh Water Fish Commission, Tallahassee.

Ogilvie, V. E. 1966a. Report on the peacock bass project including Venezuelan trip report and a description of five *Cichla* species. Florida Game and Fresh Water Fish Commission, Tallahassee.

Ogilvie, V. E. 1966b. Peacock bass—preliminary report, revised. Florida Game and Fresh Water Fish Commission, Tallahassee.

Robins, C. R., and six coauthors. 1991. Common and scientific names of fishes from the United States and Canada, 5th edition. American Fisheries Society Special Publication 20.

Rutledge, W. P., and B. W. Lyons. 1978. Texas peacock bass and nile perch: status report. Proceedings of the Annual Conference Southeastern Association of Fish and Wildlife Agencies 30(1976):18–23.

Shafland, P. L. 1982. Recommendation for removing peacock bass from Florida's restricted fish list. Florida Game and Fresh Water Fish Commission, Tallahassee.

Shafland, P. L. 1984. A proposal for introducing peacock bass (*Cichla* spp.) in southeast Florida canals. Florida Game and Fresh Water Fish Commission, Tallahassee.

Shafland, P. L. 1986. A review of Florida's efforts to regulate, assess and manage exotic fishes. Fisheries (Bethesda) 11(2):20–25.

Shafland, P. L. 1989. Florida's peacock bass program. American Fisheries Society, Introduced Fish Section Newsletter 9(3):24–27.

Shafland, P. L. 1990. Peacock bass investigations. First annual performance report. Florida Game and Fresh Water Fish Commission, Tallahassee.

Shafland, P. L. 1993a. An overview of Florida's introduced butterfly peacock bass (*Cichla ocellaris*) sportfishery. Natura (Venezuala) 96:26–29.

Shafland, P. L. 1993b. Peacock bass investigations. Annual performance report. Florida Game and Fresh Water Fish Commission, Tallahassee.

Shafland, P. L., and B. D. Hilton. 1985. Introduction of peacock bass (*Cichla* spp.) in southeast Florida canals. First annual performance report. Florida Game and Fresh Water Fish Commission, Tallahassee.

Shafland, P. L., and B. D. Hilton. 1986. Introduction and evaluation of peacock bass (*Cichla* spp.) in southeast Florida canals. Second annual performance report. Florida Game and Fresh Water Fish Commission, Tallahassee.

Shafland, P. L., B. D. Hilton, and R. J. Metzger. 1985. Fishes of Black Creek Canal. Completion report for study XII:1981–1985. Florida Game and Fresh Water Fish Commission, Tallahassee.

Shafland, P. L., and J. M. Pestrak. 1982. Lower lethal temperatures for fourteen non-native fishes in Florida. Environmental Biology of Fishes 7(2):149–156.

Swingle, H. A. 1967. Temperature tolerance of the peacock bass and a pond test of its value as a piscivorous species. Proceedings of the Annual Conference Southeastern Association of Fish and Wildlife Agencies 20(1966):297–299.

VanArman, J., and ten coauthors. 1984. South Florida water management district. Pages 138–157 *in* E. A. Fernald and D. J. Patton, editors. Water resources atlas of Florida. Florida State University, Gainesville.

Verani, J. R., M. de A. Marins, A. B. da Silva, and A. C. Sobrinho. 1983. Population control in intensive fish culture associating *Oreochromis* (*Sarotherodon*) *niloticus* with the natural predator *Cichla ocellaris*—quantitative analysis. Pages 580–587 *in* L. Fishelson and Z. Yaron, compilers. Proceedings of the International Symposium on Tilapia in Aquaculture. Tel Aviv University, Tel Aviv, Israel.

Welcomme, R. L. 1988. International introductions of inland aquatic species. FAO (Food and Agriculture Organization of the United Nations) Fisheries Technical Paper 294.

Zaret, T. M. 1980. Life history and growth relationships of *Cichla ocellaris*; a predatory South American cichlid. Biotropica 12(2):144–157.

Zaret, T. M., and R. T. Paine. 1973. Species introduction in a tropical lake. Science 182:449–455.

Establishment and Expansion of Redbelly Tilapia and Blue Tilapia in a Power Plant Cooling Reservoir

JOHN U. CRUTCHFIELD, JR.

Carolina Power and Light Company, 412 South Wilmington Street
Raleigh, North Carolina 27601, USA

Abstract.—After inadvertent introduction in 1984 redbelly tilapia *Tilapia zilli* and blue tilapia *T. aurea* established populations in Hyco Reservoir, North Carolina, a power plant cooling lake with elevated selenium concentrations. Redbelly tilapia became the fourth most abundant species in the reservoir within 3 years after introduction, whereas blue tilapia remained a minor component of the fish community. Feeding by redbelly tilapia eliminated all aquatic macrophytes from the reservoir within a 2-year period that coincided with declines in populations of golden shiner *Notemigonus crysoleucas*, eastern mosquitofish *Gambusia holbrooki*, and green sunfish *Lepomis cyanellus*. Populations of gizzard shad *Dorosoma cepedianum*, satinfin shiner *Cyprinella analostana*, common carp *Cyprinus carpio*, channel catfish *Ictalurus punctatus*, flat bullhead *Ameiurus platycephalus*, and white catfish *Ameiurus catus* remained unchanged or showed no consistent trends after the introduction of the tilapias. A tilapia fishery was established in the reservoir; both tilapias constituted 19% by number and 13% by weight of the fish harvested by anglers during 1992. Introductions of redbelly tilapia and blue tilapia should be carefully evaluated because of tilapias' opportunistic feeding strategies, their adaptability to a variety of environmental conditions, and the ability of redbelly tilapia to alter habitat through feeding.

Redbelly tilapia *Tilapia zilli* and blue tilapia *T. aurea*, cichlids native to Africa and the Middle East, have been introduced throughout the United States primarily for biological control of aquatic macrophytes and algae and for aquacultural food production (Shelton and Smitherman 1984; Legner et al. 1975; Shireman 1984). Both species have established reproducing populations where winter water temperatures are above 10°C (e.g., Florida, Texas, and California) and in waters with heated power plant discharges or thermal springs (Courtenay et al. 1984). Documented effects of both species include habitat alteration, such as removal of aquatic macrophytes (mainly redbelly tilapia) and algae (Habel 1975; Hauser 1975) and changes in water quality from feeding activities (Taylor et al. 1984), and negative effects on fish populations (particularly clupeids and centrarchids) through competition for spawning sites and food sources or from predation on eggs and larvae (Buntz and Manooch 1969; Noble et al. 1976; Hendricks and Noble 1980; Shafland and Pestrak 1984; Noble and Germany 1986).

Few studies have specifically evaluated the effects of redbelly tilapia on native fish communities in large reservoirs in the United States (Taylor et al. 1984). Furthermore, very few studies in this country have documented a simultaneous introduction of redbelly tilapia and blue tilapia in a large reservoir. The objectives of this paper are to (1) document the establishment of redbelly tilapia and blue tilapia populations in a North Carolina power plant cooling impoundment, Hyco Reservoir, over a 10-year period; (2) describe changes in the native fish community following introduction of these tilapias; (3) document the development of a recreational fishery for both tilapias; and (4) discuss the influential factors in the establishment of both tilapias.

Redbelly tilapia and blue tilapia were inadvertently introduced into Hyco Reservoir from an on-site aquacultural study during 1984. The exact number of tilapias released is not known; but it is estimated that no more than 100 fish of both species combined escaped from a holding cage located in the heated discharge area (W. W. Hassler, AB Ltd., personal communication). Both species quickly established reproducing populations within the reservoir, and the feeding activities of redbelly tilapia resulted in complete removal of all submersed and floating macrophytes (including a 57-ha infestation of *Egeria densa*) in the reservoir within a 2-year period. Evidence that redbelly tilapias were responsible for macrophyte removal included vegetation present in their stomachs and visual observations of feeding (Crutchfield et al. 1992). Prior to the tilapias' introduction, the aquatic macrophyte community consisted of *Nitella* sp., *Potamogeton crispus*, *P. diversifolius*, *Najas guadalupensis*, *Najas minor*, *Eleocharis baldwinii*, *Egeria densa*, and *Hydrilla verticillata*. Changes in water quality variables in the reservoir were minimal after elimination of the macrophytes, and nutrients in the water column did not increase (Crutchfield et al. 1992).

During 1975 to 1982, prior to the tilapias' introduction, continuous ash-pond discharge into Hyco Reservoir from the power plant resulted in selenium bioaccumulation in the biota and subsequent reproductive impairment of largemouth bass *Micropterus salmoides* and bluegill *Lepomis macrochirus* (Gillespie and Baumann 1986; Woock et al. 1987). Reproductive impairment was also suspected for redear sunfish *L. microlophus*, pumpkinseed *L. gibbosus*, warmouth *L. gulosus*, crappie *Pomoxis* spp., and yellow perch *Perca flavescens* (CP&L 1984). By 1982, the fish community was dominated by fishes that appeared to tolerate high levels of selenium—green sunfish *L. cyanellus*, gizzard shad *Dorosoma cepedianum*, satinfin shiner *Cyprinella analostana*, and eastern mosquitofish *Gambusia holbrooki*. From 1982 to 1989, mean selenium concentrations (± SD) in reservoir surface waters at the power plant discharge ranged from 7 ± 0.9 μg/L in 1983 to 14 ± 2.4 μg/L in 1986. During 1990, the Carolina Power and Light Company (CP&L) installed a dry fly ash disposal system at the power plant to reduce selenium inputs into the reservoir and induce recovery of the affected fish species. Since the system began operating, selenium concentrations in reservoir surface waters have ranged from less than 1 to 4 μg/L and have remained below the North Carolina water quality standard of 5 μg/L (CP&L 1993).

Study Area

Hyco Reservoir is a 1,760-ha impoundment located in north-central North Carolina (Figure 1). The reservoir was constructed by CP&L in 1964 to provide condenser cooling water and receiving waters for coal-ash pond effluent for the four-unit Roxboro Steam Electric Plant. The power plant has 2,462 MW of generating capacity and yields year-round heated discharge into Hyco Reservoir. Discharge area surface water temperatures for the December through March overwintering period ranged from 14.0° to 24.8°C during 1984 to 1993.

The reservoir is mesotrophic, is dendritic in morphology, and has four major tributaries: Cane Creek complex, Cobb's Creek, North Hyco River, and South Hyco River. Heated effluent enters the reservoir at the confluence of the North Hyco and South Hyco rivers (Figure 1). Six transects were sampled for fish during 1982 to 1993: Transect 1 (Cobb's Creek), Transect 2 (North Hyco), Transect 3 (South Hyco), Transect 4 (discharge area), Transect 6 (dam area), and Transect 9 (Cane Creek complex) (Figure 1).

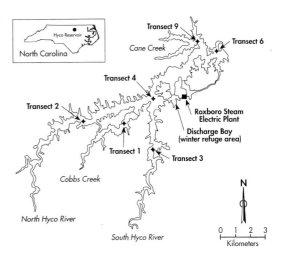

FIGURE 1.—Location of sampling transects and thermal discharge overwintering area for redbelly tilapia and blue tilapia in the Hyco Reservoir, North Carolina.

Methods

Fish Population Assessment

The relative abundance of both tilapias and native fish species was determined with annual cove rotenone sampling during August. Sampling was conducted at Transects 1, 2, 3, 4, 6, and 9, although the number of sampled transects varied by year (Table 1). The surface area of sampled coves ranged from 0.2 to 0.5 ha. Maximum depths in the coves ranged from 2.3 to 4.9 m. Cove sampling procedures generally followed those outlined by Grinstead et al. (1978); fish were collected for 3 d. The block nets employed in sampling had 5-mm-square mesh. All fish were identified, measured to the nearest millimeter (total length), and weighed to the nearest gram. The most abundant fish taxa were subsampled for length and weight measurements. The data were not adjusted for nonrecovery of fish.

Electrofishing was conducted quarterly (March, June, September, and December) at two stations each at Transects 3, 4, and 9 (Table 1). Each station was sampled for 15 min by use of pulsed DC current (280 V at approximately 3 A).

Rotenone coves and shoreline electrofishing stations varied in macrophyte composition and relative abundance depending upon the respective transect. Prior to the tilapias' introduction (1984), the coves of Transects 1, 2, and 3 supported locally abundant stands of naiad *Najas minor* and *Nitella* sp., the coves of Transects 4 and 6 supported sparse (primarily *Najas minor*) or no macrophytes, and the

TABLE 1.—Schedule for annual cove rotenone (R) and quarterly electrofishing (E) sampling by transect at Hyco Reservoir, 1982–1993.

Year	Transect					
	1	2	3	4	6	9
1982	R	R	R,E	R,E		R,E
1983	R	R	R,E	R,E		R,E
1984			R,E	R,E		R,E
1985	R	R	R,E	R,E		R,E
1986			R,E	R,E		R,E
1987	R	R	R,E	R,E		R,E
1988			E	E		E
1989		R	R,E	R,E	R	
1990			E	E		
1991	R	R	R,E	R,E	R	
1992			R,E	R,E	R	
1993	R	R	R,E	R,E	R	

Transect 9 cove was heavily infested with *Egeria densa*. Shoreline electrofishing stations at Transect 3 supported locally abundant stands of *Najas minor*, *Nitella* sp., and *E. densa*; Transect 4 had sparse stands of *Najas minor*; and Transect 9 had a heavy infestation of *E. densa*. Although macrophyte composition and abundance varied by transect, all vegetation disappeared at a similar rate from sampled transects within a 2-year period after the tilapias' introduction.

A one-way analysis of variance (ANOVA) was used to test annual differences in native fish abundance in mean cove rotenone and electrofishing samples from 1982 to 1993. Fisher's protected least-significant-difference test was used to separate annual means for a species if there was a significant difference among means. Data from 1984 to 1986 were not included in the ANOVA because these years were considered a transitional period for the fish community between the presence (before 1985) and complete absence (after 1986) of aquatic macrophytes. The years 1982 to 1983 correspond to macrophytes present and tilapias absent, and the years 1987 to 1993 correspond to macrophytes absent and tilapia present. Species whose reproduction was known or suspected of being affected by selenium bioaccumulation—bluegill, most *Lepomis* spp. (except green sunfish), largemouth bass, crappies, and yellow perch—were not subjected to statistical testing of changes in abundance. Electrofishing data at Transects 3, 4, and 9 were tested separately for differences in annual mean catches of selected species; at Transect 9 only 1982, 1983, 1987, and 1988 electrofishing annual mean data were tested for differences. Months were used as a blocking factor in the electrofishing ANOVA. All data were normalized with a $\log_e$ transformation prior to statistical testing. Statistical tests were performed using the General Linear Models Procedure of the Statistical Analysis System (SAS Institute 1990). A type I error rate of 5% ($\alpha = 0.05$) was used to judge the significance of the tests.

Recreational Fishery

A nonuniform probability creel survey (Malvestuto et al. 1978) was conducted during 1992 to provide estimates of catch and harvest of tilapias and native sport fish by bank and boat anglers. The survey was divided into two periods and two areas. One period (January, February, and December) evaluated the winter fishery in the heated discharge area, and the other period evaluated the spring through fall (March through November) fishery for the entire reservoir. The winter creel survey was conducted only in that part of the reservoir with water temperatures 10°C or greater (522 ha). The entire reservoir was divided into four zones for the spring through fall survey. Sampling probability for each zone was based on surface area.

Sampling was stratified by morning and afternoon-evening hours and by weekday and weekend day. No sampling was conducted at night, and holidays were treated as weekend days. One weekday and one weekend day were randomly selected for each week during the March through November period for a total of 78 sampling periods. Two weekdays and two weekend days each month were sampled during the winter fishery period for a total of 12 periods.

Results

Tilapia Abundance

Thirty-one species were collected by means of cove rotenone and electrofishing sampling during 1982 to 1993 (Table 2). Redbelly tilapia was first collected in 1984; at that time, it was the eighth most abundant species in the fish community and had a mean density of 203 fish/ha and a mean standing crop of 2.5 kg/ha (Table 3). Redbelly tilapia was the fourth most abundant species by density and standing crop 3 years after its introduction. The density and standing crop of redbelly tilapia peaked in 1991, 5 years after macrophyte disappearance. Blue tilapia was not collected in Hyco Reservoir until 1985, at which time the mean density was 17 fish/ha and mean standing crop was 0.5 kg/ha. Although blue tilapia gradually increased in abundance from 1984 to 1993, it remained a minor component of the fish community, constituting 5% or less of the total fish density and standing crop (Table 3).

Densities and standing crops of most fish collected in cove rotenone samples did not differ over time. Significant declines in golden shiner, eastern mosquitofish, and green sunfish occurred after macrophyte removal (Table 4). Although there were not always significant differences in the statistical ranking of the mean densities and standing crops between years when macrophytes were present and years when macrophytes were absent, these three species had significantly declined below their 1982 to 1983 abundance by the end of the study period (i.e., 1992 and 1993). The decline in green sunfish mean densities after vegetation removal was directly related to reduced abundance of age-0 fish (Figure 2). There were significant differences in mean density of satinfin shiner ($P = 0.03$) and mean standing crop of channel catfish ($P = 0.03$); however, the ranking of means did not show any consistent trend in abundance for these species in years when vegetation was absent (1987 to 1993).

Electrofishing Catch Rates

Electrofishing catch rates of golden shiner significantly declined at Transects 3, 4, and 9 after the

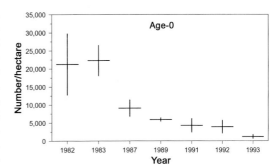

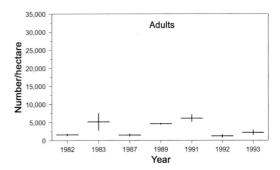

FIGURE 2.—Mean densities (number/ha) ± SE of age-0 (≤50 mm) and adult (>50 mm) green sunfish collected during annual cove rotenone sampling at Hyco Reservoir, 1982–1993. Samples were omitted for 1984 to 1986, and no samples were collected during 1988 and 1990.

TABLE 2.—Common and scientific names of fish species collected in Hyco Reservoir, 1982–1993 (taxonomic nomenclature follows Robins et al. 1991).

Common name	Scientific name
Gizzard shad	*Dorosoma cepedianum*
Satinfin shiner	*Cyprinella analostana*
Common carp	*Cyprinus carpio*
Creek chubsucker	*Erimyzon oblongus*
Silver redhorse	*Moxostoma anisurum*
Shorthead redhorse	*Moxostoma macrolepidotum*
V-lip redhorse	*Moxostoma pappillosum*
Unidentified redhorse	*Moxostoma* spp.
Bluehead chub	*Nocomis leptocephalus*
Golden shiner	*Notemigonus crysoleucas*
Spottail shiner	*Notropis hudsonius*
Snail bullhead	*Ameiurus brunneus*
White catfish	*Ameiurus catus*
Yellow bullhead	*Ameiurus natalis*
Brown bullhead	*Ameiurus nebulosus*
Flat bullhead	*Ameiurus platycephalus*
Channel catfish	*Ictalurus punctatus*
Margined madtom	*Noturus insignis*
Redfin pickerel	*Esox americanus americanus*
Eastern mosquitofish	*Gambusia holbrooki*
Redbreast sunfish	*Lepomis auritus*
Green sunfish	*Lepomis cyanellus*
Pumpkinseed	*Lepomis gibbosus*
Warmouth	*Lepomis gulosus*
Bluegill	*Lepomis macrochirus*
Redear sunfish	*Lepomis microlophus*
Hybrid sunfish	*Lepomis* spp.
Largemouth bass	*Micropterus salmoides*
White crappie	*Pomoxis annularis*
Black crappie	*Pomoxis nigromaculatus*
Yellow perch	*Perca flavescens*
Blue tilapia	*Tilapia aurea*
Redbelly tilapia	*Tilapia zilli*

tilapias' introduction (Table 5). Significantly lower electrofishing catch rates of eastern mosquitofish occurred after tilapia introduction at Transect 9, where large amounts of *E. densa* were present at the sampling stations prior to the introduction. Low catch rates (<2 fish/h) of eastern mosquitofish at Transects 3 and 4 precluded statistical analyses of these data. There were significant differences in annual mean electrofishing catch rates of green sunfish and satinfin shiner at all transects among years; however, the ranking of mean catch rates for both species at all transects did not show any clear trends after macrophyte removal (Table 5). Electrofishing catch rates of gizzard shad significantly declined at Transect 4 when macrophytes were absent, but no temporal trend was evident at Transect 9.

The lack of agreement in trends of green sunfish abundance between the rotenone and electrofishing sampling methods may have been related to gear-type bias and the species' behavior. Several electrofishing stations included riprap areas and beaver lodges where green sunfish congregated. In the absence of macrophytes green sunfish may have sought other cover such as riprap or brush. Consequently, the elec-

TABLE 3.—Mean density (number/ha) and mean standing crop (kg/ha) of major fish taxa collected by cove rotenone sampling at Hyco Reservoir, 1982–1993.

Taxa[a]	1982 Density	1982 Standing crop	1983 Density	1983 Standing crop	1984 Density	1984 Standing crop	1985 Density	1985 Standing crop	1986 Density	1986 Standing crop
Gizzard shad	3,118	35.2	6,069	44.2	8,691	97.6	3,975	64.8	872	29.4
Golden shiner	1,024	3.8	937	6.3	94	1.4	34	0.5	5	0.1
Satinfin shiner	9,752	4.7	3,812	1.5	3,516	1.8	2,163	2.5	4,563	3.4
Common carp	17	11.3	13	30.8	9	32.3	5	4.4	6	9.6
Moxostoma spp.	0	0	54	5.3	11	1.1	61	15.1	2	0.6
Channel catfish	54	24.4	71	31.2	77	4.4	151	27.3	58	28.4
White catfish	73	3.7	68	3.5	110	7.5	25	2.8	9	0.7
Flat bullhead	69	4.0	192	4.7	192	7.1	205	9.9	39	2.5
Ameiurus spp.	160	4.0	332	3.5	65	0.4	141	3.6	14	0.3
Eastern mosquitofish	9,158	2.8	3,079	0.8	322	0.1	489	0.2	306	0.2
Green sunfish	22,842	64.5	27,444	56.1	23,802	52.9	10,997	56.4	3,793	24.3
Bluegill	5,884	10.9	6,373	32.2	1,154	13.3	3,184	29.3	389	10.7
Other *Lepomis* spp.	163	1.8	653	4.0	1,144	6.6	1,383	11.8	473	6.4
Largemouth bass	12	0.3	6	0.6	6	<0.1	25	1.0	13	0.3
Pomoxis spp.	18	0.6	18	0.3	1	<0.1	17	0.5	0	0
Redbelly tilapia	0	0	0	0	203	2.5	1,080	16.6	783	15.9
Blue tilapia	0	0	0	0	0	0	17	0.5	29	1.4
Yellow perch	43	0.7	224	1.0	336	1.0	286	1.2	15	0.2
Other taxa	12	0.1	12	0.8	2	0.2	4	<0.1	2	<0.1
Total	52,399	172.8	49,357	226.8	39,735	230.2	24,242	248.4	11,371	134.4
Sample size	5		5		3		5		3	

(*Continued on next page*)

trofishing data may have reflected a redistribution of green sunfish, particularly adults, into these areas or better sampling efficiency without obstructive vegetation. Similarly, the loss of vegetation would improve fish collection efficiency during rotenone sampling. For these reasons, the rotenone data probably reflected changes in green sunfish abundance after macrophyte removal better than did electrofishing data.

Recreational Fishery

Creel survey data collected during 1992 indicated that a recreational fishery for both tilapia species was established in Hyco Reservoir (Table 6). An estimated 17,248 tilapias (both species combined) weighing 2,425 kg were harvested from Hyco Reservoir during 1992, 8 years after their introduction. After *Lepomis* spp. and crappies, tilapia was the third most prevalent fish group caught and harvested by anglers during the spring through fall. Of the tilapias, anglers harvested (kept) 70% of the redbelly tilapia and 98% of the blue tilapia caught during that period. Tilapia constituted only 6% and 5% of the total number of fish caught and harvested, respectively, by anglers in the winter fishery.

Discussion

Redbelly tilapia and blue tilapia quickly established reproducing populations in Hyco Reservoir after their introduction in 1984. Redbelly tilapia became abundant within 2 years after the introduction and remained the fourth most numerically dominant species from 1987 to 1993. In contrast to introductions of this species in other U.S. reservoirs (Buntz and Manooch 1969; Noble et al. 1976), blue tilapia remained a minor component of the fish community.

The dominance of redbelly tilapia over blue tilapia may have been related to a dissimilarity in reproductive sensitivity to selenium bioaccumulation or to the relative differences in reproductive strategies and aggressiveness between both species. Redbelly tilapia are nest-guarding spawners with the ability to mature sexually at a small size and produce multiple broods within a spawning season (Lowe-McConnell 1982). Blue tilapia are mouthbrooders that produce fewer eggs per spawn but invest more energy (i.e., larger egg size and higher yolk content) and care in their young than do redbelly tilapia. In Lake Victoria, Africa, introduced redbelly tilapia supplanted the mouthbrooder *Tilapia variabilis* that spawned in the same areas (Lowe-McConnell 1982). This displacement was attributed to the aggressive nature of redbelly tilapia, which dominated the spawning areas. Noakes and Balon (1982) suggested that due to their higher fecundity and earlier sexual maturity, tilapia spawners which guard nests, like redbelly tilapia, would be better suited for initial invasions into lakes compared with mouthbrooder tilapias. Populations of mouth-

TABLE 3.—Extended.

Taxa[a]	1987 Density	1987 Standing crop	1989 Density	1989 Standing crop	1991 Density	1991 Standing crop	1992 Density	1992 Standing crop	1993 Density	1993 Standing crop
Gizzard shad	5,946	62.6	10,695	81.0	11,853	94.6	26,788	256.6	5,820	71.2
Golden shiner	63	0.2	3	<0.1	10	<0.1	11	<0.1	6	<0.1
Satinfin shiner	5,973	3.0	4,385	2.5	2,656	1.5	514	0.7	1,203	0.7
Common carp	3	6.7	20	38.9	1	2.3	2	4.7	7	15.2
Moxostoma spp.	6	0.7	9	5.8	5	1.7	3	2.0	7	6.5
Channel catfish	41	16.9	30	14.4	44	3.9	8	2.2	27	5.8
White catfish	52	3.9	428	29.5	152	11.6	182	22.5	42	3.2
Flat bullhead	104	2.3	104	5.5	127	4.2	98	2.8	151	4.2
Ameiurus spp.	68	0.5	66	0.1	148	0.8	95	0.7	26	0.6
Eastern mosquitofish	1,232	0.4	424	0.2	607	0.2	104	<0.1	201	0.1
Green sunfish	10,578	23.9	10,514	54.3	10,439	32.7	5,122	21.5	3,383	16.0
Bluegill	481	4.1	565	5.8	729	6.8	1,771	17.2	2,543	13.5
Other Lepomis spp.	191	3.3	375	3.3	283	3.5	1,206	9.4	910	9.3
Largemouth bass	3	<0.1	3	0.2	9	0.8	197	2.2	107	4.0
Pomoxis spp.	1	<0.1	12	1.0	4	0.1	82	1.8	16	1.0
Redbelly tilapia	1,694	11.9	1,996	25.8	6,192	44.3	1,564	23.2	2,159	20.5
Blue tilapia	45	1.2	121	3.4	335	3.9	163	15.2	290	9.9
Yellow perch	44	0.1	160	0.7	500	2.4	714	5.3	830	4.7
Other taxa	44	0.2	9	<0.1	4	2.2	0	0	14	0.1
Total	26,569	141.9	29,919	272.4	34,098	217.5	38,624	388.0	17,742	186.5
Sample size	5		4		5		3		5	

[a] Moxostoma spp. includes silver redhorse, shorthead redhorse, V-lip redhorse, and unidentified redhorse; Ameiurus spp. includes snail bullhead, brown bullhead, and yellow bullhead; other Lepomis spp. includes hybrid Lepomis, pumpkinseed, redbreast sunfish, redear sunfish, warmouth, and unidentified Lepomis spp.; Pomoxis spp. includes black and white crappies; and other taxa include redfin pickerel, spottail shiner, bluehead chub, creek chubsucker, and margined madtom.

brooders would be favored in lakes over longer periods of time. In the Teso dams' reservoirs in Uganda, redbelly tilapia and the mouthbrooder *Tilapia leucostictus* were stocked simultaneously, and redbelly tilapia predominated in the first few years after the introduction. However, *T. leucostictus* became the dominant tilapia species after several years. The decline of redbelly tilapia in these reservoirs was attributed to a reduction of macrophytes by redbelly tilapia and subsequent lack of food for this species. A similar species shift has not occurred in Hyco Reservoir. Redbelly tilapia remained the dominant tilapia even after its preferred food source, macrophytes, was eliminated from the reservoir. In Hyco Reservoir redbelly tilapia switched to a diet dominated by organic detritus and filamentous algae (Crutchfield et al. 1992).

Elimination of aquatic macrophytes by redbelly tilapia coincided with observed declines in popula-

TABLE 4.—Changes in density (number/ha) and biomass (kg/ha) of fish collected in cove rotenone samples in Hyco Reservoir, 1982–1983 and 1987–1993.

Species	P-value	Mean rank[a]						
Density								
Golden shiner	<0.01	1982z	1983z	1987y	1992yx	1991yx	1993yx	1989x
Satinfin shiner	0.03	1982z	1987z	1989z	1983zy	1991zy	1993yx	1992x
Eastern mosquitofish	<0.01	1982z	1983zy	1987y	1991yx	1989yx	1993xw	1992w
Green sunfish	<0.01	1983z	1982zy	1987zy	1989y	1991yx	1992xw	1993w
Biomass								
Golden shiner	<0.01	1983z	1982z	1987y	1992y	1991y	1993y	1989y
Eastern mosquitofish	<0.01	1982z	1983y	1987y	1989yx	1991yx	1993x	1992x
Channel catfish	0.03	1983z	1982z	1987z	1989z	1993zy	1991zy	1992y
Green sunfish	<0.01	1983z	1982z	1989z	1991zy	1987yx	1992yx	1993x

[a] The ranking of annual means are shown in decreasing order. Mean ranks within a row sharing a common letter do not differ ($P > 0.05$, Fisher's protected least-significant-difference test). See Table 3 for nontransformed annual means for each species.

TABLE 5.—Changes in catch rates (number/h; given in parentheses) of species collected by electrofishing at Transects 3, 4, and 9 in Hyco Reservoir, 1982–1983 and 1987–1993.

Species	P-value	Mean rank[a]								
Transect 3										
Golden shiner	<0.01	1983z (2.3)	1982z (1.5)	1989y (0.9)	1987yx (0.5)	1988x (0.4)	1990x (0.4)	1993x (0.4)	1991x (0.2)	1992x (0)
Satinfin shiner	<0.01	1988z (3.7)	1989zy (2.5)	1983yx (2.0)	1987yx (1.9)	1982x (1.9)	1990x (0.9)	1992xw (0.8)	1991x (0.5)	1993w (0)
Green sunfish	0.02	1989z (5.7)	1990zy (5.6)	1983yx (5.2)	1991yx (5.1)	1987yx (5.0)	1992yx (4.8)	1988yx (4.7)	1982x (4.6)	1993w (3.7)
Transect 4										
Gizzard shad	<0.01	1983z (4.3)	1982y (2.6)	1987yx (2.5)	1990yx (2.4)	1989yx (2.3)	1988yx (2.1)	1991yx (1.6)	1992yx (1.6)	1993x (1.1)
Golden shiner	<0.01	1983z (1.2)	1982z (0.6)	1987y (0)	1988y (0)	1989y (0)	1990y (0)	1991y (0)	1992y (0)	1993y (0)
Satinfin shiner	<0.01	1988z (3.6)	1990z (3.4)	1987z (3.1)	1989z (2.9)	1982z (2.5)	1991zy (2.4)	1983zy (2.2)	1992yx (1.0)	1993x (0)
Green sunfish	<0.01	1989z (5.4)	1990z (5.4)	1983zy (5.0)	1988zy (5.0)	1987yzx (4.4)	1991zyx (4.2)	1992yx (3.7)	1982x (3.5)	1993x (2.9)
Transect 9										
Gizzard shad	<0.01	1983z (4.2)	1987y (2.0)	1982y (1.9)	1988y (1.6)					
Golden shiner	<0.01	1983z (2.1)	1982y (0.8)	1987x (0)	1988x (0)					
Satinfin shiner	0.02	1983z (4.2)	1988z (4.0)	1987z (4.0)	1982y (2.5)					
Eastern mosquitofish	0.01	1983z (2.6)	1982z (2.0)	1987y (0)	1988y (0)					
Green sunfish	<0.01	1983z (5.8)	1988y (4.7)	1987y (4.6)	1982y (4.2)					

[a] Mean ranks within a row sharing a common letter do not differ ($P > 0.05$, Fisher's protected least-significant-difference test).

tions of golden shiner, eastern mosquitofish, and green sunfish. These species prefer dense aquatic vegetative cover for either spawning or protection against predation (Forney 1957; Miley 1978; Ware and Gasaway 1978). Bettoli et al. (1993) noted declines in several *Lepomis* spp. after vegetation removal by grass carp *Ctenopharyngodon idella* in Lake Conroe, Texas. Reduced density of age-0 green sunfish in Hyco Reservoir also suggested possible competitive interactions with tilapias during green sunfish spawning or early life stages.

It seems unlikely that elevated selenium concentrations were responsible for the decline in golden shiner, eastern mosquitofish, and green sunfish populations, because these species were reproducing under elevated selenium conditions during the early 1980s prior to the tilapias' introduction. The abundance of these species also remained unchanged or decreased during 1990 to 1993, when selenium concentrations decreased in the reservoir following implementation of the dry fly ash disposal system. Cumbie and Van Horn (1979) reported eastern mosquitofish and flat bullhead populations persisted in Belews Lake, North Carolina, a nearby reservoir where fish reproduction also was impaired by selenium bioaccumulation. Cherry et al. (1976) documented an eastern mosquitofish population surviving and reproducing in a coal ash pond that had elevated water temperatures and a mean selenium concentration in water of 107 µg/L.

Populations of gizzard shad, satinfin shiner, common carp, flat bullhead, channel catfish, and white catfish either showed no significant changes or no consistent trends in abundance after macrophyte removal. Fluctuations in gizzard shad abundance were related to relative year-class strength and cove to cove variation. Bettoli et al. (1993) also found no significant change in gizzard shad and common carp populations after vegetation removal in Lake Conroe, Texas.

Several factors likely contributed to the successful establishment of both tilapias in Hyco Reservoir. First, and most importantly, the power plant's thermal discharge provided water temperatures suitable for overwintering. There have been no documented winterkills of tilapias in the reservoir discharge area since the introduction of both species. Both tilapias are able to migrate, reproduce, and repopulate areas in the reservoir each spring at distances of up to 12 km from the discharge area.

Second, both tilapias appear tolerant of a wide range of environmental conditions. Both species reproduced and expanded their populations despite elevated selenium concentrations within the reservoir. Both tilapias can tolerate a wide variety of environmental conditions including high water temperatures, fluctuating pH and dissolved oxygen concentrations, elevated ammonia concentrations, brackish to saline waters, and oligotrophic to hypereutrophic productivity conditions (Shelton and Smitherman 1984). Both tilapia species have flexible feeding strategies and can use organic detritus, filamentous algae, zooplankton, and benthic invertebrates when preferred macrophyte and algal food sources are scarce or absent (Hauser 1975; Spataru 1978; Hendricks and Noble 1980; Fitzpatrick 1981; Mallin 1986; Khallaf and Alne-na-ei 1987). This ability to tolerate different environmental conditions would be advantageous to a species when colonizing new habitats or maintaining populations during periods of less than optimal living conditions.

A final factor contributing to the success of the tilapias was the altered nature of the fish community and the low abundance of the major native piscivore, largemouth bass, due to reproductive impairment from selenium bioaccumulation (Gillespie and Baumann 1986). Largemouth bass constituted less than 1% of the total standing crop in Hyco Reservoir cove rotenone samples prior to and after the tilapias' introduction. Largemouth bass typically constitute approximately 5% of the total standing crop in southeastern U.S. mainstream reservoirs (Davies et al. 1982). Hauser (unpublished data) documented heavy predation on redbelly tilapia by adult largemouth bass and channel catfish and suggested predation lowered the redbelly tilapia population and subsequently reduced its effect on macrophytes. Lowe-McConnell (1982) suggested that the rapid expansion of redbelly tilapia in Lake Victoria, Africa, resulted from the species' high fecundity and the comparative absence of predators. Bickerstaff et al. (1984) reported increased predation on redbelly tilapia fry by adult bluegill when the amount of protective vegetation cover was reduced in experimental enclosures. The investigators speculated that introductions of redbelly tilapia would likely fail under conditions of high predator densities, especially adult bluegill and largemouth bass.

A tangible benefit of the tilapias' introduction was the development of a hook-and-line tilapia fishery. Tilapias were acceptable to anglers as a food fish as demonstrated by their ranking among all fish harvested by anglers from the reservoir. Most tilapias were caught and harvested during warmer months when the fish were distributed throughout the reservoir rather than the overwintering period when fish were congregated in the discharge area. Similar to their relative abundances in the reservoir, redbelly tilapia was also the more often caught and harvested tilapia. Interviews with anglers indicated most tilapias were caught with earthworms and crickets.

Development of tilapia fisheries in other U.S. reservoirs, especially of redbelly tilapia, has been infrequently documented by fishery managers. Buntz and Manooch (1969) reported that blue tilapia constituted approximately 17% of the total number of fish harvested in a 14-d creel at Lake

TABLE 6.—Estimated total number (± approximate SE) of fish caught, number harvested, and weight (kg) harvested by taxa for March–November (entire reservoir) and January, February, and December (heated discharge area only) at Hyco Reservoir, 1992.

Taxa[a]	Number caught	Number harvested	Weight harvested
March–November			
Channel catfish	1,073 ± 495	986 ± 508	495 ± 205
White catfish	4,462 ± 952	3,878 ± 829	852 ± 211
Ameiurus spp.	7,960 ± 1,582	2,708 ± 836	302 ± 92
Creek chubsucker	103 ± 109	0	0
Common carp	683 ± 519	672 ± 514	849 ± 647
Bluegill	965 ± 343	518 ± 260	32 ± 16
Green sunfish	23,449 ± 9,666	10,837 ± 3,360	669 ± 234
Lepomis spp.	45,989 ± 14,502	10,663 ± 4,233	1,044 ± 427
Largemouth bass	3,244 ± 863	947 ± 577	698 ± 431
Pomoxis spp.	25,606 ± 7,403	18,365 ± 6,482	4,925 ± 1,816
Yellow perch	633 ± 469	0	0
Blue tilapia	1,326 ± 962	1,302 ± 945	404 ± 297
Redbelly tilapia	18,816 ± 6,734	13,241 ± 5,049	1,165 ± 519
Tilapia spp.	3,835 ± 1,597	1,516 ± 754	467 ± 232
Total	138,144	65,633	11,902
January, February, and December			
Channel catfish	241 ± 123	112 ± 14	27 ± 6
White catfish	156 ± 19	69 ± 10	26 ± 8
Ameiurus spp.	1,198 ± 814	1,022 ± 671	158 ± 105
Bluegill	98 ± 111	286 ± 100	24 ± 8
Green sunfish	61 ± 116	102 ± 48	10 ± 5
Lepomis spp.	971 ± 286	355 ± 150	44 ± 22
Largemouth bass	208 ± 181	75 ± 65	25 ± 21
Pomoxis spp.	23,655 ± 4,656	22,765 ± 4,632	5,501 ± 1,389
Yellow perch	49 ± 6	49 ± 6	7 ± 9
Blue tilapia	131 ± 82	131 ± 82	45 ± 28
Redbelly tilapia	600 ± 166	207 ± 81	49 ± 43
Tilapia spp.	1,054 ± 110	851 ± 81	295 ± 33
Total	28,422	26,024	6,211

[a] *Ameiurus* spp. includes snail bullhead, brown bullhead, and yellow bullhead; *Lepomis* spp. includes hybrid *Lepomis*, pumpkinseed, redbreast sunfish, redear sunfish, warmouth, and unidentified *Lepomis* spp.; *Pomoxis* spp. includes black and white crappies; and *Tilapia* spp. includes blue tilapia and redbelly tilapia.

Parker, Florida. In Crenshaw Lake, Arkansas, blue tilapia, blackchin tilapia *T. melanotheron*, and Mozambique tilapia *T. mossambica* represented 25% of the number and 72% of the weight of all fish harvested by anglers with hook-and-line and dip nets from 1972 to 1973 (Habel 1975). Commercial fisheries for blue tilapia have occurred at several Florida lakes (Langford et al. 1978). Hook-and-line fisheries for *T. mossambica* and redbelly tilapia have been reported for the Gila River and associated irrigation canals in Arizona and in the Salton Sea in California (Radonski et al. 1984).

With the reduction of selenium inputs into Hyco Reservoir during 1990, reproduction of bluegill, warmouth, pumpkinseed, redear sunfish, largemouth bass, and yellow perch has resumed at low levels (Table 2; CP&L 1993). The black crappie population has rapidly expanded. The actual effects that redbelly tilapia or blue tilapia populations will have on the long-term recovery of these species or other species affected by selenium toxicity are not presently known. Although both tilapias have become established in the reservoir, they have not completely dominated the fish community. The limited size of the overwintering area (approximately 30% of the total reservoir area), the length of the growing season (210 d) for growth and reproduction in unheated areas, and the quantity and quality of food sources (i.e., no macrophytes and moderate algal productivity) present at Hyco Reservoir may prevent these species from reaching nuisance levels in future years.

Management Implications

In areas suitable for overwintering, the use of redbelly tilapia and blue tilapia should be restricted to tightly controlled, closed systems to prevent unwanted escapement of these two species. Both species can tolerate a wide variety of environmental conditions, which increases the likelihood of successful establishment of populations that can persist through time. Introductions of both species are also more likely to be successful in systems with low or depauperate predator populations. Fishery managers need to consider carefully whether redbelly tilapia and blue tilapia should be introduced into reservoirs for aquatic macrophyte and algae control. Both species can shift to alternate food sources in the absence or reduced abundance of preferred food sources and still maintain viable, reproducing populations. The ability of redbelly tilapia to eliminate all aquatic macrophytes in Hyco Reservoir within a 2-year period also demonstrates how quickly this species can alter habitat in large reservoirs.

Acknowledgments

I wish to thank Carolina Power and Light Company for the financial support to conduct this study. Special thanks to Kyle Martin, who conducted the creel survey, and to various members of the Biological Assessment and Environmental Assessment Units who helped with field sampling. Ann Harris assisted with the statistical analyses of the data.

References

Bettoli, P. W., M. J. Maceina, R. L. Noble, and R. K. Betsill. 1993. Response of a reservoir fish community to aquatic vegetation removal. North American Journal of Fisheries Management 13:110–124.

Bickerstaff, W. B., C. D. Ziebell, and W. J. Matter. 1984. Vulnerability of redbelly tilapia fry to bluegill predation with changes in cover availability. North American Journal of Fisheries Management 4:120–125.

Buntz, J., and C. S. Manooch III. 1969. *Tilapia aurea* (Steindächner), a rapidly spreading exotic in south central Florida. Proceedings of the Annual Conference Southeastern Association of Game and Fish Commissioners 22(1968):495–501.

Cherry, D. S., R. K. Guthrie, J. H. Rodgers, Jr., J. Cairns, Jr., and K. L. Dickson. 1976. Responses of mosquitofish (*Gambusia affinis*) to ash effluent and thermal stress. Transactions of the American Fisheries Society 105:686–694.

Courtenay, W. R., Jr., D. A. Hensley, J. N. Taylor, and J. A. McCann. 1984. Distribution of exotic fishes in the continental United States. Pages 41–77 *in* W. R. Courtenay, Jr. and J. R. Stauffer, Jr., editors. Distribution, biology, and management of exotic fishes. Johns Hopkins University Press, Baltimore, Maryland.

CP&L (Carolina Power and Light Company). 1984. Roxboro Steam Electric Plant 1983 environmental monitoring report. CP&L, New Hill, North Carolina.

CP&L (Carolina Power and Light Company). 1993. Roxboro Steam Electric Plant 1992 environmental monitoring report. CP&L, New Hill, North Carolina.

Crutchfield, J. U., Jr., D. H. Schiller, D. D. Herlong, and M. A. Mallin. 1992. Establishment and impact of redbelly tilapia in a vegetated cooling reservoir. Journal of Aquatic Plant Management 30:28–35.

Cumbie, P. M., and S. L. Van Horn. 1979. Selenium accumulation associated with fish reproductive failure. Proceedings of the Annual Conference Southeastern Association of Fish and Wildlife Agencies 32(1978):612–624.

Davies, W. D., W. L. Shelton, and S. P. Malvestuto. 1982. Prey-dependent recruitment of largemouth bass: a conceptual model. Fisheries (Bethesda) 7(6):12–15.

Fitzpatrick, L. A., B. W. Rickel, M. O. Saeed, and C. D. Ziebell. 1981. Factors influencing the effectiveness of *Tilapia zilli* in controlling aquatic weeds. Arizona Cooperative Fisheries Unit, Report 81-1, Tucson.

Forney, J. L. 1957. Raising bait fish and crayfish in New York ponds. Cornell Extension Bulletin 986:3–30, Cornell, New York.

Gillespie, R. B., and P. C. Baumann. 1986. Effects of high tissue concentrations of selenium on reproduction of bluegills. Transactions of the American Fisheries Society 115:208–213.

Grinstead, B. G., R. M. Gennings, G. R. Hooper, C. A. Schultz, and D. A. Whorton. 1978. Estimation of standing crop of fishes in predator-stocking evaluation reservoirs. Proceedings of the Annual Conference Southeastern Association of Fish and Wildlife Agencies 30(1976):120–130.

Habel, M. 1975. Overwintering of the cichlid, *Tilapia aurea*, produces fourteen tons of harvestable sized fish in a south Alabama bass-bluegill public fishing lake. Progressive Fish-Culturist 37:31–32.

Hauser, W. J. 1975. *Tilapia* as biological control agents for aquatic weeds and noxious aquatic insects in California. Proceedings of the Annual Conference of the California Mosquito Control Association, Inc. 43:51–53.

Hendricks, M. K., and R. L. Noble. 1980. Feeding interactions of three planktivorous fishes in Trinidad Lake, Texas. Proceedings of the Annual Conference Southeastern Association of Fish and Wildlife Agencies 33(1979):324–330.

Khallaf, E. A., and A. A. Alne-na-ei. 1987. Feeding ecology of *Oreochromis niloticus* (Linnaeus) & *Tilapia zilli* (Gervais) in a Nile canal. Hydrobiologia 146:57–62.

Langford, F. H., F. J. Ware, and R. D. Gasaway. 1978. Status and harvest of introduced *Tilapia aurea* in Florida lakes. Pages 102–108 *in* W. L. Shelton and J. H. Grover, editors. Culture of exotic fishes; symposium proceedings. Fish Culture Section, American Fisheries Society, Bethesda, Maryland.

Legner, E. F., W. J. Hauser, T. W. Fisher, and R. A. Medved. 1975. Biological aquatic weed control by fish in the lower Sonoran Desert of California. California Agricultural Newsletter 29:8–10.

Lowe-McConnell, R. H. 1982. Tilapias in fish communities. Pages 83–114 *in* R. S. V. Pullin and R. H. Lowe-McConnell, editors. The biology and culture of tilapias. International Center for Living Aquatic Resources, Manila, Philippines.

Malvestuto, S. P., W. D. Davies, and W. L. Shelton. 1978. An evaluation of the roving creel survey with nonuniform probability sampling. Transactions of the American Fisheries Society 107:255–262.

Mallin, M. A. 1986. The feeding ecology of the blue tilapia (*T. aurea*) in a North Carolina reservoir. Proceedings of the Annual Conference and International Symposium of the North American Lake Management Society 5:323–326.

Miley, W. W., II. 1978. Ecological impact of the pike killifish, *Belonesox belizanus* Kner (Poeciliidae), in southern Florida. Master's thesis. Florida Atlantic University, Boca Raton.

Noakes, D. L. G., and E. K. Balon. 1982. Life histories of tilapias: an evolutionary perspective. Pages 61–82 *in* R. S. V. Pullin and R. H. Lowe-McConnell, editors. The biology and culture of tilapias. International Center for Living Aquatic Resources, Manila, Philippines.

Noble, R. L., R. D. Germany, and C. R. Hall. 1976. Interactions of blue tilapia and largemouth bass in a power plant cooling reservoir. Proceedings of the Annual Conference Southeastern Association of Game and Fish Commissioners 29:247–251.

Noble, R. L., and R. D. Germany. 1986. Changes in fish populations of Trinidad Lake, Texas, in response to abundance of blue tilapia. Pages 455–461 *in* R. H. Stroud, editor. Fish culture in fishery management. American Fisheries Society, Fish Culture Section and Fisheries Management Section, Bethesda, Maryland.

Radonski, G. C., N. S. Prosser, R. G. Martin, and R. H. Stroud. 1984. Exotic fishes and sport fishing. Pages 313–321 *in* W. R. Courtenay, Jr. and J. R. Stauffer, Jr., editors. Distribution, biology, and management of exotic fishes. Johns Hopkins University Press, Baltimore, Maryland.

Robins, C. R., and six coauthors. 1991. Common and scientific names of fishes from the United States and Canada, 5th edition. American Fisheries Society Special Publication 20.

SAS Institute. 1990. SAS/STAT user's guide, version 6, 4th edition, volume 1. SAS Institute, Inc., Cary, North Carolina.

Shafland, P. L., and J. M. Pestrak. 1984. Predation of blue tilapia by largemouth bass in experimental ponds. Proceedings of the Annual Conference Southeastern Association of Fish and Wildlife Agencies 35(1981):443–448.

Shelton, W. L., and R. O. Smitherman. 1984. Exotic fishes in warmwater aquaculture. Pages 262–301 *in* W. R. Courtenay, Jr. and J. R. Stauffer, Jr., editors. Distribution, biology, and management of exotic fishes. Johns Hopkins University Press, Baltimore, Maryland.

Shireman, J. V. 1984. Control of aquatic weeds with exotic fishes. Pages 302–312 *in* W. R. Courtenay, Jr. and J. R. Stauffer, Jr., editors. Distribution, biology, and management of exotic fishes. Johns Hopkins University Press, Baltimore, Maryland.

Spataru, P. 1978. Food and feeding habits of *Tilapia zilli* (Gervais) (Cichlidae) in Lake Kinneret (Israel). Aquaculture 14:327–338.

Taylor, J. N., W. R. Courtenay, Jr., and J. A. McCann. 1984. Known impacts of exotic fishes in the continental United States. Pages 322–373 *in* W. R. Courtenay, Jr. and J. R. Stauffer, Jr., editors. Distribution, biology, and management of exotic fishes. Johns Hopkins University Press, Baltimore, Maryland.

Ware, F. J., and R. D. Gasaway. 1978. Effects of grass carp on native fish populations in two Florida lakes. Proceedings of the Annual Conference Southeastern Association of Fish and Wildlife Agencies 30(1976):324–335.

Woock, S. E., W. R. Garrett, W. E. Partin, and W. T. Bryson. 1987. Decreased survival and teratogenesis during laboratory selenium exposures to bluegill, *Lepomis macrochirus*. Bulletin of Environmental Contamination and Toxicology 39:998–1005.

Hatcheries, Habitat, and Regulations:

Past and Future Uses

in Fisheries Management

A Common Sense Protocol for the Use of Hatchery-Reared Trout

ROBERT W. WILEY

Wyoming Game and Fish Department
528 South Adams, Laramie, Wyoming 82070, USA

Abstract.—Fish hatcheries are vital to fisheries management, maintenance of high-quality angling, and restoration of Endangered fishes. However, people tend to expect too much from hatcheries and rely on stocking to provide more fish than lakes and streams can sustain. In the Rocky Mountains, salmonids were imported for rearing in hatcheries and stocking to supplement native fish for sport angling, commercial use, and food. Fish were stocked in any water that looked suitable. No one understood that natural waters have productive limits, introduced fishes might extirpate native fishes, and fish stocks might be adapted to specific stream and lake conditions. People believed that rearing and stocking fish was necessary to continued good angling. Without evidence to the contrary, fisheries biologists and fish culturists thought one fish was just as good as another. The idea that differences in fish stocks could be hereditary, adaptive, and result from local evolution was slowly recognized.

In western streams in seven ecoregions, trout biomass averaged 67 kg/ha or less for 56–96% of stocks measured. Salmonids occupy stream habitat to its potential carrying capacity, so survival of trout stocked in streams beyond this relatively small biomass is low. Survival is usually higher in lakes because salmonids may not fill lake habitat to its potential capacity. Regulations are used to control fishing effort and harvest. Regulations could also be used to govern the use of hatchery fish by specifying rearing and stocking objectives based on environmental requirements of the species to be stocked, as dictated by habitat condition and productivity of the receiving water. For successful fish stocking, habitat must be in good condition, the species must be physiologically and behaviorally capable of surviving where stocked, and biologists must be sure that stocking solves the real problem.

Stocking fish according to regional fish-stocking protocols would improve consistency in the use of fish. Foresighted fisheries management should (1) be based on drainage surveys that document habitat conditions and natural limits of production; (2) determine genetic strengths of broodstocks and stock hatchery fish where best suited; (3) manage for native or wild fish first; (4) establish priorities for fish stocking in standing waters; and (5) understand public desires.

Fish hatcheries are vital to fish management, maintenance of quality angling, and restoration of Threatened and Endangered species. However, the wisdom of fish stocking has been questioned (Behnke 1991; White 1992), and some suggest that the waters of North America are polluted with hatchery fish (Beckstrom 1991). Rearing and stocking hatchery salmonids has become an incendiary issue in the Pacific Northwest; emotion also runs high in the Rocky Mountain region. The controversy about fish stocking may be attributable to the variety of stocking practices in the western United States. If there were regionally accepted protocols for fish culture and fish stocking based on biological need and the fish production capacity of watershed habitats, much of the debate would disappear.

The purpose of this paper is to describe the history of fish culture in Wyoming, explain its role in fisheries management, define the role of habitat in sustaining fish stocks, and provide a biologically based protocol to fish rearing and stocking. Historical fisheries literature and fish stocking data from Wyoming and other regions of the western United States were reviewed to develop an overall view of fish culture and fisheries management in the region. The resulting perspective was used to recommend a common-sense fisheries management strategy.

Artificial rearing and stocking of fish was known in Europe in 1763 (Netboy 1973). By the late 1800s, trout hatcheries were common in North America. Since then, fish have been stocked, sometimes promiscuously and without regard for habitat or native fishes, to establish fishing (Stone 1872; Wales 1939); stimulate economies (Barkwell 1883); supplement populations that had been depleted by overfishing, habitat destruction, or other maladies (Barkwell 1883; Hedgepeth 1941; Wiley et al. 1993); and protect or restore extirpated stocks of native, Threatened, and Endangered species (Rinne et al. 1986).

Wyoming Angling, Fish Culture, and Fish Management

Fishing in Wyoming Territory was evidently very good. While camped for 3 weeks in the Tongue River valley (northeastern Wyoming) in July 1876, General Crook's soldiers caught an estimated 15,000 trout (Bourke 1891). By 1882, laws governing propagation and culture of fish were enacted,

TABLE 1.—Catchable- and subcatchable-size trout stocked in Wyoming, 1987 through 1992.[a] Values in parentheses are trout stocked per licensed angler.

Year	Total number of trout stocked	Catchable-size trout	Subcatchable-size trout
1987	9,631,681 (35)	1,773,596	7,858,085
1988	7,872,300 (29)	1,287,503	6,584,797
1989	9,558,946 (36)	1,074,484	8,484,462
1990	8,510,770 (32)	1,037,070	7,473,700
1991	6,633,232 (23)	835,764	5,797,468
1992	11,478,864 (36)	952,458	10,526,406
Mean	8,947,632 (32)	1,160,146	7,787,486

[a] Data are from fish stocking records, Wyoming Game and Fish Department, 1987-1992.

and the first fish hatchery was commissioned (Barkwell 1883).

The first Wyoming hatchery opened in 1884. Many waters were either devoid of desirable sport fish (the Platte River drainage had no salmonids) or the stocks had been extirpated by dams without fishways, unscreened water diversions, the use of explosives or poisons, or overfishing. Interest in fish culture and fish stocking also was stimulated by increased fishing-based tourism (Barkwell 1883). There were 4 hatcheries by 1900, 9 by 1950, and 12 today. About 9 million trout (32 per licensed angler) were stocked yearly from 1987 through 1992 (Table 1). They were produced from 19 broodstocks of 9 trout species or subspecies: brook trout *Salvelinus fontinalis*, brown trout *Salmo trutta*, cutthroat trout *Oncorhynchus clarki* (four subspecies), golden trout *O. aguabonita*, lake trout *Salvelinus namaycush*, and rainbow trout *O. mykiss*.

The course for Wyoming fish culture and fish management in the early 1900s was set by Louis Miller and Gustave Schnitger, the third and fourth State Fish Commissioners. They believed good fishing and fish stocking were obligate partners, and annual stocking, like the annual reseeding of agricultural crops, was essential (Miller 1890). Schnitger (1896) estimated that there were 259,000 ha of water in Wyoming that would sustain 1,000 trout per hectare. If each trout weighed 0.45 kg, Schnitger thought Wyoming waters could sustain 720 million kilograms of trout. Trout were stocked as though productive capacity was virtually unlimited, and by 1918 it was estimated that all waters in the state had been stocked (Lenihan 1918).

Reports of success from stocking salmonids were common. Rainbow trout stocked in the Laramie River (without trout before 1885) produced 240,000 eggs yearly for the state fish hatchery in the late-nineteenth century (Miller 1890). Native cutthroat trout were moved east of the Continental Divide to establish new fisheries (Barkwell 1883). In the 1930s, people believed that the numerous rearing ponds located around the state were producing many 15-cm trout, 90% of which survived to stocking (Cook 1936). The ponds were operated cooperatively with many Wyoming angling groups, but were abandoned in 1950 because actual production was much less than anticipated (Greene 1950). There were no reliable records of angler catch in waters that were stocked, only the notion that angling was very good. By 1940, although fish stocking was believed beneficial, much of it was considered unnecessary (Simon 1940).

Simon (1940) was concerned about native trout; he deplored promiscuous fish stocking and identified it as one cause for extirpation of native trout from some Wyoming waters. In 1940, he proposed a systematic, statewide fisheries management plan for Wyoming that had management activities based on watershed characteristics, native fish stocks, and need for stocked fish. It specified that no fish would be stocked in virgin waters (those with no fish) before those waters were surveyed, only native trout would be stocked into high lakes, no trout would be stocked in places where they were not the predominant species, and no trout would be stocked when there was uncertainty about which species to stock.

The statewide fisheries management plan also provided a salmonid stocking protocol (Simon 1940). Golden trout were to be stocked in highest elevation lakes, brook trout in the next lower elevation waters, and cutthroat trout, rainbow trout, and brown trout would be stocked at progressively lower elevations. The plan further specified that Arctic grayling *Thymallus arcticus* would be stocked with no other species, lake trout with no other game fish, brown trout with only forage fish, and golden trout alone. No rainbow trout were to be stocked with native cutthroat trout. The protocol was far-sighted considering it predated fisheries management crews by 10 years.

Simon's (1940) fisheries management protocol was not followed until 1950, when the Dingell–Johnson Act (16 U.S.C.A. §§ 777 to 777K) was passed and sport-fish restoration funds were available to hire fishery management crews. By that time, anglers expected fish stocking, water storage projects were becoming common, and natural habitat quality was being compromised.

The recently hired biologists began watershed surveys to measure species' distribution and abundance to deduce the need for fish stocking. They slowly realized that distinct stocks of trout adapted to specific drainages required careful study before being transplanted elsewhere or supplemented with hatchery trout. Transplants worked best when receiving waters were most like those of home drainages. If trout were simply moved around, failure was common (Wiley et al. 1993). Slowly, fisheries management plans for watersheds (including lakes and reservoirs) developed, and native fishes received special attention (Binns 1977, 1978, 1981; Kiefling 1978). However, stocking nonnative species with endemic cutthroat trout still continued until the mid-1980s. In 1993, a survey for native, nonsport fish was initiated, and the Wyoming Game and Fish Department designated all streams for wild trout management. Where stream stocking continues, fisheries managers must justify the need. Meanwhile, angler support for stocking hatchery trout, including catchable-size fish, continues because people expect fish to be stocked (Wiley 1989).

Return of Hatchery Trout to Anglers

Miller (1890), Schnitger (1892), and Cook (1936) thought 80–90% of the trout stocked would survive and return to anglers. Analysis of available data for 1953–1989 showed 80–90% return rates (percent of number stocked that were caught) were rare (Wiley et al. 1993). Return rates were usually greater for catchable-size trout ($\geq$21 cm) than they were for subcatchable-size trout (<21 cm). Return rate of catchable-size trout stocked in spring, the time of greatest angling effort, was highest, and there was virtually no carryover to the next angling season (Wiley et al. 1993). Returns of catchable-size trout can exceed 90% where angling effort is high (Wiley et al. 1993).

Best returns of subcatchable-size trout stocked into streams occurred if fish were stocked in spring, hatchery and receiving waters were of similar quality, water temperature and flows were not limiting, and few competing fish were present. Return rates of subcatchable-size fish were greater for lakes than for streams (Wiley et al. 1993). Return rates are best for lakes and reservoirs in which water is productive, number of competing planktivores or piscivores is low, and water level fluctuates little (Chamberlain 1993).

Hatchery Broodstocks

Maintaining broodstocks of trout (native and nonnative) and raising them in hatcheries has benefited management. Annual production of about 2.4 million native cutthroat trout has enabled reintroduction of these fish to parts of their original range and has made them available for other suitable waters.

Genetic analysis of 13 of the 19 Wyoming broodstocks provided information on the best uses of progeny in five fisheries management categories; supplementing natural recruitment, restoring native populations, establishing nonnative stocks, maintaining put-grow-and-take fisheries, and maintaining catchable-size trout fisheries (Alexander 1993). Knowing the genetic composition of broodstocks allows better judgements about the best use of progeny from the broodstocks and helps safeguard native and wild stocks.

Productive Capacity

It is axiomatic that native fishes and most sport fishes, both native and nonnative, thrive in undisturbed habitats or habitats that most closely resemble those in which they evolved (Li and Moyle 1993). Platts and McHenry (1988) concluded that trout occupy a habitat to its potential carrying capacity, at least in the absence of human intervention (including habitat perturbations and intensive harvest). They found that in most undisturbed, lightly fished streams in seven western ecoregions, standing stocks of trout were 67 kg/ha or less (56–96% of observations). About 55% of the measured trout stocks in Wyoming streams were 68 kg/ha or less, and almost 80% were 112 kg/ha or less (Table 2).

Many streams in Wyoming and other western states have been changed by logging, timber transport, gold dredging, water storage or diversion, livestock use, and other anthropogenic activities (Sedell et al. 1991; Platts 1991; Wesche 1993). The changes reduce the complexity of natural stream systems (Griffith 1993). Salmonid populations in streams usually decline in response to reduced habitat complexity (Hicks et al. 1991). Despite the reduced productive capacity of these altered salmon streams, managers all too often look to stocking salmonids to offset the loss of habitat. They forget about the loss of productive capacity, are perplexed by the low return, and may further respond by stocking still more fish.

Angler Attitudes

In 1975, 1980, and 1988, surveys of the attitudes of anglers and other interested people revealed interests changing from a utilitarian (harvest-oriented) to a nonconsumptive use (catch and release or just observing fish) (Anderson et al. 1990). In 1988, people were asked to participate in fisheries man-

TABLE 2.—Standing stocks of salmonids in Wyoming streams[a] and streams in seven western ecoregions[b] (RF is Rocky Mountain Forest; SF is Sierra Forest; PF is Pacific Forest; CF is Columbia Forest; IS is Intermountain Sage; CP is Colorado Plateau; GM is Gila Mountain). Number of streams surveyed in Wyoming and seven ecoregions is included in parentheses.

Standing stock range (kg/ha)	Percent of streams							
	Wyoming (1,037)	RF (62)	SF (70)	PF (53)	CF (42)	IS (22)	CP (15)	GM (9)
0–22	21.8	22.6	20.0	60.4	54.8	50.0	6.7	11.1
23–45	20.3	21.0	20.0	26.4	14.3	13.6	33.3	33.3
46–68	13.1	19.4	18.6	9.4	16.6	13.6	20.0	11.1
69–90	11.1	8.1	11.4	1.9	4.7	13.6	13.3	11.1
91–112	7.3	9.7	7.1	1.9	2.4		26.7	22.2
113–134	4.9	6.4	10.0		2.4	9.1		
135–157	3.9	6.4	1.4		2.4			
158–179	2.9	1.6						
180–202	3.9	1.6	4.3					
203–224	1.4		1.4					
225–246	1.9	1.6				2.4		11.1
247–269	0.9		1.4					
270–291	1.1		2.9					
292–314	1.1							
315–336	1.0							
337–359	0.5							
≥360	2.9	1.6	1.4					

[a] Data from Wyoming Game and Fish Department files.
[b] Data from Platts and McHenry (1988).

agement decisions by expressing their preferences. Fisheries managers listened, and implemented management options that were sociologically desirable, economically feasible, and biologically reasonable.

Stocking trout can create demand for continued stocking (Hobbs 1948; Clawson 1963), even though changes in habitat conditions may suggest reduced or no stocking. For example, when zooplankton was depressed in Flaming Gorge Reservoir by competing game and nongame fish, and the number of large, piscivorous trout increased as the reservoir aged, the return of stocked trout (75–125 mm) declined. In 1990, fishery managers thought that stocking larger trout (200 mm) for similar return rates (30% of number stocked) might not be economical and did not meet management objectives; however, they continued stocking because the public expected it. I suggest that fisheries agencies created the expectation by stocking fish.

Regulations

The primary purpose of regulations in sport-fishery management has been to govern fishing effort, limit fish harvest, and maximize enjoyment among as many anglers as possible (Noble and Jones 1993; Griffith 1993). Varley (1984) pointed out that the amount of fishing effort required to remove about half the legal-size trout stock varied among species in Yellowstone Park. He estimated that about 35 angling hours per hectare per year would remove half the legal-size cutthroat trout, but 285 hours were necessary for rainbow trout, and over 400 hours were necessary for brown trout. Therefore, need for harvest regulation varies among species.

In Wyoming, bag limits for fish have become more stringent since 1899 (9.1 kg/d); the 1994 general regulation is six fish in possession, only one of which may be over 50 cm. With the perceived decline in fishing success (relative to expectations), increased human population, increased economic development, construction of reservoirs, depressed habitat, and so on, the number of fish hatcheries also increased after 1900.

Goodman (1990) expressed concern that unregulated fish stocking would depress genetic diversity of salmonid stocks and recommended federal regulation of hatchery practices, partly because of low returns from stocking fish. Goodman's plea for hatchery regulation makes some sense. But, federal intervention would be intrusive to every state conservation agency. The states can implement watershed-based fisheries and hatchery management practices that promote the genetic conservation which Goodman (1990) suggested.

Interstate cooperative management, at least in the West, is not without precedent. For example, the Colorado River Fish and Wildlife Council recommended fish health management guidelines and inspection protocols for salmonids produced or imported into the Colorado River basin. In 1973 the seven basin states ratified the guidelines and protocols and have cooperated on all fish health issues since. In addition, stocking nonnative fish in the drainage required the approval of all seven states. The states recognized that fish do not respect geographic boundaries and that native fish in the system could be protected only by managing fisheries based on a river basin focus.

Cooperation among states in the Rocky Mountains on fish stocking might mean following similar stocking protocols. States have fish-stocking guidelines, but they vary widely. Most states recommend acceptable return rates for stocked trout, usually 50% or greater return of number of catchable-size trout stocked or 100% or greater return of weight of subcatchable-size trout stocked.

Common Sense Protocol for Modern Fish Management

All management decisions are value judgements requiring choices among various and sometimes opposing options. In trout management, there are three general options: complete dependence on wild fish, judicious use of hatchery and wild fish, or virtually complete dependence on hatchery trout. The first two are preferred even though there are difficult social and biological choices in balancing the wild (natural) with the hatchery (artificial) product.

Choices between what can be stocked safely and cost effectively are not solely dependent on responding to public desires. Stocking can generate demand (Hobbs 1948; Behnke 1989), which results in increased and unnecessary dependence on hatchery trout, because people learn to expect trout stocking. Successful management programs address public preferences as well as the biology of the fish so that angler expectations are at least partly met (Wiley 1989) by foresighted fishery management programs (McFadden 1969).

The following protocol is recommended to meet the needs of anglers, nonanglers, and fish through fish management and fish culture. The ideas are not new. Simon (1940) suggested some, but there were no fisheries managers to implement them.

1. Thoroughly inventory biological and habitat characteristics of watersheds because streams, lakes, and their fish stocks are dependent on the integrity of habitats in watersheds. Solve habitat problems before stocking; hatchery fish survive no better than do wild fish in limited habitat.
2. Understand the genetics and the strengths and weaknesses of broodstocks in hatcheries. Stock progeny best suited to the receiving environments.
3. Manage for native fish or wild fish wherever possible; they are best suited to natural habitat because it sustains them without stocking. Also, conservation agencies must understand that their fishery resource stewardship includes native nongame fish as copartners with game species.
4. Manage according to the fish production limits of waters. Stocking fish can meet natural production limits but cannot raise those limits. Continually increasing fish stocking is not the solution for increased angling, limited natural production, nor the larger problem of degraded habitat.
5. Stock fish in standing waters in which salmonid reproduction is often limited or absent.
6. Develop fisheries management plans by drainage basin, identifying limiting factors and recommending specific management strategies to fit drainage conditions.
7. Establish fish management programs based on recognition of all public desires for fisheries, not just the utilitarian aspect. Fisheries management agencies are responsible for public perception of the resource because management actions largely create that perception (Clawson 1963). The public participates in fisheries management by expressing its desires, and conservation agencies are obligated to listen sincerely and to work to satisfy those requests which are biologically sound, economically feasible, and sociologically desirable.

I suggest that adoption of this or similar fisheries management protocols among states in a river basin, or larger area, would assuage much of the controversy about fish stocking practices and sooth concerns about conserving the genetic diversity of trout. The specter of federal intrusion into state fishery management prerogatives would also be reduced by such voluntary cooperation among states.

References

Alexander, C. B. 1993. History and genetic variation of salmonid broodstocks, Wyoming. Master's thesis. University of Wyoming, Laramie.

Anderson, D., C. Phillips, and T. Krehbiel. 1990. Wyoming angler attitudes and preferences; application of strategic choice modeling. University of Wyoming, Laramie.

Barkwell, M. C. 1883. Report of the Board of Fish Commissioners for the two years ending December 31, 1883. Wyoming Game and Fish Department, Cheyenne.

Beckstrom, K. 1991. The right fish. North American Fisherman, February–March 1991.

Behnke, R. J. 1989. Summary of progress in wild trout management: 1974–1989. Pages 12–17 *in* F. Richardson and R. Hamre, editors. Wild trout IV. Trout Unlimited, Vienna, Virginia.

Behnke, R. J. 1991. From hatcheries to habitat, look again. Trout (Autumn) 1991:55–58.

Binns, N. A. 1977. Present status of indigenous populations of cutthroat trout, *Salmo clarki*, in southwest Wyoming. Wyoming Game and Fish Department Fisheries Technical Bulletin 2, Cheyenne.

Binns, N. A. 1978. Habitat structure of Kendall Warm Springs with reference to the endangered Kendall Warm Springs dace, *Rhinichthys osculus thermalis*. Wyoming Game and Fish Department Fisheries Technical Bulletin 4, Cheyenne.

Binns, N. A. 1981. Bonneville cutthroat trout, *Salmo clarki utah*, in Wyoming. Wyoming Game and Fish Department Fisheries Technical Bulletin 5, Cheyenne.

Bourke, J. G. 1891. On the border with Crook. Charles Scribners' Sons, New York.

Chamberlain, C. B. 1993. Evaluation criteria for trout in the lakes and reservoirs of Wyoming. Master's thesis. University of Wyoming, Laramie.

Clawson, M. 1963. Land and water for recreation. Rand McNally and Company, Chicago.

Cook, F. 1936. Report of the division of fish culture. Pages 15–19 *in* A. J. Martin, Biennial report of the State Game and Fish Commissioner of Wyoming, 1935–1936. Wyoming Game and Fish Department, Cheyenne.

Goodman, M. L. 1990. Preserving the genetic diversity of salmonid stocks: a call for federal regulation of hatchery programs. Environmental Law 20(83):111–167.

Greene, A. F. C. 1950. Report of the Fish Division. Pages 30–45 *in* Annual report of the Wyoming Game and Fish Commission, 1950. Wyoming Game and Fish Department, Cheyenne.

Griffith, J. S. 1993. Coldwater streams. Pages 405–426 *in* C. Kohler and W. A. Hubert, editors. Inland fisheries management in North America. American Fisheries Society, Bethesda, Maryland.

Hedgepeth, J. 1941. Livingston Stone and fish culture in California. California Fish and Game 25(3):126–148.

Hicks, B. J., J. D. Hall, P. A. Bisson, and J. R. Sedell. 1991. Responses of salmonids to habitat changes. Pages 483–518 *in* W. R. Meehan, editor. Influences of forest and rangeland management on salmonid fishes and their habitats. American Fisheries Society, Bethesda, Maryland.

Hobbs, D. F. 1948. Trout fisheries in New Zealand, their development and management. New Zealand Marine Department Fisheries Bulletin 9. Wellington, New Zealand.

Kiefling, J. W. 1978. Studies on the ecology of the Snake River cutthroat trout. Wyoming Game and Fish Department Fisheries Technical Bulletin 3, Cheyenne.

Lenihan, J. J. 1918. Sixth annual report of the State Fish Commissioner, 1918. Wyoming State Archives, Cheyenne.

Li, H. W., and P. B. Moyle. 1993. Management of introduced fishes. Pages 287–308 *in* C. Kohler and W. A. Hubert, editors. Inland fisheries management in North America. American Fisheries Society, Bethesda, Maryland.

McFadden, J. T. 1969. Trends in freshwater sport fisheries of North America. Transactions of the American Fisheries Society 98:136–150.

Miller, L. 1890. Report of Louis Miller, State Fish Commissioner of Wyoming for the year ending September 30, 1890. Wyoming Game and Fish Department, Cheyenne.

Mullan, J. W., K. R. Williams, G. Rhodus, T. W. Hillman, and J. D. McIntyre. 1992. Production and habitat of salmonids in mid-Columbia River tributary streams. U.S. Fish and Wildlife Service Monograph 1, Fishery Assistance Office, Leavenworth, Washington.

Netboy, A. 1973. The salmon, their fight for survival. (Not seen; cited in Goodman 1990).

Noble, R. L., and T. W. Jones. 1993. Managing fisheries with regulations. Pages 383–402 *in* C. Kohler and W. A. Hubert, editors. Inland fisheries management in North America. American Fisheries Society, Bethesda, Maryland.

Platts, W. S. 1991. Livestock grazing. Pages 389–424 *in* W. R. Meehan, editor. Influences of forest and rangeland management on salmonid fishes and their habitats. American Fisheries Society, Bethesda, Maryland.

Platts, W. S., and M. L. McHenry. 1988. Density and biomass of trout and char in western streams. U.S. Forest Service General Technical Report INT-241.

Rinne, J. H., J. E. Johnson, B. L. Jensen, A. W. Regier, and R. Sorensen. 1986. The role of fish hatcheries in the management of threatened and endangered species. Pages 271–286 *in* R. Stroud, editor. Fish culture in fisheries management. American Fisheries Society, Fish Culture Section and Fisheries Management Section, Bethesda, Maryland.

Schnitger, G. 1892. Annual report of the State Fish Commissioner of Wyoming for the year 1892. Wyoming State Archives, Cheyenne.

Schnitger, G. 1896. Biennial report of the State Fish Commissioner of Wyoming for the years 1895 and 1896. Wyoming Game and Fish Department, Cheyenne.

Sedell, J. R., F. N. Leone, and W. S. Duval. 1991. Water transportation and storage of logs. Pages 325–368 *in* W. R. Meehan, editor. Influences of forest and rangeland management on salmonid fishes and their habitats. American Fisheries Society, Bethesda, Maryland.

Simon, J. R. 1940. Report of the Fish Division. Pages 31–41 *in* R. Grieve, editor. Biennial report of the Wyoming Game and Fish Commission, 1939–1940. Wyoming Game and Fish Department, Cheyenne.

Stone, L. 1872. Trout culture. Proceedings of the American Fish Culturists' Association 1:46–56.

Varley, J. D. 1984. The use of restrictive regulations in managing wild trout in Yellowstone National Park with particular reference to cutthroat trout, *Salmo clarki*. Pages 145–146 *in* J. M. Walton and D. B.

Houston, editors. Proceedings of the Olympic Wild Fish Conference, Port Angeles, Washington.

Wales, J. H. 1939. General report of investigations on the McCloud River in 1938. California Fish and Game 25(4):272–309.

Wesche, T. A. 1993. Watershed management and land use practices. Pages 181–204 *in* C. Kohler and W. A. Hubert, editors. Inland fisheries management in North America. American Fisheries Society, Bethesda, Maryland.

White, R. J. 1992. Why wild fish matter: a biologist's view. Trout (Summer) 1992:25–33, 44–50.

Wiley, R. W. 1989. Anglers, common sense, and fishery management. Pages 193–196 *in* F. Richardson and R. W. Hamre, editors. Wild Trout IV. Trout Unlimited, Vienna, Virginia.

Wiley, R. W., R. A. Whaley, J. B. Satake, and M. Fowden. 1993. Assessment of stocking hatchery trout, a Wyoming perspective. North American Journal of Fisheries Management 13:160–170.

Development of an Optimal Stocking Regime for Walleyes in East Okoboji Lake, Iowa

JOE G. LARSCHEID

Iowa Department of Natural Resources, 611 252nd Avenue
Spirit Lake, Iowa 51360, USA

Abstract.—East Okoboji Lake is part of an interconnected chain of glacial lakes along the Iowa–Minnesota border. East Okoboji Lake supports an important walleye *Stizostedion vitreum* fishery and is a source of broodstock walleye for the Spirit Lake Fish Hatchery. Natural reproduction of walleye in East Okoboji Lake is limited, and the fishery has been sustained by stocking sac-fry, small (50 mm) walleye fingerlings, and large (100–150 mm) intensively reared (reared in raceways and fed dry prepared food) and extensively reared (reared in nursery lakes and fed minnows) walleye fingerlings. The objective of this study was to determine the optimal stocking strategy to meet the goal of 25 yearling walleyes per hectare. Each type of walleye stocked into East Okoboji Lake was unique with respect to its first-year survival and production costs. These variables were used in a linear programming model to determine the optimal stocking regime necessary to meet study objectives, while remaining within the production and budgetary constraints of the Spirit Lake Fish Hatchery. Prior to stocking, all fingerlings were marked with a freeze brand. Unmarked fingerlings were assumed to have originated from sac-fry stockings. Percent contributions, densities, and survival rates of stocked walleyes were estimated from population surveys conducted with boat-mounted AC electrofishing gear. Sac-fry were the most cost-effective type of walleye to stock into East Okoboji Lake. Small walleye fingerlings were the next most cost-effective option followed by large intensively reared and then the large extensively reared walleye fingerlings. Stocked large fingerlings contributed significantly to the young-of-year walleye populations in East Okoboji Lake; however, most of these fish perished within 2–5 weeks after they were stocked. The critical size for assuring at least short-term survival of stocked large walleye fingerlings appears to be 130 mm. First-year survival of stocked large fingerlings was lower than expected (1–6%) and may be one reason why densities of walleye have not been increasing in these lakes despite increased stocking efforts. Larger fall walleye fingerlings (200 mm) need to be stocked to assure greater survival and to improve the benefit-to-cost ratio of these fingerlings.

Spirit Lake and East and West Okoboji lakes are part of an interconnected chain of glacial lakes known as the Iowa Great Lakes. The Iowa Great Lakes region is a popular vacation area that receives over one million visitors each year. Walleye *Stizostedion vitreum* is a popular game fish in these lakes, and each year 30–50% of Iowa Great Lakes anglers specifically fish for walleyes even though walleyes constitute less than 1% of the total fish harvested (Larscheid 1993). Natural reproduction of walleyes in these lakes is limited (McWilliams 1976; McWilliams and Larscheid 1992), and the fisheries have been sustained by annual stockings of sac-fry and fingerlings. Despite these stockings, population densities and harvests of walleyes in these lakes have declined. In addition to being important walleye sport fisheries, these lakes are the main source of broodstock walleyes (≥430 mm) for the Spirit Lake Hatchery. In recent years, population densities of broodstock walleyes in these lakes were not adequate to meet the demands of the Spirit Lake Hatchery and additional eggs were obtained from other natural lakes.

In response to the declining numbers of walleyes, the Iowa Department of Natural Resources initiated an investigation (McWilliams 1990; McWilliams and Larscheid 1992) to determine the causes of the decline and to develop a management plan to increase walleye densities in the Iowa Great Lakes. The initial findings of this study determined overwinter survival of stocked fingerlings was lower than that of fingerlings originating from sac-fry stockings. Also, the quality of hatchery-produced fingerlings varied greatly during this study, but this variation was not quantified. Therefore, definite explanations for the observed differential mortality were not possible.

The objective of the study was to determine the optimal stocking strategy to meet the goal of 25 yearling walleyes per hectare. This number was based on estimates of survival typical of 2–7-year-old walleyes in these lakes and in the literature (Colby et al. 1979). With typical survival, recruitment of 25 yearlings per hectare would sustain densities of at least 5 broodstock walleyes per hectare, which is the density necessary to meet the needs of the Spirit Lake Fish Hatchery (Larscheid 1991).

This fishery has been sustained by stocking four types of walleyes: sac-fry in April, small (50 mm) fingerlings in June, and large (100–150 mm) intensively reared (reared in raceways and fed dry prepared food) and extensively reared (reared in nursery lakes and fed minnows) fingerlings in September. Each type of walleye stocked was unique with respect to its first-year survival in East Okoboji Lake and its initial production costs. These variables were used in a linear programming model to determine the optimal stocking regime—one that meets study objectives while remaining within the production and budgetary constraints of the Spirit Lake Hatchery. Factors affecting the initial poststocking mortality of large walleye fingerlings stocked into East Okoboji Lake were also examined.

Methods

Study Site

East Okoboji Lake is located in Dickinson County, Iowa, and is part of an interconnected chain of glacial lakes along the Iowa–Minnesota border (Figure 1). East Okoboji Lake is a long, narrow, shallow (average depth = 3 m), eutrophic lake that usually has extensive algae blooms late in the summer. The lake has about 742 surface hectares and a maximum depth of 7 m. East Okoboji Lake has extensive muck flats (particularly in the northern portions of the lake), few rocky reefs, and is connected to West Okoboji Lake by a narrow channel in the southern portion of the lake. Dense stands of curlyleaf pondweed *Potamogeton crispus*, sago pondweed *Potamogeton pectinatus*, northern watermilfoil *Myriophyllum exalbescens*, and coontail *Ceratophyllum demersum* are present throughout the lake. Currently there are at least 38 species of fish in the lake with excellent fisheries for walleye, smallmouth bass *Micropterus dolomieu*, largemouth bass *Micropterus salmoides*, northern pike *Esox lucius*, muskellunge *Esox masquinongy*, channel catfish *Ictalurus punctatus*, black bullhead *Ictalurus melas*, bluegill *Lepomis machrochirus*, and yellow perch *Perca flavescens* (Larscheid 1993).

Walleye Production

Sac-fry were produced at the Spirit Lake Hatchery from adult walleyes collected from Spirit Lake and East and West Okoboji lakes (Jorgensen 1993). Small (50 mm) walleye fingerlings were extensively reared at Welch and Sunken lakes (Figure 1). Large fingerlings (100–150 mm) were intensively reared in cement raceways at the Spirit Lake Hatchery and

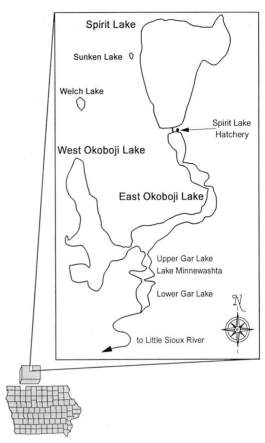

FIGURE 1.—The study area comprised East Okoboji Lake, the Spirit Lake Fish Hatchery, and the Welch Lake and Sunken Lake walleye nurseries. The location of the study area in relation to Iowa is shown in the left corner.

extensively reared at the Welch and Sunken walleye nursery lakes (Jorgensen 1993). Intensively reared walleyes were raised on a dry-pellet feed; extensively reared walleyes were fed minnows (primarily *Pimephales promelas*). Sac-fry were stocked at 8,800–13,500/ha in East Okoboji Lake (Table 1). Small extensively reared fingerlings were stocked at 60–68/ha, and large fingerlings were stocked at 9–46/ha. Sunken Lake was the primary source of the small extensively reared fingerlings and Welch Lake was the primary source of the large extensively reared fingerlings stocked into East Okoboji Lake.

Production costs included all expenditures related to the production and distribution of walleyes including labor, chemicals, fertilizer, food, station and equipment maintenance relating to walleye culture, and vehicle expense (depreciation, overhead, insurance, and maintenance). Not included in the

TABLE 1.—The number, average length, and date walleye were stocked into East Okoboji Lake, 1991–1993.

Year	Type of walleye	Mean length (mm)	Date(s) stocked	Number stocked	Density stocked (#/ha)
1991	Sac-fry	8.7	Apr 25–May 1	10,000,000	13,466
	Small extensive	66	Jun 25–27	44,224	60
	Large intensive	132.1	Sept 18–19	10,003	13
	Large extensive	124.2	Sept 23–25	21,121	27
1992	Sac-fry	9.3	May 5–8	6,500,000	8,753
	Small extensive	55.9	Jun 19–24	50,555	68
	Large intensive	132.1	Oct 2	10,000	13
	Large extensive	91.4	Oct 9	11,915	16
1993	Sac-fry	9.2	May 10–11	6,500,000	8,753
	Small extensive[a]	61.0	Jun 28–Jul 2	47,191	64
	Large intensive	129.8	Sept 16–22	6,526	9
	Large extensive[b]	124.4	Sept 29–30	33,920	46

[a] 43% of these fish were not marked.
[b] 58% of these fish were not marked.

production costs are depreciation on the station, capital improvements, and those expenditures related to maintaining a district office or public information and education programs.

Freeze Branding

All fingerlings were marked with a freeze brand (LaJeone and Bergerhouse 1991) to differentiate type of fingerling stocked (small extensively reared, large intensively reared, and large extensively reared), and year stocked. Small fingerlings were held on the brand for 0.5–1 s, and large fingerlings were held on the brand for 2–4 s. Short- (0–10 d) and long-term (120 d) mortality due to branding was assessed by comparing the observed mortality of branded and unbranded walleyes reared at the Spirit Lake Hatchery. Unbranded fingerlings were used as the control for this experiment and were brought into the hatchery the same day as the branded walleyes. This mortality figure was used to adjust the actual stocking rates of the walleye fingerlings. A linear multinomial analysis of variance (ANOVA) model was used to test for significant differences between the observed mortality rates and to compute the power $(1-\beta)$ of these tests (Woodward et al. 1990). A minimal trivial difference (Woodward et al. 1990) of 5% was used and all power testing was conducted at the $\beta = 0.10$ level. Short- (10 d) and long-term (up to 357 d) brand retention was assessed by holding branded walleyes in raceways and ponds at the Spirit Lake Hatchery.

Experienced workers were able to effectively brand between 1,000 and 1,200 small walleye fingerlings per hour and between 500 and 800 large walleye fingerlings per hour. All fish were branded indoors at the Spirit Lake Hatchery immediately after seining. Maximum water temperatures varied from 27°C (while branding the small fingerlings) to 12°C (while branding the large fingerlings).

Sampling Fingerling and Yearling Walleyes

In October, young-of-year and yearling walleyes were sampled at night with boat-mounted AC electrofishing gear. An attempt was made to sample the entire shoreline each night, but this was not always possible. All walleyes captured were examined for brands, measured to the nearest tenth of an inch (2.5 mm), weighed to the nearest hundredth of a pound (4.5 g), marked with a caudal fin punch, and released near the point of capture. A Schumacher–Eschmeyer multiple mark–recapture model was used to estimate the densities of yearling and young-of-year walleyes both in 1992 and 1993; population estimates were not conducted in 1991. Sampling was continued until the 95% confidence intervals (CI) converged to yield stable population estimates. The underlying assumptions of this model were examined by plotting the cumulative number of marked fish in the population versus the observed proportion of marked fish in each sample. Although not conclusive, a linear relationship was evidence that the underlying assumptions (i.e., closed population, equal catchability) were satisfied (Krebs 1989).

Percent contributions of stocked walleyes to year-class abundance were estimated from the proportions of each group of walleyes observed in the electrofishing samples. All unmarked fish were assumed to have originated from stocked sac-fry. McWilliams and Larscheid (1992) found no evidence for natural reproduction of walleyes during 1984–1989 in East and West Okoboji lakes. The percent contribution estimates were biased due to lost marks. Nearly unbiased contribution estimates were obtained by applying an error matrix correction procedure (Cook and Lord 1978). The error matrix was composed of data from the brand-retention experiments (short- and long-term retention). The nearly unbiased contribution estimates were constrained to be between 0 and 100 (Cook 1983). The variance and 95% CI of these estimates were calculated using the method of Pella and Robertson (1979); Cook (1983) found this method provided statistically valid, conservative confidence intervals. The population estimates were divided into the various origins of walleye (i.e., sac-fry stockings, small fingerlings, large intensively and extensively reared fingerlings) by multiplying these population estimates by the nearly unbiased percent contribution estimates. These estimates were divided by the number of walleye fry or fingerlings originally stocked to derive estimates of poststocking survival.

Growth of walleye fingerlings as a function of temperature was predicted by

$$G = 1.98 + 0.117X - 0.0218X^2 - 0.0017X^3;$$

G = specific growth rate (% change in weight per day);
X = water temperature − 23°C (Hokanson and Koenst 1986).

The changes in weight were converted to length using tables in Piper et al. (1986).

Assessment of Health and Condition of Fingerlings

The Health Condition Profile (HCP) was used to assess the health and condition of walleye fingerlings (Goede and Barton 1990). The HCP of stocked large fingerlings was assessed immediately prior to stocking, whereas the HCP of fingerlings originating from the sac-fry plants and from the small extensively reared fingerlings were assessed the day after they were collected from electrofishing samples. Standard errors and 95% confidence limits were calculated for each parameter of the HCP. The following equations of the cumulative binomial distribution were used to solve iteratively for the exact lower (P_L) and the exact upper (P_U) 95% confidence limits for each proportion of the HCP.

$$\sum_{j=0}^{X} \left(\frac{N!}{j!(N-j)!}\right)(P_U)^j(1-P_U)^{N-j} = 0.025,$$

$$\sum_{j=X}^{N} \left(\frac{N!}{j!(N-j)!}\right)(P_L)^j(1-P_L)^{N-j} = 0.025;$$

X = number of normal fish,
N = sample size,
$j = 0 .. N$.

This procedure is recommended for small sample sizes (Johnston and Kotz 1969) over more conventional approximation methods (e.g., Gustafson 1988).

Relative weight (Anderson 1980) was calculated using the standard weight equation for walleyes proposed by Murphy et al. (1990). Since each fish species varies in the way it stores mesenteric fat (Goede and Barton 1990), we developed separate criteria for assessing mesenteric fat deposition in walleyes. Fat deposition was characterized as the relative coverage of mesenteric fat on the pyloric ceca where: (0) = no fat present, (1) = 0–25% coverage, (2) = 25–50% coverage, (3) = 50–75% coverage, (4) = 75–100% coverage, and (5) = 100% coverage.

Benefit–Cost Analysis of Stocking

The following linear programming model (Goldstein and Schneider 1984) was used to analyze the cost differential of the various types of walleye stocked into these lakes and to determine the optimal stocking strategy to reach study objectives.

$$N = S_1X + S_2Y + S_3Z_1 + S_4Z_2;$$

N = number of yearling walleye;
X = number of sac-fry to stock;
Y = Number of small fingerlings to stock;
Z_1 = number of intensively reared to stock;
Z_2 = number of extensively reared to stock;
S_1 = 18-month survival of sac-fry plants;
S_2 = 16-month survival of small fingerlings;
S_3 = 12-month survival of intensively reared;
S_4 = 12-month survival of extensively reared.

This model was used to solve for N with the following constraints:

Space and time (how many walleyes can be physically produced, allocated, and stocked into East Okoboji Lake);

$$X \leq 10,000,000,$$
$$Y \leq 100,000,$$
$$Z_1 \leq 20,000,$$
$$Z_2 \leq 40,000.$$

TABLE 2.—The estimated contribution of stocked sac-fry, small extensively reared fingerlings, and large intensively and extensively reared fingerlings to the fall young-of-year walleye population in East Okoboji Lake, 1991–1993.

Year	Type of walleye	N	Nearly unbiased contribution (%)	SD (%)	CV (%)	Simultaneous 95% CI
1991	Sac-fry	409	50.29	2.48	4.93	44.09–56.48
	Small extensive	56	10.38	1.45	13.96	6.75–14.00
	Large intensive	132	19.74	1.77	8.95	15.32–24.16
	Large extensive	131	19.59	1.76	8.99	15.19–24.00
	Totals	728	100.00			
1992	Sac-fry	846	59.06	1.36	2.31	55.26–62.85
	Small extensive	164	12.09	0.90	7.43	9.59–14.60
	Large intensive	201	14.40	0.95	6.58	11.76–17.05
	Large extensive	180	13.27	0.94	7.06	10.66–15.88
	Totals	1,391	98.82[a]			
1993	Sac-fry	939	92.49	1.08	1.17	89.79–95.19
	Small extensive	22	4.04	0.87	19.00	1.87–6.20
	Large intensive	27	2.75	0.52	21.47	1.44–4.05
	Large extensive	3	0.73	0.42	57.89	0.00–1.78
	Totals	991	100.00			

[a] An incidental stocking of pond-reared walleye fingerlings accounts for 1.18% of the total.

Production costs (assuming an annual stocking budget of US$20,000):

$$XC_1 + YC_2 + Z_1C_3 + Z_2C_4 \leq \$20,000;$$

C_1 = Costs to produce and stock sac-fry;
C_2 = Costs to produce and stock small walleye fingerlings;
C_3 = Costs to produce and stock intensively reared walleye fingerlings;
C_4 = Costs to produce and stock extensively reared walleye fingerlings.

Unless otherwise noted, all significance testing was conducted at the $\alpha = 0.05$ level and all power testing was conducted at the $\beta = 0.10$ level.

Results

Freeze Branding

Each year, over a 10-d period, between 11 and 15% of the small walleye fingerlings died due to the branding process; however, there was no mortality of large fingerlings attributed to branding ($\beta < 0.001$). Brand retention of the small fingerlings was poor (74%) the first year, but improved (95%) in subsequent years due to better branding techniques. Nearly 100% of the large fingerlings retained their brands (up to 357 d).

Contribution of Stocked Walleyes to Young-of-Year Abundance

Each year, 50–92% of the total young of year collected in the fall originated from the sac-fry stockings (Table 2). Stocked 50-mm walleye fingerlings contributed substantially to the young-of-year walleye population in 1991 and 1992, but not in 1993 (Table 2). During 1991 and 1992, stocked large fingerlings contributed 28–39% of the young-of-year walleye population in East Okoboji Lake, but this contribution decreased to only 3% in 1993.

The mark–recapture plots (cumulative number of marked walleyes versus the proportion of marked walleyes in each sample) in 1992 ($r^2 = 0.8900$) and 1993 ($r^2 = 0.7443$) were significant and linear, which was evidence that the underlying assumptions of the mark–recapture models were satisfied. In 1992, 1,153 young-of-year walleyes were caught (28 were recaptures), and 1,023 were caught in 1993 (36 were recaptures). The 1992 population estimate was 19,365 (95% CI 15,466–25,893); the 1993 estimate was only 11,242 (95% CI 6,914–30,048). The 95% confidence intervals were asymmetric and relatively narrow around the 1992 estimate (20–34% of the estimate) but were wide for the 1993 estimate (38–167% of the mean). The 1992 estimate was nearly twice as large as the 1993 estimate; however, a conclusive statement about this difference was not possible due to the wide intervals around the 1993 estimate. These population estimates corresponded to about 26 young-of-year fingerlings per hectare in 1992, and 15 young-of-year fingerlings per hectare in 1993 (Table 3). The densities of young-of-year walleyes originating from the sac-fry plants were very similar in 1992 and 1993; however, the densities of young-of-year walleyes originating from the

TABLE 3.—Densities (number per hectare) and survival (%) of young-of-year walleyes stocked into East Okoboji Lake, 1992–1993. Values in parentheses are 95% confidence intervals.

Type of walleye	Density	Poststocking survival	Number of days at large
1992 Year-Class			
Sac-fry	16.41 (12.38–23.12)	0.19 (0.14–0.26)	171–174
Small extensive	5.38 (3.69–8.21)	9.02 (6.19–13.76)	124–129
Large intensive	3.98 (2.66–6.20)	29.57 (19.75–46.01)	10–24
Large extensive[a]			0–12
Totals[b]	25.77 (18.73–37.53)		
1993 Year-Class			
Sac-fry	14.00 (8.36–38.52)	0.16 (0.10–0.44)	196–197
Small extensive	0.61 (0.17–2.51)	1.03 (0.29–4.22)	117–121
Large intensive	0.41 (0.13–1.64)	4.94 (1.59–19.45)	35–41
Large extensive	0.11 (0.00–0.72)	0.24 (0.00–1.58)	27–28
Totals	15.13 (8.66–43.39)		

[a] The stocked extensively reared walleye fingerlings were not included in the multiple mark–recapture experiment because 80% of these fish were stocked on 14 October 1992 (2 d after sampling began).
[b] These totals do not include a small incidental stocking of pond-reared walleye fingerlings.

fingerling stockings were eight times greater in 1992 than in 1993.

The survival of the sac-fry plants were nearly identical in 1992 and 1993 (Table 3). Assuming no natural reproduction, about 0.19% of the sac-fry that were stocked in May 1992 survived to October 1992, and about 0.16% of the sac-fry that were stocked in May 1993 survived to October 1993. The survival of stocked walleye fingerlings, however, was much greater in 1992 than in 1993. The survival of stocked small fingerlings was nine times greater in 1992 than in 1993 (Table 3).

Most of the large fingerlings that were stocked into East Okoboji Lake perished within 2–5 weeks of stocking (Table 3). In 1992, nearly 70% of the large intensively reared fingerlings perished within 24 d after stocking. The initial poststocking mortality was even more severe in 1993 when nearly all (95–99%) of the large fingerlings perished within 41 d after stocking (Table 3).

Health, Condition, and Growth of Fall Young-of-Year Walleyes

Overall, all the walleye fingerlings were similar in terms of health and condition (Table 4). However, deposition of mesenteric fat was significantly higher for intensively reared fingerlings than for any other group of walleyes. Bile and relative weight varied each year and no consistent trends were evident (Table 4).

Each year, the average length of the large fingerlings stocked into East Okoboji lake was significantly smaller than the fingerlings originating from the stocked sac-fry or from the small fingerlings stocked in June (Table 5). Although significantly different, the difference in length of fingerlings orig-

TABLE 4.—Mean health and condition indices of walleye fingerlings originating from stocked sac-fry, small fingerlings, and the large intensively and extensively reared fingerlings. Values in the same column and year, with the same letter were not significantly different.

Year	Origin of fall fingerlings	Relative weight (%)	Bile rating	Mesenteric fat rating	Tissue and organ normality (%)
1991	Sac-fry	NA	0.75 z	0.30 z	97 z
	Small extensive	NA	1.05 z	0.67 z	100 z
	Large intensive	NA	0.33 y	1.85 y	98 z
	Large extensive	NA	1.58 z	0.50 z	95 z
1992	Sac-fry	88 z	0.65 z	1.45 z	97 z
	Small extensive	85 z	0.81 z	1.67 z	98 z
	Large intensive	84 z	0.85 z	3.50 y	99 z
	Large extensive	98 y	0.70 z	1.30 z	98 z
1993	Sac-fry[a]	86 z	0.35 zy	1.90 z	99 z
	Small extensive	NA	NA	NA	NA
	Large intensive	92 y	0.25 y	3.40 y	96 z
	Large extensive	82 x	0.70 z	1.60 z	98 z

[a] About 2% of these fish were unmarked large extensively reared walleye.

TABLE 5.—Total lengths in October of walleye fingerlings originating from sac-fry stocked in April, small fingerlings stocked in June, and large intensively and extensively reared walleye fingerlings stocked each fall into East Okoboji Lake, 1991–1993. Values in the same column and year with the same letter were not significantly different.

Year	Type of walleye	N	Mean length (mm)	95% CI	%
1991	Sac-fry	429	157z	155–158	1
	Small extensive	56	149y	146–153	6
	Large intensive	355	133x	131–135	36
	Large extensive	314	124w	123–126	64
1992	Sac-fry	846	159z	158–160	1
	Small extensive	264	161y	160–163	0
	Large intensive	396	131x	129–132	45
	Large extensive	150	92w	89–94	96
1993	Sac-fry[a]	939	144z	143–145	9
	Small extensive	22	152y	147–157	0
	Large intensive	273	132x	130–133	31
	Large extensive	225	124w	123–125	62

[a] About 2% of these fish were unmarked large extensive walleyes.

inating from the stocked sac-fry or from the small fingerlings stocked in June was small (Table 5). Fewer than 10% of fingerlings that originated from the sac-fry plants or from the small fingerlings stocked in June were less than 130 mm; whereas, 31–96% of the large intensively and extensively reared fingerlings were less than 130 mm in length (Table 5).

In 1991 and 1992, the average length of intensively reared fingerlings recaptured in East Okoboji Lake were significantly greater than the average length of these fish immediately prior to stocking (Figure 2). Prior to stocking, 36–45% of the intensively reared walleye fingerlings were less than 130 mm in length. However, when these fish were recaptured in East Okoboji Lake 2–3 weeks later, only 2–6% were less than 130 mm in length, which is an 87–94% reduction in the number of these small walleyes (Figure 2). The same trend was noted for extensively reared walleye fingerlings in 1991. Prior to stocking 64% of these fish were less than 130 mm in length. When these fish were recaptured 2–3 weeks later, only 33% of the fish were less than 130 mm in length, a 48% reduction in the number of these small fingerlings (Figure 2). In 1992, there was not a significant shift in the length-frequency distribution of extensively reared walleye fingerlings after these fish were stocked into East Okoboji Lake (Figure 2). However, 80% of these fish were stocked 2 d after sampling began, so there was not sufficient time for any shifts to occur. The sample sizes in 1993 were not adequate to evaluate these shifts in the length-frequency distributions (Table 2).

Contribution of Stocked Walleyes to Yearling Abundance

Sac-fry contributed significantly to the yearling walleye densities with over 50% of the yearlings originating from stocked sac-fry (Table 6). Stocked 50-mm fingerlings contributed 13–18% of the yearling walleye population, and stocked large fingerlings contributed between 15 and 38% of the yearling walleye population. The contribution of sac-fry and the small walleye fingerlings to the 1991 and 1992 year-classes was not significantly different; however, the contribution of the large fingerlings was significantly different between years (Table 6).

The mark–recapture plots (cumulative number of marked walleye versus the proportion of marked walleye in each sample) in 1992 ($r^2 = 0.8974$), and 1993 ($r^2 = 0.9101$) were significant and linear, which was evidence that the underlying assumptions of the mark–recapture models were satisfied. In 1992, 574 yearling walleyes were caught (40 were

TABLE 6.—The estimated percent contribution of stocked sac-fry, small extensively reared fingerlings, and large intensively and extensively reared fingerlings to the fall yearling walleye population in East Okoboji Lake, 1992–1993.

Year-class	Type of walleye	N	Nearly unbiased contribution	SD	CV	Simultaneous 95% CI
1991	Sac-fry	382	48.91	2.57	5.25	42.49–55.33
	Small extensive	66	12.90	1.67	12.93	8.73–17.00
	Large intensive	119	18.78	1.76	9.36	14.38–23.18
	Large extensive	123	19.41	1.79	9.22	14.94–23.88
	Totals	690	100.00			
1992	Sac-fry	428	65.58	1.92	2.92	59.75–71.40
	Small extensive	112	18.01	1.56	8.67	13.27–22.76
	Large intensive	46	7.19	1.02	14.23	4.08–10.30
	Large extensive	48	7.50	1.04	13.91	4.33–10.67
	Totals	634	98.28[a]			

[a] An estimated 1.72% of the population originated from an incidental stocking of pond-reared walleye fingerlings or were fish that immigrated from nearby West Okoboji Lake.

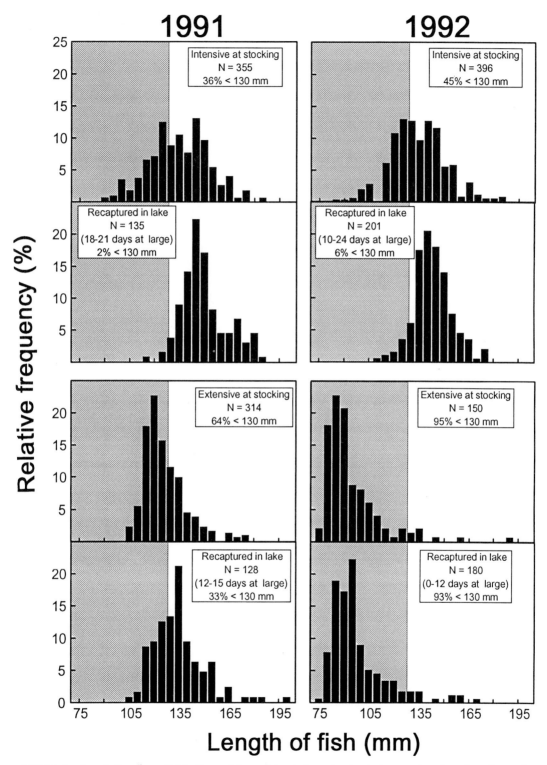

FIGURE 2.—Length-frequency distributions of large intensively and extensively reared walleye fingerlings when stocked and when recaptured in East Okoboji Lake, October 1991 and October 1992.

TABLE 7.—Densities (number per hectare) of yearling walleye and survival (%) of young-of-year walleye stocked into East Okoboji Lake, and recaptured 1 year later as yearling fish, 1992–1993. Values in parentheses are 95% confidence intervals.

Type of walleye	Density	Poststocking survival
1992 Year-Class		
Sac-fry	2.08 (1.43–3.11)	0.02 (0.0107–0.0232)
Small extensive	0.54 (0.30–0.96)	1.08 (0.59–1.91)
Large intensive	0.79 (0.49–1.31)	5.90 (3.61–9.70)
Large extensive	0.82 (0.49–1.33)	3.03 (1.87–4.97)
Totals	4.23 (2.71–6.71)	
1993 Year-Class		
Sac-fry	4.75 (3.17–8.15)	0.05 (0.0362–0.0931)
Small extensive	1.30 (0.70–2.60)	2.18 (1.17–4.34)
Large intensive	0.52 (0.22–1.18)	3.87 (1.61–8.73)
Large extensive	0.54 (0.23–1.22)	3.39 (1.43–7.59)
Totals	7.11 (4.32–13.15)	

TABLE 8.—Comparison of 1-year survival of fall walleye fingerlings originating from the sac-fry or the small fingerling stockings, to the fingerling walleyes stocked in the fall in East Okoboji Lake, 1992 year-class. Values in the same column with the same letter were not significantly different.

Type of walleye	Poststocking survival (%)	95% CI
Sac-fry	28.95 z	19.32–49.66
Small extensive	24.21 z	13.07–48.22
Large intensive	3.87 y	1.61–8.73
Large extensive	3.39 y	1.43–7.59

recaptures), and 664 were caught in 1993 (31 were recaptures). The 1992 population estimate was 3,141 (95% CI 2,513–4,186), and the 1993 estimate was 5,378 (95% CI 3,939–8,475). The 95% confidence intervals were asymmetric and were relatively tight around the 1992 estimate (20–33% of the estimate), as well as the 1993 estimate (27–58% of the mean). These estimates were significantly different. These population estimates correspond to about four yearling walleyes per hectare in 1992, and seven yearling walleyes per hectare in 1993 (Table 7). The densities of the yearling walleyes originating from the stocked fingerlings were similar in 1992 and 1993; however, the densities of yearling walleyes originating from the sac-fry plants in 1993 were more than twice the density in 1992.

The survival of sac-fry stocked in May 1992 and recaptured in October 1993 was significantly higher than the survival of sac-fry stocked in April 1991 and recaptured in October 1992 (Table 7). Survival rates of stocked walleye fingerlings were similar each year and no significant differences were noted (Table 7). Only about 1–2% of the small fingerlings that were stocked each June survived to be yearling fish. Twelve-month survival of the stocked large fingerlings varied from 3 to 6%. In 1991, the 12-month survival of intensively reared walleyes was greater than the 12-month survival of extensively reared walleye fingerlings, but the magnitude of this difference was small (Table 7). Fall young-of-year walleyes originating from the sac-fry plants or from the small fingerlings stocked in June survived to be yearling fish at a rate seven times greater than the large fingerlings that were stocked each fall (Table 8).

Benefit–Cost Analysis of Stocking

The most cost-effective type of walleye to stock is sac-fry, followed by small fingerlings, large intensively reared fingerlings, and finally large extensively reared fingerlings (Table 9). Initially, intensively reared walleyes were more expensive to produce than the extensively reared walleyes. However, more of the intensively reared fingerlings survived 1 year after stocking than did the extensively reared fingerlings.

Discussion

Sac-fry were the most cost-effective stocking option and contributed more than 50% to each year-class. Rose (1955), Carlander et al. (1960), and Carlander and Payne (1977) documented significant contributions of sac-fry stockings. However, as found in this and other studies (Mitzner 1992; McWilliams and Larscheid 1992) the success of sac-fry stocking is extremely variable among years and these stockings alone cannot be relied upon to develop uniform year-classes of walleye.

Small walleye fingerlings were the next most cost-effective stocking option. Fielder (1992) and Kop-

TABLE 9.—The cost of producing and stocking young-of-year and yearling walleyes in East Okoboji Lake, 1991–1992.

Year	Type of walleye	Initial cost to produce and stock one walleye	Final cost of one yearling walleye
1991	Sac-fry	$0.00032	$2.08
	Small extensive	$0.051	$4.70
	Large intensive	$0.32	$5.38
	Large extensive	$0.25	$8.35
1992	Sac-fry	$0.00032	$0.59
	Small extensive	$0.034	$1.74
	Large intensive	$0.28	$7.13
	Large extensive	$0.25[a]	$7.48

[a] 1991 production costs.

pelman et al. (1992) found that stocked small fingerlings contributed more to year-class abundance of young-of-year walleyes than stocked sac-fry. Stocked 50-mm walleyes in nearby West Okoboji Lake had low survival (McWilliams and Larscheid 1992) but were effective in establishing year-classes of walleyes in other lakes and impoundments in Iowa (e.g., Wahl and Kalishek 1985).

Most of the large fingerlings perished within 2–5 weeks after they were stocked into East Okoboji Lake. This high poststocking mortality was most likely size related. Each year, fewer walleyes less than 130 mm were recaptured than expected. This shift in the length-frequency distributions could have been caused by lower vulnerability of the smaller fish to capture, significant poststocking growth, poorer condition of the smaller stocked fish, or size-related predation.

In 1992, 95% of the extensively reared walleye fingerlings stocked into East Okoboji Lake were less than 130 mm in length. Most of these fish (80%) were stocked 2 d after sampling began, so there was very little time for any shifts in the length-frequency distributions to occur. We collected many fish less than 130 mm in length, which illustrates that we effectively sampled the smaller walleyes. Therefore, size-related differences in vulnerability to capture alone cannot explain the shifts in the length-frequency distributions.

Fingerling walleye growth is optimal at temperatures around 23°C (Hokanson and Koenst 1986), and is virtually nonexistent at 10°C (Forney 1966). To achieve the observed shift in the length-frequency distribution in 1991, the intensively reared walleye fingerlings had to grow an average of 25 mm within 21 d from the time they were stocked to the end of the sampling period. Optimal growth (food not limiting) and water temperatures exceeding 23°C would have been necessary to accomplish this. Similarly, the intensively reared walleye fingerlings stocked in 1992 would have had to grow 15 mm to cause the observed shift in the length-frequency distributions. Water temperatures exceeding 18°C and optimal growth over the 24-d period would have been necessary. The same trend was noted for extensively reared walleye fingerlings; growth of these fingerlings would have had to exceed 10 mm over the 15-d period in order to mimic the observed shift in the length-frequency distribution. Water temperatures exceeding 16°C and optimal growth over the 15-d period would have been necessary to cause this shift. Each year, the large intensively and extensively reared walleye fingerlings were stocked in late September or early October when the water temperatures were 12–14°C, and the water temperatures continually cooled to around 10°C at the end of the sampling period in late October. Hence the conditions for this rapid growth were not available, and growth of the stocked walleye fingerlings was not a plausible reason for the observed shifts in the length-frequency distributions.

All fingerlings appeared to be in good condition and few differences among groups were noted except that intensively reared fingerlings had significantly greater amounts of mesenteric fat than the other groups of fingerlings. Adequate amounts of fat are necessary for fish to overwinter (Newsome and Leduc 1975; Adams et al. 1985; Goede and Barton 1990); therefore, higher ratings of mesenteric fat deposition should correspond to higher survival. Still, most of these fish perished within 2–5 weeks after they were stocked into East Okoboji Lake. Therefore, the condition of the stocked fingerlings does not explain the observed shifts in the length-frequency distributions.

The observed shifts in the length-frequency distributions were most likely due to size-related predation. Santucci and Wahl (1993) found that predation by largemouth bass was significantly greater for 132–145-mm than for 186–216-mm walleyes. Size-related mortality of walleye fingerlings was also noted by Forney (1966) and Toneys and Coble (1979).

Each year, 12-month survival (survival of age-0 to age-1) of stocked fingerlings was lower than that of fingerlings originating from sac-fry in East Okoboji Lake. McWilliams and Larscheid (1992) and Mitzner (1992) also concluded that fingerlings originating from sac-fry stockings survived better than stocked fingerlings. Mraz (1968) found that survival of native fingerlings was 50 times that of stocked fingerlings. Most of the mortality of the stocked fingerlings occurred within 5 weeks of stocking. The first-year survival of stocked large walleye fingerlings was lower than expected. Originally, we speculated that 20–30% of the large fingerlings stocked each fall survived at least 1 year in East Okoboji Lake. However, actual survival was only 3–6%. This may be one reason why densities of walleye have not been increasing in these lakes despite increased stocking efforts.

Even if all the walleyes allocated for East Okoboji Lake (10 million sac-fry, 100,000 small fingerlings, 20,000 intensively reared fingerlings, and 40,000 extensively reared fingerlings) were stocked, only four yearlings per hectare would be produced, which is still short of the objective. This is because of the low 12-month survival of the large fingerlings stocked into East Okoboji Lake. If the 12-month survival of the large fingerlings could be increased to equal the 12-month survival of fingerlings originating from the sac-fry stockings, then the objective of 25 yearling walleyes per hectare is obtainable. For instance, if the 12-month survival of the large intensively and extensively reared walleye fingerlings was increased to 30%, then the objective of 25 yearling walleyes per hectare would be reached by stocking 10 million sac-fry, 100,000 small fingerlings, 20,000 intensively reared fingerlings, and 24,500 extensively reared walleye fingerlings, at a total cost of $19,575.

The size of the large fingerlings would undoubtedly have to be increased to increase the 12-month survival rate of these fish, which would increase the initial production costs. However, the higher survival rate would ultimately decrease the final costs of these fish. For instance, if the 12-month survival of intensively reared walleye fingerlings stocked in 1991 was increased to 30%, the initial costs of producing these fish could have been doubled, and the fish would still have been as cost-effective as the sac-fry stocked in 1991. Santucci and Wahl (1993) found that the initial costs of producing 200-mm walleyes was nearly twice as great as the cost of producing 150-mm walleye fingerlings. However, the 200-mm fingerlings were the most cost-effective stocking option when the first-year survival of these fish was considered.

Larger fall fingerlings (200 mm) need to be stocked to assure greater survival and to increase the benefit-to-cost ratio of the large walleye fingerlings. Reducing the mortality rate of the stocked fingerlings is necessary to meet the study objective of 25 yearling walleyes per hectare. These stocked fish would be significantly larger than the fall resident fingerlings that originated from the sac-fry stockings, and therefore should be less vulnerable to any size-related mortality. Mitzner (1992) found that large intensively reared (up to 200 mm) walleye fingerlings stocked in Lake Rathbun survived as well or even better than the fingerlings originating from the sac-fry plants. Also, Santucci and Wahl (1993) found that the first-year survival of stocked 200-mm walleyes was 31%, whereas the first-year survival of stocked 150-mm walleyes was only 7%. Therefore, stocking larger fall fingerlings in East Okoboji Lake may produce similar results. If the Spirit Lake Hatchery can produce and stock 200-mm walleye fingerlings, and these fish survive as well as the fingerlings originating from the sac-fry stockings, then the objective of 25 yearling walleyes is obtainable and feasible. This objective is not obtainable by stocking the current size of fall fingerling walleyes in East Okoboji Lake.

It is important to note that it was not feasible to reach this objective by stocking only one type of walleye. For instance, even though sac-fry are the most cost-effective type of walleye to stock, it would take an estimated stocking of 52 million sac-fry in East Okoboji Lake in order to reach the objective of 25 yearlings per hectare. This stocking would consume 65% of the production capability of the Spirit Lake Hatchery, and would eliminate walleye stocking in many other areas in Iowa. Alternatively, 1.1 million small fingerlings would be needed to reach this objective, which is nearly four times the current production of the Spirit Lake Hatchery. Similarly, it would take approximately 375,000 intensively reared walleye fingerlings or 572,000 extensively reared walleye fingerlings to reach this objective, which far exceeds the production capabilities of the Spirit Lake Hatchery. However, the objective of 25 yearlings per hectare can be met by stocking a combination of these types of walleye, while remaining within the budgetary and production constraints of the Spirit Lake Hatchery.

Acknowledgments

I would like to thank Wally Jorgenson and Jim Christianson for their cooperation during all aspects of this study. Tom Gengerke and Jim Christianson provided valuable technical advice on various phases of this study. Ed Thelen, Cindy Martens, Mark Sexton, Brian Bristow, Tom Gengerke, Jim Christianson, Jim Berquist, Lannie Miller, Don Herrig, and many others helped collect the field data. Don Bonneau, Brian Bristow, Tom Gengerke, Robert Wiley, Michael Staggs, and an anonymous individual critically reviewed this document. This research was supported by the Federal Aid in Sport Fish Restoration, Project F-135-R, U.S. Fish and Wildlife Service, and by the Iowa Department of Natural Resources.

References

Adams, S. M., J. E. Breck, and R. B. MClean. 1985. Cumulative stress-induced mortality of gizzard shad in a southeastern U.S. reservoir. Environmental Biology of Fishes 13:103–112.

Anderson, R. O. 1980. Proportional stock density (PSD) and relative weight (W_r): interpretative indices for fish populations and communities. Pages 27–33 in S. Gloss and B. Shupp, editors. Practical fisheries management: more with less in the 1980's. American Fisheries Society, New York Chapter, Ithaca.

Carlander, K. D., and P. M. Payne. 1977. Year-class abundance, population and production of walleye

(*Stizostedion vitreum vitreum*) in Clear Lake, Iowa 1948–74, with varied stocking rates. Journal of the Fisheries Research Board of Canada 34:1792–1799.

Carlander, K. D., R. R. Whitney, E. B. Speaker, and K. Madden. 1960. Evaluation of walleye fry stocking in Clear Lake, Iowa, by alternate year planting. Transactions of the American Fisheries Society 89:249–254.

Colby, P. J., R. E. McNichol, and R. A. Ryder. 1979. Synopsis of biological data on the walleye, *Stizostedion v. vitreum* (Mitchill 1818). FAO (Food and Agriculture Organization of the United Nations) Fisheries Biology Synopsis 119.

Cook, R. C. 1983. Simulation and application of stock composition estimators. Canadian Journal of Fisheries and Aquatic Sciences 40:2113–2118.

Cook, R. C., and G. E. Lord. 1978. Identification of stocks of Bristol Bay sockeye salmon *Oncorhynchus nerka*, by evaluating scale patterns with a polynomial discriminant function. Fishery Bulletin 76:415–423.

Fielder, D. G. 1992. Evaluation of stocking walleye fry and fingerlings and factors affecting their success in lower Lake Oahe, South Dakota. North American Journal of Fisheries Management 12:336–345.

Forney, J. L. 1966. Factors affecting first-year growth of walleyes in Oneida Lake, New York. New York Fish and Game Journal 13:147–167.

Goede, R. W., and B. A. Barton. 1990. Organismic indices and an autopsy-based assessment as indicators of health and condition of fish. American Fisheries Society Symposium 8:93–108.

Goldstein, L. J., and D. I. Schneider. 1984. Finite mathematics and its applications, 2nd edition. Prentice-Hall, Englewood Cliffs, New Jersey.

Gustafson, K. A. 1988. Approximating confidence intervals for indices of fish population size structure. North American Journal of Fisheries Management 8:139–141.

Hokanson, K. E. F., and W. M. Koenst. 1986. Revised estimates of growth requirements and lethal temperature limits of juvenile walleyes. Progressive Fish-Culturist 48:90–94.

Johnston, N. J., and S. Kotz. 1969. Discrete distributions. Houghton Mifflin Company, Boston.

Jorgensen, W. 1993. Production report: Spirit Lake Hatchery. Iowa Department of Natural Resources, Des Moines.

Koppelman, J. B., K. P. Sullivan, and P. J. Jeffries, Jr. 1992. Survival of three genetically-marked walleyes stocked into two Missouri impoundments. North American Journal of Fisheries Management 12:291–298.

Krebs, C. J. 1989. Ecological methodology. Harper Collins Publishers, Inc. New York.

LaJeone, L. J., and D. L. Bergerhouse. 1991. A liquid nitrogen freeze-branding apparatus for marking fingerling walleyes. Progressive Fish-Culturist 53:130–133.

Larscheid, J. G. 1991. Contribution of stocked walleye and population dynamics of adult walleye in Spirit and East and West Okoboji Lakes. Iowa Department of Natural Resources, Federal Aid in Sport Fish Restoration, Project F-135-R-1, Annual Performance Report, Des Moines.

Larscheid, J. G. 1993. Contribution of stocked walleye and population dynamics of adult walleye in Spirit and East and West Okoboji Lakes. Iowa Department of Natural Resources, Federal Aid in Sport Fish Restoration, Project F-135-R-3, Annual Performance Report, Des Moines.

McWilliams, R. H. 1976. Larval walleye and yellow perch population dynamics in Spirit Lake and the contribution of stocked sac-fry to the larval walleye density. Iowa Conservation Commission, Technical Series 76-1, Des Moines.

McWilliams, R. H. 1990. Large natural lakes. Iowa Department of Natural Resources, Federal Aid in Sport Fish Restoration, Project F-95-R, Completion Report, Des Moines.

McWilliams, R. H., and J. G. Larscheid. 1992. Assessment of walleye fry and fingerling stocking in the Okoboji Lakes, Iowa. North American Journal of Fisheries Management 12:329–335.

Mitzner, L. 1992. Evaluation of walleye fingerling and fry stocking in Rathbun Lake, Iowa. North American Journal of Fisheries Management 12:321–328.

Mraz, D. 1968. Recruitment, growth, exploitation and management of walleyes in a southeastern Wisconsin lake. Wisconsin Conservation Department Technical Bulletin 40.

Murphy, B. R., M. L. Brown, and T. A. Springer. 1990. Evaluation of the relative weight (W_r) index, with new applications to walleye. North American Journal of Fisheries Management 10:85–97.

Newsome, G. E., and G. Leduc. 1975. Seasonal changes of fat content in the yellow perch (*Perca flavescens*) of two Laurentian lakes. Journal of the Fisheries Research Board of Canada 32:2214–2221.

Pella, J. J., and T. L. Robertson. 1979. Assessment of composition of stock mixtures. Fishery Bulletin 77:387–398.

Piper, R. G., I. B. McElwain, L. E. Orme, J. P. McCraren, L. G. Fowler, and J. R. Leonard. 1986. Fish hatchery management. U.S. Fish and Wildlife Service, Washington, DC.

Rose, E. T. 1955. The fluctuation in abundance of walleye in Spirit Lake, Iowa. Proceedings of the Iowa Academy of Science 62:567–575.

Santucci, V. J., Jr., and D. H. Wahl. 1993. Factors influencing survival and growth of stocked walleye (*Stizostedion vitreum*) in a centrarchid-dominated impoundment. Canadian Journal of Fisheries and Aquatic Sciences 50:1548–1558.

Toneys, M. L., and D. W. Coble. 1979. Size-related first winter mortality of freshwater fishes. Transactions of the American Fisheries Society 108:415–419.

Wahl, J., and B. Kalishek. 1985. Evaluation of walleye fry and fingerling stocking regimes. *In* Fishery management investigations. Iowa Department of Natural Resources, Des Moines.

Woodward, J. A., D. G. Bonett, and M-L. Brecht. 1990. Introduction to linear models and experimental design. Harcourt Brace Jovanovich, San Diego, California.

Evaluation of the Florida Largemouth Bass in Texas, 1972–1993

ALLEN A. FORSHAGE

Texas Parks and Wildlife Department
11942 FM 848, Tyler, Texas 75707, USA

LORAINE T. FRIES

Texas Parks and Wildlife Department
PO Box 947, San Marcos, Texas 78667, USA

Abstract.—The creation of lentic habitats in Texas by reservoir construction provided the impetus to introduce a fish that was assumed to be better adapted to the new conditions. Based on reported faster growth, greater maximum size, and potentially increased fitness in lentic habitats of the Florida largemouth bass *Micropterus salmoides floridanus* compared with the native largemouth bass *M. salmoides salmoides*, the Texas Parks and Wildlife Department has stocked Florida largemouth bass into selected Texas waters since 1972. The overall goal of these stockings was to maximize quality recreational fishing for largemouth bass.

Population genetics studies conducted in 1991 through 1993 on 126 reservoirs stocked with Florida largemouth bass suggest that introductions have had varying success. Electrophoresis of loci exhibiting fixed differences between the subspecies ($AAT-B^*$ and $sIDHP^*$) indicated an average of 36.3% occurrence of Florida largemouth bass alleles. Ninety-one reservoirs had a frequency of Florida largemouth bass alleles 20% or greater; six had 0% occurrence. In seven reservoirs, Florida largemouth bass virtually replaced northern largemouth bass with almost complete fixation (>80%) of the Florida largemouth bass alleles. Linear models suggest Florida largemouth bass stocking success is lower in older reservoirs, reservoirs in northern latitudes, and larger reservoirs. Establishment of Florida largemouth bass alleles is increased through repeated annual stockings.

Florida largemouth bass stockings also have affected largemouth bass angling. Since their introduction, the state record largemouth bass increased from 6.12 to 8.25 kg. Mean weight for largemouth bass submitted to the Big Fish Awards program increased from a low of 3.93 kg in 1977 to a high of 5.41 kg in 1991. The number of reservoirs yielding trophy largemouth bass (≥4.54 kg) increased from 2 in 1974 to 35 in 1993. The increase in trophy largemouth bass catches is considered positive by Texas anglers and fishery managers.

Texas covers an extensive geographic area that includes five major physiographic regions of North America. In the early 1900s, freshwater fishery resources in Texas primarily were limited to streams and rivers. Texas had only 12 public lakes with a combined area of 12,111 ha; Lake Caddo constituted 85% of this area. Extensive impoundment of lotic habitats began in the 1920s, increased in subsequent decades until reaching a peak in the 1960s, and declined after the 1970s. Today, there are over 620 public freshwater reservoirs with a combined area of 677,941 ha.

Northern largemouth bass *Micropterus salmoides salmoides* were native to Texas rivers and streams before impoundments were built (Philipp et al. 1983). In the late 1800s federal hatcheries cultured and stocked largemouth bass throughout the United States. Early stocking records from the U.S. Commission of Fish and Fisheries (USCFF) (1895, 1896, 1898a, 1898b, 1899, 1900) indicate some stockings of Texas waters included black bass *Micropterus* spp. from Virginia, Illinois, and Missouri. In addition, Texas hatcheries have cultured northern largemouth bass since 1940 using locally procured broodstock and have stocked over 46 million fish throughout the state.

In 1972 Florida largemouth bass *M. s. floridanus* were imported into Texas to improve largemouth bass fishing in terms of size and provide more trophy fish. The rationale used to justify this introduction was based on the reported superiority of Florida largemouth bass in growth rate, longevity, and possibly lower angling vulnerability that would provide larger fish than did existing largemouth bass stocks (Pelzman 1980). Bottroff and Lembeck (1978) reported that Florida largemouth bass stocked in San Diego County, California, grew faster than did northern largemouth bass stocked in the same locations after their first year of life. Bottroff and Lembeck (1978) speculated that the Florida largemouth bass outperformed northern largemouth bass because the southern California environment more closely resembled Florida than did the Illinois environment, the origin of the stocked northern largemouth bass. Largemouth bass native to Texas were likely adapted to the lotic

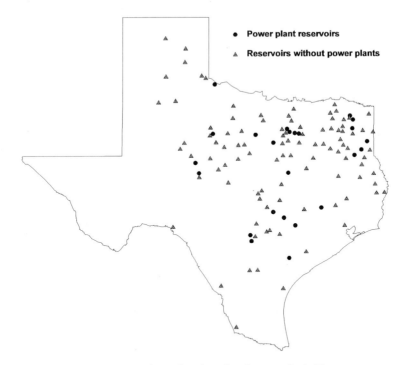

FIGURE 1.—Approximate location of study reservoirs in Texas.

conditions prevalent prior to extensive impoundment; however, Florida largemouth bass evolved in natural lake systems. The creation of lentic habitats in Texas provided an impetus to introduce a fish that was assumed to be better adapted to the new conditions.

From 1972 through 1993, over 66 million Florida largemouth bass have been stocked into 454 different Texas reservoirs. This widespread stocking of Florida largemouth bass into Texas waters with existing northern largemouth bass populations has resulted in the potential for interaction between two previously isolated gene pools.

Evaluations of the Florida subspecies and its effects on Texas' fisheries have been conducted since stocking began. Growth studies (Inman et al. 1978) indicated the Florida largemouth bass and F_1 female Florida largemouth bass × male northern largemouth bass hybrids grew faster than did northern largemouth bass during a 3-year period. Studies also evaluated catchability (Inman et al. 1978), temperature tolerance (Guest 1985), food habits (Ryan 1987), and genetic effects of stocking (Kulzer et al. 1987). The objectives of the present study were to evaluate the effects of Florida largemouth bass stocking as measured by allele frequency change and trophy bass catches in Texas reservoirs.

Methods

Largemouth bass were collected by electrofishing from 126 reservoirs (Figure 1) in the fall or spring of 1991 through 1993. Most reservoirs ($N = 81$) were sampled only one time, but some were sampled two ($N = 36$) or three times ($N = 9$). In cases where reservoirs were sampled more than once, data from each sampling event were included in the analyses. The sampling goal at each reservoir was to obtain 30 age-0 fish. In all cases, samples were collected from different areas of the reservoir. Actual sample size ranged from 7 to 107 (mean, 30.5; SE, 0.75). Age-1 largemouth bass also were collected to provide adequate sample size when sufficient numbers of age-0 fish were unavailable. Fish were placed on ice immediately after collection and either processed at the reservoir or transported to laboratory facilities. Total length (mm) and weight (g) were recorded for each fish. Otoliths were removed from each fish for age determination. Liver tissue was excised from each fish, placed in prenumbered vials, frozen, and stored at $-80°C$.

We assumed that essentially no Florida largemouth bass genetic influence was present in Texas reservoirs prior to introductions by the Texas Parks and Wildlife Department. Because of contaminated broodstock at state hatcheries (Harvey et al. 1980), some Florida largemouth bass stockings prior to 1980 may have included intergrade fish.

Samples were electrophoretically analyzed using the procedures described by Philipp et al. (1983), except 2% agarose (weight per volume) gels cast on acetate support films were substituted for starch gels. For each population, allele frequencies were calculated for the diagnostic loci *sIDHP** (which codes for isocitrate dehydrogenase, enzyme number 1.1.1.42 [IUBNC 1984]) and *AAT-B** (aspartate aminotransferase, 2.6.1.1). The frequency of Florida largemouth bass alleles was determined by totaling their occurrence and dividing by the total number of alleles scored for each reservoir population. Although most of the reservoirs contained intergrade populations, genotypes were used to infer the heritage of individual fish. Therefore, fish with Florida largemouth bass genotypes were considered Florida largemouth bass and those exhibiting northern largemouth bass genotypes were considered northern largemouth bass. Fish heterozygous for northern and Florida largemouth bass alleles were designated as F_1 intergrades. Fish with any other genotype were considered F_x intergrades. Variables that could influence Florida largemouth bass stocking success also were considered: reservoir size (range, 18–185,000 ha; mean, 13,614 ha; SE, 2,178); stocking density (range, 0.016–485 fish/ha; mean, 40.5 fish/ha; SE, 4.30); number of years stocked (range, 1–9 years; mean, 2.5 years; SE, 0.13); age of impoundment when first stocked (range, 5–71 years; mean, 20.7 years; SE, 1.31); latitude (range, 26.28–35.5°N; mean, 31.73°N; SE, 0.123); longitude (range, 93.34–101.5°W; mean, 97.09°W; SE, 0.163); and whether or not the reservoir was heated (power plant reservoir).

Allele and genotypic frequencies were arcsin transformed to normalize the distribution (Wilkinson 1990a). Multiple linear regressions (Wilkinson 1990b) using the transformed data related the frequency of occurrence of Florida largemouth bass alleles to reservoir and stocking conditions. Dependent variables were percent Florida largemouth bass alleles in the population, percent Florida largemouth bass genotypes, percent F_1 genotypes, percent F_x genotypes, percent of fish with Florida largemouth bass alleles, and the percent northern largemouth bass genotypes. Independent variables were reservoir and stocking characteristics. Separate regressions were analyzed for the entire data set, power plant reservoirs, and non-power plant reservoirs. Regression formulas were considered significant when *P* was 0.05 or less.

Trends in occurrence of trophy bass from Texas reservoirs following introduction of Florida largemouth bass were evaluated using two angler recognition programs: the State Record Fish program and Big Fish Awards program. The State Record Fish program awards a State Record Fish Certificate to an angler for the catch and proper documentation of the largest (in weight) individual of a species caught in Texas. The Big Fish Awards program provides a certificate for the catch and proper documentation of a largemouth bass larger than 3.6 kg. Information from largemouth bass State Record Fish and Big Fish Awards certified since 1970 and 1974, respectively, was analyzed. State Record Fish program data coupled with genotype identification information were used to evaluate effects of Florida largemouth bass stocking on the maximum size of largemouth bass caught in Texas waters. Big Fish Awards data were tabulated by weight, date of capture, and water body. Simple linear regression was used to determine if mean weight increased over time. Stocking results were correlated with information obtained from Big Fish Awards to infer stocking effects.

Results and Discussion

Stocking Evaluation

The 126 reservoirs had an average of 36.3% occurrence of Florida largemouth bass alleles, and Florida largemouth bass alleles were found in more than 20% of fish sampled in 91 reservoirs (Table 1). Six reservoirs contained only northern largemouth bass genotypes; conversely only one reservoir contained exclusively Florida largemouth bass genotypes. However, seven reservoirs showed almost complete fixation (>80%) for Florida largemouth bass alleles. Florida largemouth bass allele frequencies were greater than 50% in 28% of the reservoirs surveyed. These data demonstrate the Florida largemouth bass genome has become successfully established in some Texas reservoirs.

Regression analysis for all reservoirs combined indicated the variables stocking frequency, reservoir size, latitude, and age accounted for relatively large and significant amounts of the variation in percent Florida largemouth bass alleles and genotypes, percent fish with Florida largemouth bass alleles, and percent northern largemouth bass genotypes (Table 2). Similar results were obtained for non-power plant reservoirs. In general, reservoir age, latitude, and size were the most influential variables in these models.

Reservoir age when first stocked with Florida

TABLE 1.—Mean (SE) percent northern largemouth bass (NLMB), Florida largemouth bass (FLMB), F_1, and F_x genotypes and FLMB alleles of largemouth bass collected during 1991–1993 from 126 Texas reservoirs. Reservoir groupings are based on the percent of FLMB alleles.

Percent of FLMB alleles	Number of lakes	Genotype				FLMB alleles
		NLMB	FLMB	F_1	F_x	
0	6	100.00	0.00	0.00	0.00	0.00
		(0.00)	(0.00)	(0.00)	(0.00)	(0.00)
≤20	29	74.65	0.23	6.56	18.53	9.16
		(3.01)	(0.16)	(1.90)	(2.30)	(1.08)
21–40	40	32.85	6.02	18.24	42.77	31.96
		(2.38)	(1.14)	(1.68)	(2.60)	(0.82)
41–60	29	14.38	11.00	25.29	50.47	49.29
		(2.45)	(0.97)	(1.92)	(2.19)	(1.06)
61–80	15	5.78	32.43	20.43	41.35	68.05
		(1.59)	(3.13)	(2.25)	(3.12)	(1.48)
>80	7	0.00	75.16	10.77	14.07	91.47
		(0.00)	(7.29)	(5.19)	(5.95)	(2.36)

largemouth bass was significantly negatively correlated with percent Florida largemouth bass alleles in the population, percent Florida largemouth bass genotypes, and percent of fish with Florida largemouth bass alleles in all regressions (Table 2). Conversely, this independent variable was positively related to percent northern largemouth bass alleles. This finding contrasts with Kulzer et al. (1987) who found no significant correlation between impoundment age and percent intergradation in a study of Florida largemouth bass introductions in 19 Texas reservoirs. However, Gilliland and Whitaker (1991) found reservoir age positively correlated with percent F_1 genotypes, but not with percent introgression in Oklahoma reservoirs. Our finding suggests that it is difficult to establish Florida largemouth bass alleles in older reservoirs. According to Boxrucker (1986), Florida largemouth bass stocking influence is greatest in reservoirs that are stocked

TABLE 2.—Correlation coefficients between largemouth bass population genetic variables and reservoir and stocking variables for 126 Texas reservoirs in which Florida largemouth bass (FLMB) were stocked into existing northern largemouth bass (NLMB) populations. Coefficients of determination from regression analysis used all independent variables. Reservoir and stocking characteristics are stocking frequency and density and reservoir size, latitude, longitude, and age of impoundment at first stocking (reservoir age). Asterisks denote significance at $P < 0.05$.

Independent variable	Dependent variable					
	Percent FLMB alleles	Percent FLMB genotypes	Percent F_1 genotypes	Percent F_x genotypes	Percent fish with FLMB alleles	Percent NLMB genotypes
All reservoirs, $N = 170$						
Stocking frequency	0.169*	0.110	0.091	0.204*	0.203*	−0.156*
Reservoir size	−0.286*	−0.211*	−0.157	−0.213*	−0.328*	0.321*
Reservoir latitude	−0.311*	−0.346	−0.093	−0.027	−0.264*	0.346*
Reservoir longitude	0.040	−0.063	0.034	−0.017	−0.047	−0.067
Stocking density	0.122	0.011	0.008	0.041	−0.098	−0.111
Reservoir age	−0.415*	−0.400*	−0.098	−0.087	−0.331*	0.375*
Coefficient of determination	0.360*	0.326*	0.007	0.031	0.288	0.361*
Non-power plant reservoirs, $N = 134$						
Stocking frequency	0.253*	0.198*	0.146	0.209	0.253*	−0.212*
Reservoir size	−0.240*	−0.155	−0.192	−0.248	−0.285*	0.271*
Reservoir latitude	−0.353*	−0.391*	−0.136	−0.118	−0.301*	0.383*
Reservoir longitude	0.234*	0.220*	0.122	0.065	0.212*	−0.233*
Stocking density	0.116	0.114	0.071	0.078	0.118	−0.111
Reservoir age	−0.320*	−0.270*	−0.170	−0.208*	−0.286*	0.307*
Coefficient of determination	0.388*	0.290*	0.058*	0.087*	0.282*	0.333*
Power plant reservoirs, $N = 36$						
Stocking frequency	0.032	0.036	−0.048	0.180	0.117	−0.030
Reservoir size	−0.171	−0.054	−0.325	−0.100	−0.346*	0.318*
Reservoir latitude	0.010	−0.097	−0.209	0.181	−0.008	0.089
Reservoir longitude	0.103	0.152	−0.040	−0.091	0.107	−0.169
Stocking density	0.174	0.057	−0.183	0.056	0.120	−0.171
Reservoir age	−0.760*	−0.657*	0.180	0.195	−0.605*	0.630*
Coefficient of determination	0.589*	0.437*	0.000	0.000	0.499*	0.598*

when first impounded. Reduced competition with native largemouth bass in expanding populations allows greater recruitment of the stocked fish, avoiding the high mortality often associated with supplemental stocking of largemouth bass into existing populations. In Texas, older reservoirs generally have decreased largemouth bass productivity, which may decrease recruitment and stocking success. It also is possible that some older reservoirs are at carrying capacity for largemouth bass, and stocking any largemouth bass in such reservoirs is less successful.

Latitude usually had the second highest correlation coefficients and was negatively correlated with percent Florida largemouth bass alleles in the population, percent Florida largemouth bass genotypes, and percent of fish with Florida largemouth bass alleles; it was positively correlated to percent northern largemouth bass alleles (Table 2). The notable exceptions were regressions for power plant reservoirs, where latitude was not significant. Dunham et al. (1992) reported no significant correlation between latitude and percent introgression of Florida largemouth bass into native populations in Alabama. However, Gilliland and Whitaker (1991) reported cold temperatures in northern latitudes of Oklahoma were negatively correlated with total Florida largemouth bass alleles. They reported Florida largemouth bass were less tolerant of rapid water temperature drops and prolonged cold periods seen in much of northern Oklahoma. Cichra et al. (1982) and Guest (1985) also reported that the Florida subspecies is less cold tolerant than is northern largemouth bass. Texas covers an extensive geographic area, and cooler water temperatures associated with northern reservoirs could affect Florida largemouth bass stocking success. A z-test for sample means indicated that the mean latitudes for non-power plant (mean, 31.81°N; range, 26.28–35.5°N) and power plant reservoirs (mean, 31.44°N; range, 28.43–34.2°N) were not significantly different ($P > 0.05$). The failure of latitude to be significant in linear models for power plant reservoirs supports the conclusion that cooler temperatures in northern latitudes may be cause for a decreased incidence of Florida largemouth bass alleles.

Reservoir size generally had the third highest correlation coefficients and was negatively correlated with percent Florida largemouth bass alleles in the population, percent Florida largemouth bass genotypes, and percent of fish with Florida largemouth bass alleles; it was positively correlated to percent northern largemouth bass alleles (Table 2). Likely this variable reflected stocking density because large reservoirs are stocked at lower densities than are smaller reservoirs. Reservoir area was not significantly correlated with Florida largemouth bass introgression in Oklahoma (Gilliland and Whitaker 1991), but Dunham et al. (1992) and Kulzer et al. (1987) reported a positive correlation between stocking density and Florida largemouth bass introgression in Alabama and Texas, respectively.

Stocking frequency had the fourth highest correlation coefficients and was positively correlated to percent Florida largemouth bass alleles in the population, percent Florida largemouth bass genotypes, and percent of fish with Florida largemouth bass alleles; it was negatively correlated with percent northern largemouth bass alleles (Table 2). This result is different from those of Kulzer et al. (1987), who reported stocking frequency as having the greatest effect of variables tested on the frequency of Florida largemouth bass alleles in Texas reservoirs, and Dunham et al. (1992) who found the number of times a lake had been stocked as the best one-variable, stepwise model for predicting percent introgression in Alabama lakes.

Overall, the best linear models in this study were for power plant reservoirs (Table 2). These models accounted for more variation ($R^2 \geq 0.44$) in percent Florida largemouth bass alleles, percent Florida largemouth bass genotypes, percent fish with Florida largemouth bass alleles, and percent northern largemouth bass genotypes than did models for all reservoirs or models for non-power plant reservoirs. In power plant reservoirs, reservoir age was the most influential independent variable. Reservoir size was the only other variable with significant correlations. These results suggest that among power plant reservoirs, Florida largemouth bass introductions will be most successful in younger and smaller impoundments. It is probable that the regression models for power plant reservoirs were stronger because, in general, impoundments in this category tended to have more in common (i.e., reservoir age and size) than did the other, broader categories.

In summary, the linear models for all reservoirs combined and for non-power plant reservoirs suggest Florida largemouth bass introgression is more successful when this subspecies is stocked into newer reservoirs, reservoirs in southern latitudes, and smaller reservoirs. In older and larger reservoirs and those at more northern latitudes (<35.5°N), managers may increase Florida largemouth bass alleles through repeated annual stockings.

TABLE 3.—Texas largemouth bass state records, 1943–1992.

Date	Weight (kg)	Identification
Jan 16, 1943	6.12	Northern largemouth bass[a]
Feb 2, 1980	6.39	Florida largemouth bass[b]
Jan 10, 1981	6.45	F_1 intergrade largemouth bass[b]
Feb 7, 1981	7.03	F_1 intergrade largemouth bass[c]
Feb 16, 1986	7.67	Florida largemouth bass[c]
Nov 26, 1986	8.02	Florida largemouth bass[c]
Jan 24, 1992	8.25	Florida largemouth bass[c]

[a] Genetic identification inferred from historical data.
[b] Genetic identification based upon meristic data.
[c] Genetic identification based upon electrophoresis.

Angling Quality

An expected benefit of introducing Florida largemouth bass was increased numbers of trophy bass and subsequently increased maximum size of largemouth bass for Texas anglers. Anderson (1975) and Weithman and Anderson (1978) identified the importance of trophy fish to the angler's perception of the quality of a fishery. They reported that one large fish contributes more to personal gratification and memories, and therefore to fishing quality, than does an equal weight of small fish. The first Texas record largemouth bass, 6.12 kg, was caught in 1943. This record stood for 37 years until a 6.39-kg Florida largemouth bass was caught from Lake Monticello. This was the first of six new records in a 12-year time span (Table 3). All new records have been identified as Florida largemouth bass or F_1 intergrade genotypes and were caught in reservoirs previously stocked with Florida largemouth bass. This is very similar to results from Oklahoma (Horton and Gilliland, in press), California (Dennis P. Lee, California Department of Fish and Game, personal communication), and Louisiana (Gary A. Tilyou, Louisiana Department of Wildlife and Fisheries, personal communication). Each of these states reported an increase in state record largemouth bass after introduction of Florida largemouth bass.

In addition to increases in weight of state record largemouth bass, mean weight of largemouth bass certified through the Big Fish Awards program increased significantly ($N = 20$; $P < 0.001$) from 1974 to 1993. Mean weight increased from 3.98 kg in 1974 to a high of 5.41 kg in 1991 (Figure 2). Prior to the introduction of Florida largemouth bass, fish larger than 4.54 kg were rarely caught in Texas. Big Fish Awards program applications were certified for only seven largemouth bass larger than 4.54 kg from 1974 through 1978; however, from 1979 through 1993, 525 largemouth bass larger than 4.54

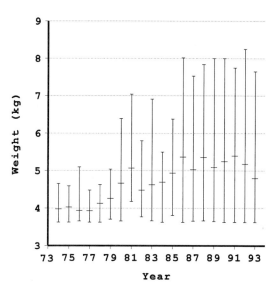

FIGURE 2.—Mean weight (kg) of largemouth bass certified for the Big Fish Awards program in Texas, 1974–1993. Vertical bars represent ranges of weights.

kg were certified. The number of different reservoirs yielding these trophy fish increased from 2 in 1974 to 35 in 1993. Although genetic analyses generally were not performed on individual fish submitted to the Big Fish Awards program, there was genetic information on the reservoir populations. Significant positive correlation was found between the number of largemouth bass 4.54 kg or larger submitted to the Big Fish Awards program and the percent of fish with Florida largemouth bass alleles (Forshage et al. 1991). The increased frequency in Big Fish Awards for largemouth bass followed the introduction of Florida largemouth bass by about 7 to 9 years, which corresponds to the time it would take for larger individuals to appear in recipient populations.

It has been postulated that there could be negative consequences resulting from stocking Florida largemouth bass outside its native range. Philipp (1991) and Philipp and Whitt (1991) documented inferior growth and survival of introgressed largemouth bass populations in central Illinois; this reduced performance was attributed to outbreeding depression. The results of our study imply that Florida largemouth bass and their intergrades would not perform well in Illinois because, in Texas, Florida largemouth bass alleles were negatively correlated with latitude. In situations of radically altered habitat, managers should consider stocking fish that are genetically fit to the new conditions (Maceina and

Murphy 1992). This approach has been used in Texas with the Florida largemouth bass. The results of the present study indicate successful introgression of Florida largemouth bass (Florida largemouth bass alleles 20% or greater) in 97 out of 126 Texas reservoirs. The virtual replacement of northern largemouth bass alleles with Florida largemouth bass alleles in some Texas reservoirs implies increased fitness of the Florida subspecies. Furthermore, the apparent effect of Florida largemouth bass stockings on trophy largemouth bass fishing are viewed by many anglers and fisheries managers as positive.

Acknowledgments

The authors acknowledge and greatly appreciate the assistance of Joe N. Fries, National Biological Service, San Marcos, Texas, who helped with the statistical analysis. Thanks to Frederick Janssen for assistance with manuscript preparation. We also thank Dick Luebke, Gary Garrett, Dennis Lee, Harold Schramm, Jr., and two anonymous reviewers for their comments and suggestions on this manuscript. Funding for this research was provided by the Federal Aid in Sport Fish Restoration Project F-30-R of the Texas Parks and Wildlife Department.

References

Anderson, R. O. 1975. Factors influencing the quality of largemouth bass fishing. Pages 183–194 *in* R. H. Stroud and H. Clepper, editors. Black bass biology and management. Sport Fishing Institute, Washington, DC.

Bottroff, L. J., and M. E. Lembeck. 1978. Fishery trends in reservoirs of San Diego County, California following introductions of Florida largemouth bass, *Micropterus salmoides floridanus*. California Fish and Game 64:4–23.

Boxrucker, J. 1986. Evaluation of supplemental stocking of largemouth bass as a management tool in small impoundments. North American Journal of Fisheries Management 6:391–397.

Cichra, C. E., W. H. Neil, and R. L. Noble. 1982. Differential resistance of northern and Florida largemouth bass to cold shock. Proceedings of the Annual Conference Southeastern Association of Fish and Wildlife Agencies 34(1980):19–24.

Dunham, R. A., C. S. Turner, and W. C. Reeves. 1992. Introgression of the Florida largemouth bass genome into native populations in Alabama public lakes. North American Journal of Fisheries Management 12:494–498.

Forshage, A. A., P. P. Durocher, M. A. Webb, and D. G. Lewis. 1991. Management application of angler recognition program data. Proceedings of the Annual Conference Southeastern Association of Fish and Wildlife Agencies 43(1989):36–40.

Gilliland, E. R., and J. Whitaker. 1991. Evaluation of genetic status of largemouth bass populations. Proceedings of the Annual Conference Southeastern Association of Fish and Wildlife Agencies 43(1989):182–190.

Guest, W. C. 1985. Survival of adult Florida and northern largemouth bass subjected to cold temperature regimes. Proceedings of the Annual Conference Southeastern Association of Fish and Wildlife Agencies 36(1982):332–339.

Harvey, W. D., I. F. Greenbaum, and R. L. Noble. 1980. Electrophoretic evaluation of five hatchery stocks of largemouth bass in Texas. Annual Proceedings of the Texas Chapter, American Fisheries Society 3:49–63.

Horton, R. A., and E. R. Gilliland. In press. Monitoring trophy largemouth bass in Oklahoma using a taxidermist network. Proceedings of the Annual Conference Southeastern Association of Fish and Wildlife Agencies 47(1993).

Inman, C. R., R. C. Dewey, and P. P. Durocher. 1978. Growth comparison and catchability of three largemouth bass strains. Proceedings of the Annual Conference Southeastern Association of Fish and Wildlife Agencies 30(1976):40–47.

IUBNC (International Union of Biochemistry, Nomenclature Committee). 1984. Enzyme nomenclature. Academic Press, San Diego, California.

Kulzer, K. E., R. L. Noble, and A. A. Forshage. 1987. Genetic effects of Florida largemouth bass introductions into selected Texas reservoirs. Proceedings of the Annual Conference Southeastern Association of Fish and Wildlife Agencies 39(1985):56–64.

Maceina, M. J., and B. R. Murphy. 1992. Stocking Florida largemouth bass outside its native range. Transactions of the American Fisheries Society 121:686–691.

Pelzman, R. J. 1980. Impact of Florida largemouth bass, *Micropterus salmoides floridanus*, introduction at selected northern California waters with a discussion of the use of meristics for detecting introgression and for classifying individual fish of intrograded populations. California Fish and Game 66(3):132–162.

Philipp, D. P. 1991. Genetic implications of introducing Florida largemouth bass, *Micropterus salmoides floridanus*. Canadian Journal of Fisheries and Aquatic Sciences 48(Supplement 1):58–65.

Philipp, D. P., W. F. Childers, and G. S. Whitt. 1983. A biochemical genetic evaluation of the northern and Florida subspecies of largemouth bass. Transactions of the American Fisheries Society 112:1–20.

Philipp, D. P., and G. S. Whitt. 1991. Survival and growth of northern, Florida, and reciprocal F_1 hybrid largemouth bass in central Illinois. Transactions of the American Fisheries Society 120:58–64.

Ryan, M. R. 1987. Food habits of largemouth bass in two heated east Texas reservoirs. Proceedings of the Annual Conference Southeastern Association of Fish and Wildlife Agencies 39(1985):166–175.

Weithman, A. S., and R. O. Anderson. 1978. A method

of evaluating fishing quality. Fisheries (Bethesda) 3(3):6–10.
Wilkinson, L. 1990a. SYGRAPH: the system for graphics. SYSTAT, Inc., Evanston, Illinois.
Wilkinson, L. 1990b. SYSTAT: the system for statistics. SYSTAT, Inc., Evanston, Illinois.
USCFF (U.S. Commission of Fish and Fisheries). 1895. Report of the commissioner for the year ending June 30, 1893, part XIX. USCFF, Washington, DC.
USCFF (U.S. Commission of Fish and Fisheries). 1896. Report of the commissioner for the year ending June 30, 1894, part XIX. USCFF, Washington, DC.
USCFF (U.S. Commission of Fish and Fisheries). 1898a. Report of the commissioner for the year ending June 30, 1896, part XXII. USCFF, Washington, DC.
USCFF (U.S. Commission of Fish and Fisheries). 1898b. Report of the commissioner for the year ending June 30, 1897, part XXIII. USCFF, Washington, DC.
USCFF (U.S. Commission of Fish and Fisheries). 1899. Report of the commissioner for the year ending June 30, 1898, part XXIV. USCFF, Washington, DC.
USCFF (U.S. Commission of Fish and Fisheries). 1900. Report of the commissioner for the year ending June 30, 1899, part XXV. USCFF, Washington, DC.

Accomplishments and Roadblocks of a Marine Stock Enhancement Program for White Seabass in California

DONALD B. KENT AND MARK A. DRAWBRIDGE

Hubbs-Sea World Research Institute
2595 Ingraham Street, San Diego, California 92109, USA

RICHARD F. FORD

Department of Biology, 5300 Campanile Avenue
San Diego State University, San Diego, California 92182, USA

Abstract.—The Ocean Resources Enhancement and Hatchery Program (OREHP) was established in 1983 to evaluate the feasibility of using cultured fish to enhance depleted populations of white seabass *Atractoscion nobilis* in southern California. This paper describes some of the operational and political aspects associated with developing a marine stock enhancement program. Initial OREHP research was directed toward developing culture protocols, evaluating tagging techniques, identifying population characteristics of wild stocks (including genetics), investigating patterns of postrelease survival, and using computer models to conduct cost-benefit analyses of the stocking program. In 1991 OREHP began an expansion to incorporate a production-scale marine fish hatchery and satellite cage rearing facilities. Operational funds have been secured through extended state legislation and construction funds from a mitigation requirement and private donations. A site for the hatchery has been identified adjacent to a coastal lagoon. All required preconstruction permit applications have been approved. An overlap in the authority of two state agencies with regard to coastal resource issues is being resolved through a memorandum of agreement between these agencies. This agreement stipulates the dedication of money for postrelease and genetic assessments, and the drafting of a written plan describing how fish are to be cultured, distributed, tagged and released.

Based on the experiences of OREHP, we recommend that each species be evaluated individually to determine the efficacy of stocking. The evaluation should consider the biological characteristics and management history of each species. If possible, operational funds should be secured through a dedicated account so the program can be evaluated for several years and so funds cannot be diverted into other unrelated programs. Both scientific and user group advisors should be involved when establishing the goals and oversight responsibilities for the program, and lines of authority for the program should be established early in its development. A high-profile review process should be maintained, and postrelease and genetic assessments should be incorporated into the program as early as possible.

The harvest of many fishery resources at or above maximum sustainable levels worldwide (NMFS 1992) has increased the need for research and development of stock enhancement. The Food and Agriculture Organization of the United Nations (FAO) has reported that approximately 33% of the 200 fisheries it monitors are depleted or overexploited (FAO 1988). The National Marine Fisheries Service (NMFS) has similarly reported that fisheries resources of the United States were fully utilized (26%) or overutilized (28%), and the status of 34% of the stocks is unknown (NMFS 1992). By comparing worldwide population growth with the reported diminishing yields from harvest fisheries, New (1991) concluded that there will be an annual aquaculture production deficit of 50 million tons by the year 2025. Because New (1991) reports current production at around 11 million tons, this more than fourfold expanded production in just over 3 decades will require that the aquaculture industry try to meet this goal through development of three as yet underexploited production methods: (1) the expanded use of inland farms that integrate aquaculture and agriculture, (2) the increased use of off shore sites where the development of protected embayments is not in conflict with aquaculture, and (3) the further development of aquaculture-based fisheries (i.e., stock enhancement) that will augment the natural production capabilities of wild populations. If this expanded food production capability is to be met, it is clear that future fisheries management plans will have to evaluate stock enhancement as a means of helping to maintain or increase food resources and overall resource diversity. This paper describes the case history of a marine stock enhancement program for white seabass *Atractoscion nobilis* in southern California and emphasizes the political and operational aspects of establishing such a program.

Project Background

Marine coastal fisheries provide significant economic value through both commercial and recreational harvests. Venrick (1985) estimated that marine sportfishing contributed more than US$2 billion to the California economy in 1983. Because coastal fisheries represent such a significant resource to California, many sportfishing groups have expressed serious concerns about the declines of key species. In 1982 both an organization of anglers, the National Coalition for Marine Conservation—Pacific Region, and an affiliation of commercial sportfishing vessel owners, the California Sportfishing Association, suggested evaluating the potential for marine fish hatcheries to augment depleted populations of fish in southern California.

In 1983, the California legislature established the Ocean Resources Enhancement and Hatchery Program (OREHP) under the direction of the California Department of Fish and Game (CDF&G) to conduct "basic and applied research on the artificial propagation and distribution of adversely affected marine fish species . . ." (California State Assembly 1983). The legislation established $1 sportfishing and $10 commercial marine fishing stamps to fund this program. It also mandated the formation of an advisory panel to oversee the program with members representing the commercial and sportfishing industries (Sportfishing Association of California and California Gillnetters Association), a conservationist group (National Coalition for Marine Conservation), the aquaculture industry (California Aquaculture Association), the scientific community (University of California and California State University), and CDF&G.

The OREHP advisory panel identified the white seabass as the most appropriate species for use in an experimental stocking program. White seabass is an important sport and commercial species, and catches have declined to low levels. Regulations to manage the white seabass fishery have been in place since 1931 and continue to the present day with some modifications. The regulations include a minimum size limit (711 mm total length [TL]), closed seasons, bag limits, and gear restrictions. Despite the regulations, commercial and recreational fisheries catches have declined (Vojkovich and Reed 1983; Vojkovich, CDF&G, personal communication).

In order to obtain definitive answers regarding the efficacy of stocking marine fish, the advisory panel established specific goals for the program that were used to solicit and evaluate research proposals. These goals were directed toward (1) developing culture techniques, (2) assessing natural population characteristics and postrelease survival, (3) evaluating genetic characteristics of wild and hatchery stocks, and (4) determining the economic feasibility of marine stock enhancement.

FIGURE 1.—Map of southern California showing culture, release, and assessment areas for white seabass. Key to sites: gill-net sampling (G); cage rearing and release (P); and recapture (R).

Because the funding source for the operation of OREHP is derived from fishers north of the Mexican border and south of Point Arguello, the culture, release, and assessment work has also been confined to this region (Figure 1).

Work Accomplished to Date (1983–1993)

Culture

Culture and stocking research was conducted at the marine laboratory on Mission Bay, California, jointly maintained by Hubbs-Sea World Research Institute and San Diego State University. White seabass broodfish were obtained from several sources, primarily commercial sportfishing vessels. As of this writing the hatchery maintains a breeding population of 33 wild-caught white seabass (18 kg average weight). Effort is being dedicated to increase the total broodfish spawning population to 200 animals—well over the number (i.e., 75) recommended by Bartley and Kent (1990) to ensure the released fish will have minimal genetic effect on the wild population.

Broodfish are divided into separate recirculating-water pools where they are maintained under controlled temperature and photoperiod regimes to induce spawning throughout the year. Eggs are collected and reared through the larval, postlarval, and juvenile stages in culture pools until reaching a size of approximately 65 mm TL (60 days). During this culture sequence, juveniles are weaned from a diet of live and frozen crustacean food to a commercially available pellet. After weaning the juvenile white seabass are transported to cages where they are held for an additional 6–7 months prior to release.

Pre- and Postrelease Assessment

Initial ecological surveys and subsequent attempts to recapture hatchery-reared white seabass used different gear types, including beach seines, beam and otter trawls, experimental gill nets, and hook and line. Experimental gill nets have been the most effective gear because they catch a wide size range of fish (200–850 mm TL), can be used in a diversity of habitats (kelps beds, embayments, and rock reefs) and have relatively high catch rates. Since the gill-net sampling program was initiated in 1988, the majority of effort has been focused within Mission Bay, the primary release site, and along the adjacent open-ocean coast in kelp beds. A hook and line sampling program was initiated in 1992 through the efforts of anglers aboard commercial sportfishing vessels. This program is relatively inexpensive and samples a wide area. Due to the current size limit of 711 mm TL, this method provides data for only larger white seabass.

Prior to release, all hatchery-reared fish are marked for future identification. Oxytetracycline was initially used to mark hatchery-produced fish. This mark was found ineffective for our purposes because it did not last for more than 4 years. Coded wire tags have been used since 1990. This tagging system has enabled identification of the release group to which recaptured fish belong and provided more accurate estimates of growth and patterns of migration.

Genetic Assessment

The genetic diversity of white seabass in southern California has been measured (Bartley and Kent 1990). From genetic analysis of cultured white seabass the program concluded (1) there are no measurable temporal, clinal, or geographic components to the genetic diversity of the white seabass population studied; (2) the genetic diversity of cultured fish from a single spawn is less than that of the wild population; (3) the genetic diversity observed between multiple spawns of cultured broodfish approaches that of the wild population; and (4) a spawning group of at least 75 broodfish is necessary to provide the rarest alleles (approximately 2%) observed in the wild population (Bartley et al. 1995, this volume).

Bioeconomic Assessment

A computer model was developed by L. W. Botsford and R. C. Hobbs (unpublished data) that provides a standard method for evaluating new culture techniques and for estimating the costs to produce fish of different ages prior to release. These culture cost estimates are then used in combination with estimates of postrelease survival to predict the benefits and costs of the program. By use of a calculated, theoretical curve that defines the relationship between the size at release and postrelease survival, the optimal size at release was found to be 210 mm TL when culture costs were considered. This bioeconomic model will be updated as new growth and survival data are gathered.

Planning for Expansion (1991–1994)

Beginning in 1991, OREHP initiated an expansion of the stock enhancement program. A review of the work performed from 1983 to 1991 allowed the advisory panel to recommend to the CDF&G an increase in the size of the experimental rearing and release program. To expand the program in a cost-effective manner, a logistical decision was made to centralize the hatchery operation and to decentralize the grow-out culture by using cage systems operated throughout the southern California range of the experiment. The planning process involved developing funding sources for operational and capital expenses, evaluating cage rearing sites and operators, obtaining prerelease baseline data on the abundance of white seabass, and identifying other potential release areas. A location for a full-scale hatchery was identified and a preliminary design for it developed.

Funding

Operational funds to support the hatchery and assessment work had to be secured to expand OREHP. Support from the local fishing community and legislators resulted in reauthorization of the original legislation and extended the fishing stamp for an additional 10 years to 2003 (California State Assembly 1993).

In the summer of 1991 OREHP representatives approached the California Coastal Commission about including a marine hatchery as part of a mitigation plan for the San Onofre Nuclear Generating Station immediately south of San Clemente, California. Following a 2-year review of the viability of stock enhancement programs by Coastal Commission staff and scientific advisors, the Coastal Commission agreed to release $1.2 million to support the cost of hatchery construction. The cost estimate was based on a preliminary hatchery design, which in turn was based on production capabilities that could be supported by the available operational funds.

Site Selection for Central Hatchery and Cages

During the time funding sources were being sought, potential sites for the main hatchery and cages were being reviewed. It soon became evident that availability of sites within embayments for small-scale cage systems was not as limiting as was the availability of undeveloped land adjacent to a clean seawater supply along the southern California coast for the hatchery.

A suitable site for the hatchery was selected in Carlsbad, California (Figure 1) on property owned by the local utility company, San Diego Gas and Electric Co. (SDG&E), also part owner of the San Onofre Nuclear Generating Station. Through a license agreement with SDG&E, the property was made available virtually free of cost. It is situated adjacent to the outer basin of Agua Hedionda Lagoon, which receives tidal flushing from a coastal inlet, located approximately 300 m away. The site was specifically designated for aquaculture use by a local coastal plan.

Efforts to incorporate cage culture into the overall program for white seabass were initiated in 1991. The cage systems are located in various southern California embayments and are owned and operated by volunteer groups of anglers that have incorporated as nonprofit entities. These systems allow not only an expansion of the culture program by providing more fish of a larger size, but also an expansion of the release program from just San Diego to the entire southern California Bight (Figure 1).

Gill-Net Surveys

In preparation for large-scale releases of white seabass, the gill-net survey was expanded to include embayments where white seabass were cultured in cages, as well as other potential sites in southern California (Figure 1). The primary objective of the expanded gill-net survey is to collect prerelease baseline data on the relative abundances of white seabass and other sympatric fish species in these areas. Areas inhabited by wild white seabass of the same age-class as released fish should represent the most suitable areas for release. This information will also be used to help determine if wild white seabass or other species are being displaced or consumed by stocked fish.

Implementing the Expansion (1993)

Securing Capital Funds

Due to the comprehensive nature of the expanded program, the involvement of coastal resources, and the need to assess the mitigation value of the program for the San Onofre Nuclear Generating Station, several organizational tasks and a series of assurances were required by the Coastal Commission before mitigation moneys could be released.

1. A memorandum of agreement (MOA) must be developed between the two state agencies (CDF&G and California Coastal Commission) that outlines the regulatory authority of the agencies in management of the joint research and mitigation missions of the hatchery program.
2. A panel comprised of representatives from the CDF&G, the California Coastal Commission, the OREHP advisory panel, the Southern California Edison Company, the National Marine Fisheries Service, and the University of California must be formed. The responsibilities of this joint panel are stated in detail in the MOA.
3. A comprehensive hatchery plan must be prepared that details the operational methods by which the goals of the MOA will be accomplished.
4. A coastal development permit must be issued by the California Coastal Commission, permitting the construction of the hatchery facility.

The MOA identified all of the parties involved and the purpose of the agreement. It provided a description of the project and responsibilities for planning and oversight, including the composition of the joint panel. Assurances to be made regarding environmental quality were described as they relate to hatchery and cage system operations. Requirements for a postrelease evaluation program and a genetic quality assurance program were described in detail, including the minimum annual funding requirements to be dedicated to each. Finally, procedures manuals were required for both the hatchery and cage system operations.

TABLE 1.—Permits required for construction of the OREHP hatchery in Carlsbad, California.

Permit	Responsible agency	Time to process	Review and monitoring requirements	Fee
Conditional use	City of Carlsbad Planning Department	8 months	Seventy-three special conditions; annual review of compliance	$3,600
Grading	City of Carlsbad Engineering Department	2 months	Inspection and bonds required by city; approval prior to processing building permit	$6,000
Building[1]	a) City of Carlsbad Building Department b) Carlsbad Unified School District	3 months	City inspection; certificate of occupancy prior to occupying facility	$53,000 $870
Wastewater discharge	Carlsbad Municipal Water District	1 month	Periodic monitoring	$500
Coastal development	California Coastal Commission	10 months	Nineteen special conditions	$4,000
NPDES[2]	Environmental Protection Agency permit issued by California Regional Water Quality Control Board	1.5 years	Waived in lieu of periodic monitoring	$0
404	U.S. Army Corps of Engineers	1.3 years	Three special conditions; postconstruction eelgrass impact report	$500
Total Cost				$68,470

[1] Building permit fee includes $22,000 for sewer and water service; $21,000 for public and community facilities; and $500 for traffic and development impact fees.
[2] National Pollutant Discharge Eliminations System Permit.

The comprehensive hatchery plan addresses the initial objectives for culture, stocking and assessment of white seabass, and included the following:

1. defined enhancement objective or endpoint in units biomass or catch contributed;
2. culture protocols for producing white seabass with a minimum effect on the wild population's genetic variability;
3. methods for tagging fish and managing the resulting database;
4. procedures for juvenile culture and release;
5. methods for transporting the fish from the hatchery to cage systems and from cage systems to release sites;
6. standards for measuring the success of the hatchery;
7. budget and schedule for hatchery construction;
8. procedures manual for cage systems; and
9. provisions for revising the hatchery plan after the first year and biennially thereafter.

The comprehensive hatchery plan is important for several reasons. First, it acknowledges that stock enhancement programs, especially those in their infancy, are part of a dynamic process. Second, it provides a common framework from which to direct research effort. This is especially critical when many organizations and agencies have a vested interest in helping to establish the objectives and assess the results.

Permits and Approvals for Hatchery Construction

The permit requirements for development in California's coastal zone (Table 1) entail a very involved process and often require expert consultation with outside resources. Also, the permit process is site- and project-specific.

On a local level, permits were required that reviewed and approved the use of the site in addition to construction permits (i.e., grading, construction, and tenant improvements). Because the site was previously undeveloped, a Conditional Use Permit was required, which allows the local government to apply special requirements that are tailored to fit the proposed project and thus avoid problems that may be associated with the particular type of use (California Permit Handbook 1992). As part of the Conditional Use Permit, an environmental impact report may be required if a mitigated negative declaration is not found to be sufficient.

At the state level, a Coastal Development Permit was required from the California Coastal Commission. The Coastal Commission retains permit authority over tidelands, submerged lands, and certain lands held in the public trust. Because OREHP is administered by the CDF&G, no formal permit was required by this state agency to culture or release fish.

At the federal level, applications for permits were required by the Regional Water Quality Control

Board and the Army Corps of Engineers. A discharge permit, formally referred to as a National Pollutant Discharge Eliminations System Permit (NPDES), is required by the owner or operator of any facility that discharges waste into any surface waters of the state (California Permit Handbook 1992). Because our anticipated annual production level falls well below the federal requirements (9,090 kg) for a concentrated animal holding facility, we were given a waiver from this permit but were conditioned to conduct periodic monitoring of both hatchery discharge and stormwater runoff. Because the Army Corps of Engineers maintains jurisdiction over all navigable waters, a permit was required to install the seawater intake structure for the hatchery.

Discussion

Because many questions still need to be addressed, the decision to increase the scale of the OREHP experiment has gone through significant critical review. The mission early in the program to address not only the culture problems inherent to enhancement but also the economic and ecological impacts has allowed the program to carefully scale up to its proposed level. However, even with acceptance of the experimental concept and a clear mandate to proceed, OREHP's expansion has been dramatically slowed by the inertia inherent in development projects in the California coastal zone. In fact, the debate over the value of the proposed OREHP hatchery pales in comparison with the effort needed to obtain all of the permits required for construction and operation of the proposed facility.

The procedural hurdles associated with obtaining the necessary permits and approvals to construct and operate the enhancement hatchery are a major deterrent. There are multiple agencies at the federal, state, and local levels with overlapping jurisdictions and different permitting and reporting requirements. Without the help of a professional development consultant, it would be nearly impossible for an organization operated by scientists to identify the numerous agencies from which permits must be obtained The requirements imposed by different agencies on the hatchery project are duplicative and sometimes contradictory. The permitting and reporting requirements of many individual agencies are burdensome and time consuming, and agency staff have little or no apparent incentive to process permit applications on a timely basis.

The experience of OREHP in its effort to develop this relatively modest culture facility has been that even with an overall consensus to construct the hatchery, there exists a bureaucratic logjam. Agency regulations and requirements are designed to allow managed growth, but in their application, they became a discouragement to the execution of the program. The permitting process for the hatchery facility began in July 1991, yet at this writing there are still preconstruction permit applications pending review and approval. The permits acquired thus far have involved two public hearings, an appeal to city council that required an additional public hearing, and a lawsuit that is under review for appeal. In an effort to provide fair public review of development projects, California law has, in effect, allowed single individuals to cause significant expenditures in time and funding to delay and, in some cases by attrition, halt projects that have an overwhelming majority of public support.

In addition to this permitting impasse, pressure continues to be exerted from both sides of the stock enhancement question. Some feel that the maintenance of fishing yields should be accomplished only through the informed management of the existing stock. Many in this group also hold that stock enhancement simply represents a seemingly attractive technical solution to very complicated environmental resource problems, and that its appeal to user groups in the short term cannot really balance the need to correct the underlying causes of the diminished harvest. In addition, because of the significant number of hatcheries built in the Pacific Northwest for salmon *Oncorhynchus* spp. in response to logging and hydroelectric projects, a fear exists that the acceptance of marine stock enhancement as a viable resource management tool might result in its common use for mitigation, allowing further degradation of the coastal environment.

On the other side of the argument, user groups of both recreational and commercial fishers have stated the opinion that the concerns raised by resource managers and the scientific community have little practical merit, and that costs associated with scientific investigations can be eliminated in lieu of supporting increased hatchery production. It can be extremely difficult to convince the lay person that concerns as seemingly esoteric as the genetic variability of the progeny produced in the hatchery may in some way diminish the viability of the wild population.

These on-going arguments often serve only to polarize the debate further on the usefulness of enhancement hatchery programs, ultimately toward limiting the ability to actually test their efficacy. It is incumbent upon the fisheries management community to resist these pressures and apply the best scientific procedures in testing the real potential for fisheries enhancement. If this does not occur, then user groups that control or strongly influence the economic and political resources supporting enhancement programs will cause projects to be performed that lack the scientific structure required to allow adequate assessment of positive or negative impacts.

Based on the experiences of OREHP, we recommend that each species be evaluated individually to determine the efficacy of stocking. The evaluation should consider the biological characteristics and management history of each species. If possible, operational funds should be secured through a dedicated account so the program can be evaluated over several years, or at least until the stocked fish are recruited into the fishery, and so funds cannot be diverted into unrelated programs. Both scientists and user groups should be involved when establishing the goals and oversight responsibilities for the program, and lines of authority for the program should be established early in its development. A high-profile review process should be maintained, and postrelease and genetic assessments should be incorporated into the program as early as possible.

In the case of California's OREHP, we feel that an excellent working relationship has been developed between the scientists conducting the research, the management agencies responsible for the resource, and the user groups providing the funding. With continued scientific review and barring further permitting constraints, the OREHP hopes to further the goal of adequately testing marine fisheries enhancement as a responsible resource management tool.

References

Bartley, D. M., and D. B. Kent. 1990. Genetic structure of white seabass population from the southern California Bight region: applications to hatchery enhancement. California Cooperative Oceanic Fisheries Investigations Report 31:97–105.

Bartley, D. M., D. B. Kent, and M. A. Drawbridge. 1995. Conservation of genetic diversity in a white seabass hatchery enhancement program in southern California. American Fisheries Society Symposium 15:249–258.

California Permit Handbook. 1992. Governor's office of planning and research, office of permit assistance, Sacramento, California.

California State Assembly. 1983. Ocean fishery research. By Larry Stirling. Bill AB 1414, Chapter 982. Sacramento.

California State Assembly. 1993. California ocean resources enhancement and hatchery program. By Dede Alpert. Bill AB 960, Chapter 987. Sacramento.

Food and Agriculture Organization of the United Nations (FAO). 1988. Review of the state of world fishery resources. FAO Fisheries Circular 710, Revision 7.

New, M. B. 1991. Turn of the millennium aquaculture: navigating troubled waters or riding the crest of the wave? World Aquaculture 22(3):28–49.

NMFS (National Marine Fisheries Service). 1992. Our living oceans: report on the status of U.S. living marine resources, 1992. NOAA (National Oceanic and Atmospheric Administration) Technical Memorandum, NMFS F/SPO-2, Silver Spring, Maryland.

Venrick, E. L. 1985. Marine recreational fishing. Pages 38–42 in The oceans and the economy of San Diego. University of San Diego, Scripps Institution of Oceanography, San Diego, California.

Vojkovich, M., and R. J. Reed. 1983. White seabass, Atractoscion nobilis, in California-Mexico waters: status of the fishery. California Cooperative Oceanic Fisheries Investigations Report XXIV:79–83.

Using Genetics in the Design of Red Drum and Spotted Seatrout Stocking Programs in Texas: A Review

TIM L. KING

*National Biological Service, Leetown Science Center
Aquatic Ecology Laboratory, 1700 Leetown Road
Kearneysville, West Virginia 25430, USA*

ROCKY WARD, IVONNE R. BLANDON, AND ROBERT L. COLURA

*Perry R. Bass Marine Fisheries Research Station
Texas Parks and Wildlife Department, HC 02, Box 385
Palacios, Texas 77465, USA*

JOHN R. GOLD

*Department of Wildlife and Fisheries Sciences
Texas A&M University, College Station, Texas 77843, USA*

Abstract.—In response to stock identification and hatchery-product evaluation concerns, the Texas Parks and Wildlife Department conducted extensive genetic surveys on populations of red drum *Sciaenops ocellatus* and spotted seatrout *Cynoscion nebulosus* to estimate baseline levels of genetic variation, delineate population structure, and identify rare genes that could be useful as genetic markers for assessing supplemental stocking programs. Significant clinal variation at the locus that codes for aspartate aminotransferase-2 and in average individual heterozygosity was observed in spotted seatrout. Allozyme analyses suggest that both spotted seatrout and red drum within the areas surveyed are single, randomly mating populations. Rare alleles that could function as genetic markers were identified in both species. These genetic surveys have led to modifications of the respective species' supplemental stocking programs. For example, presence of clinal genetic variation in spotted seatrout allele frequencies has resulted in the development of three distinct broodstocks representative of the upper, middle, and lower Texas coasts. To evaluate stocking success, marker alleles were chosen for red drum and spotted seatrout. Research was conducted to test inheritance and viability of the selected alleles. No differences in growth were observed among fish of various ages possessing all possible marker allele genotypes, and the alleles appeared to be inherited according to simple Mendelian expectations.

Prior to 1981, red drum *Sciaenops ocellatus* in Texas supported commercial and recreational fisheries. Declining populations in the late 1970s, attributable to growth and recruitment overfishing, resulted in a ban on commercial sale of red drum and strict harvest regulations on sportfishing (TPWD 1985). Additional regulations were imposed throughout the 1980s to increase population abundance in the coastal bays and to augment recruitment into the Gulf of Mexico spawning population.

The spotted seatrout *Cynoscion nebulosus* is a vital recreational resource along much of the U.S. Atlantic coast and throughout the Gulf of Mexico. In addition to experiencing high fishing harvest, geographic populations throughout the spotted seatrout's range are susceptible to environmental perturbations. Recently, spotted seatrout abundance on the Texas coast has been reduced by severe freezes (McEachron et al. 1984) and red tide *Ptychodiscus brevis* (Hammerschmidt 1987). These incidents, combined with increased recreational fishing demand (Meador and Green 1986), have led to changes in management policies, such as restricted size, bag, and possession limits (TPWD 1985).

The Texas Parks and Wildlife Department (TPWD) has initiated marine stocking programs to reverse the decline of red drum and spotted seatrout populations. Many of the immediate effects of the red drum stocking program have been assessed by TPWD (Bumguardner et al. 1990). Survival of stocked red drum has been evaluated by stocking out of season and tagging (Matlock et al. 1984). Due to intrinsic variability in growth, the former method is useful in identifying the contribution of hatchery-produced individuals for only about 1 year. The latter method is generally cost prohibitive because of low recapture rate and poor tag retention (Bumguardner et al. 1990).

An effective marking method is needed to determine the contribution of the stocked fish to the red drum population and recreational fishery; this tool also will be needed to determine the contribution of stocked spotted seatrout. Furthermore, extensive knowledge of the population structure of red drum and spotted seatrout inhabiting the Texas coast, including the identification of stocks, is needed to facilitate prudent long-range management decisions.

Gene marking, which has proven successful in evaluating stocking programs of reservoir walleye *Stizostedion vitreum* (Murphy et al. 1983) and anadromous salmonids (Seeb et al. 1986) may be a viable method for assessing red drum and spotted seatrout stocking programs. Using allozymes as genetic markers can provide all the information (e.g., growth, movement, survival, population estimation, and stock contribution) commonly obtained using artificial and other biological tagging methods. Furthermore, selectively bred genotypes (i.e., fish with genetics markers) mark a large number of individuals without the handling stress and mortality associated with other tagging methods. In addition, the mark will endure for the life of the fish and constitutes a heritable trait. Therefore, gene marking also allows estimation of reproductive contributions of stocked fish.

In response to stock identification and hatchery-product evaluation concerns, TPWD conducted extensive allozyme surveys on red drum and spotted seatrout populations. Ancillary to determining stock structure, the investigations were designed to estimate baseline allele frequencies and to identify rare alleles that could be used as genetic markers for assessing supplemental stocking programs. The objectives of this paper are to (1) review the population structure of red drum and spotted seatrout inhabiting the Texas Gulf Coast as suggested by allozyme analysis, (2) review the gene marking experimentation, and (3) briefly summarize the influence of genetic analyses on present management strategies such as broodstock management and on the development of stock assessment programs using gene marking.

Allozyme Surveys

Red Drum

An extensive survey of allozyme variation in adult (offshore) and juvenile (inshore) red drum from North Carolina to southern Texas suggested that genetically differentiated subpopulations occur in the northern Gulf of Mexico and along the southern Atlantic coast (Gold et al. 1993). Gold et al. (1994) analyzed allele frequencies for nine polymorphic loci by year-class and reported an absence of temporal and spatial allelic heterogeneity among inshore and offshore year-classes and among all geographic populations surveyed in the northern Gulf of Mexico. Similarly, five geographic populations of juvenile (1986 and 1987 year-classes) red drum from Texas exhibited an effective absence of genetic subdivision, a high degree of genetic similarity, and homogeneous allele frequencies.

Spotted Seatrout

Spotted seatrout from Texas bays and estuaries demonstrated levels of genetic variability, gene flow, and population differentiation similar to that reported in other portions of the species range (King and Pate 1992; Ramsey and Wakeman 1987). Examination of spotted seatrout allelic and genotypic frequencies at seven polymorphic loci revealed low levels of genetic variability and clinal patterns of geographic variation in *sAAT-2** (which codes for aspartate aminotransferase, enzyme number 2.6.1.1. [IUBNC 1984]) allele frequencies and average individual heterozygosity (King and Pate 1992). A statistically significant association (cline) existed in the frequency of the *sAAT-2*80* allele with respect to degrees north latitude and west longitude. Frequency of the *sAAT-2*80* allele increased from 0.9% in fish from east Texas to 17.1% in fish from Mexico. A statistically significant association also existed in average individual heterozygosity with respect to degrees north latitude and west longitude. Tests of heterogeneity indicated allele counts at *sAAT-2** were distributed heterogeneously among the 12 localities.

Gene Marking

Both allozyme surveys identified gene loci with allele frequencies favorable for producing a selectively bred gene mark. Moreover, selected loci were resolvable using noninvasive sampling (e.g., fin tissue). The frequency of the selected alleles was sufficiently low that individuals homozygous for the marker would be rare in the wild population. Based on these criteria, the *PEPB*85* allele (tripeptide aminopeptidase, 3.4.-.-), which averaged 7% coastwide, was chosen for the spotted seatrout gene marking study. In red drum, four loci (*ESTD** [esterase-D, 3.1.-.-], *PEPB**, *PEPD** [proline dipeptidase, 3.4.13.9], and *PEPS** [peptidase-S, 3.4.-.-]) were identified as potential markers. The dimeric esterase locus *ESTD**, was chosen primarily because stain ingredients are less toxic, less costly, and the products can be resolved much more quickly than can peptidase stains. The *ESTD*95* allele averaged 6% along the Texas Gulf Coast. Both target loci were di-allelic (*PEPB*−100* and **−85*; *ESTD*100* and **95*).

From fall 1991 to spring 1993, more than 700 red drum and 600 spotted seatrout from the entire Texas coast were screened for rare alleles. Approximately 50 mature *ESTD*100/95* and one mature *ESTD*95/95* red drum were collected. Nearly 120 mature *PEPB*100/85* and 3 mature *PEPB*85/85* spotted seatrout were obtained. These fish were placed into large indoor tanks and subjected to a temperature and photoperiod cycle that induced red drum spawns beginning in late June 1992 and spotted seatrout spawns in August 1992. Research was initiated at the onset of spawning to test basic biological assumptions concerning inheritance and viability of the selected allele.

Progeny of heterozygous red drum broodfish ($N = 7$ spawns) were cultured to 30 d of age to detect fitness differences among genotypes and to confirm Mendelian inheritance in the F_1 fish (*viz* 1:2:1 ratio of **100/100:*100/95:*95/95*). Samples of F_1 progeny ($N = 100$) from these trials revealed little growth differential (as measured by total length in millimeters) among red drum possessing any of the three genotypes (TPWD, unpublished data). Inheritance (i.e., survival) among the three genotypes was variable; however, there was no trend suggesting decreased survival of fish possessing the rare **95* allele in either the homozygous or heterozygous genotype (TPWD, unpublished data). This preliminary investigation suggests the *ESTD*95* is a biologically sound gene marker in that red drum possessing the rare allele appear to perform equally well as fish possessing the common allele. See King et al. (1993) for an overview of the Texas red drum gene marking program. Fish sampled from spotted seatrout pond culture trials are awaiting electrophoretic analysis.

Management Implications

Genetic evidence presented in this study suggests that the biological and physical processes necessary for formation of discrete stocks are not present in red drum or spotted seatrout inhabiting the Texas Gulf Coast, and gene flow is sufficient to prevent genetic partitioning. These findings suggest that each of these species should be regulated as a single stock. However, spatial heterogeneity and clinal variation in allele frequencies at the *sAAT-2** locus and in average individual heterozygosity were observed in spotted seatrout. Regardless of the mechanism(s) creating the clinal variations, the finding may indicate genetic or nonheritable variation in physiology and should be incorporated into any long-term management strategy of spotted seatrout populations in Texas.

In light of the clinal variation observed at the *sAAT-2** locus and in individual heterozygosity, and given the limited vagility of spotted seatrout, stocking success may be augmented if broodstock were collected at sites environmentally similar to those being rehabilitated. The TPWD's recently established (spring 1992) spotted seatrout population enhancement program has selected broodstock that will assist in preserving the genetic variability present. Specifically, three separate broodstocks from the upper, middle, and lower Texas coasts are maintained at TPWD's Gulf Coast Conservation Association–Central Power and Light–Marine Development Center near Corpus Christi, Texas. Every effort has been made to stock fingerlings produced by these broodstocks in bays and estuaries contiguous to broodstock collection sites. In addition to the stock identification efforts, research has identified an allozyme marker to assist in assessing the success of the supplemental stocking program. Broodstock homozygous for the rare *PEPB*85* allele have been spawned and are being reared to maturity. Genetically marked spotted seatrout could be stocked by fall 1994.

A red drum gene marking study, based on a rare allozyme, was implemented in fall 1993. Approximately 250,000, 30-d-old fingerlings were stocked into East Matagorda Bay, Texas. A choice was made by TPWD to stock offspring from heterozygous crosses rather than awaiting the development of homozygous rare (marked) fish. This plan consists of monitoring changes in *ESTD*95* frequency (and the absolute number of **95/*95* individuals) at the stocking site and adjacent bay systems. The recovery phase, which began immediately, consists of the removal of fin tissue from all fish collected in East Matagorda Bay bag-seine and gill-net samples by TPWD Resource Program personnel. Fin tissue will also be taken from fish during harvest surveys.

Gene marking projects using selectively bred characters raise concerns about potential deleterious effects on the red drum population. Texas Parks and Wildlife Department has gone to unprecedented lengths to assure the utility of the marker alleles and the success of the assessment procedures. The potential benefits of a successful stock assessment should overshadow negative effects the selectively bred fish might impart on the natural population (King et al. 1993).

In summary, allozyme surveys of spotted seatrout

and red drum populations have allowed identification of broodstock that better represent the extant genetic variation in targeted stocking areas. The same surveys have provided a robust mechanism (*viz*, gene mark) for direct estimates of stocking success, abundance, natural mortality, movement, recruitment into the natural population, and reproduction.

Acknowledgments

The U.S. Fish and Wildlife Service provided partial funding under the Federal Aid in Sport Fish Restoration Act (Project F-36-R). Work was also supported by the Coastal Fisheries Branch of the Texas Parks and Wildlife Department, the Marine Fisheries Initiative (MARFIN) Program of the U.S. Department of Commerce (grants NA89-WC-H-MF025 and NA90AA-H-MF107), and the Texas A&M University Sea Grant College Program (grants NA85AA-D-SG128 and NA89AA-D-SG139). The authors are indebted to Paul Hammerschmidt, whose insight and persistence has given the Perry R. Bass Marine Fisheries Research Station an opportunity to prove that gene marking is a viable assessment method for the red drum stocking program.

References

Bumguardner, B. W., R. L. Colura, A. F. Maciorowski, and G. C. Matlock. 1990. Tag retention, survival, and growth of red drum fingerlings marked with coded wire tags. American Fisheries Society Symposium 7:286–292.

Gold, J. R., L. R. Richardson, T. L. King, and G. C. Matlock. 1993. Temporal stability of nuclear gene (allozyme) and mitochondrial DNA genotypes among red drums from the Gulf of Mexico. Transactions of the American Fisheries Society 122:659–668.

Gold, J. R., T. L. King, L. R. Richardson, D. A. Bohlmeyer, and G. C. Matlock. 1994. Genetic studies in marine fishes, part 7. Allozyme differentiation within and between red drum (*Sciaenops ocellatus*) from the Gulf of Mexico and Atlantic Ocean. Journal of Fish Biology 44:567–590.

Hammerschmidt, P. C. 1987. The tide of the red death. Texas Parks and Wildlife Magazine 45:10–15.

IUBNC (International Union of Biochemistry, Nomenclature Committee). 1984. Enzyme nomenclature. Academic Press, San Diego, California.

King, T. L., and H. O. Pate. 1992. Population structure of spotted seatrout inhabiting the Texas Gulf coast: an allozymic perspective. Transactions of the American Fisheries Society 121:746–756.

King, T. L., R. Ward, and I. R. Blandon. 1993. Gene marking: a viable assessment method. Fisheries (Bethesda) 18(2):4–5.

MacLean, J. A., D. O. Evans, N. V. Martin, and R. L. DesJardine. 1981. Survival, growth, spawning distribution, and movements of introduced and native lake trout (*Salvelinus namaycus*) in two inland Ontario Lakes. Canadian Journal of Fisheries and Aquatic Sciences 38:1685–1700.

Martin, J., J. Webster, and G. Edwards. 1992. Hatcheries and wild stocks: are they compatible? Fisheries (Bethesda) 17(1):4.

Matlock, G. C., B. T. Hysmith, and R. L. Colura. 1984. Returns of tagged red drum stocked into Matagorda Bay, Texas. Texas Parks and Wildlife Department, Coastal Fisheries Branch, Management Data Series 63, Austin.

McEachron, L. W., G. Saul, J. Cox, C. E. Bryan, and G. C. Matlock. 1984. Winter's arctic weather takes a heavy toll on marine life. Texas Parks and Wildlife Magazine 43:10–13.

Meador, K. L., and A. W. Green. 1986. Effects of a minimum size limit on spotted seatrout recreational harvest. North American Journal of Fisheries Management 6:509–518.

Murphy, B. R., L. A. Neilsen, and B. J. Turner. 1983. Use of genetic tags to evaluate stocking success for reservoir walleyes. Transactions of the American Fisheries Society 112:457–463.

Ramsey, P. R., and J. M. Wakeman. 1987. Population structure of *Sciaenops ocellatus* and *Cynoscion nebulosus* (Pisces:Sciaenidae): biochemical variation, genetic subdivision, and dispersal. Copeia 1987:682–695.

Rutledge, W. P., and G. C. Matlock. 1986. Mariculture and fisheries management—a future cooperative approach. Pages 119–127 *in* R. H. Stroud, editor. Fish culture in fisheries management. American Fisheries Society, Fish Culture Section and Fisheries Management Section, Bethesda, Maryland.

Seeb, J. E., L. W. Seeb, and F. M. Utter. 1986. Use of genetic marks to assess stock dynamics and management programs for chum salmon. Transactions of the American Fisheries Society 115:448–454.

TPWD (Texas Parks and Wildlife Department). 1985. Saltwater finfish research and management in Texas: report to the Governor and the 69th Legislature. Texas Parks and Wildlife Department, Report 3000-154, Austin.

History, Genetic Variation, and Management Uses of 13 Salmonid Broodstocks Maintained by the Wyoming Game and Fish Department

CHARLES B. ALEXANDER AND WAYNE A. HUBERT

National Biological Service, Wyoming Cooperative Fish and Wildlife Research Unit
University of Wyoming, Laramie, Wyoming 82071, USA

Abstract.—The Wyoming Game and Fish Department maintains 19 unique broodstocks of salmonids from eight different species to supply the Department's needs for cultured salmonids. We assessed the history and current genetic variation of 13 of these broodstocks in 1992–1993. The length of time that the various broodstocks had been cultured ranged from less than a decade to over a century. Most of the broodstocks had experienced genetic bottlenecks at the time of their first removal from natural populations or upon transfer to new locations. Purposeful selection for features such as spawning time, rapid growth, and survival in the hatchery had occurred in most stocks prior to initiation of broodstock management programs in the 1980s. Analysis of genetic variation (proportion of polymorphic loci and heterozygosity) indicated that most of the broodstocks were above average compared with similar hatchery stocks in North America. The past history of genetic bottlenecks and purposeful selection for specific traits may have reduced genetic variability of most of the 13 broodstocks compared with the stocks from which they were established. Information on the genetic variability of not only cultured broodstocks, but also on populations with which stocked fish may interbreed, is needed when planning to supplement natural recruitment of native or nonnative populations, re-establish wild stocks of native salmonids, create nonnative populations, or maintain put-grow-and-take and put-and-take fisheries where cultured fish may escape to interbreed with existing populations.

Genetic variation is important to the long-term survival of fish stocks because it gives them evolutionary flexibility to adapt to changing environmental conditions (Soule 1983; Krueger et al. 1981; Philipp et al. 1986; Danzmann et al. 1989). Reduced genetic heterozygosity in hatchery broodstocks of trout can reduce the ability of their progeny to compete and lower their progeny's survival (Miller 1958; Allendorf and Phelps 1980; Allendorf and Leary 1986; Wydoski 1986). Stocking fish with substantial genetic variation decreases the chances of diluting gene pools and lowering fitness of wild stocks (Kapuscinski and Jacobson 1987). The genetic variability of wild stocks from which hatchery broodstocks were developed is generally unknown, so assessment of genetic variation in a broodstock often depends upon comparison with other broodstocks and introduced populations of wild fish.

The genetic variation that naturally occurs in native stocks of fish can be reduced during the development of broodstocks. A broodstock consists of mature males and females that supply gametes for the production of fish in a hatchery. The founders (fish obtained from a wild population) used to start a broodstock embody the upper limit of genetic variation available for future generations (Altukhov and Salmenkova 1981; Allendorf and Ryman 1987).

Effective population size is the number of broodfish contributing to the next generation and is defined by Tave (1986) to be (4 × number of females × number of males)/(number of females + number of males). Recommendations for population sizes of broodstocks range from 50 (FAO 1981) to 1,000 broodfish (Eschmeyer and Harris 1984) to maintain genetic diversity. However, the recommendations for a minimum effective population size for broodstocks to be used in stocking programs range from 200 (Kincaid 1976) to 500 (FAO 1981) fish.

Hatchery practices can have negative effects on the genetic variation in stocks (Vuorinen 1984). In the past, hatcheries were generally operated under economic constraints, and culturists were asked to meet production demands for specific stocks; little concern was given to the population genetics of the stocks (Allendorf and Ryman 1987). Cultured broodstocks have experienced a wide range of human intervention, from those broodstocks that have had minimal human influence to highly domesticated broodstocks that are unique to individual hatcheries.

Stocking of hatchery trout is an important component of the Wyoming Game and Fish Department's (WGFD) fisheries management program (Wiley et al. 1993). The WGFD has used hatchery trout to replace wild stocks lost through disaster, restore waters after chemical removal of fish, sup-

TABLE 1.—Salmonid broodstocks maintained by the Wyoming Game and Fish Department that were assessed in this study.

Common and scientific name	Stock name	Location of the broodstock in Wyoming
Rainbow trout *Oncorhynchus mykiss*	Kamloops	Ten Sleep/Wigwam Hatchery, Ten Sleep
	Fall	Boulder Rearing Station, Boulder
	Eagle Lake	Tillett Springs Rearing Station, Lovell
	Eagle Lake	Lake DeSmet, Johnson County
Cutthroat trout *O. clarki* subsp.	Snake River	Auburn Hatchery, Auburn
	Snake River	Ten Sleep/Wigwam Hatchery, Ten Sleep
Colorado River cutthroat trout *O. c. pleuriticus*	Colorado River	Daniel Hatchery, Daniel
Bonneville cutthroat trout *O. c. utah*	Bear River	Daniel Hatchery, Daniel
California golden trout *O. aguabonita aguabonita*[a]	Surprise Lake	Surprise Lake, Sublette County
Brook trout *Salvelinus fontinalis*	Soda Lake	Soda Lake, Sublette County
Lake trout *Salvelinus namaycush*	Jenny Lake	Story Hatchery, Story
Brown trout *Salmo trutta*	Soda Lake	Soda Lake, Sublette County
Arctic grayling *Thymallus arcticus*	Meadow Lake	Meadow Lake, Sublette County

[a] Called *Oncorhynchus mykiss aguabonita* in Behnke (1992).

plement natural recruitment, and establish populations where none exist (Wiley et al. 1993). The WGFD maintains eight salmonid species that comprise 19 unique genetic broodstocks (Alexander 1993). These broodstocks have supplied eggs to over 20 states, so these hatchery stocks are important to Wyoming and other states that use them.

Wiley et al. (1993) stated that the WGFD has no formal policy to identify when stocking of hatchery trout is appropriate, but fisheries managers are admonished to see that Wyoming waters are properly stocked and to guard against overstocking or otherwise negatively affecting existing fish stocks.

The WGFD recognized the value of genetic diversity and initiated procedures in the 1980s to maintain and enhance diversity within its salmonid broodstocks (S. Sharon, WGFD, personal communication). The goal of its broodstock management program is to provide a diverse array of fish that is applicable for all uses by fisheries managers in Wyoming. Annual spawning practices to maintain the genetic diversity of each broodstock include one-to-one pairings of at least 250 pairs from two or more year-classes, collection of eggs a minimum of four times throughout at least 60% or more of the spawning period, and equal representation of eggs from each female in the cohort. Despite the extensive effort to maintain genetic diversity within WGFD broodstocks since the 1980s, many of the broodstocks were established much earlier and have experienced purposeful and inadvertent selection during their management.

The purpose of our study was to define and evaluate the histories and genetic backgrounds of 13 WGFD salmonid broodstocks to better ensure effective management of the broodstocks in the future. Our objectives were to (1) describe the origin, breeding, and management histories of 13 salmonid broodstocks maintained by the WGFD (listed in Table 1), (2) determine the current genetic variability of the broodstocks, and (3) assimilate the information into recommendations on the use of these stocks by fisheries managers. Among the broodstocks were those that have very limited or no genetically pure wild stocks (Kamloops rainbow trout, Eagle Lake rainbow trout, Surprise Lake California golden trout, and Jenny Lake lake trout; Table 1), those that are native species or subspecies in Wyoming (Snake River cutthroat trout, Colorado River cutthroat trout, Bonneville cutthroat trout, and Meadow Lake Arctic grayling), and those that supply a large proportion of the cultured fish used by fisheries managers in Wyoming (Fall rainbow trout, Soda Lake brook trout, and Soda Lake brown trout; Alexander 1993).

Methods

History

Information on origin, breeding, and management of broodstocks was obtained from several sources. We reviewed WGFD and U.S. Fish and Wildlife Service (USFWS) file material, including administrative reports, stocking records, and technical bulletins, and we interviewed current and retired fish culturists and managers. A questionnaire soliciting information on first founding of each broodstock, founding of Wyoming's broodstocks, breeding by the WGFD, hatchery practices, current use of each broodstock, and the sources of information for each broodstock was completed by the hatchery managers responsible for the broodstocks.

Records of and personal communications with personnel of the USFWS, U.S. Forest Service, and other state fisheries agencies were also used. For each species we reviewed published literature and reports of the Wild Trout and Salmon Genetics Laboratory at the University of Montana to obtain information on specific stocks and hatchery practices that have been applied.

Genetics

A random sample of 50 fish was taken from each broodstock and sent to the Wild Trout and Salmon Genetics Laboratory for analysis. Horizontal starch gel electrophoresis was performed on muscle, liver, and eye tissues to evaluate genetic traits at a varying number of loci (Leary and Booke 1990): 40 in Arctic grayling, 54 in brook trout, 41 in brown trout, 62 in California golden trout, 55 in lake trout, 47 in Fall and Kamloops rainbow trout, 45 in Eagle Lake rainbow trout, 64 in Colorado River cutthroat trout, 68 in Bonneville cutthroat trout, and 64 in Snake River cutthroat trout.

Specific alleles were identified by the absolute or relative electrophoretic distances traveled by the polypeptide they encode. An estimate of the allele frequencies was calculated for each sample and used to determine the proportion of polymorphic loci and the mean frequency of heterozygous loci per individual. The mean proportion of polymorphic loci (PP; number of genetically variable loci divided by the total number analyzed) and mean expected heterozygosity (H; mean number of heterozygous genes for all individuals in the sample divided by the total number of loci analyzed) were used as estimates of the amount of genetic variation in each sample. The estimates for the WGFD stocks were compared with estimates for other hatchery and wild stocks previously analyzed by the Wild Trout and Salmon Genetics Laboratory (Leary and Allendorf 1992, 1993; file documents submitted to the Wyoming Game and Fish Department and Wyoming Cooperative Fish and Wildlife Research Unit).

Results

History

We summarized available information on the date, number of broodfish, and number of eggs obtained for the initial establishment (founding) of each stock and for the later establishment of the WGFD stocks (Table 2). Among the 13 broodstocks, information was available on the number of fish taken from native stocks to establish the initial broodstock only for Ten Sleep/Wigwam Hatchery Snake River cutthroat trout and Kamloops rainbow trout stock (13 females and 5 males). The number of fish from hatchery sources used to establish subsequent broodstocks from the first-founded Wyoming broodstock frequently was small. For example, the current stocks of Surprise Lake California golden trout (13 fish) and Ten Sleep/Wigwam Hatchery Kamloops rainbow trout (10 pairs) had progenitor broodstocks established from small numbers of fish.

Information on the number of eggs used to establish broodstocks was found for 4 of the 13 stocks (Table 2). The number of eggs used to establish broodstocks initially or upon transfer to a new location was less than 10,000 in several cases: Boulder Rearing Station Fall rainbow trout (300), Tillett Springs Rearing Station Eagle Lake rainbow trout (2,000), Lake DeSmet Eagle Lake rainbow trout (2,000), Ten Sleep/Wigwam Hatchery Kamloops rainbow trout (3,827), Meadow Lake Arctic grayling (9,429), and Surprise Lake California golden trout (9,600).

The number of generations that a broodstock has been cultured depends on the date of initial founding, as well as the age at which stocks reach sexual maturity. The minimum number of generations that Wyoming's stocks have been cultured ranged from three (Ten Sleep/Wigwam Hatchery Snake River cutthroat trout) to over 25 (Boulder Rearing Station Fall rainbow trout, Surprise Lake California golden trout, Soda Lake brown trout, and Soda Lake brook trout).

We found evidence of intentional or unintentional selection for specific traits for all 13 broodstocks. The Fall rainbow trout has been subjected to selection since the 1800s, and the broodstock's spawning time has been set back from April to October. The spawning time of Auburn Snake River cutthroat trout has been moved back from March to November–December. Intentional selection for characteristics such as spawning time, early maturation, fecundity, growth rate, disease resistance, body condition, coloration, or longevity has occurred among the various broodstocks.

Eleven of the 13 broodstocks have been maintained in a hatchery at some time since their removal from the wild. Two broodstocks, Surprise Lake California golden trout and Meadow Lake Arctic grayling, have been maintained in lakes throughout their management histories. Other WGFD broodstocks held in lakes are Eagle Lake rainbow trout at Lake DeSmet and brook and brown trout at Soda Lake.

TABLE 2.—History and management of 13 salmonid broodstocks maintained by the Wyoming Game and Fish Department.

Broodstock	Original founding			Founding in Wyoming		
	Year	Number of broodfish	Number of eggs taken	Year	Number of broodfish	Number of eggs taken
Ten Sleep/Wigwam Kamloops rainbow trout	1970	13 females 5 males	unknown	1987	262 pairs	3,827
Boulder Fall rainbow trout	1870s	unknown	unknown	1956	unknown	50,848
Tillett Eagle Lake rainbow trout	1950	unknown	2,000	1981–1983	unknown	109,612
Lake DeSmet Eagle Lake rainbow trout	1950	unknown	2,000	1981–1983 1984	unknown 718 females	109,612 unknown
Auburn Snake River cutthroat trout				1953	unknown	29,000
Ten Sleep/Wigwam Snake River cutthroat trout				1987–1990	200 pairs	283,731
Daniel Colorado River cutthroat trout				1973 1982–1989	unknown 50 pairs	unknown 50,000
Daniel Bear River Bonneville cutthroat trout				1977	unknown	unknown
Suprise Lake California golden trout	1976	unknown	unknown	1920s 1949	unknown unknown	unknown 9,600
Soda Lake brook trout	unknown	unknown	unknown	1920s	unknown	unknown
Story Jenny Lake lake trout	1884	unknown	30,000	1964	unknown	unknown
Soda Lake brown trout	unknown	unknown	unknown	1889–1890	unknown	unknown
Meadow Lake Arctic grayling	unknown	unknown	unknown	1921	unknown	1,000,000

Genetic Diversity

Most of the 13 broodstocks that we assessed had above average genetic diversity (PP and H) compared with other broodstocks and wild stocks that have been assessed, but the number of stocks to which comparisons could be made was very limited in several cases (Table 3). No data were available for comparison of the Bear River Bonneville cutthroat trout broodstock with other stocks of Bonneville cutthroat trout, and only one sample outside the WGFD hatchery system was available for comparison of Snake River cutthroat trout and Jenny Lake lake trout broodstocks.

Kamloops rainbow trout.—The Ten Sleep/Wigwam Hatchery broodstock had genetic variation greater than that of a stock maintained at the Ennis National Fish Hatchery in Montana (Table 3). The Kamloops rainbow trout broodstock probably lost genetic diversity upon initial founding of the stock in 1970 when 13 females and 5 males yielded a total of 37 offspring. The broodstock probably experienced additional loss of genetic variation when founded in Wyoming; spawning of 262 pairs of fish yielded only 3,827 eggs. Kamloops rainbow trout have genetic variability lower than that of all of the other rainbow trout stocks (Table 3) due to limited diversity within the native stock or founder effects.

Fall rainbow trout.—This broodstock of rainbow trout initially was taken from the wild in California in 1872 and has experienced at least 40 generations in culture and substantial purposeful selection for several traits. The genetic diversity of the WGFD's Fall rainbow trout was high compared with other domesticated broodstocks (Table 3), including other fall-spawning broodstocks. However, it is very likely that the Fall rainbow trout broodstock possesses less genetic diversity than it had upon founding due to selection for earlier spawning time.

Eagle Lake rainbow trout.—The Tillett Springs Rearing Station and Lake DeSmet broodstocks had genetic variation similar to that of broodstocks maintained at Creston and Ennis National Fish Hatcheries in Montana (Table 3). It is likely that genetic variation was lost upon the initial establishment of the broodstock from the native stock and upon establishment in Wyoming. In 1950, an unknown number of native broodfish yielded 2,000 eggs from which this broodstock was established. Wyoming received eggs from California in 1981, 1982, and 1983. In 1983, a shipment of about 30,000 eggs were hatching upon arrival in Wyoming and only 844 embryos survived a Betadine treatment. It is not known how many embryos survived to become broodfish. This incident may have selected fish with later hatching time and reduced the genetic variation of the broodstock.

Snake River cutthroat trout.—The 1992 sample from the Auburn Hatchery was similar to a 1985

TABLE 3.—Proportion of polymorphic loci (*PP*) and heterozygosity (*H*) of salmonid broodstocks maintained by the Wyoming Game and Fish Department and other stocks (comparison stocks) that have been assessed by the Wild Trout and Salmon Genetics Laboratory, University of Montana.

Wyoming stock	PP	H	Comparison stocks	N	PP Mean	PP Range	H Mean	H Range
Ten Sleep/Wigwam Kamloops rainbow trout	0.167	0.038	Kamloops rainbow trout	1		0.095		0.035
Boulder Fall rainbow trout	0.333	0.093	Domesticated rainbow trout	7	0.284	0.146–0.372	0.083	0.071–0.117
Tillett Eagle Lake rainbow trout	0.243	0.094	Eagle Lake rainbow trout	2	0.257	0.243–0.270	0.101	0.094–0.108
Lake DeSmet Eagle Lake rainbow trout	0.270	0.094	Eagle Lake rainbow trout	2	0.257	0.243–0.270	0.101	0.094–0.108
Auburn Snake River cutthroat trout	0.067	0.021	Snake River cutthroat trout	2	0.084	0.050–0.117	0.019	0.016–0.021
Ten Sleep/Wigwam Snake River cutthroat trout	0.133	0.019	Snake River cutthroat trout	2	0.084	0.050–0.117	0.019	0.016–0.021
Daniel Colorado River cutthroat trout	0.033	0.008	Colorado River cutthroat trout	7	0.069	0.017–0.119	0.010	0.002–0.022
Daniel Bear River Bonneville cutthroat trout	0.111	0.024	Bonneville cutthroat trout	0				
Surprise Lake California golden trout	0.018	0.002	California golden trout	4	0.099	0.036–0.143	0.025	0.018–0.033
Soda Lake brook trout	0.306	0.069	Brook trout	7	0.226	0.175–0.273	0.072	0.050–0.089
Jenny Lake lake trout	0.137	0.044	Lake trout	1		0.118		0.037
Soda Lake brown trout	0.243	0.057	Brown trout	6	0.186	0.118–0.235	0.046	0.031–0.063
Meadow Lake arctic grayling	0.061	0.030	Arctic grayling	17	0.073	0.030–0.121	0.022	0.006–0.033

sample from the same location, indicating no substantial loss of genetic variation since 1985. The Auburn broodstock was established from native fish in 1953 and the Ten Sleep/Wigwam Hatchery broodstock was founded in 1987. The Auburn broodstock has been cultured for a longer period of time than has the Ten Sleep/Wigwam broodstock and has experienced intentional selection for spawning time, size, longevity, and color prior to initiation of the WGFD's broodstock management practices in the 1980s.

Colorado River cutthroat trout.—The Daniel Hatchery broodstock had low genetic variation relative to other stocks (Table 3). Five of the seven stocks to which the Daniel broodstock was compared were native fish from tributaries to the Little Snake River in southeastern Wyoming, and all five stocks had greater genetic diversity. The Daniel Hatchery stock was founded from a different enclave of Colorado River cutthroat trout in central Wyoming in 1973 and 1984. It is possible that the native stock from which the Daniel broodstock was obtained had lower genetic diversity than that of fish in the Little Snake River drainage. However, small numbers of fish were obtained to establish the Daniel broodstock in 1973 (160 fingerlings) and 1984 (50 pairs produced 50,000 eggs), and founder effects most likely have contributed to the limited diversity.

Bear River Bonneville cutthroat trout.—The broodstock of Bonneville cutthroat trout at the Daniel Hatchery was established with fish collected from three tributaries to the Bear River in 1977. No genetic data were available from other stocks of Bear River cutthroat trout to enable comparison.

California golden trout.—The Surprise Lake broodstock of California golden trout has very low genetic diversity compared with fish sampled from Golden Trout Creek, the initial source of the broodstock in California, and other native populations. The broodstock from which Wyoming obtained fish was established initially in California in 1876 and has been cultured for over 40 generations. The Surprise Lake broodstock probably lost genetic variation upon establishment in Wyoming. A 1920s shipment of eggs destined to the eastern United States was found to be in poor condition when inspected by a forest ranger in western Wyoming, and the ranger stocked the surviving embryos in Cooks Lake, Wyoming. The Cooks Lake stock was the source of 9,600 fry used to start the Surprise Lake broodstock in 1949.

Brook trout.—The genetic diversity of the broodstock in Soda Lake is high relative to seven other stocks that have been assessed (Table 3). The initial broodstock was established in 1869 and has been cultured for over 35 generations. The WGFD broodstock was established in Shoshone Lake, Fremont County, Wyoming, in the 1920s, and transferred to Soda Lake, Sublette County, Wyoming, in 1951.

Lake trout.—Wyoming's Jenny Lake broodstock of lake trout maintained at the Story Hatchery has genetic diversity similar to the Jenny Lake stock once held at the Jackson National Fish Hatchery (Table 3). The Jackson stock had been founded from the Story Hatchery broodstock in 1973.

Brown trout.—The Soda Lake broodstock of brown trout showed high genetic diversity compared with seven other stocks (Table 3). The brown trout was introduced to the United States in the 1880s, and the original genetic variation is unknown.

Arctic grayling.—The Meadow Lake broodstock of Arctic grayling has genetic variation that is similar to other stocks that have been assessed (Table 3). The diversity was less than that of the Red Rocks Lake population (where the broodstock originated), but it was almost identical to the Grebe Lake stock (established from Red Rocks Lake broodstock) in Yellowstone National Park, which served as the source of 9,429 fry used to establish the WGFD Meadow Lake broodstock in 1949.

Discussion

Management Uses

Maintenance of genetic diversity in hatchery stocks is important (Ryman and Stahl 1980; Allendorf and Phelps 1980; Hynes et al. 1981; Krueger et al. 1981). Reduction in genetic diversity of hatchery broodstocks can reduce the ability of stocked progeny to compete and survive (Miller 1954, 1958; Allendorf and Phelps 1980). Interbreeding of stocked and wild fish of the same species may reduce desirable traits in wild stocks (Reisenbichler and McIntyre 1977). However, fishery managers cannot predict if changes in gene pools will be advantageous, deleterious, or benign (Nicholas et al. 1978). Elimination of unique genomes of native fishes, through hybridization with hatchery fish, may result in the loss of residual traits that are critical to the long-term survival of native stocks (Marnell 1986).

Despite the research that has been conducted, there is no information from which to predict the genetic influences of hatchery fish on native stocks. A protocol for use of hatchery stocks by fisheries managers that acknowledges the existence of unique genomes among native stocks, as well as wild stocks of nonnative species, and strives to assure that hatchery stocks will not genetically impact wild populations is needed. The genetic background of hatchery fish must be considered prior to stocking (Murphy and Kelso 1986). Despite the potential impact of genetic traits of stocked fish, few states include these factors among prestocking considerations (Murphy and Kelso 1986).

Fish produced from salmonid broodstocks can be used to (1) supplement natural recruitment of wild stocks, (2) restore stocks of fish native to a drainage, (3) establish populations of nonnative fish, and (4) maintain put-and-take or put-grow-and-take fisheries where natural reproduction is lacking. We provide recommendations on the use of Wyoming's broodstocks for each of these management purposes based on our knowledge of the history and genetic diversity of the broodstocks and literature in the area.

Supplement natural recruitment.—In those cases in which interbreeding of wild and hatchery-reared fish may occur, there may be a reduced fitness of progeny in terms of survival and natural reproduction (Reisenbichler and McIntyre 1977; Ryman and Stahl 1981; Allendorf 1991; Waples 1991). The chance of loss of fitness from interbreeding would be increased if the stocked fish were progeny of broodstock with lower genetic diversity than that of the recipient stock. Based on our results, stocking cutthroat trout from WGFD's broodstocks, with the possible exception of the Ten Sleep/Wigwam Hatchery Snake River cutthroat trout and Bear River Bonneville cutthroat trout, may be conducive to loss of genetic diversity. This may also be the case for nonnative salmonid populations established in Wyoming; there is no information on the genetic diversity of wild stocks except for native cutthroat trout. Knowledge of the genetic diversity of both hatchery broodstocks and wild stocks is needed to evaluate the possible genetic consequences of supplemental stocking programs.

Restore native populations.—Among the 13 broodstocks that we assessed, four (Snake River cutthroat trout from the Auburn and Ten Sleep/Wigwam hatcheries, and Colorado River cutthroat trout and Bear River Bonneville cutthroat trout from the Daniel Hatchery) originated from native subspecies of cutthroat trout in Wyoming. Because many native stocks of cutthroat trout have been affected by anthropogenic activities, there is interest in the restoration of the native subspecies in many drainages. Our assessment indicates that genetic variation of broodstocks may be less than that of native stocks in two (Auburn Snake River cutthroat trout and Daniel Colorado River cutthroat trout) of the four broodstocks. The Snake River cutthroat trout at the Ten Sleep/Wigwam Hatchery appears to have nearly the same genetic diversity as does the native stock. Insufficient information on native stocks of Bonneville cutthroat trout prevents comparison with WGFD's broodstock. Use of broodstocks with limited genetic diversity may hinder the long-term success of programs to re-establish native subspecies of cutthroat trout.

The WGFD also maintains a broodstock of Arctic grayling in Meadow Lake. The genetic diversity of this stock seems to be similar to other native and wild stocks that have been assessed, including the Red Rocks Lake stock from which Wyoming's broodstock originated. Therefore, the Meadow Lake broodstock should be useful for restoration purposes.

Establish nonnative populations.—Rainbow trout, golden trout, brown trout, brook trout, and lake trout are species that have been introduced into Wyoming. The WGFD's broodstocks of these species have been cultured for several generations, genetic bottlenecks are known to have occurred, and active selection for specific traits has occurred in most of the broodstocks. However, Wyoming's Soda Lake brown trout and brook trout broodstocks have as much genetic diversity as any broodstocks for these species available in the United States; consequently, the chances for survival of introduced populations in the wild may be as good for fish from these broodstocks as any that are available.

The Fall rainbow trout also has high genetic diversity, but it has been selected to become a fall-spawning broodstock. The fall spawning period of this broodstock will likely hinder reproduction in the wild due to winter incubation and, therefore, limit its use in the establishment of wild stocks.

The California golden trout from Surprise Lake had very low genetic diversity compared with native populations in California. This broodstock has probably lost substantial genetic diversity due to founder effects; the long-term effect can not be predicted.

There is insufficient information on stocks of Kamloops rainbow trout to predict use of WGFD's broodstock for establishing wild populations. The history of the Kamloops rainbow trout broodstock suggests that genetic variability has been lost; however, the measured genetic diversity was greater than that of the only comparable stock. We cannot predict how this amount of genetic variation may influence the establishment of wild stocks.

The four broodstocks of cutthroat trout could be used to establish populations outside the natural range of cutthroat trout. The genetic variation of Ten Sleep/Wigwam Hatchery Snake River cutthroat trout and Bear River Bonneville cutthroat trout broodstocks was similar to that of native stocks. Therefore, these stocks have potential for establishing introduced populations.

Maintain put-and-take or put-grow-and-take fisheries.—There is often no natural recruitment by trout in lakes, reservoirs, or in streams such as tailwaters downstream from reservoirs. Hatchery-reared trout are stocked into these waters to create put-and-take or put-grow-and-take fisheries. The histories and genetic variation of broodstocks used for these purposes are of little concern as long as the stocked fish are pleasing to anglers and return to the creel is at a satisfactory rate. Genetic variation may be important for survival, growth, longevity, and catchability of stocked fish, but performance in individual situations is the primary concern of managers (Ayles and Baker 1983; Bailey and Saunders 1984; Wydoski 1986). All of WGFD's broodstocks are probably suitable for these two management applications. However, if there is a possibility that stocked fish will emigrate and mix with wild populations, then the genetic concerns are identical to those associated with supplemental stocking, and genetic information for both the hatchery broodstock and the receiving populations should be assessed prior to initiation of a stocking program.

Recommendations

The maintenance of genetic variation is important for the long-term survival of fish populations (Danzmann et al. 1989) and is crucial for broodstocks (Tave 1986). The use of protein electrophoresis to assess genetic variation of broodstocks is relatively inexpensive. In this study, each sample of 50 fish was analyzed for US$1,250–1,500. Comparisons of genetic variation and other genetic features can be made among hatchery broodstocks and wild populations using protein electrophoresis, but the technique has limitations. The existing database for comparison is limited for many stocks, so there is little comparative information for use in an assessment. Protein electrophoresis evaluated 40–68 loci among the species in this study, which is a minute portion of the entire genome (Powell 1983). Consequently, this technique may give indications of differences in genetic variation, but too little of the genome may be assessed to provide conclusive information about a stock's overall genetic variation. Continued use of protein electrophoresis to assess genetic variation should be done with this limitation in mind.

One of our objectives was to assimilate the findings of this case study into recommendations to the WGFD. Given the founder effects and purposeful selection that has occurred among many of the broodstocks in Wyoming prior to the 1980s, we recommend a conservative approach by fisheries managers in the use of these broodstocks to prevent damage to wild populations.

1. Identify the genetic diversity of wild populations.
2. Stock cultured fish only when their genetic variation exceeds or closely resembles that of populations with which they may interbreed to avoid diluting gene pools and lowering the fitness of wild stocks.
3. Refrain from stocking hatchery-reared fish that may be partially hybridized (such as rainbow trout × cutthroat trout) into waters supporting wild stocks of pure species with which they may interbreed.
4. Refrain from stocking of hatchery-reared cutthroat trout into drainages containing native stocks of cutthroat trout until it is determined that genetic diversity and subsequent fitness of native stocks will not be damaged by interbreeding.
5. Increase understanding of genetic conservation among fisheries managers so that they will use cultured fish in a manner that does not increase the chances of eliminating unique genomes of native and wild stocks.
6. Continue to maintain a rigorous hatchery program to assure that genetic diversity is maintained in broodstocks by means of preliminary genetic analyses of fish stocks that are proposed for artificial propagation, use of large numbers of founders and parents in successive generations, and periodic genetic analyses after development of broodstocks (Ryman and Stahl 1980).

Acknowledgments

We thank R. F. Leary, Wild Trout and Salmon Genetics Laboratory, University of Montana, for the genetic analysis and comparative data on fish stocks; T. Annear, S. Facciani, W. A. Gern, C. Hunter, J. Johnson, J. Larscheid, H. Schramm, S. Sharon, M. D. Stone, and R. W. Wiley for guidance and criticism; S. Sharon and other WGFD personnel for their direct assistance; and retired WGFD employees J. Conley, W. Huggins, O. Landen, and C. Raper for information. The work was supported by the WGFD. The Wyoming Cooperative Fish and Wildlife Research Unit is jointly supported by the University of Wyoming, Wyoming Game and Fish Department, and National Biological Service.

References

Alexander, C. B. 1993. History and genetic variation of salmonid broodstocks, Wyoming. Master's thesis. University of Wyoming, Laramie.

Allendorf, F. W. 1991. Ecological and genetic effects of fish introductions: synthesis and recommendations. Canadian Journal of Fisheries and Aquatic Sciences 48:178–181.

Allendorf, F. W., and R. F. Leary. 1986. Heterozygosity and fitness in natural populations of animals. Pages 57–76 in M. E. Soule, editor. Conservation biology: the science of scarcity and diversity. Sinauer Associates, Sunderland, Massachusetts.

Allendorf, F. W., and S. R. Phelps. 1980. Loss of genetic variation in a hatchery stock of cutthroat trout. Transactions of the American Fisheries Society 109:537–543.

Allendorf, R. W., and N. Ryman. 1987. Genetic management of hatchery stocks. Pages 141–159 in N. Ryman and F. Utter, editors. Population genetics and fishery management. University of Washington Press, Seattle.

Altukhov, Y. P., and E. A. Salmenkova. 1981. Applications of the stock concept to fish populations in the USSR. Canadian Journal of Fisheries and Aquatic Sciences 38:1591–1600.

Ayles, G. B., and R. F. Baker. 1983. Genetic differences in growth and survival between strains and hybrids of rainbow trout (*Salmo gairdneri*) stocked in aquaculture lakes in the Canadian prairies. Aquaculture 33:269–280.

Bailey, J. K., and R. L. Saunders. 1984. Returns of three year-classes of sea-ranched Atlantic salmon of various river strains and strain crosses. Aquaculture 41:259–270.

Behnke, R. J. 1992. Native trout of western North America. American Fisheries Society Monograph 6.

Danzmann, R. G., M. M. Ferguson, and F. W. Allendorf. 1989. Genetic variability and components of fitness in hatchery strains of rainbow trout. Journal of Fish Biology 35(Supplement A):313–319.

Eschmeyer, P. H., and D. K. Harris, editors. 1984. Fisheries and wildlife research development 1983. U.S. Fish and Wildlife Service, Denver.

FAO (Food and Agriculture Organization of the United Nations). 1981. Conservation of the genetic resources of fish: problems and recommendations. Report of the expert consultation on the genetic resources of fish, Rome, 1980. FAO Fisheries Technical Paper 217.

Hynes, J. D., E. H. Brown, J. H. Helle, N. Ryman, and D. A. Webster. 1981. Guidelines for the culture of fish stocks for resource management. Canadian Journal of Fisheries and Aquatic Sciences 38:1867–1876.

Kapuscinski, A. R., and L. D. Jacobson. 1987. Genetic guidelines for fisheries management. University of Minnesota, Saint Paul.

Kincaid, H. L. 1976. Effects of inbreeding on rainbow trout populations. Transactions of the American Fisheries Society 105:273–280.

Krueger, C. C., A. J. Gharrett, T. R. Dehring, and F. W. Allendorf. 1981. Genetic aspects of fisheries rehabilitation programs. Canadian Journal of Fisheries and Aquatic Sciences 38:1877–1881.

Leary, R. F., and F. W. Allendorf. 1992. Genetic analysis of four populations of Eagle Lake rainbow trout. University of Montana, Wild Trout and Salmon Genetics Laboratory Report 92/2, Missoula.

Leary, R. F., and F. W. Allendorf. 1993. California golden trout, founder effects, and lost genetic diversity. University of Montana, Wild Trout and Salmon Genetics Laboratory Report 93/2, Missoula.

Leary, R. F., and H. E. Booke. 1990. Starch gel electrophoresis and species distinctions. Pages 141–170 in C. B. Schreck and P. B. Moyle, editors. Methods for fish biology. American Fisheries Society, Bethesda, Maryland.

Marnell, L. F. 1986. Impacts of hatchery stocks on wild fish populations. Pages 339–347 in R. H. Stroud, editor. Fish culture in fisheries management. American Fisheries Society, Fish Culture Section and Fisheries Management Section, Bethesda, Maryland.

Miller, R. B. 1954. Comparative survival of wild and hatchery-reared cutthroat trout in a stream. Transactions of the American Fisheries Society 83:120–130.

Miller, R. B. 1958. The role of competition in the mortality of hatchery trout. Journal of the Fisheries Research Board of Canada 15:27–45.

Murphy, B. R., and W. E. Kelso. 1986. Strategies for evaluating fresh-water stocking programs: past practices and future needs. Pages 303–313 in R. H. Stroud, editor. Fish culture in fisheries management. American Fisheries Society, Fish Culture Section and Fisheries Management Section, Bethesda, Maryland.

Nicholas, J. W., R. R. Reisenbichler, and J. D. McIntyre. 1978. Genetic implications of stocking hatchery trout on native trout populations. Pages 189–192 in J. R. Moring, editor. Proceedings of the wild trout-catchable trout symposium. Oregon Department of Fish and Wildlife, Corvallis.

Philipp, D. P., J. B. Koppelman, and J. L. Van Orman. 1986. Techniques for identification and conservation of fish stocks. Pages 323–337 in R. H. Stroud, editor. Fish culture in fisheries management. American Fisheries Society, Fish Culture Section and Fisheries Management Section, Bethesda, Maryland.

Powell, J. R. 1983. Molecular approaches to studying founder effects. Pages 229–240 in C. Schonewald-Cox, S. Chambers, B. MacBryde, and W. Thomas, editors. Genetics and conservation: a reference for managing wild animal and plant populations. Benjamin Cummins Publishing Company, Menlo Park, California.

Reisenbichler, R. R., and J. D. McIntyre. 1977. Genetic differences in growth and survival of juvenile hatchery and wild steelhead trout, *Salmo gairdneri*. Journal of the Fisheries Research Board of Canada 34:123–128.

Ryman, N., and G. Stahl. 1980. Genetic changes in hatchery stocks of brown trout (*Salmo trutta*). Canadian Journal of Fisheries and Aquatic Sciences 37:82–87.

Ryman, N., and G. Stahl. 1981. Genetic perspectives of identification and preservation of Scandinavian stocks of fish. Canadian Journal of Fisheries and Aquatic Sciences 38:1562–1575.

Soule, M. E. 1979. Heterozygosity and developmental stability: another look. Evolution 33:396–401.

Soule, M. 1980. Thresholds for survival: maintaining fitness and evolutionary potential. Pages 151–169 in M. E. Soule and B. A. Wilcox, editors. Conservation biology: an evolutionary-ecological perspective. Sinauer Associates, Sunderland, Massachusetts.

Soule, M. E. 1983. What do we really know about extinction? Pages 111–124 in C. Schonewald-Cox, S. Chambers, B. MacBryde, and W. Thomas, editors. Genetics and conservation: a reference for managing wild animal and plant populations. Benjamin Cummings Publishing Company, Menlo Park, California.

Tave, D. 1986. Genetics for fish hatchery managers. AVI Publishing Company, Westport, Connecticut.

Vuorinen, J. 1984. Reduction of genetic variability in a hatchery stock of brown trout, *Salmo trutta*. Journal of Fish Biology 24:339–348.

Waples, R. S. 1991. Genetic interactions between hatchery and wild salmonids—lessons from the Pacific Northwest. Canadian Journal of Fisheries and Aquatic Sciences 48:124–133.

Wiley, R. W., R. A. Whaley, J. B. Satake, and M. Fowden. 1993. Assessment of stocking hatchery trout: a Wyoming perspective. North American Journal of Fisheries Management 13:160–170.

Wydoski, R. S. 1986. Information needs to improve stocking as a cold-water fisheries management tool. Pages 41–57 in R. H. Stroud, editor. Fish culture in fisheries management. American Fisheries Society, Fish Culture Section and Fisheries Management Section, Bethesda, Maryland.

The Roles of Hatcheries, Habitat, and Regulations in Wild Trout Management in Idaho

ALLAN R. VAN VOOREN

Idaho Department of Fish and Game
600 South Walnut Street, Post Office Box 25, Boise, Idaho 83707, USA

Abstract.—Both special regulations and hatcheries play important roles in wild trout management in Idaho. Over 7,000 miles of streams are managed with special regulations to limit exploitation in unproductive waters, manage for quality and trophy fisheries, and segregate harvest and non-harvest-oriented anglers. Public misconception and misuse of regulations has potentially threatened habitat restoration and wild trout management efforts. Stocking trout has been an important tool in wild trout management to redirect and absorb consumptive effort, educate the public on the importance of habitat, and recruit supporters for habitat management. The demand on fisheries management budgets and the misconceptions reinforced by trout stocking have had negative effects on wild trout management. Since 1985, Idaho's stream stocking program has reduced by one-third numbers of fish stocked and by one-half stream miles stocked. Policies have been adopted that use habitat quality to dictate management for wild trout or with stocked trout. Hatchery trout need to remain a tool for fish managers, but must be used in ways that do not reinforce the misconception that hatcheries are a cure-all.

Beginning with its first 15-year policy plan in 1975, and continuing into the current 1991–2005 plan, the Idaho Department of Fish and Game (Department) has had an expressly stated policy that wild native salmonids would be given priority consideration in all management decisions. Even prior to 1975, recognizing the need to protect wild trout, Idaho led the way on special regulations in the West with the adoption of catch-and-release regulations on Kelly Creek in 1970 (Ball 1971).

At the same time, Idaho has had one of the most intensive trout stocking programs in the West, annually stocking over 3 million harvestable-size trout in lakes, streams, and reservoirs statewide for 400,000 licensed anglers. As recently as 1985, over 2,000 of the state's 26,000 miles of fishable streams (streams that provide habitat for adult sport fishes throughout at least most of the year) were regularly stocked with nearly 1 million catchable-size trout per year.

In the Department's most recent statewide angler opinion survey (Reid 1987), 67% of anglers who purchased a license to fish in Idaho in 1987 indicated they would like to see increased emphasis on protection of wild trout, and 72% wanted increased emphasis on habitat protection and enhancement. In the same survey, 60% of the anglers also wanted increased emphasis on hatchery trout programs. Although there seems to be inherent contradiction in these concurrent desires, these programs do, in several ways, complement each other.

The purpose of this paper is to describe and offer personal perspectives on (1) how Idaho's special trout regulation program and put-and-take trout stocking program have evolved, and (2) on how each affects wild trout management. I will compare the outcomes of these programs with the importance of habitat conditions and describe how the special regulations, trout stocking programs, and the ultimate success of wild trout management are driven by habitat conditions. Lastly, I will discuss the Department's current policies on wild trout management and put-and-take trout stocking and offer a challenge to fishery managers.

Special Regulations

Special regulations are synonymous with wild trout management in Idaho. Special regulations, or regulations that are more restrictive than the general six-trout limit with no size restriction, are an important tool to (1) protect wild trout from overharvest in unproductive waters; (2) restore overharvested populations; (3) provide quality and trophy fisheries; and (4) recognize a stream's wild trout value and redirect consumptive angling effort to other waters.

Ball (1971) recognized that marked declines in populations of westslope cutthroat trout *Oncorhynchus clarki lewisi* soon followed increased angler access to an area. This finding led to the adoption of catch-and-release regulations on Kelly Creek. Evolution of cutthroat trout in relatively sterile streams led to aggressive feeding behavior and higher susceptibility to angling than is exhibited by other salmonid species, even in the same waters (Lewynsky

1986). MacPhee (1966) documented 50% exploitation on a cutthroat trout population in a small stream from only 13 h of angling effort per mile. Rieman and Apperson (1989) reviewed information on westslope cutthroat trout in Idaho and found that 78% of Idaho's strong cutthroat trout populations occurred in areas under special regulations, primarily catch and release. They concluded that "restrictive regulations may be necessary to sustain most cutthroat populations."

Rainbow trout *O. mykiss* are the predominant native salmonid in south-central and southwestern Idaho. Where they occur in relatively sterile watersheds, they fill a niche similar to westslope cutthroat trout and have also developed aggressive feeding behavior and susceptibility to overharvest (Rohrer 1990). Under these conditions, wild rainbow trout populations not only are susceptible to overharvest but also show quick response to restrictive harvest regulations. Within 2 years after adoption of a 14-in minimum size and two-fish bag limit, Rohrer (1991) found wild rainbow trout densities increased 55% and wild rainbow trout catch rates increased from 0.73 to 1.97 fish/h.

In the more productive waters of south-central and southeastern Idaho, populations of rainbow trout and Yellowstone cutthroat trout *O. c. bouvieri* have more resiliency to harvest. Here, special regulations are not necessary to perpetuate wild trout, but have been an important tool used to provide quality (enhanced catch rate and size) and trophy (greater than 20% of the catch is greater than 16 in) fisheries (Moore et al. 1979), even where use of bait is not restricted. Regulations include catch and release and slot and minimum-size limits.

The most recent use of special regulations for wild trout management in Idaho is the adoption of a two-fish bag limit on designated wild trout streams. In the same survey of Idaho anglers that showed a desire for increased emphasis on protection of wild trout (Reid 1987), anglers were asked which of several management options they would prefer if harvest restrictions were necessary. The option most frequently selected (31%) was a reduction in bag limit. The Department recognized that relatively few anglers harvest a limit and that a reduction in bag limit alone would not result in reduced harvest of wild trout. However, the Department anticipated that angler recognition of a stream as managed for wild trout, and the availability of alternate fishing sites with a six-fish limit, would direct consumptive-oriented anglers to streams with more liberal regulations and somewhat partition consumptive and nonconsumptive angling effort.

Elle (1993) evaluated the wild trout population and fishery the first year following adoption of this two-fish wild trout limit on the South Fork Payette River. Although there were no recent data for comparison, he found nearly identical fishing effort in the two-trout and six-trout areas (59 h/acre versus 55 h/acre), yet exploitation of wild trout over 10 in was 5% in the two-trout zone and 44% in the six-trout zone. Differences in access and other variables preclude definitive conclusions, but the results suggest that harvest-oriented anglers were more likely to fish the six-trout area, and anglers fishing the two-trout area were less inclined to harvest fish.

The success of special regulations in improving catch rates and population size structure, however, have misled some anglers to think that special regulations can be a cure-all and overcome the effects of degraded habitat. For example, angler groups have urged the Department to enact special regulations on the middle Snake River and create a trophy fishery, despite extensive news coverage and obvious symptoms of acute water quality problems. This misconception of the utility of special regulations can divert attention and energy from addressing the root factors affecting a fishery.

Organized angler groups have lobbied the Idaho Fish and Game Commission for total catch-and-release and bait restrictions in streams for which Department research indicated such restrictions were not necessary to achieve catch-rate and fish-size objectives. These actions by organized anglers are viewed by other anglers as an effort to exclude them from the fishery. Gigliotti and Peyton (1993) provided an excellent description of such a situation on the AuSable River in Michigan, where the Natural Resources Commission established catch-and-release regulations proposed by vocal organized anglers. Strong local opposition from anglers not affiliated with any organized group led to a restraining order against the Michigan Natural Resources Commission.

A nearly identical scenario on the Big Wood River near Sun Valley, Idaho, led to an order restraining the Idaho Fish and Game Commission from enacting catch-and-release regulations. The Big Wood River case led to the introduction of a bill in the Idaho Legislature that would have prohibited the Commission from enacting any special regulations without conclusive evidence that such regulations were biologically necessary to perpetuate the population. This bill would have taken away the Department's ability to manage for quality or trophy fishing opportunities. Although the bill did not pass, the example illustrates the potential neg-

TABLE 1.—Miles and percent of Idaho streams under special trout regulations. These data do not include regulations for finespotted Snake River cutthroat trout *O. clarki* subsp. (no harvest of 8–16 in fish) in all streams throughout southeastern Idaho.

Regulation and totals	Stream under regulation	
	Miles	Percent
Trophy (2 fish; 20 in minimum or 12–20 in slot)	186	0.7
Quality (2 fish; 14 in minimum or 12–16 in slot)	486	1.9
Catch and release	1,405	5.5
Wild trout (2 fish)	3,495	12.7
Other (species- or habitat-specific regulations for species preservation)	1,792	5.7
Subtotal	7,395	26.5
General (6 fish)	18,240	73.5
Total fishable stream miles	25,604	100.0

ative effect that politically driven regulations could have on wild trout management.

The Department has taken an approach of educating anglers on the desirability of self-sustaining wild trout populations, promoting wild trout management, and implementing bag-limit restrictions to reduce, without prohibiting, consumptive fishing on wild trout. Since 1991, nearly 3,500 stream miles have been placed under the two-fish wild trout regulation. Idaho now has over 7,000 of its approximately 26,000 miles of fishable streams under special regulations for wild trout, including 1,400 miles under catch-and-release (Table 1). The regulations have been well received and, as people see the results of special regulations, more are being requested.

Stocking

Historically, the U.S. Bureau of Fisheries and the Idaho Department of Fish and Game stocked hatchery trout into nearly every Idaho trout stream accessible by vehicle. White (1992) has reviewed the potential negative effects of stocking cultured trout on wild trout. In contrast to these potential negative effects, stocking trout has had several positive benefits on wild trout management in Idaho.

Hatchery trout have been used extensively for educational purposes, such as helping students and adults learn about the need for clean gravel and good water quality. Over 300,000 people per year view hatchery trout and incubating trout eggs in a living stream display at the Department's Morrison-Knudsen Nature Center in Boise.

Put-and-take hatchery trout programs purportedly recruit anglers. These anglers may then become important constituents in support of policies and legislation to improve water quality and minimum stream flows. Over 58,000 angler-hours are expended in a put-and-take fishery on an 11-mi reach of the Boise River in Boise (Reid and Mabbot 1987). During the winter of 1992–1993, when the Bureau of Reclamation cut river flows to less than half the previously agreed to 150 ft^3/s minimum flow, the Department publicized that trout stockings were being terminated because of low flows. Numerous television and newspaper reports and editorials explained the need to provide adequate flows for fish. As a result, a bill to allow the conversion from consumptive to nonconsumptive instream water rights was introduced in the Idaho Legislature. The bill did not pass, but its introduction reflects the potential influence that participants in an urban stocked-trout fishery can have on environmental legislation.

Hatchery trout fisheries also play an important role in redirecting harvest-oriented fishing effort that might otherwise be directed to wild trout fisheries. I asked attendees at a sport show in southwest Idaho and a fair in southeast Idaho to complete a questionnaire in which they were given the opportunity to spend 8 h fishing in any combination of eight fishing ponds. Individual ponds contained fish of 10, 13, 15, or 18 inches. Catch rates approximated those in actual harvest and release fisheries for fish of those sizes, ranging from 0.1 to 1.0 fish/h in ponds open to harvest and 0.5 to 7.0 fish/h in ponds where all fish had to be released. Forty percent of the total effort was spent in ponds where fish could be harvested, despite lower catch rates. Because of this desire to harvest fish, fisheries of even moderate effort (30–60 h/acre) can not be sustained without restrictive regulations in most wild trout streams in Idaho, where standing stocks are generally less than 25 lb/acre and frequently less than 5 lb/acre (Elle 1993).

Hatchery-supported trout fisheries in lakes, ponds, and reservoirs can be more productive and better suited to provide consumptive opportunity, but 56% of Idaho anglers prefer to fish in streams (Reid 1987). Providing hatchery-supported fisheries in designated streams or stream reaches where habitat is insufficient to support a healthy wild trout population can absorb a great deal of this desire for harvesting trout in a stream environment.

Providing hatchery-supported fisheries in proximity to wild trout fisheries effectively partitions consumptive and nonconsumptive effort. In 1988, estimated fishing effort in a 40-mi reach of the Middle Fork of the Boise River was 8,727 h when it was stocked with trout. In 1990, fish were stocked into only the lower 11 mi, and a two-trout, none-under-14-in, artificial-flies-and-lures-only restriction was placed on the upper reach. Total fishing effort was almost the same (8,680 h); however, proportion of fishing effort in the lower reach increased from 62% in 1988 to 90% in 1990 (Rohrer 1991). Angler attitudes were not surveyed to determine the reason for this shift in fishing effort; however, harvest orientation of the anglers is a likely explanation.

Concentrating trout stocking in designated areas responds to the anglers' desire for increased emphasis on hatchery trout programs (Reid 1987) and results in increased catch rates and greater return rates of stocked fish (Mauser 1994). Department proposals for restrictive wild trout regulations on the Big Wood River and lower Selway River had greater acceptance after stocked trout fisheries were provided nearby.

The potential detrimental biological effects of stocking trout are well documented (e.g., White 1992). However, the Department's trout stocking program has additional negative effects that may transcend the biological effects. One obvious effect is the high cost of using hatchery trout. Not counting capital costs, Mauser (1994) estimated the Department's average cost of rearing, transporting, and stocking a put-and-take trout was US$0.62. Even at the current goal of a 40% minimum return, each fish in the angler's creel would cost $1.50, one-tenth the cost of an annual license. The Department currently spends 43% of its discretionary (license and Sport Fish Restoration Act) funds generated from fishing for put-and-take trout stocking.

A second adverse effect of using hatchery fish is reinforcement of misconceptions that harvest is independent of the production capacity of natural systems, and that stocking is an acceptable substitute for harvest restraint. Fish and game agencies have often responded to overharvest-related declines in fisheries with supplemental stocking. Indiscriminant supplemental stocking undermines efforts to manage wild trout fisheries within production capabilities.

I believe the greatest adverse effect associated with using hatchery trout, however, is reinforcement of the misconception that hatcheries are an acceptable alternative to habitat protection or restoration. Fish and game agencies often have responded to habitat-related declines in fisheries with supplemental stocking. Although hatcheries may play an important role in some fisheries restoration efforts, these uses of stocking and insufficient information and education on habitat and harvest-related limitations on fisheries have given the public the message for decades that hatcheries can fix all fisheries problems.

To get a preliminary measure of Idaho anglers' attitudes about the role of stocking in fisheries management, I sent a postcard questionnaire to 200 randomly selected resident licensed anglers in each of the Department's seven regions. The questionnaire asked anglers to allocate the $15 they spent on a fishing license among six major fisheries-related programs: access, law enforcement, habitat protection and enhancement, information and education, stocking, and management surveys and research. Although the survey had a low response rate (36%), the 506 respondents allocated 32% of their license money to stocking compared with 21% to habitat, 14% to enforcement, 13% to access, 7% to surveys and research, 4% to information and education, and 10% to miscellaneous activities.

It is no wonder that developers propose, and much of the general public finds it acceptable, to use hatcheries as mitigation for damming rivers, changing flows, and otherwise altering wild trout habitat. When mitigation by stocking hatchery-produced fish is less expensive than avoiding habitat impacts, the pressure from development interests and public acceptance of stocking presents major challenges to fisheries managers to maintain habitat conditions and perpetuate wild trout.

Since 1985, Idaho has reduced the number of put-and-take trout stocked in streams from 960,000 to 658,000. More significantly, the miles of streams stocked have been reduced from over 2,000 mi in 1985 to 853 mi in 1993. The Department did this while still meeting the anglers' desires for increased emphasis on hatchery programs by stocking a greater density, stocking more frequently, stocking in higher use areas with greater return, and informing anglers of stocking locations. Most stocking reductions have been made without major deprivations to anglers. Reductions were made where return rates were low due to low fishing effort, rapid dispersal of stocked fish, or poor survival of stocked fish. Reductions in stream stocking were often also offset with increases in stocking in adjacent ponds. The effect on wild trout has been reduced in direct interactions with hatchery trout and reduced harvest on wild trout waters.

Habitat

As important as hatcheries and special regulations are, their influence on wild trout management pales by comparison with the limitations of habitat productivity and the impacts of habitat alteration. Elle (1993) reviewed data on 22 wild trout fisheries in Idaho and found the estimated biomass at less than 15 lb/acre for about one-third, 15–50 lb/acre for another third, and 50–150 lb/acre for the other third. The Department has management control on a very limited portion of Idaho's trout habitat, but it consults with land management agencies and provides technical guidance and carries out cooperative habitat management projects with private landowners to benefit wild trout populations.

Habitat conditions drive both wild trout and hatchery trout programs. In their 1991–1995 Fish Management Plan, the Department adopted a policy of managing only for wild trout where habitat is capable of supporting an acceptable fishery. (In its initial version, this policy included a criterion of minimum catch [not harvest] rate of 0.3 fish/h for trout greater than 8 in; a criteria of acceptable without specific definition was adopted in the final version to allow flexibility in unique circumstances.) Hatchery trout will be stocked only where habitat can not support an acceptable wild trout fishery and where a minimum 40% return of stocked fish to the creel is obtainable. Achieving a 40% return is a function of angler effort, water quality, and habitat that will support adult trout.

To see how our practices measured up to our policies, I compared our current hatchery trout program and our special regulation program with a Northwest Rivers Inventory habitat quality classification of Idaho streams (Nellis and Allen 1986). The classification effort relied on both measured variables and professional judgement of state, federal, and tribal biologists to classify habitat quality of Idaho's 26,000 mi of fishable streams as either low, intermediate, high, or unknown quality. Under this classification, 3,530 mi were designated as low-quality habitat, 7,580 mi were designated as intermediate, 6,769 mi were designated as high, and 7,805 mi were designated as unknown (Table 2). The last category is largely smaller, remote streams that would likely have high-quality habitat. Of Idaho's 7,364 mi of special regulation trout waters, only 1% are designated low-quality habitat and 93% are designated high-or unknown-quality habitat. Clearly, wild trout management with special regulations is focused on streams with high-quality habitat.

TABLE 2.—Miles of Idaho streams stocked with put-and-take trout and under special trout regulations by low-, intermediate-, high-, and unknown-quality habitat (from Nellis and Allen 1986).

Streams	Low	Intermediate	High and unknown	Total
Statewide	3,530	7,580	14,574	25,684
Stocked	166	370	321	857
Special regulations	51	430	6,880	7,364

Of the 857 stream miles stocked with hatchery trout, 19% are designated low-quality habitat, 43% intermediate, and 38% high or unknown (Table 2). The relatively small proportion of low-quality streams receiving put-and-take stocking is related to achieving the 40% return rate—the streams must have high fishing effort and suitable adult trout rearing habitat. Although the habitat quality classification process did not specifically address the natural trout production capacity of stream reaches, the majority of put-and-take stocking occurs in low- and intermediate-quality streams, those which, at present, are likely to have less potential for self-sustaining trout populations (i.e., wild trout management) than are high-quality streams. Not all of the streams ranked as high quality support trout reproduction. Some of the put-and-take stocking occurs in high-quality streams that do not support natural reproduction of trout; however, some stocking does occur in streams with natural reproduction. The Department must continue to evaluate put-and-take stocking locations.

Summary

State fishery managers have at their disposal the tools of stocking, special regulations, and habitat management. State managers have direct management control of very little fishery habitat. Special regulations and stocking are controlled directly by states and, when used properly, benefit sportfishing and wild trout management. Both can be used to manage exploitation of wild trout and both can be used to educate the public on the influence and importance of habitat.

Because stocking fish is one of the relatively few quantifiable, tangible, and easily measured accomplishments of state fisheries agencies, it has often been the first thing held up to the public in media contacts, agency annual reports, and other accounting of agency accomplishments and actions to enhance fisheries. In so doing, fisheries agencies have reinforced a major misconception that stocking solves most fishery problems and undermined support for wild trout harvest restrictions and habitat management. We must resist using hatchery trout in a manner that reinforces these misconceptions. We need to measure our fish management contributions and successes in terms of fishery management objectives met, not numbers of fish stocked.

Major changes have occurred in trout stocking programs to minimize effects on wild trout and efforts are being taken to educate the public on the importance of habitat.

The challenge to fisheries managers and the professional fisheries community is to continue these beneficial programs without creating another misconception—that the use of hatchery trout has no role in wild trout management.

References

Ball, K. W. 1971. Evaluation of catch-and-release regulations on cutthroat trout in the North Fork of the Clearwater River. Idaho Department of Fish and Game, Federal Aid in Sport Fish Restoration, Project F-59-R-2, Job 1, Job Completion Report, Boise.

Elle, S. 1993. Wild trout investigations: South Fork Payette River studies. Idaho Department of Fish and Game, Federal Aid in Sport Fish Restoration, Project F-73-R-15, Study IV, Job Performance Report, Boise.

Gigliotti, L. M., and R. B. Peyton. 1993. Values and behaviors of trout anglers, and their attitudes toward fishery management, relative to membership in fishing organizations: a Michigan case study. North American Journal of Fisheries Management 13:492–501.

Lewynsky, V. A. 1986. Evaluation of special angling regulations in the Coeur d'Alene River trout fishery. Master's thesis. University of Idaho, Moscow.

MacPhee, C. 1966. Influence of differential angling mortality and stream gradient on fish abundance in a trout-sculpin biotope. Transactions of the American Fisheries Society 95:381–387.

Mauser, G. 1994. Hatchery trout evaluations. Idaho Department of Fish and Game, Federal Aid in Sport Fish Restoration, Project F-73-R-15, Job 1, Job Performance Report, Boise.

Moore, V. K., D. R. Cadwallader, and S. M. Mate. 1979. South Fork of the Boise River creel census and fish population studies. Annual Report to the Bureau of Reclamation. Idaho Department of Fish and Game, Boise.

Nellis, C. H., and S. Allen. 1986. Pacific Northwest rivers study, 1986 Final Report: Idaho. Idaho Department of Fish and Game, Boise.

Reid, W. W. 1987. A survey of 1987 Idaho anglers' opinions and preferences. Idaho Department of Fish and Game, Federal Aid in Sport Fish Restoration, Project F-35-R-13, Job 8, Job Completion Report, Boise.

Reid, W. W., and B. Mabbot. 1987. Regional fishery management investigations: Region 3 rivers and streams investigations. Idaho Department of Fish and Game, Federal Aid in Sport Fish Restoration, Project F-71-R-11, Job 3-b, Job Performance Report, Boise.

Rieman, B. E., and K. A. Apperson. 1989. Status and analysis of salmonid fisheries: westslope cutthroat trout synopsis and analysis of fishery information. Idaho Department of Fish and Game, Federal Aid in Sport Fish Restoration, Project F-73-R-11, Job 1, Job Performance Report, Boise.

Rohrer, R. L. 1990. Upper Boise River basin fisheries investigations. Idaho Department of Fish and Game, Federal Aid in Sport Fish Restoration, Project F-73-R-12, Study I, Job 1, Job Performance Report, Boise.

Rohrer, R. L. 1991. Upper Boise River basin fisheries investigations. Idaho Department of Fish and Game, Federal Aid in Sport Fish Restoration, Project F-73-R-13, Study I, Job 1, Job Performance Report, Boise.

White, R. J. 1992. Why wild fish matter: a biologist's view. Trout (Summer 1992):24–50.

Use of Cultured Fish for Put-Grow-and-Take Fisheries in Kentucky Impoundments

BENJY KINMAN

Kentucky Department of Fish and Wildlife Resources
1 Game Farm Road, Frankfort, Kentucky 40601, USA

Abstract.—Twenty large (340–160,300 acres) impoundments and 160 small impoundments are important fisheries resources in Kentucky; 55% of Kentucky anglers prefer to fish in these impoundments. Put-grow-and-take stocking programs were established to create recreational angling opportunities that otherwise would not exist in these altered aquatic environments. Progeny of native muskellunge *Esox masquinongy* are stocked into three large impoundments that inundated riverine muskellunge habitat. Native walleye *Stizostedion vitreum* was extirpated as a result of dam construction, and a lake-spawning strain has been used to reintroduce this species into seven large impoundments. Channel catfish *Ictalurus punctatus* is extensively stocked into small impoundments. Striped bass *Morone saxatilis* was introduced as a pelagic predator into one large impoundment with available coolwater habitat. Hybrid striped bass *Morone saxatilis* ♀ × *M. chrysops* ♂ has been introduced into six large impoundments to create a pelagic fishery. Two warmwater fish hatcheries owned and operated by the Kentucky Department of Fish and Wildlife Resources allocated approximately 90% of their fish production (biomass) for these stockings. Rainbow trout *Oncorhynchus mykiss* was introduced into seven small impoundments and three large impoundments with coldwater habitats. A federal hatchery in Kentucky dedicates 35% of annual production to provide rainbow trout for these stockings. These put-grow-and-take stocking programs have achieved the annual yield (weight harvested) objective of at least 1 lb/acre or a 10% addition in the total yield for all species except walleye. Maintenance of these fisheries has successfully supplemented and diversified angling opportunities, possibly dispersed fishing effort, and established trophy fisheries for striped bass and muskellunge.

Kentucky's fishery resource consists of 20 large impoundments (340–160,300 acres), more than 160 small impoundments, 18,500 mi of perennial streams, and more than 50,000 ponds (private farm ponds, strip pits, and floodplain lakes). Among Kentucky anglers, 32% prefer to fish large impoundments (including their tailwaters), 27% prefer to fish streams, 23% small impoundments, and 16% ponds (Hale et al. 1992). Warmwater fishing opportunities predominate; however, fisheries for rainbow trout *Oncorhynchus mykiss* and brown trout *Salmo trutta* are available in 71 streams, including several tailwaters, encompassing 312 mi. The Kentucky Department of Fisheries and Wildlife Resources' (KDFWR) fishery management efforts primarily have been directed toward impoundments, except for a farm pond stocking and a technical guidance program. Stream management could be characterized as protectionist and the primary management activity is a statewide stream inventory to document distribution and relative abundance of fish.

Demand for new fishing opportunities in Kentucky in 1960–1980 was largely offset by the construction of new impoundments. Fishery management of these systems primarily has consisted of stocking and regulation. Regulation of the fishery resource has vacillated as described by Redmond (1986); currently, emphasis is on quality fishing experiences, and harvest regulations are more restrictive than those in place a few years ago. Stocking programs began in Kentucky with stockings of walleye *Stizostedion vitreum*, common carp *Cyprinus carpio*, and Pacific salmon *Oncorhynchus* spp. in the 1880s (U.S. Commission of Fish and Fisheries 1884, 1885). Stocking practices have changed from indiscriminate stocking to decisions based on habitat suitability and fish community compatibility; however, these stocking practices need further refinement.

The KDFWR has actively used maintenance stockings for put-grow-and-take fisheries for striped bass *Morone saxatilis*, hybrid striped bass *Morone saxatilis* ♀ × *M. chrysops* ♂ (and the reciprocal cross), muskellunge *Esox masquinongy*, walleye, channel catfish *Ictalurus punctatus*, and rainbow trout (Table 1). Two warmwater fish hatcheries operated by KDFWR are dedicated to culturing these fish. Of these fish, only walleye have annual recruitment. All salmonids are provided by the Wolf Creek National Fish Hatchery in Kentucky. The objective for put-grow-and-take fisheries is a minimum 1 lb/acre annual yield (weight harvested) or a 10% addition to the total annual yield by the stocked fish. A secondary objective for put-grow-and-take rainbow trout fisheries is a minimum return of 50% by weight of the annual stocking (Wilkins et al. 1967).

TABLE 1.—Current stocking of cultured fish to maintain put-grow-and-take fisheries in Kentucky. The minimum current stocking sizes are given for rainbow trout and channel catfish.

Species and water body	Size (acres)	Initial stocking year	Annual stocking rate (number/acre)	Current stocking size (in)
Striped bass				
Lake Cumberland	50,250	1979[a]	5	1.5–2.0
Muskellunge				
Buckhorn Lake	1,230	1980	0.50–1.00	8.0–14.5
Cave Run Lake	8,270	1973	0.23–0.49	13.2–14.6
Green River Lake	8,210	1982	0.23–0.43	12.7–14.6
Rainbow trout				
Bert Combs Lake	36	1969	42[b]	9
Beulah Lake	87	1977	29[b]	9
Cannon Creek Lake	243	1977	25[b]	9
Cranks Creek Lake	219	1985	27[b]	9
Fishpond Lake	32	1979	47[b]	9
Greenbo Lake	181	1985	83	9
Laurel River Lake	6,060	1974	21	9
Mill Creek Lake	41	1963	24[b]	9
Paintsville Lake	1,139	1985	31	9
Wood Creek Lake	672	1976	21	9
Hybrid striped bass				
Barren River Lake	10,000	1979	20	1.0–2.0
Dewey Lake	1,100	1993	20	1.0–2.0
Fishtrap Lake	1,143	1990	20	1.0–2.0
Herrington Lake	2,610	1979	20	1.0–2.0
Taylorsville Lake	3,050	1989	20	1.0–2.0
Guist Creek Lake	317	1992	20	1.0–2.0
Channel catfish				
Marion County Lake	27	1962	50	8
Spurlington Lake	36	1965	50	8
Mauzy Lake	27	1969	50	8
Carpenter Lake	64	1962	50	8
64 public lakes	1–826		25	8
20 public lakes	3–158		50	8
Walleye				
Carr Fork Lake	710	1984	50	1.0–1.5
Martins Fork Lake	334	1979	50	1.0–1.5
Nolin River Lake	5,790	1974	50	1.0–1.5
Lake Cumberland	50,250	1973	7	1.0–1.5
Laurel River Lake	6,060	1985	50	1.0–1.5
Green River Lake	8,210	1993	50	1.0–1.5
Paintsville Lake	1,139	1984	50	1.0–1.5

[a] First year of consistent stocking rate.
[b] Receives additional fish in April, May, and October.

Creel surveys have been periodically conducted on large impoundments (minimum once per 5 years) and nonroutinely on small impoundments. Generally, roving creel surveys were used, except some nighttime rainbow trout fisheries were surveyed by the access-site method. Total catch (numbers and weight), catch rate, and length frequency by species were estimated along with the collection of angler characteristics.

This paper explains Kentucky's management strategies regarding cultured fish in put-grow-and-take fisheries and synthesizes available creel data to examine the relative importance and contribution of these cultured fish to impoundment fisheries.

Striped Bass

Striped bass was stocked in many large impoundments in Kentucky before its thermal preferences and tolerances were fully understood (Coutant 1985). Presently, only Lake Cumberland (50,250 acres) and the Ohio River (664 mi along Kentucky's northern border) are stocked with striped bass. Lake Cumberland has been the priority stocking site and has received 5 fingerlings (1.5–2.0 in) per acre since 1984. If hatchery production is sufficient approximately 5 fingerlings/acre are stocked into the navigation pools of the Ohio River. The striped bass fishery on the Ohio River mainly is confined to the tailwaters below the navigation dams; creel surveys have not been conducted in these areas since the fishery has developed.

Lake Cumberland is an oligotrophic, multipurpose impoundment containing the largest volume of oxygenated coolwater habitat of any impoundment in the state. Striped bass stockings began in 1969 with low annual stocking densities (3 fingerlings/acre) that continued until 1978 (4.5 fingerlings/acre). In 1979 and 1981 the lake was stocked at the recommended rate of 10 fingerlings/acre (Striped Bass Committee, Southern Division American Fisheries Society, personal communication). In 1984, stocking rates were reduced to 5 fingerlings/acre after growth rates were determined to be faster at lower stocking rates (Kinman 1988). The early stockings remained unexploited, which later created a regionally renowned trophy fishery. Heightened interest in striped bass fishing resulted in a 15-fold increase in fishing effort from 1983–1986. A lucrative fishing guide industry has now developed as a direct result of this fishery.

Recent creel surveys have been confined to the lower half of the impoundment, where the majority of striped bass fishing occurs. Striped bass was the dominant fishery in the lower lake; fishing effort averaged 5 h/acre, or 32% of the total effort; and harvest averaged 61% of the total yield (all species) by weight (Table 2). Harvested striped bass averaged 27 in and 7.8 lb during the four creel-survey years, but the average size has decreased during the last three surveys. Concomitant with this decline has been a decline in total numbers harvested despite relatively stable fishing effort. Yield declined from 4.4 (1988) and 5.7 (1990) lb/acre to 2.3 lb/acre in 1992.

Striped bass was regulated with a 15-in-minimum

length limit and a 5 fish/d creel limit until 1989 when the daily creel was reduced to 3 fish. This fishery has an insignificant catch-and-release component except for 15–20-in striped bass; a 20-in striped bass has arbitrarily been established as keeper size by anglers at Lake Cumberland. Catch rates of striped bass larger than 30 in have steadily declined since 1988, indicating the fishery is approaching overharvest. Anglers have expressed a desire to maintain the trophy aspect of this fishery. Regulatory alternatives to protect a trophy fishery are very limited with an open-access fishery and relatively high mortality (38% overall and 58% with live bait) of released striped bass (Hysmith et al., in press). A 24-in-minimum-length limit and a two-fish-daily creel limit became effective in March 1994. A 24-inch striped bass in Lake Cumberland is age 3; the 24-in-length limit protects striped bass from harvest 1.5 years longer than did the previous 15-in-length limit.

Muskellunge

Muskellunge is stocked in three major impoundments and 14 streams in Kentucky. Stream stockings are classified as supplemental stockings because limited natural reproduction occurs in these native muskellunge streams (Brewer 1980). The put-grow-and-take impoundment fisheries are Cave Run (8,270 acres), Green River (8,210 acres) and Buckhorn (1,230 acres) lakes; all are U.S. Army Corps of Engineers impoundments that inundated native muskellunge stream habitat. Muskellunge fisheries are regulated with a 30-in-minimum-length limit and a 2 fish/d creel limit. Muskellunge in impoundments attain legal size during their third growing season (age 2).

Stocking strategies for muskellunge have changed significantly in the last 20 years. A muskellunge fishery was successfully established with 6-in fingerlings stocked at approximately 1 fish/acre in Cave Run Lake following impoundment in 1973 (Axon 1978). Poor survival of similar-size muskellunge stocked into Green River Lake in 1977–1978 prompted a shift to stocking larger fingerlings (8-in average) in Cave Run Lake. Stocking began in Buckhorn Lake in 1980 using 9-in muskellunge. In 1982, an experimental stocking at a lower density (0.4 fish/acre) of 13-in muskellunge in alternate years was initiated in Green River Lake; the off years received a 9-in-average-size muskellunge at 1 fish/acre. Hatchery growth rates produce a 9-in muskellunge in late July–early August, which requires stocking when water temperatures are 80°F.

After 6 years of stocking, a significantly higher survival of 13-in versus 9-in muskellunge was observed, presumably due to lower predation and cooler water temperatures during the fall stocking of the larger fish (Kinman 1989). Hatchery rearing patterns have now shifted toward producing a smaller number of larger muskellunge (13-in) for both Cave Run and Green River lakes; Buckhorn Lake has received 9-in muskellunge with some surplus 13-in fish. Different stocking densities (0.2–0.4 fish/acre) of 13-in muskellunge are now being examined in Cave Run and Green River lakes to match hatchery capabilities, economic feasibility, and sportfishing objectives.

Trophy muskellunge fisheries are unique for this latitude and display some distinct differences among the three impoundments. Cave Run Lake has the oldest (1974) muskellunge fishery and subsequently has a better reputation. This is reflected in the higher fishing effort of 6.8 h/acre (32% of total effort) compared with 2.6 h/acre (9% of total effort) at Green River Lake (Table 2). Fishing effort was higher for muskellunge than for any other species in Cave Run Lake in 1993. Fishing effort for other warmwater fishes, such as black bass *Micropterus* spp., crappies *Pomoxis* spp., and white bass *Morone chrysops*, exceed that for muskellunge at Green River Lake. Muskellunge harvest at Green River Lake has averaged 0.65 lb/acre or 9% of the total yield compared with 0.81 lb/acre or 22% of the total yield at Cave Run Lake. Muskellunge at Buckhorn Lake represented only 1% of the total fishing effort; however, the yield was 1 lb/acre or 9% of the total yield. This fishery significantly differs from both Cave Run and Green River lakes where 30 and 41% of the legal-size muskellunge were released; no legal-size muskellunge were released at Buckhorn Lake. Whereas many muskellunge anglers fishing Cave Run and Green River lakes are members of the Silver Musky Club and Muskie, Inc., most muskellunge at Buckhorn Lake are caught incidently by anglers seeking other species. The density of muskellunge at Buckhorn Lake is probably negatively affected by the extreme reservoir drawdown and exchange rate, which contribute to a significant loss of muskellunge in the water released from the reservoir; muskellunge harvest is reported to be good in the tailwaters following major discharges.

Rainbow Trout

Rainbow trout is stocked in many streams and tailwaters on a put-and-take basis; however, only 10

TABLE 2.—Creel survey estimates for put-grow-and-take fisheries in Kentucky from 1977 to 1993. (Standard error is included in parentheses.)

Lake and sample size	Harvest			Effort				
	Number harvested	Pounds per acre harvested	Percent of total weight (all species)	Fishing effort (h/acre)	Percent of total effort (all species)	Harvest rate (fish/h)	Average length (in)	Average weight (lb)
Striped bass								
Lake Cumberland	10,912	3.5	61[a]	5.0[a]	32[a]	0.09	27.0	7.8
(N = 4)	(2,618)	(0.9)	(10)	(1.4)	(1)	(0.01)	(0.5)	(0.5)
Muskellunge								
Cave Run Lake	599	0.8	22	6.8	32	0.01	35.5	11.2
(N = 6)	(96)	(0.14)	(4)	(0.6)	(2)	(<0.01)	(0.6)	(0.7)
Green River Lake	512	0.7	9	2.6	9	0.02	35.1	10.7
(N = 6)	(117)	(0.13)	(2)	(0.4)	(1)	(0.002)	(0.7)	(0.8)
Buckhorn Lake	137	1.0	9	1.5	1	0.02	35.1[b]	10.9[b]
(N = 2)	(35)	(0.2)	(0.2)	(1.4)	(1)	(0.002)		
Rainbow trout[c]								
Paintsville Lake	15,589	6.1	63	23.0	37	0.41	10.7	0.4
(N = 1)								
Greenbo Lake	1,593	7.0	52	7.9	9	0.51	12.5	0.8
(N = 1)								
Woods Creek Lake	694	0.8	31	3.7	17	0.26	12.3	0.7
(N = 2)	(54)	(0.05)	(10)	(0.6)	(4)	(0.06)	(0.6)	(0.1)
Laurel River Lake	15,402	1.8	65	5.7	79	0.35	12.9	0.8
(N = 4)	(7,802)	(0.8)	(17)	(1.8)	(8)	(0.09)	(0.4)	(0.6)
Hybrid striped bass								
Herrington Lake	5,534	3.0	8	4.8	7	0.53	13.2	1.2
(N = 4)	(544)	(0.5)	(1)	(2.2)	(0.05)	(0.03)	(0.8)	(0.2)
Barren River Lake	14,987	2.0	17	4.1	11	0.37	13.4	1.2
(N = 4)	(5,281)	(0.9)	(5)	(1.5)	(2)	(0.05)	(0.6)	(0.2)
Taylorsville Lake	2,999	2.3	13	9.1	7	0.10	16.4	2.2
(N = 2)	(1,099)	(1.0)	(5)	(3.2)	(2)	(0)	(0.7)	(0.3)
Channel catfish								
Marion County Lake	656	27.8	20	92.4	12	0.13	13.7	0.8
(N = 2)	(269)	(14.3)	(3)	(46.2)	(3)	(0.04)	(0.5)	(0.08)
Spurlington Lake	807	23.7	39	34.9	14	0.32	14.6	1.0
(N = 2)	(81)	(4.5)	(9)	(3.0)	(0.05)	(0.15)	(0.7)	(0.13)
Lake Mauzy	3,664	28.9	34	97.4	25	0.24	12.7	0.7
(N = 2)	(1,178)	(7.3)	(4)	(18.2)	(0.4)	(0.04)	(0.4)	(0.07)
Carpenter Lake	4,108	62.7	56	84.2	19	0.57	14.4	1.0
(N = 2)	(1,506)	(26)	(8)	(1.7)	(4)	(0.21)	(0.3)	(0.05)
Walleye								
Lake Cumberland	6,136	0.3	6	0.6	4	0.16	18.3	2.0
(N = 1)								
Nolin River Lake	1,224	0.4	3	0.2	<1	0.07	17.0	1.6
(N = 1)								
Carr Fork Lake	170	0.5	3	1.4	2	0.06	17.9	1.8
(N = 2)	(58)	(0.2)	(1)	(0.8)	(1)	(0.003)	(0.3)	(0.1)
Laurel River Lake	1,596	0.4	13	2.1	8	0.09	17.0	1.6
(N = 1)								

[a] Based on 3 years of data.
[b] Based on 1 year of data.
[c] Based on nocturnal creel data (May–September).

lakes are managed for put-grow-and-take fisheries (Table 1). These lakes maintain an oxygenated hypolimnion with sufficient cold water to support rainbow trout in the summer. The basic management strategy is to stock 9-in (minimum) rainbow trout in January–February, a period of low fishing effort and low predation by warmwater predators, particularly black bass. This early stocking also allows some growth prior to the beginning of the summer fishery. Five of these 10 lakes are stocked only in January or February; the remainder receive additional stockings in both April and May (prestratification) and October (poststratification). The statewide creel limit is 8 fish/d. These liberal regulations coupled with the 9-in-minimum stocking size allow for immediate harvest; therefore, without a closed season there is a put-and-take component of the rainbow trout fisheries at these lakes.

The success of the put-grow-and-take rainbow trout fisheries has been examined only at 4 of the 10 lakes. These fisheries occur at night during the major recreational season (mid-May–mid-September); therefore, a nocturnal creel survey is necessary. The nocturnal surveys have been an access-site survey or a limited roving survey in the lower lake because the rainbow trout population and fishery is concentrated near the dam where the largest volume of coldwater habitat occurs. Warmwater fisheries in the upper regions of these lakes were not surveyed at night by roving creel surveys and were unequally sampled compared with lower lake access-site surveys; therefore, estimated percentages of total yield and fishing effort (Table 2) are probably biased toward the rainbow trout fisheries.

Laurel River Lake, by reputation and size, is considered our best rainbow trout fishery; however, Greenbo and Paintsville lakes have higher yields. The Paintsville Lake creel survey was conducted in 1988, 4 years after impoundment. Subsequent changes in the dissolved oxygen–temperature profile have decreased the volume of suitable coldwater habitat. The high yield of 6.1 lb/acre and fishing effort of 23 h/acre have subsequently declined due to the reduction of coldwater habitat and the increased harvest shortly after the annual stocking. Greenbo Lake had the highest yield (7 lb/acre); this lake annually receives the highest stocking at 83 rainbow trout per acre compared with 21–31 rainbow trout per acre in the other lakes surveyed. The Laurel River Lake data reflect a 4-year average of three successive years between 1977 and 1979 and a recent survey in 1993. The 1993 creel survey revealed a return of 38,251 rainbow trout and a yield of 4.2 lb/acre. All rainbow trout lakes displayed a high harvest rate (0.26–0.51 fish/h), and fish averaged 10.7–12.9 in. Prior-year stockings constitute a minor contribution to the total rainbow trout harvest, although quality-size (>13 in) rainbow trout contributed 29% to the nighttime harvest in Laurel River Lake in 1993. Essentially these rainbow trout fisheries are only the product of the annual stocking and the limited growth that occurs within the same year. Except for Wood Creek Lake, all rainbow trout lakes are achieving a 50% return by weight in the harvest.

Hybrid Striped Bass

Hybrid striped bass was first introduced in Kentucky in 1979 after poor results with striped bass stocking were obtained in two Kentucky impoundments (Axon 1979). Positive results had been attained in other southern impoundments with this fish (Williams 1971; Ware 1975). All hybrid striped bass are received as fry from other southern states or purchased from a private producer and reared in state hatcheries to fingerling size (1–2 in). Presently, six impoundments are receiving hybrid striped bass and the stocking rate has been standardized at 20 fingerlings/acre/year (Table 1).

Hybrid striped bass stocking success in Herrington Lake was examined from the initial introduction (1979) through 1986 (Kinman 1987). The original regulation for hybrid striped bass was a 15-in-minimum-length limit and 5 fish/d creel limit. Beginning in 1982 the regulation was changed due to angler misidentification of hybrid striped bass as white bass. The new regulation was a 20 fish/d aggregate limit for hybrid striped bass and white bass, with no more than 5 fish exceeding 15 in. The harvest objective of 1 lb/acre was achieved every year following the liberalization of creel and length limits. The fishing effort (4.8 h/acre) represented 7% of the total fishing effort with a mean harvest rate of 0.53 fish/h (Table 2). This fishery has a low live-release rate (<13%), and the incidence of trophy-size (>23 in) fish was low in 1992 (<3%). The average size hybrid striped bass in this fishery is 13.2 in or an age-1 fish.

The harvest of hybrid striped bass in Barren River Lake has consistently exceeded 10% of the total yield (mean = 17%) and has averaged 1.96 lb/acre in four creel survey years (Table 2). Barren River Lake was formerly regulated with a five-fish-daily creel limit and a 15-in length limit before the regulation was changed in 1989 to the aggregate limit previously described for Herrington Lake due to angler misidentification of hybrid striped bass as white bass. The mean size hybrid striped bass in the harvest decreased from 19.6 in (4.0 lb) with the former regulation to 13.4 in (1.2 lb) with the aggregate limit. Fishing effort (4 h/acre) for hybrid striped bass has averaged 11% of the total fishing effort, and the harvest rate has averaged 0.37 hybrid striped bass per hour. Only 1% of the hybrid striped bass larger than 12 in were released in the 1992 creel survey.

Hybrid striped bass was first introduced in 1989 in Taylorsville Lake; therefore, only a limited number of year-classes was available for harvest during the 1991–1992 creel survey years. At the time of the introduction, white bass were absent from the system, which would alleviate any angler misidentification. A 15-in length limit on all *Morone* species and a 5 fish/d creel limit was implemented to create a higher quality fishery. Hybrid striped bass yield has averaged 2.3 lb/acre or 13% of the total yield. The hybrid striped bass fishing effort (9.1 h/acre) was double the rate of the other two lakes and averaged 7% of the total effort. In 1992 the fishing effort for hybrid striped bass ranked third among the fishing effort for other sport fishes. The higher length limit produced a lower harvest rate (0.10 fish/h) but larger fish (average 16.4 in or 2.2 lb). In 1992, 7% of all creeled hybrid striped bass were larger than 20 in.

Channel Catfish

Channel catfish stocking in small impoundments is one of the oldest fish stocking programs in Kentucky. Maintenance of channel catfish fisheries has been considered paramount because, collectively, catfishes *Ictalurus* spp. are the third most popular group of fish in Kentucky (Hale et al. 1991) and natural reproduction has been considered minimal in small impoundments (Marzolf 1957). Eighty-four smaller state-owned or state-managed impoundments receive annual stockings of channel catfish 9–10 in long; large impoundments are not stocked with channel catfish. Historical stocking rates have been highly variable with little rationale other than allocating the available production. In 1990, stocking rates were standardized; state-owned lakes were given stocking priority and stocked at 50 channel catfish per acre, and other public-owned impoundments were stocked at 25 channel catfish per acre if fish were available. There are presently no creel or size limits on channel catfish.

Creel surveys were conducted in 1991 and 1992 at four state-owned lakes specifically to examine the contribution of channel catfish to the fishery. Average channel catfish yield for these lakes ranged from 23.7 to 62.7 lb/acre or 20–56% of the total yield. Average fishing effort ranged from 34.9 to 97.4 h/acre or 12–25% of the total fishing effort. Harvest rates varied between years and among lakes, but ranged from 0.26 to 0.51 channel catfish per hour. The average length harvested (12.7–14.6 in) indicated that stocked channel catfish are being harvested soon after stocking. A survey of catfish anglers on these four lakes indicated that most anglers (86%) would keep channel catfish smaller than 15 in. Similarly, based on a statewide angler survey (Hale et al. 1991), 74% of the anglers preferred catfish smaller than 15 in.

Walleye

The KDFWR began a limited stocking program with northern (lake) strain walleye (Hackney and Holbrook 1978) in the 1960s–1970s in several impoundments. The original goal was to establish self-sustaining populations following introductory stockings. Walleye stocking philosophy changed in the 1980s, and KDFWR decided to concentrate on put-grow-and-take fisheries because walleye fisheries had not developed by natural reproduction. Also, stocking would be limited to impoundments with coolwater habitat (oxygenated water <75°F in the summer growing season). Presently, KDFWR is stocking seven impoundments with fingerling (1.0–2.0 in) walleye at 50 fingerlings/acre except for Lake Cumberland. Natural reproduction in this lake in 2 of 3 years contributed significantly to the fish population (Kinman 1990); therefore, the stocking number was reduced to 7 fingerlings/acre. The statewide regulations for walleye are a 15-in-minimum-length limit and a 10 fish/d creel limit.

Walleye fisheries have failed to meet harvest objectives; however, there is strong public support for this fish in Kentucky. The best recorded yields were 0.42 lb/acre (13% of the total yield) from Laurel River Lake and 0.63 lb/acre (5% of the total yield) at Carr Fork Lake in 1993. Lake Cumberland, by reputation, is our best walleye lake but has not been surveyed since 1988 when the yield was 0.29 lb/acre (6% of the total yield). The best recorded harvest rate of 0.16 walleyes per hour also occurred at Lake Cumberland. Walleye fishing effort (2.1 h/acre, 8% of the total fishing effort) was highest at Laurel River Lake. Walleye fishing effort at the remaining lakes was less than 1.5 h/acre and less than 5% of the total fishing effort. The average length walleye in the harvest was similar (17–18.3 in) among the lakes; fish of this length are age-3 fish. Walleyes enter the legal fishery at age 2. The harvest in all these lakes principally occurs in May–June prior to thermal stratification and in October following destratification. Many walleye are harvested by black bass and white bass anglers while night fishing. Walleye fishing has increased in those impoundments where walleye spawning runs into headwater tributaries occur in the spring.

Discussion

Fishing is a major outdoor recreational activity in Kentucky; combined resident and nonresident participation rates are above the national average (Adams et al. 1993). Demand for fishing opportunities is difficult to assess with the continuum of angler desires and motivations; however, we do know that 80% of the angler-days in Kentucky are devoted to black bass, crappies, sunfishes *Lepomis* spp., and catfishes, in that order (Hale et al. 1991). These populations are self sustaining, except channel catfish, in small impoundments; therefore, our fishery management options are limited to regulation in allocating these renewable resources. The KDFWR has emphasized the use of cultured fish in its sport-fish management program, and put-grow-and-take fisheries represent a major component. The importance of these put-grow-and-take fisheries is lower than that of most self-sustaining fisheries on a statewide basis; however, on a lake-specific basis, creel survey results have demonstrated that several of these fisheries are major contributors to both yield and fishing effort. These fisheries have also possibly reduced fishing effort on other popular species.

A cost-benefit analysis of these fisheries is difficult because annual hatchery production and transportation costs can not easily be partitioned for put-grow-and-take stocking and data on annual use of these fisheries are not available. Collectively, the operation of the two KDFWR hatcheries and a fish transportation section represented approximately 35% (1991) of the Fisheries Division US$3 million operating budget. However, in 1990, 78% of the total weight of sport-fish production at Minor Clark Fish Hatchery was fish to maintain put-grow-and-take fisheries, and 84% (by weight) of the Frankfort Fish Hatchery sport-fish production was devoted to the channel catfish program. A cost of $873,000 was estimated from these percentages and the total budget of the fish transportation sections. Based on $22/d trip-related expenditures in Kentucky (USDI 1993) and estimated angler use from creel surveys, annual trip-related expenses for muskellunge, striped bass, and hybrid striped bass were $1,466,500. Similarly, annual trip-related expenditures to fish for catfish (based only on lakes stocked with channel catfish) were $1,284,600. These four fishes accounted for $2.7 million in trip-related expenditures and an estimated 3:1 benefit-to-cost ratio. Walleyes were not included in this estimate because the contribution of natural reproduction cannot be quantified. Also rainbow trout was omitted because they are produced at a federal hatchery.

The future use of these fisheries will require the development of lake-specific management objectives and consideration of genetic conservation for some species. The KDFWR is developing lake management plans that will address lake-specific objectives. A draft strategic plan for the KDFWR addresses the need to consider the effects of our management practices on natural diversity.

Presently, no immediate plans exist for drastic changes to the put-grow-and-take fisheries for either channel catfish or rainbow trout. The KDFWR's concern for both these fisheries is the level of immediate harvest, because both are stocked with catchable-size fish and no size limits are in place for protection. Both fisheries employ the use of live bait, and potential problems exist with catch-and-release mortality if a length limit is implemented. Although the growth potential of these fish has not been fully achieved, there is an acceptable level of angler satisfaction with the current management regime.

Genetic contamination of native channel catfish stocks due to escapement from our stocking practices or via the aquaculture industry remains a possibility. No genetic examination of channel catfish stocks (except for hatchery broodstock) has been conducted.

Striped bass fisheries in Kentucky will not be expanded, but will be refined with different stocking and regulatory strategies. The Lake Cumberland striped bass fishery has exceeded all expectations with the additional benefit of a significant winter fishery. The effect of catch-and-release mortality will possibly be examined following the recent length limit change. The amount of natural reproduction in the Ohio River will be examined in the future by use of marked hatchery fingerlings. Striped bass telemetry studies have been conducted in the midsection of the Ohio River (Henley 1993) and lower river (in progress). One objective of these studies is to provide anglers information they can use to improve their fishing success for this species when fish are not concentrated below navigation dams.

Muskellunge was propagated to maintain the genetic integrity of the native stock; plans are to continue this stocking program at the current level. Muskellunge remains KDFWR's most expensive cultured fish because it is extensively reared on minnows and goldfish. Efforts are in progress to refine stocking density of 13-in muskellunge to achieve the greatest survival and return at the least cost in the two significant fisheries at Green River and Cave Run lakes. Current management objectives are based on yield (weight harvested). Lake-specific objectives will be redefined to incorporate the high catch-and-release rate of legal-size muskellunge.

Walleye was native to several major river systems in Kentucky (Everman 1918); however, native river populations were effectively extirpated by impoundments. Native southern strain walleyes failed to sustain themselves following river impoundment and most impoundment walleye fisheries have been developed by introducing a northern (lake) strain that is considered a more successful lake spawner (Hackney and Holbrook 1978). Kentucky has used a walleye strain of Lake Erie origin. Walleye escapement and migration outside the state could genetically contaminate other walleye stocks. No strategies have been discussed to address this potential problem.

Walleye fisheries have been the least successful put-grow-and-take fisheries in Kentucky based on total return, catch rate, and percent of fishing effort, but the reasons for this are not completely understood. Fall gill netting generally reflects good density and size distribution (McLemore et al. 1992). Summer harvest of walleye in Kentucky is affected, probably negatively, by the light-sensitive nature of walleye (Ryder 1977) coupled with the structure and bottom orientation of this fish. These two factors in our deep, clear storage impoundments place these fish in inaccessible locations for conventional fishing methods in Kentucky. Education efforts have increased angler knowledge, and telemetry studies (in progress) on two impoundments will assist anglers to locate walleye better. Future plans also include mass marking of hatchery fish to examine the level of natural reproduction and possibly reduce stocking numbers. Stocking time will be examined to ascertain the level of synchrony with the spawning of gizzard shad *Dorosoma cepedianum* (Ney and Orth 1986). An additional impoundment, Green River Lake (8,210 acres), was added to the walleye stocking program in 1993; however, no plans exist for any further expansion of this program.

Hybrid striped bass has been the most recent addition to our list of cultured fish, and it produced the most predictable result with relatively high catch rates and significant contributions to the yield. One reason for the large contribution to yield is hybrid striped bass provides a nearly year-round fishery in Kentucky impoundments. Similar positive results have been documented by Axon and Whitehurst (1985). The trophy aspect of this fishery has not met our expectations; average-size hybrid striped bass are in the 12–17-in length range, despite growth capabilities to 20 lb. Misidentification of smaller hybrid striped bass as white bass by both anglers and creel clerks may be responsible for underestimating hybrid striped bass harvest and reducing densities of trophy-size fish in the population. The regulation strategy at Taylorsville Lake may improve densities of trophy-size hybrid striped bass following the recruitment of more year-classes. However, current angler attitudes favor harvesting smaller hybrid striped bass; therefore, angler satisfaction may be met without the trophy aspect of this fishery. Regulating hybrid striped bass fisheries remains a problem due to anglers' inabilities to distinguish white bass from hybrid striped bass correctly. Muoneke (1987) has recommended managing both hybrid striped bass and white bass as a single fishery with combined regulations when both occur in the same lake. Based on the current success of hybrid striped bass fisheries, the KDFWR plans to expand the stocking program, and new water bodies are under consideration. The inability to obtain *Morone* broodfish consistently has prevented KDFWR production of hybrid striped bass fry. Limited genetic monitoring in two lakes has not located any hybrid striped bass backcrossing with the native white bass; however, the potential exists in both stocked impoundments and contiguous waters.

The use of cultured fish has created significant additional angling opportunities in impoundments in Kentucky. The artificiality of these systems must be considered in the evolution of ecosystem management and the protection of biodiversity. These highly altered systems, rivers and reservoirs on these rivers, exist for a variety of reasons, and there are no plans to restore these systems to their natural state. Many voids in knowledge remain regarding fish community analysis in reservoirs, including ones receiving cultured fishes. The cost of improving this knowledge base must be balanced in limited state budgets with other existing operational activities, other new programs, and development projects.

References

Adams, C. E., J. K. Thomas, and W. R. Knowles, Jr. 1993. Explaining differences in angling rates in United States. Fisheries (Bethesda) 18(4):11–17.

Axon, J. R. 1978. An evaluation of the muskellunge fishery in Cave Run Lake, Kentucky. American Fisheries Society Special Publication 11:328–333.

Axon, J. R. 1979. An evaluation of striped bass introductions in Herrington Lake. Kentucky Department of Fish and Wildlife Resources Bulletin 62, Frankfort.

Axon, J. R., and D. K. Whitehurst. 1985. Striped bass management in lakes with emphasis on management problems. Transactions of the American Fisheries Society 114:8–11.

Brewer, D. K. 1980. A study of native muskellunge populations in eastern Kentucky streams. Kentucky Department of Fish and Wildlife Resources Bulletin 64, Frankfort.

Coutant, C. C. 1985. Striped bass, temperature, and dissolved oxygen: a speculative hypotheses for environmental risks. Transactions of the American Fisheries Society 114:31–61.

Everman, B. W. 1918. The fishes of Kentucky and Tennessee, a distribution catalogue of the known species. Bulletin of U.S. Bureau of Fisheries 35:295–368.

Hale, S. R., M. Price, and E. Schneider. 1992. 1991 Kentucky angler survey. Urban Research Institute Bulletin, University of Louisville, Louisville, Kentucky.

Hackney, P. A., and J. A. Holbrook, II. 1978. Sauger, walleye, and yellow perch in the southeastern United States. American Fisheries Society Special Publication 11:74–81.

Henley, D. T. 1993. Seasonal movement and distribution of striped bass. Proceedings of the Annual Conference Southeastern Association of Fish and Wildlife Agencies 45(1991):370–384.

Hysmith, B. T., J. H. Moczygembia, and G. R. Wilde. In press. Hooking mortality of striped bass in Lake Texoma, Texas–Oklahoma. Proceedings of the Annual Conference Southeastern Association of Fish and Wildlife Agencies.

Kinman, B. T. 1987. Evaluation of hybrid striped bass introductions in Herrington Lake. Kentucky Department of Fish and Wildlife Resources Bulletin 82, Frankfort.

Kinman, B. T. 1988. Evaluation of striped bass introductions in Lake Cumberland. Kentucky Department of Fish and Wildlife Resources Bulletin 83, Frankfort.

Kinman, B. T. 1989. Evaluation of muskellunge introductions in Green River Lake. Kentucky Department of Fish and Wildlife Resources Bulletin 85, Frankfort.

Kinman, B. T. 1990. Evaluation of walleye introduction in Lake Cumberland. Kentucky Department of Fish and Wildlife Resources Bulletin 88, Frankfort.

Marzolf, R. C. 1957. The reproduction of channel catfish in Missouri ponds. Journal of Wildlife Management 21:22–28.

McLemore, W. N., and six coauthors. 1992. Annual performance report for statewide fisheries management project. Subsection 1: lakes and tailwaters research and management. Kentucky Department of Fish and Wildlife Resources, Frankfort.

Muoneke, M. I. 1987. Population dynamics of white bass and striped bass × white bass hybrids in Lake Carl Blackwell, Oklahoma. Doctoral dissertation. Oklahoma State University, Stillwater.

Ney, J. J., and D. J. Orth. 1986. Coping with future shock: matching predator stocking programs to prey abundance. Pages 81–92 in R. H. Stroud, editor. Fish culture in fisheries management. American Fisheries Society, Fish Culture Section and Fisheries Management Section, Bethesda, Maryland.

Redmond, L. C. 1986. The history and development of warmwater fish harvest regulation. Pages 186–195 in G. E. Hall and M. J. Van Den Avyle, editors. Reservoir fisheries management. American Fisheries Society, Southern Division, Reservoir Committee, Bethesda, Maryland.

Ryder, R. A. 1977. Effects of ambient light variations on behavior of yearling, subadult, and adult walleyes (*Stizostedion vitreum vitreum*). Journal of the Fisheries Research Board of Canada 34(10):1481–1491.

Ware, F. J. 1975. Progress with *Morone* hybrids in freshwater. Proceedings of the Annual Conference Southeastern Association of Fish and Wildlife Agencies 28(1974):48–54.

Wilkins, P., L. Kirkland, and A. Hulsey. 1967. The management of trout fisheries in reservoirs having a self-sustaining warm water fishery. Pages 444–452 in Reservoir fishery resources symposium. American Fisheries Society, Southern Division, Reservoir Committee, Bethesda, Maryland.

Williams, H. M. 1971. Preliminary studies of certain aspects of the life history of the hybrid (striped bass × white bass) in two South Carolina reserves. Proceedings of the Annual Conference Southeastern Association of Game and Fish Commissioners 24(1970):424–431.

U.S. Commission of Fish and Fisheries. 1884. Report to the Commissioner for 1881, part IXI. Government Printing Office, Washington, DC.

U.S. Commission of Fish and Fisheries. 1885. Report of the Commission for 1883, part XI. Government Printing Office, Washington, DC.

USDI (U.S. Department of the Interior). 1993. 1991 national survey of fishing, hunting, and wildlife associated recreation: Kentucky. USDI, Washington, DC.

Better Roles for Fish Stocking in Aquatic Resource Management

RAY J. WHITE

320 12th Avenue North, Edmonds, Washington 98020, USA

JAMES R. KARR

Institute for Environmental Studies, University of Washington, Seattle, Washington 98195, USA

WILLA NEHLSEN

2100 Southeast Hemlock Avenue, Portland, Oregon 97214, USA

Abstract.—Artificial propagation and stocking have been a central feature of North American fishery management since its inception well over a century ago. Despite questions that have long been voiced about the effectiveness of fish stocking, citizens, fishery agencies, and legislators often pin hopes for the future of fisheries on stocking programs. Evidence that poststocking performance (survival and reproduction) of hatchery-produced fish is inferior to that of wild conspecifics is abundant, as is evidence that stocked fish harm wild fish. Poststocking behavioral problems of hatchery fish appear at least as numerous and damaging as physiological and anatomical problems but receive less attention in culture programs; they may also be less solvable. Severity of culture-induced problems increases with the length of the period under artificial propagation; raising larger fish and more generations of captive-bred fish increase the likelihood of problems. The problems with stocking are worse in streams, in part because streams tend to present harsher environments than do standing waters. Stocking is least successful where a reproducing population of the same species already exists, so stocking programs should be separated from healthy and potentially healthy wild stocks. Supplementation, stocking of progeny of wild broodstock into the parent population, is especially likely to be counterproductive. Better evaluation of poststocking performance and of effects on wild populations may improve programs, if agencies act on the results. Evaluation of stocking programs should go beyond accounting for operational costs and monetary value of harvested fish; it should recognize full costs and benefits in economic, ecological, and social contexts. Analysis of the environmental impacts of stocking should be done for each species and water body. State, provincial, tribal, and federal programs should use fish stocking only within ecologically sound policies. The top priority of fishery programs should be protection of wild, native fish stocks and the environments upon which they depend.

North American freshwater and anadromous fisheries management began essentially as the artificial spawning and hatching of fish to produce large numbers of young for release in water bodies—an innovation over ancient culture of self-reproducing fish, such as common carp *Cyprinus carpio* in China (Landau 1992). Public enthusiasm for such new technology, in keeping with the nineteenth century spirit of spreading agriculture across the continent, led to massive expansion of government hatchery and stocking programs (Bowen 1970; USCFF 1873–1994; Bottom, in press), an emphasis that continues today.

The legacy of early enthusiasm for fish stocking can be seen in current agency budgets. For example, the Washington Department of Fisheries (WDF, no date) spent US$31.3 million for salmon culture,[1] the largest category (35%) in a 1991–1992 fiscal year operating budget of $88.7 million. (Culture and stocking of steelhead *Oncorhynchus mykiss* and freshwater fishes were conducted by another state agency.) Incorporation of other assignable costs for administration, planning, research, and amortized capital expenditure would raise percentage spent for the salmon culture even higher. Similarly, Oregon's $90.6 million 1993–1995 biennial budget for fisheries included 42.5% to propagate fish but only 3% to manage for natural production (ODFW 1993), and Pennsylvania's $24.8 million 1991–1992 fiscal year fisheries budget showed 37% for fish propagation (PFBC 1992).

Some fishery authorities began to recognize failure of major stocking programs soon after they started, and others expressed serious fundamental criticism. Spencer F. Baird, the first U.S. Fish Commissioner and founder of the federal hatchery program, concluded that East Coast stocking of Atlantic salmon *Salmo salar* and Pacific salmon *Oncorhynchus* spp. was not working and halted it in the 1880s (Foster 1991). He apparently tried but was largely unsuccessful in warding off nationwide indiscriminate stocking (Bo-

---

[1] Direct operational cost plus what we estimate to be the proportion of within-salmon-program administrative cost attributable to culture. This does not include a culture-attributable amount for general agency administration.

wen 1970). Cobb (1917) warned that "in some sections an almost idolatrous faith in the efficacy of artificial culture of fish for replenishing the ravages of man and animals is manifested, and nothing has done more harm than the prevalence of such an idea ... the consensus of opinion is that artificial culture does considerable good, yet the very fact that this cannot be conclusively proven ought to be a warning to all concerned not to put blind faith in it alone." Koelz (1926) presented data and emphatic arguments on the futility of stocking coregonid fry in the Great Lakes; confirmation followed (Miller 1946, 1952b; Christie 1963), and the program was discontinued in 1960 (Todd 1986).

Despite longstanding expressions of concern about stocking, reactions to declining fish abundance have continued to center on stocking (or demanding it), and traditionally less emphasis is placed on dealing with causes of decline. Early fish culturists were confident that stocking would "negate the need for regulations on commercial fishing" (Todd 1986), and it was commonly thought stocking could offset fish losses from damming, forest clearance, agriculture, and other anthropogenic degradations of habitat (Bowen 1970; Foster 1991). This belief still resounds among the public and some fisheries scientists when problems of overharvest or fish habitat destruction arise, and some major stocking programs expanded tremendously from the late 1950s into the 1980s. For example, Blumm and Simrin (1991) described how agencies in the Columbia River basin tried unsuccessfully to reverse salmon decline by concentrating on hatchery construction and stocking, not on treating causes. New public hatchery systems are still being planned; for example, one on Washington's Skagit River (WDW 1994) and a $40.5 million complex in the Yakima and Klickitat river basins of Washington consisting of several central hatcheries, several more satellite facilities, and perhaps as many as 33 acclimation and release sites (BPA 1990).

But countering views have increased. Thompson (1965) warned eloquently about ways in which hatchery programs violate basic biological principles. In the last 18 years, investigation and critical analysis of stocking and stocked hatchery fish has intensified in the form of at least 72 articles.[2] As the critical aspect of the articles increased, hatchery-based management was defended (Gladson 1990; Martin et al. 1992; SFI 1992; Daley 1993; Nickum 1993; Kochman 1994). Also, recent articles have explained stocking program benefits (Bachman 1994; Chan 1994; Fowden 1994; Schmidt 1994; Whaley 1994).

Our goal here is to review evidence on effects of fish stocking and suggest better roles for stocking. We conclude that stocking can be appropriate in a narrow range of conditions. Failure to assess those conditions in designing fisheries management programs, and thus stocking under the wrong circumstances, is wasteful and can contribute to decline of wild[3] fish stocks. Some fishery agencies recognize these realities, but for various reasons stocking persists as the primary management tool in many areas where wild fish populations thrive or where they could thrive if properly fostered. Even when scientific evidence is substantial that stocking is unneeded, or will fail or cause harm, many agencies allow political pressure for stocking to determine their activities. Protecting fish stocks to benefit future generations requires change from emphasis on stocking for single-species goals and adoption of more integrative and ecologically sound fisheries resource management.

Stocking Programs—An Unsettled History with Uncertain Results

North American public stocking programs were begun by the Canadian government (USCFF 1873–

---

[2] Reisenbichler and McIntyre 1977, 1986; Harris 1978; Laarman 1978; Johnson 1978; Hosmer et al. 1979; Sholes and Hallock 1979; Sosiak et al. 1979; Allendorf and Phelps 1980; Ryman and Stahl 1980; Krueger et al. 1981; Hynes et al. 1981; Wright 1981, 1993; Brown 1982; Dickson and MacCrimmon 1982; Cross and King 1983; Doyle 1983; Ersback and Haase 1983; Bachman 1984; Leider et al. 1984, 1986, 1990; O'Grady 1984; Reisenbichler 1984, 1988; Vuorinen 1984; Campton and Johnston 1985; Smith et al. 1985; Chilcote et al. 1986; Levings et al. 1986; Marnell 1986; Moyle et al. 1986; Nickelson et al. 1986; Todd 1986; Allendorf and Ryman 1987; Lichatowich and McIntyre 1987; Vincent 1987; Petrosky and Bjornn 1988; Vespoor 1988; Behnke and Johnson 1989; White 1989; Francis 1990; Goodman 1990; Jonsson et al. 1990, 1991; Miller et al. 1990; Steward and Bjornn 1990; Behnke 1991; Egidius et al. 1991; Evans and Willox 1991; Hindar et al. 1991; Stevens 1991; Waples 1991; SAC 1991; Hard et al. 1992; Hilborn 1992; Meffe 1992; White 1992a, 1992b; Dickson 1993; Fleming and Gross 1993; Hilborn and Winton 1993; Walters and Juanes 1993; Washington and Koziol 1993; Wiley et al. 1993a, 1993b; Bakke 1994; Griffith 1994; Van Vooren 1994; Winton and Hilborn 1994.

[3] We define a wild fish as one that completes its life cycle without direct help from humans after hatching in a largely natural environment from an egg deposited there by its mother. Wild fish rear to maturity, spawn, and have offspring to produce the next generation of wild fish.

1994) and by eastern states of the United States in the 1860s as efforts to create more sport fishing and offset fish population declines that resulted from overharvest and habitat destruction (Bowen 1970). The overall effort accelerated in the United States after establishment of the U.S. Fish Commission in 1872 (Bowen 1970). The first U.S. federal fish hatchery was built in 1871; the number rose to 77 in 1918 and then underwent a fluctuating phase having a slow upward trend until the mid-1930s, when another rapid growth period began due to a job-creation effort during the Great Depression. After reaching an all-time high of 122 federal hatcheries in 1943, the number declined to the present 74 (USCFF 1873–1994). Often, the federal government established hatcheries and later transferred them to state agencies.

The amount of fish distributed from federal hatcheries grew during 1871–1919 to a fluctuating plateau of about 1 million kg annually during 1920–1930 (USCFF 1873–1994). A decline began and continued through 1948, when output was about 250,000 kg, despite the increase in federal hatchery building of the late 1930s and early 1940s. As in-hatchery disease and nutrition problems were solved in the 1950s and, particularly, the 1960s (Lichatowich and McIntyre 1987), the federal output grew again. It persists today at somewhat less than 3 million kg per year (USCFF 1873–1994). Individual federal stocking programs grew, plateaued, waned, and regrew or were halted within this overall pattern. We did not systematically investigate fish stocking trends in state and Canadian agencies, but it appears that many of these also underwent initial growth phases in the late 1800s, then diverged from the U.S. federal pattern, but also tended to expand production from the 1950s into the 1980s.

Even when stocking programs ended in an area or a fish hatchery closed, the human impulses and political pressures that led to those programs lived on and reasserted themselves. Many defunct programs resumed after public and institutional "forgetting periods," often involving retirements or deaths of the persons who stopped the programs. For example, Foster (1991) reported that U.S. Commissioner of Fisheries Baird suspended stocking for the Connecticut River's unsuccessful Atlantic salmon restoration in 1881, that he halted transfer of Pacific salmon to East Coast streams in 1883, declaring it a proven failure, and that by 1890, "even the states had given up on restoring salmon to the Connecticut River." Baird died in 1887. In six of the years between 1894 and 1902, Atlantic salmon eggs from a federal hatchery were sent to a Connecticut State station on the Connecticut River. Intermittently after Baird's death, entirely unsuccessful stocking of Pacific salmon was done in East Coast states (USCFF 1873–1994).

Alaskan salmon hatcheries provide a major example of halt and resumption. In 1891, Alaskan cannery operators began to develop hatcheries mainly for sockeye salmon $O.$ $nerka$, soon after which federal hatcheries were also built. At the peak of this combined effort, 220 million sockeye salmon eggs were taken at 6 facilities in 1910, resulting in 180 million fry released in 1911 (Roppel 1982). After that, biological and funding problems caused closing of hatcheries until two federal and one private facility remained in 1933, when the U.S. Commissioner of Fisheries closed the two federal hatcheries (the private one lasted until 1936), because he considered them a waste of public funds and an unjustified subsidy to a special industry. The Commissioner advocated maintaining runs by restricting harvest to let enough adults reach spawning grounds (Roppel 1982). Thereafter, the only Alaskan hatcheries were four research stations (one converted to stocking of pink salmon $O.$ $gorbuscha$ in 1965) until the 1970s, when the state initiated a new system of public and private hatcheries, which concentrated on pink salmon but also built up a sizable sockeye salmon component and stocked fewer of other salmonids (McNair and Holland 1993).

The efficacy of the renewed Alaskan hatchery program is in question. After being depressed in the late 1950s, Alaskan salmon catches began a trend of general increase, which Royce (1989) attributed largely to scientifically based harvest control. After a 1972–1976 period of poor runs due to unfavorable weather, the trend rose sharply to a record catch in 1987, to which the Alaskan hatchery effort contributed about 20% by weight (Royce 1989). Beamish and Bouillon (1993) analyzed North Pacific Ocean climate, salmon management, and catches of pertinent countries and concluded that the trends in salmon catch did not result primarily from fishing effort, harvest management, or stocking, but from climate trends. That conclusion may warrant modification in view of later knowledge about shifts in Pacific Ocean circulations (D. M. Eggers, Alaska Department of Fish and Game [ADFG], personal communication). Beamish and Bouillon (1993) suggested reducing stocking during periods of decreasing marine survival and increasing it to boost catch during periods of climate-related high marine survival of salmon. But predicting climate far enough in advance to adjust hatchery production would be difficult (T. Ellison, ADFG, personal communication), and the value of increased stocking when ocean conditions enable better survival might be negated because prices received by commercial fishers fall as the harvests, which are primarily wild fish, increase (Boyce et al. 1992). Benefit-cost analysis of 16 alternatives for changing the Alaskan salmon stocking program indicated that, over the next 30 years, the greatest positive net benefit (financial return to commercial fishers minus their fishing costs and the cost of operating the hatchery program) would come from eliminating the hatchery program, that the greatest net detriment would derive from increasing the whole program by 15% (the largest increase examined), and that intermediate measures (increasing or decreasing hatchery production of different species within the program) would have intermediate effects (Boyce et al. 1992). The analysis did not consider sport fishers, other users, or noneconomic benefits and costs.

Another example of stopping and starting is the Canadian stocking of Pacific salmon. Based on poor results from sockeye salmon stocking (Foerster 1931, 1936), Canada closed all Pacific salmon hatcheries after the 1935 hatching season (Larkin 1970) but between 1977 and 1988 redeveloped a massive Salmonid Enhancement Program (SEP). The SEP consisted of 28 major salmon hatcheries (Shepherd 1991), about 270 smaller facilities and spawning channels, and a habitat restoration effort. The primary goal of SEP is to double the number of salmon available for catch in British Columbia (Hilborn and Winton 1993). But as increasing amounts of fish were stocked, the catches of the main species stocked, chinook *O. tschawytsha* and coho salmon *O. kisutch*, declined. By the late 1980s SEP's contribution of all species to the annual landed value of Canadian fisheries barely equaled the program's annual cost and was only 64% of the same period's value of wild Fraser River sockeye salmon, which came almost entirely from a watershed having little SEP activity (Hilborn and Winton 1993).

The extreme example of a recently expanded stocking program may be Washington State. In 1960, state and federal hatcheries in Washington stocked less than 0.5 million kg of salmon (Fletcher et al. 1976; USCFF 1873–1994) in addition to an unknown weight of steelhead and other trout. By the early 1980s, a system of state, federal, tribal, and cooperative facilities was annually releasing 2.7 to 3.3 million kg of salmon (Hill 1984) in addition to the other salmonids. In 1990, 121 state, federal, and tribal fish hatcheries and over 140 smaller satellite and volunteer-operated facilities in Washington released 358 million salmon and trout (WDF 1993a; K. Watson, State of Washington House of Representatives Natural Resources Committee, unpublished data). The biomass of hatchery salmonids stocked in Washington that year was 4.67 million kg (USCFF 1873–1994; WDF 1993a; J. Kerwin, Washington Department of Fish and Wildlife, Olympia, personal communication; K. Phillipson, Northwest Indian Fisheries Commission, Olympia, personal communication). This was 61% more than the 2.9-million-kg U.S. federal hatchery output of all species, nationwide, in 1990 (USCFF 1873–1994).

Despite, and perhaps partly because of (Goodman 1990; Waples 1991; Wright 1993), such substantial annual stocking in Washington, and other large programs in Oregon and Idaho, the salmon declines continued. By 1990, at least 94 of Washington's wild anadromous salmonid stocks were considered at risk of extirpation, among 214 at risk in Washington, Oregon, Idaho, and California (Nehlsen et al. 1991). Since then several Northwest salmon stocks have been listed as Threatened or Endangered under the U.S. Endangered Species Act of 1973 (16 U.S.C.A. §§1531 to 1544), and in 1994, for the first time in history, ocean fishing for chinook and coho salmon was closed along most of the Washington and Oregon coast.

Stocking programs can support harvest in the short term, and many fisheries have depended on such government support. Common examples are put-and-take recreational fisheries for legally catchable-size trout, put-and-grow fisheries in suitable lakes, and some commercial salmon fisheries in parts of the Pacific Northwest where overfishing and habitat destruction have seriously reduced wild stocks. However, salmon stocking itself has contributed to declines of wild stocks. We question the economic efficacy of intense artificial support of fisheries, as well as the biological and social effects. Full biological, social, and economic evaluation of a stocking program probably has never been done. A few stocking programs have been studied in terms of the immediate biological and social objectives of increasing the abundance of fish and the amount harvested by anglers. Wiley et al. (1993b) observed that for more than 100 years low return rates to fisheries from stocking less than catchable-size hatchery trout have been reported.

Poor poststocking survival is common (Shetter and Hazzard 1941; Needham and Slater 1944; Smith and Smith 1945; Schuck 1948; Miller 1954; Wales 1954; Brynildson and Christenson 1961; Reimers 1963; Mason et al. 1967; Kanid'yev et al. 1970), and stocking hatchery fish can depress trout abundance (Thuemler 1975; Vincent 1987; Evans and Willox 1991). Several studies have shown a genetic basis for poor poststocking survival of hatchery fish (Reisenbichler and McIntyre 1977; Chilcote et al. 1981). There may be only one study (Nielsen et al. 1957) in which hatchery trout stocked in a stream often survived as well as wild trout of the same size, although some of the study's results were at odds with this conclusion, especially during severe winters. Todd (1986) reviewed the long history of poor success in stocking coregonids in the Great lakes. Lichatowich and McIntyre (1987) described the collapse of the Oregon coho salmon fishery despite massively increased stocking. Even where survival after stocking in initial years is satisfactory, the results of continued programs typically deteriorate and, after a few years, no longer meet biological or economic objectives (Hilborn 1992; Hilborn and Winton 1993).

Stocked hatchery fish may generally survive less well in streams than in lakes and deep ponds. Streams require fish to spend energy in currents and, especially where shallow, tend to present harsher environments in terms of small and varying volume, rapidly fluctuating temperature, and depth-related exposure to avian, mammalian, and reptilian predators. The difference between survival of stocked trout in flowing- and in standing-water environments is in part the basis for Alberta's (Nelson and Paetz 1992) and Montana's (Wells 1985) policies of stocking trout only in lakes, reservoirs, and ponds.

Effects of Hatchery and Stocking Programs

Fish produced in hatcheries exhibit poor poststocking survival and reproduction because of many morphological, physiological, and behavioral problems (Table 1), which are often interrelated. Hatchery fish also can have multiple adverse influences on wild fish (Table 2). In addition, hatchery facilities and operations can harm wild fish and biotic communities via environmental effects (Table 2). Most importantly, perhaps, stocking programs can detrimentally influence people's attitudes and expectations (Wiley et al. 1993a), especially by instilling a false sense of security that a solution to fishery decline is at hand (Goodman 1990). Some of the poststocking problems tend to occur only in gross mismanagement and can be solved or reduced by commonly applied hatchery and stocking methods; e.g., proper timing of release can obviate the problem of inappropriate physiological state when fish are stocked, waste treatment can remedy contamination of streams with hatchery effluent, and strictly enforced policies against stocking nonnative fishes can avoid the problems of introduced species. Other problems will be much harder to solve, and some will probably be intractable. Notable among the difficult-to-unsolvable problems are those that involve fish behavior.

Differences from wild fish arise in hatchery conspecifics by genetic divergence and by induction of the hatchery environment, as Swain et al. (1991) demonstrated for morphological differences in coho salmon. Learning of behaviors would be another form of environmental induction in hatcheries. Altered fish behavior is probably the most ignored effect of hatcheries. Of 27 ways that hatchery programs can decrease the survival or reproductive success of stocked fish (Table 1), 14 are primarily behavioral. Of 10 ways that hatchery fish can harm wild fish (Table 2), at least 8 are primarily behavioral. The behavioral problems we list are complexes of many interrelated subbehaviors and often also involve anatomical and physiological problems. For example, the high vulnerability of hatchery-reared fish to predators has many contributory behaviors —e.g., conspicuous hyperactivity (Bachman 1984), near-surface swimming (Vincent 1960; Moyle 1969), weak concealment behavior (Raney and Lachner 1942; Vincent 1960; Roadhouse et al. 1986), lack of overhead fright response (Ritter and MacCrimmon 1973), inappropriate willingness to risk predation in favor of feeding (Johnsson and Abrahams 1990)—some of these inferable from altered behavior of wild fish in captivity—and various behaviors listed in Table 1. The greater predator vulnerability of hatchery fish than of wild conspecifics must also derive from most of the anatomical and physiological problems that hatchery fish can have (Table 1).

Fish reared in hatcheries acquire behaviors that are inappropriate to life in many natural environments, and they lose adaptive behaviors. This is harmful in the wild because, in nature, fish must make complicated decisions in order to survive and reproduce. For example, under changing conditions a fish must decide how simultaneously to maximize energy intake, minimize energy expenditure, and minimize risk of predation (Dill 1987; Walters and Juanes 1993).

Fish culturists typically have a physiological, not behavioral, focus. They manipulate water quality, nutrition, disease control, and genetics for better physiological performance (primarily health and growth) of fish under hatchery conditions; the small attention they give fish behavior also typically per-

TABLE 1.—Forms of poor (or potentially disadvantageous) performance of hatchery-reared fish relative to wild fish.

Trait or process	References or clarification
Primarily morphological	
Abnormal gross body proportions	Swain et al. 1991
Deformity	Aulstad and Kittelsen 1971; Kincaid 1983; Bouck and Smith 1979
Unnatural coloration	Hager 1964; Maynard 1995, this volume
Hyperbuoyancy	Sosiak 1982
Primarily physiological	
Rapid decline of energy stores	Klak 1941; Miller 1952a, 1954, 1958; Reimers 1963; Ersbak and Haase 1983
Stress response	Stress in handling limits survival; Barton et al. 1986; Woodward and Strange 1987
Poor stamina	Miller et al. 1959; Vincent 1960; Green 1964; Bams 1967; Horak 1972; Thomas and Donahoo 1977
Inappropriate physiologic state when stocked	If rearing and release are poorly coordinated
Inability to cope with frequent natural variations in environmental conditions, e.g., thermal fluctuation	Vincent 1960
Inability to cope with infrequent natural extremes, e.g., thermal, drought	Thompson 1965
High pathogenic disease incidence	Goede 1986, 1994
Nutritional disease	If the fish were not properly fed
Delayed onset of breeding in females	Fleming and Gross 1993
Primarily behavioral	
Vulnerability to predators	Mead and Woodall 1968; Fraser 1974; Olla and Davis 1989; Olla et al. 1991; Jarvi and Uglem 1993
Difficulty satisfying nutritional needs	Miller 1958; Reimers 1963; Kanid'yev et al. 1970; Gillen et al. 1981; Ersbak and Haase 1983; Bachman 1984
Inferior ability to cope with winter conditions	Reimers 1963
Inappropriate habitat occupation	Vincent 1960; Dickson and MacCrimmon 1982; Petrosky and Bjornn 1988
Energy waste in fast water	Bachman 1984
Stress in unfamiliar surroundings and social disorientation	Miller 1958; Bachman 1984
Lack of social hierachy	Bachman 1984
Weak territorial behavior	Norman 1987
Increased or decreased aggressiveness	Swain and Riddell 1990, 1991; Mesa 1991; Ruzzante 1994
Inappropriate timing and duration of migration and spawning	Washington and Koziol 1993
Less precise homing	Bams 1976; Stabell 1981, 1984
Tameness (reduced fear of humans and animals resembling them)	Doyle and Talbot 1986
Vulnerability to angling	Greene 1952; Flick and Webster 1962
Reduced breeding success	Chilcote et al. 1986; Leider et al. 1990; Campton et al. 1991; Fleming and Gross 1993

tains only to hatchery life. Selecting fish for docility and tameness is done intentionally and unintentionally in fish culture. Hatchery personnel consider raising wild fish bothersome because the fish often become frantic and jump out of containers (R. J. White, personal observation), a situation which itself must select for docility. Feeding behavior of fish obviously is also of interest in fish culture. Selection for docility and certain feeding behaviors, and ignoring other kinds of changes in behavior, can alter the way fish act after release and, therefore, negatively influence the fish's success in poststocking life and contribution to the fishery.

In many ways, performance of fish in hatchery conditions bears little resemblance to what they must do in nature. Hatcheries obviously lack many hazards and other conditions that confront the fish when stocked in open waters, so hatchery life may induce behaviors that are disadvantageous after release. These include inappropriate habitat occupation (Vincent 1960; Dickson and MacCrimmon 1982), weak territoriality (Norman 1987), and waste of energy in fast water (Bachman 1984). The longer the period that a fish is artificially reared beyond hatching, the more likely it is to acquire behaviors that are inappropriate for life after release. In regard to duration of hatchery influence on the individual fish, we emphasize learned behaviors as opposed to innate behaviors, which are genetically acquired (Alcock 1975).

Ruzzante (1994) reviewed the evidence that domestication of fish can genetically change aggressive (agonistic) and schooling behavior. He concluded that when fish culturists select for increased growth,

TABLE 2.—Adverse effects of hatcheries and stocking on wild fish.

Adverse effect on wild fish	References or clarification
Effects of hatchery facilities and operations	
Chemical contamination (pollution)	Organic wastes and disinfectant chemicals can flow from hatcheries
Pathogenic contamination	Disease organisms can flow from hatcheries
Flow reduction in stream	Diverting water from streams (or from their groundwater aquifers) to supply hatcheries reduces instream flow.
Destruction of structural habitat in channel and riparian zone	Building water diversion structures and fish traps in streams; building hatcheries, associated roads, and other appurtenances in riparian zones
Physical interference with migration and spawning of wild fish	Craig and Townsend 1946
Wild spawner robbing for use in hatcheries	Hilborn 1992
Up-river nutrient depletion	Where hatcheries block migration or adult salmon are used as broodstock and the carcasses are not deposited in upstream areas
Effects of the stocked fish	
Presence of exotic species or varieties	Magnuson 1976
Genetic contamination	Waples et al. 1990; Waples 1991; Hindar et al. 1991
Predation	Sholes and Hallock 1979
Competition for food and favorable space	Bjornn 1978; Bachman 1984; Nickelson et al. 1986
Disruptive behavior	Bachman 1984
Stimulation of wild presmolts to migrate prematurely (Pied-Piper effect)	Kuehn and Schumacher 1957; Hansen and Jonsson 1985; Hillman and Mullan, in press
High density of stocked fish (crowding)	Juvenile salmonid mortality is strongly density-dependent (Mills 1969; Fraser 1969)
Mixed-stock exploitation problems	Wright 1981; McIntyre and Reisenbichler 1986
Predator attraction	Beamish et al. 1992
Disease transmission	Goede 1986, 1994
Effects via human perceptions and resource management	
False sense of security	Goodman 1990
Diversion of funds from more appropriate programs	Functional groups within agencies compete for funds

indirect selection for either increased or decreased agonistic behavior can result, depending on the amount of food available to the fish group. We hypothesize that besides this genetic concern, changed levels of aggressiveness may also be induced by learning in hatcheries by fish that are reared there long enough for learning to occur.

The need to make hatcheries less harmful to poststocking behavior seems urgent, and work toward improvements in that regard should be done. But the complexity of natural environments will make it hard to devise "training" and genetic procedures to remedy the adverse postrelease behaviors. Wiley et al. (1993b) extensively reviewed literature on the needs and possibilities for altering hatchery facilities and operations to prepare artificially reared salmonids better for life in streams. Despite work on these problems (Maynard 1995, this volume), contrasts between wild and practical hatchery environments and the information presented in Wiley et al. (1993b) lead us to believe that, economically, production hatcheries could not develop the poststocking behaviors needed in fish that are released into most streams (and some lakes), particularly those water bodies having significant numbers of predators or competitors. Stocking programs should not proceed on the assumption that the behavioral problems will be solved; the problems should be solved before major programs continue and before new ones are developed.

The Roles of Stocking in Aquatic Resource Management

Fish culture will probably serve society best when it is a component of ecologically sound aquatic resource management. Fish culture should not drive management programs, as has been a strong tendency in the past. Instead, stocking should derive from needs within ecological approaches to fishery management, besides serving economic and social objectives. Hatchery reform to improve poststocking results is essential but too narrow. The broader need is stocking programs operating within ecologically, economically, and socially responsible management of watersheds and river systems.

Ecological approaches are needed because fisheries management will be most successful if geared to the productive capacities of water bodies, rather than trying to exceed them, and to sustaining the native (locally normal) biodiversity and health of ecosystems, rather than extirpating species and destroying ecological functions of biotic communities. Conforming to productive capacity of systems seems self-evidently wise from biological and economic standpoints, is a major reason for detailed premanagement examination of water bodies, has a theoretical basis in ecology (Warren et al. 1979), and is explicitly called for in various guidelines and plans for stocking and other management (e.g., RASP 1992). Stocking programs entail risks of upsetting community (Magnuson 1976) and genetic (Nelson and Soulé 1987) diversity. In reviewing literature on biodiversity in relation to fisheries science, Carlson and Muth (1993) made a strong case for the conservation principles promoted by preserving the integrity of natural communities. Maximizing diversity per se is not sound management; the native diversity of a natural community, however, holds great value for the sustainability of aquatic ecosystems and is embodied in the concept of biological integrity (Karr 1991; Angermeier and Karr 1994).

For fisheries management to serve human objectives, it must accommodate resource users' outlooks, including ethical and aesthetic as well as economic values. In recreational fisheries, many anglers do not care whether the fish they catch originated from a hatchery or are wild fish (Hummon 1992), whereas, other anglers value wild fish, consider stocked fish objectionable or less valuable and feel recreational quality infringed upon if hatchery-produced fish are released into streams or lakes where wild populations are expected (Taylor 1992; Hummon 1992; Belsey 1994; Craig 1994). In that respect, Leopold's (1933) theorem that "the recreational value of a head of game is inverse to the artificiality of its origin, and hence in a broad way to the intensiveness of the system of game management which produced it" holds major implication for stocking programs.

Human values and desires in regard to fisheries can become unreasonable when those desires require manipulating resources in ways that would destroy resource sustainability. When people request, even on the basis of admirable values or desires, that stocking be done that is incompatible with productive capacity of an aquatic ecosystem or that would harm it ecologically or genetically, the request should be denied. Wiley et al. (1993a) advised against overresponse to public demands for stocking, saying that "trout stocking programs can generate further pernicious demand, resulting in increased and unnecessary dependence on hatchery trout, because people come to expect planted trout." Jackson (1989) defined good management in part as "giving people what they want to the extent that the ecosystem can support it." We would alter this to define good management as providing people opportunity to use natural resources to the extent that the ecosystem can support and withstand that use in the long term.

Stocking should be restricted to certain, more specific situations than in the past. As the histories of the Atlantic salmon, Great Lakes, and Pacific Northwest fisheries indicate, and as Eschmeyer (1955) vividly described, much past stocking was overdone and poorly focused. This may still be so wherever fishery agencies treat fish propagation and stocking more as objectives than as means to better fishing and ecosystem health.

Numbers of fish stocked are inadequate measures of management success, just as numbers of NPDES (National Pollution Discharge Elimination System) permits issued and numbers of trees planted are. Appropriately sizing and directing fish stocking requires awareness of ecological reality. White (in press) suggested that (a) if a water body has a healthy population of a given fish species, then stocking more of it is unlikely to increase the population and will often harm it; (b) if a stream or lake and its connected waters do not provide for all life stages of a fish species, then stocking it there cannot create a self-sustaining population; (c) if an ecosystem is suitable for the whole life cycle of a species but has none of that species, then starting a population there should require stocking only once or a few times; and (d) if a species is not native to a region, stocking it there is unlikely to establish it, and if it does become established, it will probably harm the native biotic community. Of the 25 fishes introduced into North America from other continents for management purposes since the 1870s, only 12 became established, and only brown trout *Salmo trutta* and common carp are widespread (Courtenay and Kohler 1986). Where introduced species persist long enough to affect the native biota, the harm that can ensue is well known (Magnuson 1976). Fishery scientists know such ecological realities; they must educate the public and policymakers about them.

In view of the above, we (White et al., in press) suggest that fish stocking be limited to temporary programs of three kinds and prolonged programs of two kinds. The temporary programs include stocking (a) to recolonize native species into waters from which human activities eliminated them, after reducing or eliminating the abuses that caused the loss (e.g., habitat destruction, exotic introductions, or excessive harvest); (b) to create populations in new waters; e.g., artificial lakes and ponds, or tailwaters (we recommend against colonizing naturally fishless natural water bodies, which harms established natural communities); and (c) to sustain a presently overharvested local fishery through a planned program of downsizing and transition to other employment or from reliance on stocked fish to reliance on wild fish. In the last type, the stocking usually should be decreased, gradually so as to wean the fishery from it with minimal social and economic dislocation.

Prolonged stocking programs, generally undesirable from ecologic, genetic, and economic standpoints, may be justified (a) to augment weak stocks (by put-and-grow stocking) in waters having little or no reproductive habitat but substantial productive capacity (particularly artificial water bodies, such as reservoirs and reservoir tailwaters), but only where this will not harm the indigenous biota; and (b) to provide highly artificial opportunities for recreation where fishing will be intensive (e.g., put-and-take stocking of catchable-size fish, for quick and easy catch in urban ponds). A caveat is that put-and-take stocking can instill human outlooks on resources (e.g., that fishing is an instant-gratification sport) that interfere with other management objectives in the future. Also, as is obvious to people who handle wild and hatchery fish, newly stocked fish usually have strikingly unnatural appearances such as deformities (crooked fins and abraded noses) and discoloration (Hager 1964), from hatchery life (Table 1). Such appearances can mislead anglers about characteristics of wild conspecific fish and about aesthetics of fishing. In extensive areas where pond conditions favor it, put-and-take trout stocking can be avoided by introducing panfishes, such as bluegill *Lepomis macrochirus*, which would reproduce and need no further stocking.

Fish stocking has often been categorized as introductory (introducing a species or genetic variety that is not native to the recipient water body but that is expected to establish a self-sustaining population there), as maintenance (repeated annual stocking to create a population of fish that could not reproduce in the water body), or as supplemental (augmenting a self-reproducing population by stocking fish of the same species). Referring to the kinds of stocking we previously described, introductory stocking comprises temporary types a and b, and maintenance stocking equates to prolonged type a. In reviewing literature on coolwater fish stocking, Laarman (1978) found that introductory stocking generally succeeded much better than did maintenance stocking, and the latter better than did supplemental stocking. Evans and Willox (1991) concluded from results of supplementally stocking lake trout *Salvelinus namaycush* in two Ontario lakes and from simulation modeling based on those data, that stocking hatchery-reared lake trout causes loss of native populations. Also, after supplemental stocking of rainbow trout (nonanadromous *O. mykiss*) in two Montana streams, abundances of conspecifics and of brown trout declined (Vincent 1987). Supplemental stocking seems inadvisable also because, by involving the same or potentially interbreeding species, it can have harmful genetic effects on the recipient population, such as contamination with nonadaptive genes, breakdown of natural systems of semi-isolated populations, and interspecific hybridization (Allendorf et al. 1987).

A variant of supplemental stocking is the supplementation concept recently promoted as a possible solution to decline of Pacific anadromous salmonids. Supplementation has been defined in diverse ways, but the most prevalent basic concept involves increasing "natural production" in an existing wild stock by using wild adults as broodstock and releasing their offspring back into the wild population (RASP 1992). The intent is to create a larger spawning population of adults and provide for increased harvest. Not only is supplementation inconsistently defined, but so many methods and objectives are espoused as to make the term almost meaningless. For example, RASP (1992) characterizes it as encompassing "a wide range of management" within four general objectives: (a) restoration of a native species to habitats from which it was extirpated, (b) introduction of a species into habitat where it was not native, (c) rearing augmentation in underused habitat, and (d) harvest augmentation.

The supplementation concept seems out of keeping with ecological reality and rational management. If habitat were protected (or restored toward a previously better state) and harvest properly managed, why would the wild, already naturally reproducing stock not increase on its own? If the habitat is not protected or restored, will the stocked fish survive any better than do the wild ones? Supplementation also is at variance with the principle of separating hatchery-produced fish from wild populations (see "Recent Reform—Wild Fish Policies"). Bowles (1995, this volume) of the Idaho Fish and Game Department observes that little is yet known about supplementation, urges a cautious approach, foresees very limited application, and considers that, at least for anadromous salmonids that migrate relatively far in the Columbia River system, it cannot succeed as a substitute for solving the basic causes of stock decline and therefore should not be undertaken as a long-term program.

A different view is conveyed by Cuenco et al. (1993) from the Columbia Inter-Tribal Fish Commission, which represents Native American commercial fishing interests and is a proponent of supplementation. Although Cuenco et al. (1993) advocate proceeding "cautiously so that productivity of supplemented stocks can be tested," they assert that a "holistic rehabilitation plan, given the constraints imposed upon the Columbia River fish production system by human activities, requires the effective use of supplementation in conjunction with [other measures]," and they present an outline of theoretical, almost undocumented possibilities for salmon supplementation. They promote optimism via such remarks as "The supplementation scheme described . . . seeks to eliminate any divergence between the hatchery fish and the wild fish by using representative samples of the wild population as hatchery broodstock, by avoiding any artificial selection, and by minimizing the difference between the natural and the hatchery environments." Yet they do not say (and we believe no one else has) how truly representative samples of eggs can be secured from wild populations under practical conditions (short of taking almost all spawners in low-population crises), and other material in their paper indicates that methods for avoiding selection and for minimizing differences between natural and hatchery environments are far from realization. We believe those aspirations to be unattainable. Cuenco et al. (1993) also state that "supplementation may be unnecessary in the long term if other factors limiting populations in the basin are corrected," but in describing six types of depressed-stock situations in which supplementation might be used, for only one (restoration of extirpated stocks) do they mention stopping the program, and then only if it is successful. They provide only two examples of existing, purportedly successful supplementation efforts. Cuenco (1994) further discussed supplementation based on modeling but based his models on the assumption "that artificial propagation does not alter important genetic traits affecting stock productivity"; we suggest this assumption is unrealistic.

In supplementation lies another danger. The word can easily mislead. It expresses the intent behind an as yet undeveloped procedure, but it is undoubtedly misunderstood by many to represent a fairly certain result. In the last few years, advocates of hatchery-based management have also commonly applied to hatchery and stocking programs another potentially misleading term: enhancement. From that has come the expression "enhanced fish," now sometimes used to describe hatchery-produced fish (e.g., in Boyce et al. 1992, and at the "International Symposium on Biological Interaction of Enhanced and Wild Salmonids," held in Nanaimo, British Columbia, June 1991). An implication, intended or not, is that the fish have been improved by culturing them, which is ironic in view of the evidence (Table 1). We maintain that such terms are unwarranted, and that they risk promoting mismanagement and should be used sparingly and only when clarified by appropriate adjectives.

Stocking and other aspects of fish culture usually should be separated from wild populations. Because hatchery fish pose ecological, genetic, and disease risks for wild fish (Table 2), waters containing healthy wild stocks should be managed for those fish and not be stocked. Likewise, impoverished wild stocks that are restorable should be allowed to recover via habitat restoration and harvest control whenever possible; they should not be subjected to hazards of captive breeding or augmentation with hatchery fish. Reisenbichler and McIntyre (1986) examined the possibilities and complications of integrating natural and artificial production of anadromous salmonids and concluded that procedures to mitigate the adverse effects of stocked fish on wild populations involved separating hatchery and wild fish by releasing the hatchery fish as smolts, thus sending them quickly to sea and reducing their interactions with wild fish in rearing streams. However, problems of residual hatchery smolts (those that fail to migrate) can occur (Pearsons et al. 1993). Reisenbichler and McIntyre (1986) warned about the genetic and ecological dangers of outplanting (release of hatchery fish that will rear or spawn in streams, as opposed to those that will spend little juvenile time in streams and will home to a hatchery) and reported that "apparently, most outplanting programs have been unsuccessful." Belsey (1994) also strongly advocated separation of wild and hatchery-based management. The principle of separation is embodied in the management policies of Alberta and Montana, which no longer stock trout in streams but only in standing waters (Nelson and Paetz 1992; Wells 1985), and to a less striking extent in modern management by other agencies (see "Recent Reform—Wild Fish Policies").

Only as a last and temporary resort should managers try to restore wild stocks by fish culture. Geneticists advocate that fish culture be done in ways that minimize gene flow to wild populations, that fish culture facilities be located and designed to minimize escape of fish, that sites of rearing and stocking be those having minimal effect on natural populations and the waters they inhabit, that waters be reserved for complete protection from fish cultural influences, and that transport of fish between areas be restricted (Ryman 1981; Hindar et al. 1991). Therefore, fish culture for direct marketing (aquaculture) should be restricted to aquaculture-dedicated ponds or small lakes from which the fish cannot emigrate or, in larger aquatic systems, to enclosures from which fish cannot escape. Also, aquatic ecosystems should be protected from the effects of such feedlot operations on water quality.

Evaluation of Stocking Programs

Poststocking results should be monitored, and performance of stocking programs evaluated against the objectives of the programs. Depending on management objectives, evaluation should include information about (a) performance of the stocked fish (survival, body growth, and reproductive success); (b) effects on other fish populations (hybridization and ecological effects, e.g., competition, predation, habitat alteration, and disease transmission); (c) effects on other members of the native biota; and (d) effects on humans (catch rate and other measures of fishing quality; economic, aesthetic, and social effects; and ramifications for public understanding of and attitudes toward resource issues). Weithman (1986) suggested including angler acceptance in evaluating stocking for sport fisheries. If the objective is to increase the reproductive capacity of a population, then reproductive success and long-term fitness of the fish should also be assessed.

Weithman (1986) recommended that all important stocking programs be evaluated on an economic basis. He later (Weithman 1993) expanded the concepts in terms of socioeconomic benefits, acknowledging coordination with biological benefits. We suggest that besides direct production and fish transport costs, full costs of stocking programs would include matters that have usually been ignored or externalized, such as stocking-related administration, the determination of where and when to stock (e.g., field surveys, planning, public meetings, and consultations with politicians), amortization of designing and building new hatcheries and of disposing of old ones, environmental damage by hatcheries (e.g., pollution), harm to wild stocks by hatchery-reared fish, evaluation of stocking programs, the administration involved in all the proceeding, and negative social effects. Full accounting of benefits should include the values of harvest, recreation, and other pay-off to anglers and to state and local economies (Weithman 1986), as well as biological and social benefits. We found no such comprehensive economic or socioeconomic analysis of a stocking program. Most economic analyses have dealt only with direct costs of hatchery operation and stocking, but some have included hatchery maintenance and amortization of capital costs (Weithman 1986).

We caution against evaluating on economic and commodity bases alone. Failure to include ecological and social aspects (some of them intangible) in evaluations of stocking and other managements will result in ignoring or belittling such benefits as the parent–child bonding that can derive from fishing together, and fishing for aesthetically pleasing fish in pleasant environments, thereby enriching the human spirit with greater appreciation of the natural world.

When economic analyses of stocking programs are negative, considering social benefits may at least partially offset the deficit; but environmental detriments may, in turn, cancel the social benefits. For four SEP chinook salmon hatcheries in British Columbia, operating costs alone ranged from about Can$45 to $380 per adult fish that survived to catch or escapement (Winton and Hilborn 1994); if hatchery construction costs were added, the estimated per-fish cost was $50 to $435, and the estimated per-kilogram cost was $7 to $62 (Winton 1991).[4] If discounting of capital costs were done to account for the time value of money, and such other costs as administration, evaluation, and environmental harm were added, the cost per fish or kilogram would rise. The cost per fish contributed by the stocking program to the catch (the adult run minus escapement) would be even higher. Benefit-cost analysis could not be conducted, even on the basis of just hatchery operation costs, because the

---

[4]For comparison, average per-fishing-trip expenditure by U.S. Pacific salmon anglers was US$62.60 (ca. Can$77.50) in 1987, during the period covered in the above estimates (Huppert and Fight 1991). The in-season retail market price in Seattle for gutted whole-carcass chinook salmon fluctuated within the approximate range of $5.50 to $18 (Can$7 to $22) per kilogram in the last decade (Survey of fish markets).

SEP did not sufficiently measure biological results; after over 15 years of operation, information is still inadequate to compare the program's (or any hatchery's) performance against its objectives (Winton and Hilborn 1994). Progress toward meeting secondary SEP goals of economically stimulating underdeveloped areas via jobs in hatcheries and via involving the public (Hilborn and Winton 1993) might increase the benefit, but environmental detriments, such as adverse effects on wild fish, that remain unmeasured would also have to be taken into account.

Ratios of economic benefit to production cost in Oregon salmon hatcheries range from near zero to 8.5, averaging 1.3 for summer chinook salmon, 1.9 for fall chinook salmon, and 1.6 for coho salmon; these ratios are below those for steelhead and trout stocking (ODFW 1992b). Benefit-cost ratios would be lower—and far more often under the 1.0 breakeven point—if full, long-term costs were considered.

Stocking programs should undergo environmental impact analysis and other regulation. This should be done on a site- and stock-specific basis, not solely on a broader, programmatic basis. In view of the adverse effects of hatcheries and stocking (Table 2), government agencies having hatchery programs should reduce hatchery impacts, and regulatory bodies, such as the U.S. Environmental Protection Agency, should act on pertinent problems. Goodman (1990) argued for federal regulation of hatcheries.

Recent Reform—Wild Fish Policies

Various U.S. states and Canadian provinces have made major advances in ecologically sound fisheries resource management, often explicitly based on wild fish policies. We suspect these advances have been made in part because biologists trained in ecology and genetics increasingly direct fisheries management and base it on field study of waters and their biota. A trend toward separating hatchery-based management from waters that support wild fish began in Alberta in the mid-1950s. Based on findings that hatchery-produced cutthroat trout *O. clarki* survived poorly in streams containing wild trout (Miller 1952a, 1954, 1958), Alberta discontinued stream stocking and restricted stocking of hatchery fish to lakes, reservoirs, and some beaver ponds (Nelson and Paetz 1992).

In 1956, Yellowstone National Park halted its stocking program (Varley 1981) and in the 1970s set special angling regulations to foster wild trout populations (Varley and Schullery 1983). In the 1960s, Michigan (H. A. Tanner, Michigan Department of Natural Resources, personal communication) and Wisconsin (WDNR 1966) surveyed and classified trout streams according to trout population characteristics, decreased use of hatchery trout in those streams that supported natural trout reproduction, and increased protection and restoration of trout habitat. In the 1970s, based on Vincent's (1975, 1987) findings and the obvious presence of thriving wild trout in many streams, Montana switched a substantially stream-oriented trout stocking program to one of stocking reservoirs, lakes, and ponds almost exclusively (Anonymous 1973; Wells 1985).

Pennsylvania, which traditionally had stocked trout extensively, began biological surveys of all its stocked streams in 1976, and, based on early results, planned a stream classification system to include wild trout water in which no hatchery trout would be stocked (Graff 1986). In 1981, the Pennsylvania Fish and Boat Commission (PFBC) adopted a "resource first" policy of managing trout streams biologically, not politically (Nale 1988). In 1983, management in which about 5% or 623 km (R. T. Green, PFBC, personal communication) of the previously stocked streams were designated for wild trout because they supported substantial self-reproducing populations of brown trout or brook trout *Salvelinus fontinalis* ($\geq$40 kg/ha brown trout or mixed brown-and-brook trout or $\geq$30 kg/ha brook trout; Graff 1986). These wild trout streams were managed under five kinds of angling regulations (PFBC 1987). Under the new management, trout populations increased markedly in regularly monitored streams (White 1992b), and others (Green, personal communication). Apparently many anglers perceived this improvement and were enthusiastic (Nale 1988). Stocking of hatchery trout, now focused on waters that do not support wild trout well, also increased (D. R. Graff, PFBC, personal communication). In 81% of the stream lengths first designated for wild trout, trout populations did not change or increased. Stream lengths in which wild trout populations decreased were reclassified to lower wild trout status and stocking was permitted. Also, more streams have been added to the wild trout class, and the amount under such management is now 1,517 km (Green, personal communication).

In the mid-1980s, the Maryland Department of Natural Resources also established a policy recognizing opportunities to manage for wild trout, made field surveys identifying self-sustaining populations, developed management plans that separated stocked trout from substantial wild populations, improved habitat, and tailored angling regulations to capacities of waters and to anglers' diverse values (Bachman 1994). The field surveys revealed a far greater length of trout streams than previously recognized. Most became managed for wild trout, which rose in abundance; at the same time, put-and-take stocking of lakes and streams not suitable for wild trout was also greatly increased (Bachman 1994).

Ontario and British Columbia place high priority on wild fish and concentrate stocking of hatchery fish in waters best suited to that method. Ontario set forth an aquatic ecosystem approach, in which fishery management is increasingly oriented toward conserving naturally reproducing fish communities, particularly native species, and in which use of hatchery fish and nonnative species is based on analysis of long-term ecological, social, and economic benefits and costs (OMNR 1992). Since 1990, Ontario's hatchery production has declined due to budget constraints. In this reduction, two stocking programs considered to be of low priority (poor return rates to fisheries) were discontinued: coho salmon stocking in Lake Ontario (where plantings of five other salmonid species continue) and muskellunge *Esox masquinongy* stocking in inland lakes. Also, brook trout plantings in some southern Ontario streams were halted after habitat restoration increased wild populations (B. A. Potter, Ontario Ministry of Natural Resources, personal communication).

British Columbia, which manages inland fishes and steelhead (marine fish and salmon are in the federal domain), developed a 1991–1995 strategic plan, putting top priority on conserving wild stocks (BCFB, no date). The plan emphasizes protecting and managing fish habitat according to the needs of individual water bodies. Most stocking is to augment populations of trout in lakes, fewer nonanadromous trout are stocked in streams, and many steelhead are stocked in rivers. Stocking programs are designed to avoid jeopardizing wild populations. Each lake and stream of the province will be classified (H. Andrusak, British Columbia Fisheries Branch, personal communication) according to the target fish stock or community of stocks that it can support and will be managed according to the priority (1) wild indigenous (significant, self-sustaining populations of native fishes—no stocking, managed by angling regulations and habitat protection and improvement); (2) wild naturalized (nonnative species that are self-sustaining—managed as for category 1 but stocked under special circumstances); (3) augmented (insufficient habitat to sustain reproduction in wild stocks, thus requiring stocking to maintain abundance—hatchery broodstocks of recent wild origin from nearby waters that are habitat adapted will usually be used); and (4) hatchery (closed systems unsuitable for stocks of the above kinds—usually stocked with hatchery fish from broodstock developed for many generations in hatcheries).

Oregon has established an elaborate set of policies pertaining to wild fish composed of general fish management goals and policies on natural production, wild fish management, wild and hatchery fish gene resource conservation, and transgenic fish (ODFW 1992a). In Washington, state and tribal fishery agencies have enunciated a wild stock restoration initiative to "maintain and restore healthy wild salmon and steelhead stocks and their habitats to support the region's fisheries, economies, and other societal values" (WDF 1993b). The state has begun to develop a wild salmonid policy that involves aspects of harvest management, fish culture and stocking, and habitat protection and restoration (WDFW 1994).

Conclusions

Primary needs in reforming the use of stocking in fishery management are to (1) subordinate stocking programs to ecologically sound fishery management, (2) devise fish culture methods that are less domesticating, (3) stock as few fish as possible in as few waters as possible for as few years as possible, (4) monitor and evaluate stocking programs rigorously, and (5) act on the results.

The ability of humans to alter their environment by applying increasingly sophisticated technologies often leads to unexpected, negative results. Fish hatcheries and stocking programs exemplify a technology whose benefits seem obvious to many people but conceal an ominous future. As our analysis of the literature suggests, the good intentions behind fish stocking often went astray. The public, decision-makers, and many fisheries professionals perpetuated oversimplified, confused views about the roles of culture and stocking in natural resource management. Misperceptions were fostered by naive optimism, exaggerated claims, misleading terminology, and poorly defined objectives. Many important fisheries in North America and the world are in crisis, due primarily to overfishing and habitat destruction. Artificial propagation and stocking usually cannot offset those problems but can exacerbate them and create further problems. Fishery management must focus on the basic resources and be made more effective. Discrepancies between the assumed benefits of fish stocking programs and their environmental, social, and economic costs should no longer be ignored.

Acknowledgments

Oregon Trout, under funding by the Northwest Area Foundation, supported research and writing for this article. We incorporated comments by staff of the Washington Department of Fish and Wildlife Hatchery Program, which were made on a related, unpublished report.

References

Alcock, J. 1975. Animal behavior: an evolutionary approach. Sinauer Associates, Sunderland, Massachusetts.

Allendorf, F. W., and S. R. Phelps. 1980. Loss of genetic variation in a hatchery stock of cutthroat trout. Transactions of the American Fisheries Society 109:537–543.

Allendorf, F. W., and N. Ryman. 1987. Genetic management of hatchery stocks, p. 141–159 in N. Ryman and F. Utter, editors. Population genetics & fishery management. University of Washington Press, Seattle.

Allendorf, F. W., N. Ryman, and F. M. Utter. 1987. Genetics and fishery management. Pages 1–19 in N. Ryman and F. Utter, editors. Population genetics & fishery management. University of Washington Press, Seattle.

Angermeier, P. L., and J. R. Karr. 1994. Biological integrity versus biological diversity as policy directives: protecting biotic resources. BioScience 44:690–697.

Anonymous. 1973. Minutes of the December 6, 1973, meeting of the Montana Fish and Game Commission (item 34), Helena.

Aulstad, D., and A. Kittelsen. 1971. Abnormal body curvatures of rainbow trout (*Salmo gairdneri*) in inbred fry. Journal of the Fisheries Research Board of Canada 28:1918–1920.

Bachman, R. A. 1984. Foraging behavior of free-ranging wild and hatchery brown trout in a stream. Transactions of the American Fisheries Society 113:1–32.

Bachman, R. A. 1994. Restoring wild trout resources: how hatchery trout fit into the equation. Pages 26–37 in D. Zaldokas and P. A. Slaney, editors. Proceedings of the clean water-wild trout international conservation symposium. British Columbia Ministry of Environment, Lands and Parks, Fisheries Project Report RD40, Vancouver.

Bakke, B. 1994. A naturalist's view of hatcheries. Osprey (Steelhead Committee of the Federation of Fly Fishers) 21:8–10.

Bams, R. A. 1967. Differences in the performance of naturally and artificially propagated sockeye salmon migrant fry as measured with swimming and predation tests. Journal of the Fisheries Research Board of Canada 24:1117–1153.

Bams, R. A. 1976. Survival and propensity for homing as affected by presence or absence of locally adapted paternal genes in two transplanted populations of pink salmon (*Oncorhynchus gorbuscha*). Journal of the Fisheries Research Board of Canada 33:2716–2725.

Barton, B. A., C. B. Schreck, and L. A. Sigismondi. 1986. Multiple acute disturbances evoke cumulative physiological stress responses in juvenile chinook salmon. Transactions of the American Fisheries Society 115:245–251.

BCFB (British Columbia Fisheries Branch). No date. Conserving our resource, Fisheries Program strategic plan, 1991–1995. Fisheries Program, British Columbia Environment, Victoria.

Beamish, R. J., and D. R. Bouillon. 1993. Pacific salmon production trends in relation to climate. Canadian Journal of Fisheries and Aquatic Sciences 50:1002–1016.

Beamish, R. J., B. L. Thompson, and G. A. McFarlane. 1992. Spiny dogfish predation on chinook and coho salmon and the potential effects on hatchery-produced salmon. Transactions of the American Fisheries Society 121:444–455.

Behnke, R. J. 1991. From hatcheries to habitat? Look again. Trout (Autumn):55–58.

Behnke, R. J., and D. Johnson. 1989. The role of catchable trout in a state fishery program: how much is enough? Pages 92–99 in Proceedings, 24th Annual Meeting, Colorado-Wyoming Chapter, American Fisheries Society.

Belsey, J. 1994. A non-scientific observation promoting the separation of wild and domestic trout. Pages 182–185 in R. W. Wiley and W. A. Hubert, editors. Wild trout and planted trout: balancing the scale. Wyoming Game and Fish Department, Laramie.

Bjornn, T. C. 1978. Survival, production, and yield of trout and chinook salmon in the Lemhi River, Idaho. University of Idaho, Forest, Wildlife and Range Experiment Station Bulletin 27, Moscow.

Blumm, M., and A. Simrin. 1991. The unraveling of the parity promise: hydropower, salmon, and endangered species in the Columbia basin. Environmental Law 21:657-254.

Bottom, D. L. In press. To till the water: a history of ideas in fisheries conservation. In D. J. Stouder, P. A. Bisson, and R. J. Naimau, editors. Pacific salmon and their ecosystems: status and future options. Chapman and Hall, New York.

Bouck, G. R., and S. D. Smith. 1979. Mortality of experimentally descaled smolts of coho salmon (*Oncorhynchus kisutch*) in fresh and salt water. Transactions of the American Fisheries Society 108:67–69.

Bowen, J. T. 1970. A history of fish culture as related to the development of fishery programs. American Fisheries Society Special Publication 7:71–93.

Bowles, E. C. 1995. Supplementation: panacea or curse for the recovery of declining fish stocks? American Fisheries Society Symposium 15:277–283.

Boyce, J., M. Herrmann, D. Bischak, and J. Greenberg. 1992. A benefit-cost analysis of the Alaska salmon enhancement program. Alaska Senate Special Committee on Domestic and International Commercial

Fisheries Legislative Review of the Alaska Salmon Enhancement Program, Juneau.

BPA (Bonneville Power Administration). 1990. Preliminary design report for the Yakima/Klickitat Production Project. BPA Division of Fish and Wildlife, Portland, Oregon.

Brown, B. 1982. Mountain in the clouds: a search for the wild salmon. Macmillan, New York.

Brynildson, O. M., and L. M. Christenson, 1961. Survival, yield, growth and coefficient of condition of hatchery-reared trout stocked in Wisconsin waters. Wisconsin Conservation Department, Miscellaneous Research Report 3 (Fisheries), Madison.

Campton, D. E., F. W. Allendorf, R. J. Behnke, and F. M. Utter. 1991. Reproductive success of hatchery and wild steelhead. Transactions of the American Fisheries Society 120:816–822.

Campton, D. E., and J. M. Johnston. 1985. Electrophoretic evidence for a genetic admixture of native and nonnative rainbow trout in the Yakima River, Washington. Transactions of the American Fisheries Society 114:782–793.

Carlson, C. A., and R. T. Muth. 1993. Endangered species management. Pages 355–381 in C. C. Kohler and W. A. Hubert, editors. Inland fisheries management in North America. American Fisheries Society, Bethesda, Maryland.

Chan, B. M. 1994. Wild trout and augmented trout: the British Columbia experience. Pages 140–144 in R. W. Wiley and W. A. Hubert, editors. Wild trout and planted trout: balancing the scale. Wyoming Game and Fish Department, Laramie.

Chilcote, M. W., S. A. Leider, and R. Jones. 1981. Kalama River salmonid studies. Washington State Game Department, Fishery Research Report 81-11, Olympia.

Chilcote, M. W., S. A. Leider, and J. J. Loch. 1986. Differential reproductive success of hatchery and wild summer-run steelhead under natural conditions. Transactions of the American Fisheries Society 115:726–735.

Christie, W. J. 1963. Effects of artificial propagation and the weather on recruitment in the Lake Ontario whitefish fishery. Journal of the Fisheries Research Board of Canada 20:597–638.

Cobb, J. N. 1917. Pacific salmon fisheries. Appendix 3 to the report of U.S. Commissioner of Fisheries for 1916. U.S. Department of Commerce, Bureau of Fisheries Document 839, Washington, DC.

Courtenay, W. R., and C. C. Kohler. 1986. Exotic fishes in North American fisheries management. Pages 401–413 in R. H. Stroud, editor. Fish culture in fisheries management. American Fisheries Society, Fish Culture Section and Fisheries Management Section, Bethesda, Maryland.

Craig, J. A., and L. D. Townsend. 1946. An investigation of fish maintenance problems in relation to the Willamette Valley Project. U.S. Fish and Wildlife Service, Special Scientific Report 33, Chicago.

Craig, J. S. 1994. Wild trout—a non-native perspective emphasizing resource vs. recreation differentiation. Pages 134–138 in R. W. Wiley and W. A. Hubert, editors. Wild trout and planted trout: balancing the scale. Wyoming Game and Fish Department, Laramie.

Cross, T. F., and J. King. 1983. Genetic effects of hatchery rearing in Atlantic salmon. Aquaculture 33:33–40.

Cuenco, M. L. 1994. A model of an internally supplemented population. Transactions of the American Fisheries Society 123:277–288.

Cuenco, M. L., W. H. Backman, and P. R. Mundy. 1993. The use of supplementation to aid in natural stock restoration. Pages 269–293 in J. G. Cloud and G. H. Thorgaard, editors. Genetic conservation of salmonid fishes. Plenum, New York.

Daley, W. J. 1993. The use of hatcheries: polarizing the issue. Fisheries (Bethesda) 18(3):4–5.

Dickson, T. 1993. To stock or not? Minnesota Volunteer (May-June 1993):41–49.

Dickson, T. A., and H. R. MacCrimmon. 1982. Influence of hatchery experience on growth and behavior of juvenile Atlantic salmon (*Salmo salar*) within allopatric and sympatric stream populations. Canadian Journal of Fisheries and Aquatic Sciences 39:1453–1458.

Dill, L. M. 1987. Animal decision making and its ecological consequences: the future of aquatic ecology and behaviour. Canadian Journal of Zoology 65:803–811.

Doyle, R. W. 1983. An approach to the quantitative analysis of domestic selection in aquaculture. Aquaculture 33:167–185.

Doyle, R. W., and A. J. Talbot. 1986. Artificial selection on growth and correlated selection on competitive behaviour in fish. Canadian Journal of Fisheries and Aquatic Sciences 43:1059–1064.

Egidius, E., L. P. Hansen, B. Jonsson, and G. Naevdal. 1991. Mutual impact of wild and cultured Atlantic salmon in Norway. Journal du Conseil International pour l'Exploration de la Mer 47:404–410.

Ersbak, K., and B. L. Haase. 1983. Nutritional deprivation after stocking as a possible mechanism leading to mortality in stream-stocked brook trout. North American Journal of Fisheries Management 3:142–151.

Eschmeyer, R. W. 1955. Fish conservation fundamentals. Sport Fishing Institute Bulletin 38(January):79–109.

Evans, D. O., and C. W. Willox. 1991. Loss of exploited, indigenous populations of lake trout, *Salvelinus namaycush*, by stocking of non-native stocks. Canadian Journal of Fisheries and Aquatic Sciences 48 (Supplement 1):134–147.

Fleming, I. A., and M. R. Gross. 1993. Breeding success of hatchery and wild coho salmon (*Oncorhynchus kisutch*) in competition. Ecological Applications 3:230–245.

Fletcher, V., B. Kiser, B. Rogers, and B. Foster. 1976. 1975 hatcheries statistical report of production and plantings. Washington Department of Fisheries, Olympia.

Flick, W. A., and D. A. Webster. 1962. Problems in sampling wild and domestic stocks of brook trout. Transactions of the American Fisheries Society 91:140–144.

Foerster, R. E. 1931. A comparison of the natural and artificial propagation of salmon. Transactions of the American Fisheries Society 61:121–130.

Foerster, R. E. 1936. An investigation of the life history

and propagation of the sockeye salmon (*Oncorhynchus nerka*) at Cultus Lake, British Columbia, Number 5. The life history cycle of the 1926 year class with artificial propagation involving the liberation of free-swimming fry. Journal of the Biological Board of Canada 2:311–333.

Foster, C. H. W. 1991. Yankee salmon: the Atlantic salmon of the Connecticut River. CIS, Cambridge, Massachusetts.

Fowden, M. 1994. What are the fisheries benefits from current stocking practices in rivers, streams and tailwaters? Pages 126–133 *in* R. W. Wiley and W. A. Hubert, editors. Wild trout and planted trout: balancing the scale. Wyoming Game and Fish Department, Laramie.

Francis, R. C. 1990. Fisheries science and modeling: a look to the future. Natural Resource Modeling 4:1–10.

Fraser, F. J. 1969. Population density effects on survival and growth of juvenile coho salmon and steelhead trout in experimental stream channels. Pages 253–265 *in* T. G. Northcote, editor. Symposium on salmon and trout in streams. University of British Columbia, H. R. MacMillan Lectures in Fisheries, Vancouver.

Fraser, J. M. 1974. An attempt to train hatchery-reared brook trout to avoid predation by the common loon. Transactions of the American Fisheries Society 103:815–818.

Gillen, A. L., R. A. Stein, and R. F. Carline. 1981. Predation by pellet-reared tiger muskellunge on minnows and bluegills in experimental systems. Transactions of the American Fisheries Society 110:197–209.

Gladson, J. 1990. Fish hatcheries: part of the solution or the problem? Oregon Wildlife, November-December 1990.

Goede, R. W. 1986. Management considerations in stocking of diseased or carrier fish. Pages 349–355 *in* R. H. Stroud, editor. Fish culture in fisheries management. American Fisheries Society, Bethesda, Maryland.

Goede, R. W. 1994. Aquaculture/disease/wild fish. Pages 176–180 *in* R. W. Wiley and W. A. Hubert, editors. Wild trout and planted trout: balancing the scale. Wyoming Game and Fish Department, Laramie.

Goodman, M. L. 1990. Preserving the genetic diversity of salmonid stocks: a call for federal regulation of hatchery programs. Environmental Law 20:111–166.

Graff, D. R. 1986. The politics of wild trout. Trout 1986(Winter): 12–19.

Green, D. M. 1964. A comparison of stamina of brook trout from wild and domestic parents. Transactions of the American Fisheries Society 93:96–100.

Greene, C. W. 1952. Results from stocking brook trout of wild and hatchery strains at Stillwater Pond. Transactions of the American Fisheries Society 81:43–52.

Griffith, J. 1994. Native Rocky Mountain trout and stocked trout: can we have both? Pages 20–39 *in* R. W. Wiley and W. A. Hubert, editors. Wild trout and planted trout: balancing the scale. Wyoming Game and Fish Department, Laramie.

Hager, F. 1964. Ortstreue Regenbogenforellen in den Wildbächen des Innviertels. Österreichs Fischerei 17:189–190.

Hansen, L. P., and B. Jonsson. 1985. Downstream migration of hatchery-reared smolts of Atlantic salmon (*Salmo salar* L.) in the River Imsa, Norway. Aquaculture 45:237–248.

Hard, J. J., R. P. Jones, Jr., M. R. Delarm, and R. S. Waples. 1992. Pacific salmon and artificial propagation under the Endangered Species Act. National Marine Fisheries Service, Technical Memorandum NMFS-NWFSC-2, Washington, DC.

Harris, G. S., editor. 1978. Salmon propagation in England and Wales. Report by the Association of River Authorities/National Water Council Working Party, National Water Council, London.

Hilborn, R. 1992. Hatcheries and the future of salmon in the Northwest. Fisheries (Bethesda) 17(1):5–8.

Hilborn, R., and J. Winton. 1993. Learning to enhance salmon production: lessons from the salmonid enhancement program. Canadian Journal of Fisheries and Aquatic Sciences 50:2043–2056.

Hilborn, R. 1992. Hatcheries and the future of salmon in the Northwest. Fisheries (Bethesda) 17(1):5–8.

Hilborn, R., and J. Winton. 1993. Learning to enhance salmon production: lessons from the salmonid enhancement program. Canadian Journal of Fisheries and Aquatic Sciences 50:2043–2056.

Hill, P. M. 1984. A detailed listing of the liberations of salmon into the open waters of the state of Washington during 1983. Washington Department of Fisheries, Progress Report 210, Olympia.

Hillman, T. W., and J. W. Mullan. In press. The effect of hatchery releases on the abundance and behavior of naturally produced juvenile salmon and steelhead. North American Journal of Fisheries Management.

Hindar, K., N. Ryman, and F. Utter. 1991. Genetic effects of cultured fish on natural fish populations. Canadian Journal of Fisheries and Aquatic Sciences 48:945–957.

Horak, D. L. 1972. Survival of hatchery-reared rainbow trout (*Salmo gairdneri*) in relation to stamina tunnel ratings. Journal of the Fisheries Research Board of Canada 29:1005–1009.

Hosmer, M. J., J. G. Stanley, and R. W. Hatch. 1979. Effects of hatchery procedures on later return of Atlantic salmon to rivers in Maine. Progressive Fish Culturist 41:115–119.

Hummon, N. P. 1992. The 1991 trout angler telephone survey. Pennsylvania Fish Commission, Harrisburg.

Huppert, D. D., and R. D. Fight. 1991. Economic considerations in managing salmonid habitats. American Fisheries Society Special Publication 19:559–585.

Hynes, J. D., E. H. Brown, J. H. Helle, N. Ryman, and D. A. Webster. 1981. Guidelines for the culture of fish stocks for resource management. Canadian Journal of Fisheries and Aquatic Sciences 38:1867–1876.

Järvi, T., and I. Uglem. 1993. Predator training improves anti-predator behavior of hatchery reared Atlantic salmon (*Salmo salar*) smolt. Nordic Journal of Freshwater Research 68:63-71.

Jackson, R. M. 1989. Why, why, why? The human dimensions of trout angling motivations and satisfactions. Pages 186–192 *in* F. Richardson and R. H. Hamre, editors. Wild Trout IV. Trout Unlimited, Vienna, Virginia.

Johnson, L. D. 1978. Evaluation of esocid stocking program in Wisconsin. Pages 298–301 in R. J. Kendall, editor. Selected coolwater fishes of North America. American Fisheries Society Special Publication 11, Washington, DC.

Johnsson, J. I., and M. V. Abrahams. 1990. Domestication increases foraging under threat of predation in juvenile steelhead trout (Oncorhynchus mykiss)—an experimental study. Canadian Journal of Fisheries and Aquatic Sciences 48:243–247.

Jonsson, B. N., N. Jonsson, and L. P. Hansen. 1990. Does juvenile experience affect migration and spawning of adult Atlantic salmon? Behavioral Ecology and Sociobiology 26:225–230.

Jonsson, B. N., N. Jonsson, and L. P. Hansen. 1991. Differences in life history and migratory behaviour between wild and hatchery-reared Atlantic salmon in nature. Aquaculture 98:69–78.

Kanid'yev, A. N., G. M. Kostyunin, and S. A. Salmin. 1970. Hatchery propagation of the pink and chum salmons as a means of increasing the salmon stocks of Sakhalin. Journal of Ichthyology 10:249–259.

Karr, J. R. 1991. Biological integrity: a long-neglected aspect of water resource management. Ecological Applications 1:66–84.

Kincaid, H. L. 1983. Inbreeding in fish populations used for aquaculture. Aquaculture 33:215–227.

Klak, G. E. 1941. The condition of brook and rainbow trout from four eastern streams. Transactions of the American Fisheries Society 70:282–289.

Kochman, E. 1994. Sportfish management in Colorado with emphasis on hatchery-reared fish. Pages 12–19 in R. W. Wiley and W. A. Hubert, editors. Wild trout and planted trout: balancing the scale. Wyoming Game and Fish Department, Laramie.

Koelz, W. 1926. Fishing industry of the Great Lakes. Appendix XI to the report of the U. S. Commissioner of Fisheries for 1925. U.S. Bureau of Fisheries Document 1001, Washington, DC.

Krueger, C. C., A. J. Gharrett, T. R. Dehring, and F. W. Allendorf. 1981. Genetic aspects of fisheries rehabilitation programs. Canadian Journal of Fisheries and Aquatic Sciences 38:1877–1881.

Kuehn, J. H., and R. E. Schumacher. 1957. Preliminary report on a two-year census on four southeastern Minnesota streams. Minnesota Department of Conservation, Investigational Report 186, Division of Game Fisheries, St. Paul.

Laarman, P. W. 1978. Case histories of stocking walleyes in inland lakes, impoundments, and the great Lakes—100 years with walleyes. American Fisheries Society Special Publication 11:254–260.

Landau, M. 1992. Introduction to aquaculture. Wiley, New York.

Larkin, P. A. 1970. Management of Pacific salmon of North America. American Fisheries Society Special Publication 7:223–236.

Leider, S. A., M. W. Chilcote, and J. J. Loch. 1984. Spawning characteristics of sympatric populations of steelhead trout (Salmo gairdneri): evidence for partial reproductive isolation. Canadian Journal of Fisheries and Aquatic Sciences 41:1454–1462.

Leider, S. A., M. W. Chilcote, and J. J. Loch. 1986. Comparative life history characteristics of hatchery and wild steelhead trout (Salmo gairdneri) of summer and winter races in the Kalama River, Washington. Canadian Journal of Fisheries and Aquatic Sciences 43:1389–1409.

Leider, S. A., P. L. Hulett, J. J. Loch, and M. W. Chilcote. 1990. Electrophoretic comparison of the reproductive success of naturally spawning transplanted and wild steelhead trout through the returning adult stage. Aquaculture 88:239–252.

Leopold, A. 1933. Game management. Charles Scribner's Sons. New York.

Levings, C. D., C. D. McAllister, and B. D. Chang. 1986. Differential use of the Campbell River estuary, British Columbia, by wild and hatchery-reared juvenile chinook salmon (Oncorhynchus tschawytsha). Canadian Journal of Fisheries and Aquatic Sciences 43: 1386–1397.

Lichatowich, J. A., and J. D. McIntyre. 1987. Use of hatcheries in the management of Pacific anadromous salmonids. American Fisheries Society Symposium 1:131–136.

Magnuson, J. J. 1976. Managing with exotics—a game of chance. Transactions of the American Fisheries Society 105:1–9.

Marnell, L. F. 1986. Impacts of hatchery stocks on wild fish populations. Pages 339–347 in R. H. Stroud, editor. Fish culture in fisheries management. American Fisheries Society, Fish Culture Section and Fisheries Management Section, Bethesda, Maryland.

Martin, J., J. Webster, and G. Edwards. 1992. Hatcheries and wild stocks: are they compatible? Fisheries (Bethesda) 17(1):4.

Mason, J. W., O. M. Brynildson, and P. E. Degurse. 1967. Comparative survival of wild and domestic strains of brook trout in streams. Transactions of the American Fisheries Society 96:313–318.

Maynard, D., T. A. Flagg, and C. V. W. Mahnken. 1995. A review of seminatural culture strategies for enhancing the postrelease survival of anadromous salmonids. American Fisheries Society Symposium 15:307–314.

McIntyre, J. D., and R. R. Reisenbichler. 1986. A model for selecting harvest fraction for aggregate populations of hatchery and wild anadromous salmonids. Pages 179–189 in R. H. Stroud, editor. Fish culture in fisheries management. American Fisheries Society, Fish Culture Section and Fisheries Management Section, Bethesda, Maryland.

McNair, M., and J. S. Holland. 1993. FRED 1992 Annual Report to the Alaska state legislature. Alaska Department of Fish and Game, Juneau.

Mead, R. W., and W. Woodall. 1968. Comparison of sockeye salmon fry produced by hatcheries, artificial channels, and natural spawning areas. Progress Report 20, International Pacific Salmon Fisheries Commission, New Westminster, British Columbia.

Meffe, G. K. 1992. Techno-arrogance and halfway technologies: salmon hatcheries on the Pacific Coast of North America. Conservation Biology 6:350–354.

Mesa, M. G. 1991. Variation in feeding, aggression, and position choice between hatchery and wild cutthroat trout in an artificial stream. Transactions of the American Fisheries Society 120:723–727.

Miller, R. B. 1946. Effectiveness of a whitefish hatchery. Journal of Wildlife Management 10:316–322.

Miller, R. B. 1952a. Survival of hatchery-reared cutthroat trout in an Alberta stream. Transactions of the American Fisheries Society 81:35–42.

Miller, R. B. 1952b. The relative strengths of whitefish year classes as affected by egg plantings and weather. Journal of Wildlife Management 16:39–50.

Miller, R. B. 1954. Comparative survival of wild and hatchery-reared trout in a stream. Transactions of the American Fisheries Society 83:120–130.

Miller, R. B. 1958. The role of competition in the mortality of hatchery trout. Journal of the Fisheries Research Board of Canada 15:27–45.

Miller, R. B., A. C. Sinclair, and P. W. Hochachka. 1959. Diet, glycogen reserves and resistance to fatigue in hatchery rainbow trout. Journal of the Fisheries Research Board of Canada 16:321–328.

Miller, W. H., T. C. Coley, H. L. Burge, and T. T. Kisanuki. 1990. Analysis of salmon and steelhead supplementation: emphasis on unpublished reports and present programs. Report to Bonneville Power Administration (Project 88–100), Portland, Oregon.

Mills, D. H. 1969. The survival of juvenile salmon and brown trout in some Scottish streams. Pages 217–228 in T. G. Northcote, editor. Salmon and trout in streams. H. R. MacMillan Lectures in Fisheries, University of British Columbia, Vancouver.

Moyle, P. B. 1969. Comparative behavior of young brook trout of domestic and wild origin. Progressive Fish-Culturist 31:51–56.

Moyle, P. B., H. W. Li, and B. A. Barton. 1986. The Frankenstein effect: impact of introduced fishes on native fishes in North America. Pages 415–426 in R. H. Stroud, editor. Fish culture in fisheries management. American Fisheries Society, Fish Culture Section and Fisheries Management Section, Bethesda, Maryland.

Nale, M. A. 1988. The wild trout of Penn's woods. Trout (Summer)1988:12–24.

Needham, P. R., and D. W. Slater. 1944. Survival of hatchery-reared brown and rainbow trout as affected by wild trout populations. Journal of Wildlife Management 8:22–36.

Nehlsen, W., J. E. Williams, and J. A. Lichatowitch. 1991. Pacific salmon at the crossroads: stocks at risk from California, Oregon, Idaho, and Washington. Fisheries (Bethesda) 16(2):4–21.

Nelson, J. S., and M. J. Paetz. 1992. The fishes of Alberta, 2nd edition. University of Alberta Press, Edmonton.

Nelson, K., and M. Soulé. 1987. Genetic conservation of exploited fishes. Pages 345–368 in N. Ryman and F. Utter, editors. Population genetics & fishery management. University of Washington Press, Seattle.

Nickelson, T. E., M. F. Solazzi, and S. L. Johnson. 1986. The use of hatchery coho salmon (*Oncorhynchus kisutch*) presmolts to rebuild wild populations in Oregon coastal streams. Canadian Journal of Fisheries and Aquatic Sciences 43:2443–2449.

Nickum, J. 1993. The role of fish hatcheries in the 21st century. Fisheries (Bethesda) 18(2):2.

Nielsen, R. S., N. Reimers, and H. D. Kennedy. 1957. A six-year study of the survival and vitality of hatchery-reared rainbow trout of catchable size in Convict Creek, California. California Fish and Game 43:5–42.

Norman, L. 1987. Stream aquarium observations of territorial behavior in young salmon (*Salmo salar* L.) of wild and hatchery origin. Salmon Research Institute, Älvkarleby, Report 1987-2. (In Swedish, English summary).

ODFW (Oregon Department of Fish and Wildlife). 1992a. Division 7 Oregon administrative rules: fish management and hatchery operation (in part). Oregon Administrative Rule 635-07-501, Portland.

ODFW (Oregon Department of Fish and Wildlife). 1992b. Evaluation of salmonids released from ODFW hatcheries (sorted by cost per fish recovered in Oregon). ODFW internal report, Portland.

ODFW (Oregon Department of Fish and Wildlife). 1993. 1993–1995 Governor's recommended budget, Department of Fish and Wildlife. ODFW, Portland.

O'Grady, M. F. 1984. Observations on the contribution of planted brown trout (*Salmo trutta* L.) to spawning stocks in four Irish lakes. Fisheries Management 15:117–122.

Olla, B. L., and M. W. Davis. 1989. The role of learning and stress in predator avoidance of hatchery-reared coho salmon (*Oncorhynchus kisutch*) juveniles. Aquaculture 76:209–214.

Olla, B. L., M. W. Davis, and C. H. Ryer. 1991. Foraging and predator avoidance in hatchery-reared Pacific salmon achievement of behavioral potential. Pages 5–12 in J. E. Thorpe and F. A. Huntingford, editors. The importance of feeding behavior for the efficient culture of salmonid fishes. World Aquaculture Society, Baton Rouge, Louisiana.

OMNR (Ontario Ministry of Natural Resources). 1992. Strategic plan for Ontario fisheries—SPOF II—an ecosystem approach to managing fisheries. OMNR, Toronto.

Pearsons, T. N., G. A. McMichael, E. L. Bartrand, M. Fisher, J. T. Monahan, and S. A. Leider. 1993. Yakima species interactions study annual report 1992. Washington Department of Wildlife report to Bonneville Power Administration (Contract DE-B179–89BP01483) Portland, Oregon.

Petrosky, C. E., and T. C. Bjornn. 1988. Response of wild rainbow trout (*Salmo gairdneri*) and cutthroat trout (*S. clarki*) to stocked rainbow trout in fertile and infertile streams. Canadian Journal of Fisheries and Aquatic Sciences 45:2087–2105.

PFBC (Pennsylvania Fish and Boat Commission). 1987. Management of trout fisheries in Pennsylvania waters, 2nd edition. PFBC, Bellefonte.

PFBC (Pennsylvania Fish and Boat Commission). 1992. Pennsylvania Fish and Boat Commission Annual Report, Fiscal Year 1991–92. PFBC, Harrisburg.

Raney, E. C., and E. A. Lachner. 1942. Autumn food of recently planted young brown trout in small streams of central New York. Transactions of the American Fisheries Society 71:106–111.

RASP (Regional Assessment of Supplementation Project). 1992. Supplementation in the Columbia basin. Report (Contract DE-AC06-75RL01830) to Bonneville Power Administration, Portland, Oregon.

Reimers, N. 1963. Body condition, water temperature and over-the-winter survival of hatchery-reared trout in Convict Creek, California. Transactions of the American Fisheries Society 92:39–46.

Reisenbichler, R. R. 1984. Outplanting: potential for harmful genetic change in naturally spawning salmonids. Pages 33–39 in J. M. Walton and D. B. Houston, editors. Proceedings of the Olympic wild fish conference. Peninsula College, Port Angeles, Washington.

Reisenbichler, R. R. 1988. Relation between distance transferred from natal stream and recovery rate for hatchery coho salmon. North American Journal of Fisheries Management 8:172–174.

Reisenbichler, R. R., and J. D. McIntyre. 1977. Genetic differentiation in growth and survival of juvenile hatchery and wild steelhead trout (*Salmo gairdneri*). Journal of the Fisheries Research Board of Canada 34:2333–2337.

Reisenbichler, R. R., and J. D. McIntyre. 1986. Requirements for integrating natural and artificial production of anadromous salmonids in the Pacific Northwest. Pages 365–374 in R. H. Stroud, editor. Fish culture in fisheries management. American Fisheries Society, Fish Culture Section and Fisheries Management Section, Bethesda, Maryland.

Ritter, J. A., and H. R. MacCrimmon. 1973. Influence of environmental experience on response of yearling rainbow trout (*Salmo gairdneri*) to a black and white substrate. Journal of the Fisheries Research Board of Canada 30:1740–1742.

Roadhouse, S., M. J. Saari, D. Roadhouse, and B. A. Pappas. 1986. Behavioral and biochemical correlates of hatchery rearing methods on lake trout. Progressive Fish-Culturist 48:38–42.

Roppel, P. 1982. Alaska's salmon hatcheries 1891–1959. Alaska Department of Fish and Game, Division of Fisheries Rehabilitation, Enhancement and Development, Juneau.

Royce, W. F. 1989. Managing Alaska's salmon fisheries for a prosperous future. Fisheries (Bethesda) 14(2):8–13.

Ruzzante, D. E. 1994. Domestication effects on aggressive and schooling behavior in fish. Aquaculture 120:1–24.

Ryman, N., editor. 1981. Fish gene pools. Ecological Bulletin (Stockholm) 34.

Ryman, N., and G. Stahl. 1980. Genetic changes in hatchery stocks of brown trout (*Salmo trutta*). Canadian Journal of Fisheries and Aquatic Sciences 37:82–87.

SAC (The Salmon Advisory Committee). 1991. Assessment of stocking as a salmon management strategy. Report to the Ministry of Agriculture, Fisheries and Food, London.

Schmidt, B. R. 1994. Wild trout vs. stocked trout: a state management agency perspective. Pages 80–85 in R. W. Wiley and W. A. Hubert, editors. Wild trout and planted trout: balancing the scale. Wyoming Game and Fish Department, Laramie.

Schuck, H. A. 1948. Survival of hatchery trout in streams and possible methods of improving the quality of hatchery trout. Progressive Fish-Culturist 10:3–14.

SFI (Sport Fishing Institute). 1992. A bucket of fish: successes and failures of stocking. SFI Bulletin No. 432:1–2, Washington, DC.

Shepherd, B. G. 1991. On choosing well: bioengineering reconnaissance of new hatchery sites. American Fisheries Society Symposium 10:354–364.

Shetter, D. S., and A. S. Hazzard. 1941. Results from plantings of marked trout of legal size in streams and lakes of Michigan. Transactions of the American Fisheries Society 70:446–466.

Sholes, W. H., and R. J. Hallock. 1979. An evaluation of rearing fall-run chinook salmon, *Oncorhynchus tshawytscha*, to yearlings at Feather River Hatchery, with a comparison of returns from hatchery and downstream releases. California Fish and Game 64:239–255.

Smith, E. M., B. A. Miller, J. D. Rodgers, and M. A. Buckman. 1985. Outplanting anadromous salmonids. A literature survey. Bonneville Power Administration Project No. 85-68, Portland, Oregon.

Smith, L. L., and B. S. Smith. 1945. Survival of seven-to-ten-inch planted trout in two Minnesota streams. Transactions of the American Fisheries Society 73:108–116.

Sosiak, A. J. 1982. Buoyancy comparisons between juvenile Atlantic salmon and brown trout of wild and hatchery origin. Transactions of the American Fisheries Society 111:307–311.

Sosiak, A. J., R. G. Randall, and J. A. McKenzie. 1979. Feeding by hatchery-reared and wild Atlantic salmon (*Salmo salar*) parr in streams. Journal of the Fisheries Research Board of Canada 36:1408–1412.

Stabell, O. B. 1981. Homing of Atlantic salmon in relation to olfaction and genetics. Pages 238–246 in E. L. Brannon and E. O. Salo, editors. Salmon and trout migratory behavior symposium. University of Washington School of Fisheries, Seattle.

Stabell, O. B. 1984. Homing and olfaction in salmonids: a critical review with special reference to the Atlantic salmon. Biological Review 59:333–388.

Stevens, W. K. 1991. Hatched and wild fish: clash of cultures. New York Times, July 23:C1 and C10.

Steward, C. R., and T. C. Bjornn. 1990. Supplementation of salmon and steelhead stocks with hatchery fish: a synthesis of published literature. Technical Report 90-1, Bonneville Power Administration, Portland, Oregon.

Swain, D. P., and B. E. Riddell. 1990. Variation in agonistic behavior between newly emerged juveniles from hatchery and wild populations of coho salmon, *Oncorhynchus kisutch*. Canadian Journal of Fisheries and Aquatic Sciences 47:566–571.

Swain, D. P., and B. E. Riddell. 1991. Domestication and agonistic behavior in coho salmon: reply to comment. Canadian Journal of Fisheries and Aquatic Sciences 48:520–522.

Swain, D. P., B. E. Riddell, and C. B. Murray. 1991. Morphological differences between hatchery-reared and wild populations of coho salmon (*Oncorhynchus kisutch*): environmental versus genetic origin. Canadian Journal of Fisheries and Aquatic Sciences 48: 1783–1791.

Taylor, D. S. 1992. Why wild fish matter: an angler's view. Trout (Summer):52–58.

Thomas, A. E., and M. J. Donahoo. 1977. Differences in swimming performance among strains of rainbow trout (*Salmo gairdneri*). Journal of the Fisheries Research Board of Canada 34:304-306.

Thompson, W. F. 1965. Fishing treaties and the North Pacific. Science 150:1786–1789.

Thuemler, T. 1975. Fish the blue ribbon streams. Wisconsin Conservation Bulletin 40(3):16–17.

Todd, T. N. 1986. Artificial propagation of coregonines in the management of the Laurentian Great Lakes. Archiv für Hydrobiologie, Beiheft Ergebnisse der Limnologie 22:31–50.

USCFF (United States Commissioner of Fish and Fisheries [and successor positions]). 1873–1994. Reports of the Commissioner [usually annual]. USCFF [and successor agencies, e.g., U.S. Bureau of Fisheries, U. S. Fish and Wildlife Service], Washington, DC.

Van Vooren, A. 1994. State perspectives on the current role of trout culture in the management of trout fisheries. Pages 100–108 *in* R. W. Wiley and W. A. Hubert, editors. Wild trout and planted trout: balancing the scale. Wyoming Game and Fish Department, Laramie.

Varley, J. D. 1981. A history of fish stocking activities in Yellowstone National Park between 1881 and 1980. U.S. National Park Service, Information Paper 35, Yellowstone National Park, Wyoming.

Varley, J. D., and P. Schullery. 1983. Freshwater wilderness: Yellowstone fishes and their world. Yellowstone Library and Museum Association, Yellowstone National Park, Wyoming.

Verspoor, E. 1988. Reduced genetic variability in first-generation hatchery populations of Atlantic salmon (*Salmo salar*). Canadian Journal of Fisheries and Aquatic Sciences 45:1686–1690.

Vincent, E. R. 1975. Effect of stocking catchable trout on wild populations. Pages 88–91 *in* W. King, editor. Wild trout management: proceedings of the wild trout management symposium at Yellowstone National Park. Trout Unlimited, Denver, Colorado.

Vincent, E. R. 1987. Effects of stocking catchable-size hatchery rainbow trout on two wild trout species in the Madison River and O'Dell Creek, Montana. North American Journal of Fisheries Management 7:91–105.

Vincent. R. E. 1960. Some influences of domestication upon three stocks of brook trout (*Salvelinus fontinalis* Mitchell). Transactions of the American Fisheries Society 89:35–52.

Vuorinen, J. 1984. Reduction of genetic variability in a hatchery stock of brown trout, *Salmo trutta*. Journal of Fish Biology 24:339–348.

Wales, J. H. 1954. Relative survival of hatchery and wild trout. Progressive Fish-Culturist 16:125–127.

Walters, C. J., and F. Juanes. 1993. Recruitment limitation as a consequence of natural selection for use of restricted feeding habitats and predation risk taking by juvenile fishes. Canadian Journal of Fisheries and Aquatic Sciences 50:2058–2070.

Waples, R. S. 1991. Genetic interactions between hatchery and wild salmonids: lessons from the Pacific Northwest. Canadian Journal of Fisheries and Aquatic Sciences 48(Supplement 1):124–133.

Waples, R. S., G. A. Winans, F. M. Utter, and C. Mahnken. 1990. Genetic approaches to the management of Pacific salmon. Fisheries (Bethesda) 15(5):19–25.

Warren, C. E., M. Allen, and R. E. Thompson. 1979. Conceptual frameworks and the philosophical foundations of general living systems theory. Behavioral Science 24:296–310.

Washington, P. M., and A. M. Koziol. 1993. Overview of the interactions and environmental impacts of hatchery practices on natural and artificial stocks of salmonids. Fisheries Research 18:105–122.

WDF (Washington Department of Fisheries). No date. 1991–1993 Biennial Report. WDF, Olympia.

WDF (Washington Department of Fisheries). 1993a. A detailed listing of the liberations of salmon into the open waters of the state of Washington during 1990. WDF Salmon Culture Division, Olympia.

WDF (Washington Department of Fisheries). 1993b. 1992 Washington state salmon and steelhead inventory. WDF (and Washington Department of Wildlife and Western Washington Treaty Indian Tribes), Olympia.

WDFW (Washington Department of Fish and Wildlife). 1994. Draft scoping paper for a wild salmonid policy. WDFW, Olympia.

WDW (Washington Department of Wildlife). 1994. Final environmental impact statement for Grandy Creek Trout Hatchery. WDW, Olympia.

WDNR (Wisconsin Department of Natural Resources). 1966. Wisconsin trout streams. WDNR Publication 213-66, Madison.

Weithman, A. S. 1986. Economic benefits and costs associated with stocking fish. Pages 357–363 *in* R. H. Stroud, editor. Fish culture in fisheries management. American Fisheries Society, Fish Culture Section and Fisheries Management Section, Bethesda, Maryland.

Weithman, A. S. 1993. Socioeconomic benefits of fisheries. Pages 159–177 *in* C. C. Kohler and W. A. Hubert, editors. Inland fisheries management in North America. American Fisheries Society, Bethesda, Maryland.

Wells, J. 1985. Wild trout management: commitment to excellence. Montana Outdoors (July/August):19–22.

Whaley, R. A. 1994. What are the fisheries benefits from current stocking practices in lakes and reservoirs? Pages 118–124 *in* R. W. Wiley and W. A. Hubert, editors. Wild trout and planted trout: balancing the scale. Wyoming Game and Fish Department, Laramie.

White, R. J. 1989. We're going wild: a 30-year transition from hatcheries to habitat. Trout (Summer):15–23, 26–35, 38–49.

White, R. J. 1992a. Why wild fish matter: a biologist's view. Trout (Summer):25–32, 44–50.

White, R. J. 1992b. Why wild fish matter: balancing ecological and aquacultural fishery management. Trout (Autumn):16–33, 44–48.

White, R. J. In press. Hatchery versus wild salmon. Proceedings of the New England Atlantic salmon management conference. New England Salmon Association, Danvers, Massachusetts.

White, R. J., W. Nehlsen, and J. R. Karr. In press. Why wild trout matter. In R. H. Hamre, editor. Wild Trout V. Trout Unlimited, Vienna, Virginia.

Wiley, R. W., R. A. Whaley, J. B. Satake, and M. Fowden. 1993a. Assessment of stocking hatchery trout: a Wyoming perspective. North American Journal of Fisheries Management 13:160–170.

Wiley, R. W., R. A. Whaley, J. B. Satake, and M. Fowden. 1993b. An evaluation of the potential for training trout in hatcheries to increase poststocking survival in streams. North American Journal of Fisheries Management 13:171–177.

Winton, J. N. 1991. Supplementation of wild salmonids: management practices in British Columbia. Master's thesis, University of Washington, Seattle.

Winton, J. N., and R. Hilborn. 1994. Lessons from supplementation of chinook salmon in British Columbia. North American Journal of Fisheries Management 14:1–13.

Woodward, C. C., and R. J. Strange. 1987. Physiological stress responses in wild and hatchery-reared rainbow trout. Transactions of the American Fisheries Society 116:574–579.

Wright, S. 1981. Contemporary Pacific salmon fisheries management. North American Journal of Fisheries Management 1:29–40.

Wright, S. 1993. Fishery management of wild Pacific salmon stocks to prevent extinctions. Fisheries (Bethesda) 18(5):3–4.

Poster Presentations

Use of Gene Marking to Assess Stocking Success of Red Drum in Texas Bays

ROCKY WARD

Texas Parks and Wildlife Department, Perry R. Bass Marine Fisheries Research Station
HC 2 Box 385, Palacios, Texas 77465, USA

TIM L. KING

National Biological Service, Aquatic Ecology Branch
1700 Leetown Road, Kearneysville, West Virginia 25430, USA

IVONNE R. BLANDON

Texas Parks and Wildlife Department, Perry R. Bass Marine Fisheries Research Station

LAWRENCE W. MCEACHRON

Texas Parks and Wildlife Department
702 Navigation Circle, Rockport, Texas 78382, USA

Abstract.—Over 200 million eggs, fry, and fingerlings of red drum *Sciaenops ocellatus* have been stocked into Texas coastal waters since the inception of the Texas red drum enhancement program. The Coastal Fisheries Branch of Texas Parks and Wildlife Department (TPWD) has instituted a program that uses a gene marker to assess the success of the red drum stocking effort and the stocking program's contribution to the recreational fishery. Preliminary surveys of population structure identified an uncommon allele of dimeric esterase-D (ESTD) as a suitable gene marker. This allele occurs at a low frequency (0.05 to 0.08), and there is little evidence for nonrandom changes in frequency along the Texas coast. Individuals heterozygous for this allele have been collected and maintained as broodfish. The expected frequency of the marker allele among the progeny of these heterozygous fish is 0.5, and the ratio of the three phenotypes is expected to be 1:2:1.

Ten to 12 broodfish were placed in large tanks (efforts were made to place equal numbers of males and females in each tank), and spawning was stimulated by temperature and photoperiod manipulations. Ten separate spawns from these tanks were maintained in 0.1-ha ponds for 30 d, and one pond was maintained for 300 d. At the end of this period, the weight and length of 100 individuals from each pond were measured and the ESTD phenotype of each individual was determined. No differences in weight or length at 30 d of age among the three phenotypes of the progeny were found in 9 of 10 spawns. We also found no differences in weight or length at 300 d. This finding does not prove that the ESTD* allelic variability is selectively neutral but does suggest that growth differences between the phenotypes are minor.

A variable number of fingerlings (250,000 to 1,500,000), spawned by parents heterozygous for the uncommon allele, will be stocked into East Matagorda Bay each year for 4 years. Every red drum collected by routine TPWD sampling from this bay and adjacent areas of neighboring bays will be surveyed to determine ESTD phenotype. This survey will continue for 4 years after the last stocking of genetically marked fish. The stocking program will be considered successful (i.e., stocked fish will have contributed to the population) if the frequency of the uncommon ESTD allele or the percentage of individuals homozygous for the uncommon allele increases above baseline values (0.07 and 0.005, respectively).

Multivariate Analysis of Red Drum Stocking in Texas Bays

ROCKY WARD

Texas Parks and Wildlife Department, Perry R. Bass Marine Fisheries Research Station
HC 2 Box 385, Palacios, Texas 77465, USA

LAWRENCE W. MCEACHRON AND BILLY E. FULS

Texas Parks and Wildlife Department, Rockport Marine Laboratory
702 Navigation Circle, Rockport, Texas 78382, USA

MAURICE MUOENKE

Texas Parks and Wildlife Department, Operational Services Branch
4200 Smith School Road, Austin, Texas 78744, USA

Abstract.—Exploratory multivariate analyses were used in a preliminary assessment of the success of stocking red drum *Sciaenops ocellatus* into Texas coastal waters to enhance the abundance of this recreationally important species. Regression analyses on the relationship between stocking rate and subsequent relative abundance of red drum (as measured by gill-net harvest) suggested stocking may have contributed to changes in abundance in the lower Laguna Madre, East Matagorda Bay, San Antonio Bay, and possibly Matagorda Bay. Little or no evidence of stocking success was found in other bays.

Principal components analysis (PCA) and stepwise regression analyses were then used to examine the ability of a set of physical variables (bay area and mean, upper and lower salinity, turbidity, and temperatures for the fall and spring gill-net seasons) to explain variation in stocking success among eight bays during the years 1975 through 1992. The PCA was applied to generate a reduced number of orthogonal factors to which a specific amount of variance may be attributed and to combat multicolinearity. Five factors generated by the PCA were retained that had eigenvalues greater than 1.0. The first factor was loaded most heavily by variables related to salinity, the second and third factors were loaded most heavily by variables related to turbidity, the fourth factor was loaded heavily by spring temperature, and the fifth factor was loaded most heavily by bay area and fall turbidity.

Based on the inability of stepwise regression analyses to develop satisfactory explanatory models, the variables employed in the present study failed to explain the success or failure of stocking. Further research is needed to identify variables that better predict success or failure of stock enhancement efforts.

Predation and Cannibalism on Hatchery-Reared Striped Bass in the Patuxent River, Maryland

LINDA L. ANDREASEN

U.S. Fish and Wildlife Service, Maryland Fisheries Resource Office
1825 Virginia Street, Annapolis, Maryland 21401, USA

Abstract.—Hatchery-reared striped bass *Morone saxatilis* have been stocked into the Patuxent River, Maryland, since 1985 as part of a multiagency striped bass restoration effort. Fingerlings tagged with binary-coded wire tags (CWTs) are released as phase I fish (35–50 mm) in July and as phase II fish (150–200 mm) in November. Mortality rates have been reported for hatchery fish in the Patuxent River; however, the initial loss of hatchery fish to predation is unknown. This study attempts to (1) characterize initial predation on hatchery fish (species and size distribution of predators, and selectivity by predators for certain attributes of hatchery fingerlings) and (2) estimate initial loss of hatchery fish to predation. The second objective (estimation of loss) is ongoing and therefore will not be reported here.

Phase II striped bass fingerlings were released into the Patuxent River on four consecutive days. Fingerlings were marked daily prior to release with a uniquely coded CWT. Each day, prior to release, a subsample was additionally marked with Floy tags (Floy Tag Co., Seattle, Washington). Predators were sampled after peak feeding on each day of release. Prior to predator sampling, an otter trawl was used to establish the ratio of wild to hatchery fish in the river. Stomach contents were examined for number of wild versus hatchery striped bass, percent Floy-tagged hatchery fish, and size distribution of hatchery fish.

Striped bass was found to be the primary predator species at the sampling site (Patuxent River, river mile 24). Other potential predator species (channel catfish *Ictalurus punctatus*, white catfish *Ameiurus catus*, and, sporadically, Atlantic croaker *Micropogonias undulatus*) were determined to be unimportant. Of the predator striped bass captured with identifiable prey, the average number of hatchery fingerlings per predator stomach was 1.43. Few cases of cannibalism appear in the literature; however, striped bass are known to be opportunistic

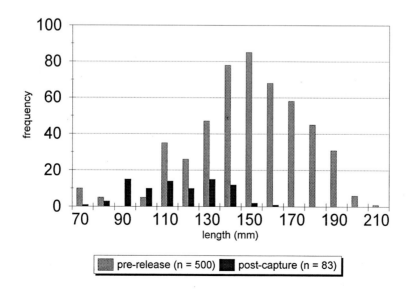

FIGURE 1.—Fingerling length distribution (prerelease vs. postcapture).

feeders. Cannibalism also has been shown to be a mechanism for density-dependent regulation of fish populations.

Length-frequency distributions of prerelease versus postcapture (in predator stomachs) hatchery striped bass (tagged with CWTs only) were compared (Figure 1) and shown to be significantly different (95% confidence interval). Predators appear to be selecting smaller hatchery fish. Similarly, there was a significant difference between the percentage of Floy-tagged fish to hatchery fish without Floy tags in prerelease versus postcapture samples (5% versus 25%, respectively), indicating that predators may be selecting Floy-tagged fish. Floy-tagged fish and small hatchery fish are possibly more vulnerable than other fingerlings, due to more awkward or slower swimming. Ratios of wild to hatchery fish in trawl samples were not found to be significantly different from the ratio of wild to hatchery fish found in predator stomachs.

The Use of Hatchery-Produced Striped Bass for Stock Restoration and Validation of Stock Parameters

STEVEN P. MINKKINEN, CHARLES P. STENCE, AND BRIAN M. RICHARDSON

Maryland Department of Natural Resources
580 Taylor Avenue, Annapolis, Maryland 21401, USA

Abstract.—Since 1985, a joint restoration program by the Maryland Department of Natural Resources (MDNR) and U.S. Fish and Wildlife Service has stocked 6,457,000 striped bass *Morone saxatilis* into Maryland tributaries of the Chesapeake Bay. All were tagged with binary-coded wire tags (Northwest Marine Technology, Shaw Island, Washington). Two different stocking strategies were used: (1) stock 50-mm phase-I fish during June and July and (2) stock 100- to 200-mm phase-II fish in the fall. Cage holding experiments were conducted over 2 weeks to determine poststocking tag loss and mortality. The number of fish available for recapture was adjusted using these estimates. A survey was conducted to assess the contribution of hatchery-produced striped bass to the juvenile population. Sampling was conducted with a 3.1-m × 1.2-m beach seine and a 4.9-m semiballoon trawl.

The Patuxent River has received 2,816,000 hatchery fish which constituted 30, 72, 85, 48, and 5% of the juveniles collected from the 1989 to 1993 year-classes, respectively (40% overall; Figure 1). Hatchery and wild fish were collected in the same areas and habitats. Peterson mark–recapture estimates in December each year indicated recruitment of 586,000; 125,000; 39,000; 413,000; and 2,575,000 wild striped bass from 1989 to 1993, respectively. Juvenile recruitment indices, which are collected by MDNR to monitor striped bass reproductive success, also demonstrated two levels of production. During 1993, a dominant year-class was indicated by the survey with an index that was an order of magnitude higher than those recorded previously (1983–1992; Figure 2). These dominant year-classes have historically been the primary contributors to the stock (Rago, P. J., and C. P. Goodyear. 1987. Recruitment mechanisms of striped bass and Atlantic salmon: comparative liabilities of alternative life histories. American Fisheries Society Symposium 1:402–416).

Hatchery fish represented 5.3% of the juvenile striped bass collected in the Choptank River during 1991. Six million wild juvenile striped bass were present after phase-I stocking (July) and 1,000,000 wild fish remained after phase-II stocking (December).

In the Nanticoke River hatchery fish constituted 18.1% of the 1992 year-class of striped bass collected in summer and 68% of those collected in winter. During 1993, hatchery striped bass constituted 5.6 and 63.7% of the fish collected in the summer and winter

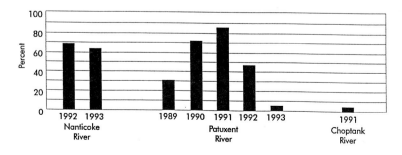

FIGURE 1.—Percentage of hatchery-produced juvenile striped bass by river system and year from winter trawl surveys.

surveys, respectively. Abundance estimates for wild striped bass during 1992 were 227,000 in July and 90,000 in December. In 1993, abundance estimates were 3,540,000 (July) and 250,000 (December).

Instantaneous mortality varied tenfold (between 0.002 and 0.02) in all three rivers from 1991 to 1993, influencing eventual recruitment to the adult population (Figure 3). Factors that control these mortality rates are being investigated.

Hatchery efforts have contributed significantly to juvenile abundance in these systems. These fish should contribute to the adult spawning stock when mature. The use of mass marking to measure absolute abundance validated juvenile abundance indices that predict recruitment of striped bass along the Atlantic Coast.

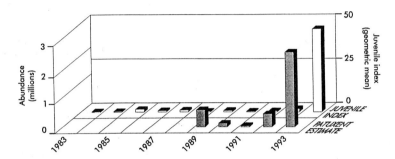

FIGURE 2.—Peterson mark–recapture estimates in the Patuxent River for 1989 to 1993 and the geometric mean juvenile index from a juvenile recruitment seine survey conducted by the Maryland Department of Natural Resources to measure relative year-class success.

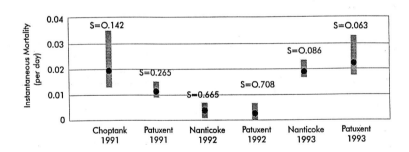

FIGURE 3.—Instantaneous rates of mortality (S) and 95% confidence interval for juvenile striped bass (July to December) calculated using Ricker's method (Ricker, W. E. 1975. Computation and interpretation of biological statistics of fish populations. Bulletin of the Fisheries Research Board of Canada 191:382).

American Shad Handling, Rearing, and Marking Trials

BRIAN M. RICHARDSON AND STEVEN P. MINKKINEN

Maryland Department of Natural Resources
580 Taylor Avenue, Annapolis, Maryland 21401, USA

Abstract.—During the 1960s, the American shad *Alosa sapidissima* in Chesapeake Bay declined drastically (Figure 1). Since that time, there has been some increase in the upper bay population due to larval stocking efforts, fish passage facilities, and a complete harvest moratorium. However, American shad runs in other rivers are still severely depressed. We began a program, in cooperation with the U.S. Fish and Wildlife Service and Susquehanna River Anadromous Fish Restoration Committee, to restore viable American shad runs in the Patuxent River. This Chesapeake Bay tributary historically had viable American shad runs but no juvenile American shad have been captured in the Patuxent River since 1960. An experimental pilot study was initiated in 1993 to assess the potential for rearing, handling, and marking American shad in a hatchery setting. The study has three components: (1) incubate strip-spawned fertilized eggs and grow larvae to juvenile size for stocking; (2) transport, handle, and naturally spawn American shad adults collected at Conowingo Dam fish lift; and (3) investigate different methods of marking juvenile American shad.

Strip spawning is the traditional method used in American shad culture. This method usually entails nighttime gill netting on the spawning grounds to collect male and ripe female American shad. All fish captured are sacrificed as eggs and milt are extracted. The process is limited by the difficulty of obtaining an ample supply of males and ripe females. In addition, American shad are sequential spawners so stripping the fish uses only a portion of the potential fecundity. Strip-spawned fertilized eggs were incubated and hatched, and larvae were grown to juvenile size.

Because American shad are sequential spawners, we surmised that the use of natural

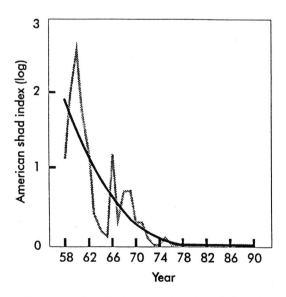

FIGURE 1.—Juvenile index for American shad at the head of the Chesapeake Bay. Index is catch-per-unit-effort of summer seine survey from 1958 to 1990.

TABLE 1.—Results of a 10-d tag retention and mortality study in hatchery tanks. Two trials of fish were given four different treatments. Control fish had no treatment. Anesthetized fish were dosed with MS-222 (tricaine methanesulfonate). Horizontal-tag-insertion fish were implanted with CWTs laterally through the body. Vertical-tag-insertion fish were implanted with CWTs dorsally through the body. Abbreviations are RI, round 1; RII, round 2; and NT, no tags implanted.

Variables	Control		Anesthetized		Horizontal tag insertion		Vertical tag insertion	
	RI	RII	RI	RII	RI	RII	RI	RII
Fish in trial (N)	100	100	100	100	100	100	100	101
Survival %	94	99	89	99	85	97	91	100
Tag retention %	NT	NT	NT	NT	80	90	85	96

spawn techniques could increase production without sacrificing the adults. American shad are known to be sensitive to handling stress. Because all hatchery culture work requires some handling of the fish, tolerance to this type of stress had to be determined.

Seventy-seven adult American shad were collected at Conowingo Dam fish lift on the Susquehanna River and transported to Manning Hatchery in Maryland. American shad were equally distributed among four systems that allowed collection of naturally spawned fertilized eggs without disturbing the adults.

Otoliths of strip-spawned juveniles were marked with tetracycline (TC). This method is a practical system for marking large numbers of small fish but lacks a discrete tag for identification purposes; furthermore, the animal must be sacrificed to establish a positive identification. Otolith examination is tedious and requires expensive microscopic equipment.

Naturally spawned juveniles were marked with binary-coded wire tags (CWT; Northwest Marine Technology, Shaw Island, Washington). This type of mark allows identification of discrete stocks, and a positive tag can be established without sacrificing the animal; however, fish must be larger in size. A CWT loss and mortality study was performed in

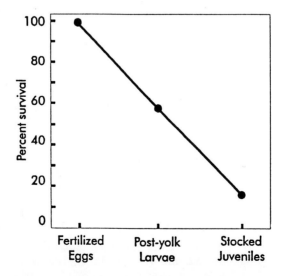

FIGURE 2.—Survival of strip-spawned American shad at each early life stage. Survival from fertilized egg to post-yolk larvae was 58%, and survival from post-yolk larvae to stocked juvenile was 29%. Overall survival from egg to juvenile was 16.7%.

hatchery tanks. Fish were given four different treatments, in two rounds of 10-d trials (Table 1).

Approximately 93,000 strip-spawned American shad juveniles with TC marks were stocked in the Patuxent River. Survival of American shad at each early life stage was excellent (Figure 2). Of the 77 American shad adults collected for natural spawning, 2 died in transport, 45 died at the hatchery, and 30 were released into the Patuxent River after spawning. Approximately 1,000 naturally spawned American shad juveniles with CWTs were stocked into the Patuxent River. A winter trawl survey recovered five American shad juveniles in the Patuxent River. These will be examined for otolith marks. Prior to this experiment a seine survey conducted in this river since 1983 had not captured any juvenile American shad.

Initial success at handling and spawning American shad demonstrates the feasibility of using natural spawning techniques for American shad production. During 1994, eggs from strip-spawned fish will be incubated and stocked as larvae and juveniles in the Patuxent River. Natural spawning will be attempted using hormones to stimulate egg maturation and spawning. We believe it will be possible to use coded wire tags on American shad juveniles to aid in population assessment.

Long-Term Survival of Three Size-Classes of Striped Bass Stocked in the Savannah River, Georgia–South Carolina

JULIE E. WALLIN AND MICHAEL J. VAN DEN AVYLE

National Biological Service, Georgia Cooperative Fish and Wildlife Research Unit
Daniel B. Warnell School of Forest Resources
University of Georgia, Athens, Georgia 30602, USA

Abstract.—Hatchery-reared fingerlings of striped bass *Morone saxatilis* have been stocked annually since 1990 to restore the depleted population in the Savannah River. We compared the effectiveness of stocking three size-classes of fish: phase I (15–25 mm total length), advanced phase I (50–80 mm), and phase II (180–225 mm). Fish were stocked in freshwater or at brackish water sites (0–7‰ salinity). All fish were marked by means of micromagnetic coded wire tags, anchor tags, or immersion in oxytetracycline. We calculated the number of fingerlings effectively stocked each year based on estimates of survival rates at 48 hours poststocking (Wallin and Van Den Avyle, 1995, this volume). Electrofishing was used during January–April to monitor abundance of age-1 and older striped bass.

Relative recruitment rates of phase-II fish were 7–52 times greater than for advanced phase-I fish at age 1 and 7–10 times greater at age 2, suggesting a substantially improved survival rate for the larger fish. Relative recruitment of phase-I fish at age 1 was only about 10% of that observed for advanced phase-I fingerlings. Long-term survival of both advanced phase-I and phase-II fish was 3–11 times greater for fish stocked at freshwater sites than for those stocked at brackish water sites.

Rearing fish to the larger sizes and stocking at freshwater sites appears to maximize long-term survival of striped bass stocked in the Savannah River estuary. Data on the relative survival of phase-II and advanced phase-I fish show that stocking phase-II fish would be economically efficient if the cost per fish did not exceed about seven times the cost of producing advanced phase-I fingerlings.

The Development of PCR-Based Genomic DNA Assays to Assess Genetic Variation in the Striped Bass Santee–Cooper River Population, South Carolina

Marilyn Diaz, Gilles M. LeClerc, and Bert Ely

*University of South Carolina, Department of Biological Sciences
Columbia, South Carolina 29208, USA*

James S. Bulak

*South Carolina Wildlife and Marine Resources Department, Fisheries Division
Columbia, South Carolina 29202, USA*

Abstract.—The Santee–Cooper River population of striped bass *Morone saxatilis* has undergone extensive stocking since 1984. The Wateree, the Congaree, and the Santee are the major rivers that define this landlocked system. To maintain any unique genetic adaptations that the Santee–Cooper population has evolved in response to its habitat, it is imperative that the fish used to generate broodstock be representative of the entire system. The parents used as broodstock fish for the stocking program are usually taken from the Santee River. Previous studies using allozyme or mitochondrial DNA analysis have had little success in detecting genetic variation within striped bass populations. Therefore, we have developed a set of nuclear genetic markers that detect variation within populations of striped bass to evaluate whether fish taken from the Santee River are representative of the entire Santee–Cooper system (Figure 1). These markers were identified from genomic clones that either yielded restriction fragment length polymorphisms (RFLPs) or contained simple sequence repeats (SSRs; microsatellites). Our goal was to design polymerase chain reaction (PCR)-based assays for each of the markers. To date, we have identified two polymorphic loci with this approach: RFLP 10-59 and SSR 83. Both of these markers can be used to screen the population with PCR-based assays. Data from SSR 83 suggest little or no difference in the allele distribution among the three rivers and are consistent with the hypothesis that Santee River fish are representative of the entire system. However, these results must be confirmed with additional markers before firm conclusions can be drawn.

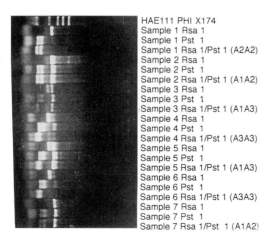

Figure 1.—A rapid PCR-based assay for genetic variation in striped bass using a double digest of a simple sequence repeat. The results demonstrate the presence of three of the four possible alleles. *A1* is the allele which does not have a recognition site for either *RSA-1* or *PST-1*. *A2* contains the *PST-1* site but not the *RSA-1* site. *A3* contains sites for both enzymes. We screened the striped bass population of the Santee–Cooper system with these alleles.

Development of a Hatchery Broodstock for Recovery of Green Lake Strain of Lake Trout

HAROLD L. KINCAID

National Biological Service, National Fishery Research and Development Laboratory
Rural Delivery 4, Box 63, Wellsboro, Pennsylvania 16901, USA

Abstract.—The Green Lake (Wisconsin) strain of lake trout *Salvelinus namaycush*, the last remnant of historic southern Lake Michigan deepwater spawning strains, was discontinued as a hatchery broodstock after the 1975 year-class was stocked into Lake Michigan in 1976. In 1982, a decision was made to preserve the gene pool of the Green Lake strain in a captive broodstock developed from feral fish recaptured from Black Can Reef and Julian's Reef in southern Lake Michigan. Between 1986 and 1988, 24 females and 31 males

TABLE 1.—Source of broodstock used to produce the 1984 to 1988 lots of the Green Lake strain.

			Parental source[a]		
Group	Year-class	Hatchery location[b]	Domestic brood	Julian's Reef	Black Can Reef
84-DOM	1984	Charlevoix	44/39		
86A-WILD	1986	Charlevoix			3/6
86B-WILD	1986	Jake Wolfe		3/5	
86C-WILD	1986	Charlevoix		7/10	5/12
87-WILD	1987	Charlevoix		3/4	2/2
88-WILD	1988	Charlevoix			6/6

[a]Number of fish (female/male) spawned to produce the group from each source. Both reefs are located in Lake Michigan.

[b]Hatcheries rearing individual groups were the Charlevoix State Fisheries Station, Charlevoix, Michigan; and the Jake Wolfe Memorial State Fish Hatchery, Manito, Illinois.

were captured from the 1976 stocking and spawned in five progeny groups (Table 1). A sixth group was produced from a remnant of the Green Lake broodstock at the Genoa (Wisconsin) National Fish Hatchery.

Broodfish from four of the six progeny groups were transferred to the Iron River National Fish Hatchery in 1991, where a two-stage breeding plan was initiated to develop a composite Green Lake broodstock. In 1991, a diallel mating design (Table 2) was imposed to equalize matings among groups and to equalize the genetic contribution of each group. At the eyed egg stage, 110 eggs were removed from each family and placed in a pool for rearing to maturity. This procedure was repeated in 1992, except lots were sampled at

TABLE 2.—Number of single-family matings produced for each group combination in fall 1991 and 1992 for pooling into the new Green Lake broodstock. One female and one male from each group was used to produce each family. A total of 67 families (35 in 1991 and 32 in 1992) were produced in the two year-classes.

Female groups	Male groups					
	84-DOM		86-WILD[a]		87-WILD	
	1991	1992	1991	1992	1991	1992
84-DOM	3	2	4	4	4	2
86-WILD[a]	4	4	4	4	4	4
87-WILD	4	4	4	4	4	4

[a] 86-WILD is a composite of 86A-WILD and 86C-WILD. 86B-WILD was dropped from the program.

the swim-up stage, and 3–100 fish from each lot were pooled. When the fish reach maturity in 1998, a reciprocal-crossing design will require that a minimum of 100 matings of 1991 year-class males with 1992 year-class females and 100 matings of 1991 year-class females with 1992 year-class males be used to construct the final broodstock. The breeding plan used to develop the new Green Lake broodstock will not recreate the original Green Lake strain, but it will preserve the remaining genetic variation from this unique gene pool.

Hidden Falls Hatchery Chum Salmon Program

BRUCE A. BACHEN AND TIM LINLEY

Northern Southeast Regional Aquaculture Association
1308 Sawmill Creek Road, Sitka, Alaska 99835, USA

Abstract.—Reacting to low salmon abundance in the 1970s, Alaska used some of its oil wealth to establish a public and private nonprofit salmon enhancement program to provide more stable returns. Comprehensive salmon plans that identified harvest goals and enhancement opportunities were developed for specific areas of Alaska. Chum salmon *Oncorhynchus keta* were identified in the Southeast Regional Plan as the preferred species for major hatchery production from a fishery management perspective. Summer stocks were specifically identified as desirable, provided adequate terminal harvest areas could be found.

The Hidden Falls State Fish Hatchery (SFH) was constructed in 1978. By 1993, approximately 7.2 million summer chum salmon had returned to the facility, including 1.8 million in 1993. The total value of the commercial chum salmon harvest through 1993 was US$22.1 million (Figure 1). In addition, enough fish are sold each year to generate about $1 million to pay for hatchery operations. The chum salmon return to Hidden Falls SFH in 1993 was the largest ever to a North American hatchery (Washington Department of Fisheries and Wildlife. 1993. Puget Sound run reconstruction database for fall chum. Washington Department of Fisheries and Wildlife, Olympia; McNair, M., and J. Holland. 1994. Alaska fisheries enhancement program. 1993 Annual Report, Alaska Department of Fish and Game, Juneau; E. Perry, personal communication). In 1993, Hidden Falls SFH produced 22.8% of the southeast Alaska commercial chum salmon catch and helped to make the catch the highest in nearly 70 years.

The success of the Hidden Falls SFH program is partially due to the consideration of interactions between returning hatchery-produced adults and wild stocks. The broodstock selected for use at Hidden Falls SFH returns midsummer, and chum salmon are harvested prior to the time when most pink salmon *O. gorbuscha* migrate through the terminal harvest area. The temporal separation between hatchery returns and most wild stocks allows aggressive harvest on hatchery fish without apparent harm to wild stocks. Hidden Falls SFH returns allow fishery managers to schedule commercial openings throughout southeast Alaska knowing that the hatchery harvest will usually attract between 20 and 80% of the boats and thereby reduce fishing pressure in wild stock harvest areas.

An important factor in the decision to site a hatchery at Hidden Falls was proximity to salmon-producing streams. The Hidden Falls area has little wild production nearby due to the limited freshwater habitat available. The nearest chum salmon-producing stream to Hidden Falls SFH is Clear River, about 13 water miles away. Return per spawner data for

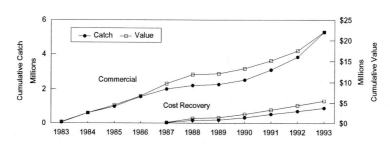

FIGURE 1.—Cumulative catch and values of commercial and cost-recovery harvests of Hidden Falls chum salmon, 1983–1993.

pink and chum salmon to Clear River show no significant correlation with chum salmon production at Hidden Falls (Figure 2). By locating a hatchery away from wild stocks, adverse effects of juvenile competition between hatchery and wild stocks as well as from excessive harvest of returning wild stock adults are minimized.

By incorporating wild stock concerns in the design of this enhancement project, managers have greater flexibility to use harvest strategies that generate maximum public benefit. The relatively large terminal harvest area, available because of the temporal separation between hatchery and wild stocks, allows chum salmon to be harvested while most are still bright and of high value. Hidden Falls chum salmon have a reputation for being high quality, a characteristic that is increasingly important in the competitive world of salmon marketing.

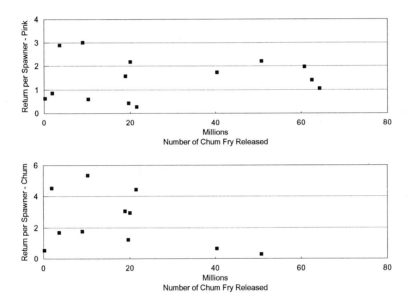

FIGURE 2.—The relationship between releases of chum salmon fry from Hidden Falls SFH and the return per spawner for Clear River pink and chum salmon. Returns are peak escapement counts. The correlations are $r = 0.07$ and $r = 0.39$, respectively, for pink and chum salmon.

Separation of Hatchery-Reared from Wild Red Drum Based on Discriminant Analysis of Daily Otolith Growth Increments

J. Jeffery Isely

National Biological Service, South Carolina Cooperative Fish and Wildlife Research Unit
Department of Aquaculture Fisheries and Wildlife, Clemson University
Clemson, South Carolina 29634, USA

Churchill B. Grimes and Andrew W. David

National Marine Fisheries Service
3500 Delwood Beach Road, Panama City, Florida 32408, USA

Abstract.—Identification of hatchery-reared fish after stocking is essential for the evaluation of stock enhancement programs. Fish typically are individually or mass marked with physical or chemical markers that can be identified upon recapture. Growth of hatchery fish can be significantly different from growth of wild fish because hatchery conditions generally do not mimic natural conditions. This difference in growth is recorded in the otolith microstructure. We developed a procedure to discriminate hatchery-reared from wild red drum *Sciaenops ocellatus* by quantifying the history of growth recorded in the otoliths. We measured asteriscus radius from the primordia to each successive daily increment for the first 25 increments in 114 hatchery and 292 wild fish. We chose the asteriscus because a consistent axis of growth could be identified and increments were visible throughout the otolith. Change in radius, equal to the increment width, was used as an index of daily growth. Independent linear discriminant equations using the first 25 otolith increment widths as independent variables were used to classify juvenile red drum into hatchery and wild groups in duplicate trials. The two-group discriminant functions correctly classified 100% of hatchery fish and 100% of wild fish in each trial. When duplicate trials were combined in a single four-group analysis based on production pond for hatchery fish and collection date for wild fish, the rule developed from training samples correctly classified 100% of hatchery fish into production pond and 92–96% of wild fish into collection date. When the four-group rule was applied to samples of unknown origin in a blind test, 62–70% of fish were correctly classified to hatchery production pond or wild collection date. Misclassifications generally occurred within similar groups. When the classification matrix was collapsed into two major groups, wild and hatchery stocks, 87 and 85% of fish, respectively, were correctly classified. We conclude that discriminant analysis of otolith banding patterns can provide an accurate means of evaluating stock enhancement programs in the future without additional chemical or physical markers.

A Captive Broodstock Approach to Rebuilding a Depleted Chinook Salmon Stock

James B. Shaklee, Carol Smith, Sewall Young, Christopher Marlowe, Chuck Johnson

Washington Department of Fish and Wildlife
Post Office Box 43151, Olympia, Washington 98504, USA

Brad B. Sele

Jamestown S'Klallam Tribe
305 Old Blyn Highway, Sequim, Washington 98382, USA

Abstract.—Many wild stocks of Pacific salmon *Oncorhynchus* sp. have declined severely in Washington, Oregon, Idaho, and California. Lost and damaged habitat, barriers to freshwater migration, and overfishing probably caused most of these declines. Low population numbers diminish harvest opportunities, threaten the genetic integrity of stocks, and increase the risk of population extinction due to episodic environmental fluctuation and catastrophe.

The Washington Department of Fish and Wildlife, the Jamestown S'Klallam Tribe, the U.S. Fish and Wildlife Service, and local volunteer groups have joined forces to rebuild the critically depressed stock of wild chinook salmon *O. tshawytscha* in the Dungeness River of Washington's Olympic Peninsula. Several other government agencies are contributing to this effort. Success will require increasing fish numbers, restoring habitat, and controlling harvest.

Based on logistical constraints and genetic considerations, our approach is to use a captive broodstock derived from wild fry to provide large numbers of progeny for future release into the river while allowing continued natural production. We are using both pre-emergent fry collected from redds by hydraulic pumping and postemergent fry collected by electroshocking and seining. Given the recent average estimated escapements of adults to the river of less than 200 individuals and a predominantly 4-year life cycle, our annual goal is to sample at least 25 redds (=25 families) per year and to obtain a representative sample from the postemergent fry population in order to capture the genetic characteristics of the wild population in the captive broodstock. We aim to keep the effective population size of the captive broodstock at approximately 200 fish and to restrict inbreeding. In 1993, we successfully collected 3,872 pre-emergent fry from 14 redds and 1,598 postemergent fry. Although hydraulic sampling proved an effective method of collecting pre-emergent fry, we are concerned about the possible detrimental effects this sampling method may have on the survival of the fry remaining in the redds.

The families (groups) are being reared separately until the fish can be marked with group-specific tags. The tags will allow us to pool the groups for subsequent rearing and yet avoid within-group spawning (e.g., avoid brother–sister matings) and to monitor family performance and mortality. We will rear one-half of the broodstock entirely in freshwater and one-half in marine net pens, both to reduce the risk of complete loss of the broodstock due to site-specific mortality and to allow us to assess the relative merits of these two different approaches. The progeny of the captive broodstock will be 100% tagged and released as presmolts into the Dungeness River to allow a detailed assessment of harvest effects on this stock. The intended duration of captive broodstock production is 8 years (two 4-year brood cycles). It will take an additional 8 years to complete evaluation of the program.

A survey of the basin to identify limiting habitat factors is planned. Although habitat restoration and harvest management offer the only basis for sustained stock restoration, the captive broodstock program should protect the stock from the potentially devastating genetic effects of a severe population bottleneck and provide large numbers of fry with which to initiate rapid population rebuilding.

The Assessment of Marine Stock Enhancement in Southern California: A Case Study Involving the White Seabass

MARK A. DRAWBRIDGE AND DONALD B. KENT

Hubbs-Sea World Research Institute
2595 Ingraham Street, San Diego, California 92109, USA

MICHAEL A. SHANE AND RICHARD F. FORD

San Diego State University, Department of Biology
San Diego, California 92182, USA

Abstract.—The feasibility of enhancing depleted populations of white seabass *Atractoscion nobilis* has been investigated since 1983. During 1986–1993, more than 153,000 tagged juvenile white seabass were released into Mission Bay, California. The size of fish released varied from 31 to 317 mm total length (TL); the majority (55%) were between 80 and 120 mm TL. Oxytetracycline was used initially to mark fish prior to release. This marker was easy to administer but did not identify individual release batches and failed to persist for more than 4 years. Coded wire tags have been used since 1990. We found white seabass had a high rate and long duration of tag retention (>90% after 300 d).

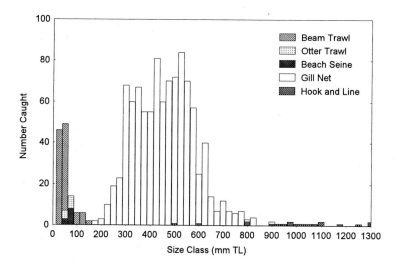

FIGURE 1.—Catch of white seabass using different gear types (not adjusted for effort).

Efforts to recapture stocked white seabass have employed different gear types, including 15.2-m beach seines, 1.6-m beam and 7.6-m otter trawls, 45.7-m gill nets, and hook and line (Figure 1). Sampling effort was focused within the Mission Bay release area and along the open coast to the north and south of the entrance to the bay. Gill nets were the most versatile gear type in that they could sample all white seabass habitats (including kelp beds). Each net consisted of replicate panels of different mesh sizes (12.7-, 19.0-, 25.4-, 31.7-, 38.1-, and 50.8-mm stretch mesh), which caught a wide size range of white seabass (150–850 mm TL, ages 1–6). Stocked white seabass recaptured in gill nets ($N = 39$) ranged from less than 1.0 to 4.5 years old, and periods at liberty extended over 4 years. The proportion of stocked white seabass collected in the Mission Bay gill-net survey has increased from 1.1 (1991) to 11.8% (1993), which demonstrates the cumulative effect of multiple releases and supports the idea that white seabass remain resident within the embayments for several years. The majority (79%) of these recaptured fish were originally released during April–July, the period of lowest stocking (Figure 2). This preliminary finding suggests that season of release may be an important factor in determining the rate of postrelease survival.

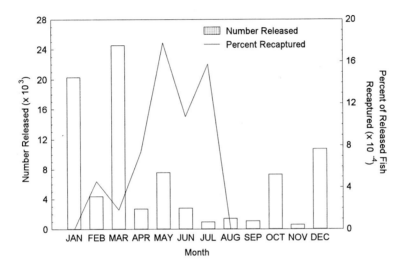

FIGURE 2.—Total number of white seabass released per month (bar) and the percent recapture rate of the total number released each month (line). Release and recapture data for each month are pooled for years 1990–1993.

An expanded release program was initiated in 1992 to stock greater numbers of larger (approximately 200 mm TL) white seabass over a greater geographic range from San Diego to Santa Barbara, California. A major objective of this expanded program is to identify the optimal size of fish at release and the optimal release sites and seasons. Coded wire tags will be used to mark fish, and gill nets and hook and line will be the primary collection gear.

Broodstock Development Plan for the Fluvial Arctic Grayling in Montana

WILLIAM P. DWYER

U.S. Fish and Wildlife Service
4050 Bridger Canyon Road, Bozeman, Montana 59715, USA

ROBB F. LEARY

The University of Montana, Division of Biological Sciences
Missoula, Montana 59812, USA

Abstract.—The Big Hole River fluvial Arctic grayling *Thymallus arcticus* is the last surviving remnant population of a fish that was once found throughout the Missouri River drainage upstream from Great Falls, Montana. The number of Arctic grayling in the Big Hole River has declined drastically in the last 10–20 years. One of the primary elements in the recovery process is to develop and maintain a captive broodstock at the U.S. Fish and Wildlife Service's Fish Technology Center in Bozeman, Montana. In order to accomplish this, wild fish have been captured by electrofishing at spawning time in May. Mature fish are then spawned, and the eggs taken to the Fish Technology Center for hatching and rearing. Development of an Arctic grayling broodstock will require specific safeguards in order to represent the entire population genetically and avoid a bottleneck that would decrease genetic diversity. Precautions being taken to prevent problems and to provide direction for the program include the following: (1) Big Hole River Arctic grayling will be spawned until an effective founding size of at least 50 fish have contributed to the captive reserve; (2) eggs from each mating will be incubated separately so that the number of parents successfully contributing to a year-class is known; (3) the contribution of each successful mating in a year-class should be kept nearly equal if possible; (4) different year-classes will be crossed in order to convert year-class genetic differences into within-population genetic diversity; and (5) to prevent broodstock from becoming highly adapted to the hatchery environment, gametes from Big Hole River fish will be infused at least every 10 years for 3 successive years.

In an attempt to meet the overall goals of this project, we have successfully spawned fish in the Big Hole River during the annual spring run in 1988 and 1990–1992. At the present time we are holding broodstock from each of these year-classes. All year-classes will be crossed in 1994 and 1995 (Table 1). We will attempt to get at least 100 spawning pairs. Each cross (year-class) should have an equal contribution to the final broodstock. Progeny from these crosses will represent a genetically complete broodstock that may be used to produce offspring to establish new populations, supplement the existing population, and serve as a genetic reserve.

TABLE 1.—In 1994–1995, the modified diallel crosses will be made using the 1988 year-class from 1992 matings and the 1990, 1991, and 1992 year-classes from the captive broodstock.

Male year-class	Female year-class			
	1988[a]	1990	1991	1992
1988[a]	0	X	X	X
1990	X	0	X	X
1991	X	X	0	X
1992	X	X	X	0

[a]The 1988 year-class, which was stocked into Axolotl Lake, Montana, is represented by a 1992 spawning of those fish. A portion of the progeny of 50 pairs has been retained as broodstock.

Interactive Effects of Stocking-Site Salinity and Handling Stress on Short-Term Survival of Striped Bass Stocked in the Savannah River, Georgia–South Carolina

JULIE E. WALLIN AND MICHAEL J. VAN DEN AVYLE

National Biological Service, Georgia Cooperative Fish and Wildlife Research Unit
Daniel B. Warnell School of Forest Resources
University of Georgia, Athens, Georgia 30602, USA

Abstract.—Efforts to restore populations of striped bass *Morone saxatilis* along the Atlantic and Gulf coasts have included stocking in estuaries, which requires consideration of salinity regimes when selecting stocking sites. The documented benefits of adding salts to reduce handling stress suggest that there could be variation in poststocking survival when fish are stocked in waters with different salt content. This variation could be erroneously attributed to differences in food availability, predation, or other stocking-site features, thereby leading to inaccurate conclusions regarding long-term survival rates. Our study was conducted to evaluate the effects of handling stress and stocking-site salinity on survival of striped bass fingerlings during the first 48 hours after stocking.

Experiments were conducted at freshwater and brackish water (0–7‰ salinity) sites and three size-classes of cultured fingerlings were evaluated: phase I (15–25 mm total length [TL]), advanced phase I (50–80 mm TL), and phase II (180–225 mm TL). Fish were held in floating cages at the stocking sites, and survival was determined at 48 hours. Survival was evaluated for fish that experienced three different prestocking handling procedures: minimal handling, routine handling, and microtagging. Minimally handled fish were transported from undrained ponds directly to the stocking sites at low densities (0.6–20 g/L). Routinely handled fish were transported from drained ponds to hatchery raceways, held 12–36 hours, loaded onto trucks (densities 60–90 g/L), and transported to stocking sites. Microtagged and anchor-tagged fish were handled like routine fish except for tag application. All fish were transported in 5 g NaCl/L and 0.1 g Ca^{++}/L. Stress levels of advanced phase-I and phase-II fish were evaluated prior to stocking by measuring serum cortisol concentrations. Phase-I fish were too small for cortisol evaluation.

Stocking-site salinity and handling prior to stocking significantly affected survival of striped bass. For the two larger size-classes, minimally handled fish had survivals ranging from 97–100% at both sites, which was significantly higher than survival for routinely handled fish. Routinely handled fish and tagged fish had relatively high survival when stocked at brackish water sites (survival 84–98%) but significantly lower survival when stocked at freshwater sites (survival 15–50%). For the smallest size-class, survival was significantly higher for fish stocked in brackish water (survival 80–98%) than for fish stocked in freshwater (survival 1–9%), regardless of handling prior to stocking. Prior to stocking, minimally handled advanced phase-I fish had lower cortisol levels (mean = 22.0 μg/dL) than did fish that were routinely handled or microtagged (mean = 35.1 μg/dL), indicating higher stress levels among tagged fish. For phase-II fish, there was no significant difference in cortisol levels between handling treatments (means = 9–11 μg/dL).

Short-term survival of stocked fish could be enhanced by stocking at brackish sites or by minimizing handling prior to stocking. Stocking densities should be adjusted to account for short-term mortality in order to accurately evaluate stocking success.

Restoration of the Savannah River Striped Bass Population

MICHAEL J. VAN DEN AVYLE AND JULIE E. WALLIN

National Biological Service, Georgia Cooperative Fish and Wildlife Research Unit
Daniel B. Warnell School of Forest Resources
University of Georgia, Athens, Georgia 30602, USA

CARL HALL

Georgia Department of Natural Resources, Wildlife Resources Division
Richmond Hill, Georgia 31324, USA

Abstract.—The abundance of adults of striped bass *Morone saxatilis* and number of eggs spawned in the Savannah River, Georgia–South Carolina, declined by about 95% in the 1980s because of habitat alterations at the estuarine spawning grounds. Operation of a tide gate and attendant channel modifications increased salinity at striped bass spawning sites and accelerated the seaward transport of eggs and larvae to areas with toxic salinity levels. In 1989, a goal of reestablishing a self-maintaining striped bass stock was established, and programs aimed at restoring suitable spawning habitat and a self-sustaining population were implemented. The tide gate was taken out of operation in 1991, and a canal that diverted eggs and larvae to saline habitats was blocked in 1992.

Striped bass fingerlings have been released annually since 1990 at estuarine stocking sites. Fish were cultured using wild broodstock from the adjacent Ogeechee River, Georgia, because adequate numbers of broodfish could not be obtained in the Savannah River. Size-classes and numbers of fingerlings stocked annually in 1990–1993 were (1) phase I (15–30 mm total length [TL]), 56,000–196,000; (2) advanced phase I (50–80 mm TL), 40,000–143,000; and (3) phase II (175–250 mm TL), 14,000–35,000. All were marked with micromagnetic coded wire tags or internal anchor tags or by immersion in an oxytetracycline solution to allow discrimination between stocked and wild fish.

Effects of stocking efforts were monitored annually by electrofishing at the spawning grounds during January–April and trawling for juveniles during July–September. Sampling effort consisted of about 100 hours of electrofishing (pedal time) and 64–148 trawl samples each year. Catch per unit effort (CPUE) with each gear was determined separately for naturally spawned and stocked year-classes.

Striped bass abundance increased following the stocking program. Electrofishing CPUE of subadult and adult fish (>1 kg) increased from about 0.2 fish/h in 1989 to 2.0 fish/h in 1992–1993, approaching levels similar to those recorded in the early 1980s. Stocked fish represented 77–94% of all juvenile striped bass collected in trawls since 1990 and 62% of 582 fish captured by electrofishing in 1993, indicating that the present population is composed of mainly stocked fish.

Restoration will be considered successful when survival of naturally spawned year-classes is sufficient to provide enough spawners to maintain the population. Stocked fish are beginning to reach sexual maturity and are expected to be capable of spawning beginning in 1994. Egg production and recruitment will be monitored to determine if successful reproduction occurs. Annual stocking efforts will continue until rates of natural recruitment at age 2 equal or exceed those resulting from stocking efforts.

Genetic Contribution of Hatchery Fish to Walleye Stocks in Saginaw Bay, Michigan

THOMAS N. TODD

National Biological Service
Great Lakes Center, Ann Arbor, Michigan 48105, USA

ROBERT C. HAAS

Department of Natural Resources
333135 South River Road, Mt. Clemens, Michigan 48045, USA

Abstract.—Stocks of walleye *Stizostedion vitreum* were severely depressed in Saginaw Bay in the 1970s. In 1979, the Michigan Department of Natural Resources began intensive stocking of walleye fingerlings to bolster fish populations (Figure 1). Subsequent to stocking, the walleye fishery has recovered. The study objective was to determine if recovery was due to the stocking program or natural reproduction. Inherent genetic differences between hatchery fish and endemic walleyes were used to determine the effect and contribution of hatchery fish to Saginaw Bay. Horizontal starch gel electrophoresis of muscle and liver tissue was used to assay the genotypes of walleyes sampled during spawning runs in the Tittabawassee River tributary, as well as the genotypes of pond-reared walleye fingerlings derived from eggs obtained from spawning walleyes in the Muskegon River tributary of Lake Michigan. Four of the 21 scored loci exhibited polymorphisms: *ADH-1* (which codes for the enzyme alcohol dehydrogenase); *sIDHP* (isocitrate dehydrogenase); *sMDH-3* (malate dehydrogenase); and *PROT-3* (general [unidentified] protein). Sampling was conducted during the spawning runs of 1983–1985, 1987, and 1988.

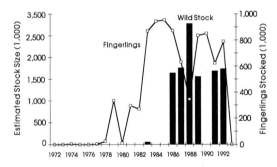

FIGURE 1.—Walleye fingerlings stocked in Saginaw Bay and estimates of the number of adult walleyes in Saginaw Bay, 1972–1993.

Allele frequencies of walleyes in the 1983 spawning run were similar, but not identical, to those of the pond-reared walleyes that had been stocked in Saginaw Bay, and the frequencies of alleles at *sMDH-3** and *PROT-3** deviated significantly ($P < 0.05$) from random mating expectations (Table 1). Allele frequencies were increasingly less like those of stocked walleyes in subsequent spawning runs and varied from year-class to year-class as well. The deviations from random mating expectations in walleyes from the 1983 spawning run resulted from a mixture of hatchery and endemic walleyes. Only the 1983 spawning run, which was dominated primarily by 1979 year-class walleyes on their first spawning run, was composed of almost entirely first-generation stocked walleyes as well as wild walleyes endemic to the system. Subsequent spawning runs potentially were constituted of not only stocked and endemic walleyes but also natural hybrids and backcrosses between the two. Such a genetic mixture of walleyes has undoubtedly obscured the contribution of first-generation stocked walleyes to spawning runs, but the rapid return of allele frequencies to random mating expectations after 1983 suggests that extensive natural reproduction has occurred. Stocked fish contributed a proportionally large number of walleyes to spawning runs in the first few years of the stocking program, but the trends in allele frequencies indicate that natural reproduction by walleyes with genotypes differing from those of stocked fish was probably outstripping input from the stocking program. Although stocked walleyes contributed little to the spawning runs, they might have contributed more to the fishery than the data suggest if, for example, they did not run up the Tittabawassee River in proportion to their abundance or did not reproduce as successfully as did wild fish.

TABLE 1.—Frequency of the common allele at four polymorphic loci in walleyes from the Muskegon and Tittabawassee rivers, Michigan.

Locus	Muskegon River	Tittabawassee River				
		1983	1984	1985	1987	1988
ADH-1*	0.96	0.89	0.84	0.85	0.90[a]	0.90
sIDHP-1*	0.63	0.78	0.75	0.74	0.77	0.76
sMDH-3*	0.73	0.81[a]	0.68	0.76	0.74	0.72[a]
PROT-3*	0.11	0.15[a]	0.19	0.20	0.26	0.23

[a] Denotes significant departure from random mating expectations ($P < 0.05$).

Homing Propensity in Transplanted and Native Chum Salmon

WILLIAM W. SMOKER

University of Alaska, Juneau Center for Fisheries and Ocean Science
11120 Glacier Highway, Juneau, Alaska 99801, USA

FRANK P. THROWER

National Marine Fisheries Service, Auke Bay Laboratory
11305 Glacier Highway, Juneau, Alaska 99801, USA

Abstract.—Homing to the natal stream and the converse, straying (which may lead to gene migration), are important to the adaptedness and fitness of local salmon populations. Homing is the basis of reproductive isolation; it enables salmon populations to adapt to particular local environments and to maximize evolutionary fitness and economic productivity. Homing is thought to be in part genetically based. Bams (Bams, R. A. 1976. Survival and propensity for homing as affected by presence or absence of locally adapted paternal genes in two transplanted populations of pink salmon [*Oncorhynchus gorbuscha*]. Journal of the Fisheries Research Board of Canada 33:2716–2725) observed survival and homing of transplanted and of hybrids between transplanted and native pink salmon *Oncorhynchus gorbuscha*. They survived oceanic life equally well but some transplanted pink salmon apparently failed to enter the natal river, and a disproportionate number of hybrids failed to transit the natal river completely.

Transplanted salmon produced by enhancement programs would therefore be expected to stray more readily than do native salmon. Hybrids of native and nonnative salmon, either purposefully or inadvertently produced by straying of hatchery salmon into local populations, would be expected to stray more readily. Hybridization, therefore, could have deleterious effects, presumably by disrupting adaptive gene complexes.

We tested the hypothesis of increased straying by transplanted fish in chum salmon *O. keta* near Juneau, Alaska, by searching for strays in streams in the region of their natal stream. We transplanted gametes and spawned native, transplanted, and hybrid groups (Figure 1). We increased the power of Bams' design by transplanting from two stocks, by replicating the experiment in two brood years, and by observing native as well as transplanted and hybrid fish. We tagged (coded microwire tags) and released approximately 20,000 age-0 fry in each group each year, brood years 1981 and 1982.

Straying in transplanted chum salmon could not be detected. We recovered no strays among transplanted chum salmon and only one among hybrid chum salmon (Table 1; Figure 1). Our criterion for identifying a stray was liberal, requiring only presence in the stream, not actual spawning. Recovery efforts were intense. Three likely destinations of strays were monitored by weirs at which every returning chum salmon was examined in all 3 years of return. At the natal stream there was exhaustive sampling, and at a fourth likely destination of strays more than 20% of the run was directly examined. Relative survival (recovery) rates of groups are not pertinent to our conclusion that straying could not be detected. These rates varied significantly between groups, but the two brood years' experiments had different patterns. Survivals were probably affected by uncontrolled differences in the treatment of the groups and aren't readily interpretable as reflecting effects of transplantation or hybridization. (In a brood year 1983 experiment, two replicate groups similar to those reported here and treated alike with the exception of different tag codes were recovered at significantly different rates, suggesting that large observational error was associated with estimates of treatment effects on survival in the 1981 and 1982 brood years.) Differential recovery is probably not explained by differential straying outside our recovery region: (1) in each cohort relative recovery rates in local fishing districts did not vary from recoveries at Salmon Creek, and (2) there were no recoveries from distant fishing districts.

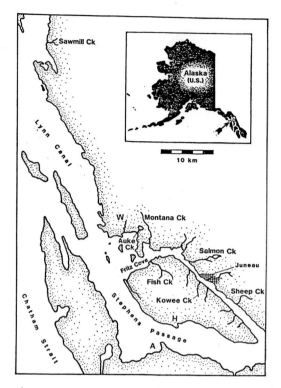

FIGURE 1.—Chum salmon gametes from Sawmill, Montana, and Salmon creeks were spawned and released from Salmon Creek in two brood years (1980, 1981). Returning salmon migrate north via Chatham Strait, southeast via Stephens Passage to Fritz Cove, then either east or counterclockwise around Douglas Island to Salmon Creek. Adults were examined (1) at weirs (every fish) at Auke, Kowee, and Sheep creeks, (2) by intense random sampling at Fish and Salmon creeks, and (3) by carcass surveys at Sawmill, Montana, Wadleigh (W), Hilda (H), and Admiralty (A) creeks.

That hybrids of native and transplanted salmon did not stray is similar to Bams' finding: pink salmon hybrids homed successfully to the mouth of the natal stream. Bams, however, estimated that 75% of transplanted pink salmon (no native parentage) strayed to other streams. If the number of transplanted salmon recovered at Salmon Creek had represented only 25% of returns, we would have expected over 700 to have strayed into local streams; we found none. We transplanted chum salmon about 65 km, and Bams transplanted pink salmon more than 200 km, so, perhaps, we did not transplant chum salmon far enough to disrupt their homing propensity. However, it seems more likely that homing is a complex ability that varies among species and populations and according to environmental circumstances, and that even transplantation does not necessarily disrupt it.

TABLE 1.—Tagged chum salmon recovered over three return years (1984, 1985, 1986) in random samples and carcass surveys at Salmon Creek from a total return over 10,000. Numbers recovered should not be interpreted to infer relative survival of native, transplanted, or hybrid salmon. No coded microwire tags were observed among carcasses surveyed at Sawmill Creek (120 surveyed), Montana Creek (64), Wadleigh Creek (160), Hilda Creek (1,670), or Admiralty Creek (2,620). No tags were observed among returns to weirs at Auke Creek (5,161 total return), Kowee Creek (2,001), or Sheep Creek (1,939). One tagged chum salmon, a hybrid between Salmon Creek and Sawmill Creek parents, was recovered from a sample of 4,548 at Fish Creek.

Origin	Number recovered
Salmon Creek native	200
Montana Creek transplant	103
Sawmill Creek transplant	139
Hybrids	169

Advances to Increase Restocking of Native Fishes in Lake Patzcuaro, Mexico

CATALINA ROSAS-MONGE

Dirección General de Pesca
Gobierno del Estado de Michoacán, Michoacán, México

ARTURO CHACON-TORRES

Universidad Michoacana de San Nicolas de Hidálgo, Escuela de Biología
Michoacán, México

Abstract.—Because of its ecological, historical, cultural, and economic values, Lake Patzcuaro represents one of the most important water bodies in Mexico. For centuries local fisheries have been based upon capture of the pez blanco *Chirostoma estor* (a Mexican silverside), a fish with the highest market value in the country (up to US$25/kg). The charales (several species of the same genus but smaller in size) and the acumara *Algansea lacustris* are other commercially important components of the unique and endemic fish fauna, as is the amphibian achoque *Ambystoma dumerilii*. However, the fishery has deteriorated as a result of deforestation of the watershed, cultural eutrophication, overfishing, and introduction of nonnative species, including largemouth bass *Micropterus salmoides*, common carp *Cyprinus carpio*, grass carp *Ctenopharyngodon idella*, and tilapias *Tilapia* spp. Similar circumstances are increasingly apparent throughout Mexico and other Latin American countries.

The progressive decline in catch rates of native fishes prompted several attempts to spawn and culture them under controlled conditions. These efforts were unsuccessful. In 1985, a small area (approximately 5 ha) of the lake was set aside as a natural reserve. This area is closed to public access and allows investigation of processes necessary for ecological recovery and provides a site for culturing pez blanco and acumara. The culture method includes egg collection, fertilization, and incubation and fry rearing. Egg incubation and fry rearing are done in net cages suspended in the lake. The fry are reared at a density of 18,000/m^3 to a size of 5 cm. We obtained 80% hatching success and 90% juvenile survival during the first 2 years of work. We will continue and refine the initial egg incubation and fry rearing practices; fish will be reared to larger size (10–15 cm) in earthen channels adjacent to the lake and will be fed live and prepared feeds.

Is Genetic Change from Hatchery Rearing of Anadromous Fish Really a Problem?

REGINALD R. REISENBICHLER

National Biological Service, Northwest Biological Science Center
6505 NE 65th Street, Seattle, Washington 98115, USA

GAYLE BROWN

National Biological Service, Washington Cooperative Fish and Wildlife Unit, School of Fisheries WH-10
University of Washington, Seattle, Washington 98195, USA

Abstract.—Data suggest that genetic change caused by artificial propagation has the potential to reduce the fitness of naturally spawning (wild) populations of anadromous salmonids when hatchery fish interbreed with wild fish, even when hatchery programs are designed to cause minimum genetic difference between hatchery and wild fish. Several studies show domestication selection in the hatchery (natural selection for fish that do well in the hatchery environment) and a coincident loss in genetic fitness for rearing in natural streams and suggest that only one generation in the hatchery environment may be prob-

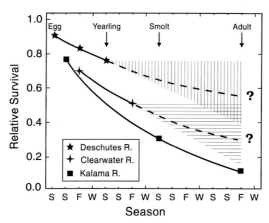

FIGURE 1.—Survival of hatchery steelhead *Oncorhynchus mykiss* relative to wild steelhead when reared together in natural streams. Data are from three different studies in three different river systems: Deschutes River, Oregon (Reisenbichler, R. R., and J. D. McIntyre. 1977. Genetic differences in growth and survival of juvenile hatchery and wild steelhead trout, *Salmo gairdneri*. Journal of the Fisheries Research Board of Canada 34:123–128; Kalama River, Washington (Leider, S. A., P. L. Hulett, J. J. Loch, and M. W. Chilcote. 1990. Electrophoretic comparison of the reproductive success of naturally spawning transplanted and wild steelhead trout through the returning adult stage. Aquaculture 88[3–4]:239–252); Clearwater River, Idaho (Reisenbichler 1994, unpublished data). The data from the Kalama River are averages from four year-classes and the two methods of calculating relative survival. Curves were fitted by eye and the shaded areas represent boundaries for reasonable estimates of survival to maturity where actual data were not available.

lematic (Figure 1). Other studies, and simple observation, indicate that environmental conditions in hatcheries fall far outside the range of conditions encountered in natural streams. On this basis, we suggest that hatchery rearing over several generations also may disrupt coadapted genetic systems for rearing in natural streams (Figure 2). When such hatchery fish interbreed with wild fish, reduced fitness (outbreeding depression) results and may persist for many generations. A simple, conceptual model illustrates how even modest reductions in fitness can lead to extirpation of wild populations, especially those also

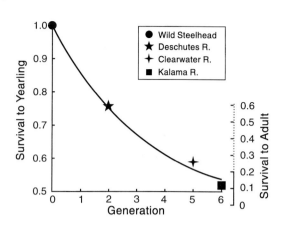

FIGURE 2.—Survival of hatchery steelhead relative to wild steelhead when reared together in natural streams, versus generations the hatchery population has been in the hatchery. Values for relative survival to yearling were from Reisenbichler and McIntyre (1977; Deschutes River, Oregon) or were interpolated from Leider et al. (1990; Kalama River, Washington) and Reisenbichler (1994, unpublished data; Clearwater River, Idaho). Values for survival to returning adult were from Leider et al. (Kalama River), or were extrapolated from the other two studies (see Figure 1). The value for zero generations is 1.0 by definition.

affected by environmental degradation or intensive fishing. Existing data, taken with the mandate to conserve populations of wild fish, are sufficient to require a cautious approach to artificial propagation in fisheries management. Such an approach should include protection of a substantial proportion of the viable wild populations from interbreeding with hatchery fish.

Development of a Regional Policy for the Prevention and Control of Nonnative Aquatic Species: The Chesapeake Basin Experience

DANIEL E. TERLIZZI

National Oceanic and Atmospheric Administration
University of Maryland Sea Grant Extension, Chesapeake Bay Program Office
410 Severn Avenue, Suite 107-A, Annapolis, Maryland 21401, USA

RONALD J. KLAUDA AND FRANCES P. CRESSWELL

Maryland Department of Natural Resources
Chesapeake Bay Research and Monitoring Division, Tawes State Office Building,
580 Taylor Avenue, Annapolis, Maryland 21401, USA

Abstract.—The discovery of veligers of zebra mussel *Dreissena polymorpha* in the upper Susquehanna River, the major tributary to Chesapeake Bay, the increased use of triploid grass carp *Ctenopharyngodon idella* in some states of the Chesapeake basin, and the initiation of *in situ* experiments with the nonnative Pacific oyster *Crassostrea gigas* were major factors stimulating the development of a basinwide policy for prevention and control of nonnative aquatic species. We present the policy developed by the Exotic Species Workgroup (EXSWG) of the Chesapeake Bay Programs Living Resources Subcommittee in response to the potential environmental threat posed by nonnative aquatic species. Workgroup members include scientists, policy staff, and resource managers from the Chesapeake Bay Program jurisdictions (Maryland, Pennsylvania, Virginia, and the District of Columbia) with representation from the nonsignatory basin states (Delaware, New York, and West Virginia). The Chesapeake Bay Policy for the Introduction of Nonindigenous Aquatic Species was approved by the policy advisors to the jurisdictional leaders and forwarded to the jurisdictions for signature on 2 December 1993. This regional policy was adopted and is now being used as a guide for intentional and nonintentional introductions.

This policy includes three important provisions. First, species approved by individual jurisdictions at the time of adoption for aquaculture or stocking are grandfathered by the policy and their introduction can continue as part of jurisdictional fisheries management. Other species will be considered first-time introductions. First-time introductions include (1) species that are nonnative or not naturalized; (2) species that a jurisdiction has not previously regulated; or (3) previously approved species for which changes in scope, including changes in stocking rates and locations or culture methods, might significantly increase risk of escapement. A second provision requiring applications for first-time introductions into a jurisdiction will be submitted to multijurisdictional review by an ad hoc Technical Review Panel consisting of one representative each from the signatory jurisdictions plus two additional members appointed by the EXSWG after consultation with the Technical Advisory Committee (Figure 1). The role of the ad hoc panel is advisory. After deliberation, the recommendation of the panel as well as dissenting views are presented. Third, all jurisdictions agree to participate in development of an implementation plan to be completed by the EXSWG by 30 September 1994 that specifies protocols for introduction by different pathways (aquaculture, fisheries management, biological controls, research, and ballast water discharge), prevention, education, and control.

The implementation plan for the research introduction pathway has been addressed, in part, through the development of research protocols for the handling and containment of dreisseinid mussels. These protocols contain a list of information items that a researcher should provide to the appropriate jurisdictional agency as part of a permit application process. A risk assessment questionnaire should be completed by the state agency to determine if the risk of introduction is acceptable and if a permit can be issued to allow importation of live specimens for research purposes. The research protocols were developed to allow modification for use with other nonnative aquatic species. Recently, research

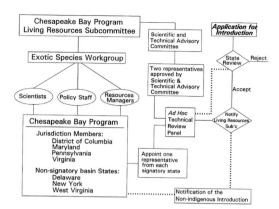

FIGURE 1.—Flow chart illustrating the *ad hoc* technical review process. Dashed lines indicate procedure following technical review.

protocols were employed to issue permits to several Maryland laboratories to conduct studies on key environmental parameters and behavioral characteristics of dreisseinid larvae.

AFS Section Position Statements

At the opening symposium session, representatives from eight of the nine sponsoring American Fisheries Society Sections delivered statements regarding the uses and effects of cultured fishes in aquatic ecosystems from the unique perspective of the discipline they each represented. As new information was presented during the week, the position statements evolved and were refined. At the closing symposium session, Section representatives presented their revised position statements. These documents were taken back to the Sections, presented to Section members, and approved after a comment period.

FISH CULTURE SECTION

Presented by Kirby D. Cottrell, Section Delegate

The Fish Culture Section includes a broad variety of fisheries professionals from public, private, private nonprofit, tribal, and educational programs. Hatchery production is a tool used by fisheries management entities as an alternative for portions of the natural life cycle of fishes and as a means to contribute to management goals for a wide variety of users. Fish are reared for release in a variety of ecosystems including freshwater, estuarine, and marine environments which range from aquaria in public schools to a hatchery for chum salmon *Oncorhynchus keta* above the Arctic Circle. The Fish Culture Section promotes and advocates the propagation of the highest quality of fishes possible and development of advanced technologies to meet regulatory, legislative, environmental, socioeconomic, and fisheries management expectations. Fish production is guided by information regarding species, strain, size, sex, body condition, health, genetics, physiology, and ecology.

Presentations during this symposium have demonstrated value and wise use of fish hatcheries and their products.

- In areas where suitable environmental conditions are restored and recovery of wild or natural populations is a goal, artificial propagation may be used in conjunction with other rehabilitation measures to assist in recovery.
- In areas where habitat has been permanently lost, environmental conditions cannot sustain natural populations, or the cultured fish will not substantively interact with natural populations, artificial propagation may be used to establish and maintain fisheries or replacement populations.
- In areas where wild stocks are healthy, cultured fish should be used only if the program is designed to have negligible effects on healthy wild stocks.
- Fisheries enhancement can establish new fishing opportunities in certain areas when such enhancement is consistent with overall fisheries management policies regarding native stocks.
- Overharvest of natural stocks in the presence of abundant hatchery fish is due either to improper release locations or production levels or to our inability to design and implement a selective harvest strategy to discriminate hatchery fish from wild fish. We must continue to provide means to identify hatchery-produced fish and differentiate them from wild stocks.
- All recommended guidelines for appropriate stocking should also apply in aquaculture situations where escapes are likely.
- We must meet the needs of the target species while safeguarding nontarget species associated with the aquatic ecosystem.
- Fish culture facilities and their production have intrinsic value in educational outreach and as sites where scientific knowledge about fish can be enhanced and promoted.
- Private aquaculture produces at least 70% of the fish cultured in the United States. We must continue to provide guidance to the aquaculture community through Fish Culture Section policy, technical information, and activities.

Hatchery policies should provide guidance for hatchery operation once production numbers are established. Production operations must support a comprehensive plan, completed by fisheries managers, for natural and hatchery fish production.

The Fish Culture Section is proud of our members' technological and philosophical achievements in producing a wide variety of fish culture products. We accept the continuing challenge to evolve our science to produce cultured fish that survive and contribute to our fishery user requirements without adversely affecting the aquatic ecosystem. The science of fish culture through its ability to solve technical problems is becoming more goal oriented. The Fish Culture Section does not advocate indiscriminate release or use of cultured fishes. We strongly support the recognition of and management for wild stocks as part of a balanced fisheries management program.

The Fish Culture Section is supportive of the effort for total ecosystem management and believes that we, together with our fellow professional disciplines, must develop compatible goals and objectives for the production and use of fish culture products. All fisheries disciplines must fully understand the goals and objectives related to the use and production of cultured fishes. We also recognize that public and resource shareholders have every right to make value judgements of where and where not to release fish.

Given our current knowledge, the role of fish culture can be succinctly summarized as tell us

- what you want,
- how you want it to look,
- how you want it to act,
- how you want it constructed (genetics),
- where you want it put,
- when you want it put there, and we'll do it!

We commend the organizers and presenters of this symposium and applaud their ideas, insights, and efforts to foster a strong union among the various fisheries disciplines.

FISHERIES MANAGEMENT SECTION

Presented by Harold L. Schramm, Jr., President

Fisheries managers are charged with stewardship of aquatic resources. Accomplishing this responsibility requires making decisions that are biologically sound, economically desirable, and socially acceptable. The difficulty lies in balancing biologic, economic, and sociologic ideals with ethical and moral stewardship responsibilities. As stewards of aquatic resources, we have responsibilities to maintain, enhance, and conserve resources that supersede short-term social and economic values. However, we also have responsibilities to recognize social needs and economic realities.

We need agreed-upon criteria to identify situations in which stock conservation is appropriate, as well as circumstances for which it is appropriate to stock nonnative fishes. These criteria must then be translated into operational fisheries management guidelines.

There are numerous goals for managing aquatic resources. For each system, fisheries managers are in a position to identify those goals and choose the appropriate tools to be used to achieve the objectives. We suggest that identification of the most appropriate goals can be done best at the local level; there is no one management goal that is best for all habitats or even all habitats of a given type. In prioritizing the management objectives and selecting the best one, we must always consider the ecological risks and the societal benefits. As we manage the aquatic resources to achieve the specified objectives, we must continually evaluate the effects, both positive and negative, of the management activity and always be ready to make appropriate changes.

Meanwhile, on the eve of the Denver meeting, the Fisheries Management Section strives to forge criteria that can be adopted to guide our actions and the actions of conservation agencies. We do recognize the following principles: (1) healthy habitat is key to maintaining diverse and productive fish stocks; (2) fish and wildlife agencies must work to ensure that habitat and all that depends on it is not sacrificed on the altar of money and politics; (3) stock conservation concepts are an appropriate component of management plans; (4) fish live where they do because of hereditary, biological, and environmental influences; and (5) we have to understand all public desires, instill a sound resource philosophy in the public, and then integrate that public awareness with biological technology to move forward in fisheries management. The challenge for those who meet in Denver will be to translate these principles into suggested protocols to apply to uses of cultured fish in successful fisheries management.

Genetics Section

Presented by Fred M. Utter, President

Perceptions of the appropriate role of hatcheries and cultured fishes have shifted from time to time over the past century in fisheries management. The attitude that hatcheries were the primary solution to problems of maintaining productivity in the face of continually declining natural populations, faded when better understandings of stock structures coincided with aroused concerns for identifying and preserving natural populations. Conflicting goals of cultural and conservation interests resulted in a polarization that lingers to some degree into the present and is represented by "hatchery bashers" versus "fish dumpers" at the extremes.

This symposium reflects an emerging middle ground based to a large extent on a broadly shared concern for better understanding and application of genetic and ecological principles required for realistically dealing with the inherent incompatibilities reflected in the polarized extremes. The extensive genetic motivation underlying this moderated perspective is apparent in the titles of many of the presentations offered during this symposium. The days of lip service to the black box of issues within the vague and general framework of genetics appears to be passing. Informed fisheries personnel are becoming increasingly interested in, and armed with, tools and theory that were largely unavailable just a few years ago. As a result, they are capable of taking actions on problems that were previously unsolvable.

Given this apparent enlightenment, what role can hatcheries play in optimizing ecosystem yield over the long term, in a manner that is not detrimental to valuable and irreplaceable native populations and the genetic resources they represent? The key lies in sequential categories of activity that require joint participation of fish culture and conservation interests. The native populations must be identified in order to conserve them adequately. Fulfillment of this requirement depends on different kinds of information. Field studies are needed to verify the existence and ecological parameters of natural populations. In addition, laboratory studies are necessary to identify the geographic patterns of indigenous ancestral groupings and the presence of exogenous or introgressed populations resulting from populations introduced or altered through cultural or other anthropogenic activities. The genetic and ecological relationships among hatchery, native, and other naturally reproducing populations must also be determined. Information from these studies will permit development of strategies to optimize the protection of native populations while supporting fisheries through hatchery operations.

No stocking should be permitted without adequate biological assessments followed by development of appropriate management and protective strategies to establish the necessary foundation for subsequent activities. The potential range of such activities is extensive and largely situation dependent. At one extreme, maintenance of large hatchery broodstocks to support extensive releases of exogenous hatchery fish may continue if monitoring—a critical and difficult element of fisheries management to protect resources—failed to detect measurable effect on healthy native populations of a region. Even here, the development of broodstocks from fish maturing naturally under local conditions is desirable to maximize productivity.

At the other extreme, an irreplaceable native population found to be increasingly imperiled by past and existing hatchery practices may require that all releases be stopped and, where possible, a program to eradicate established nonnative fish undertaken. The replacement of nonnative hatchery populations with those derived from indigenous fish may be desirable provided guidelines for hatchery maintenance are followed to prevent excessive genetic losses through such processes as inbreeding and domestication. Assurances based on adequate empirical tests are needed so that any new activity will not have negative genetic effects on native or naturally reproducing populations.

The membership of the Genetics Section reflects a diversity of backgrounds and interests. The Section supports a cooperative rather than a polarized approach to the interfacing of conservation and cultural activities based on a commitment to shared responsibilities. Cooperation coupled with adequate and timely knowledge of underlying genetic issues will permit effective responses to the inevitable problems that will arise in the future.

Early Life History Section

Presented by Fred P. Binkowski, Section Delegate

The Early Life History Section (ELHS) strongly supports continued and expanded fundamental and strategic research on the uses and effects of cultured fishes in aquatic ecosystems. The Section also supports the use of scientific findings to improve the understanding of associated mechanisms and ecological issues and to solve problems associated with this topic.

In principal, the ELHS agrees that the stocking of hatchery fishes can be an effective tool for the enhancement, rehabilitation, and restoration of fish stocks; but such efforts must consider all the essential biological and physical factors, and specific objectives must be formulated within the context of the aquatic ecosystem into which the fishes are stocked. In certain situations, benefits of stockings to support high-demand urban recreational fisheries in park ponds and lagoons may overshadow possible risks to fish stocks already effected. In these cases put-and-take fishing can be an appropriate objective. In situations with more pristine wild stocks, no stockings may be the best policy. Between these extreme situations, each specific situation can fit into a tiered approach to the use of stocking for enhancement and restoration.

Moving toward less reliance on stocking for enhancement will be appropriate in some of these situations. Promoting other management strategies such as catch and release, habitat improvement, pollution control, stricter regulation, and public education can help decrease reliance on stockings. Another approach would be to deemphasize use of wild fisheries for human consumption and replace or supplement fishes for human consumption with aquaculture products. Improvement in the quality of stocked fishes through modifications in rearing techniques may improve adaptability of cultured fishes to wild systems.

In cases where rehabilitation and restoration are attempted through the use of introductions of fishes in early life stages, the ELHS concerns are what effect will these (eggs, sac-fry, or larvae) have on the wild resource and community structure and how can the methods and results of such programs be evaluated?

The stocking of fishes in early life stages generally has fallen out of favor as an enhancement strategy, but it is sometimes advocated for introduction or restoration situations, especially when important early life history events such as olfactory imprinting to natal grounds are considered. Whether or not such stockings can benefit recruitment has been difficult to establish. Some of the symposium presentations addressed such problems. Recent developments in otolith analysis and marking offer greater hope for evaluation of early life stage stockings and their possible influences on wild stock recruitment. Most of the rearing mortality in a cohort of fish intended for rehabilitation or restoration usually occurs during early life stages. Accordingly, the greatest influence of unintended selection due to the rearing environment may occur during these phases of domestication in cultured fishes. A better understanding of the types of selective mortality that occur during the early life stages of fishes in culture situations as compared with wild systems is needed. Is it possible that this early effect has as much or more unintentional selection due to broodstock selection and population size? How do survival conditions for early life history stages in hatcheries restrict and how do they relax the expression of genetic variability in the parental stocks into the next generation? How can rearing conditions result in propagation of individuals that have fitness in aquatic ecosystems and still produce enough numbers to restore or rehabilitate stocks? Much remains to be learned about the process of domestication, even for fishes that have had long histories of cultivation. For certain of our Threatened and Endangered fishes, we are totally dependent on wild broodfish, and techniques for intensive hatchery rearing of large numbers of fish are yet to be developed and applied. We believe it would be unreasonable not to explore and develop such techniques, although some might prefer to rely solely on habitat management and protection for restoration and rehabilitation efforts. The application of sound broodstock management will be an important component of the development of such techniques.

The brunt of concern at this symposium has focused on broodstock, genetics, population size, and management, and the early life history aspects of the rearing process have received less emphasis. The ELHS believes that further examination of factors influencing early life history survival and fitness in both hatchery situations and natural aquatic ecosystems will be required in order to construct effective restoration and rehabilitation strategies.

FISHERIES ADMINISTRATORS SECTION

Presented by Fred Harris, Section Delegate

The Fisheries Administrators Section is composed of administrators of federal, state, provincial, and private fisheries programs. Section members administer a variety of fisheries programs, many of which involve the use of cultured fish.

The Section views cultured fish as one component of fisheries management. As such, stocked fish are a management tool. Effective use of this tool requires selection of the right species for introduction into the right water to accomplish the right management objective. In all instances, fish stocking practices should be consistent with resource stewardship responsibilities.

The science associated with rearing and using cultured fish is advancing. Fisheries agencies have changed fish culture and fish stocking practices in response to advances in knowledge. Additional changes will occur as more knowledge becomes available. Currently, the full magnitude of issues associated with the use of cultured fish is undefined. As a result, the development of a comprehensive nationwide policy on the use of cultured fish is probably premature.

An immediate action that should be accomplished is the identification of issues associated with the use of cultured fish for which sufficient data exist to reach general agreement on appropriate practices. The development of resultant policies or guidelines should incorporate representatives of affected user groups as well as management agencies. Close cooperation and clear communication among government agencies, user groups, universities, and other interested parties will be required for the successful accomplishment of this task.

A longer-term need is the incorporation of additional data into the body of knowledge used to make decisions concerning the use of cultured fish. In some cases, data may be available but not widely known or accepted. In other cases, data may be lacking. The American Fisheries Society, using its Fisheries Action Network, is capable of effectively addressing both cases and should initiate actions to do so.

This symposium on Uses and Effects of Cultured Fishes in Aquatic Ecosystems represented a valuable contribution toward the definition of the proper roles of cultured fish in management programs. The Fisheries Administrators Section extends its appreciation to all who were involved.

INTRODUCED FISH SECTION

Presented by Dennis R. Lassuy, President

As noted by Gil Radonski in the keynote address, most fishes used for introductions are from cultured sources. The perspective of the Introduced Fish Section (IFS) simply reiterates existing American Fisheries Society (AFS) guidelines: When cultured fishes are introduced (i.e., released, directly into open ecosystems or into facilities from which they may subsequently escape) beyond their natural range, the existing position statement of the American Fisheries Society should be followed in its entirety. The AFS position on introductions of aquatic species (Kohler, C. C., and W. R. Courtenay, Jr. 1986. American Fisheries Society position on introduction of aquatic species. Fisheries [Bethesda] 11[2]:34–38) is available from the American Fisheries Society.

During the course of the conference, two caveats developed. First, we recognize that "what's here is here," and, for the purposes of this conference only, limit the recommendations to introductions of "new" species (i.e., those not already present in the receiving ecosystem). Second, we recognize that other valid protocols exist that may be followed. As long as those protocols encompass the policy and decision-making approach of the existing AFS protocol, use of such protocols may also be acceptable.

The IFS also recognizes that many fisheries managers have been making an effort to incorporate the approach represented by the AFS protocol, but full implementation depends on recognition of the protocol's importance and validity by the public and their governing bodies who set the context in which fisheries managers work. The importance of such buy-in is part and parcel of the existing AFS protocol in its iterative calls for publicity and open review.

The IFS emphasizes that an understanding of the full life history of the species being considered for introduction and an assessment of the implications of the presence of such life stages on the receiving ecosystem is integral to proper implementation of the AFS protocol. In the interim between now and the Denver meeting, we will seek to identify a success story to exemplify what adherence to the AFS protocol means. The work of Paul Shafland in Florida is instructive in exemplifying what adherence to the AFS protocol means.

Although the IFS also recognizes the concerns expressed at this conference regarding the potential genetic implications of some introductions, we were not comfortable incorporating a specific comment on its importance into this statement of perspective and defer to the existing AFS position statement on transgenic fishes (Kapuscinski, A. R., and E. M. Hallerman. 1990. AFS position statement: transgenic fishes. Fisheries [Bethesda] 15[4]:2–4) and the deliberations of the Genetics Section at this conference for more detailed advice in this arena.

This IFS perspective represents the consensus of those IFS members who participated in deliberations at the Albuquerque meeting.

BIOENGINEERING SECTION

Presented by Susan Baker, President

The role of the Bioengineering Section is to provide a forum for American Fisheries Society members to exchange information, express views, report on activities, and discuss solutions to problems related to processes, procedures, techniques, and effects of fisheries bioengineering research and practice. The goal of the Section is to promote excellence in the field of fisheries bioengineering through improved communications between biologists, engineers, and other professionals. The Section provides the opportunity for tracking the latest developments in aquaculture systems, habitat protection, hydro mitigation, equipment development, and project design and operation.

The majority of Section members are professionals who work or are interested in one of the following areas of expertise:

- hatchery and aquaculture technologies;
- habitat protection and restoration;
- hydro mitigation and fish passage;
- water treatment and recirculation;
- pond culture and reservoir fisheries;
- mariculture and net-cage systems;
- research laboratories and aquariums;
- fishing gear and fish processing equipment and facility design; and
- related training and education programs.

On the whole, the Section is technology based, not policy based. However, as a group of professionals, we will work closely with our scientific peers to ensure that the physical facilities we design and build will integrate the necessary requirements to insure the wise use of cultured fishes. The hatchery facilities will provide the hatchery manager with the capability to manage in a manner that ensures the product meets the designated fisheries management needs. Over the last few days we have achieved a much better understanding of resource needs. Toward this end, the Bioengineering Section is willing and ready to accept the challenge of meeting these needs.

MARINE FISHERIES SECTION

Presented by Churchill B. Grimes, President

Artificial enhancement of marine fish stocks is not a new venture. In the United States and other countries, stocking to augment depleted fish stocks was commonly practiced without demonstrable success from about 1880–1950. During the last decade, Norway and several states—most notably California, Florida, and Texas—have implemented, or are contemplating, marine stock enhancement programs. The Marine Fisheries Section perception of stock enhancement has been changed by the proceedings of this conference from healthy skepticism to cautious optimism. If enhancement can be practiced with economic feasibility and ecological responsibility, it could have great practical value in managing marine fishery resources. With particular reference to the marine environment, the debate on uses and effects is simpler for those of us concerned with marine rather than freshwater and anadromous fishes, because it is mostly scientific and technical, at least at this time. Discussion is concentrated on if and how marine stocks can be enhanced to any practically useful degree. In contrast, in the freshwater environment managers must deal not only with the scientific and technical issues surrounding uses and effects, but the debate is very much exacerbated because it involves complex and difficult to resolve societal value issues that were discussed at some length during the meeting. The extreme points of view are represented, on one side, by the idea that policy on the use of cultured fishes should be guided by the desires of resource users and, on the other side, the idea that the policy should be determined by what is best for resources and ecosystems. It is difficult to balance these competing values. If stock enhancement becomes widespread in the marine environment, we will be involved in the same debate in the future. I hope we can profit from the freshwater and anadromous experiences. I will restrict our comments to the scientific and technical issues.

The large variations in annual recruitment characteristic of many marine fish populations, and the open nature of the systems they exist in, suggest that there may be considerable scope in ecosystems to allow the level of recruitment to be enhanced, at least during years of poor recruitment. However, carrying capacity is probably best viewed as the trophic capacity of an ecosystem to support a certain biomass that may be partitioned into various community structures. Thus, if the biomass of a species in the system is drastically reduced via exploitation, other elements may simply increase to assume any unused trophic capacity of the system with some new community structure. The ecosystem may no longer possess the capacity to support stocked individuals of the depleted species, at least not at their original population level. There may be physiological and behavioral deficits in hatchery-produced fish that will severely limit their survival to enter fishable stocks. There are legitimate concerns about the effects of stocking on biodiversity and the ability of enhanced populations to survive in the long term. Competitive interactions of hatchery-produced fish and wild conspecifics, as well as with other elements of the fish community, are not understood. Habitat relationships and requirements of many species also are unknown. These are just some of the more obvious information gaps that must be bridged, perhaps repeatedly, to predict how and if stock enhancement schemes will have any practical effect on the fishable stock.

As demonstrated at this symposium, there have been many significant accomplishments. Great advances have been made in rearing techniques so that it is possible to produce enormous numbers of healthy progeny for stocking. Marking techniques such as fed and injected chemicals, genetic markers, otolith microstructure measures, and coded wire tags, which are rapid, economical, and long-term techniques for estimating stocked fish survival, have been developed.

The bioeconomic feasibility of stock enhancement in marine environments remains an open question. We have heard promising results from Texas, Florida, Hawaii, and California. Especially relevant questions here are the trophic carrying capacity of the system, competition of hatchery fish with wild conspecifics, and replacement of wild with hatchery fish. For example, several claims were made at this conference that the catch of some species consisted of a significant percentage of hatchery fish. This percentage is rather meaningless unless we know if it is new production or is simply subtracted from wild production and total production remains the same. The results of research on striped mullet *Mugil cephalus* in Hawaii has yielded good evidence that hatchery production is not at the expense of wild. This is a good example of the research that must be done but may be a special situation because the striped mullet fishery, like most

Hawaiian coastal fisheries, is very depressed and well below production capacity. Also, Hawaiian ichthyofauna is relatively depauperate and, therefore, contains few potential competing species.

The genetic risks associated with stocking are equally relevant to marine, freshwater, and anadromous situations. Because marine fish generally are genetically more homogeneous, due presumably to the greater homogeneity of their environment (*viz*, the existence of fewer barriers and good mechanisms, such as ocean currents, to promote gene flow), genetic risks may be lower in the marine environment. Regardless, any plan for hatchery introductions should take appropriate safeguards as are Texas and Florida with red drum *Sciaenops ocellatus* and is California with white seabass *Atractoscion nobilis*. The hatchery environment has lead to artificial selection for undesirable traits, such as aggressive behavior. Hatchery practices should be improved to reduce these problems.

The use of nonnative fishes is not as likely to be a problem in marine situations. However, we generally oppose the introduction of nonnatives. There is no need to repeat the mistakes of the past, and marine faunas are usually not depauperate, which was suggested during this meeting to justify introduction of nonnative species. Production of nonnative species likely will be at the expense of the native fish community.

There are legitimate concerns about physiological and behavioral deficits of hatchery-reared fish. For example, salmon feeding behavior, predator avoidance, agonism, schooling, and shelter seeking, were all negatively influenced in hatcheries. It was suggested that some of these problems have a genetic basis, and that hatchery practices have selected for them. These issues are just as relevant to marine as freshwater and anadromous fish culture.

In conclusion, there are large potential benefits from successful stock enhancement. However, we strongly emphasize that enhancement must be a part of an integrated program that focuses primarily on conservation of natural stocks and the ecosystems that produce them. We should not lead environmental and resource policymakers to believe that hatcheries are a technological fix for irresponsible management of natural stocks and ecosystems. We recommend Blankenship and Lebers' 10-point plan as presented at this meeting as a thoughtful and responsible approach. We would add the preamble that the approach should be stepwise. First should come a rigorous research program to demonstrate the bioeconomic feasibility and investigate the underlying biological questions. Only after successful completion of the research phase should implementation begin.

Symposium Summary

Melding Science and Values in Pursuit of Ethical Guidelines for the Uses of Cultured Fishes in Aquatic Ecosystems

PAUL BROUHA

American Fisheries Society
5410 Grosvenor Lane, Suite 110, Bethesda, MD 20814, USA

I congratulate everyone involved with this symposium since its inception. We all recognize the potential far-reaching significance of our efforts to produce guidelines for the use of cultured fishes in aquatic ecosystems, and everyone is working hard as a result. Although it is too much to expect that criticisms will not be leveled at our efforts as being biased some way or other, I think we have largely succeeded in providing an open, objective forum in which to present available scientific information from all disciplines and perspectives of fisheries science. Every participant I have talked to has indicated the format of discipline-oriented presentations followed by point-counterpoint sessions has been instructive and fruitful in increasing understanding across the disciplines and in reducing the adversarial and emotion-laden context that has attended past discussions regarding use of cultured fishes.

Much scientific information has been presented to support the thesis that cultured fishes can continue to be an important tool in fisheries management and conservation. Within the hatchery environment it is clear we are applying more scientific information to culture fishes to mimic wild stocks than is widely realized. We can put hatcheries and nature into a common selection system to ensure fitness of cultured fishes. It is also clear that we know a great deal more in the area of genetics than is being applied, yet at this point we are peering into the fish genome with a set of tools that reveals images of uncertain focus and significance. We are akin to a blind man trying to appreciate an elephant using only the sense of touch, but that situation is changing rapidly as genetic research technology evolves. In addition, the symposium has not explored in depth the status of our knowledge in other areas of culturing and stocking fishes—fish health, fish diseases, behavioral training, and optimal fish size and timing of introductions. One often-repeated bit of advice has been to prepare and execute a plan to monitor the fate and effects of cultured fish stocked into the environment.

Moving beyond the hatchery environment, fisheries scientists and ecologists enter a very difficult world. As we have seen during the symposium, determining the effects of cultured fishes in aquatic ecosystems is problematic. Many assertions of loss of fitness, outbreeding depression, introduction of diseases in wild populations, and ecological effects have not been supported by scientific evidence. The scientific method, however, is an exacting taskmaster that requires devotion of resources beyond those historically available to yield definitive information. Just because we have not proved that we have reduced the fitness or population size of a particular fish stock because of genetic pollution, introduction of disease agents, increased predation, behavioral changes, increased harvest, "broodstock mining," or increased competition does not mean it is not happening. Conversely, asserting that some or all of these factors are occurring on the basis of theoretical dogma is equally unsupportable. Especially in areas where habitats have been substantially altered by humans, a "Mother Nature knows best" attitude may yield less sustainable, diverse, and productive fish communities than would using our fisheries management knowledge to its best effect. Optimally performing native fish species or assemblages of them may not be available in altered environments. Perhaps in these environments we can select and culture stocks of fishes on the basis of a suite of performance objectives that will be more fit and productive than those of native stocks.

Even as we recognize that hatcheries historically have not been operated responsibly from a genetics perspective and that many mistakes have been made during more than a century of stocking, let us move forward to fully realize the potential of cultured fishes in ecosystem and fisheries management.

As we move forward, values must shape the applications of our science. Symposium participants recognized that fact and suggested such values even as they objectively presented scientific information. We were cautioned that the use of cultured fishes must not continue to be a politically expedient alternative to ecosystem restoration and harvest management. It was suggested, because of our state of knowledge, that we manage native species in natural ecosystems to the extent possible. We should at least avoid making things worse and making irreversible resource commitments. Conversely, it was asserted that the public has a right to make value judgements within biological limits to maximize public benefits derived from aquatic habitats. The steady state of healthy ecosystems was challenged, and the term "ecosystem health" was judged by our keynote speaker, Gil Radonski, to be anthropocentric. The terms "biodiversity" and "ecosystem integrity" were questioned on the basis of the time baseline to which the present and the desired future should be compared. Finally, continuing human-induced ecological effects and the certainty that they will continue to increase in scope and magnitude was acknowledged. With a rapidly expanding population of nearly 300 million humans in North America, we agreed it is unrealistic to suggest we can return ecosystems to presettlement conditions. However, we also agreed that where pristine habitats and fish stocks exist, we must preserve them.

Just as Gil Radonski opened the symposium by referring to Bowen's (1970) maxim regarding the ethical use of cultured fishes, let me now close it by suggesting a somewhat different ethical model for the use of cultured fishes in ecosystem and fisheries management. Leopold's Land Ethic may be stated as "A thing is right when it tends to preserve the integrity, stability, and beauty of the biotic community. It is wrong when it tends otherwise" (Leopold 1949). Callicott (1991) expands on this statement by offering several commandments, three of which are especially appropriate to developing guidelines for the use of cultured fishes in aquatic ecosystems: (1) Thou shalt not exterminate species or render them extinct, (2) Thou shalt exercise great caution in introducing exotic species into local ecosystems, and (3) Thou shalt exercise great caution in damming and polluting water courses.

Clearly, this ethical model is appropriate when considering management alternatives in pristine habitats where native fish stocks exist—we must preserve their integrity, stability, and beauty. It is the right thing to do. Where fisheries managers have the authority and all the "cogs and wheels" (Leopold 1953:147) of the functional ecosystem remain or are recoverable, managers must work to restore them.

As Fred Harris stated during the opening session of this symposium on behalf of the Fisheries Administrators Section, the fisheries manager only rarely has authority to manage habitats. Far more frequently the challenge of making the best of a habitat altered or polluted by human activities must be faced. In such cases the fisheries manager is more like the auto mechanic at the corner garage working on an old jalopy. The fisheries manager is doing well just to get that aquatic ecosystem to "run," let alone be beautiful or stable. Similar to the demands on the auto mechanic, economics and politics dictate the manager use the full toolbox of fisheries management tools, as necessary, to "get the jalopy back on the road and running along, good as new." The public and the manager's fishing constituents, even though they or their forbearers may have acquiesced to the decision to alter or pollute the habitat, expect no less.

In such altered aquatic habitats I would propose modification of Leopold's Land Ethic to "A thing is right when it tends to preserve the integrity, stability, [productivity], and beauty of the biotic community." Where cogs and wheels have been discarded through mindless tinkering, and aquatic ecosystems are impaired or dysfunctional as a result, we must be able to use all our ecological knowledge and fisheries management tools to recraft and introduce cogs and wheels to replace those lost and thus recreate a stable and productive biotic community that is as nearly natural as possible. As humans continue their anthropogenic modification of the world's ecosystems, such corner garage miracles must increasingly become the norm. The use of cultured fishes must remain in our local fisheries manager's toolbox, and the scientific and academic research community must continue to work to understand the problems, refine the tools, and support their application in the context of this modified ethical model. Presentation of our technical information and the ensuing discussions during this symposium has been a remarkable step in the right direction. Now our challenge is to continue to work together to produce guidelines for the use of cultured fishes in aquatic ecosystems.

Following this symposium, even as peer review and editing of the proceedings is progressing, executive summaries of each paper will be compiled. In addition, participants will again be requested to suggest guidelines for use of cultured fishes in various aquatic habitat classes. Representatives from each fisheries management jurisdiction in North America will be invited along with American Fisheries Society (AFS) Section representatives to a facilitated workshop to be held in Denver, Colorado, 28–29 July 1994. Invitees will be sent the executive summaries along with your suggested guidelines compiled from your responses. The invitees will be asked to study and distill this information in preparation for a productive workshop.[1]

Our products will be the proceedings of this symposium and a guidelines document that will represent the best collective application of fisheries science in the context of professional and land ethics which we are capable of developing. Once completed, these products will be offered to policymakers and administrators of North American fisheries management jurisdictions with an invitation to consider our symposium and guidelines as they revise or develop policies on the use of cultured fishes in aquatic ecosystems.

---

[1] A facilitated workshop was held 29–30 July in Denver, Colorado. The workshop report and results are presented on page 601.

On behalf of AFS, thank you for your cooperation and commitment to producing these products—they are truly worthwhile, and I am sure will be viewed in retrospect as being tremendously significant.

References

Bowen, J. T. 1970. A history of fish culture as related to the development of fishery programs. Pages 71–72 *in* N. G. Benson, editor. A century of fisheries in North America. American Fisheries Society Special Publication 7.

Callicott, J. B. 1991. Conservation ethics and fishery management. Fisheries (Bethesda) 16(2):22–28.

Leopold, A. 1949. A Sand County almanac: and sketches here and there. Oxford University Press, New York.

Leopold, L. B. 1953. Round River: from the journals of Aldo Leopold. Oxford University Press, New York.

WORKSHOP RESULTS

Considerations for the Use of Cultured Fishes in Fisheries Resource Management

Considerations for the Use of Cultured Fishes in Fisheries Resource Management

Seventy-one individuals representing 41 fisheries resource management agencies (Appendix I) attended a facilitated workshop 29–30 July 1994, in Denver, Colorado. Collectively, the workshop participants manage or influence the management of a diversity of resources and have a broad range of responsibilities. To ensure that workshop recommendations were developed based on the most recent advances in fisheries resource management, participants were provided with abstracts of peer-reviewed presentations from the symposium "Uses and Effects of Cultured Fishes in Aquatic Ecosystems" before the workshop. To further prepare the participants, all were mailed two documents designed to focus their thoughts on the task of identifying recommendations for the use of cultured fishes. One of these documents was a matrix that listed types of fishery habitats and possible uses of cultured fishes. The second document described 17 fictitious, but realistic, fisheries management scenarios. Each scenario provided information about aquatic habitat, social, and fishery conditions. The participants, working in small groups of 6–10 people, associated an assigned scenario with a cell or several cells in the matrix and developed recommendations for the use of cultured fishes in the situation described by the scenario. Where consensus for specific recommendations could not be reached, minority opinions were recorded and carried forward to the next step of the process. Two small groups worked independently on each scenario.

After this process was completed, the two groups that independently developed stocking recommendations for the same scenario were combined and asked to blend their ideas into one set of recommendations. These blended recommendations (with minority views where necessary) were presented to all workshop participants for clarification and questions. The participants then either accepted the recommendations as written or identified possible changes. Where necessary, changes were addressed by the same working groups that developed the recommendations and final wording was presented to all participants for final approval.

Most, if not all, fisheries management agencies develop fisheries management (or equivalent) plans for management of waters in their jurisdiction. The workshop participants agreed that (1) fish stocking is a viable strategy in modern fisheries management, (2) use of cultured fishes should be addressed within an overall fisheries management plan prior to stocking, and (3) a checklist of considerations for the use of cultured fishes would be a beneficial addition to fisheries management plans when fish stocking is a management option. As the checklist developed, participants were cognizant of their colleagues needs and concerns. The resulting considerations for the use of cultured fishes recognize the need for dynamism and allow for jurisdictional flexibility.

The considerations for the use of cultured fishes identified at this workshop result from integration of current information about the uses and effects of cultured fishes, past and present management programs and experiences, and the philosophies of many fisheries management agencies. In time, additional information about the uses and effects of cultured fishes will be acquired and fisheries management agencies will change their programs and philosophies. Through implementation, these considerations for the use of cultured fishes can be tested. In the future (possibly 5–10 years), we encourage fisheries managers to re-evaluate these considerations in light of changes in fisheries information, management programs, and management philosophies in order to achieve the best possible management of fisheries resources.

Considerations for the Use of Cultured Fishes in Fisheries Resource Management

Biological Feasibility

Decisions to stock cultured fish should be based on evaluations that determine whether the environment can support the cultured fish and whether stocking will achieve positive management objectives. The following recommendations will help determine whether the management objectives will be achieved using cultured fishes.

1. Complete watershed surveys and assessments to determine the status of fish populations and environmental conditions.
2. Evaluate the status and trends of existing fish populations to determine whether there is a need for stocking.
3. Determine the carrying capacity of the aquatic system to predict stocking rates.
4. Determine opportunities for habitat restoration to sustain native and naturalized fishes as well as to ensure successful introduction of compatible species.[1]
5. Compare environmental requirements of cultured species considered appropriate candidates for stocking with habitat conditions (including fish populations) to predict the suitability of the habitat for the candidate species.
6. Assess hatchery production capability to meet fishery management objectives.
7. Conduct pilot studies or review data from comparable stocking programs to evaluate survival, growth, and reproduction of stocked fish, to test beneficial or harmful effects of stocking, and to develop stocking protocols.

Effects Analysis

Evaluations should be conducted to determine what effects stocked fish may have on the environment, native and naturalized biota, and humans. The following recommendations may be useful in determining the potential positive or negative effects of stocking cultured fishes.

1. Consider all possible beneficial or harmful effects on diversity of native and naturalized fishes with particular emphasis on threatened and endangered species.
2. Identify and evaluate any potential beneficial or harmful genetic effects on native and naturalized fish if interbreeding of cultured fish with native or naturalized fish is possible.
3. Evaluate potential beneficial or harmful effects of cultured fishes on population abundance and population variables, such as size structure, growth rate, recruitment rate, and mortality rate, of native and naturalized fishes.
4. Evaluate the history of other transplants into the fishery to determine the genetic history of existing fish populations.
5. Determine the potential for introduction of diseases to native and naturalized fishes from cultured fishes.

---

[1]Throughout this document, "species" should be interpreted as a biological unit that can be managed; therefore, "species" may be a species, a subspecies, or a stock.

6. Evaluate potential interspecific and intraspecific behavioral interactions (e.g., competition, predation, changes in reproductive behavior) that would have significant adverse effects on native and naturalized fishes.
7. Determine the potential for stocked fish to invade nontarget areas or expand their range to nontarget habitats.
8. Identify potential beneficial or harmful environmental effects of fish culture (e.g., water discharge, broodfish collection, fish escapement) on the local aquatic community.
9. Evaluate potential beneficial or harmful effects of fish hatchery construction and operation on humans (e.g., aesthetics, traffic, local economy).
10. Consider potential beneficial or harmful effects of increased and directed public use of aquatic environments on biotic communities.
11. Evaluate public health issues related to hatchery operations and the use of cultured fish.
12. Evaluate the potential for the stocked fish to persist and flourish without continued stocking.
13. If the cultured fish will be a previously untried introduction, the American Fisheries Society position statement on introduction of aquatic species (Kohler and Courtenay 1986) should be consulted.
14. Consider the potential for introduction of nontarget species (i.e., native, nonnative, or exotic species brought in with shipments of target species).
15. Develop monitoring activities and continue to evaluate effects after stocking.

To reduce the potential for adverse ecological effects, five specific recommendations for stocking fish are provided.

1. Remote isolated aquatic habitats managed for preservation of native or naturalized species should not be stocked for put-and-take fisheries. If put-and-take fisheries management is used, the stocking of cultured fishes should have no negative impact on established fishes.
2. When fish are stocked into altered habitats for the purpose of creating put-and-take fisheries, the species of fish used should be compatible with the physical, chemical, and biotic conditions of the altered habitat. In determining the species to be stocked:
    - Primary consideration should be given to native or naturalized species;
    - If no native or naturalized species can meet management goals, select the nonnative species best suited to use the productivity of the altered habitat;
    - If nonnative species are used, impacts on existing native or naturalized species should be considered.

    When nonnative species are stocked into altered habitats as part of a put-grow-and-take program, effects on native and naturalized species and reversibility of the stocking program should be considered.
3. When a species is stocked that has potential to interbreed with native or naturalized populations, appropriate genetic analyses of existing populations should be conducted.
    - If the extant population was founded from native or naturalized fish, native or naturalized broodfish should be used.

- If the extant population was not founded from native or naturalized fish, alternate broodfish may be used.
4. Stocked fish should be derived from hatcheries with broodfish performance plans and appropriate disease certifications.[2]
5. Stocking should be discontinued if self-sustaining populations, at levels sufficient to meet management goals, are achieved.

Economic Evaluation

Fisheries management strategies have benefits and costs. Benefits are often those to society such as angling days, fish yield, and public access, but also include ecosystem function, stability, and productivity. Costs include operations, staff, and capital investment; depending on the impacts of stocked fish, costs for biological, habitat, or social effects may also be incurred. Although some measurements remain difficult (e.g., what a "species" is worth), benefits and costs should be comprehensively evaluated.

1. Analyze benefits and costs for the stocking of cultured fish.
2. Analyze benefits and costs for fish culture operations.

Public Involvement

The public has the right and responsibility to make value judgements concerning fisheries management and, along with fisheries managers, has the right and responsibility to protect, enhance, conserve, and enjoy fisheries resources. As such, they have a role in decisions to use (or not use) cultured fishes for fisheries resource management. The following recommendations are intended to help maintain or increase public acceptance of stocking cultured fishes as a management tool.

1. Keep the public informed and encourage dialogue.
2. Make efforts to educate the public about the biological and social benefits of stocking or not stocking fishes.
3. Determine public consumptive and nonconsumptive uses of the fishery. If fishes are stocked for recreational fishing, then angler preferences, angler accessibility, uniqueness of the fishing opportunity, and potential changes in public use should be considered.
4. Develop and encourage opportunities for public participation in fish stocking programs.

Interagency Cooperation

Stocked fishes, or the fish culture operations that produce fish for stocking, can have direct and indirect biological, social, and economic effects on other political or geographic jurisdictions. The following recommendations may help develop cooperative management strategies and contribute to achieving management objectives when such cases occur.

1. The parties responsible for stocking cultured fishes should inform and seek concurrence from other fisheries management jurisdictions where resources may

---

[2]Broodfish performance plans should address fish origins, appropriate population size, maintenance methods, breeding plans to minimize inbreeding, and production schedules.

be impacted by stocked fish. Determination of jurisdictions solicited for concurrence should consider possible migration of the stocked species.
2. When the stocked fishes are introduced species, the American Fisheries Society position statement on introduction of aquatic species should be consulted (Kohler and Courtenay 1986).

Administrative Considerations

The above considerations identify specific biological, social, economic, and political issues that should be considered in decisions to stock fish. The following considerations may be useful addenda to fisheries management plans when stocking cultured fish is part of the management strategy.

1. State specific management objectives to be accomplished by stocking cultured fishes, identify criteria to determine when objectives are achieved, and specify when stocking will be evaluated.
2. Identify additional regulations that may be needed. If new regulations are required, the enforcement of these regulations and their compatibility with other management objectives should be considered.
3. Develop and maintain operational guidelines for fish stocking.
4. Consider agency strategic plans, guidelines, policies, regulations, and laws.

Reference

Kohler, C. C., and W. R. Courtenay, Jr. 1986. American Fisheries Society position on introductions of aquatic species. Fisheries (Bethesda) 11(2):39–42.

Appendix
Workshop Participants

Donald L. Archer
Utah Wildlife Resources

Gary L. Armstrong
Indiana Department of Natural Resources

Elliott Atstupenas
U.S. Fish and Wildlife Service, North Carolina

Dale Best
U.S. Fish and Wildlife Service, Wisconsin

Cliff Bengston
The Tulalip Tribes, Washington

Richard L. Berry
Oregon Fish and Wildlife

Paul Brouha
American Fisheries Society, Maryland

Robert D. Burkett
Alaska Department of Fish and Game

Brian C. Cates
U.S. Fish and Wildlife Service, Washington

Chris Christianson
Columbia Basin Fish and Wildlife Authority, Oregon

Lawrence E. Claggett
Wisconsin Department of Natural Resources

Glen Contreras
Virginia

Kirby D. Cottrell
Illinois Department of Conservation

Thomas A. Curtis
South Carolina Wildlife and Marine Resources

Wayne J. Daley
Steering Committee

Michael Donofrio
Keweenaw Bay Indian Community, Michigan

Paul Dorn
Pacific Northwest Tribes, Washington

Philip P. Durocher
Texas Parks and Wildlife Department

Gary Edwards
U.S. Fish and Wildlife Service, Washington, D.C.

John M. Epifanio
Point-Counterpoint Organization, Michigan

Kim E. Erickson
Oklahoma Department of Wildlife

Lee Evenhuis
Squaxin Island Tribe, Washington

Stephen Facciani
Wyoming Game and Fish Department

Benjamin M. Florence
Maryland

Delano R. Graff
Steering Committee

Churchill B. Grimes
Steering Committee

Donald Horak
Colorado Division of Wildlife

Ronald G. Howey
U.S. Fish and Wildlife Service, Massachusetts

Alan Huff
Florida Department of Environmental Protection

Philip J. Hulbert
New York Department of Environmental Conservation

Martin J. Jennings
Wisconsin Department of Natural Resources

Harold L. Kincaid
Steering Committee

Kenneth M. Leber
Hawaii Department of Aquatic Resources

Conrad Mahnken
National Marine Fisheries Service, Washington

Martin T. Marcinko
Steering Committee

James A. Marshall
Ohio Division of Wildlife

Mallory G. Martin
North Carolina Wildlife Resources Commission

Gary C. Matlock
National Marine Fisheries Service, Maryland

Franklin T. McBride
North Carolina Wildlife Resources Commission

C. Eugene McCarty
Texas Parks and Wildlife Department

Duncan McInnes
New Hampshire Fish and Game

Donald D. MacKinlay
Department of Fisheries and Oceans, British Columbia

Roy W. Miller
Delaware Division of Fish and Wildlife

Brenda Mitchell
Bureau of Land Management, Colorado

Vincent A. Mudrak
Steering Committee

John G. Nickum
U.S. Fish and Wildlife Service, Washington, D.C.

Douglas D. Nygren
Kansas Wildlife and Parks

Frank M. Panek
National Park Service, Washington, D.C.

Ronald D. Payer
Minnesota Department of Natural Resources

Vern Pepper
Department of Fisheries and Oceans, Newfoundland

Ted Perry
Department of Fisheries and Oceans, British Columbia

Peter W. Pfeiffer
Kentucky Department of Fish and Wildlife Resources

David P. Philip
Illinois Natural History Survey

Terry Radcliffe
Facilitator – AFS, Montana

Dennis C. Ricker
Steering Committee

Harold L. Schramm, Jr.
Steering Committee

Jill C. Silvey
Bureau of Land Management, Idaho

R. Z. Smith
National Marine Fisheries Service, Oregon

Robert H. Soldwedel
New Jersey Fish, Game and Wildlife

Ronald Southwick
Virginia Game and Fish

Michael D. Spencer
Georgia Fish and Game

Beth D. Staehle
American Fisheries Society, Maryland

Fred M. Utter
American Fisheries Society Genetics Section, Washington

Allan R. Van Vooren
Idaho Fish and Game

David A. Watsjold
U.S. Fish and Wildlife Service, Alaska

Robert J. Wattendorf
Florida Game and Fresh Water Fish

Joe Webster
U.S. Fish and Wildlife Service, Colorado

Arthur M. Williams
Louisiana Department of Wildlife and Fish

Jack E. Williams
Bureau of Land Management, Washington, D.C.

J. Holt Williamson
Steering Committee

Terry E. Wright
Northwest Indian Fish Commission, Washington